NAYFEH—Perturbation Methods
NAYFEH and MOOK—Nonlinear Oscillations
PETRICH—Inverse Semigroups
PIER—Amenable Locally Compact Groups
PRENTER—Splines and Variational Methods
RAO—Measure Theory and Integration
RICHTMYER and MORTON—Difference Methods for Initial-Value Problems, 2nd Edition
ROCKAFELLAR—Network Flows and Monotropic Optimization
SCHUMAKER—Spline Functions: Basic Theory
SHAPIRO—Introduction to the Theory of Numbers
SIEGEL—Topics in Complex Function Theory
        Volume 1—Elliptic Functions and Uniformization Theory
        Volume 2—Automorphic Functions and Abelian Integrals
        Volume 3—Abelian Functions and Modular Functions of Several Variables
STAKGOLD—Green's Functions and Boundary Value Problems
STOKER—Differential Geometry
STOKER—Nonlinear Vibrations in Mechanical and Electrical Systems
TURÁN—On A New Method of Analysis and Its Applications
WHITHAM—Linear and Nonlinear Waves
ZAUDERER—Partial Differential Equations of Applied Mathematics

NONCOMMUTATIVE
NOETHERIAN RINGS

NONCOMMUTATIVE NOETHERIAN RINGS

J. C. McCONNELL AND J. C. ROBSON
University of Leeds

with the cooperation of **L. W. Small**
University of California at San Diego

A Wiley–Interscience Publication

JOHN WILEY & SONS

Chichester · New York · Brisbane · Toronto · Singapore

Copyright © 1987 by John Wiley & Sons Ltd.

All rights reserved.

No part of this book may be reproduced by any means, or transmitted, or translated into a machine language without the written permission of the publisher

Library of Congress Cataloging-in-Publication Data:
McConnell, J. C. (John C.)
  Noncommutative Noetherian rings.
  (Wiley series in pure and applied mathematics)
  'A Wiley–Interscience publication.'
  Bibliography: p.
  Includes indexes.
  1. Noetherian rings. 2. Noncommutative rings.
I. Robson, James Christopher, 1940–   II. Small,
Lance W., 1941–   III. Title. IV. Series: Pure
and applied mathematics (John Wiley & Sons)
QA251.4.M38 1988   512'.4   87-8153

ISBN 0 471 91550 5

British Library Cataloguing in Publication Data:
McConnell, J. C.
  Noncommutative Noetherian rings.—
  (Wiley series in pure and applied
  mathematics).
  1. Noetherian rings
  I. Title II. Robson, J. C. III. Small,
  L. W.
  512'.4   QA251.4

ISBN 0 471 91550 5

Typeset by Thomson Press (India) Limited, New Delhi
Printed and bound in Great Britain by
Anchor Brendon Ltd, Tiptree, Essex

Contents

Preface   xi

Notation   xv

0. Preliminaries   1

    §1   Chain conditions   1
    §2   Prime radical   3
    §3   Jacobson radical   5

PART I   BASIC THEORY

1. Some Noetherian rings   11

    §1   Matrices   11
    §2   Skew polynomial rings   15
    §3   Weyl algebras   18
    §4   Skew power series and Laurent polynomials   21
    §5   Group rings and generalizations   22
    §6   Skew polynomial rings in several variables   25
    §7   Enveloping algebras and their generalizations   29
    §8   Further simple rings   33
    §9   Additional remarks   37

2. Quotient rings and Goldie's theorem   40

    §1   Right quotient rings   40
    §2   Uniform dimension   49
    §3   Goldie's theorem   55
    §4   Additional remarks   60

3. Structure of semiprime Goldie rings — 61

- §1 Orders in quotient rings   61
- §2 Minimal primes   68
- §3 Right ideals in semiprime rings   71
- §4 Endomorphism rings   75
- §5 Morita equivalence   80
- §6 Morita contexts and prime rings   85
- §7 Additional remarks   91

4. Semiprime ideals in Noetherian rings — 92

- §1 Reduced rank and applications   92
- §2 Artin–Rees property and localization   100
- §3 Localization and prime ideals   106
- §4 Affiliated primes and regular elements   116
- §5 Additivity   120
- §6 Patch continuity   124
- §7 Additional remarks   130

5. Some Dedekind-like rings — 132

- §1 Maximal orders   133
- §2 Asano and Dedekind prime rings   137
- §3 Classical orders   142
- §4 Hereditary Noetherian rings   149
- §5 Idealizer rings   152
- §6 Hereditary Noetherian prime rings   156
- §7 Modules over Dedekind prime rings   161
- §8 Additional remarks   169

PART II   DIMENSIONS

6. Krull dimension — 173

- §1 Deviation of a poset   174
- §2 Krull dimension of modules   180
- §3 Krull dimension in rings   187
- §4 Prime ideals and FBN rings   191
- §5 Bounds on Krull dimensions   196
- §6 Calculation of Krull dimension   199
- §7 Stable bounds on generators   206
- §8 Quotient rings and localization   209
- §9 Skew polynomials over commutative rings   218
- §10 Additional remarks   227

7. Global dimension   230

§1   Preliminaries   230
§2   Change of rings   234
§3   Factor rings   237
§4   Localization   244
§5   Estimates of global dimension   245
§6   Filtered and graded modules   254
§7   Regular rings   259
§8   Fixed rings   261
§9   Skew polynomial rings   266
§10 Skew polynomials over commutative rings   273
§11 Simple Dedekind domains   274
§12 Additional remarks   277

8. Gelfand–Kirillov dimension   278

§1   Definition and examples   279
§2   Dimensions of related algebras   285
§3   Module theory   291
§4   Almost commutative algebras and Hilbert polynomials   299
§5   Applications   303
§6   Somewhat commutative algebras   308
§7   Additional remarks   316

PART III   EXTENSIONS

9. The Nullstellensatz   321

§1   Algebras over a field   321
§2   Algebras over rings   326
§3   Generic flatness   327
§4   Constructible algebras   331
§5   Finite dimensional endomorphism rings   336
§6   Polynomials over division rings   338
§7   Additional remarks   341

10. Prime ideals in extension rings   343

§1   Finite extensions and chain conditions   343
§2   Prime ideals   347
§3   Quotient rings and closure   352
§4   Incomparability   357

viii   Contents

§5   Crossed products and fixed rings   363
§6   Primes in polynomial rings   369
§7   Additional remarks   375

11. Stability   377

§1   Stably free modules   377
§2   Stably free nonfree modules   382
§3   Stable and elementary ranks   386
§4   Cancellation of modules   391
§5   Ranks of certain rings   393
§6   Local information   400
§7   Stability and cancellation   407
§8   Additional remarks   413

12. K_0 and extension rings   415

§1 K_0 of a ring   416
§2   Projective-graded modules   419
§3   Filtrations and the syzygy theorem   422
§4 K_0 of module categories   425
§5   Skew Laurent extensions   429
§6   Filtered rings   432
§7   Applications to simple rings   436
§8   Additional remarks   441

PART IV   EXAMPLES

13. Polynomial identity rings   445

§1   Polynomial identities   445
§2   Nilpotence   451
§3   Central simple algebras   454
§4   Embeddings and matrix rings   456
§5   Central polynomials   459
§6   Semiprime rings and central polynomials   464
§7   Prime ideals and Azumaya algebras   468
§8   Integral extensions and prime rings   473
§9   The trace ring and maximal orders   479
§10 Affine k-algebras   483
§11 Additional remarks   488

14. Enveloping algebras of Lie algebras — 490

§1 Basics   491
§2 Prime ideals   495
§3 Eigenvalues and prime ideals   498
§4 Primitive ideals   501
§5 The solvable case   506
§6 When eigenvectors are central   509
§7 When g is algebraic   513
§8 The simple algebras $\mathcal{A}(V, \delta, \Gamma)$ 515
§9 The general case   520
§10 Additional remarks   526

15. Rings of differential operators on algebraic varieties — 527

§1 Algebras over a ring   528
§2 Affine algebras over a field   534
§3 Dimensions   540
§4 Further properties   546
§5 Rings of differential operators   550
§6 Additional remarks   554

References — 556

Index of notation — 579

Index — 583

Preface

'Do you see this ring?'
    Robert Browning (*The Ring and the Book*)

This book describes an abstract theory whose elegance and strength will be recognized easily by the reader. However, as the quotation indicates, it is examples which suggest and require theory. Therefore the book starts and ends with examples of rings arising from matrices, differential operators, groups, Lie algebras, etc.; if these are not familiar, a glance though Chapter 1 will help introduce them.

The theory itself is concerned essentially with noncommutative Noetherian rings although, when the methods apply more generally, as often is the case, the class of rings discussed is widened accordingly. The theory can reasonably be viewed as completing a picture which already contains the two rich, well-developed bodies of results concerning commutative Noetherian rings and noncommutative Artinian rings.

The problems involved in completing such a picture are clearly illustrated by considering the elementary but powerful technique of localization of a commutative ring at a prime ideal. For a noncommutative ring this is not, in general, possible, even at the zero ideal of an integral domain. The key to the theory presented here is Goldie's theorem which demonstrates that localization is possible at the zero ideal of any Noetherian prime ring. The additional fact that the resulting localized ring is simple Artinian points to the link with the Artinian theory.

The book is organized into four parts. After a chapter summarizing certain prerequisites, Part I starts with examples. Goldie's theorem, for prime and semiprime rings, takes up Chapter 2, and the succeeding chapter details the structure of these rings. Then Chapter 4 shows, in a more tentative way, what can be deduced about the structure of prime and semiprime ideals in Noetherian rings. This part ends with an illustration of the theory so far, an account of rings akin to commutative Dedekind domains.

Part II introduces three dimensions, Krull, global and Gelfand–Kirillov, each of which can be viewed as a measure of how far a ring, or module, is from having some desirable property. The properties concerned are, respectively, being Artinian, semisimple, or a finite dimensional algebra.

Part III then goes in some depth into certain special aspects, with Chapters 9, 10, 12 all focused on the preservation of properties under appropriate ring extension. Thus Chapter 9 concentrates on simple modules and includes several results related to the Nullstellensatz; Chapter 10 deals with prime ideals and the 'Krull relations' like 'lying over'; and Chapter 12 concerns the Grothendieck group K_0. This is linked with Chapter 11 which studies stability of modules over noncommutative Noetherian rings, culminating in Stafford's remarkably complete version of the Bass–Serre–Swan theory.

Finally, Part IV comprises three chapters, each of which concentrates on a particular class of rings and applies to it the preceding theory. The three classes concerned are polynomial identity (PI) rings, enveloping algebras of Lie algebras, and rings of differential operators on algebraic varieties. None of these chapters is encyclopaedic but each is, rather, an introduction to that area from this perspective. One further class, group rings, although studied throughout the book, is covered less fully. We refer to [Passman **77, 84**] for further details.

This book aims to be an accessible source for the material it contains and, whenever possible, worthwhile progress is made using elementary tools. For example, in Chapter 12 there are three versions, of increasing generality and difficulty, of the syzygy theorem. It does not aim to be a history of the subject although the final section of each chapter provides attributions and indicates other sources. The preparation of such sections is rightly recognized as a thankless and time-consuming task; cf. [Jacobson **64**, Preface]. We trust the reader will forgive the many imperfections and accept them as an approximate guide only.

There are, of course, threads connecting all the topics described here. Nevertheless, we have attempted to make the chapters, from Chapter 4 onwards, relatively independent. Thus, if the reader is willing to accept the occasional result quoted from another chapter, several chapters can be read almost in isolation. In particular, the experienced ring theorist will be able to dip into the book as required (but is warned that standing conventions are announced from time to time, usually at the start of a chapter or section).

For the less knowledgeable reader, for whom this book is written, we can offer some more detailed advice. Note first that the examples and techniques in Chapter 1 may seem difficult. All that is required initially is to browse through the easier portions of it; the remainder can be digested when needed later. When combined with Chapters 2, 3 and Chapter 4, §1, this makes up a **basic course**.

Once that has been covered there are several alternatives. The most obvious is to complete the reading of Part I, with or without the later sections of

Chapter 4; and another is to examine the dimensions by reading Chapter 6, §§1–6, Chapter 7, §§1–5, and Chapter 8, §§1–3.

The other possible reading courses we suggest are aimed at one or other of the later chapters. Each follows on from the basic course.

Affine algebras: Chapter 8, §§1–3; Chapter 9.

Prime ideals: Chapter 7, §8; Chapter 10; Chapter 14, §2. (But possibly omitting Chapter 10, §§3, 4.)

Stability: Chapter 6, §§1–3, 7; Chapter 11.

K_0: Chapter 7, §§4, 6, 7; Chapter 12.

PI rings: Chapter 8, §§1–3; Chapter 9, §§1, 2; Chapter 13.

Enveloping algebras: Chapter 14.

Differential operators: Chapter 6, §§1, 2, 5; Chapter 7, §§1–5, 9; Chapter 8, §§1, 2, 6; Chapter 9; Chapter 15.

We now have the happy task of thanking all who have helped in producing this book. We start with Alfred Goldie whose definitive research unlocked the door to this theory. Over the years he has been supervisor, colleague, friend and adviser to both authors and, some years ago, sowed the seed of this book. Too much is owed to express here. We also thank Lance Small not only for cooperating with us throughout the writing but also for help and friendship over many years. We thank all those who have helped us by discussing, explaining, correcting and advising.

In particular we mention A. Braun, K. A. Brown, P. M. Cohn, S. C. Coutinho, K. R. Goodearl, R. Hart, A. G. Heinicke, M. P. Holland, E. E. Kirkman, S. Montgomery, K. R. Pearson, L. H. Rowen, S. P. Smith, J. T. Stafford and R. B. Warfield. We thank A. Landford and M. Williams for typing so much, so well, P. Boyes for duplicating draft chapters, and the staff of John Wiley for their care and helpfulness.

Finally, our greatest debt. We have learnt, whilst writing, why authors so often thank their wives so fulsomely: it is they who have sacrificed most! Thank you Doreen and Lyn.

JOHN MCCONNELL
CHRIS ROBSON
University of Leeds

Notation

Notation will be introduced as required, and an index of notation can be found at the end of the book. However, we note here some, largely standard, notation and conventions to be used throughout.

A ring R will mean a nonzero associative noncommutative ring, usually assumed to have a 1 although, occasionally, it will be specified otherwise. Ring homomorphisms will be expected to preserve the 1. $M_n(R)$ denotes the ring of all $n \times n$ matrices over R, $Z(R)$ the centre of R and R^+ the underlying additive group.

The notation M_R indicates a right R-module which, when $1 \in R$, is assumed to be unital (i.e. to have the property that $m1 = m$ for each $m \in M$). Likewise $_RM$ denotes a left R-module. $\operatorname{Hom}(M, N)$ denotes the set of R-module homomorphisms between two modules and, if further emphasis is needed, the notation $\operatorname{Hom}_R(M, N)$ or $\operatorname{Hom}(M_R, N_R)$ is used. In particular $\operatorname{End} M$ denotes the ring of endomorphisms of M. This is viewed as acting on the left of M thus making M a left $\operatorname{End} M$-module. The symbols $\hookrightarrow$ and $\twoheadrightarrow$ denote homomorphisms which are, respectively, injective and surjective.

A submodule N of M is indicated by writing $N \triangleleft M$. Also $I \triangleleft R_R$ or $I \triangleleft_r R$ means that I is a right ideal of R, and $I \triangleleft {}_R R$ or $I \triangleleft_l R$ that I is a left ideal. The notation $I \triangleleft R$ is reserved for ideals, i.e. two-sided ideals. The image of $m \in M$ in M/N is denoted by $m + N$, $[m + N]$ or $\bar{m}$; and the lattice of submodules of M by $\mathscr{L}(M)$.

The symbol $\subset$ is reserved for proper inclusion with $\subseteq$ indicating conditional inclusion. A chain of $n+1$ subsets $A_0 \supset A_1 \supset \cdots \supset A_n$ is said to have **length** n.

As usual, $\mathbb{N}, \mathbb{Z}, \mathbb{Q}, \mathbb{R}, \mathbb{C}$ represent the sets of natural numbers, integers, rational, real and complex numbers. For p a prime number, $\mathbb{Z}_{p^\infty}$ denotes the abelian group $\{q \in \mathbb{Q} \mid qp^n \in \mathbb{Z} \text{ for some } n \in \mathbb{N}\}/\mathbb{Z}$. For $a \in \mathbb{R}$, $\lceil a \rceil$ denotes the least integer $n \geq a$; and the notation $m \gg 0$, for $m \in \mathbb{N}$, means 'for all sufficiently large m'.

Cross-referencing within a chapter makes use of the appropriate paragraph number, e.g. 2.3; and to another chapter, gives first the chapter number, e.g. 5.2.3.

Chapter 0
PRELIMINARIES

Bearing in mind our experience that one person's well-known result is another's dim memory, this brief chapter aims to describe some of the basic objects and results which underpin the remainder of the book. It contains three short sections. The first covers chain conditions and Artinian rings, and the other two deal with the prime radical and the Jacobson radical.

§1 Chain Conditions

1.1 The basic facts regarding chain conditions and the structure of Artinian rings are described first. These can be found in many texts—e.g. [Cohn **77**], [Jacobson **80**], [Lambek **66**]. All proofs are omitted here.

1.2 If $M_R \neq 0$ and M has exactly two submodules, namely M and 0, then M is a **simple**, or **irreducible**, module. A module which is a direct sum of simple modules is called **semisimple**; and if the simple modules are pairwise isomorphic, it is called **isotypic**.

If a module M_R has the property that each descending chain
$$M = M_0 \supset M_1 \supset M_2 \supset \cdots$$
must terminate after a finite number of steps, then M satisfies the **descending chain condition**, or **d.c.c.**, and then M is called an **Artinian** module.

The dual concept, using ascending chains of submodules of M, is the **ascending chain condition**, or **a.c.c.** Modules satisfying the a.c.c. are called **Noetherian**.

Note that if $N \triangleleft M$ then M is Noetherian or Artinian if and only if both N and M/N have that property.

1.3 We now summarize some basic results.

Theorem. (Jordan–Hölder) (*a*) *A module M_R satisfies both the a.c.c and the d.c.c. if and only if there is an upper bound, n say, on the lengths of chains of submodules of M.*

(b) *If that is so, then every chain of submodules of M can be refined to one of length n, say*

$$M = M_0 \supset M_1 \supset \cdots \supset M_n = 0.$$

The composition factors M_i/M_{i+1} are simple and are uniquely determined up to order and isomorphism. □

1.4 Proposition. *The following conditions on a semisimple module M_R are equivalent:*
 (i) *M_R satisfies the a.c.c.;*
 (ii) *M_R satisfies the d.c.c.;*
 (iii) *M_R has finite length.* □

1.5 Proposition. *The following conditions on a module M_R are equivalent:*
 (i) *M_R is Noetherian;*
 (ii) *each submodule of M_R is finitely generated;*
 (iii) *every non-empty set of submodules of M_R has a maximal member.* □

1.6 Next we turn to the ring R itself. If R_R is Noetherian then R is a **right Noetherian ring**. Similarly, one defines **left Noetherian, right Artinian** and **left Artinian** rings. Then a **Noetherian**, or **Artinian**, ring is one which has both the right- and left-hand properties. The structure of Noetherian rings is the object of study in this book. The structure of Artinian rings, which has been well understood for many years, is an important ingredient of this study.

1.7 Application of 1.5 to R_R gives

Corollary. *The following conditions on a ring R are equivalent:*
 (i) *R is right Noetherian;*
 (ii) *R satisfies the a.c.c on right ideals;*
 (iii) *every right ideal of R is finitely generated;*
 (iv) *every non-empty set of right ideals of R has a maximal member.* □

1.8 In the special case when every right ideal of R is **principal** (or **cyclic**), R is called a **principal right ideal ring**, or **pri-ring**. The left-handed version is a **pli-ring**; and a **principal ideal ring** is one which is both a pri-ring and a pli-ring.

1.9 The next few facts connect module properties with ring properties. They comprise the **Artin–Wedderburn theory**.

Lemma. (Schur's lemma) *If M_R is simple then $\text{End}(M_R)$ is a division ring.* □

1.10 A ring R is **simple** if R has precisely two ideals, 0 and R. This is equivalent to asserting that 0 is the unique maximal ideal of R.

Theorem. *The following conditions on a ring R are equivalent:*
 (i) *R is a simple right Artinian ring;*
 (ii) *$R \simeq M_n(D)$ for some n and some (uniquely determined) division ring D;*
 (iii) *$R \simeq \operatorname{End} M_S$, where M is a semisimple isotypic S-module of length n over some ring S.* □

Note that (ii) is symmetric. So the left-hand versions of (i) and (iii) are also valid, and hence R is Artinian.

1.11 An ideal A of a ring R is **nilpotent** if $A^n = 0$ for some n.

Theorem. *The following conditions on R are equivalent:*
 (i) *R is a finite direct product of simple Artinian rings;*
 (ii) *R_R is semisimple;*
 (iii) *every right R-module is semisimple;*
 (iv) *R is right Artinian and has no nilpotent ideals;*
 (v) *R is right Artinian and the intersection of its maximal ideals is 0.* □

Note again the symmetry, provided here by (i) and 1.10. Such a ring R is called a **semisimple (Artinian) ring.**

1.12 The **radical** $N(R)$ of a right Artinian ring R is the intersection of its maximal ideals.

Theorem. *If R is right Artinian then $N(R)$ is nilpotent and is the largest nilpotent ideal of R. Also $R/N(R)$ is a semisimple ring.* □

1.13 A straightforward application of 1.4 to the factors $N(R)^m/N(R)^{m+1}$ in 1.12 leads easily to Hopkins' theorem.

Corollary. *If R is right Artinian then R is right Noetherian.* □

1.14 The apparent duality between a.c.c. and d.c.c. makes 1.13 surprising. However, it is the presence of $1 \in R$ (which is in fact needed for 1.12 and 1.13) which produces this asymmetry. It does indicate, however, that Artinian rings are a special type of Noetherian ring. Consequently the structure of Noetherian rings can be expected to be rather more complex.

§2 Prime Radical

2.1 For arbitrary non-Artinian rings there are two main candidates for the role played by $N(R)$ in 1.12, namely the prime radical and the Jacobson radical. Each reflects a different feature and both coincide with $N(R)$ when R is right Artinian.

This section describes the prime radical. Since this is less widely known, yet is particularly relevant to this book, proofs are given. Moreover, since these results will, occasionally, be applied to rings without 1, it is not assumed in this section that the rings concerned have a 1.

2.2 A ring R is called an **integral domain** if the product of nonzero elements is always nonzero; and an ideal A of any ring R is **completely prime** if R/A is an integral domain. Whilst such examples abound, they do not play a major role in the structure theory.

2.3 Instead the main role belongs to the rings characterized next.

Proposition. *The following conditions on a ring R are equivalent*:
 (i) *if $0 \neq a$, $b \in R$ then $aRb \neq 0$*;
 (ii) *if $0 \neq A$, $B \triangleleft R_R$ then $AB \neq 0$*;
 (iii) *if $0 \neq A$, $B \triangleleft R$ then $AB \neq 0$.*

Proof. (i)$\Rightarrow$(ii) Choose $0 \neq a \in A$, $0 \neq b \in B$. Then $aRb \neq 0$ and so $AB \neq 0$.
 (ii)$\Rightarrow$(iii) Trivial
 (iii)$\Rightarrow$(i) Note first that if $C = \{c \in R \mid RcR = 0\}$ then $C \triangleleft R$ and $RCR = 0$ and hence $C = 0$. Thus if $A = RaR$ and $B = RbR$ then $AB \neq 0$ and so $aRb \neq 0$. □

Such a ring R is called a **prime ring**. It has the property, clear from (iii), that if $0 \neq A$, $B \triangleleft R$ then $A \cap B \neq 0$. An ideal A of any ring R is called **prime** if R/A is a prime ring. The set of all prime ideals of R is denoted by Spec R, the **prime spectrum** of R.

2.4 The **prime radical** of R is the intersection of all prime ideals of R. It is clear that it contains all nilpotent ideals; and if R is right Artinian this coincides with $N(R)$. The same notation, $N(R)$, will be retained.

In non-Artinian rings, even in the commutative case, $N(R)$ need not be nilpotent. But it does have some properties of that type, as is shown next.

2.5 An element of a ring R is **nilpotent** if $a^n = 0$ for some n; and if each element of a subset of R is nilpotent the subset is **nil**.

A less obvious concept is that of an element a being **strongly nilpotent**. This requires that any sequence $a = a_0, a_1, a_2, \ldots$ such that $a_{n+1} \in a_n R a_n$ is ultimately zero. Of course, every strongly nilpotent element is nilpotent; and each element in a nilpotent right ideal is strongly nilpotent.

2.6 Theorem. *The prime radical $N(R)$ is precisely the set of strongly nilpotent elements of R. In particular $N(R)$ is nil.*

Proof. Suppose $a \in R$ and $a \notin N(R)$. Therefore $a \notin P$ for some $P \in \operatorname{Spec} R$. Hence $aRa \not\subseteq P$ and so $ara \notin P$ for some $r \in R$. Let $a_0 = a$ and $a_1 = ara$. Repetition of this process leads to $a_2, a_3, \ldots$ none of which belongs to P. Hence none is zero and a is not strongly nilpotent.

Conversely, suppose a is not strongly nilpotent and an infinite sequence $\mathscr{S}$ of nonzero elements a_i exists as described. Using Zorn's lemma, we may choose an ideal P of R maximal with respect to having $P \cap \mathscr{S} = \varnothing$. Now suppose $B, C \triangleleft R$ with $B \supset P$, $C \supset P$. Then $B \cap \mathscr{S} \neq \varnothing$ and $C \cap \mathscr{S} \neq \varnothing$ and so $a_i \in B$, $a_j \in C$ for some i, j. If $k = \max\{i, j\}$ it follows that $a_{k+1} \in BC$ and yet $a_{k+1} \notin P$. This demonstrates that P is prime and so $a \notin N(R)$. □

2.7 Corollary. *The following conditions on a ring R are equivalent:*
(i) *R has no nonzero nilpotent right ideal;*
(ii) *R has no nonzero nilpotent ideal;*
(iii) *$N(R) = 0$.*

Proof. (iii)⇒(i)⇒(ii). Trivial.
 (ii)⇒(iii) Evidently the ideal $C = \{c \in R \mid RcR = 0\}$ is nilpotent and hence zero. Thus, if $0 \neq a \in R$, then RaR is nonzero and so not nilpotent. Therefore $aRa \neq 0$ and so $ara \neq 0$ for some $r \in R$. Setting $a_0 = a$ and $a_1 = ara$, and then repeating the process, we see that a is not strongly nilpotent. Hence $N(R) = 0$. □

These properties characterize **semiprime rings**; and an ideal A of a ring R is **semiprime** if R/A is a semiprime ring.

2.8 The prime radical, $N(A)$, of an ideal A is defined to be the intersection of those prime ideals which contain A. Evidently $N(R/A) = N(A)/A$ and $N(A)$ is a semiprime ideal. Indeed, $A = N(A)$ if and only if A is semiprime.

A **minimal prime** of A is a prime ideal minimal with respect to containing A. Zorn's lemma can be used to demonstrate that each prime ideal containing A contains a minimal prime of A. Thus $N(A)$ is also the intersection of the minimal primes of A.

2.9 It is clear from 2.7 that if a ring R has a nonzero nilpotent right ideal then it also has a nonzero nilpotent ideal. The **Köthe conjecture** is that this is true with 'nilpotent' replaced by 'nil'. It will be seen, in 2.3.7, that, in Noetherian rings, nil right ideals are nilpotent; so the conjecture is valid there. Another special case is dealt with in 13.2.5.

§3 Jacobson Radical

3.1 The radical $N(R)$ of a right Artinian ring R has the property that $R/N(R)$ is semisimple. Thus $N(R)$ is the intersection of the maximal right (or left) ideals.

It is this aspect which is preserved, for arbitrary rings, in the Jacobson radical.

Once again, the theory here is valid for rings without 1. However, we restrict our account to rings with 1, referring to [Jacobson **64**] for a more complete account. Since this theory is rather well known, we omit proofs.

3.2 A **faithful** module M_R is one such that $Mr = 0$ only if $r = 0$, for any $r \in R$. If a ring R has a faithful simple module M_R then R is **(right) primitive**. Whilst this is not left–right symmetric, the word 'right' is usually omitted. Of course each simple module $M \simeq R/I$ for some maximal right ideal I. Therefore R is primitive if and only if R has a maximal right ideal I which contains no nonzero ideal.

3.3 Lemma. (i) *A simple ring is primitive.*
(ii) *A primitive ring is prime.* □

3.4 An ideal A of a ring R is called **(right) primitive** if R/A is a primitive ring. Thus by 3.3, each maximal ideal is primitive and each primitive ideal is prime.

3.5 There is a structure theorem for primitive rings. It involves a left vector space, V say, over a division ring D. Let $E = \text{End}(_D V)$, the ring of D-linear transformations of V, and let R be a subring of E. We say R is **dense** in E if, given any $\alpha \in E$ and any finite dimensional subspace W of V, there exists $r \in R$ such that $W(\alpha - r) = 0$; i.e. restricted to W, r and α coincide. Of course, if V itself is finite dimensional then this would entail R equalling E.

3.6 Theorem. (Density theorem) *Let R be primitive, M_R be a faithful simple module, and $D = \text{End } M_R$. Then M is a left D-vector space and R is dense in $E = \text{End }_D M$.* □

3.7 The dichotomy between $_D M$ being finite or infinite dimensional gives

Corollary. *If R is primitive and D is as above then either $R \simeq M_n(D)$ or else, for each natural number t, there is a subring R_t of R and a surjective ring homomorphism $R_t \to M_t(D)$.* □

3.8 The **Jacobson radical**, $J(R)$, of R is defined to be the intersection of all (right) primitive ideals. The symmetry of (iii) below shows that this can also be defined via left primitive ideals.

Theorem. *The following conditions on an ideal A of R are equivalent:*
(i) $A = J(R)$;
(ii) *A is the intersection of the maximal right ideals of R;*
(iii) *A is the largest ideal such that $\quad - a$ is a unit for each $a \in A$.* □

3.9 If $J(R) = 0$ then R is **semiprimitive.** An ideal A of R is **semiprimitive** if R/A is a semiprimitive ring.

3.10 One remarkably useful fact is noted next.

Lemma. (Nakayama's lemma) *If M_R is finitely generated and nonzero, then $M \neq MJ(R)$.* □

Part I
BASIC THEORY

Chapter 1
SOME NOETHERIAN RINGS

This chapter demonstrates how certain Noetherian rings arise in particular contexts. These rings will serve, in later chapters, to illustrate the theory with both examples and counter-examples. The contexts considered involve matrices, groups of automorphisms and Lie algebras of derivations. In each case the construction of the ring is given, together with a description of its basic properties including, in particular, a demonstration that it is Noetherian.

A complete understanding of these examples is not needed for a first reading of the theory of later chapters; but it is required for assessing the significance of the theory.

§1 Matrices

This section describes some rings associated with matrices. Despite being rather elementary, these examples have particular importance.

1.1 As in linear algebra, if R is a ring (with 1) and R^n is the free right R-module of rank n (whose elements are viewed as column vectors over R) then End $R^n \simeq M_n(R)$, the ring of all $n \times n$ matrices over R. As usual, the matrices act on R^n via left multiplication. There is a standard embedding of R into $M_n(R)$ given by $r \mapsto (a_{ij})$ where $a_{ii} = r$ for each i, and $a_{ij} = 0$ for $i \neq j$. It is convenient to identify R with its image under this map.

Similarly, if $I \triangleleft_r R$, then $M_n(I)$ is defined, and evidently $M_n(I) \triangleleft_r M_n(R)$.

1.2 Proposition. *$M_n(R)$ is right Noetherian if and only if R is right Noetherian.*

Proof. As an R-module, $M_n(R) \simeq R^{n^2}$, so if R is right Noetherian then $M_n(R)$ is Noetherian as a right R-module and so, *a fortiori*, as a right $M_n(R)$-module. Conversely, if $\{I_j | j = 1, 2, \ldots\}$ is a strictly ascending chain of right ideals of R, then $\{M_n(I_j)\}$ is a similar chain in $M_n(R)$. □

1.3 Whenever a subring R of a ring S is being considered, S can be considered as a right R-module, S_R. The argument above can be used to prove

Lemma. *If R is a subring of S, with S_R finitely generated and R right Noetherian, then S is right Noetherian.* □

1.4 Subrings of $M_n(R)$ arise naturally, particularly in the context of PI rings (see Chapter 13). The ring of all upper triangular matrices

$$T_n(R) = \{(a_{ij}) \mid a_{ij} \in R, a_{ij} = 0 \text{ for } i > j\}$$

is one obvious example. For such rings, the proof of 1.2 is easily extended, to give the next result.

Corollary. *If $R \subseteq S \subseteq M_n(R)$ are rings then S is right Noetherian if and only if R is right Noetherian.* □

1.5 A rather different type of 'matrix' construction is also common. It is helpful to consider first a special case which arises whenever R is a ring and M is a right R-module. If one lets $M^* = \mathrm{Hom}(M, R)$ and $S = \mathrm{End}\, M_R$ then $_S M_R$ and $_R M_S^*$ are bimodules. Moreover, given $m, n \in M$ and $\alpha \in M^*$, there are products $\alpha \cdot m \in R$, $m \cdot \alpha \in S$ given by $\alpha \cdot m = \alpha(m)$, and $m \cdot \alpha$ is the map $n \mapsto m \cdot \alpha(n)$. Thus $MM^* \subseteq S$, $M^*M \subseteq R$. One can check that the array

$$\begin{bmatrix} R & M^* \\ M & S \end{bmatrix} = \left\{ \begin{bmatrix} r & \alpha \\ m & s \end{bmatrix} \middle| r \in R, s \in S, m \in M, \alpha \in M^* \right\},$$

with the formal operations of 2×2 matrices, is indeed a ring.

1.6 More generally, suppose that R, S are rings, $_R V_S$, $_S W_R$ are bimodules and $\theta: V \otimes_S W \to R$, $\psi: W \otimes_R V \to S$ are bimodule homomorphisms. The array

$$T = \begin{bmatrix} R & V \\ W & S \end{bmatrix}$$

can be given the formal operations of 2×2 matrices again, using θ and ψ in defining multiplication. If θ, ψ satisfy the associativity conditions required to make T a ring then the collection $(R, S, V, W, \theta, \psi)$ is called a **Morita context**, and T is the **ring of the Morita context**.

1.7 Proposition. *The ring T is right Noetherian if and only if R_R, S_S, V_S and W_R are right Noetherian.*

Proof.

$$T = \begin{bmatrix} R & V \\ 0 & 0 \end{bmatrix} \oplus \begin{bmatrix} 0 & 0 \\ W & S \end{bmatrix} = A \oplus B \text{ say.}$$

So T_T is Noetherian if and only if A_T and B_T are Noetherian. Note that A is also a right $R \oplus S$-module and any T-submodule of A is also an $R \oplus S$-sub-

1.1.11 module. Thus $\mathscr{L}(A_T) \hookrightarrow \mathscr{L}(A_{R\oplus S})$, where $\mathscr{L}$ denotes the lattice of submodules. Hence if R_R and V_S are Noetherian, so too is A_T. Similar arguments deal with B_T.

Conversely, there are embeddings $\mathscr{L}(R_R) \hookrightarrow \mathscr{L}(A_T)$ and $\mathscr{L}(V_S) \hookrightarrow \mathscr{L}(A_T)$ given, for $I \triangleleft R_R$, $V' \triangleleft V_S$, by

$$I \mapsto \begin{bmatrix} I & IV \\ 0 & 0 \end{bmatrix}, \quad V' \mapsto \begin{bmatrix} V'W & V' \\ 0 & 0 \end{bmatrix}.$$

These, together with similar comments for W, S, prove the converse. □

1.8 In the special case when $W = 0$, the maps θ, ψ are of course zero, and no associativity is involved. The resulting ring

$$T = \begin{bmatrix} R & V \\ 0 & S \end{bmatrix}$$

is frequently used to provide interesting counter-examples. Here the ideal

$$\begin{bmatrix} 0 & V \\ 0 & 0 \end{bmatrix}$$

is nilpotent, of index 2.

1.9 For example, if $R = \mathbb{Z}$ and $V = S = \mathbb{Q}$ then

$$T = \begin{bmatrix} \mathbb{Z} & \mathbb{Q} \\ 0 & \mathbb{Q} \end{bmatrix}$$

is a right Noetherian ring but is not left Noetherian since $_{\mathbb{Z}}\mathbb{Q}$ is not Noetherian. Indeed the nilpotent radical of T is simple as a right module, but not Noetherian as a left module.

1.10 Similarly, the ring

$$\begin{bmatrix} \mathbb{Q} & \mathbb{R} \\ 0 & \mathbb{R} \end{bmatrix}$$

is right Artinian but not left Artinian.

1.11 There is a Morita context arising naturally in connection with any right ideal A of a ring R; viz.

$$\begin{bmatrix} \mathbb{I}(A) & A \\ R & R \end{bmatrix},$$

where $\mathbb{I}(A) = \{r \in R | rA \subseteq A\}$. The ring $\mathbb{I}(A)$ is called the **idealizer of** A, and is easily seen to be the largest subring of R containing A as a two-sided ideal. The ring $\mathbb{I}(A)/A$ is called the **eigenring** of A. This acts, by left multiplication, on the module R/A; and it can be checked that $\mathbb{I}(A)/A \simeq \text{End}(R/A)$. These rings are discussed in more detail in Chapter 5, §5. For now it will be sufficient to consider the case when A is a maximal right ideal.

1.12 Theorem. *If A is a maximal right ideal of R, then*:
(i) *R is a finitely generated right $\mathbb{I}(A)$-module*;
(ii) *R/A, viewed as a right $\mathbb{I}(A)$-module, has a composition series of length 1 if $RA = A$, and of length 2 otherwise*;
(iii) *$\mathbb{I}(A)$ is right Noetherian if and only if R is right Noetherian*;
(iv) *$\mathbb{I}(A)$ is right Artinian if and only if R is right Artinian*.

Proof. Note first that if $RA = A$ (i.e. if A is a two-sided ideal of R) there is nothing to prove. Suppose, therefore, that $RA \neq A$, and hence $RA = R$.

(i) $1 = \sum_{i=1}^{n} r_i a_i$ for some $r_i \in R$, $a_i \in A$. Since $A \subseteq \mathbb{I}(A)$ it follows that $\{r_i\}$ generates R over $\mathbb{I}(A)$.

(ii) Let B/A be a proper $\mathbb{I}(A)$-submodule of R/A. Then $(BA + A)/A \subseteq B/A$ and is a proper R-submodule of R/A, and hence is zero. Thus $BA \subseteq A$ and $B \subseteq \mathbb{I}(A)$. However, $\mathbb{I}(A)/A \simeq \text{End}(R/A)$ which is a division ring. Therefore either $B/A = \mathbb{I}(A)/A$ or $B/A = 0$.

(iii) If $\mathbb{I}(A)$ is right Noetherian then, by (i), R is a right Noetherian $\mathbb{I}(A)$-module, and hence a right Noetherian R-module.

Conversely, if R is right Noetherian, let B be any right ideal of $\mathbb{I}(A)$. Now $BR \supseteq B \supseteq BA$, with both BR and BA being finitely generated as right ideals of R and hence, by (i), as right $\mathbb{I}(A)$-modules. Furthermore, it follows that BR/BA is a homomorphic image of $(R/A)^n$ for some n. This, by (ii), has a finite composition series over $\mathbb{I}(A)$; so B/BA has too, and thus is finitely generated. Hence B is finitely generated; and so $\mathbb{I}(A)$ is right Noetherian.

(iv) This follows easily from the arguments of (iii). □

1.13 Several specific examples will be obtained later using this technique—see 3.10 for example. For now, we note that, if A is a maximal right ideal of a right Noetherian ring R, then the ring

$$T = \begin{bmatrix} \mathbb{I}(A) & A \\ R & R \end{bmatrix}$$

is also right Noetherian. This follows directly from 1.7 and 1.12(iii). However, T is also the idealizer of the maximal right ideal

$$\begin{bmatrix} A & A \\ R & R \end{bmatrix}$$

of $M_2(R)$. So it also follows just from 1.12(iii).

§2 Skew Polynomial Rings

2.1 This section concerns polynomials over a ring R in a variable x which is not assumed to commute with the elements of R. It is desired, however, that each polynomial should be expressible uniquely in the form $\sum x^i a_i$ for some $a_i \in R$. This applies, of course, to the elements ax, for any $a \in R$; but, in order that degrees behave appropriately, (i.e. $\deg(f(x)g(x)) \leq \deg f(x) + \deg g(x)$), it is required that $ax \in xR + R$, $ax = x\sigma(a) + \delta(a)$ say. Under these conditions it is apparent that σ, δ are endomorphisms of the underlying additive group R^+ of R. Moreover,

$$(ab)x = x\sigma(ab) + \delta(ab)$$

and

$$a(bx) = x\sigma(a)\sigma(b) + \delta(a)\sigma(b) + a\delta(b).$$

Thus σ is a ring endomorphism of R and δ satisfies

$$\delta(ab) = \delta(a)\sigma(b) + a\delta(b)$$

which is the defining property of a **σ-derivation**. Note in particular that $\sigma(1) = 1$ and $\delta(1) = 0$.

2.2 Example. Let k be a field, $R = k[x]$ be the commutative polynomial ring and σ any endomorphism of R which is the identity on k. Then, for any choice of $f \in R$, there is a σ-derivation δ defined by $\delta(x) = f$.

2.3 Given a ring R, an endomorphism σ and a σ-derivation δ, it is possible to construct a polynomial ring as described above. First consider the ring E of abelian group endomorphisms of $R^{\mathbb{N}}$. Note that $R \hookrightarrow E$, acting by right multiplication. Also there is an element $x \in E$ defined by $(r_i)x = (\sigma(r_{i-1}) + \delta(r_i))$, where $(r_i) = (r_0, r_1, \ldots) \in R^{\mathbb{N}}$ and $r_{-1} = 0$. In the subring S of E generated by R and x one can check, for $a \in R$, that $ax = x\sigma(a) + \delta(a)$. It follows easily that every element of S can be written in the form $\sum x^i a_i$. Moreover, since

$$(1, 0, 0, \ldots)\sum x^i a_i = (a_i),$$

this expression is unique.

The ring S thus constructed is called a **skew polynomial ring**, and is denoted by $R[x; \sigma, \delta]$. If $\delta = 0$ this is written as $R[x; \sigma]$; and, if $\sigma = 1$, as $R[x; \delta]$.

2.4 The ring $R[x; \sigma, \delta]$ can also be described as being the ring T generated freely over R by an element x subject only to the relation $ax = x\sigma(a) + \delta(a)$ for each $a \in R$. To see this, note that each element of T can be written in the form $\sum x^i a_i$ and that there is an obvious surjection $T \to R[x; \sigma, \delta]$ (using the freeness of T). Since the x^i are R-independent in $R[x; \sigma, \delta]$, they are also R-independent in T. Hence $T \simeq R[x; \sigma, \delta]$.

2.5 Clearly $R[x;\sigma,\delta]$ has the universal property that if $\psi:R\to S$ is a ring homomorphism, and $y\in S$ has the property that

$$\psi(a)y = y\psi(\sigma(a)) + \psi(\delta(a)) \qquad \text{for all } a\in R$$

then there exists a unique ring homomorphism $\chi:R[x;\sigma,\delta]\to S$ such that $\chi(x)=y$ and the diagram

$$\begin{array}{ccc} R & \longrightarrow & R[x;\sigma,\delta] \\ & \psi \searrow \swarrow \chi & \\ & S & \end{array}$$

is commutative.

2.6 One could follow an alternative convention, forming a skew polynomial ring using the relation $xa = \sigma(a)x + \delta(a)$ and writing elements in the form $\sum a_i x^i$. In the case when σ is an automorphism the ring obtained is simply $R[x;\sigma^{-1},-\delta\sigma^{-1}]$. This alternative convention will be used when convenient.

2.7 It is clear that $R[x;\sigma,\delta]/xR[x;\sigma,\delta] \simeq R$ as a right R-module. Hence $R[x;\sigma,\delta]$ has a representation in $\operatorname{End} R^+$ with elements of R acting by right multiplication and x acting as δ since $ax = x\sigma(a) + \delta(a) \equiv \delta(a)$.

Similarly, $R[x;\sigma,\delta]/(x-1)R[x;\sigma,\delta] \simeq R$, and this gives a representation in which x acts as $\sigma + \delta$.

2.8 The **degree** of an element $\sum_{i=0}^{n} x^i a_i$ is defined to be n, provided $a_n \neq 0$; and then a_n is called the **leading coefficient** of the element. Conventionally the zero polynomial has degree $-\infty$ and leading coefficient 0.

Note that, for $a\in R$,

$$ax^n = x^n \alpha_n(a) + x^{n-1}\alpha_{n-1}(a) + \cdots + \alpha_0(a),$$

where each $\alpha_i \in \operatorname{End} R^+$. So if $\alpha_n(a) \neq 0$ then ax^n has degree n and leading coefficient $\alpha_n(a)$. The general formulae for α_i in terms of σ,δ and n are rather complicated, but for $R[x;\sigma]$ and $R[x;\delta]$ they are simple. For $R[x;\sigma]$, $\alpha_n = \sigma^n$ and $\alpha_i = 0$ for $i \neq n$; and for $R[x;\delta]$,

$$\alpha_i = \binom{n}{i}\delta^{n-i}.$$

(Compare Leibnitz's formula for $\delta^n(ab)$.)

2.9 The properties of R are, of course, reflected in those of $R[x;\sigma,\delta]$ as the next result shows.

Theorem. Let $S = R[x;\sigma,\delta]$.
(i) If σ is injective and R is an integral domain, then S is an integral domain.

(ii) If σ is injective and R is a division ring, then S is a principal right ideal domain.
(iii) If σ is an automorphism and R is a prime ring, then S is a prime ring.
(iv) If σ is an automorphism and R is right (or left) Noetherian, then S is right (respectively left) Noetherian.

Proof. (i) If $f = \sum_{i=0}^{n} x^i a_i$ and $g = \sum_{j=0}^{m} x^j b_j$ with a_n, b_m nonzero, then fg has degree $n+m$ and leading coefficient $\sigma^m(a_n)b_m$, which is nonzero.

(ii) If I is a nonzero right ideal containing a nonzero element of degree n, then I contains a monic polynomial of degree n, i.e. $\sum_{i=0}^{n} x^i b_i$ with $b_n = 1$. A division algorithm shows that I is generated by the monic element of least degree belonging to I.

(iii) Let f, g be as in (i). Since σ is injective, $\sigma^m(a_n) \neq 0$ and so $\sigma^m(a_n)Rb_m \neq 0$. Since σ is surjective, each element of this set is the leading coefficient of some element of fRg. Hence $fRg \neq 0$ and so $fSg \neq 0$.

(iv) (This is a variation on a standard proof of the Hilbert basis theorem.) Suppose R is right Noetherian. Since σ is an automorphism, each element of S can be written in the alternative form $\sum b_i x^i$. If $I \triangleleft_r S$, let I_n be the set of leading coefficients, when written in this form, of elements in I of degree $\leq n$. It is clear that I_n is a right ideal of R (called the nth **leading right ideal** of I) and that $I_n \subseteq I_{n+1}$. Furthermore, if $I' \triangleleft_r S$ with $I \subseteq I'$ and with $I_n = I'_n$ for each $n \geq 0$, then $I = I'$. (To prove this, suppose otherwise. Choose an element in $I'\setminus I$ of least possible degree, m say. Then $I_m \neq I'_m$.)

Now, suppose that $L_0 \subseteq L_1 \subseteq \cdots$ is an ascending chain of right ideals of S, and denote by L_{in} the nth leading right ideal of L_i. Consider the array $\{L_{in} | i, n \geq 0\}$. Note first that $L_{ij} \subseteq L_{km}$ whenever $i \leq k$ and $j \leq m$. The ascending chain $\{L_{ii} | i \geq 0\}$ of right ideals of R stabilizes, say at L_{jj}. For each n with $0 \leq n \leq j-1$ the chain $\{L_{in} | i \geq 0\}$ stabilizes, say at k_n. Choose

$$m = \max\{j, k_0, k_1, \ldots, k_{j-1}\}.$$

Then, for all $i \geq m$ and all $n \geq 0$, $L_{in} = L_{mn}$. Thus $L_i = L_m$, and so S is right Noetherian. The left Noetherian case is similar. $\square$

2.10 The restrictions placed upon the endomorphism σ above are needed, as the examples below will help to indicate. Nevertheless, even the existence of an endomorphism is not actually required in the proof of (iv). A careful reading of the proof establishes

Theorem. *Let R be a right Noetherian ring, and S be an over-ring generated by R and an element x such that $Rx + R = xR + R$. Then S is right Noetherian.* $\square$

2.11 The following examples help to illustrate 2.9. The first shows what can happen if σ is not injective. The others show that even if σ is injective the Noetherian property can be lost on one side or on both sides.

Example. (i) Let k be a field, $R = k[y]$ and $\sigma(f(y)) = f(0)$. The ring $S = R[x;\sigma]$ is not an integral domain since $yx = x\sigma(y) = 0$. Indeed $ySx \subseteq yxS = 0$ and so $(Sxy)^2 = 0$. Therefore S is neither prime nor semiprime. The sums $\sum Sxy^i$ and $\sum x^i yS$ are both direct sums, so S is neither left nor right Noetherian.

Example. (ii) Let k be a field, $R = k(y)$, the rational function field, and $\sigma(f(y)/g(y)) = f(y^2)/g(y^2)$, where $f, g \in k[y]$. By 2.9(ii), $S = R[x;\sigma]$ is a principal right ideal domain. However S is not a principal left ideal domain—and in fact is not left Noetherian.

To see this, let $\deg(f/g) = \deg f - \deg g$. Notice that for each element of Sx, when written in the form $\sum x^i a_i$, the coefficients a_i have even degree. Thus $Sxy \cap Sx = 0$. It follows that the sum $\sum_{i=0}^{\infty} Sxyx^i$ is direct. To see this, suppose, to the contrary, there exists $m < n$, and elements $s_i \in S$ with

$$s_m xyx^m + \cdots + s_n xyx^n = 0$$

and s_m, s_n nonzero. Then

$$s_m xy = -(s_{m+1} xyx + \cdots + s_n xyx^{n-m}) \in Sxy \cap Sx = 0$$

which is a contradiction. Hence S is not left Noetherian.

Example. (iii) Let k be a field, $R = k[y]$, $\sigma(f(y)) = f(y^2)$ and $S = R[x;\sigma]$. This is, of course, a subring of the previous example. The argument above shows that $\sum Sxyx^i$ is a direct sum. Let $A_n = \sum_{i=0}^{n} x^i yS$ and note that, since $x^{n+1}y \notin A_n$, the chain of right ideals $\{A_n\}$ is strictly increasing. Thus S is neither left nor right Noetherian.

2.12 One might wonder if 2.9(iv) is valid for σ a surjection. It is—because σ would then be an automorphism! For if $\ker \sigma \neq 0$, then the infinite chain $0 \subset \ker \sigma \subset \ker \sigma^2 \subset \ldots$ would contradict R being right or left Noetherian.

§3 Weyl Algebras

3.1 This section introduces an example which arises in several guises. Let k be a field of characteristic zero. Then $A_n(k)$ denotes the k-algebra with $2n$ generators $x_1, \ldots, x_n, y_1, \ldots, y_n$ and relations

$$x_i y_j - y_j x_i = \delta_{ij}, \quad \text{the Kronecker delta,}$$

and

$$x_i x_j - x_j x_i = y_i y_j - y_j y_i = 0$$

and is called the nth **Weyl algebra** over k. This algebra first appeared in quantum mechanics as an algebra generated by position and momentum operators. The noncommutativity of the generators reflected the Heisenberg uncertainty principle. When $n = 1$, the generators are written as x, y rather than x_1, y_1.

1.3.6   Weyl algebras

3.2 An alternative description of $A_n(k)$, in terms of iterated skew polynomials, is also useful. Let $R = k[x_1,\ldots,x_n]$, the commutative polynomial ring, and consider the sequence of rings

$$R_0 = R, R_{i+1} = R_i[y_{i+1}; \partial/\partial x_{i+1}].$$

The k-algebra R_n then has generators which satisfy the relations which define the Weyl algebra. On the other hand, the generators of $A_n(k)$ satisfy the defining relations for the R_i.

Theorem. $A_n(k) \simeq R_n$.   □

3.3 The results of 3.2 and 2.3 give a 'normal form' for elements of R_n, and hence for elements of A_n. Specifically, each element can be written uniquely in the form

$$\sum a_{\alpha\beta} y^\alpha x^\beta, \quad a_{\alpha\beta} \in k,$$

where $\alpha = (m_1,\ldots,m_n)$, $\beta = (p_1,\ldots,p_n)$ and y^α denotes $y_1^{m_1} y_2^{m_2} \cdots y_n^{m_n}$ etc. As noted in 2.6, one could also write the elements (again uniquely) in the form $\sum b_{\beta\alpha} x^\beta y^\alpha$.

3.4 It is straightforward to check, for $c \in A_n$, that

$$x_i c - c x_i = \partial c/\partial y_i$$

and

$$y_i c - c y_i = -\partial c/\partial x_i.$$

with the derivatives calculated formally. (That is, $\partial(x_i^n)/\partial x_i = n x_i^{n-1}$ etc.)

3.5 Theorem. *If char $k = 0$ then $A_n(k)$ is a simple Noetherian integral domain.*

Proof. That $A_n(k)$ is a Noetherian integral domain is clear from 3.2 and 2.9(i), (iv). To see that $A_n(k)$ is simple, suppose that $0 \neq I \triangleleft A_n(k)$. Choose $0 \neq c \in I$. By 3.4, both $\partial c/\partial x_i$ and $\partial c/\partial y_i$ belong to I for $i = 1,\ldots,n$. This, together with 3.3 and induction, shows that I contains $q a_{\alpha\beta}$ for some nonzero integer q and some $a_{\alpha\beta}$, a nonzero coefficient in c. Thus $1 \in I$ and $I = A_n$.   □

3.6 In the same way as described in 2.7, $k[x_1,\ldots,x_n]$ is a right $A_n(k)$-module with $x_1,\ldots,x_n$ acting by multiplication and y_i acting as $\partial/\partial x_i$. Repeated differentiation, as in the proof of 3.5, shows that the $A_n(k)$-module so obtained is simple. This representation of $A_n(k)$ maps it on to the ring of differential operators on $k[x_1,\ldots,x_n]$ generated by $k[x_1,\ldots,x_n]$ together with the operators $\partial/\partial x_i$, $i = 1,\ldots,n$. Since $A_n(k)$ is simple the kernel of this map is zero. This proves that $A_n(k)$ is isomorphic to that '**ring of differential operators with polynomial coefficients**'.

3.7 If k is a field of finite characteristic, the construction of $A_n(k)$ can be repeated as before. However, it is no longer true that $A_n(k)$ is simple; for the element $qa_{\alpha\beta}$ might turn out to be zero. Indeed, if k has characteristic m then x_i^m is a central element and generates a nonzero ideal (x_i^m).

3.8 If R is any ring with identity then $A_n(R)$ is the ring generated by R and $x_1, \ldots, x_n, y_1, \ldots, y_n$ with the relations $x_i r - r x_i = 0 = y_j r - r y_j, r \in R$,

$$x_i y_j - y_j x_i = \delta_{ij}$$

and

$$x_i x_j - x_j x_i = y_i y_j - y_j y_i = 0.$$

Just as in 3.2, $A_n(R)$ may be described via a sequence of skew polynomial extensions.

Theorem. (i) *If R is right Noetherian or an integral domain then so also is $A_n(R)$.*
(ii) *If R is a division ring of characteristic zero then $A_n(R)$ is simple.*

Proof. Similar to the proof of 3.5. □

3.9 There is another ring closely related to $A_n(k)$. It is constructed as described in 3.2, except that R_0 is taken to be the rational function field $k(x_1, \ldots, x_n)$. The ring thus obtained is named $B_n(k)$. Note the obvious embedding $A_n(k) \hookrightarrow B_n(k)$.

Theorem. (i) *$B_n(k)$ is a Noetherian integral domain which, if k has characteristic zero, is simple.*
(ii) *$B_1(k)$ is a principal left and right ideal domain.*

Proof. (i) This is proved in the same way as 3.5.
(ii) This follows from 2.9(ii). □

3.10 One can use idealizers, as described in §1, to construct other unusual subrings of the $A_n(k)$. This is well illustrated by the following example.

Example. *If k is a field of characteristic zero then the ring $S = k + xA_1(k)$ is a Noetherian integral domain having just three ideals, viz. 0, $xA_1(k)$ and S.*

Proof. Note that each element $r \in A_1$ has the form $p + xs$ with $s \in A_1$ and $p \in k[y]$. Moreover, $xp - px = dp/dy$. It follows easily that $p + xs \in \mathbb{I}(xA_1)$ if and only if $p \in k$; so $\mathbb{I}(xA_1) = k + xA_1 = S$.

Also, using the technique of 2.7 and 3.6, it is clear that xA_1 is a maximal right ideal of A_1. Therefore, by 1.12 and 3.5, S is a right Noetherian integral domain.

1.4.5 *Skew power series and Laurent polynomials*

Symmetry shows that $A_1 x$ is a maximal left ideal of A_1, that $k + A_1 x$ is its idealizer, and so that $k + A_1 x$ is a left Noetherian integral domain. However, the inner automorphism of B_1 given by $b \mapsto x^{-1} bx$ restricts to give an isomorphism $k + xA_1 \to k + A_1 x$. Thus S is left Noetherian too.

Finally, note that if I is any nonzero ideal of S then $I \supseteq xA_1 I x A_1 = xA_1$, since A_1 is simple. But $S/xA_1 \simeq k$, and so $I = xA_1$ or S. □

3.11 As a contrast, consider the *commutative* polynomial ring $k[x, y]$. The analogous subring $R = k + xk[x, y]$ is not Noetherian; for R acts as k on $xk[x, y]/x^2 k[x, y]$ which is infinite dimensional over k.

§4 Skew Power Series and Laurent Polynomials

4.1 The construction of skew polynomials described in §2 has several useful variants. The results and arguments involved are very similar to those already given, so will be at most sketched.

First, let R be a ring and σ an endomorphism. Then $R[[x; \sigma]]$ denotes the ring of power series $\sum_{i=0}^{\infty} x^i a_i$ subject only to the relations $ax = x\sigma(a)$. This is called the **skew power series** ring.

4.2 There is a difficulty in attempting to form a power series ring involving a derivation too, since then, in multiplying two power series, one could obtain infinite sums of elements of R as coefficients of x^i. Further conditions on R, σ and δ would thus be required.

4.3 Let R be a ring and σ an automorphism. Then $R[x, x^{-1}; \sigma]$ denotes the ring of polynomials over R in x and x^{-1} subject to $ax = x\sigma(a)$. This is the ring of **skew Laurent polynomials**. Each element has a unique representation in the form $\sum_{i \in \mathbb{Z}} x^i a_i$ with all but finitely many coefficients being zero. Note that $R[x; \sigma]$ is a subring.

4.4 Given R and an automorphism σ, one can form the **skew Laurent power series** ring $R[[x, x^{-1}; \sigma]]$ in which each element has a unique representation as $\sum_{i \in \mathbb{Z}} x^i a_i$ with $a_{-n} = 0$ for almost all $n \in \mathbb{N}$ and with $ax = x\sigma(a)$.

4.5 Theorem. *Let R be a ring with an automorphism σ and let S be either $R[x, x^{-1}; \sigma]$ or $R[[x; \sigma]]$ or $R[[x, x^{-1}; \sigma]]$. Then statements (i)–(iv) of 2.9 are valid.*

Proof. The proofs for $R[x; \sigma, \delta]$ are easily modified. The leading coefficient of a power series, however, is the nonzero coefficient of the least power of x occurring. □

§5 Group Rings and Generalizations

5.1 The notion of a group ring is well known. Given a ring R and a group G the **group ring** RG is defined to be a free right R-module with the elements of G as a basis and with multiplication given by $(g_1 r_1)(g_2 r_2) = (g_1 g_2)(r_1 r_2)$ together with bilinearity.

5.2 There are embeddings $R \hookrightarrow RG$ and $G \hookrightarrow RG$. In fact RG has the following universal property: given a ring S, a ring homomorphism $\psi: R \to S$ and a group homomorphism χ from G to the group of units of S such that

$$\psi(r)\chi(g) = \chi(g)\psi(r), \qquad r \in R, \quad g \in G,$$

then there exists a unique ring homomorphism $\xi: RG \to S$ such that $\xi(r) = \psi(r)$ and $\xi(g) = \chi(g)$.

5.3 Many group rings are related to skew Laurent polynomial rings as the following examples illustrate.

Example. (i) Let G be the discrete Heisenberg group, $G = \langle x, y, z \rangle$ with relations $xyx^{-1}y^{-1} = z$, and z is central. Then

$$RG \simeq R[y, y^{-1}][z, z^{-1}][x, x^{-1}; \sigma]$$

where $\sigma(y) = z^{-1}y$, $\sigma(z) = z$.

Example. (ii) Let G be the infinite dihedral group; $G = \langle x, y \rangle$ with relations $x^2 = 1 = (xy)^2$. Then

$$RG \simeq R[y, y^{-1}][x; \sigma]/(x^2 - 1),$$

where $\sigma(y) = y^{-1}$. (Here $(x^2 - 1)$ is the ideal generated by the central element $x^2 - 1$.)

5.4 It is convenient for our purposes to extend the idea of a group ring by allowing the group elements to have some action on the ring of scalars. More precisely, let R be a ring, G a group and φ a homomorphism from G to Aut R, the group of ring automorphisms of R. For $g \in G$, $r \in R$ the image of r under $\varphi(g)$ will be denoted by r^g.

The **skew group ring** $R\#G$ (read 'R smash G') is then the free right R-module with elements of G as a basis and with multiplication defined by $(hr)(gs) = (hg)(r^g s)$ for $g, h \in G$, $r, s \in R$. Thus each element of $R\#G$ has a unique expression as $\sum_{g \in G} g r_g$ with $r_g = 0$ for all but finitely many $g \in G$. (It also has a unique representation in the form $\sum_{g \in G} s_g g$.)

Evidently $R\#G$ contains G as a subgroup of its group of units, and R as a subring. The ordinary group ring is obtained when $\varphi(g) = 1$ for all g.

5.5 Example. (i) If $G = \{x^n | n \in \mathbb{Z}\} \simeq \mathbb{Z}$ and $\varphi(x) = \sigma$ then $R\#G = R[x, x^{-1}; \sigma]$, the skew Laurent polynomial ring.

Example. (ii) A classical example is $K\#G$, where K is a finite Galois extension of a field k, and G is the Galois group.

5.6 Just like RG, the skew group ring has a universal property, the relation for $R\#G$ being

$$\psi(r)\chi(g) = \chi(g)\psi(r^g).$$

In particular there is a homomorphism ξ from $R\#G$ to $S = \operatorname{End} R^+$ given by $g \mapsto \varphi(g)$, $r \mapsto$ 'right multiplication by r'. Whilst ξ is not, in general, injective, it is when restricted to R. By analogy with differential operators (see 3.6) one can view $R\#G$ as a 'ring of formal "automorphic" operators'.

5.7 There is a connection with semidirect products of groups. Let N, H be groups and $\varphi: H \to \operatorname{Aut} N$ a group homomorphism. For $n \in N$, $h \in H$ the image of n under $\varphi(h)$ will be written as n^h. The corresponding **semidirect product** G is $H \times N$ with the multiplication $(f, n)(h, m) = (fh, n^h m)$. In fact, G being a semidirect product of N by H is equivalent to there being a split short exact sequence

$$1 \to N \to G \to H \to 1.$$

Now φ can be extended to $H \to \operatorname{Aut} RN$, this being the group of ring automorphisms of RN, by letting H act as the identity on R. This way the ordinary group ring RG can be identified with $RN\#H$. The next more general definition will handle nonsplit sequences.

5.8 Let R be a ring and G a group. Let S be a ring containing R and containing a set of units $\bar{G} = \{\bar{g} | g \in G\}$ isomorphic *as a set* to G such that
(i) S is free as a right R-module with $\bar{G}$ as a basis and $\bar{1}_G = 1_S$; and
(ii) for all $g_1, g_2 \in G$, $\bar{g}_1 R = R \bar{g}_1$ and $\bar{g}_1 \bar{g}_2 R = \overline{g_1 g_2} R$.
Then S is called a **crossed product** of R by G, written $R * G$.

It follows from (ii) that S is also freely generated as a left R-module.

If in addition $\bar{g}r = r\bar{g}$ for all $r \in R$, $g \in G$, S is sometimes called a **twisted group ring**.

5.9 Corresponding to a subgroup N of G, there is clearly a subring $R * N$ of $R * G$.

Lemma. (i) *Let $X \subseteq G$ be a set of representatives of the cosets of G modulo N. Then $R * G$ is freely generated as an $R * N$-module by $\bar{X}$.*
(ii) *If N is a normal subgroup of G, and $H = G/N$, then $R * G \simeq (R * N) * H$.*

Proof. Straightforward. □

This, of course, extends 5.7 to cover nonsplit extensions. For if $1 \to N \to G \to H \to 1$ is an extension of N by H, then (ii) above shows that $RG \simeq RN * H$. Likewise $R\#G \simeq (R\#N) * H$.

5.10 Example. (i) Let G be the Heisenberg group, as in 5.3(i), and $N = \langle z \rangle$. Then $H = G/N \simeq \mathbb{Z}^2$ and so $kG \simeq k\langle z \rangle * \mathbb{Z}^2$; indeed, one can choose $x^m y^n$ as the representative, in kG, of $(m, n) \in \mathbb{Z}^2$.

Example. (ii) Given a positive integer n, choose $\binom{n}{2}$ nonzero elements λ_{ij} of a field k, where $1 \leqslant i < j \leqslant n$, and then set $\lambda_{ii} = 1, \lambda_{ji} = \lambda_{ij}^{-1}$. Consider the k-algebra $S = k[x_1, x_1^{-1}, \ldots, x_n, x_n^{-1}]$ with relations $x_i x_j = \lambda_{ij} x_j x_i$. Then, as in (i), $S = k * \mathbb{Z}^n$ with $(m_1, \ldots, m_n)$ being represented by $x_1^{m_1} \cdots x_n^{m_n}$. Alternatively, S may be viewed as an n-fold iterated skew Laurent polynomial ring over k,

$$S = k[x_1, x_1^{-1}][x_2, x_2^{-1}; \sigma_2] \cdots [x_n, x_n^{-1}; \sigma_n].$$

5.11 Next we turn to the Noetherian property for such rings. Since both group rings and skew group rings are examples of crossed products, it will be enough to consider crossed products.

Lemma. *Let R be a ring, G a group with a subgroup N of finite index. If $R * N$ is right Noetherian then so too is $R * G$.*

Proof. This is an easy consequence of 1.3 and 5.9(i). □

Proposition. *Let R be a ring, G a group with a normal subgroup N such that G/N is infinite cyclic. Then:*
(i) $R * G \simeq (R * N)[x, x^{-1}; \sigma]$ *for some automorphism σ of $R * N$; and*
(ii) *if $R * N$ is right Noetherian then so too is $R * G$.*

Proof. (i) Choose $x \in G$ such that xN generates G/N, and write $\bar{x}$ for its image in $R * G$. By 5.8(ii), $\bar{x}^n R = \overline{x^n} R$ for each n. It follows from 5.9(i) that the set $\{\bar{x}^n | n \in \mathbb{Z}\}$ generates $R * G$ freely as an $R * N$-module. Since $xN = Nx$, it follows from 5.8(ii) that $\bar{x}(R * N) = (R * N)\bar{x}$. Let σ be the automorphism of $R * N$ given by $\sigma(y) = (\bar{x})^{-1} y \bar{x}$. Then $R * G \simeq (R * N)[\bar{x}, (\bar{x})^{-1}; \sigma]$.
(ii) This follows directly from (i) and 4.5(iv). □

5.12 Recall that a **poly-(cyclic or finite)** group is a group G with a finite chain

$$1 = G_0 \triangleleft G_1 \triangleleft \cdots G_n = G$$

where G_i is normal in G_{i+1}, and each factor G_{i+1}/G_i is either infinite cyclic or finite. It is easy to show that, by choosing a different chain, one can arrange that the only finite factor is the last, G_n/G_{n-1}. This explains the more usual name, **polycyclic by finite** group, which will be used henceforth.

An easy induction argument, using Lemma 5.11 and Proposition 5.11(ii), gives the next result.

Theorem. *If R is right Noetherian and G is a polycyclic by finite group, then $R*G$ is right Noetherian.* □

It is still an open question as to when a group ring is Noetherian and, in particular, as to whether $\mathbb{Z}G$ Noetherian implies that G is polycyclic by finite.

§6 Skew Polynomial Rings in Several Variables

This section concerns various rings which, like the Weyl algebra $A_n(k)$, resemble polynomial rings in several variables. The techniques used will involve 'filtered and graded' arguments; so first these terms are introduced.

6.1 A **filtered ring** is a ring S with a family $\{F_n, n = 0, 1, 2, \ldots\}$ of subgroups of S^+ such that
(i) for each i, j, $F_i F_j \subseteq F_{i+j}$,
(ii) for $i < j$, $F_i \subseteq F_j$, and
(iii) $\bigcup F_n = S$.
The family $\{F_n\}$ is called a **filtration** of S.

For example, viewing $A_1(k)$ as $k[x][y; d/dx]$, let F_n be the $k[x]$-submodule of $A_1(k)$ generated by $\{1, y, \ldots, y^n\}$.

6.2 More generally, suppose that S is a ring with a subring R, and that S is generated, as a ring, by R together with elements $\{x_i | i \in I\}$. Here S is called a **ring extension** of R, and a **finitely generated ring extension** if I is finite. An element of S of the form

$$r_0 x_{i_1} r_1 x_{i_2} \cdots x_{i_n} r_n,$$

with $r_i \in R$, is called a **word** of length n. Define F_j to be the additive subgroup of S generated by all words of length j or less, with $F_0 = R$. This gives the **standard filtration** of S with the respect to those generators.

One could view $A_1(k)$ as being generated by the subring k, together with x, y. The standard filtration this time has F_j being the k subspace generated by all $x^m y^n$ with $m + n \leqslant j$.

6.3 A **graded ring** is a ring T together with a family $\{T_n, n = 0, 1, 2, \ldots\}$ of subgroups of T^+ such that
(i) $T_i T_j \subseteq T_{i+j}$, and

(ii) $T = \oplus_n T_n$, as an abelian group.

The family $\{T_n\}$ is called a **grading** of T; and a nonzero element of T_n is said to be **homogeneous** of degree n. For homogeneous elements it follows that $\deg ab = \deg a + \deg b$ provided $ab \neq 0$.

For example, $T_n = x^n R$ gives a grading of $R[x; \sigma]$.

6.4 Any graded ring T has a natural filtration $\{F_n\}$ with $F_n = T_0 \oplus \cdots \oplus T_n$. On the other hand, from any filtered ring S one can construct a graded ring T. We set $T_n = F_n/F_{n-1}$ and $T = \oplus T_n$. To define multiplication in T it is enough to consider homogeneous elements. If $a \in F_n \backslash F_{n-1}$, then a is said to have **degree** n and $\bar{a} = a + F_{n-1} \in T_n$ is the **leading term** of a. Suppose c has degree m. Then $\bar{a}\bar{c}$ is defined to be $ac + F_{m+n-1} \in T_{m+n}$. (Note that if $ac \in F_{m+n-1}$ then $\bar{a}\bar{c} = 0$, otherwise $\bar{a}\bar{c} = \overline{ac}$.) This well-defined multiplication makes T into a ring called gr S, the **associated graded ring** of S.

For example, if a graded ring T is given the natural filtration mentioned above, then gr $T \simeq T$. As another example, if $S = A_1(k)$ is filtered in either of the ways described above then $\bar{x}\bar{y} = \bar{y}\bar{x}$. Indeed gr $S = k[\bar{x}, \bar{y}]$, the commutative polynomial ring in two variables, but for the first filtration (gr $S)_n = k[\bar{x}]\bar{y}^n$ while, for the second, gr$(S)_n$ is the space of homogeneous polynomials, in $\bar{x}$ and $\bar{y}$, of degree n.

6.5 It will be useful to note the effect of this process when S is an extension of a subring R, with generators $\{x_i | i \in I\}$ and with the standard filtration $\{F_n\}$ described in 6.2. It is convenient, and causes no loss of generality, to insist that no $x_i \in R$.

Lemma. gr S is a ring extension of R with generators $\{\bar{x}_i | i \in I\}$; and each $\bar{x}_i$ is homogeneous of degree 1.

Proof. Let $0 \neq z \in (\text{gr } S)_j$; so $z = \bar{y}$ for some $y \in F_j \backslash F_{j-1}$ with y being a finite sum of words $x_0 x_{i_1} r_1 \cdots x_{i_j} r_j$, each belonging to $F_j \backslash F_{j-1}$. The leading term of such a word is $\overline{r_0 x_{i_1} r_1 \cdots x_{i_j} r_j}$; i.e. $\bar{r}_0 \bar{x}_{i_1} \bar{r}_1 \cdots \bar{x}_{i_j} \bar{r}_j$. Thus $z = \bar{y}$ is a finite sum of such words and so gr S is generated, over R, by the $\bar{x}_i$. □

6.6 Naturally there is some connection between the properties of a filtered ring S and its associated graded ring gr S.

Proposition. (i) *If* gr S *is an integral domain then* S *is an integral domain.*
(ii) *If* gr S *is prime, then* S *is prime.*

Proof. (i) Let $a, c \in S$ be nonzero elements of degrees m, n respectively. Then $\bar{a}, \bar{c}$ are nonzero elements of gr S and so $\bar{a}\bar{c} \neq 0$. By definition, this means that $ac \notin F_{m+n-1}$ and so $ac \neq 0$.

1.6.9   Skew polynomial rings in several variables   27

(ii) If $a, c \in S$ are nonzero then $\bar{a} \operatorname{gr} S \ \bar{c} \neq 0$. Thus there is a homogeneous element $\bar{s}$ of gr S with $\bar{a}\bar{s}\bar{c} \neq 0$. If s is a representative of $\bar{s}$ in S, then $asc \neq 0$; and so $aSc \neq 0$. □

The converses are not true. For example if $S = k[x]$ and F_n is the set of all polynomials of degree at most $2n$, then $\bar{x}\bar{x} = 0$ in gr S. So gr S is not an integral domain (nor prime, of course).

6.7 Next we compare right ideals of a filtered ring S and its associated graded ring, with the Noetherian property in mind.

First consider a graded ring $T = \bigoplus_n T_n$. If $t \in T$, then $t = t_0 + t_1 + \cdots + t_m$ for some $t_i \in T_i$. The element t_i is called the ith **homogeneous component** of t; and a right ideal A of T is called a **graded right ideal** if all the homogeneous components of each element of A also belong to A.

Next suppose that S is a filtered ring and take $T = \operatorname{gr} S$. Associated with any right ideal I of S, there is a graded right ideal gr I of gr S which is defined by setting $(\operatorname{gr} I)_n = (I + F_{n-1}) \cap F_n / F_{n-1} \subset F_n / F_{n-1}$, and

$$\operatorname{gr} I = \bigoplus_n (\operatorname{gr} I)_n.$$

(Note that $(I + F_{n-1}) \cap F_n / F_{n-1} \simeq I \cap F_n / I \cap F_{n-1}$.) If $\bar{a} \in (\operatorname{gr} I)_n$ and $\bar{r} \in (\operatorname{gr} S)_m$ then

$$\bar{a}\bar{r} = \overline{ar} + F_{m+n-1} \in (I + F_{m+n-1}) \cap F_{m+n}/F_{m+n-1}.$$

This shows that gr I is indeed a right ideal of gr S.

Proposition. *The map $I \mapsto \operatorname{gr} I$ is a partially ordered set map $\mathscr{L}(S_S) \to \mathscr{L}(\operatorname{gr} S_{\operatorname{gr} S})$ which is injective on chains.*

Proof. It is enough to show that if $I \subseteq J$ are right ideals of S with $I \subseteq J$ and gr $I = \operatorname{gr} J$ then $I = J$; and this is done as in the proof of 2.9(iv). □

6.8 As a simple example, take $S = R[x]$, the polynomial ring over R in a central indeterminate. The standard filtration gives $\operatorname{gr} S = R[\bar{x}] \simeq S$. Let I be a right ideal of S and $I_0, I_1, \ldots$ its leading right ideals, as in the proof of 2.9(iv). Then $\operatorname{gr} I = I_0 \oplus I_1 \bar{x} \oplus I_2 \bar{x}^2 \oplus \cdots$.

6.9. Theorem. *If S is a filtered ring and gr S is right Noetherian then S is right Noetherian.*

Proof. Since gr S has no infinite strictly ascending chain of right ideals neither, by 6.7, has S. □

As with 6.6, the converse is false. For example if $S = k[x]$ and $F_j = k[x]$ for $j > 0$, $F_0 = k$, then gr S is not Noetherian.

6.10 Having established the 'filtered and graded' method, we now aim to apply it to rings 'resembling skew polynomial rings'. More specifically, suppose that S is a finitely generated extension of a subring R, with generators $x_1,\ldots,x_n$ which satisfy the conditions, for all i,j,
(i) $Rx_i + R = x_i R + R$, and
(ii) $x_i x_j - x_j x_i \in \sum_{f=1}^{n} x_f R + R$.
Then S is called an **almost normalizing extension** of R. (As before, we will suppose $x_i \notin R$). This name will appear more natural after encountering almost commutative algebras in Chapter 8, §4 and normalizing extensions in Chapter 10.

6.11 A number of examples considered earlier can be viewed this way.
(i) $S = R[x; \sigma, \delta]$, where σ is an automorphism. In this case take $x_1 = x$.
(ii) $S = R[x, x^{-1}; \sigma]$, with $x_1 = x$, $x_2 = x^{-1}$.
(iii) $S = A_n(k)$, the Weyl algebra, with $R = k$ and with generators $\{x_1,\ldots,x_n, y_1,\ldots,y_n\}$.
(iv) $S = A_n(k)$ again, with $R = k[x_1,\ldots,x_n]$ and with generators $\{y_1,\ldots,y_n\}$.
(v) As a new example, let S be the k-algebra $k[x, y, z]$ with relations $xy - yx = z$, $yz - zy = x$, $zx - xz = y$.
(This is the universal enveloping algebra of a three-dimensional simple Lie algebra. Other such algebras will be described in §7). Here we take $R = k$ and generators $\{x, y, z\}$.

6.12 For almost normalizing extensions the standard filtration described in 6.2 gives associated graded rings of a rather special form.

Proposition. *If S is an almost normalizing extension of R then* $\operatorname{gr} S = R[\bar{x}_1,\ldots,\bar{x}_n]$ *with*
(i) $R\bar{x}_i = \bar{x}_i R$, *and*
(ii) $\bar{x}_i \bar{x}_j = \bar{x}_j \bar{x}_i$.

Proof. The form of $\operatorname{gr} S$ is established in 6.5. If $r \in R$ and $i \in \{1,\ldots,n\}$ then, by definition, $rx_i = x_i r' + r''$. Note that $r \in F_0$ and $x_i \in F_1$. So
$$\bar{r}\bar{x}_i = rx_i + F_0 = x_i r' + r'' + F_0$$
$$= x_i r' + F_0 = \bar{x}_i \bar{r}'.$$
Hence $R\bar{x}_i = \bar{x}_i R$; and (ii) is proved similarly. □

6.13 To illustrate this, consider again the rings described in 6.11, with standard filtrations.
(i) When $S = R[x; \sigma, \delta]$ then $\operatorname{gr} S = R[\bar{x}; \sigma]$.
(ii) When $S = R[x, x^{-1}; \sigma]$, then $\operatorname{gr} S = R[\bar{x}_1, \bar{x}_2]$ with relations $r\bar{x}_1 = \bar{x}_1 \sigma(r)$, $r\bar{x}_2 = \bar{x}_2 \sigma^{-1}(r)$, and $\bar{x}_1 \bar{x}_2 = \bar{x}_2 \bar{x}_1 = 0$. (The last equation indicates that, in general, care is required in passing from S to $\operatorname{gr} S$).

(iii) Here $S = A_n(k)$ and gr $S = k[\bar{x}_1,\ldots,\bar{x}_n,\bar{y}_1,\ldots,\bar{y}_n]$, the commutative polynomial ring in $2n$ variables, graded by total degree in the generators.
(iv) Here $S = A_n(k)$ again, and gr $S = k[x_1,\ldots,x_n][\bar{y}_1,\ldots,\bar{y}_n]$ the commutative polynomial ring in $2n$ variables, graded by total degree of the $\bar{y}_i$'s.
(v) Here gr S is the commutative polynomial ring in the three indeterminates $\bar{x}, \bar{y}, \bar{z}$, as will be shown in Corollary 7.5.

6.14 Theorem. *Let S be an almost normalizing extension of a right Noetherian ring R with the standard filtration of 6.2. Then both S and gr S are right Noetherian rings.*

Proof. Here 6.9 shows that it is enough to prove that gr S is right Noetherian. The description of gr S in 6.12 shows that gr S has a chain of subrings

$$R \subseteq R[\bar{x}_1] \subseteq R[\bar{x}_1, \bar{x}_2] \subseteq \cdots \subseteq \text{gr } S;$$

and then 2.10 can be applied at each step. □

§7 Enveloping Algebras and Their Generalizations

In this section enveloping algebras of Lie algebras are described, discussed and generalized in a way parallel to that for group rings in §5.

7.1 A Lie algebra g over a field k, or k-**Lie algebra** for short, is a k-vector space equipped with a **Lie product**, i.e. a k-bilinear map $g \times g \to g$, $(x, y) \mapsto [x, y]$ such that $[x, y] = -[y, x]$, $[x, x] = 0$, and satisfying the **Jacobi identity**

$$[x, [y, z]] + [y, [z, x]] + [z, [x, y]] = 0.$$

Note that if A is an associative k-algebra one can define a k-Lie algebra structure on A by setting $[a, b] = ab - ba$, this product being called the **bracket product**. A **representation** of a k-Lie algebra g is defined to be a Lie algebra homomorphism from g to any such A. In fact each finite dimensional k-Lie algebra can be represented as a subalgebra of $M_n(k)$ for some n (Ado and Iwasawa's theorem; see [Jacobson **62**]).

7.2 If g is a k-Lie algebra then the **universal enveloping algebra** of g is an (associative) k-algebra $U = U(g)$ together with a representation $\theta: g \to U$ which is universal; i.e. given any (associative) k-algebra A and representation $\varphi: g \to A$, there exists a unique algebra homomorphism $\psi: U \to A$ such that $\psi\theta = \varphi$. As usual with universal objects, $U(g)$ is uniquely determined up to isomorphism.

7.3 If $\{x_i | i \in I\}$ is a k-basis for g, then U may be described as the associative k-algebra generated by elements $\{x_i | i \in I\}$ with the relations $x_i x_j - x_j x_i = [x_i, x_j]$.

In other words, $U(g)$ is the factor of the free k-algebra F on the set $\{x_i\}$ by the ideal generated by the elements $x_i x_j - x_j x_i - [x_i, x_j]$.

For example, the ring described in 6.11(v) is $U(g)$ with g being the three-dimensional Lie algebra with basis $\{x, y, z\}$ and products

$$[x, y] = z, \qquad [y, z] = x, \qquad [z, x] = y.$$

7.4 The description of $U(g)$ in 7.3 makes $U(g)$ an extension of the subring k, with generators $\{x_i | i \in I\}$. The standard filtration described in 6.2 leads to an associated graded ring $\mathrm{gr}(U(g))$, as in 6.5.

Proposition. $\mathrm{gr}\, U(g)$ is a commutative k-algebra, generated over k by the set $\{\bar{x}_i | i \in I\}$.

Proof. The relation $x_i x_j - x_j x_i = [x_i, x_j]$ shows that

$$\bar{x}_i \bar{x}_j - \bar{x}_j \bar{x}_i = [x_i, x_j] + F_1$$

in F_2/F_1. But $[x_i, x_j] \in g \subset F_1$ and so $\bar{x}_i \bar{x}_j = \bar{x}_j \bar{x}_i$. The rest is clear from 6.5. □

Corollary. *If g is finite dimensional then $U(g)$ is right and left Noetherian.*

Proof. The proposition shows that $\mathrm{gr}\, U$ is Noetherian, and so 6.9 applies. □

It is in fact the case that $\mathrm{gr}\, U(g)$ is a polynomial ring, with the $\bar{x}_i$ as indeterminates. This depends upon the ideas discussed next.

7.5 Suppose that $\{x_i | i \in I\}$ is a k-basis for g and let I be totally ordered by $\leq$. Now if $i, j \in I$ with $i > j$, one can write

$$x_i x_j = x_j x_i + [x_i, x_j]$$

and here, of course, $j < i$. This, together with induction, makes it clear that $U(g)$ is generated, as a k-module, by the **standard monomials** $\{x_{i_1} x_{i_2} \cdots x_{i_n} | i_t \in I, i_1 \leq i_2 \leq \cdots \leq i_n\}$. It does not, however, make it clear that the standard monomials are k-independent. This is the content of the next result.

Theorem. (Poincaré–Birkhoff–Witt theorem) *The standard monomials are a k-basis for $U(g)$.*

The (nontrivial) proof is omitted–see [Dixmier 77] or [Humphreys 80] for details. □

It has the following interesting consequences.

Corollary. (i) *The canonical mapping $g \to U(g)$ is an embedding.*
(ii) $\mathrm{gr}\, U(g)$ *is a polynomial ring over k in the commuting indeterminates $\{\bar{x}_i | i \in I\}$.*

1.7.8   Enveloping algebras and their generalizations

(iii) $U(g)$ is an integral domain.
(iv) If $g \neq 0$, then $U(g)$ is not a simple ring.

Proof. (i) (ii). These are obvious from the theorem.
(iii) This follows from (ii), using 6.6.
(iv) Here g has a trivial representation in k via the zero map. This gives a k-algebra homomorphism $U(g) \to k$ whose kernel is the ideal generated by the $\{x_i\}$. □

7.6 For some Lie algebras the truth of Theorem 7.5 is evident. For example, if g is an abelian Lie algebra (meaning that $[x,y] = 0$ for all $x, y \in g$) then the relations in $U(g)$ simply state that its generators commute. Thus $U(g)$ is a commutative polynomial ring with $\{x_i\}$ as generators, and $U(g) \simeq \operatorname{gr} U(g)$.

As another example, let g be the two-dimensional Lie algebra with basis $\{x, y\}$ and product $[x, y] = y$. Then $U(g) = k[x, y]$ subject to $xy - yx = y$; and this is simply the skew polynomial ring $k[y][x; -y\, d/dy]$. Alternatively, the relation $xy - yx = y$ can be written as $xy = y(x+1)$; and then $U(g)$ becomes identified with $k[x][y; \sigma]$, where σ is the k-automorphism of $k[x]$ defined by $x \mapsto x + 1$.

In fact, one can view Theorem 7.5 as showing for any g that $U(g)$ is a form of noncommutative polynomial ring over k. But note that it cannot always be built up by adjoining generators one at a time.

7.7 We now turn to generalizations of enveloping algebras analogous to the generalizations of group algebras described in §5. This introduces new examples and, perhaps more important, gives new ways of looking at old examples. In particular we will see further descriptions of the Weyl algebra $A_n(k)$—which by Corollary 7.5(iv) is not an enveloping algebra. It is, however, a homomorphic image of the universal enveloping algebra of the $2n+1$ dimensional Heisenberg Lie algebra g. This has generators $x_1, \ldots, x_n, y_1, \ldots, y_n, z$; and products $[x_i, y_i] = z$ for $i = 1, \ldots, n$, all other products being zero. Evidently $A_n(k) \simeq U(g)/(z-1)$.

7.8 First consider extending the scalars involved. It is possible to proceed in the same way as in 7.4 and 7.5 whenever k is a commutative ring and g is a k-**Lie algebra**, by which is meant a free k-module (with basis $\{x_i | i \in I\}$) with a Lie product as in 7.1. Theorem 7.5 is again valid, and so too is its corollary provided, for (iii), that k is an integral domain. Furthermore, if k is Noetherian and I is finite, then $U(g)$ is right and left Noetherian.

For example, one could form the $\mathbb{Z}$-algebra on $2n + 1$ generators satisfying the relations described in 7.7. The enveloping algebra obtained is then a Noetherian integral domain.

The reader might wonder why g is restricted above to be a free k-module. If g were not free then $U(g)$ could still be constructed as a homomorphic image of the tensor algebra of the k-module g; but there is no analogue of the

Poincaré–Birkhoff–Witt (PBW) theorem. Indeed, the canonical map $g \to U(g)$ need not even be injective (see [Cohn **63**] for example).

7.9 More generally if k is a commutative ring, g, as in 7.8, is a k-Lie algebra and R is any noncommutative k-algebra containing k, the ring $R \otimes_k U(g)$ exists, and can be viewed as being obtained by extending scalars.

7.10 A further generalization allow an action of g on the coefficient k-algebra R. Whereas, in §5, elements of G acted as automorphisms, here elements of g act as derivations. By a k-**derivation** of R is meant a k-module homomorphism $d: R \to R$ which is also a derivation (i.e. $d(ab) = d(a)b + ad(b)$). The set of all k-derivations of R is denoted by $\text{Der}_k R$. It is a k-module and a Lie algebra over k under the bracket product.

The action to be considered here is a k-Lie algebra homomorphism $\varphi: g \to \text{Der}_k R$. If $x \in g$ and $r \in R$, let $x(r)$ denote $\varphi(x)(r)$. The **skew enveloping algebra** $R \# U(g)$ is now defined to be the k-algebra generated by R and $\{x_i | i \in I\}$ with the relations, for all $i, j \in I$,
(i) $x_i r - r x_i = x_i(r)$, and
(ii) $x_i x_j - x_j x_i = [x_i, x_j]$.
As for $U(g)$ it can be shown (see 9.9) that $R \# U(g)$ is a free right (and left) R-module with the standard monomials as a basis. Also one can describe $R \# U(g)$ by means of a universal property. Note that if $\varphi(x_i) = 0$ for all $i \in I$, then $R \# U(g) \simeq R \otimes U(g)$, as in 7.9.

7.11 Examples. (i) Let $R = k[y_1, \ldots, y_n]$, the commutative polynomial ring over k, and let g be the abelian k-Lie algebra with basis $x_1, \ldots, x_n$. Let $\varphi: g \to \text{Der}_k R$ be defined by $x_i \mapsto \partial/\partial y_i$. Then $R \# U(g) \simeq A_n(k)$.

(ii) If k is taken to be a field, $R = k(y_1, \ldots, y_n)$, the rational function field, and g, φ are as in (i) then $R \# U(g) \simeq B_n(k)$, as defined in 3.9.

(iii) Let R be a k-algebra, with $k \subseteq R$, $\delta \in \text{Der}_k R$ and g the abelian k-Lie algebra with basis $\{x\}$. Suppose $\varphi: g \to \text{Der}_k R$ maps x to δ. Then $R \# U(g) \simeq R[x; -\delta]$.

(iv) Suppose that k is a field, and g is a k-Lie algebra which is a semidirect product of an ideal n and a subalgebra h; that is, there is an exact sequence of Lie algebras
$$0 \to n \to g \to h \to 0$$
which splits. Then $U(g) \simeq U(n) \# U(h)$.

7.12 Finally, we describe a crossed product, which will, in particular, deal with nonsplit exact sequences of Lie algebras.

Suppose that R is a k-algebra, with $k \subseteq R$, and that g is a k-Lie algebra with basis $\{x_i | i \in I\}$. A k-algebra S containing R is called a **crossed product** of R by $U(g)$ (and written $R * U(g)$) provided there is a k-module embedding of g into S, $x \mapsto \bar{x}$, such that

(i) $\bar{x}r - r\bar{x} \in R$ and $r \mapsto \bar{x}r - r\bar{x}$ is a k-derivation of R,
(ii) $\bar{x}\bar{y} - \bar{y}\bar{x} \in \overline{[x,y]} + R$ for all $x, y \in g$, and
(iii) S is a free right (and left) R-module with the standard monomials in $\{\bar{x}_i\}$ as a basis.

7.13 Examples. (i) Any $R \otimes U(g)$ as in 7.9 and any $R \# U(g)$ as in 7.10 is a crossed product of R by $U(g)$:
(ii) If g is the abelian Lie algebra of dimension $2n$ over a field k, then

$$A_n(k) \simeq k * U(g)$$

under the obvious embedding of g.
(iii) Given a short exact sequence of Lie algebras

$$0 \to n \to g \to h \to 0$$

then $U(g) \simeq U(n) * U(h)$.

7.14 It is not difficult now to check

Proposition (i) $\mathrm{gr}(R * U(g))$ is a polynomial ring over R in central variables $\{\bar{x}_i | i \in I\}$.
(ii) If R is an integral domain, so is $R * U(g)$.
(iii) If R is prime, then $R * U(g)$ is prime.
(iv) If I is finite, then $R * U(g)$ is an almost normalizing extension of R. If, further, R is right Noetherian then $R * U(g)$ is also right Noetherian. □

7.15 As a final illustration, recall that $A_2(k) \simeq R \# U(g)$, where $R = k[y_1, y_2]$, g is the two-dimensional abelian Lie algebra with basis x_1, x_2 and x_i acts as $\partial/\partial y_i$. If this action of g on R is kept fixed, then every crossed product $R * U(g)$ is isomorphic to $A_2(k)$. For $R * U(g)$ will be $R[\bar{x}_1, \bar{x}_2]$ with relations
(i) $\bar{x}_i r - r\bar{x}_i = \partial r/\partial y_i$, and
(ii) $\bar{x}_1 \bar{x}_2 - \bar{x}_2 \bar{x}_1 = f$, for some $f \in R$.
If $g \in R$ is chosen so that $\partial g/\partial y_1 = f$, then, in $A_2(k)$, $x_1(x_2 + g) - (x_2 + g)x_1 = f$. It follows that the map $R * U(g) \to A_2(k)$ given by $\bar{x}_1 \mapsto x_1$, $\bar{x}_2 \mapsto x_2 + g$ is a ring isomorphism.

§8 Further simple rings

8.1 The fact that Weyl algebras may be viewed as crossed products by enveloping algebras (7.11 (i)) shows that such rings can be simple. This section discusses when crossed products, by enveloping algebras or by groups, are simple, concentrating mainly upon the easiest cases, namely skew polynomials and skew Laurent polynomials.

8.2 First we introduce a little terminology. Given any element a of a ring R there is a derivation of R, namely $r \mapsto ra - ar = [r, a]$; this is called an **inner derivation** and denoted by ad a. For example, if $R = S[x; \delta]$ then ad x is an inner derivation of R which, since it extends δ from S to R, is denoted by δ.

Similarly, given any unit a of R there is an automorphism given by $r \mapsto a^{-1}ra$, and this is called an **inner automorphism**. The inner automorphism of $R = S[x, x^{-1}; \sigma]$ induced by x extends σ from S to R, and is denoted by σ.

An ideal I of a ring R is said to be **stable** under a set of derivations if $\delta(I) \subseteq I$ for each derivation δ in the set; and I is **stable** under a set of automorphisms if $\sigma(I) = I$ for each automorphism.

8.3 Let R be a ring, $S = R * U(g)$, $T = R * G$ for some Lie algebra g and group G. The elements of g induce inner derivations of S which restrict to derivations of R. So it makes sense to describe ideals of R as stable with respect to g. Similarly, elements of G give automorphisms of R and ideals may be stable with respect to G.

Lemma. (i) *If $R * U(g)$ is simple, then R has no proper nonzero ideal which is stable with respect to g.*

(ii) *If $R * G$ is simple, then R has no proper nonzero ideal which is stable with respect to G.*

Proof. (i) If $0 \neq I \triangleleft R$ and I is stable with respect to g then $SI = IS$ is a proper ideal of S.

(ii) Likewise. □

The next two results show that the conditions here are not far from being sufficient in the special cases considered.

8.4 Theorem. *Let R be a $\mathbb{Q}$-algebra and $S = R[x; \delta]$. Then S is a simple ring if and only if R has no proper nonzero δ-stable ideals and δ is not an inner derivation of R.*

Proof. If δ were an inner derivation, say $\delta = \text{ad } a$ for some $a \in R$, then $S = R[x - a; 0]$ which is not simple. This, together with 8.3, shows that the two conditions are necessary.

Conversely, suppose the conditions hold and that $0 \neq I \triangleleft S$. It is readily checked that the leading right ideals I_n of I (as in 2.9) are actually δ-stable ideals of R. Choose the least n with $I_n \neq 0$, and so $I_n = R$ by hypothesis. If $n = 0$ then $I = S$. If $n > 0$ then there is a monic polynomial

$$f = x^n + x^{n-1}r_{n-1} + \cdots + r_0 \in I.$$

For any $r \in R$,
$$fr - rf = x^{n-1}(r_{n-1}r - rr_{n-1} - n\delta(r)) + \cdots.$$
The minimality of n shows that $r_{n-1}r - rr_{n-1} - n\delta(r) = 0$ for each $r \in R$; thus $\delta = \mathrm{ad}(-r_{n-1}/n)$, an inner derivation. This shows that $I = S$ and so S is simple. □

8.5 Theorem. *Let R be a ring and $T = R[x, x^{-1}; \sigma]$. Then T is a simple ring if and only if R has no proper nonzero σ-stable ideals and no power of σ is an inner automorphism of R.*

Proof. If I is a σ-stable ideal of R, then I is also G-stable where G is the group generated by σ. So 8.3 shows that the first condition is necessary for T to be simple.

Next, suppose σ^n is inner, induced by $a \in R$, say; so $\sigma^n(r) = a^{-1}ra$. However, σ^n is also induced by x^n. It follows that a^{-1} and x^n commute with each other and so $a^{-1}x^n$ is central in $R[x^n, x^{-n}; \sigma^n]$. The same is true of $\sigma^i(a^{-1}x^n)$ for each i. It is now easily checked that if $z = a^{-1}x^n$ and
$$w = (1-z)(1-\sigma(z))\cdots(1-\sigma^{n-1}(z))$$
then $\sigma(w) = w$. Hence w is central in T. Now $R[x^n, x^{-n}; \sigma^n] = R[z, z^{-1}; 1]$. Therefore $1 - z$ is neither a unit nor a zero divisor in $R[x^n, x^{-n}; \sigma^n]$ and it follows easily that the same is true of w. However, T is a free module over this ring; so w is neither a unit not a zero divisor in T. Hence wT is a proper nonzero ideal of T. Thus the second condition is necessary for T to be simple.

Conversely, suppose the conditions both hold and let $0 \neq I \triangleleft T$. We define $\{I_n\}$ to be the leading ideals of $I \cap R[x; \sigma]$, and note that then I_n is a σ-stable ideal of R. Suppose n is minimal such that $I_n \neq 0$, and so $I_n = R$. If $n = 0$ then $I = T$ as required. If $n > 0$ then there is a monic polynomial
$$g = a_0 + a_1 x + \cdots + a_{n-1} x^{n-1} + x^n \in I.$$
Note that $a_0 \neq 0$, or else $gx^{-1} = a_1 + \cdots + x^{n-1} \in I$, contradicting the minimality of n. Note also that $g - x^{-1}gx \in I$ yet has degree $n-1$ or less; hence $\sigma(a_i) = a_i$ for each i. This shows that $Ra_i R$ is a σ-stable ideal and, in particular, $Ra_0 R = R$. Thus $\sum p_j a_0 q_j = 1$ for some $p_j, q_j \in R$.

Let $f = \sum p_j g q_j - \sum p_j \sigma^{-n}(q_j)g$. Since $\deg f < n$ and $f \in I$, it follows that $f = 0$. However, the degree 0 coefficient of f is $1 - \sum p_j \sigma^{-n}(q_j) a_0$, hence a_0 has a left inverse in R. Symmetry shows that a_0 is a unit. A similar argument shows that $rg - g\sigma^n(r) = 0$ for each $r \in R$, and the degree 0 coefficient shows that $ra_0 = a_0 \sigma^n(r)$. Thus σ^n is inner, contrary to the hypothesis. □

8.6 We now consider some specific examples, first some involving a derivation.

Example. Let k be a field of characteristic zero, and
$$R = k[y_1, y_1^{-1}, y_2, y_2^{-1}, \ldots, y_n, y_n^{-1}],$$
the commutative Laurent polynomial ring. Choose $\lambda_1, \ldots, \lambda_n \in k$ to be linearly independent over $\mathbb{Q}$, and define δ by $\delta(y_i) = \lambda_i y_i$; so $y_i x - x y_i = \lambda_i y_i$. Then the ring $S = R[x; \delta]$ is a simple Noetherian integral domain.

Proof. Here 2.9 shows all except the simplicity of S. Note that if $a = y_1^{m_1} \cdots y_n^{m_n}$ then $\delta(a) = (\lambda_1 m_1 + \cdots + \lambda_n m_n) a$. The independence of the λ's shows that these eigenvalues are all distinct, and it follows that any δ-stable k-subspace of R contains a monomial in the y's. Hence there are no proper nonzero δ-stable ideals and so, by 8.4, S is simple. □

When $n = 1$, one can replace x by $\lambda^{-1} x$ getting the ring $S = k[y, y^{-1}][x; \delta]$ with $yx - xy = y$. Then $y(y^{-1}x) - (y^{-1}x)y = 1$, and so $A_1(k) \hookrightarrow S$. Indeed $S \simeq k[x, x^{-1}][y; d/dx]$ this latter ring being denoted by $A_1'(k)$. More generally,
$$A_n'(k) = k[x_1, x_1^{-1}, \ldots, x_n, x_n^{-1}][y_1; \partial/\partial x_1] \cdots [y_n; \partial/\partial x_n].$$
The same arguments as in 3.5 show $A_n'(k)$ to be a simple Noetherian integral domain.

8.7 Finally, we discuss examples involving an automorphism.

Example. (i) The ring $k[y][x, x^{-1}; \sigma]$ with $\sigma(y) = y + 1$, where char $k = 0$, is isomorphic to A_1', as is easily checked. So it is simple. □

Example. (ii) For any field k, let $R = k[y_1, y_1^{-1}, \ldots, y_n, y_n^{-1}]$ and choose $\lambda_1, \ldots, \lambda_n \in k \backslash \{0\}$ in such a way that the subgroup of $k \backslash \{0\}$ generated by them has rank n. Define σ by $\sigma(y_i) = \lambda_i y_i$ and let $T = R[x, x^{-1}; \sigma]$. Then T is a simple Noetherian integral domain.

Proof. By 4.5, it is enough to check that T is simple. The independence of the λ_i shows, as in 8.6, that any nonzero σ-stable ideal contains a monomial, and so equals R. Thus 8.5 applies. □

One possible choice of the λ's, when k has characteristic zero, is any set of distinct primes.

The ring T described above is denoted by $P_{\lambda_1, \ldots, \lambda_n}(k)$. In the case where $n = 1$, the condition on λ is simply that it is not a root of unity. The relation in the ring $P_\lambda(k)$ is then $yx = \lambda xy$, which, as for Weyl algebras, is well known in quantum mechanics. The analogy with Weyl algebras will become stronger in later chapters.

§9 Additional Remarks

9.0 (= General comments on chapter) Since most of this chapter is rather basic and well known, it is difficult to give many precise attributions. The major classes of examples described in this chapter will be studied more deeply later in the book. The index can be used to discover the results pertaining to each such class.

9.1 (= Notes on §1) (a) The idea of Morita contexts arises from that of Morita equivalence; see Chapter 3, §§5, 6. The ring of a Morita context comes from [Sands **73**].

(b) The triangular examples in 1.9, 1.10 appear in [Small **65**]. Similar examples will recur later.

(c) Idealizers and eigenrings were introduced in [Ore **32**] with the isomorphism $\mathbb{I}(A)/A \simeq \mathrm{End}(R/A)$ being noted in [Fitting **35**]; and 1.12 comes from [Robson **72**]. Idealizers play an important part in Chapter 5, §§5, 6.

9.2 (a) Rings $R[x; \sigma, \delta]$ are often called **Ore extensions** of R. Their systematic study began with [Ore **33**] although Hilbert worked earlier with rings related to §4.

(b) The Hilbert basis theorem, 2.9(iv), is, naturally, due to Hilbert. The version in 2.10 comes from [McConnell **68**].

(c) Example 2.11(ii) comes from [Goldie **60**] where it is credited to G. Higman.

9.3 (a) The first study of Weyl algebras is in [Dirac **26**] with an account in [Weyl **28**]. Their simplicity was noted in [Littlewood **33**].

(b) The representation of $A_n(k)$ in 3.6 as differential operators acting on $k[x_1, \ldots, x_n]$ from the right appears unnatural. The more normal action from the left can also be used, provided care is taken with $\pm$ signs. In fact, in Chapter 15, where differential operators are dealt with more fully, we will prefer to use the usual left action. It is a fact that there is no 'best' side when studying noncommutative rings; indeed often both sides are used at once.

(c) 3.10 comes from [Hart and Robson **70**].

9.4 This type of ring was used in [Hilbert **03**].

9.5 (a) Skew group rings and crossed products of finite groups were used by Noether and others in considering invariant subrings. This is described in Chapter 10, §5.

(b) 5.12 comes from [Hall **54**]. We note, in connection with the open question in 5.12, that $\mathbb{Z}G$ being Noetherian implies that G has the a.c.c. for subgroups; but [Ol'šanskiĭ **79**] and E. Rips (unpublished) have given examples of groups with a.c.c. on subgroups which are not polycyclic by finite.

(c) There is an extensive literature on polycyclic by finite group rings for which [Passman **77**] is a major source. See also [Passman **84**]. This book does not aim to cover this topic.

9.6 (a) Filtered and graded techniques are clearly related to the proof of the Hilbert basis theorem. They will be used frequently in later chapters. Graded rings are discussed in [Nastasescu and Van Oystaeyen **82**].

(b) Almost normalizing extensions were introduced in [McConnell **82**]. This notion is an important ingredient of Chapter 9.

9.7 (a) Enveloping algebras are studied in Chapter 14.

(b) Skew enveloping algebras appear in [Sweedler **69**] under the name **smash products**.

(c) There are two points in §7 about which further comment is required:
(i) The proof that $R\#U(g)$ 'satisfies the PBW theorem' (i.e. has the standard monomials as a basis as in 7.5) was omitted from 7.10.
(ii) The problem of construction of crossed products $R*U(g)$ was not addressed in 7.12. These omissions are repaired in Appendix 9.9.

9.8 (a) This section is a 'tidying up' of well-known results.

(b) Example 8.6, which will reappear in Chapter 14, is due to McConnell; see [Hart 71]. Examples like 8.7(ii) are studied in [V. A. Jategaonkar 84], [McConnell and Pettit P].

(c) In the literature, there are other similar simple k-algebras with k a field of characteristic 0.

(i) If $R = k[x_1, \ldots, x_n]$ is the commutative polynomial ring and

$$\delta = \partial/\partial x_1 + \sum_{i=2}^{n} (1 + x_i x_{i-1})\partial/\partial x_i$$

then $R[x; \delta]$ is simple. (This example appears in [Jordan 81, §1] and is due to J. Archer and G. Bergman.)

(ii) If R is a regular local ring which is the localization at a maximal ideal of an affine algebra over k, then there is a k-derivation δ of R such that $R[x; \delta]$ is simple; see [Hart 75].

(iii) Simple crossed products with Lie algebras are considered in [McConnell and Sweedler 71].

(iv) Further examples can be found in [Cozzens and Faith 75, 5.21ff] and in 12.7.16.

9.9 *Appendix* We address here the points raised in 9.7(c). We make use of some terminology and facts from Chapters 14 and 15.

Let k be a commutative ring, g a k-Lie algebra and R a k-algebra. Note, by our convention, that g is free as a k-module; but R need not be. Nevertheless we wish to relax that convention in order to view R, with the product [,], as a k-Lie algebra. A **crossed product** of R by g is a k-Lie algebra h (again not necessarily free over k) together with a short exact sequence of Lie algebras

$$0 \to R \to h \to g \to 0$$

such that, for each $y \in h$, the map $R \to R$ given by $r \mapsto [r, y]$ is a k-algebra derivation; such an h is denoted by $R*g$.

For example, if $S = R*U(g)$ is a crossed product of R by $U(g)$ as in 7.12, then it is easily verified that the k-submodule $h = R1 \oplus \bar{g}$ is a crossed product of R by g. There is a converse:

Theorem. *Given a crossed product $h = R*g$ there is a crossed product $R*U(g)$ such that $R1 \oplus \bar{g}$ coincides with h, as above.*

Proof. (Sketch) In the cited proofs of 7.5 an appropriate representation of g on the symmetric algebra $S(g)$ is constructed. This proof can be modified to give a corresponding representation of the Lie algebra $R*g$ on $R \otimes_k S(g)$, and the result follows. □

This answers the question raised in 9.7(c)(ii) and also deals with (i). For, by the definition, $R*U(g)$ satisfies the PBW theorem. So if one notes that, when $g \to \text{Der}_k R$, there is a semidirect product $R\#g$ then this theorem implies that the corresponding algebra $R\#U(g)$ satisfies the PBW theorem.

1.9.9 Additional remarks

Finally, we show how one can deduce the theorem above from 7.5 in the special case that R is a free k-module.

Proof. (Special case) Note first that if $\boldsymbol{h} = R*\boldsymbol{g}$ then $\boldsymbol{h}$ is free over k and so 7.5 applies to $U(\boldsymbol{h})$. Also, by 7.13(iii), $U(\boldsymbol{h}) \simeq U(R)*U(\boldsymbol{g})$. The universal property shows that the identity map $\iota: R \to R$ from k-Lie algebra to k-algebra induces a k-algebra homomorphism ϑ

$$\begin{array}{ccc} R & \hookrightarrow & U(R) \\ \iota \downarrow & \swarrow \theta & \\ R & & \end{array}$$

Let $I = \ker \vartheta$; so $R \simeq U(R)/I$. Evidently I is generated by

$$\{1_{U(R)} - 1_R\} \cup \{r_1 \cdot r_2 - r_1 \times r_2 | r_1, r_2 \in R\},$$

where $.$, $\times$ represent multiplication in $U(R)$ and in the k-algebra R respectively. One can check that I is stable under the adjoint action of $\boldsymbol{h}$ and therefore, as in 14.2.4,

$$U(\boldsymbol{h})/IU(\boldsymbol{h}) \simeq (U(R)/I)*U(\boldsymbol{g}).$$

However, $U(R)/I \simeq R$ and the mappings here embed $\boldsymbol{h}$ as claimed. □

Chapter 2

QUOTIENT RINGS AND GOLDIE'S THEOREM

The first major step in a structure theory for noncommutative Noetherian rings is Goldie's theorem; and that is the essential content of this chapter.

Before describing it and the relevant background it seems helpful to consider the commutative case. There, prime ideals are crucial, and localization at a prime ideal P of a commutative ring R is a standard technique. This, of course, involves inverting the elements in $R\backslash P$, and is closely related to the formation of the field of fractions of an integral domain.

For noncommutative rings, as we shall see, much of this technique is not available. Prime ideals retain their importance, but localization at a prime ideal can be impossible. Indeed even the existence of a division ring of fractions of an integral domain is not guaranteed.

Nevertheless, for Noetherian rings, Goldie's theorem shows that the formation of a ring of fractions is always possible, not merely for an integral domain, but for any semiprime ring. Moreover, the ring of fractions is a semisimple Artinian ring. Other results in this and later chapters will show how fundamental this theorem is.

§1 Right Quotient Rings

1.1 This section discusses the general problem of constructing rings of fractions. If R is a commutative ring and $\mathscr{S}$ is a **multiplicatively closed** (**m.c.** for short) subset of R, the formation of a ring of fractions $R_\mathscr{S}$ of R with respect to $\mathscr{S}$ can be viewed as a two-step operation. First an ideal I is factored out to make the image $\bar{\mathscr{S}}$ of $\mathscr{S}$ consist of nonzero divisors of $\bar{R} = R/I$. Then $R_\mathscr{S}$ is constructed as equivalence classes in $\bar{R} \times \bar{\mathscr{S}}$.

For noncommutative rings both of these steps have attendant difficulties and only under appropriate hypotheses can they both be carried out.

1.2 First, a brief discussion of m.c. sets in a noncommutative ring R. Of course, for $x \in R$, the set $\{x^n | n \in \mathbb{N}\}$ is a m.c. set. Likewise if P is a completely prime ideal, i.e. R/P is an integral domain, then $R\backslash P$ is a m.c. set. However, for a general prime ideal, R/P may have zero divisors and then $R\setminus P$ is not multiplicatively closed. There is a natural m.c. set associated with any ideal, but that requires some terminology.

An element $x \in R$ is **right regular** if $xr = 0$ implies $r = 0$ for $r \in R$. Similarly **left regular** is defined; and **regular** means both right and left regular (and hence not a zero divisor). The set, $\mathscr{C}_R(0)$, of all regular elements of R is a m.c. set of some importance. More generally, for $I \triangleleft R$, the sets

$$\mathscr{C}_R(I) = \{s \in R \mid [s+I] \text{ is regular in } R/I\},$$
$$\mathscr{C}'_R(I) = \{s \in R \mid [s+I] \text{ is right regular in } R/I\},$$

and

$$'\mathscr{C}_R(I) = \{s \in R \mid [s+I] \text{ is left regular in } R/I\}$$

are all multiplicatively closed.

1.3 Let $\mathscr{S}$ be a (non-empty) m.c. subset of a ring R, and let

$$\operatorname{ass} \mathscr{S} = \{r \in R \mid rs = 0 \text{ for some } s \in \mathscr{S}\}.$$

Then a **right quotient ring** of R with respect to $\mathscr{S}$ is a ring Q together with a homomorphism $\vartheta: R \to Q$ such that:
(i) for all $s \in \mathscr{S}$, $\vartheta(s)$ is a unit in Q;
(ii) for all $q \in Q$, $q = \vartheta(r)\vartheta(s)^{-1}$ for some $r \in R$, $s \in \mathscr{S}$; and
(iii) $\ker \vartheta = \operatorname{ass} \mathscr{S}$.
If, further, $\operatorname{ass} \mathscr{S} = 0$, one can identify R with its image under ϑ, and then each $q \in Q$ takes the form rs^{-1}.

The definition above coincides, for commutative rings, with the standard definition. Unlike that case, however, the existence of such a Q is not guaranteed, as will be seen shortly.

Other, more general, types of quotient ring will be described in Chapter 10.

1.4 Lemma. *If there exists a right quotient ring Q of R with respect to $\mathscr{S}$ then (Q, ϑ) is universal for homomorphisms $\varphi: R \to R'$ such that $\varphi(\mathscr{S})$ consists of units of R'.*

Proof. Straightforward. $\square$

This universality, together with the symmetry of the universal property, yields

Corollary. (i) *If there exists a right quotient ring Q of R with respect to $\mathscr{S}$ then it is unique up to isomorphism.*
(ii) *If R also has a left quotient ring Q' with respect to $\mathscr{S}$ then $Q \simeq Q'$.* $\square$

The latter statement shows that, when both Q and Q' exist then $\operatorname{ass} \mathscr{S}$ equals its left-hand version, namely $\{r \in R \mid sr = 0 \text{ for some } s \in \mathscr{S}\}$. The uniqueness given by (i) makes reasonable the notation $R_{\mathscr{S}}$ for Q. This ring is also called the **(right) localization** of R at $\mathscr{S}$.

1.5 The existence of a right quotient ring is often straightforward. For example, consider the case $R = A_1(k)$, the first Weyl algebra; and $\mathscr{S} = k[x]\setminus\{0\}$. Then $B_1(k)$, as in 1.3.9, is evidently $R_{\mathscr{S}}$, with ϑ the embedding $A_1 \hookrightarrow B_1$.

Similarly, if $R = A[x;\sigma]$ for some ring A and automorphism σ, and $\mathscr{S} = \{x^n | n = 1, 2, \ldots\}$ then $R_{\mathscr{S}} = A[x, x^{-1}; \sigma]$, as in 1.4.3, and again ϑ is the obvious embedding.

As an example of a different nature, consider the ring

$$R = \begin{bmatrix} \mathbb{Z} & \mathbb{Z} \\ 0 & \mathbb{Z} \end{bmatrix}$$

which, by 1.1.4, is a Noetherian ring. It has a prime ideal

$$P = \begin{bmatrix} 0 & \mathbb{Z} \\ 0 & \mathbb{Z} \end{bmatrix}$$

with $R/P \simeq \mathbb{Z}$. It is not difficult to check that, if $\mathscr{S} = R\setminus P = \mathscr{C}_R(P)$, then ass $\mathscr{S} = P$ and so $R_{\mathscr{S}} \simeq \mathbb{Q}$.

1.6 Nevertheless, in general, the existence of $R_{\mathscr{S}}$ depends upon certain properties of R and $\mathscr{S}$.

A m.c. subset $\mathscr{S}$ of R is said to satisfy the **right Ore condition** if, for each $r \in R$ and $s \in \mathscr{S}$, there exist $r' \in R$, $s' \in \mathscr{S}$ such that $rs' = sr'$.

The necessity of this condition is shown by the next result.

Proposition. *If a right quotient ring $R_{\mathscr{S}}$ exists then $\mathscr{S}$ satisfies the right Ore condition.*

Proof. Consider the element $\vartheta(s)^{-1}\vartheta(r) \in R_{\mathscr{S}}$. By definition

$$\vartheta(s)^{-1}\vartheta(r) = \vartheta(r_1)\vartheta(s_1)^{-1} \quad \text{with} \quad r_1 \in R, s_1 \in \mathscr{S}.$$

Therefore

$$\vartheta(r)\vartheta(s_1) = \vartheta(s)\vartheta(r_1) \quad \text{and so} \quad rs_1 - sr_1 \in \ker \vartheta = \text{ass } \mathscr{S}.$$

Therefore $(rs_1 - sr_1)s_2 = 0$ for some $s_2 \in \mathscr{S}$. Setting $s_1 s_2 = s'$ and $r_1 s_2 = r'$ establishes the result. □

1.7 The examples below show that the right Ore condition is not always satisfied.

Example. (i) Let $R = k\langle x, y \rangle$, the free associative algebra in two indeterminates over a field k. Let $\mathscr{S} = R\setminus\{0\} = \mathscr{C}_R(0)$. It is immediate that the right Ore condition fails, choosing, say, $r = x$ and $s = y$. This also shows, by 1.6, that $R_{\mathscr{S}}$ does not exist.

2.1.9   *Right quotient rings*

Example. (ii) Let $R = k[x, y]$ with $xy = y(x + 1)$ as described in 1.7.6. This is a Noetherian integral domain and $P = xR + yR$ is a prime ideal with $R/P \simeq k$. Let $\mathscr{S} = R\backslash P = \mathscr{C}_R(P)$. The right Ore condition again fails, choosing $r = y$ and $s = x - 1$.

To see this note that $yR = Ry$, and this too is a prime ideal of R, with $R/yR \simeq k[x]$. The equation $ya = (x - 1)b$, for $a, b \in R$, implies that $(x - 1)b \in yR$, and so $b \in yR$, $b = yc$ say. Then $ya = (x - 1)yc = yxc$ and so $a = xc \in P$. Thus $a \notin \mathscr{S}$ and hence $\mathscr{S}$ fails to satisfy the right Ore condition.

This example also shows that the standard technique of commutative ring theory, localization at a prime ideal, is not available here.

1.8 Before considering the construction of $R_{\mathscr{S}}$ it will be convenient to discuss, in the next few paragraphs, certain consequences of the right Ore condition. Note first that it can be viewed as asserting that r, s have a common right multiple of a particular form. This extends to larger numbers of elements.

Lemma. *If $\mathscr{S}$ satisfies the right Ore condition then, given $r_1, \ldots, r_n \in R$ and $s_1, \ldots, s_n \in \mathscr{S}$, there exist $r'_1, \ldots, r'_n \in R$ and $s' \in \mathscr{S}$ such that $r_i s' = s_i r'_i$ for each i.*

Proof. Suppose first that $n = 2$. The right Ore condition provides $r_1^*, r_2^* \in R$, s_1^*, $s_2^* \in \mathscr{S}$ such that $r_1 s_1^* = s_1 r_1^*$ and $r_2 s_2^* = s_2 r_2^*$. Applying the condition to s_1^* and s_2^* gives $t_1 \in \mathscr{S}$, $t_2 \in R$ such that $s_1^* t_1 = s_2^* t_2 = s$, say. Note that, since $\mathscr{S}$ is multiplicatively closed, $s \in \mathscr{S}$. Also

$$r_i s = r_i s_i^* t_i = s_i r_i^* t_i$$

for $i = 1, 2$. Setting $r'_i = r_i^* t_i$ completes the proof when $n = 2$. An easy induction extends this to the general case. $\square$

The connection between this lemma and 'common denominators' will become clear in 1.16.

1.9 Next consider the subset ass $\mathscr{S}$ introduced in 1.3. One of the conditions there is that ass $\mathscr{S} = \ker \vartheta$ and this requires that ass $\mathscr{S}$ be an ideal. Whilst it is apparent from its definition that ass $\mathscr{S}$ is closed under left multiplication by elements of R, it is not obvious that it is an ideal.

Lemma. *If $\mathscr{S}$ satisfies the right Ore condition then ass $\mathscr{S}$ is an ideal of R.*

Proof. Let $a, b \in$ ass $\mathscr{S}$ and $r \in R$. Then $as = bt = 0$ for some $s, t \in \mathscr{S}$. The right Ore condition provides t_1, s_1 such that $ss_1 = tt_1$ and r', s' such that $rs' = sr'$, where $s_1, s' \in \mathscr{S}$. Hence $(a - b)ss_1 = 0$ and $ars' = 0$; and so $a - b \in$ ass $\mathscr{S}$ and $ar \in$ ass $\mathscr{S}$. $\square$

To see that this is not true for all m.c. sets consider, as in 1.5, the ring

$$R = \begin{bmatrix} \mathbb{Z} & \mathbb{Z} \\ 0 & \mathbb{Z} \end{bmatrix},$$

with the prime ideal

$$P' = \begin{bmatrix} \mathbb{Z} & \mathbb{Z} \\ 0 & 0 \end{bmatrix}$$

and m.c. set $\mathcal{S}' = \mathscr{C}(P')$. It is easily verified that

$$\operatorname{ass} \mathcal{S}' = \{0\} \cup \left\{ \begin{bmatrix} a & b \\ 0 & 0 \end{bmatrix} \middle| a \neq 0 \right\}$$

which is not an ideal. Hence $\mathcal{S}'$ does not satisfy the right Ore condition.

This example also demonstrates an asymmetry, since, as in 1.5, one can easily check that R has a left localization at $\mathcal{S}'$ and so $\mathcal{S}'$ satisfies the left Ore condition.

1.10 The next step is to investigate the image $\bar{\mathcal{S}}$ of $\mathcal{S}$ in the ring $\bar{R} = R/\operatorname{ass} \mathcal{S}$, but first some more terminology is needed.

If X is a subset of a ring R then

$$\operatorname{r ann} X = \{r \in R \mid Xr = 0\}.$$

This is a right ideal of R, and is called an **annihilator right ideal** or, more briefly, a **right annihilator**. For example, for a single element x, $\operatorname{r ann} x = 0$ means that x is right regular. Similarly, one defines $\operatorname{l ann} X$, the **left annihilator**.

One should note that, for any m.c. set $\mathcal{S}$, $\operatorname{ass} \mathcal{S} = 0$ means that $\mathcal{S}$ consists of left regular elements.

Proposition. *Let $\mathcal{S}$ be a m.c. set satisfying the right Ore condition.*
(i) *The elements of $\bar{\mathcal{S}}$ are left regular elements of $\bar{R}$ and $\bar{\mathcal{S}}$ satisfies the right Ore condition.*
(ii) *If $\bar{R}$ satisfies the a.c.c. on right annihilators then the elements of $\bar{\mathcal{S}}$ are regular elements of $\bar{R}$.*

Proof. (i) This is immediate from the definitions.
(ii) For ease of notation, assume that $\operatorname{ass} \mathcal{S} = 0$. Suppose that $sx = 0$ with $s \in \mathcal{S}$, $x \in R$. By hypothesis

$$\operatorname{r ann} s^k = \operatorname{r ann} s^{k+1}$$

for some $k \in \mathbb{N}$. The right Ore condition provides $y \in R$ and $t \in \mathcal{S}$ with $xt = s^k y$. Now $s^{k+1} y = sxt = 0$ and so

$$y \in \operatorname{r ann} s^{k+1} = \operatorname{r ann} s^k.$$

2.1.12   *Right quotient rings*   45

Hence $xt = s^k y = 0$ and therefore $x \in \text{ass}\,\mathscr{S} = 0$. □

The following related fact will be useful later.

Lemma. *If an ideal B is a right annihilator then $'\mathscr{C}_R(0) \subseteq {}'\mathscr{C}_R(B)$.*

Proof. Suppose $c \in {}'\mathscr{C}_R(0)$ and $xc \in B = \text{r ann}\,A$. Then $Axc = 0$; hence $Ax = 0$ and $x \in B$. □

1.11 Without some extra hypothesis, like R being commutative, or that imposed in (ii) above, one cannot deduce that $\bar{\mathscr{S}}$ consists of regular elements.

Example. Let R be any ring with elements a, b such that $ab = 1$, $ba \neq 1$. (Row-finite infinite matrices over a field provide such an example, where a has 1's on the superdiagonal, and b has 1's on the subdiagonal, with zeros elsewhere.) Let $\mathscr{S} = \{a, a^2, a^3, \ldots\}$. Note that if $xa^n = 0$ then $xa^n b^n = x = 0$ and so ass $\mathscr{S} = 0$ and $\mathscr{S} = \bar{\mathscr{S}}$. Furthermore, $\mathscr{S}$ satisfies the right Ore condition since $a^n(b^n ra) = ra$ for any $r \in R$. Nevertheless

$$a^n(ba - 1) = a^{n-1}(aba - a) = 0$$

and so no element of $\mathscr{S}$ is right regular.

1.12 This example also shows that $\mathscr{S}$ satisfying the right Ore condition is not, in general, sufficient for the existence of $R_\mathscr{S}$. For evidently, in order to be units of $R_\mathscr{S}$, the elements of $\bar{\mathscr{S}}$ must be regular in $\bar{R}$.

Theorem. *Let $\mathscr{S}$ be a m.c. subset of a ring R. Then $R_\mathscr{S}$ exists if and only if $\mathscr{S}$ satisfies the right Ore condition and $\bar{\mathscr{S}}$ consists of regular elements.*

Proof. The necessity of these conditions has been shown. It remains to construct the ring $R_\mathscr{S}$. This can be done, as in the commutative case, by imposing appropriate operations on equivalence classes in $R \times \mathscr{S}$, but the verifications are arduous. For this reason, an alternative route is described below. Note first, however, that by passing to homomorphic images, it will suffice to consider the case when ass $\mathscr{S} = 0$, and then elements of $R_\mathscr{S}$ will be of the form rs^{-1}.

Construction. First consider the set $\mathscr{F}$ of those right ideals A of R such that $A \cap \mathscr{S} \neq \varnothing$. The right Ore condition makes it true, and easily verified, for A_1, $A_2 \in \mathscr{F}$ and $\alpha \in \text{Hom}(A_1, R)$ that
(i) $A_1 \cap A_2 \in \mathscr{F}$, and
(ii) $\alpha^{-1} A_2 \equiv \{a \in A_1 \,|\, \alpha(a) \in A_2\} \in \mathscr{F}$.
   Secondly, consider the set

$$\bigcup \{\text{Hom}(A, R) \,|\, A \in \mathscr{F}\}$$

together with the equivalence relation given, for $\alpha_i \in \mathrm{Hom}(A_i, R)$, by $\alpha_1 \sim \alpha_2$ if α_1 and α_2 coincide on some $A \in \mathscr{F}$, $A \subseteq A_1 \cap A_2$. Operations on equivalence classes $[\alpha_i]$ are defined by $[\alpha_1] + [\alpha_2] = [\beta]$, where β is the sum of the restrictions of α_1 and α_2 to $A_1 \cap A_2$; and by $[\alpha_1][\alpha_2] = [\gamma]$ with γ being their composition when restricted to $\alpha_2^{-1} A_1$.

It can be readily checked:
 (i) that these operations are well defined and, under them, the equivalence classes form a ring, $R_{\mathscr{S}}$ say;
 (ii) that, if $r \in R$ is identified with the equivalence class of the homomorphism $\lambda(r): R \to R$ given by $x \mapsto rx$, this embeds R in $R_{\mathscr{S}}$;
 (iii) that, under this embedding, each $s \in \mathscr{S}$ has an inverse $s^{-1} = [\alpha]$, where $\alpha: sR \to R$ is given by $sx \mapsto x$; and
 (iv) that, if $\alpha \in \mathrm{Hom}(A, R)$ with $A \in \mathscr{F}$, then $[\alpha] = as^{-1}$ where $s \in A \cap \mathscr{S}$ and $a = \alpha(s)$.

These facts complete the proof of the theorem. $\square$

1.13 A m.c. subset $\mathscr{S}$ of a ring R which satisfies the right Ore condition will be called a **right Ore set**. A right Ore set such that the elements of $\bar{\mathscr{S}}$ are regular in $\bar{R} = R/\mathrm{ass}\,\mathscr{S}$ (and so $R_{\mathscr{S}}$ exists) will be called a **right denominator set**. Some conditions for a right Ore set to be a right denominator set are easily obtained.

Lemma. *Let $\mathscr{S}$ be a right Ore set in a ring R. Then:*
 (i) *$\mathscr{S}$ is a right denominator set if and only if, when $s \in \mathscr{S}$ and $r \in R$, $sr = 0$ implies that there exists $t \in \mathscr{S}$ with $rt = 0$;*
 (ii) *if the elements of $\mathscr{S}$ are regular in R then $\mathscr{S}$ is a right denominator set;*
 (iii) *if R or $\bar{R}$ has a.c.c. on right annihilators then $\mathscr{S}$ is a right denominator set.*

Proof. (i) Exercise.
 (ii) Clear from the definitions.
 (iii) 1.10 is easily extended. $\square$

In this book we shall often be working with right Ore sets for which one or other of the hypotheses of (ii) and (iii) of the lemma hold, and in such situations this lemma will be used without further explicit mention.

1.14 The right quotient ring of a ring R with respect to $\mathscr{C}_R(0)$, the set of all regular elements, is simply called the **right quotient ring** of R, if it exists, and is denoted by $Q(R)$. For example, if R is an integral domain then it is obvious that its right quotient ring would have to be a division ring (called the **right quotient division ring** of R).

An integral domain R is called a **right Ore domain** if $\mathscr{C}_R(0)$ is a right Ore set. With this terminology, Theorem 1.12 has an immediate consequence.

2.1.16 Right quotient rings

Corollary. *An integral domain has a right quotient division ring if and only if it is a right Ore domain.* □

1.15 The connection of these results with chain conditions is demonstrated by the next result.

Theorem. *Any right Noetherian integral domain R is a right Ore domain.*

Proof. It is enough, given nonzero elements $a, b \in R$, to show that $ab' = ba' \neq 0$ for some $a', b' \in R$. However, the argument used in 1.2.11(ii) shows that either $aR \cap bR \neq 0$, as required, or else the sum $\sum b^n a R$ is direct, contradicting the hypothesis that R is right Noetherian. □

This can be applied to many of the examples described in Chapter 1. In particular, it covers the Weyl algebra $A_n(k)$, and any enveloping algebra of a finite dimensional k-Lie algebra for any field k. It thus includes the ring described in 1.7(ii) for which localization at a certain prime ideal was impossible.

It is evident that only a part of the a.c.c. was used in the proof of Theorem 1.15, namely that R contained no infinite direct sum of right ideals. This property is important for the development of our ideas, and is the topic discussed in the next section.

Note, however, that in a right Ore domain, any two nonzero right ideals intersect. Thus, for an integral domain, having no infinite direct sum implies having no direct sum at all.

1.16 Before proceeding as indicated in 1.15 it is useful to note some facts about quotient rings. Let $\mathcal{S}$ be a right denominator set in R and let $Q = R_{\mathcal{S}}$. The canonical homomorphism $\vartheta: R \to Q$ need not be injective and, to handle this, some conventions are helpful. First, we write $\vartheta(r)\vartheta(s)^{-1}$ simply as rs^{-1} or even, for the moment, as r/s. Thus ϑ sends r to $r/1$ (or to rs/s if $1 \notin \mathcal{S}$). The set $\{r \in R \mid r/1 \in B\}$ for $B \subseteq Q$ is denoted by $B \cap R$ and, for $D \subseteq R$, the right ideal of Q generated by $\{d/1 \mid d \in D\}$ is denoted by DQ.

Proposition. *Let $\mathcal{S}$ be any right denominator set in a ring R, and let $Q = R_{\mathcal{S}}$.*
  (i) *Any finite set $q_1, \ldots, q_n$ of elements of Q has a common denominator; i.e. there exist $r_1, \ldots, r_n \in R$, and $s \in \mathcal{S}$ such that $q_i = r_i s^{-1}$ for each i.*
  (ii) *Q is a flat left R-module.*
  (iii) *If $B \triangleleft_r R$, then $B \cap R \triangleleft_r R$ and $B = (B \cap R)Q$.*
  (iv) *If $D \triangleleft_r R$, then $DQ \triangleleft_r Q$, $DQ = \{ds^{-1} \mid d \in D, s \in \mathcal{S}\}$ and $DQ \cap R = \{r \in R \mid rs \in D$ for some $s \in \mathcal{S}\}$.*
  (v) *If $D_1, D_2 \triangleleft_r R$ with $D_1 \cap D_2 = 0$, then $D_1 Q \cap D_2 Q = 0$.*
  (vi) *If $I \triangleleft R$ and Q is right Noetherian, then $IQ \triangleleft Q$.*

(vii) *If R is Noetherian then there is a $(1,1)$-correspondence between $\{P \in \operatorname{Spec} R \mid P \cap \mathscr{S} = \varnothing\}$ and $\{P' \in \operatorname{Spec} Q\}$ via $P \mapsto PQ$, $P' \mapsto P' \cap R$.*
(viii) *If R is simple then $Z(R) = Z(Q)$.*

Proof. (i) If $q_i = a_i s_i^{-1}$ it is enough to find $s \in \mathscr{S}$, $b_i \in R$ such that $s_i b_i = s$ for each i. This is easily arranged using 1.8.

(ii) By a standard characterization of flatness (see 7.1.4), it is enough to check that if $I \triangleleft_r R$ then the map $\varphi : I \otimes_R Q \to R \otimes_R Q \stackrel{\sim}{\to} Q$ is injective. Suppose $\varphi(z) = 0$ with $z \in I \otimes Q$. Taking a common denominator as in (i) we may suppose that $z = i \otimes s^{-1}$ and $\varphi(z) = is^{-1} = 0$ in Q. Hence $\vartheta(i) = 0$ and so $it = 0$ in R for some $t \in \mathscr{S}$. Then $z = it \otimes t^{-1} s^{-1} = 0$.

(iii), (iv), (v) These are immediate.

(vi) If $s \in \mathscr{S}$, then $sI \subseteq I$ and so $I \subseteq s^{-1} I$. This leads to an ascending chain $\{s^{-n} IQ\}$ of right ideals of Q which must, by hypothesis, stabilize. Thus $s^{-(n+1)} IQ = s^{-n} IQ$ for $n \gg 0$ and hence $s^{-1} IQ = IQ$. It follows easily that $QIQ = IQ$ and so $IQ \triangleleft Q$.

(vii) First let $I \triangleleft R$ and consider $I' = IQ \cap R$. Since $_R I'$ is finitely generated, it follows from (iv) that $I's \subseteq I$ for some $s \in \mathscr{S}$. Applied to P, this shows that $P = PQ \cap R$. Note next that if $A, B \triangleleft Q$ with $AB \subseteq PQ$ then $(A \cap R)(B \cap R) \subseteq P$ and so, say, $A \cap R \subseteq P$. Hence $A \subseteq PQ$, by (iii), and PQ is prime. Finally it is easy to check, using (vi), that $P' \cap R$ is prime.

(viii) Evidently $Z(R) \subseteq Z(Q) \cap R$. Conversely, if $q \in Z(Q)$ then $\{r \in R \mid qr \in R\}$ is a nonzero ideal of R. Therefore $q1 = q \in R$ and $Z(Q) \subseteq Z(R)$. □

It is worth noting one consequence of (i) and its proof, namely that $a_1 s_1^{-1} = a_2 s_2^{-1}$ if and only if there exist $c_1, c_2 \in R$ such that $s_1 c_1 = s_2 c_2 \in \mathscr{S}$ and $a_1 c_1 = a_2 c_2$. This shows that Q could be described as $R \times \mathscr{S}$ modulo the equivalence relation given by the latter condition.

1.17 It is also possible, given a right denominator set $\mathscr{S}$ and a right R-module M, to construct a **module of quotients** $M_{\mathscr{S}}$. This is done as follows. First let

$$\operatorname{ass}_M \mathscr{S} = \{m \in M \mid ms = 0 \text{ for some } s \in \mathscr{S}\}.$$

One can check easily (cf. 1.9) that this is a submodule of M. It is called the **torsion submodule with respect to** $\mathscr{S}$; and if it equals M then M is called a **torsion module** with respect to $\mathscr{S}$.

Next let $\bar{M} = M/\operatorname{ass}_M \mathscr{S}$ and consider $\{\operatorname{Hom}(A, \bar{M}) \mid A \in \mathscr{F}\}$. Imposing the 'same' equivalence relation and operations as in the construction of $R_{\mathscr{S}}$ turns this into an $R_{\mathscr{S}}$-module, denoted by $M_{\mathscr{S}}$. This contains a copy of $\bar{M}$; and each of its elements has the form $\bar{m} s^{-1}$ for some $\bar{m} \in \bar{M}$ and $s \in \mathscr{S}$. One can check that the map $M \to M_{\mathscr{S}}$, $m \mapsto \bar{m}$, is universal with respect to R-homomorphisms from M to $R_{\mathscr{S}}$-modules.

Of course $M \otimes_R R_{\mathscr{S}}$ is also an $R_{\mathscr{S}}$-module.

Proposition. (i) $M \otimes_R R_{\mathscr{S}} \simeq M_{\mathscr{S}}$.
(ii) *The torsion submodule of M with respect to $\mathscr{S}$ is the kernel of the map* $M \to M \otimes_R R_{\mathscr{S}}, m \mapsto m \otimes 1$.
(iii) *M is a torsion module with respect to $\mathscr{S}$ if and only if $M \otimes_R R_{\mathscr{S}} = 0$.*

Proof. (i) The universal property of $M_{\mathscr{S}}$ gives a map in one direction; and the universal property of $\otimes$ gives one in the opposite direction.
(ii) (iii) These are now clear. □

§2 Uniform Dimension

A basic tool in the study of Noetherian rings and modules is the uniform dimension (or Goldie dimension) of a module. That is studied in this section together with the related notion of essential submodules, and an application is made to minimal prime ideals of a semiprime ring.

Throughout R will denote the ring in question.

2.1 Suppose that N is a submodule of M such that, for all nonzero submodules X of M, one has $N \cap X \neq 0$. Then N is an **essential submodule** of M, and M an **essential extension** of N. The notation $N \triangleleft_e M$ is used.

If a right ideal I is an essential submodule of R it is called an **essential right ideal**. The comments in 1.15 show that in a right Ore domain every nonzero right ideal is essential. Further examples are afforded by the following result.

Lemma. (i) *If R is a prime ring and I is a nonzero ideal then $I \triangleleft_e R_R$.*
(ii) *If N is a nilpotent ideal of a ring R, then $1 \operatorname{ann} N \triangleleft_e R_R$.*

Proof. (i) If $0 \neq X \triangleleft_r R$ then $0 \neq XI \subseteq X \cap I$.
(ii) If $0 \neq X \triangleleft_r R$, choose k such that $XN^k \neq 0$ but $XN^{k+1} = 0$. Then $XN^k \subseteq X \cap 1\operatorname{ann} N$. □

2.2 Next we establish some basic properties concerning essential submodules.

Lemma. (i) *If $N \triangleleft_e P$ and $P \triangleleft_e M$ then $N \triangleleft_e M$.*
(ii) *If $N_1 \triangleleft_e M$ and $N_2 \triangleleft_e M$ then $N_1 \cap N_2 \triangleleft_e M$.*
(iii) *If $N \triangleleft_e M$, $m \in M$ and $m^{-1}N = \{r \in R | mr \in N\}$ then $m^{-1}N \triangleleft_e R_R$.*
(iv) *If $N_i \triangleleft_e M_i$ for $i = 1, 2, \ldots, t$ then $N_1 \oplus \cdots \oplus N_t \triangleleft_e M_1 \oplus \cdots \oplus M_t$.*
(v) *If $N \triangleleft M$, there exists $N' \triangleleft M$ such that $N \cap N' = 0$ and $N \oplus N' \triangleleft_e M$.*
(vi) *M has the property that its only essential submodule is M itself if and only if M is semisimple.*

Proof. (i) (ii) Easy exercise.
(iii) If $0 \neq I \triangleleft_r R$ and $mI = 0$ then $I \subseteq m^{-1}N$. On the other hand, if $mI \neq 0$, then $mI \cap N \neq 0$ and so $I \cap m^{-1}N \neq 0$.

(iv) By induction, it is enough to consider the case $k = 2$. Let $0 \neq X \triangleleft M = M_1 \oplus M_2$. If $X \cap M_i \neq 0$ for $i = 1$ or 2, then $X \cap N_i \neq 0$ as required. Otherwise if $0 \neq m_1 + m_2 = x \in X$, then there exists $r \in R$ such that $0 \neq m_1 r \in N_1$. By assumption $m_2 r \neq 0$ also; but then there exists $s \in R$ with $0 \neq m_2 rs \in N_2$. Clearly $0 \neq xrs \in X \cap (N_1 \oplus N_2)$.

(v) One can apply Zorn's lemma to the collection of submodules X of M with $N \cap X = 0$ to obtain a maximal member, N' say. Suppose that $Y \triangleleft M$ with $Y \cap (N \oplus N') = 0$. Then $X = N' \oplus Y$ satisfies $N \cap X = 0$ and so $Y = 0$ and $N \oplus N' \triangleleft_e M$.

(vi) This is immediate from (v). □

2.3 The proof of (v) shows that given any submodule N of M there is a submodule N' of M maximal with respect to the property that $N \cap N' = 0$. This is called a **complement** to N in M; and a submodule which is a complement to some submodule of M is a **complement submodule**.

Lemma. *If N is a complement submodule in M and $N' \supset N$ then there exists $0 \neq A \triangleleft N'$ with $A \cap N = 0$.*

Proof. If N is a complement to X, take $A = N' \cap X$. □

2.4 It is worth noting that if $\mathscr{F}(R)$ denotes the set of essential right ideals of a ring R then, by 2.2(i), (ii), (iii), it satisfies conditions akin to those used in the construction of $R_\mathscr{S}$ in 1.12. One immediate consequence of this is that the set

$$\zeta(R) = \{a \in R \mid aE = 0 \text{ for some } E \in \mathscr{F}(R)\}$$

is an ideal, known as the **right singular ideal** of R.

2.5 A module U is **uniform** if $U \neq 0$ and also each nonzero submodule of U is an essential submodule.

Evidently this is equivalent to U not containing a direct sum of nonzero submodules. This yields what is called the **common right multiple property**, viz.

Lemma. *A nonzero module U is uniform if and only if, given nonzero elements $u_1, u_2 \in U$, there exist $r_1, r_2 \in R$ such that $u_1 r_1 = u_2 r_2 \neq 0$.* □

It follows immediately that, if R is an integral domain, then R_R is uniform if and only if R is a right Ore domain.

2.6 A module M is said to have **finite uniform dimension** if it contains no infinite direct sum of nonzero submodules.

This is true, of course, of any uniform module and of any Noetherian module.

2.2.10 *Uniform dimension*

Note also that if M has finite uniform dimension and $N \triangleleft M$ then N has finite uniform dimension.

The next few results study the interplay of the notions introduced above.

2.7 Lemma. *If M has finite uniform dimension and $M \neq 0$ then M contains a uniform submodule.*

Proof. Either M is uniform or else M contains a direct sum of non-zero submodules, say $M = M_0 \supseteq M_1 \oplus M_1'$. Repetition of this argument for $M_1, M_2, \ldots$ leads to the direct sum $M_1' \oplus M_2' \oplus \cdots$. This contradiction shows that the process must stop—which it does precisely when M_k is uniform. □

2.8 Lemma. *If M has finite uniform dimension then M contains an essential submodule which is a finite direct sum of uniform submodules.*

Proof. Let $M' = \bigoplus_{i=1}^n U_i$ be a direct sum of uniform submodules U_i of M. Suppose M' is not essential in M. Then there exists $0 \neq X \triangleleft M$ with $M' \cap X = 0$. By 2.7, X contains a uniform submodule U_{n+1} say; and

$$M \supseteq M' \oplus U_{n+1} = \bigoplus_{i=1}^{n+1} U_i.$$

Repetition of this process leads either to an infinite direct sum, which is precluded, or else to an essential submodule, as required. □

2.9 The next result shows that, as the name suggests, a 'dimension' exists which is an invariant of the module.

Theorem. *Let M be a module of finite uniform dimension and (cf. 2.8) let $\bigoplus_{i=1}^n U_i$ be a finite direct sum of uniform submodules of M which is essential in M. Then*
(i) *any direct sum of nonzero submodules of M has at most n summands; and*
(ii) *a direct sum of uniform submodules of M is essential in M if and only if it has precisely n summands.*

Proof. (i) (This mimics the 'exchange principle' for finite dimensional vector spaces). Let $V_1 \oplus \cdots \oplus V_k \triangleleft M$ with each $V_j \neq 0$. Set $W = V_2 \oplus \cdots \oplus V_k$, which is not essential in M. Therefore, by 2.2(iv), $W \cap U_i = 0$ for some i, say for $i = 1$. Then $V_2 \oplus \cdots \oplus V_k \oplus U_1$ is a direct sum. Repetition of this process k times shows that $k \leq n$.

(ii) This follows immediately since, if a submodule is not essential it has a complement (by 2.2(v)) which contains a uniform submodule (by 2.7). □

2.10 The nonnegative integer n given by 2.9 is called the **uniform dimension** (or

Goldie dimension) of M, and is written u dim M. If M fails to have finite uniform dimension, we write u dim $M = \infty$.

Corollary. (i) u dim $M = 1$ *if and only if M is uniform.*
(ii) u dim $M = 0$ *if and only if $M = 0$.*
(iii) *If $N \triangleleft M$ and u dim $M = n$, then u dim $N \leq n$ with equality precisely when $N \triangleleft_e M$.*
(iv) u dim $(M_1 \oplus M_2) = $ u dim $M_1 + $ u dim M_2.
(v) u dim $M < \infty$ *if and only if M has a.c.c. on complement submodules; indeed* u dim M *is the maximal length of a chain of complement submodules.*

Proof. (i)–(iv) Straightforward.
(v) Here 2.3 shows that if u dim $M < \infty$ then no chain of complement submodules will have length $> n$. On the other hand, if $M \supseteq \bigoplus_{i=1}^{t} M_i$, with t possibly infinite, then complements of the submodules $\bigoplus_{i=n}^{t} M_i$, with $n \leq t$, form an increasing chain of length t. □

2.11 When applied to the module R_R one writes r u dim R or u dim R_R.

Example. (i) Remarks in 2.5 show that, for an integral domain R, r u dim $R = 1$ if and only if R is a right Ore domain. In particular, this applies to any right Noetherian integral domain.

Example. (ii) Let k be a field, $k[x, y]$ the commutative polynomial ring, and $R = k[x, y]/(x, y)^n$. Using ¯ to indicate images in R, it is easy to check that the sum $\sum_{i=0}^{n-1} \bar{x}^i \bar{y}^{n-1-i} k$ is a direct sum of uniform submodules of R and is essential in R_R. Thus u dim $R_R = n$.

Example. (iii) Suppose that r u dim $R = n$ and consider $M_t(R)$. Recall that the **standard set of matrix units** of $M_t(R)$ is $\{e_{ij} | i, j \in (1, \ldots, t)\}$, where e_{ij} is the matrix with all entries other than the (i, j) entry being 0 and the (i, j) entry being 1. It is a nice exercise to check that u dim $e_{ii} M_t(R)$, as an $M_t(R)$-module, equals r u dim R (since submodules correspond to right ideals of R). Since $M_t(R) = \bigoplus_{i=1}^{t} e_{ii} M_t(R)$ it follows that r u dim $M_t(R) = t($u dim $R_R)$.

2.12 The way in which these properties are affected by the formation of a right quotient ring is easily deduced from 1.16.

Lemma. *Let $\mathscr{S}$ be a right Ore set of regular elements of a ring R and let $Q = R_{\mathscr{S}}$. Let $A \triangleleft_r R$ and $B \triangleleft_r Q$. Then:*
(i) $A \triangleleft_e R \Leftrightarrow AQ \triangleleft_e Q$;
(ii) $B \triangleleft_e Q \Leftrightarrow B \cap R \triangleleft_e R$;

(iii) u dim A_R = u dim AQ_Q = u dim AQ_R;
(iv) u dim B_Q = u dim B_R = u dim $(B \cap R)_R$;
(v) r u dim Q = r u dim R. □

This lemma is not true for a general right denominator set. Let $R = k[x, y]$, $x^2 = xy = 0$ with k a field. Set $P = (x)$ and $\mathscr{S} = R \backslash P$. Here $yR \cap xR = 0$ so u dim $R_R \geq 2$. But u dim $Q_Q = 1$.

2.13 The module theoretic results above could be repeated for bimodules, and applied to ideals rather than right ideals. However, such repetition is not needed. For if M is an R-bimodule, it can be viewed as a right $R \otimes R^{op}$-module. This interpretation converts sub-bimodules to right submodules. Thus, to say that $_RM_R$ has finite uniform dimension means that $M_{R \otimes R^{op}}$ has finite uniform dimension; and that means that M contains no infinite direct sum of nonzero sub-bimodules. In particular, it follows from 2.9 that u dim $_RM_R$ is an invariant of the bimodule.

2.14 This concept is of use in dealing with semiprime rings and their minimal prime ideals. Recall, from 0.2.8, that in a semiprime ring the intersection of the minimal prime ideals is zero.

Proposition. *Let R be a semiprime ring and A an ideal. Then:*
(i) *r ann A = l ann A (= ann A, say);*
(ii) *ann A is the unique complement ideal to A in R;*
(iii) *ann A is the intersection of those minimal prime ideals of R which do not contain A;*
(iv) *$_RA_R$ is uniform if and only if ann A is a minimal prime ideal;*
(v) *$A \triangleleft_e {}_RR_R$ if and only if ann A = 0;*
(vi) *if A is not contained in any minimal prime of R then $A \triangleleft_e {}_RR_R$.*

Proof. (i) $AX = 0 \Rightarrow (XA)^2 = 0 \Rightarrow XA = 0$, and vice versa.
(ii) If $A \cap C = 0$ for $C \triangleleft R$ then $AC \subseteq A \cap C = 0$; hence $C \subseteq$ ann A. On the other hand $(A \cap \text{ann } A)^2 = 0$ and so $A \cap \text{ann } A = 0$.
(iii) Let B be the intersection described; evidently $A \cap B = 0$ and so $B \subseteq$ ann A by (ii). However, A ann $A = 0$; so if P is a minimal prime and $A \not\subseteq P$ then ann $A \subseteq P$. Thus ann $A \subseteq B$.
(iv) By (ii), ann A is the unique complement ideal to A. Thus, if $_RA_R$ is uniform, then ann A = ann A' for any nonzero ideal $A' \subseteq A$. Suppose that $BC \subseteq$ ann A for some ideals B, C. Then either $AB = 0$, in which case $B \subseteq$ ann A; or else $AB \neq 0$, and then $C \subseteq$ ann AB = ann A. This shows that ann A is prime; and the minimality is a consequence of (iii).
Conversely, if ann A is a minimal prime, then ann A = ann A' for any nonzero

ideal $A' \subseteq A$ (for ann A' is also an intersection of minimal primes yet ann $A' \supseteq$ ann A). But if $A \supseteq A_1 \oplus A_2$ with A_1, A_2 nonzero ideals, then ann $A_1 \supset$ ann A. So $_RA_R$ is uniform.

(v), (vi) These are clear from (ii) and (iii). □

Corollary. *If R is a semiprime ring, then R is prime precisely when* u dim $_RR_R = 1$. □

2.15 The above corollary extends to rings of finite uniform dimension. By **annihilator ideal** is meant an ideal of the form ann A, $A \triangleleft R$.

Theorem. *The following conditions on a semiprime ring R are equivalent:*
 (i) $_RR_R$ *has finite uniform dimension;*
 (ii) *R has finitely many minimal prime ideals;*
 (iii) *R has finitely many annihilator ideals;*
 (iv) *R has a.c.c. on annihilator ideals.*

Proof. (i)⇒(ii) Say u dim $R = n$; then there are uniform ideals $U_1, \ldots, U_n$ such that $U_1 \oplus \cdots \oplus U_n \triangleleft_e {_RR_R}$. If $P_i = $ ann U_i, then by Proposition 2.14 (iv), P_i is a minimal prime. It is easy to see that $\cap P_i = 0$. These, therefore, are all the minimal primes.

(ii)⇒(iii) This is clear from Proposition 2.14(iii).

(iii)⇒(iv) Evident.

(iv)⇒(i) This is clear from 2.10(v) since, by Proposition 2.14(ii), annihilator ideals are complement ideals. □

Corollary. *If R is a semiprime ring with* u dim $_RR_R = n$ *then R has n minimal primes, 2^n annihilator ideals, any maximal chain of annihilator ideals has length n, and no minimal prime is essential in $_RR_R$.*

Proof. In the notation of the preceding proof, $P_1, \ldots, P_n$ are all the minimal primes. Indeed, $P_1 \cap \cdots \cap P_k = $ ann $(U_1 \oplus \cdots \oplus U_k)$ and so each intersection of minimal prime ideals is an annihilator. This, together with Proposition 2.14(iii), (iv) completes the proof. □

2.16 This does give some information about non-semiprime rings.

Corollary. (a) *If R is a ring with prime radical N and if* r u dim $R/N < \infty$ *then R has finitely many minimal prime ideals.*

(b) *If R is a right Noetherian ring with prime radical N then the number of minimal prime ideals is finite and equals the uniform dimension of the R-bimodule R/N.*

2.3.1 *Goldie's theorem* 55

Proof. (a) Since every prime ideal contains N it is enough to consider the case $N = 0$; but then Theorem 2.15 applies, since $\mathrm{u\,dim}\,_R R_R \leqslant \mathrm{r\,u\,dim}\,R < \infty$.
 (b) This is now clear. □

2.17 One should perhaps note that the fact, in 2.16(b), that R has finitely many minimal primes can also be deduced simply from the following result, due to Noether.

Proposition. *In a right Noetherian ring, 0 is a (finite) product of prime ideals.*

Proof. If possible, choose an ideal A of R maximal with respect to not containing a product of prime ideals. If A is prime, a contradiction occurs. So $A \supseteq BC$ with $B \supset A$, $C \supset A$. By hypothesis B, C both contain products of primes, hence so does A. This shows that no such A exists. □

§3 Goldie's Theorem

The aim now is to characterize those rings whose right quotient rings exist and are semisimple Artinian rings. One half of the process is quite easy, so that is dealt with first.

3.1 A ring R is called a **right Goldie ring** if R has finite right uniform dimension and R satisfies the a.c.c. on right annihilators.

Proposition. *Let R have a right quotient ring Q.*
 (i) *If Q is a right Goldie ring, so too is R.*
 (ii) *If Q is semisimple Artinian, then R is a semiprime right Goldie ring.*
 (iii) *If Q is simple Artinian, then R is a prime right Goldie ring.*

Proof. (i) The a.c.c. on right annihilators is a property preserved in any subring. This, together with 2.12, gives the result.
 (ii) Note first that an Artinian ring is also Noetherian; so Q is right Goldie and (i) applies. It remains to show that R is semiprime. Suppose that N is a nilpotent ideal of R. By 2.1(ii), $\mathrm{l\,ann}\,N$ is an essential right ideal of R. Therefore, by 2.12, $(\mathrm{l\,ann}\,N)Q \triangleleft_e Q$. However, since Q is semisimple, every right ideal is a direct summand of Q. It follows that $(\mathrm{l\,ann}\,N)Q = Q$ and so, using 1.16(iv), $1 = ac^{-1}$ with $a \in \mathrm{l\,ann}\,N$, and c a regular element in $\mathscr{C}_R(0)$. But then $a = c \in \mathrm{l\,ann}\,N$, so $N = 0$.
 (iii) Suppose $0 \neq A \triangleleft R$. Then $QAQ = Q$ and so $1 = \sum_i r_i c_i^{-1} a_i d^{-1}$ with $r_i \in R$, $a_i \in A$ and $c_i, d \in \mathscr{C}_R(0)$. Therefore $d \in QA$ and so $QA = Q$. Now suppose $0 \neq B \triangleleft R$. Then $QAB = QB = Q$ and so $AB \neq 0$. □

3.2 Before the main theorem, which is the converse to 3.1, some preparatory results are needed. Note that a maximal member of the set of right annihilators is called a **maximal right annihilator**.

Lemma. *Let R be a ring with a.c.c. on right annihilators.*
(i) Each maximal right annihilator has the form $\mathrm{r\,ann}\,a$ for some $a \in R$.
(ii) (Fitting's lemma) Given $b \in R$, there is an integer m such that, for all $n \geq m$, $\mathrm{r\,ann}\,b^m = \mathrm{r\,ann}\,b^n$; and then $\mathrm{r\,ann}\,b^n \cap b^n R = 0$.
(iii) Each nonzero nil left (or right) ideal of R contains a nonzero nilpotent left (resp. right) ideal.

Proof. (i) (ii) These are easily checked.

(iii) For $a \in R$, Ra is nil (or nilpotent) if and only if aR is nil (resp. nilpotent). So it is sufficient to consider a nil left ideal L, say. Choose $0 \neq a \in L$ such that $\mathrm{r\,ann}\,a$ is maximal amongst right annihilators of nonzero elements of L. If $y \in R$, then $(ya)^k = 0$ for some $k \geq 1$ and $(ya)^{k-1} \neq 0$. The maximality of $\mathrm{r\,ann}\,a$ shows that $\mathrm{r\,ann}\,(ya)^{k-1} = \mathrm{r\,ann}\,a$, hence $aya = 0$. Thus $aRa = 0$, and so Ra is nilpotent. □

3.3 Lemma. *Let R be a right Goldie ring and let $a \in R$. Then, for $n \gg 0$, $a^n R \oplus \mathrm{r\,ann}\,a^n$ is an essential right ideal of R.*

Proof. By 3.2, for $n \gg 0$, $a^n R \cap \mathrm{r\,ann}\,a^n = 0$; so the sum is direct. The result is clear now provided we show that, given $0 \neq I \triangleleft_r R$ with

$$I \cap (a^n R \oplus \mathrm{r\,ann}\,a^n) = 0,$$

then, for all t, $\sum_{k=1}^{t} a^{kn} I$ is a direct sum of nonzero right ideals. Suppose, by induction, this holds for $t-1$. Since $a^n R \cap \mathrm{r\,ann}\,a^n = 0$, it is immediate that $\sum_{k=2}^{t} a^{kn} I$ is direct. Suppose $x \in a^n I \cap \sum_{k=2}^{t} a^{kn} I$. Then $x = a^n i = a^{2n} j$ with $i \in I, j \in R$. Therefore $i - a^n j \in \mathrm{r\,ann}\,a^n$ and so

$$i \in I \cap (a^n R \oplus \mathrm{r\,ann}\,a^n) = 0.$$

Thus $x = 0$ as required. □

3.4 These results give information about the right singular ideal $\zeta(R)$, defined in 2.4, and about right regular elements.

Lemma. *If R is a ring with a.c.c. on right annihilators, then its right singular ideal $\zeta(R)$ is nilpotent. If R is also semiprime, then $\zeta(R) = 0$.*

Proof. Let $A = \zeta(R)$. For $n \gg 0$, $\mathrm{r\,ann}\,A^n = \mathrm{r\,ann}\,A^{n+1}$. Suppose $A^{n+1} \neq 0$ and choose, amongst $a \in A$ with $A^n a \neq 0$, so that $\mathrm{r\,ann}\,a$ is as large as possible. If $b \in A$, then $\mathrm{r\,ann}\,b \triangleleft_e R$, so $aR \cap \mathrm{r\,ann}\,b \neq 0$. Hence $bar = 0$ for some $r \in R$ with

$ar \neq 0$. It follows that $\operatorname{r ann} ba \supset \operatorname{r ann} a$, which contradicts the choice of a unless $A^n ba = 0$. This shows that $A^{n+1} a = 0$ and so, by the choice of n, $A^n a = 0$. Hence $A^{n+1} = 0$. □

Proposition. *If R is a semiprime ring with finite right uniform dimension and with $\zeta(R) = 0$, and if $c \in R$ is right regular, then c is regular and $cR \triangleleft_e R$.*

Proof. Since $cR \simeq R$ then $\operatorname{u dim} cR = \operatorname{u dim} R_R$. Therefore, by 2.10, $cR \triangleleft_e R$ and so $\operatorname{l ann}(cR) = 0$. Hence $\operatorname{l ann} c = 0$. □

3.5 The preceding result applies, of course, to any semiprime right Goldie ring. The same is true of the next result, which establishes a close connection between regular elements and essential right ideals.

Proposition. *Let R be a semiprime ring of finite right uniform dimension, with $\zeta(R) = 0$, and let $E \triangleleft_r R$.*
(i) *E contains an element c such that $(\operatorname{r ann} c) \cap E = 0$;*
(ii) *E is essential if and only if E contains a regular element of R.*

Proof. (i) First consider the case when E is uniform. Since $E^2 \neq 0$, then $cd \neq 0$ for some $c, d \in E$. Let $V = \operatorname{r ann} c \cap E$ and suppose $V \neq 0$. Since E is uniform, $V \triangleleft_e E$ and so, by 2.2(iii),

$$d^{-1} V = \{r \in R \mid dr \in V\} \triangleleft_e R.$$

Since $cd(d^{-1}V) = 0$ and $\zeta(R) = 0$, a contradiction has occurred. Thus $\operatorname{r ann} c \cap E = 0$.

Now consider the general case. Choose a uniform right ideal $U_1 \subseteq E$, and an element $a_1 \in U_1$ such that $\operatorname{r ann} a_1 \cap U_1 = 0$. If $\operatorname{r ann} a_1 \cap E \neq 0$, choose a uniform right ideal $U_2 \subseteq \operatorname{r ann} a_1 \cap E$, and choose $a_2 \in U_2$, with $\operatorname{r ann} a_2 \cap U_2 = 0$. Note that

$$a_1 R \oplus a_2 R \oplus (\operatorname{r ann} a_1 \cap \operatorname{r ann} a_2 \cap E) \subseteq E.$$

This process is repeated until one has, say,

$$a_1 R \oplus a_2 R \oplus \cdots \oplus a_k R \oplus (\operatorname{r ann} a_1 \cap \operatorname{r ann} a_2 \cap \cdots \cap \operatorname{r ann} a_k \cap E) \subseteq E.$$

Since $\operatorname{r u dim} R < \infty$, this process must terminate—say at this stage. This means that

$$\operatorname{r ann} a_1 \cap \operatorname{r ann} a_2 \cap \cdots \cap \operatorname{r ann} a_k \cap E = 0.$$

Let $c = a_1 + \cdots + a_k$; so $c \in E$. Since the sum $\sum a_i R$ is direct, it follows that $\operatorname{r ann} c = \cap \operatorname{r ann} a_i$. Therefore $\operatorname{r ann} c \cap E = 0$ as claimed.

(ii) If E is essential, then $\operatorname{r ann} c = 0$ and so, by Proposition 3.4, c is regular. Conversely, if $c \in E$ is regular, then $cR \triangleleft_e R$, by Proposition 3.4. Hence $E \triangleleft_e R$. □

3.6 As noted in 2.4, the set $\mathscr{F}(R)$ of essential right ideals of R satisfies the conditions which were required for the construction of $R_\mathscr{S}$ in 1.12. The connection between regular elements and $\mathscr{F}(R)$ established above is the key to the next result which includes the promised converse to 3.1.

Theorem. (Goldie's theorem) *The following conditions on a ring R are equivalent*:
(i) R is semiprime right Goldie;
(ii) R is semiprime, $\zeta(R) = 0$ and $\operatorname{r\,u\,dim} R < \infty$;
(iii) R has a right quotient ring Q which is semisimple Artinian.
Furthermore, R is prime if and only if Q is simple.

Proof. (i)$\Rightarrow$(ii) This is a consequence of Lemma 3.4.
(ii)$\Rightarrow$(iii) Let $a, d \in R$ with d regular. By Proposition 3.4, $dR \triangleleft_e R$. Hence, by 2.2(iii), $a^{-1}(dR) = \{x \in R | ax \in dR\} \in \mathscr{F}(R)$ and so, by 3.5, $a^{-1}(dR)$ contains a regular element c say. Then $ac = db$ for some $b \in R$. Thus $\mathscr{C}_R(0)$ is a right Ore set and so, by 1.12, R has a right quotient ring Q, say.
Let $J \triangleleft_e Q_Q$. By 2.12(ii), $J \cap R \triangleleft_e R$ and so contains a regular element, by 3.5 again. However, regular elements of R are units of Q; thus $J = Q$. Therefore, by 2.2(vi), Q_Q is semisimple.
(iii)$\Rightarrow$(i) This is proved in 3.1, which also shows that if Q is simple then R is prime. Finally, if R is prime, let $0 \neq X \triangleleft Q$. Then $0 \neq X \cap R \triangleleft R$. By 2.1(i), $X \cap R \triangleleft_e R$ and so, as before, $X = Q$. Thus Q is simple. $\square$

3.7 Many applications of this result appear throughout this book. For now, an immediate application to nilpotency will suffice. (The requirement of an identity element in rings is, perforce, relaxed for nil subrings.)

Theorem. *Nil subrings of a right Noetherian ring are nilpotent.*

Proof. Let N denote the prime radical of R. This is, of course, a nil ideal. The first step is to show that N is nilpotent.
To see this, let N' denote the sum of all nilpotent right ideals. Since R is right Noetherian, the sum is finite; so N' itself is nilpotent. It is easy to see that $N' \triangleleft R$; and the maximality of N' shows that R/N' can have no nilpotent right ideals. Hence, by 3.2(iii), R/N' has prime radical zero. It follows that $N' = N$.
Thus, to prove the theorem, it will suffice to consider the case when $N = 0$, and so R is semiprime. But then, by 3.6, R is a subring of a semisimple Artinian ring, for which the result is well known (see [Jacobson **64**, p. 201] for example). $\square$

3.8 So far, apart from the preceding result, rings under consideration have had an identity element. In practice, Goldie's theorem, like most of the other theorems proved so far, is true with or without a 1 in the ring. The theory can all be

2.3.9 *Goldie's theorem* 59

suitably amended (with 1.16(ii) as one exception). However, it is possible to deduce Goldie's theorem without 1, from Goldie's theorem with 1.

To this end, a 1 is adjoined to the ring R. Now the standard technique for doing this is to take the abelian group $R \oplus \mathbb{Z}$ and define multiplication by

$$(r, n)(r', n') = (rr' + rn' + nr', nn').$$

This is not the most appropriate technique here, however, since $R \oplus \mathbb{Z}$ may have a different quotient ring to R. (Try $R = 2\mathbb{Z}$ for example; or any ring R with a 1.) So a different approach is followed, which relies upon R being semiprime.

3.9 Let R be a semiprime ring. Consider the ring $\text{End}(R_R)$, and note that, since R is semiprime, $R \hookrightarrow \text{End } R_R$ with each element of R acting via left multiplication. Also, there is a 1 in $\text{End } R_R$. Let R^1 denote the subring of $\text{End } R_R$ generated by R and 1. This, then, is the ring used here.

Note, for example, that $(2\mathbb{Z})^1 = \mathbb{Z}$, and that, if $1 \in R$, then $R^1 = R$.

Lemma. (i) $R \triangleleft R^1$ and R is an essential right ideal of R^1.
(ii) R^1 is a semiprime ring.
(iii) R^1 satisfies the a.c.c. on right ideals which share the same intersection with R.

Proof. (i) It is clear that $R \triangleleft R^1$. Let $0 \neq A \triangleleft_r R^1$. Note that $AR \subseteq R$, since R^1 is generated by R and 1. Further, if $a \in A$, $r \in R$, then a and ar can both be viewed as endomorphisms of R; and ar is simply left multiplication by $a(r)$. If $a \neq 0$, then $a(r) \neq 0$ for some r; and so $AR \neq 0$. Hence R is an essential right ideal of R^1.
(ii) If $A \triangleleft_r R^1$ with $A^2 = 0$, then $(A \cap R)^2 = 0$. So R^1 is semiprime.
(iii) Suppose $A_1 \subset A_2$ yet $A_1 \cap R = A_2 \cap R$. Then

$$(A_1 + R)/R \subset (A_2 + R)/R \subseteq R^1/R;$$

but R^1/R is a factor of $\mathbb{Z}$ and so is Noetherian. □

Note next that R-submodules and R^1-submodules of R coincide. Thus, notions of uniform right ideals, u dim etc., can be extended immediately to R_R.

Proposition. *If R is semiprime right Goldie, then so too is R^1.*

Proof. Since $R \triangleleft_e R^1$ it is clear from 2.10 that $\text{u dim } R^1 = \text{u dim } R < \infty$. Next consider a right annihilator $\text{r ann } X$ in R^1. If $A \triangleleft_r R^1$ with $XA = 0$, then of course $RXA = 0$. Conversely, if $RXA = 0$, then $(R \cap XA)^2 = 0$ and so $R \cap XA = 0$. Therefore $XA = 0$, since $R \triangleleft_e R^1$. Thus $\text{r ann } X = \text{r ann } RX$ and so we may suppose $X \subseteq R$. If $I_1 \subseteq I_2 \subseteq \cdots$ is a chain of right annihilators in R^1, then $\{I_j \cap R\}$ is a chain of right annihilators in R and so $I_j \cap R = I_{j+1} \cap R$ for $j \gg 0$. By the lemma (iii) it follows that $I_k = I_{k+1}$ for $k \gg 0$. □

The final deduction, that R has a right quotient ring which is semisimple Artinian, is easy—but is best made in the language of orders. It comes in Chapter 3, §1.

§4 Additional Remarks

4.0 This account is fairly standard apart from the bimodule aspects at the end of §2 and the approach to rings without 1 at the end of §3. The content of this chapter is used throughout this text.

4.1 (a) The origins of this material, and in particular of the Ore condition, lie in [Ore **33**]. However the construction of $R_\mathscr{S}$ described in 1.12 comes from [Asano **49a**].

(b) There are other more general notions of quotient rings and localization. One of these is described in 10.3.5; but for a fuller account, including torsion theory and a categorical viewpoint, one can consult [Stenstrom **75**].

(c) The universal object defined by condition (i) of 1.3 always exists, but may well be the zero-ring. It too is sometimes denoted by $R_\mathscr{S}$.

(d) Given elements a, b with $ab = 1$, $ba \neq 1$ as in 1.11, one can construct an infinite set of orthogonal idempotents—see [Jacobson **50**]. Thus most finiteness conditions imply that right inverses are also left inverses.

(e) [Goldie **58**] gives 1.15 although this proof comes from [Lesieur and Croisot **59**]; 1.16(vi) comes from [Ludgate **72**].

4.2 (a) Uniform dimension and its basic properties originated in [Goldie **58**]. It and its variant the reduced rank, which is introduced in Chapter 4, are important throughout.

(b) The concept of the singular ideal comes from [Johnson **51**].

4.3 (a) This section is taken from [Goldie **58, 60, 69**]. However, 3.2(iii) is due to [Utumi **56**] and Lemma 3.4 comes from [Mewborn and Winton **69**].

(b) 3.7 was proved earlier for (left and right) Noetherian rings in [Levitzki **45**].

(c) Goldie's theorem is extended to non-semiprime rings in 4.1.4, and Chapter 4, §3 discusses localization at a semiprime ideal.

(d) The technique for adjoining a 1 described in 3.9 is used in [Robson **79**] in investigating whether certain types of ring, such as simple Noetherian rings, always have a 1.

Chapter 3

STRUCTURE OF SEMIPRIME GOLDIE RINGS

This chapter explores some consequences and extensions of Goldie's theorem. It starts by discussing quotient rings, but from a different viewpoint. It takes the quotient ring as fixed, and examines those subrings of which it is the right quotient ring. This is analogous to (and, in fact, includes) the notion of studying orders over integral domains in central simple algebras.

This viewpoint proves useful in the later sections. Sections 2 and 3 concentrate on the ideals, particularly the minimal primes, and the right ideals of a semiprime right Goldie ring R, and §4 analyses $\text{End } M$ for suitable modules M_R showing that $\text{End } M$ is again semiprime right Goldie. Between them these sections show that R is 'close' to a direct sum of matrix rings over right Ore domains.

Section 5 comprises a brief survey of Morita equivalence, and §6 studies Morita contexts. It describes a correspondence between certain prime ideals of the two rings making up a Morita context, and then shows when the corresponding prime factor rings share the property of being right Goldie.

§1 Orders in Quotient Rings

It is clear that different rings can have the same right quotient ring. This is true, for example, of the first Weyl algebra $A_1(k)$ and the algebra $B_1(k)$ (see Chapter 1, §3); and of $k[x;\sigma]$ and $k[x, x^{-1};\sigma]$ with σ an automorphism (see Chapter 1, §4). The study of this phenomenon and its precise connections with earlier results is the aim of this section. Its relationship with the classical theory of orders over Dedekind domains, in central simple algebras, will be discussed in Chapter 5, §3.

1.1 A ring Q is called a **quotient ring** if every regular element of Q is a unit. Examples of such rings are readily available.

Proposition. *If Q is a right Artinian ring then Q is a quotient ring. Indeed every right regular element is a unit.*

Proof. If $s \in Q$ is right regular then consider the descending chain $\{s^n Q\}$. This stabilizes with, say, $s^n Q = s^{n+1} Q$ and thus $s^n = s^{n+1} q$ for some $q \in Q$. Since s is right regular, so too is s^n; and yet $s^n(sq - 1) = 0$. Thus $sq = 1$ and, moreover, s is left regular. Finally
$$s(qs - 1) = (sq - 1)s = 0,$$
so $q = s^{-1}$. □

1.2 Given a quotient ring Q, a subring R, not necessarily containing 1, is called a **right order** in Q if each $q \in Q$ has the form rs^{-1} for some $r, s \in R$. A **left order** is defined analogously; and a left and right order is called an **order**.

Since the ring Q has a 1, there is no problem about adjoining a 1 to R if required. However, it is useful, on occasion, to allow right orders without a 1.

1.3 Lemma. *Let R_i be a subring, not necessarily with 1, of the quotient ring Q_i, for $i = 1, \ldots, n$. Then $\oplus R_i$ is a right order in $\oplus Q_i$ if and only if each R_i is a right order in Q_i.*

Proof. Straightforward. □

1.4 There is a distinction between the phrases 'R is a right order in Q' and 'Q is the right quotient ring of the ring R', over and above the convention that rings have a 1 but right orders need not; for, if R is a right order in a quotient ring Q, whilst it is clear that regular elements of R are right regular in Q, it is not clear that they are regular. However, the next result shows that this distinction vanishes when R is also a left order or Q is right Artinian.

Proposition. *Let R be a subring (with 1) of a ring Q and let $\mathscr{S} = \{$units of $Q\} \cap R$.*
(i) *If Q is the right quotient ring of R then Q is a quotient ring, R is a right order in Q, and $\mathscr{S} = \mathscr{C}_R(0)$.*
(ii) *If Q is a quotient ring and R is a right order in Q then $Q = R_\mathscr{S}$. If, further, either R is also a left order in Q or Q is right Artinian, then $\mathscr{S} = \mathscr{C}_R(0)$ and Q is the right quotient ring of R.*

Proof. (i) If $q \in Q$ is regular, with $q = rs^{-1}$ say, where $r, s \in R$, then $r = qs \in \mathscr{C}_R(0)$ and so is a unit of Q. Hence q is a unit and Q is a quotient ring. The remainder is clear.

(ii) The first claim is an immediate consequence of the definitions; and, as noted above, $\mathscr{C}_R(0) \subseteq \mathscr{C}'_Q(0)$. If R is also a left order then $\mathscr{C}_R(0) \subseteq {}'\mathscr{C}_Q(0)$ and thus $\mathscr{C}_R(0) = \mathscr{S}$. Therefore Q is the quotient ring of R.

In the other case, when Q is right Artinian, 1.1 shows that $\mathscr{C}'_Q(0) = \mathscr{C}_Q(0)$. Hence $\mathscr{S} = \mathscr{C}_R(0)$ and Q is the right quotient ring of R. □

This shows, in particular, that when Q is semisimple Artinian the distinction disappears. So 'semiprime right Goldie ring' is synonymous with 'right order, with 1, in a semisimple Artinian ring'.

1.5 Corollary. *R is a semiprime right Goldie ring with right quotient ring Q if and only if $M_n(R)$ is a semiprime right Goldie ring with right quotient ring $M_n(Q)$.*

Proof. Suppose that R is semiprime right Goldie with right quotient ring Q. Since Q is semisimple Artinian, so too is $M_n(Q)$. We will show that $M_n(R)$ is a right order in $M_n(Q)$. If $x \in M_n(Q)$ then taking a common denominator, by 2.1.16(i), one can write x in the form $(a_{ij}c^{-1})$, where $a_{ij}, c \in R$. Using the standard embedding of Q in $M_n(Q)$, this means that $x = ac^{-1}$ with $a, c \in M_n(R)$. Thus $M_n(R)$ is a right order in $M_n(Q)$ and so is semiprime right Goldie.
The converse is easily proved. □

This indirect proof of the lifting of right Goldie conditions from R to $M_n(R)$ reflects the fact that there exists a (non-semiprime) right Goldie ring R such that $M_n(R)$ is not right Goldie [Kerr 79].

1.6 We now turn to relations between right orders in the same quotient ring, starting with a very useful fact.

Lemma. *Let R be a right order in a quotient ring Q and let S be a subring of Q (not necessarily with 1). Suppose further that there are units a, b of Q such that $aRb \subseteq S$. Then S is also a right order in Q.*

Proof. Given $q \in Q$, consider the element $a^{-1}qa$. By definition, $a^{-1}qa = rt^{-1}$ for some $r, t \in R$. But then

$$q = art^{-1}a^{-1} = arb(atb)^{-1},$$

and so S is a right order in Q. □

Corollary. (i) *If R is a right order in a quotient ring Q and S is a ring (not necessarily with 1) such that $R \subseteq S \subseteq Q$ then S is a right order in Q.*
(ii) *If R is a prime right Goldie ring, $0 \neq A \triangleleft R$, and S is a subring of R with $A \subseteq S \subseteq R$ then S is a prime right Goldie ring, and has the same right quotient ring as R.*

Proof. (i) Take $a = b = 1$.
(ii) By 2.2.1, $A \triangleleft_e R_R$, and therefore, by 2.3.5, A contains a regular

element, c say, of R. Then c is a unit of the right quotient ring Q of R, and $cR \subseteq S$. □

1.7 Goldie's theorem can now be extended easily to rings without a 1.

Theorem. *A ring R, not necessarily with a 1, is a right order in a semisimple Artinian ring Q if and only if R is semiprime right Goldie.*

Proof. Suppose R is semiprime right Goldie. By Proposition 2.3.9 so too is the ring R^1, described there. Therefore, by 2.3.6, R^1 has a semisimple Artinian right quotient ring Q. Thus R^1 is a right order in Q. Now, by Lemma 2.3.9, R is an essential right ideal of R^1 and thus, by 2.3.5, contains a regular element, a say, of R^1. Since $aR^1 \subseteq R$, then Lemma 1.6 shows that R is a right order in Q.

The converse follows easily from 2.3.6 and 2.3.9. □

1.8 There is an immediate consequence which represents, perhaps, the main reason for considering rings without 1 in this context.

Corollary. *Let R be a semiprime right Goldie ring, A an essential right ideal and B a left ideal containing a regular element. Then A and AB are semiprime right Goldie rings (without 1) having the same right quotient ring as R.*

Proof. By 2.3.5 and the hypotheses, there are regular elements $a \in A$, $b \in B$. Then $aRb \subseteq AB \subseteq A \subseteq R$. The result follows from Lemma 1.6 and Theorem 1.7. □

1.9 The relationship involved in Lemma 1.6 leads to an equivalence relation on right orders R_1, R_2 in a fixed quotient ring Q; viz. $R_1 \sim R_2$ if there are units $a_1, a_2, b_1, b_2 \in Q$ such that $a_1 R_1 b_1 \subseteq R_2$ and $a_2 R_2 b_2 \subseteq R_1$. Then R_1, R_2 are termed **equivalent** right orders.

For example, the rings R and R^1 considered in 1.7 are clearly equivalent right orders. So too are the rings R, A and AB described in 1.8.

For examples in which both rings have a 1, one can cite $M_2(\mathbb{Z})$ and its subring

$$\begin{bmatrix} \mathbb{Z} & 2\mathbb{Z} \\ \mathbb{Z} & \mathbb{Z} \end{bmatrix};$$

and the first Weyl algebra $A_1(k)$ and its subring $k + xA_1(k)$ as in 1.3.10.

1.10 For right orders which are comparable, like these examples, there is a further useful fact.

Lemma. *Suppose R, S are equivalent right orders in Q with $R \subseteq S$. Then there are equivalent right orders T, T' in Q with $R \subseteq T \subseteq S$, $R \subseteq T' \subseteq S$ and units r_1,*

3.1.12 *Orders in quotient rings*

r_2 of Q contained in R such that $r_1 S \subseteq T$, $Tr_2 \subseteq R$ and $Sr_2 \subseteq T'$, $r_1 T' \subseteq R$. In particular, $r_1 S r_2 \subseteq R$.

Proof. By definition, $aSb \subseteq R$ for some units a, b of Q. Say $a = r_1 s_1^{-1}$, $b = r_2 s_2^{-1}$, with $r_i, s_i \in R$. Then

$$r_1 S r_2 \subseteq r_1 s_1^{-1} S r_2 \subseteq R s_2 \subseteq R.$$

It is readily checked that

$$T = R + r_1 S + R r_1 S \quad \text{and} \quad T' = R + S r_2 + S r_2 R$$

are as claimed. □

1.11 The study of equivalent orders is facilitated by the notion of a fractional ideal. Suppose that R is a right or left order in a quotient ring Q. Then a **fractional right R-ideal** is a submodule I of Q_R such that $aI \subseteq R$ and $bR \subseteq I$ for some units $a, b \in Q$. In a similar fashion **fractional left R-ideals** and **fractional (two-sided) R-ideals** are defined. If I is both a fractional right R-ideal and a fractional left S-ideal for some other order S, then I is called a **fractional (S, R)-ideal**.

For example, note that any essential right ideal of a semiprime right Goldie ring R is a fractional right R-ideal. Moreover, if I is any fractional right R-ideal and a is as above, then $I \simeq aI$ which is an essential right ideal of R because $ab \in aI \cap \mathscr{C}_R(0)$.

As a more specific example note that $\tfrac{1}{2}\mathbb{Z}$ is a fractional $\mathbb{Z}$-ideal and $M_2(\mathbb{Z})$ is a fractional ideal over

$$\begin{bmatrix} \mathbb{Z} & 2\mathbb{Z} \\ \mathbb{Z} & \mathbb{Z} \end{bmatrix}.$$

1.12 The **right order** and **left order** of a fractional right (or left) R-ideal I are defined respectively to be

$$O_r(I) = \{q \in Q \mid Iq \subseteq I\}$$
$$O_l(I) = \{q \in Q \mid qI \subseteq I\}.$$

For example, if

$$I = \begin{bmatrix} 2\mathbb{Z} & 2\mathbb{Z} \\ \mathbb{Z} & \mathbb{Z} \end{bmatrix} \subset Q = M_2(\mathbb{Q})$$

then

$$O_r(I) = M_2(\mathbb{Z}) \quad \text{and} \quad O_l(I) = \begin{bmatrix} \mathbb{Z} & 2\mathbb{Z} \\ \tfrac{1}{2}\mathbb{Z} & \mathbb{Z} \end{bmatrix}.$$

The basic facts about $O_r(I)$ and $O_l(I)$ are as follows.

Lemma. *Let R be a right order in Q and let I be a fractional right or left R-ideal. Then:*
(i) $O_r(I)$ and $O_l(I)$ are right orders in Q and are equivalent to R;
(ii) I is a fractional $(O_l(I), O_r(I))$-ideal.

Proof. Suppose I is a fractional right R-ideal (the alternative being dealt with symmetrically).

(i) Using the units a, b given by 1.11, note that

$$abO_r(I) \subseteq R \subseteq O_r(I),$$

that

$$aO_l(I)b \subseteq aO_l(I)I \subseteq aI \subseteq R,$$

and that $bRa \subseteq O_l(I)$, since $bRaI \subseteq bR \subseteq I$.

(ii) This is readily checked. □

1.13 Evidently a subset of a quotient ring Q can be a fractional right or left ideal over several right orders. Let I be a fractional right R-ideal. If $I \subseteq O_r(I)$ then I is an **integral right** R-**ideal**. If I satisfies the stronger condition that $I \subseteq R$ then I is a **right** R-**ideal**. (There is some confusion in the literature between these terms.)

As an example, already noted above, if R is a semiprime right Goldie ring, then the right R-ideals are precisely the essential right ideals of R.

The nature of integral right R-ideals is clarified by the next result.

Lemma. *Let R be a right or left order in a quotient ring Q and let I be a fractional right R-ideal. Then the following are equivalent:*
(i) I is an integral right $O_r(I)$-ideal;
(ii) I is an integral left $O_l(I)$-ideal;
(iii) $I^2 \subseteq I$.

Proof. Trivial. □

1.14 A fractional (R', R)-ideal I provides a link between R' and R. For equivalent (right and left) orders, there is a converse.

Proposition. *Suppose that R, R' are equivalent orders in a quotient ring Q. Then there exists a fractional (R', R)-ideal A and a fractional (R, R')-ideal B. Further, these can be chosen so that $AB \subseteq R'$ and $BA \subseteq R$, and then*

$$\begin{bmatrix} R & B \\ A & R' \end{bmatrix}$$

forms a Morita context.

3.1.16

Proof. We know that $a_1 R a_2 \subseteq R'$ for some units $a_i \in Q$. Write $a_1 = r_1 s_1^{-1}$, $a_2 = s_2^{-1} r_2$ with $r_i, s_i \in R$. Evidently $R \subseteq s_1^{-1} R s_2^{-1}$, and so $r_1 R r_2 \subseteq R'$.

Similarly, $r_1' R r_2' \subseteq R$ with $r_1', r_2' \in R'$. Set $A = R' r_2' r_1 R$, and $B = R r_2 r_1' R'$. Evidently these are fractional ideals since, for example, $r_2' r_1 R \subseteq A$ and

$$r_1' A = r_1' R' r_2' r_1 R \subseteq R r_1 R \subseteq R.$$

Moreover

$$AB = R' r_2' r_1 R r_2 r_1' R' \subseteq R' r_2' R' r_1' R' \subseteq R',$$

and likewise $BA \subseteq R$. Evidently this gives a Morita context, as claimed. $\square$

1.15 As noted in 1.4(ii), if R is a right order in a quotient ring Q then $Q = R_{\mathscr{S}}$, where $\mathscr{S} = \{\text{units of } Q\} \cap R$. In the construction of $R_{\mathscr{S}}$, in 2.1.12, each element $q \in Q$ is realized as an R-homomorphism from a right R-ideal I to R. It is a easily seen that this representation identifies $O_l(I)$ with $\text{End}(I_R)$, a fact included in (iv) of the next result.

Proposition. *Let R be a right order in a quotient ring Q, and let $I, J \triangleleft Q_R$. Then:*
(i) $I \hookrightarrow IQ \simeq I \otimes Q$;
(ii) $\text{Hom}_R(I, J) \hookrightarrow \text{Hom}_Q(IQ, JQ)$ *via* $\alpha \mapsto \alpha \otimes 1$, *this giving the unique extension of α to IQ and JQ;*
(iii) *under this embedding,* $\text{Hom}_R(I, J) = \{\beta \in \text{Hom}(IQ, JQ) | \beta I \subseteq J\}$;
(iv) *if, further, I is a fractional right R-ideal, then*

$$\text{Hom}_R(I, J) \simeq \{q \in Q | qI \subseteq J\} \quad \text{and} \quad \text{End } I \simeq O_l(I).$$

Proof. (i) This follows directly from 2.1.17 since, by 1.4(ii), $Q = R_{\mathscr{S}}$.

(ii)(iii) These are clear from (i).

(iv) If I is a fractional right R-ideal, then $IQ = Q$. Therefore $\text{Hom}(IQ, JQ) = \text{Hom}(Q, JQ)$. Since any such homomorphism is realized by left multiplication by some element of Q, the result follows easily. $\square$

It will be convenient, later, to identify $\text{Hom}_R(I, J)$, for I a fractional right R-ideal, with the subset of Q specified in (iv).

1.16 We end this section with an embedding result.

Proposition. *Let R be a prime right Goldie ring with $R \subseteq M_n(D)$ for some n and some division ring D. Then the right quotient ring Q of R embeds in $M_k(D)$ for some $k \leq n$.*

Proof. The main obstacle to be overcome is that regular elements of R may be zero divisors in the ring $S = M_n(D)$. Consequently the embedding of R in S cannot be extended to Q.

We start by viewing S as an (S, R)-bimodule. As such it has a finite composition series of length at most n, the length of $_S S$. The annihilator in R of each composition factor is easily seen to be a prime ideal; and the appropriate product of these prime ideals annihilates S_R, and so is zero. Therefore, one simple composition factor, $_S M_R$ say, has annihilator zero in R.

Let M' be the torsion submodule of M_R with respect to $\mathscr{C}_R(0)$, as described in 2.1.17. It is easy to check that M' is a sub-bimodule. Say $M' = \sum_{i=1}^{t} S x_i$. If $x_i c_i = 0$ for $c_i \in \mathscr{C}_R(0)$, $i = 1, \ldots, t$, then, using 2.1.16(i), $M'c = 0$ for some $c \in \mathscr{C}_R(0)$. Since M_R is faithful, $M' = 0$.

Note, finally, that if $c \in \mathscr{C}_R(0)$, then $_S M \simeq {}_S Mc$ and so $M = Mc$ since $_S M$ has finite length k say. Thus Q embeds in $\mathrm{End}_S M \simeq M_k(D)$. □

A similar result holds for semiprime right Goldie rings.

§2 Minimal Primes

This section compares the ideals of a semiprime right Goldie ring R with those of $Q(R)$. It is shown that R is equivalent to a direct sum of prime right Goldie rings, namely $\oplus R/P_i$, where the P_i are the minimal prime ideals. Furthermore, each prime right Goldie ring is equivalent to a matrix ring over a right Ore domain, possibly without 1.

2.1 The notation below will be kept fixed in this section. Let R be semiprime right Goldie and Q be the right quotient ring of R. By Goldie's theorem, Q is semisimple; therefore $Q = \bigoplus_{i=1}^{k} Q_i$ with each Q_i being a simple Artinian ring generated, as an ideal of Q, by a central idempotent, e_i say. Note that $e_i = 1_{Q_i}$. For each i, the ideal $M_i = \sum \{Q_j | j \neq i\}$ is a maximal ideal; and this gives all the maximal (= minimal prime) ideals of Q. Indeed each ideal of Q is the sum of a subset of the Q_i, and so is (i) an intersection of minimal prime ideals, (ii) an annihilator ideal and (iii) a complement ideal (cf. 2.2.14, 2.2.15).

Let $P_i = M_i \cap R$, $A_i = Q_i \cap R$ and $A = \oplus A_i$. It will be seen that these ideals are of special interest.

2.2 Proposition. (i) *The ideals P_i are the minimal prime ideals of R.*

(ii) *The ideals of the form $I \cap R$, with $I \triangleleft Q$, are precisely the annihilator ideals of R.*

(iii) *The A_i are the minimal nonzero annihilator ideals.*

(iv) *The ideal $A = \bigoplus_{i=1}^{k} A_i$ is an essential right ideal of R, and is a right order in Q equivalent to R.*

(v) *A_i is a right order in Q_i.*

3.2.4   *Minimal primes*

Proof. (i) Suppose $X, Y \triangleleft R$ with $XY \subseteq P_i$. By 2.1.16(vi), $XQ, YQ \triangleleft Q$; so

$$XQYQ = XYQ \subseteq P_iQ \subseteq M_i.$$

Therefore $XQ \subseteq M_i$ or $YQ \subseteq M_i$ and hence $X \subseteq P_i$ or $Y \subseteq P_i$. This shows that P_i is prime. The fact that $\bigcap P_i = 0$ shows that they are the minimal prime ideals.

(ii) If $I \triangleleft Q$ then I is an intersection of some of the M_i. Therefore $I \cap R$ is an intersection of some P_i. Now apply 2.2.14(iii), 2.2.15.

(iii) Clear from (ii).

(iv) By 2.1.16 (iii), $A_iQ = Q_i$; so $AQ = Q$. Hence by 2.2.12(i), $A \triangleleft_e R_R$. The rest follows from 1.8.

(v) This is straightforward from 1.3 and (iv) above.   ☐

Corollary. *R contains a finite direct sum of prime right Goldie rings (not necessarily with* 1) *having the same right quotient ring.*   ☐

2.3 Of course, R is not necessarily a direct sum of prime rings. It is readily checked that this is so if and only if $R = A = \oplus A_i$. This makes it clear that the semiprime Noetherian ring

$$R = \{(a,b) | a - b \in 2\mathbb{Z}\} \subseteq \mathbb{Z}^2$$

is not a direct sum of prime rings.

2.4 The next results show that R is also equivalent, as a right order, to a direct sum of prime rings with a 1, namely the rings R/P_i, with P_i a minimal prime ideal of R, as in 2.2.

Proposition. (i) $R/P_i \simeq e_iR$;
(ii) $R \subseteq R' = \bigoplus_{i=1}^{k} e_iR$;
(iii) R' *is a semiprime right Goldie ring and* $R' \sim R$;
(iv) e_iR *is a prime right Goldie ring and* $e_iR \sim A_i$;
(v) $\mathscr{C}_R(0) = \bigcap_{i=1}^{k} \mathscr{C}_R(P_i)$.

Proof. (i) M_i is the kernel of the map $q \mapsto e_iq$ from Q to e_iQ, so $P_i = M_i \cap R$ is the kernel of the restriction to R.

(ii) This is clear.

(iii) $AR' = A \subseteq R$ since $A_i = A_ie_i$. By Proposition 2.2(iv), A contains a unit, a say, of Q. Then $aR' \subseteq R \subseteq R'$ and so $R' \sim R$.

(iv) This is evident from (iii) together with 1.3.

(v) If $c \in \mathscr{C}_R(0)$ then, for each i, e_ic is a unit of e_iQ and hence is regular in e_iR. Therefore $c \in \bigcap \mathscr{C}_R(P_i)$, using (i). The argument is easily reversed.   ☐

Corollary. *R is contained in a direct sum of prime right Goldie rings (with* 1) *having the same right quotient ring.*   ☐

2.5 The fact that the R/P_i are all prime right Goldie rings has a converse. For this result only, we drop the convention that R is right Goldie.

Proposition. *Let R be a semiprime ring with finitely many minimal prime ideals $P_1,\ldots,P_k$. Then R is right Goldie if and only if R/P_i is right Goldie for each i.*

Proof. If R is right Goldie then Proposition 2.4 applies. Conversely, if each R/P_i is right Goldie then $R \hookrightarrow S = \bigoplus_{i=1}^{k} R/P_i$, a semiprime right Goldie ring with quotient ring $Q = \bigoplus_{i=1}^{k} Q_i$ say. If $A_i = Q_i \cap R$ then $A = \bigoplus_{i=1}^{k} A_i$ is an ideal of R and also an essential right ideal of S. By 1.8, $A \sim S$ so, by Corollary 1.6, R is a right order in Q. □

2.6 We now turn to the nature of a prime right Goldie ring R. We will see that R contains an equivalent right order which is a matrix ring over a right Ore domain, but not necessarily with a 1. (Later, in 4.8, it will be seen that R is equivalent to, but not necessarily comparable with, a matrix ring over a right Ore domain with a 1). First some terminology is required.

Suppose, for the moment, that $S \simeq M_n(T)$, for rings S, T with 1. Recall (from [Jacobson **64**, p. 52] for example) that a **set of matrix units** of S is a subset $\mathcal{M} = \{e_{ij} | i,j = 1,\ldots,n\}$ of S such that $\sum e_{ii} = 1$ and $e_{ij}e_{kl} = \delta_{jk}e_{il}$, where δ_{jk} is the Kronecker delta symbol. If

$$T' = \{s \in S \mid se_{ij} = e_{ij}s \quad \text{for all } i,j\},$$

the **centralizer** of $\mathcal{M}$ in S, it follows that $S \simeq M_n(T')$ with $\mathcal{M}$ corresponding to the standard set of matrix units (see 2.2.11).

If T is a division ring, then the uniqueness in the Artin–Wedderburn theory shows that $T \simeq T'$.

Theorem. (Faith–Utumi theorem) *A semiprime right Goldie ring R contains an equivalent right order which is a direct sum of matrix rings over right Ore domains (not necessarily with 1).*

Proof. Using Corollary 2.2, it is enough to consider the case when R is prime right Goldie (but not with 1) and $Q \simeq M_n(D)$ for some division ring D.

The first step is to prove that a set of matrix units $\mathcal{M}$ of Q can so be chosen that $c\mathcal{M} \subseteq R$ for some regular element $c \in R$. To see this note that, by 2.1.16(i), there will be a regular element c with $\mathcal{M}c \subseteq R$. The set

$$\mathcal{M}' = \{c^{-1}e_{ij}c \mid e_{ij} \in \mathcal{M}\}$$

can be checked to be a set of matrix units; and $c\mathcal{M}' \subseteq R$. Thus $\mathcal{M}$ can indeed be (re)chosen as desired; and we may assume that D is its centralizer in Q.

3.3.1   *Right ideals in semiprime rings*   71

It is true, as above, that $\mathcal{M}b \subseteq R$ for some other regular element b. Consider the sets

$$C = \{x \in R \mid x\mathcal{M} \subseteq R\}; B = \{x \in R \mid \mathcal{M}x \subseteq R\}.$$

Clearly $C \triangleleft_l R$ and $B \triangleleft_r R$. The multiplicative properties of $\mathcal{M}$ show that $C\mathcal{M} = (C\mathcal{M})\mathcal{M}$; hence $C\mathcal{M} = C$. Similarly, $\mathcal{M}B = B$.

Consider the subring BC of R. Since $bRc \subseteq BC \subseteq R$, Lemma 1.6 shows that BC is a right order in Q, equivalent to R. Furthermore, since $BC = \mathcal{M}BC\mathcal{M}$, then $BC = M_n(K)$ where K is the centralizer of $\mathcal{M}$ in BC. So $K \subseteq D$ and so is an integral domain. One can see from 2.2.12 together with 2.3.9 that $\text{r u dim } M_n(K) = \text{r u dim } Q = n$; and yet, as is easily checked, $\text{r u dim } M_n(K) \geq n(\text{r u dim } K)$. Therefore $\text{r u dim } K = 1$, and so K is a right Ore domain, with right quotient division ring, D_1 say. □

In fact $D_1 = D$. To see this, note that $M_n(K) \subseteq M_n(D_1) \subseteq M_n(D)$. By Corollary 1.6, $M_n(D_1)$ is a right order in $M_n(D)$. Therefore, by 1.1, $M_n(D_1) = M_n(D)$ and so $D_1 = D$.

2.7 This leads easily to an alternative characterization of the uniform dimension of a prime right Goldie ring. Recall that the **index of nilpotency** of a nilpotent element a is the least n such that $a^n = 0$.

Corollary. *If R is a prime right Goldie ring then* $\text{u dim } R_R$ *is the largest index of nilpotency of any nilpotent element of R.*

Proof. Let $Q = Q(R)$; so $Q \simeq M_n(D)$ with D a division ring and $n = \text{u dim } R_R$. By 2.6, $R \supseteq M_n(K)$ and the element $k(e_{12} + e_{23} + \cdots + e_{n-1\,n})$ with $0 \neq k \in K$ is nilpotent with index n.

On the other hand, let $a \in R$ be nilpotent. We may view a as an endomorphism of $M = D^n$, the n-dimensional right D-vector space. The chain of subspaces $M \supseteq aM \supseteq a^2M \supseteq \cdots$ must strictly decrease until it reaches 0. Hence its length, and also the index of nilpotency of a, is at most n. □

§3 Right Ideals in Semiprime Rings

Throughout this section R denotes a semiprime right Goldie ring with right quotient ring Q. The connections between right ideals of R and of Q are described, with special interest in uniform and essential right ideals. In particular, it is shown that essential right ideals are generated by regular elements.

3.1 In Q, every right ideal has the form eQ, with $e = e^2$. This is a complement right ideal (complementary to $(1-e)Q$) and is an annihilator right ideal (of $Q(1-e)$).

Proposition. (i) *A right ideal I of R is a complement right ideal if and only if $I = J \cap R$ for $J \triangleleft_r Q$.*

(ii) *An annihilator right ideal I of R has the form $J \cap R$ for $J \triangleleft_r Q$.*

(iii) *If, further, R is a left Goldie ring and $J \triangleleft_r Q$, then $J \cap R$ is an annihilator right ideal of R.*

Proof. (i) If $K \triangleleft_r R$ with $I \cap K = 0$, then one can check that $(IQ \cap R) \cap K = 0$. So if I is a complement right ideal, then $I = IQ \cap R$. The converse is clear.

(ii) Immediate.

(iii) Say $J = \mathrm{r\,ann}\, K$ with $K = \mathrm{l\,ann}\, J$. Then $K = Q(K \cap R)$, so $J = \mathrm{r\,ann}_Q(K \cap R)$ and thus $J \cap R = \mathrm{r\,ann}_R(K \cap R)$. □

The necessity of the symmetric conditions imposed in (iii) above is easily demonstrated. Let K be a right Ore domain which is not left Ore. So $Ka \cap Kb = 0$ for some nonzero $a, b \in K$. Let $R = M_2(K)$ which, by 1.5, is a prime right Goldie ring. The right ideal A of R generated by

$$\begin{bmatrix} a & 0 \\ b & 0 \end{bmatrix}$$

has $\mathrm{l\,ann}\, A = 0$. The same, therefore, is true of $AQ \cap R$. However,

$$\begin{bmatrix} a & 0 \\ b & 0 \end{bmatrix}$$

is a zero divisor; so $AQ \neq Q$ and thus $AQ \cap R \neq R$. Hence $AQ \cap R$ is not a right annihilator.

3.2 The next result indicates the importance of uniform right ideals.

Lemma. (i) *A right ideal U of R is uniform if and only if UQ is a minimal right ideal of Q.*

(ii) *If U is a uniform right ideal of R and $0 \neq u \in U$ then:*

(a) $\mathrm{u\,dim}\,(\mathrm{r\,ann}\, u) = \mathrm{u\,dim}\, R_R - 1$;

(b) *if $I \triangleleft_r R$ with $\mathrm{r\,ann}\, u \subset I$, then $I \triangleleft_e R$.*

Proof. (i) Clear from 2.2.12, since uniform right ideals of Q are minimal.

(ii) Here uQ is a minimal right ideal of Q and so $\mathrm{r\,ann}_Q u$ is a maximal right ideal. The rest follows from 3.1. □

3.3 The next few results continue our investigation of uniform right ideals.

Proposition. *Let U be a uniform right ideal of R.*

(i) *Any non-zero $\alpha \in \mathrm{Hom}\,(U, R)$ is a monomorphism.*

(ii) If $I \triangleleft_r R$ then the following are equivalent:
  (a) $\operatorname{Hom}(U, I) \neq 0$;
  (b) $IU \neq 0$;
  (c) I contains a right ideal isomorphic to U.

Proof. (i) This is clear from 1.15 (ii) since the same is true of $\operatorname{Hom}(UQ, Q)$.
  (ii) If $0 \neq \alpha \in \operatorname{Hom}(U, I)$, then $U \simeq \alpha U \subseteq I$; and $IU \supseteq \alpha(U)U = \alpha(U^2) \neq 0$. On the other hand, if $IU \neq 0$ then $xU \neq 0$ for some $x \in I$. The map $U \to xU$ is therefore a monomorphism, so $xU \simeq U \subseteq I$. □

Corollary. *If U, V are uniform right ideals of R, the following are equivalent:*
  (i) $UV \neq 0$;
  (ii) $VU \neq 0$;
  (iii) U contains an isomorphic copy of V;
  (iv) V contains an isomorphic copy of U.

Proof. $(VU = 0) \Rightarrow ((UV)^2 = 0) \Rightarrow (UV = 0) \Rightarrow ((VU)^2 = 0) \Rightarrow (VU = 0)$. This shows (i) and (ii) are equivalent, and the proposition gives the result. □

3.4 The conditions (iii) and (iv) of the corollary are sometimes summed up by saying that U, V are **subisomorphic.**

Lemma. (i) *The number of subisomorphism classes of uniform right ideals of R equals the number of minimal prime ideals.*
  (ii) *If R is prime, then all pairs of uniform right ideals are subisomorphic.*
  (iii) *If R is prime and has a minimal right ideal, then R is simple Artinian.*

Proof. (i) Both numbers can be seen to equal the number of isomorphism classes of minimal right ideals of the right quotient ring of R.
  (ii) Clear from (i).
  (iii) By 2.2.8, R contains an essential right ideal, E say, which is a direct sum of uniform right ideals. By (ii), these must all be minimal and isomorphic. Hence $\operatorname{End} E \simeq M_n(D)$ for some n and some division ring D. By 1.15(iv), $O_l(E) \simeq M_n(D)$, so is its own quotient ring; yet, by 1.12(i), R is an equivalent right order. Evidently then $R = O_l(E)$. □

3.5 It is now possible to prove the analogue, for uniform right ideals, of Schur's lemma.

Theorem. *If U is a uniform right ideal of R then $\operatorname{End} U$ is a right order in the division ring $\operatorname{End}_Q UQ$, and so $\operatorname{End} U$ is a right Ore domain.*

Proof. Since UQ is a minimal right ideal of Q by 3.2, $\operatorname{End} UQ$ is indeed a division ring, and, as noted in 1.15 (ii), End U is a subring. Let $0 \neq \alpha \in \text{End } UQ$, and consider $V = U \cap \alpha U$. This is nonzero, since UQ_R is uniform (2.2.12); and $V \triangleleft_r R$. Hence $UV \supseteq V^2 \neq 0$, and so, by Corollary 3.3, U and V are subisomorphic. Choose an isomorphic copy of U inside V, and let W be its inverse image under α. Then $W \subseteq U$ and there is an isomorphism $\beta: U \to W$; so $\beta \in \text{End } U$. Note also that $\gamma = \alpha\beta \in \text{End } U$. This shows that $\alpha = \gamma\beta^{-1}$ and so End U is a right order in End UQ. □

This result will be superseded, in 4.7, by a result which describes the endomorphism ring of any right ideal or, indeed, of any submodule of R^n.

3.6 The remainder of this section concerns generators of right ideals of R, especially essential right ideals. These all contain regular elements, by 2.3.5. It will be seen that they are generated by regular elements.

Proposition. *Let $I \triangleleft_r R$ and let $b \in R$. Then there exists $d \in I$ such that* $\text{u dim }(b + d)R = \text{u dim }(bR + I)$.

Proof. If $\text{u dim }(bR + I) = \text{u dim } bR$, one can take $d = 0$. Otherwise there exists $U \subseteq I$, a uniform right ideal such that $bR \cap U = 0$. By induction, it is enough to find $u \in U$ such that $\text{u dim }(b + u)R = \text{u dim }(bR \oplus U) = \text{u dim }(bR) + 1$.

We show next that $\text{r ann}(b) \not\subseteq \text{r ann } U$. For suppose that $\text{r ann}(b) \subseteq \text{r ann } U = A$ say, and let $A' = \text{l ann } A$. Now $A \triangleleft R$ and A' is the complement to A, by 2.2.14. Therefore $A' \simeq bA' \subseteq A'$ and so $\text{u dim } bA' = \text{u dim } A'$ and thus $bA' \triangleleft_e A'$. However, $U \subseteq A'$ and so $0 \neq bA' \cap U \subseteq bR \cap U$, a contradiction.

Therefore there is some $u \in U$ with $\text{r ann}(b) \not\subseteq \text{r ann}(u)$, and this is the required element. To see this, note that $\text{r ann}(b) + \text{r ann}(u)$ is an essential right ideal, by 3.2.(ii), and so contains a regular element c, say, with $c = x + y$ and $bx = uy = 0$. Note that $bc = (b + u)y$ and $uc = (b + u)x$. Choose any element $br + us \in bR \oplus uR$. The right Ore condition gives elements $r', s' \in R$, $c' \in \mathscr{C}(0)$ such that $rc' = cr'$ and $sc' = cs'$. But then

$$(br + us)c' = bcr' + ucs' = (b + u)(yr' + xs') \in (b + u)R.$$

Hence

$$\text{u dim }(b + u)R = \text{u dim }(bR \oplus uR) = \text{u dim } bR + 1. \qquad \square$$

3.7 Before drawing the desired consequence from 3.6, an easy result is required.

Lemma. (i) *A right ideal E of R is essential if and only if $EQ = Q$.*
(ii) *A principal right ideal cR is essential if and only if $c \in \mathscr{C}_R(0)$.*
(iii) *If R is also left Goldie, then a right ideal E of R is essential if and only if* $\text{l ann } E = 0$.

Proof. (i) This is immediate from 2.2.12(i), since Q is the only essential right ideal of Q.

(ii) If $cR \triangleleft_e R$ then $cQ = Q$. Thus, using 1.1, c is a unit of Q, and so is regular in R. The converse is now clear.

(iii) It is clear from (i) that if E is essential then $\operatorname{l ann} E = 0$. Conversely, if E is not essential then $EQ \neq Q$ and so $\operatorname{l ann}_Q EQ \neq 0$. It follows, as in the proof of 3.1(iii), that $\operatorname{l ann}_R E \neq 0$. □

Corollary. *Each essential right ideal E of R is generated by regular elements.*

Proof. By 2.3.5, there is a regular element $c \in E$. By 3.6, given any $b \in E$, there exists $d \in cR$ such that

$$\operatorname{u dim} (b + d)R = \operatorname{u dim} (bR + cR) = \operatorname{r u dim} R.$$

Therefore $(b + d)R \triangleleft_e R$ and, by the lemma, $b + d \in \mathscr{C}_R(0)$. Since $b \in (b + d)R + cR$, the result is proved. □

§4 Endomorphism Rings

Once again, **throughout the section** R denotes a semiprime right Goldie ring with quotient ring Q.

It was shown in 3.5 that the endomorphism ring of a uniform right ideal of R is a right Ore domain and, *a fortiori*, is semiprime right Goldie. This result will now be extended to describe endomorphism rings of all right ideals and certain modules.

These techniques will be used to show that R is equivalent to a direct sum of matrix rings over right Ore domains with a 1.

4.1 We start by considering End M where M is a finitely generated module over Q. This is easily described; for M decomposes as a finite direct sum of simple modules which can be grouped together in isomorphism classes. Evidently then End M is a direct sum of matrix rings over division rings; $\bigoplus_{i=1}^{k} M_{n_i}(D_i)$ say. Note that the division rings here are amongst those involved in the similar description of Q; and that $\sum_{i=1}^{k} n_i$ is the length of M (and hence the uniform dimension of M).

4.2 Next we obtain some preparatory module-theoretic results. We recall that a module M_S over any ring S is **torsionless** if given any $0 \neq m \in M$ there exists $\alpha \in M^* = \operatorname{Hom}(M, S)$ such that $\alpha(m) \neq 0$. It is easy to see that this is equivalent to saying that M embeds in some direct product of copies of S. Note also that, for any module M_S, the left S-module $_S M^*$ is torsionless since if $0 \neq \alpha \in M^*$ and $m \in M$ is such that $\alpha(m) \neq 0$ then the map $M^* \to S$ via $\beta \mapsto \beta(m)$ is as required.

A related concept, when $S = R$, is M_R being **torsion-free**. This means that its torsion submodule with respect to $\mathscr{C}_R(0)$ is zero (see 2.1.17). This is equivalent to saying that the map $M \to M \otimes_R Q$ is an embedding, since the torsion submodule is the kernel.

That a torsion-free module need not be torsionless is clearly demonstrated by $\mathbb{Q}_\mathbb{Z}$; but the next result establishes a connection between these notions.

4.3 Proposition. *Let M_R be a torsionless module. Then:*
 (i) *M_R is torsion-free and* $\operatorname{u\,dim}(M_R) = \operatorname{u\,dim}(M \otimes Q_R) = \operatorname{u\,dim}(M \otimes Q_Q)$;
 (ii) *if M_R is finitely generated then* $\operatorname{u\,dim} M_R < \infty$;
 (iii) *if $\operatorname{u\,dim} M < \infty$ then $M \hookrightarrow R^n$ for some n;*
 (iv) *if $\operatorname{u\,dim} M = t$ then $M \hookrightarrow R^t$.*

Proof. (i) With m, α as above and $c \in \mathscr{C}_R(0)$, note that $\alpha(mc) = \alpha(m)c \neq 0$; thus $mc \neq 0$. Since M is torsion-free, $M \hookrightarrow M \otimes_R Q$. As in 2.2.12, it follows that $\operatorname{u\,dim} M_R = \operatorname{u\,dim} M \otimes Q_Q$.

(ii) This is clear since $M \otimes Q_Q$ is finitely generated and Q is semisimple.

(iii) Among all maps from M to free R-modules of finite rank choose one, say $\alpha \in \operatorname{Hom}(M, R^n)$, so that $\operatorname{u\,dim}(\ker \alpha)$ is as small as possible. Suppose $K = \ker \alpha \neq 0$. Choose $0 \neq U \subseteq K$ with U uniform, and pick $0 \neq u \in U$. There exists $\beta \in \operatorname{Hom}(M, R)$ such that $\beta(u) \neq 0$. Now define $\gamma \colon M \to R^{n+1}$ by $\gamma(m) = (\alpha(m), \beta(m))$. Clearly $\ker \gamma = \ker \alpha \cap \ker \beta$.

Consider $\ker \beta \cap U = V$ say. If $V \neq 0$, then V is uniform and is essential in U. Therefore, by 2.2.2 (iii), for each $u \in U$, there exists an essential right ideal E of R with $uE \subseteq V$. Pick $c \in E \cap \mathscr{C}_R(0)$. Then $0 = \beta(uc) = \beta(u)c$, and so $\beta(u) = 0$, a contradiction. Thus $\ker \beta \cap U = 0$; but then $\operatorname{u\,dim}(\ker \gamma) < \operatorname{u\,dim}(\ker \alpha)$ which is another contradiction. Hence $\ker \alpha = 0$.

(iv) By (iii) we may suppose that $M \subseteq R^n = \sum_{i=1}^n e_i R$ say. If $n > t$, note that $M \cap e_i R = 0$ for some i, otherwise $\operatorname{u\,dim} M \geqslant \operatorname{u\,dim}(\oplus_i M \cap e_i R) \geqslant n$. But then $M \hookrightarrow R^n / e_i R \simeq R^{n-1}$ and induction completes the proof. □

Corollary. *A module M_R embeds in R^n for some n if and only if M is torsionless and $\operatorname{u\,dim} M < \infty$.* □

4.4 We now turn to the study of $\operatorname{End} M$, with M as above. The dual module $M^* = \operatorname{Hom}(M, R)$ is used to connect submodules of M with right ideals of $\operatorname{End} M$. This involves the products defined in 1.1.5.

Proposition. *Let M_R be torsionless of finite uniform dimension, and $S = \operatorname{End} M$.*
 (i) *If $N \triangleleft M$ then $\operatorname{u\,dim}(N_R) = \operatorname{u\,dim}(NM^*)_S$.*
 (ii) *If $I \triangleleft S_S$ then $\operatorname{u\,dim} I_S = \operatorname{u\,dim} IM_R$.*
 (iii) *$S \hookrightarrow \operatorname{End}_Q(M \otimes Q)$.*

3.4.6   *Endomorphism rings*   77

Proof. (i)(ii) Note that if $I \neq 0$ then $IM \neq 0$ since $I \subseteq \operatorname{End} M$. Similarly, if $NM^* = 0$ then $(M^*N)^2 = 0$ and so, R being semiprime, $M^*N = 0$. However, M is torsionless and so $N = 0$. It is now straightforward to check that if $N_1 \oplus \cdots \oplus N_k$ is a direct sum of nonzero submodules of N, then $N_1 M^* \oplus \cdots \oplus N_k M^*$ is a direct sum of nonzero right ideals of S and vice versa. So (i) and (ii) follow.

(iii) If $\alpha \in S$, then $\alpha \otimes 1 \in \operatorname{End}(M \otimes Q)$ is a unique extension. Since $M \hookrightarrow M \otimes Q$ then $S \hookrightarrow \operatorname{End}(M \otimes Q)$ via the map $\alpha \mapsto \alpha \otimes 1$. □

4.5 Theorem. *Let M_R be a torsionless module of finite uniform dimension and $S = \operatorname{End} M$. Then S is semiprime right Goldie and* $\operatorname{r u dim} S = \operatorname{u dim} M_R$.

Proof. Suppose $I \triangleleft S$ with $I^2 = 0$. Then $M^*IMM^*IM \subseteq M^*I^2M = 0$ and so M^*IM is a nilpotent ideal of R. Therefore $M^*IM = 0$, and so $IM = 0$ since M is torsionless. However, $I \subseteq S$, so $I = 0$. This shows that S is semiprime. By 4.4(i)(ii), S has finite right uniform dimension equal to that of M_R. Also, by 4.4(iii), $S \hookrightarrow \operatorname{End}(M \otimes Q)$. By 4.3(i) $(M \otimes Q)_Q$ is finitely generated, and so, as noted in 4.1, $\operatorname{End}(M \otimes Q)_Q$ is semisimple Artinian. Therefore its subring S has a.c.c. on right annihilators. This establishes that S is semiprime right Goldie. □

4.6 Note that it is not yet established that S is a right order in $\operatorname{End}(M \otimes Q)$. This will be proved after a preparatory result.

Proposition. *Let M_R be torsionless of finite uniform dimension and $S = \operatorname{End} M$. If $M' \triangleleft_e M$, then there exists $c \in M'M^* \subseteq S$ such that c is a unit of $\operatorname{End}(M \otimes Q)$.*

Proof. It follows from 4.4 (i), (ii) that $M'M^* \triangleleft_e S_S$. Therefore, since S is semiprime right Goldie, $c \in M'M^*$ for some $c \in \mathscr{C}_S(0)$. Let K be the kernel of c acting on $M \otimes Q$; so $c(K \cap M) = 0$. Then $c(K \cap M)M^* = 0$ and, since $(K \cap M)M^* \subseteq S$, it follows that $(K \cap M)M^* = 0$. By 4.4, $K \cap M = 0$ and therefore, using 4.3(i), $K = 0$. Thus c is a monomorphism on $M \otimes Q$ and so is an isomorphism. That is, c is a unit in $\operatorname{End}(M \otimes Q)$. □

Corollary. *Let M_R be torsionless of finite uniform dimension.*
(i) $\operatorname{End} M$ *is a right order in the semisimple Artinian ring* $\operatorname{End}(M \otimes Q)$.
(ii) *If R is prime then* $\operatorname{End} M$ *is prime.*
(iii) *If $N \triangleleft_e M$, then* $\operatorname{End} N \sim \operatorname{End} M$.

Proof. (i) Let $\alpha \in \operatorname{End}(M \otimes Q)$. Let M_0 be the inverse image of M under α. It is easy to see that $M_0 \triangleleft_e (M \otimes Q)_R$. Now let $M' = M_0 \cap M$, and choose c as in the proposition. Clearly $cM \subseteq M'$ and so $\alpha c = \gamma : M \to M$. Therefore $\alpha = \gamma c^{-1}$ and both $\gamma, c \in S$.

(ii) This is clear from (i) since $\operatorname{End}(M \otimes Q)$ is simple.

(iii) Choose c, a unit of $\operatorname{End}(M \otimes Q)$, so that $cM \subseteq N$. Note that since $\operatorname{u\,dim} N = \operatorname{u\,dim} M$, then $N \otimes Q = M \otimes Q$. So there is an embedding $N \hookrightarrow M \otimes Q$ and $\operatorname{End} N \hookrightarrow \operatorname{End}(M \otimes Q)$. Evidently $(\operatorname{End} N)c \subseteq \operatorname{End} M$ and $c \operatorname{End} M \subseteq \operatorname{End} N$. So $\operatorname{End} M \sim \operatorname{End} N$. □

4.7 It is natural to ask whether, if R is also left Goldie, the same holds for $\operatorname{End} M$.

Proposition. *Let R be a semiprime Goldie ring.*
(i) *If M_R is finitely generated torsion-free, then $M \hookrightarrow R^n$ and so is torsionless.*
(ii) *If M_R is torsionless of finite uniform dimension, then $\operatorname{End} M$ is semiprime Goldie.*

Proof. (i) It is clear that $M \hookrightarrow M \otimes Q \hookrightarrow Q^n$ for some n. Choose a common left denominator c for the generators in Q^n of M_R. Then $M \simeq cM \subseteq R^n$.

(ii) Note first that since $M \otimes Q_Q$ has finite uniform dimension, so too has the left Q-module $\operatorname{Hom}(M \otimes Q, Q)$. Since R is also a left order in Q, then $\operatorname{u\,dim}_R(\operatorname{Hom}(M \otimes Q, Q)) < \infty$. However, $M^* \hookrightarrow \operatorname{Hom}(M \otimes Q, Q)$. Therefore $_R M^*$ has finite uniform dimension and is torsionless. It follows from 4.5 that $\operatorname{End} M^*$ is a semiprime left Goldie ring.

Note that there is a natural embedding $\operatorname{End} M \hookrightarrow \operatorname{End} M^* \hookrightarrow \operatorname{End}(M \otimes Q)$. Also, by Proposition 4.6, there is a unit $c \in \operatorname{End}(M \otimes Q)$ with $c \in MM^*$. However,

$$c \operatorname{End} M^* \subseteq MM^*(\operatorname{End} M^*) \subseteq MM^* \subseteq \operatorname{End} M.$$

So $\operatorname{End} M \sim \operatorname{End} M^*$, and hence is left, as well as right, Goldie. □

4.8 Of course, every right ideal I of R is torsionless and of finite uniform dimension. So $\operatorname{End} I$ is also semiprime right Goldie with $\operatorname{r\,u\,dim} \operatorname{End} I = \operatorname{u\,dim} I$. In particular, if I is uniform then $\operatorname{r\,u\,dim} \operatorname{End} I = 1$, and so $\operatorname{End} I$ is, as noted in 3.5, a right Ore domain.

These notions can be used to provide another decomposition result—for orders equivalent to R. Unlike 2.6, the rings here will all have 1's, but may not be comparable to R.

Theorem. *R is equivalent to a direct sum of matrix rings over right Ore domains.*

Proof. Using 3.3 and 3.4, choose an essential right ideal I of R which is a direct sum $\bigoplus_{i=1}^{t} U_i$ of uniform right ideals so that, for each i, j, either $U_i \simeq U_j$ or else $\operatorname{Hom}(U_i, U_j) = 0 = \operatorname{Hom}(U_j, U_i)$. It is evident that $\operatorname{End} I$ is a direct sum of matrix rings over right Ore domains. By Corollary 4.6, $\operatorname{End} I \sim \operatorname{End}(R_R) \simeq R$. □

As noted earlier, the ring R as above need not itself be a direct sum of prime rings; and nor need a prime right Goldie ring be a matrix ring over a right Ore domain. To see this, note that if R is a ring (with 1) and $R \simeq M_n(K)$ then K has a 1 and R contains the idempotent elements e_{ii}. However the Noetherian prime ring

$$R = \left\{ \begin{bmatrix} a & b \\ c & d \end{bmatrix} \middle| a - d \equiv b \equiv c \equiv 0 \bmod 2 \right\} \subseteq M_2(\mathbb{Z})$$

is not a domain and has no idempotent elements $\neq 0, 1$.

4.9 There is another interesting consequence which concerns the case when R is a principal right ideal (pri) ring.

Theorem. *A semiprime pri-ring is a finite direct sum of matrix rings over right Noetherian domains.*

Proof. Applying the proof of 4.8 to this case the right ideal I obtained is principal, say $I = aR$. Since $I \triangleleft_e R$, it follows that a is a unit of Q; so $\operatorname{End} I \simeq O_l(I) = aRa^{-1} \simeq R$. Thus, R is a finite direct sum of matrix rings over right Ore domains; and since R is right Noetherian so too are the domains, by 1.1.2. □

4.10 It is not, in general, true that the domains are pri-rings. Indeed it will be shown, in 7.11.7 and 7.11.8, that although $A_1(k)$ is not a pri-ring, $M_2(A_1(k))$ is, when char $k = 0$. Nevertheless, there is a partial converse.

Proposition. *If R is a pri-ring then so too is $M_n(R)$.*

Proof. If $A \triangleleft_r M_n(R)$ let $A_i = A \cap (e_{ii} + \cdots + e_{nn})M_n(R)$, where $\{e_{ij}\}$ is a set of matrix units. Define

$$B_i = \{b \in R \mid b \text{ is an entry in row } i \text{ of some } a \in A_i\}$$

and note that $B_i \triangleleft_r R$ and that

$$B_i = \{b \in R \mid b \text{ is the } (i, i) \text{ entry of some } a \in A_i e_{ji} \text{ for some } j\}.$$

Of course, B_i is principal, say $B_i = b_{ii}R$. The latter description of B_i shows that A contains a lower triangular matrix of the form

$$a = \begin{bmatrix} b_{11} & & 0 \\ \vdots & \ddots & \\ b_{n1} & \cdots & b_{nn} \end{bmatrix}$$

and, using the former description, it is straightforward to check that $A = aM_n(R)$. □

4.11 Corollary. *If S is a simple Artinian ring then $S[x]$ is a prime principal ideal ring.*

Proof. For some division ring D, $S \simeq M_n(D)$ and then $S[x] \simeq M_n(D[x])$. By 1.2.9 $S[x]$ is prime and $D[x]$ is a principal ideal domain. Now apply 4.10. □

§5 Morita Equivalence

This section contains a brief account of the theory of Morita equivalence. In particular it shows that the property of being a semiprime right Goldie ring is preserved under Morita equivalence.

5.1 Let R be any ring and M a right R-module. Let $S = \operatorname{End} M$, and $M^* = \operatorname{Hom}(M, R)$. Then, as in 1.1.5, there is a Morita context

$$\begin{bmatrix} R & M^* \\ M & S \end{bmatrix}$$

using the multiplication maps described there. It will be seen that, under certain circumstances, this provides a very strong link between the rings R and S.

5.2 The notation of 5.1 is retained. Recall that M is **projective** if M is a direct summand of a free module.

Lemma. (Dual basis lemma) (i) M_R *is projective if and only if there exist* $m_i \in M$, $g_i \in M^*$ *such that, for each* $x \in M$, *only finitely many* $g_i x \neq 0$ *and also* $\sum m_i g_i x = x$.
(ii) M_R *is finitely generated projective if and only if* $MM^* = S$. *Then* $_R M^*$ *is also finitely generated projective and* $M \simeq \operatorname{Hom}(M^*, R) = (M^*)^*$.

Proof. (i) Suppose M_R is projective. Then $M \oplus K \simeq R^I$ for some index set I and module K_R. Let $\{e_i\}$ be a free basis for R^I, and $e_i = m_i + k_i$ with $m_i \in M$, $k_i \in K$. Let g_i be the coordinate map $R^I \to R$ corresponding to e_i, but restricted to M. Then $\{m_i\}$, $\{g_i\}$ are as claimed.
Conversely, given such 'dual bases', the epimorphism $R^I \to M$, sending each e_i to m_i, is split by the homomorphism $x \mapsto \sum e_i g_i x$. Therefore M_R is projective.
(ii) If M_R is finitely generated, then I can be picked to be finite, say $\{1, \ldots, n\}$. Then $\sum_{i=1}^n m_i g_i = 1 \in MM^*$, so $MM^* = S$. Conversely, if $MM^* = S$, then $1 = \sum_{i=1}^n m_i g_i \in S$ and then $M = \sum m_i R$.
The fact that $M \simeq M^{**}$ is verified by viewing M as a direct summand of R^n and checking the result for R itself. □

We note the following consequence for fractional ideals.

Corollary. *Let R be a right order in a quotient ring Q and let M be a fractional right R-ideal. If M_R is projective, then M_R is finitely generated.*

Proof. By 1.15, we can identify M^* with the set $\{q \in Q \mid qM \subseteq R\}$. Now M contains a unit, x say. Applying (i) shows that $g_i \neq 0$ for only finitely many i, say for $i = 1, \ldots, n$. It follows that $M = \sum_{i=1}^{n} m_i R$. □

This applies, of course, to any essential right ideal of a semiprime right Goldie ring.

5.3 A module M_R is called a **generator** if $M^*M = R$. Lemma 5.2(ii) shows that being a generator is, in a way, dual to being finitely generated projective. This duality is made clearer by the next fact.

Lemma. (i) M_R *is finitely generated projective if and only if M is isomorphic to a direct summand of R^n for some n.*
 (ii) *M_R is a generator if and only if R is isomorphic to a direct summand of M^n for some n.*

Proof. (i) Evident from above.
 (ii) If M_R is a generator then $1 = \sum_{i=1}^{n} \alpha_i x_i$ with $\alpha_i \in M^*$, $x_i \in M$. The map $M^n \to R$ via $(m_1, \ldots, m_n) \mapsto \sum \alpha_i m_i$ is thus an epimorphism; and the map splits since R_R is projective. The argument is easily reversed to give the converse. □

5.4 If M_R is both a finitely generated projective module and a generator, it is called a **progenerator**. In that case the Morita context

$$\begin{bmatrix} R & M^* \\ M & S \end{bmatrix}$$

with $S = \mathrm{End}\, M_R$ has the special property that $MM^* = S$ and $M^*M = R$. The corollary to the next result shows that this property of any Morita context characterizes progenerators. It also shows that there is a complete symmetry between R and S and between M and M^*.

Proposition. *Suppose that*

$$\begin{bmatrix} R & V \\ W & S \end{bmatrix}$$

is the ring of a Morita context and that $VW = R$. Then:
 (i) *V_S and $_S W$ are finitely generated projective;*
 (ii) *$_R V$ and W_R are generators;*
 (iii) *$V \simeq (_S W)^*$, $W \simeq (V_S)^*$ as bimodules;*
 (iv) *$R \simeq \mathrm{End}\,(V_S)$, $R \simeq \mathrm{End}\,(_S W)$;*
 (v) *the multiplication map $\mu : V \otimes W \to VW = R$ is an R-bimodule isomorphism.*

Proof. Symmetry shows that it is enough to prove the first half only of (i)–(iv). Since $VW = R$ there are $v_i \in V$, $w_i \in W$ such that $1 = \sum_{i=1}^{n} v_i w_i$.

(i) There are S-homomorphisms
$$\alpha: V \to S^n; v \mapsto (w_1 v, \ldots, w_n v)$$
$$\beta: S^n \to V; (s_1, \ldots, s_n) \mapsto \sum v_i s_i$$
and $\beta\alpha = 1_V$; so V_S is finitely generated projective.

(ii) The R-homomorphism $V^n \to R$ given by $(v'_1, \ldots, v'_n) \mapsto \sum v'_i w_i$ is surjective; so $_R V$ is a generator.

(iii) There is an (R, S)-bimodule homomorphism $\gamma: V \to (_S W)^*$ with $v \in V$ acting by right multiplication. Since $v = (\sum v_i w_i) v = \sum v_i (w_i v)$ it follows that γ is injective; and if $f \in (_S W)^*$ then
$$f(w) = f(\sum w v_i w_i) = \sum w v_i f(w_i) = wv,$$
where $v = \sum v_i f(w_i)$, and so γ is surjective.

(iv) A similar argument, with $r \in R$ acting on V_S by left multiplication, shows that $R \simeq \text{End}_S W$.

(v) Suppose $\mu(\sum_j (v'_j \otimes w'_j)) = 0$. Then
$$\sum_j (v'_j \otimes w'_j) = \sum_{i,j} (v'_j \otimes w'_j v_i w_i) = \sum_{i,j} (v'_j w'_j v_i \otimes w_i) = (\sum_j v'_j w'_j)(\sum_i v_i \otimes w_i) = 0.$$

Hence μ is an isomorphism. □

Corollary. (a) Suppose that
$$\begin{bmatrix} R & V \\ W & S \end{bmatrix}$$
is the ring of a Morita context and that $VW = R$, $WV = S$. Then:
 (i) $W_{R,S}$, $V_{S,R} V$ are progenerators;
 (ii) $V \simeq (W_R)^* \simeq (_S W)^*$ and $W \simeq (_R V)^* \simeq (V_S)^*$; and
 (iii) $S \simeq \text{End}(W_R) \simeq \text{End}(_R V)$ and $R \simeq \text{End}(_S W) \simeq \text{End}(V_S)$.
(b) Suppose that M_R is a progenerator with $S = \text{End}(M_R)$. Then:
 (i) $_S M$ is a progenerator with $R \simeq \text{End}_S M$; and
 (ii) $M \otimes_R M^* \simeq S$, $M^* \otimes_S M \simeq R$. □

5.5 Two rings are **Morita equivalent** if there is a progenerator W_R such that $S \simeq \text{End } W_R$; we write $R \overset{M}{\sim} S$. It is easy to check, with the results above, that this is indeed an equivalence relation.

By way of example, note that, for any n, $M_n(R) \overset{M}{\sim} R$. Indeed, the left ideal $M_n(R) e_{11}$ serves as a progenerator linking these two rings (being R-isomorphic to R^n, of course). Hence $M_n(R) \overset{M}{\sim} M_t(R)$ for any t.

5.6 The next result at which we aim will show that matrix rings and Morita equivalence are closely linked—but not quite as closely as this example might suggest.

Lemma. *If M_R is finitely generated projective, then there is an integer n and an idempotent $e \in M_n(R)$ such that* $\operatorname{End} M \simeq e M_n(R) e$.

Proof. There is a module K_R such that $M \oplus K \simeq R^n$ for some n (by 5.3(i)). Now $\operatorname{End} R^n$ can be identified with $M_n(R)$. Let $e \in M_n(R)$ correspond to the projection $R^n \to M \hookrightarrow R^n$. Then $e^2 = e$ and $\operatorname{End} M \simeq e M_n(R) e$. $\square$

Proposition. *Two rings R, S are Morita equivalent if and only if there exists an integer n and an idempotent element $e \in M_n(R)$ such that $S \simeq e M_n(R) e$ and $M_n(R) e M_n(R) = M_n(R)$.*

Proof. As noted in 5.5, we can identify R^n with $M_n(R) e_{11}$. So if $R \stackrel{M}{\sim} S$ and $_S M_R$ is the appropriate progenerator then, in the notation of the lemma, M can be identified with $e M_n(R) e_{11}$, R with $e_{11} M_n(R) e_{11}$ and M^* with $e_{11} M_n(R) e$. Since $R = M^* M$ then

$$e_{11} M_n(R) e_{11} = (e_{11} M_n(R) e)(e M_n(R) e_{11}).$$

It follows that $M_n(R) e M_n(R)$ is an ideal of $M_n(R)$ containing e_{11} and thus $M_n(R) e M_n(R) = M_n(R)$.

The converse is easily obtained by reversing the argument. $\square$

It is not difficult to see that Morita equivalent rings need not be matrix rings over a common subring. For example, let R be a Dedekind domain whose class group has order 2 (such as $\mathbb{Z}[\sqrt{-5}]$ say, or see 12.1.6). Then, for any ideal I of R, the ideal I^2 is principal. Choose a nonprincipal ideal A and let $S = \operatorname{End}(R \oplus A)$. Since R is a Dedekind domain, A is projective and so $R \oplus A$ is, evidently, a progenerator. Thus $S \stackrel{M}{\sim} R$. By 4.5 $\operatorname{r u dim} S = 2$; so if S is a matrix ring, it must be a 2×2 matrix ring. However, that implies that $R \oplus A \simeq B \oplus B$ for some module B_R. Then $\operatorname{u dim} B_R = 1$ and B_R is finitely generated projective. Hence, by 4.3(iv), $B \simeq C$ for some ideal C of R. Note, however, that the module theory for Dedekind domains, described in 5.7.18, shows that, since $R \oplus A \simeq C \oplus C$, then $RA \simeq C^2$. This implies that A is principal—a contradiction.

5.7 A Morita equivalence between two rings establishes a tight connection between their modules and between their ideals. This is shown in the next few results. (See [Cohn 77, p. 93] for information about category equivalence.)

Proposition. *Let R, S be Morita equivalent rings, with progenerator $_S M_R$.*
(i) The functor $N_R \mapsto (N \otimes_R M^*)_S$ provides a category equivalence between right

R-modules and right S-modules, and $\operatorname{End}(N_R) \simeq \operatorname{End}((N \otimes M^*)_S)$.

(ii) *The functor* $_R N_R \mapsto {}_S(M \otimes_R N \otimes_R M^*)_S$ *provides a category equivalence between R-bimodules and S-bimodules, and* $\operatorname{End}(_R N_R) \simeq \operatorname{End}(_S(M \otimes N \otimes M^*)_S)$.

Proof. (i) The given functor is inverse to the functor $P_S \mapsto (P \otimes_S M)_R$ by Corollary 5.4(ii). Hence the result follows.

(ii) This is proved similarly. □

5.8 The fact that the functor given by 5.7(i) yields a category equivalence, means that the modules N_R and $(N \otimes_R M^*)_S$ share many properties. Properties of a module which are preserved under a category equivalence will be termed **Morita invariant** properties. (Projective dimension and global dimension are described in Chapter 7.)

Lemma. *The following properties of a module are Morita invariant:*
  (i) *being Artinian;*
 (ii) *being Noetherian;*
(iii) *being finitely generated;*
(iv) *being projective;*
 (v) *having projective dimension k;*
(vi) *having uniform dimension k;*
(vii) *being a generator;*
(viii) *being a progenerator.*

Proof. These are all straightforward. For example, to prove (v), note that if

$$0 \to P_k \to P_{k-1} \to \cdots \to P_0 \to N \to 0$$

is a projective resolution for N_R then

$$0 \to P_k \otimes M^* \to P_{k-1} \otimes M^* \to \cdots \to P_0 \otimes M^* \to N \otimes M^* \to 0$$

is a projective resolution for $N \otimes M_S^*$. This shows that $\operatorname{pd} N \otimes M^* \leq \operatorname{pd} N$; and symmetry completes the argument. □

5.9 Theorem. *Let R, S be Morita equivalent rings with progenerator $_S M_R$.*
  (i) *The map $A \mapsto MAM^*$ gives a semigroup isomorphism between the ideals of R and those of S. In particular there is a $(1,1)$-correspondence between their prime ideals which preserves primitivity.*
 (ii) *If $A \triangleleft R$, then R/A and S/MAM^* are Morita equivalent rings.*
(iii) $Z(R) \simeq Z(S)$.

Proof. (i) This is almost immediate from 5.7.

(ii) Let $V = M^*/AM^*$ and $W = M/MA$. One can check that, under the induced multiplications, $VW = R/A$ and $WV = S/MAM^*$. Therefore, Corollary 5.4,

applied to the Morita context
$$\begin{bmatrix} R/A & V \\ W & S/MAM^* \end{bmatrix},$$
shows that $R/A \overset{M}{\sim} S/MAM^*$.

(iii) If $\vartheta: R \to R$ is an R-bimodule endomorphism then $\vartheta(1)r = \vartheta(r) = r\vartheta(1)$ and so $\vartheta(1) \in Z(R)$. It follows easily that $Z(R) \simeq \mathrm{End}\,(_R R_R)$ and therefore, by 5.7(ii), $Z(R) \simeq Z(S)$. □

5.10 For rings, as for modules, properties which are preserved under Morita equivalence are said to be **Morita invariant**.

Proposition. *The following properties of a ring are Morita invariant*:
(i) *being Artinian*;
(ii) *being Noetherian*;
(iii) *being prime*;
(iv) *being semiprime*;
(v) *being semiprime right Goldie*;
(vi) *having right global dimension k*.

Proof. (i), (ii) and (vi) are clear from 5.8, and (iii), (iv) come from 5.9(i). As for (v), note, by Corollary 4.3, that a progenerator M over a semiprime right Goldie ring is torsionless of finite uniform dimension. Hence, by 4.5, $\mathrm{End}\, M$ is semiprime right Goldie. □

Despite (v), simply being a right Goldie ring is not a Morita invariant property, as noted in 1.5.

§6 Morita Contexts and Prime Rings

6.1 The preceding two sections both include results concerning the transfer from a given ring to a related ring of the property of being prime right Goldie. Thus, from Corollary 4.6(ii), we see that if R is prime right Goldie and M_R is torsionless of finite uniform dimension, then $S = \mathrm{End}\, M_R$ also is prime right Goldie. Similarly, if R, S are Morita equivalent rings, and P, P' are prime ideals corresponding as in 5.9, then, by 5.10, R/P being right Goldie implies that S/P' is right Goldie.

These two results are both special cases of the results of this section. It is shown here that for any Morita context
$$\begin{bmatrix} R & V \\ W & S \end{bmatrix}$$
there are subsets of $\mathrm{Spec}\, R$ and $\mathrm{Spec}\, S$ in (1, 1)-correspondence and that, if P,

P' are corresponding primes, then, under mild conditions on the context, if R/P is right Goldie, so too is S/P'. This yields the quoted result from Section 4 when applied to the usual context
$$\begin{bmatrix} R & M^* \\ M & S \end{bmatrix}.$$

6.2 Theorem. *Let*
$$\begin{bmatrix} R & V \\ W & S \end{bmatrix}$$
be a Morita context. Then there is an order preserving $(1, 1)$-correspondence between the sets of prime ideals $\{P \in \operatorname{Spec} R \mid P \not\supseteq VW\}$ and $\{P' \in \operatorname{Spec} S \mid P' \not\supseteq WV\}$ given by
$$P \mapsto \{s \in S \mid VsW \subseteq P\}.$$

Proof. Let $P' = \{s \in S \mid VsW \subseteq P\}$. To see that P' is prime, suppose $AB \subseteq P'$ with $A, B \triangleleft S$. Then $VAWVBW \subseteq VABW \subseteq VP'W \subseteq P$; and $VAW, VBW \triangleleft R$. So either VAW or VBW is inside P; and hence A or B is inside P'.

Also $VW \not\subseteq P$ and so $VWVW \not\subseteq P$. Thus $WV \not\subseteq P'$,

Finally, we note that the map in the opposite direction sends P' to P. For $P' \mapsto \{r \in R \mid WrV \subseteq P'\}$; but
$$WrV \subseteq P' \Leftrightarrow VWrVW \subseteq P \Leftrightarrow r \in P$$
since $VW \not\subseteq P$. □

6.3 This result is illustrated by the following examples.

Example. (i) Let R be a semiprime right Goldie ring with right quotient ring Q and A an essential right ideal. Let $A^* = \{q \in Q \mid qA \subseteq R\}$ and $S = O_l(A)$. Then 1.14 shows that
$$\begin{bmatrix} R & A^* \\ A & S \end{bmatrix}$$
is a Morita context. The correspondence is between prime ideals of R, S not containing A^*A and AA^* respectively. Since both A and A^* contain units of Q then $A^*A \triangleleft_e R_R$ and $AA^* \triangleleft_e S_S$. Therefore no minimal prime contains one of these. Thus the correspondence includes all minimal primes of the two rings.

Example. (ii) Let R, R' be equivalent orders in a semisimple quotient ring Q. By 1.14, there is a Morita context
$$\begin{bmatrix} R & B \\ A & R' \end{bmatrix}.$$

3.6.4 Morita contexts and prime rings

Again, it is easily seen that the correspondence involves all the minimal primes of R and R'.

Example. (iii) Let R be any ring, $0 \neq A \triangleleft R$ and S a subring of R with $A \subseteq S \subseteq R$. Then

$$\begin{bmatrix} R & A \\ A & S \end{bmatrix}$$

is a Morita context. The correspondence is between $\{P \in \operatorname{Spec} R \mid P \not\supseteq A\}$ and $\{P' \in \operatorname{Spec} S \mid P' \not\supseteq A\}$ and is given by $P \mapsto P \cap S$. In particular, if R is prime, so too is S.

For a specific example, consider the rings $R = M_2(\mathbb{Z})$ and

$$S = \begin{bmatrix} \mathbb{Z} & 2\mathbb{Z} \\ \mathbb{Z} & \mathbb{Z} \end{bmatrix}.$$

Since S contains $M_2(2\mathbb{Z}) \triangleleft R$, this is covered by (iii). The only prime of R not dealt with is $M_2(2\mathbb{Z})$; whereas, in S, there are two prime ideals missed, namely

$$\begin{bmatrix} 2\mathbb{Z} & 2\mathbb{Z} \\ \mathbb{Z} & \mathbb{Z} \end{bmatrix} \text{ and } \begin{bmatrix} \mathbb{Z} & 2\mathbb{Z} \\ \mathbb{Z} & 2\mathbb{Z} \end{bmatrix}.$$

Example. (iv) If e is any nonzero idempotent element of a ring R, then

$$\begin{bmatrix} R & Re \\ eR & eRe \end{bmatrix}$$

is a Morita context. This time the correspondence is between $\{P \in \operatorname{Spec} R \mid e \notin P\}$ and $\operatorname{Spec} eRe$ and is given by $P \mapsto ePe$.

Thus if

$$R = \begin{bmatrix} \mathbb{Z} & 2\mathbb{Z} \\ \mathbb{Z} & \mathbb{Z} \end{bmatrix} \text{ and } e = \begin{bmatrix} 1 & 0 \\ 0 & 0 \end{bmatrix}$$

then $eRe \simeq \mathbb{Z}$. The correspondence covers all primes of R apart from

$$\begin{bmatrix} \mathbb{Z} & 2\mathbb{Z} \\ \mathbb{Z} & 2\mathbb{Z} \end{bmatrix}.$$

Further examples, involving idealizers, appear in 5.6.11.

6.4 Given a Morita context

$$\begin{bmatrix} R & V \\ W & S \end{bmatrix}$$

with prime ideals P of R, P' of S corresponding as in 6.2, we say the prime rings

R/P and S/P' are **context equivalent**. Note that this relation is symmetric since

$$\begin{bmatrix} S & W \\ V & R \end{bmatrix}$$

is also a Morita context.

6.5 A Morita context

$$\begin{bmatrix} R & V \\ W & S \end{bmatrix}$$

is called a **prime context** if R is prime and vW, Vw and VsW are nonzero for any $0 \neq v \in V$, $0 \neq w \in W$ and $0 \neq s \in S$. It is easy to check that this too is symmetric in that

$$\begin{bmatrix} S & W \\ V & R \end{bmatrix}$$

is also a prime context. (Indeed, the conditions are equivalent to the ring

$$\begin{bmatrix} R & V \\ W & S \end{bmatrix}$$

being prime.) An easy example is obtained by choosing any prime ring R, $0 \neq W \triangleleft_r R$, $V = W^*$, and $S = \text{End}(W_R)$.

Proposition. (i) *Two prime rings are context equivalent if and only if they belong to a prime Morita context.*
(ii) *Context equivalence preserves primitivity.*

Proof. (i)($\Rightarrow$) If

$$\begin{bmatrix} R & V \\ W & S \end{bmatrix}$$

is a Morita context, and P, P' are corresponding prime ideals, then it is clear that

$$\begin{bmatrix} R/P & V/V' \\ W/W' & S/P' \end{bmatrix}$$

is a prime context, where $W' = \{w \in W \mid Vw \subseteq P\}$ and $V' = \{v \subseteq V \mid vW \subseteq P\}$.
($\Leftarrow$) Clear, taking $P = 0$, $P' = 0$.
(ii) Let R be a primitive ring in a prime Morita context

$$\begin{bmatrix} R & V \\ W & S \end{bmatrix}.$$

Let $M = mR$ be a faithful simple module. We claim that $(V/U)_S$ is a faithful

3.6.7 *Morita contexts and prime rings* 89

simple S-module where $U = \{u \in V \mid muW = 0\}$. To check faithfulness, note that if $Vs \subseteq U$ for some $s \in S$ then $0 = mVsW = MVsW$ and so $VsW = 0$ and $s = 0$. To check simplicity, choose any $v_0 \in V \setminus U$. Now $mv_0 W \neq 0$ and so $mv_0 W = M$; hence $m = mv_0 w_0$ for some $w_0 \in W$. Finally, let $v \in V$. Then $m(v - v_0 w_0 v)W = 0$ and so $v - v_0 w_0 v \in U$. Hence $v_0 S + U = V$ and V/U is simple. □

6.6 We now wish to consider prime right Goldie rings and, in particular, aim to show that being prime right Goldie is preserved under appropriate context equivalence.

Theorem. *Let*

$$\begin{bmatrix} R & V \\ W & S \end{bmatrix}$$

be a prime Morita context in which R is prime right Goldie. Then S is prime right Goldie if and only if W_R has finite uniform dimension. In that case, $S \sim \mathrm{End}\, W_R$.

Proof. We note first that W_R must be torsionless. For if $0 \neq w \in W$, then $Vw \neq 0$ and so $vw \neq 0$ for some $v \in V$. The map $W \to vW \subseteq R$ maps w to vw which is nonzero, as required. Note also that if $sW = 0$ then $VsW = 0$ and so $s = 0$. Thus $_S W$ is faithful. Hence $S \hookrightarrow \mathrm{End}\, W_R$.

The argument of 4.4 evidently can be applied here to show that

$$\mathrm{u\,dim}\, W_R = \mathrm{u\,dim}\, WV_S = \mathrm{u\,dim}\, S_S.$$

Hence the condition on W_R is indeed necessary. Conversely, if $\mathrm{u\,dim}\, W_R < \infty$, note that $WV \triangleleft_e S_S$ and so WV contains a regular element c of S. But then

$$(\mathrm{End}\, W)c \subseteq (\mathrm{End}\, W) \cdot WV \subseteq WV \subseteq S \subseteq \mathrm{End}\, W \quad \text{and so} \quad S \sim \mathrm{End}\, W.$$

Since $\mathrm{End}\, W$ is prime right Goldie, by Corollary 4.6, then so too is S. □

6.7 Given a context

$$\begin{bmatrix} R & V \\ W & S \end{bmatrix}$$

and related prime ideals P, P' one needs, in the notation of the proof of 6.5(i), that $\mathrm{u\,dim}\, W/W' < \infty$ to obtain results like 6.6 about R/P and S/P'. The next result notes that this is the case when W_R is finitely generated.

Corollary. *Let*

$$\begin{bmatrix} R & V \\ W & S \end{bmatrix}$$

be a Morita context in which W_R is finitely generated. Let P be a prime ideal of

R such that R/P is right Goldie and $VW \not\subseteq P$, and let P' be the corresponding prime ideal of S. Then S/P' is right Goldie.

Proof. By 6.5, we may suppose that P, P' are both zero and the context is prime. Then, as in the proof of 6.6, W_R is torsionless and so, by Proposition 4.3, W_R has finite uniform dimension. Thus 6.6 applies. □

6.8 We now turn to the left-handed properties of R and S.

Corollary. *Let*

$$\begin{bmatrix} R & V \\ W & S \end{bmatrix}$$

be a prime Morita context in which W_R has finite uniform dimension and R is right and left Goldie. Then S is right and left Goldie.

Proof. As noted earlier, W_R is torsionless, and so $W \hookrightarrow W \otimes Q$ which is a finite dimensional right module over $Q = Q(R)$. It is clear that $V \hookrightarrow W^* = \mathrm{Hom}(W, R)$ and so $V \hookrightarrow (W \otimes Q)^* = \mathrm{Hom}(W \otimes Q, Q)$. This latter module also is finite dimensional over Q. Thus $V \hookrightarrow {}_RQ^n$ for some n and so ${}_RV$ has finite uniform dimension (since Q is also the left quotient ring of R).

The result now follows from the left-hand version of 6.6. □

6.9 Comparison with §5 suggests that context equivalence is a much weaker relationship than Morita equivalence. This is confirmed by the following example.

Example. Let $R = k[x, y]$, $S = k + xR$ and $A = xR$. Since

$$\begin{bmatrix} R & A \\ A & S \end{bmatrix}$$

is a prime Morita context, R and S are context equivalent. However, they cannot be Morita equivalent since R is Noetherian but S is not, as was shown in 1.3.11.

Nevertheless, there is a positive connection with Morita equivalence, via the quotient rings.

Proposition. *Let R, S be context equivalent prime right Goldie rings. Then $Q(R) \stackrel{M}{\sim} Q(S)$.*

Proof. Let
$$\begin{bmatrix} R & V \\ W & S \end{bmatrix}$$
be the appropriate Morita context. By 6.6 and its proof, W_R is torsionless and of finite uniform dimension, and $S \sim \text{End}(W_R)$. Therefore, by Corollary 4.6(i), S is a right order in $\text{End}(W \otimes Q(R))$. Since $W \otimes Q(R)$ is, perforce, a right progenerator for $Q(R)$, then $Q(R) \overset{M}{\sim} Q(S)$. □

§7 Additional Remarks

7.1 The origins of the theory of orders lie in the study of integers in algebraic number fields and of orders in central simple algebras. The arithmetic theory of orders is described in Chapter 5.

7.2/7.3 The results here are largely due to Goldie. However, 2.6 comes from [Faith and Utumi **65**], and 3.6 and Corollary 3.7 from [Robson **67**] although this proof of 3.6 follows A. Ludgate (unpublished). Some further details can be found in [Chatters and Hajarnavis **80**].

7.4 This section draws together results from [Hart **66**], [Zelmanowitz **67**] and [Jategaonkar **71**] which extended earlier work in [Levy **63**] and [Feller and Swokowski **64**]. However, 4.9 comes from [Goldie **62**].

7.5 (a) Morita equivalence arose first in terms of category equivalences, [Morita **58**]. Indeed one can define $R \overset{M}{\sim} S$ to mean there is a category equivalence between their categories of right modules; $F: \text{mod } R \to \text{mod } S$. Given such an F it is clear that $F(R)_S$ is a progenerator with $\text{End}(F(R)_S) \simeq R$. It follows that the two definitions agree. For further details, see [Anderson and Fuller **74**].

(b) Morita equivalence and progenerators are heavily involved in the arithmetic ideal theory in Chapter 5.

7.6 (a) In this text, Morita contexts are used more frequently than is usual. They were studied in [Amitsur **71**] where 6.5(ii)–6.8 were proved. The proofs here are somewhat simpler.

(b) Prime contexts were introduced by [Nicholson and Watters **79**].

(c) 6.2 is a result which, in the literature, is often reproved in special circumstances like those described in 6.3. It will be used later in 5.6.11, 10.5.14 and 13.9.7. Morita contexts are also applied in Chapter 7, §8.

Chapter 4
SEMIPRIME IDEALS IN NOETHERIAN RINGS

In this chapter the focus moves from semiprime rings to general Noetherian rings, although it does concentrate upon prime and semiprime ideals.

Let R be a Noetherian ring and A a semiprime ideal. By Goldie's theorem, R/A is an order in a semisimple Artinian ring. It is natural, in the light of the commutative theory, to hope to be able to 'localize' with respect to the set $\mathscr{C}_R(A)$—until one recalls the problems described in Chapter 2, §1. There it was shown that this was possible if and only if $\mathscr{C}_R(A)$ was a right denominator set or equivalently, since R is Noetherian, a right Ore set.

We will describe any ideal A of any ring R as being **right localizable** if $\mathscr{C}_R(A)$ is a right denominator set, and **localizable** if both left and right localizable. One of the aims of this chapter is to describe conditions under which particular ideals of a Noetherian ring are localizable, this taking up the first half of the chapter. Section 1 concentrates on the prime radical of R, §2 on ideals with the Artin–Rees property and §3 on sets of prime ideals P such that R can be localized satisfactorily with respect to $\bigcap \mathscr{C}_R(P)$. This effectively answers the question of which semiprime ideals of R are localizable.

Then §4 describes $\mathscr{C}_R(0)$ in terms of $\mathscr{C}_R(P)$ for certain prime ideals P of R. These primes are used in §5 to connect the uniform dimensions of related prime factor rings of a given Noetherian ring R and an extension S. Finally, in §6, the relationship between the uniform dimensions of different prime factors of R is studied.

Not all the topics described in this chapter can be regarded as being in a finished state. This is certainly so for §§3–6. These are intended, therefore, to be less comprehensive, concentrating on the main features.

§1 Reduced Rank and Applications

This section starts by describing the reduced rank of a module, this concept being an offshoot of the uniform dimension. There are then two applications. The first gives a characterization of right orders in right Artinian rings, and the

4.1.3 Reduced rank and applications

second is a principal ideal theorem. Some consequences regarding decompositions of rings and the Artin radical are also noted.

1.1 First comes a preliminary definition, to be subsumed in 1.2. Suppose R is semiprime right Goldie, with $Q(R) = Q$; so Q is semisimple Artinian. If M is a right R-module then $M \otimes Q$ is a semisimple Q-module whose length and uniform dimension coincide. The **reduced rank** of M is defined to be $\rho_R(M) = \operatorname{udim}(M \otimes Q)_Q$; this is finite if M_R is finitely generated, but otherwise may be infinite. It is clear from the comments in 3.4.2 that $\rho_R(M) = \operatorname{udim} M/M'$, where M' is the torsion submodule of M as described in 2.1.17; for $M/M' \hookrightarrow M/M' \otimes Q \simeq M \otimes Q$.

Note that, for a short exact sequence $0 \to A \to B \to C \to 0$ of Q-modules, one has $\operatorname{udim} B = \operatorname{udim} A + \operatorname{udim} C$ (since the sequence splits). Since $_RQ$ is flat (2.1.16(ii)), it follows that reduced rank is **additive** on short exact sequences of right R-modules; i.e. $\rho(B) = \rho(A) + \rho(C)$ if $0 \to A \to B \to C \to 0$ is a short exact sequence of right R-modules.

1.2 The preceding definition is next extended to cover non-semiprime rings. Let R be a ring with a prime radical $N = N(R)$ such that R/N is right Goldie and N is nilpotent. (This allows any right Noetherian ring, of course.) Let M be a right R-module. A **Loewy series** for M is a finite chain of submodules

$$M = M_0 \supseteq M_1 \supseteq \cdots \supseteq M_k = 0$$

such that $M_i N \subseteq M_{i+1}$ for each i; for example, take $M_i = MN^i$. The **reduced rank** of M is defined to be

$$\rho_R(M) = \sum_{i=0}^{k} \rho_{R/N}(M_i/M_{i+1}).$$

This, as it stands, is dependent upon the choice of Loewy series.

Lemma. (i) *The reduced rank of a module is independent of the choice of Loewy series.*

(ii) *Reduced rank is additive on short exact sequences.*

Proof. (i) If $M = M'_0 \supseteq M'_1 \supseteq \cdots \supseteq M'_s = 0$ is another Loewy series then the Schreier refinement theorem [Cohn **74**, p. 22] gives isomorphic refinements of the two series $\{M_i\}$ and $\{M'_i\}$. It is clear from 1.1 that the refined series yield the same reduced rank as did the original series.

(ii) This is now clear. □

Note that if M is finitely generated and R is right Noetherian, then $\rho(M) < \infty$.

1.3 We now begin the application to quotient rings.

Proposition. *Let R be a ring with $N = N(R)$ nilpotent and R/N right Goldie; and M a right R-module. Then:*
 (i) *$\rho(M) = 0$ if and only if for each $m \in M$ there exists $c \in \mathscr{C}(N)$ such that $mc = 0$.*
 (ii) *('Pseudo-Ore condition') If $a \in R$ and $c \in \mathscr{C}'(0)$ then there exists $a_1 \in R$ and $c_1 \in \mathscr{C}(N)$ such that $ac_1 = ca_1$.*
 (iii) *$\mathscr{C}'(0) \subseteq \mathscr{C}(N)$.*
 (iv) *If R has a right quotient ring Q and $N' = N(Q)$ then $N = N' \cap R$, $(N')^k = N^k Q = QN^kQ$. Also Q/N' is the right quotient ring of R/N if $\mathscr{C}_R(0) = \mathscr{C}_R(N)$.*

Proof. (i) Suppose that $\{M_i\}$ is a Loewy series for M. Then clearly $\rho(M) = 0$ if and only if $\rho(M_i/M_{i+1}) = 0$ for each i. On the other hand, since $\mathscr{C}(N)$ is multiplicatively closed, it is immediate that the second condition holds for $m \in M$ if and only if it holds for each element of M_i/M_{i+1}, for all i. Thus it is enough to consider the case when R is semiprime. Now $\rho(M) = 0$ if and only if $\rho(mR) = 0$ for each $m \in M$. So one may suppose $M = mR \simeq R/\operatorname{ann} m$; but then $M \otimes Q \simeq Q/(\operatorname{ann} m)Q$, which has rank zero precisely when $(\operatorname{ann} m)Q = Q$. However, that happens precisely when $\operatorname{ann} m$ contains a regular element.

(ii) Since $c \in \mathscr{C}'(0)$, then $R \simeq cR$ and so $\rho(R) = \rho(cR)$. Therefore $\rho(R/cR) = 0$. Therefore, by (i), there exists $c_1 \in \mathscr{C}(N)$ such that $[a + cR]c_1 = 0$; thus $ac_1 = ca_1$ for some $a_1 \in R$.

(iii) Let $c \in \mathscr{C}'(0)$ and choose $a = 1$ in (ii). Then $c_1 = ca_1$ and so $c, a_1 \in \mathscr{C}(N)$.

(iv) Let $n \in N$, $c \in \mathscr{C}(0)$; then $c^{-1}n = ad^{-1}$ for some $a \in R$, $d \in \mathscr{C}(0)$. Now $ca = nd \in N$, yet, by (iii), $c \in \mathscr{C}(N)$. Therefore $a \in N$ and so $c^{-1}n \in NQ$. It follows that $QNQ = NQ$ and thus $(NQ)^k = N^k Q$. Since N is nilpotent, so too is NQ. It follows that $NQ \subseteq N'$ and that $NQ \cap R = N$. But then Q/NQ is the right quotient ring of R/N with respect to the subset $\mathscr{S} = \{c + N \mid c \in \mathscr{C}(0)\}$ of R/N. Evidently Q/NQ is then semiprime right Goldie. It follows that $NQ = N'$. Finally, if $\mathscr{C}_R(0) = \mathscr{C}_R(N)$ then $\mathscr{S} = \mathscr{C}_{R/N}(0)$. □

1.4 Theorem. *A ring R, with prime radical N say, has a right Artinian quotient ring if and only if* (i) *R/N is right Goldie and N is nilpotent,* (ii) *$\rho(R) < \infty$ and* (iii) *$\mathscr{C}_R(0) = \mathscr{C}_R(N)$.*

Proof. Suppose (i), (ii), (iii) are given. Since $\mathscr{C}(0) = \mathscr{C}(N)$ it follows from 1.3(ii)(iii) that R satisfies the right Ore condition with respect to $\mathscr{C}_R(0)$. Therefore R has a right quotient ring Q, with prime radical N' say. By 1.3(iv), Q/N' is the right quotient ring of R/N and therefore is semisimple Artinian.

Flatness of $_RQ$, given by 2.1.16, makes it clear that $(N^k/N^{k+1}) \otimes Q \simeq N^k Q/N^{k+1}Q$. Since $\rho(R) < \infty$ so too is $\rho(N/N^2)$. Therefore N'/N'^2 has finite length and thus, by Nakayama's lemma, N' is finitely generated. It is now elementary to deduce that Q is right Artinian.

Conversely, if Q exists and is right Artinian, then N' is nilpotent, as is $N' \cap R$.

4.1.6 Reduced rank and applications

Furthermore, Q/N' is the right quotient ring of $R/N' \cap R$. Therefore $R/N' \cap R$ is semiprime right Goldie. This, together with the nilpotence of $N' \cap R$, shows that $N' \cap R = N$, the prime radical of R. It is clear, now, that $\rho(R) < \infty$. Moreover, $\mathscr{C}_R(N) \subseteq \mathscr{C}_Q(N') = \mathscr{C}_Q(0)$. Hence $\mathscr{C}_R(N) \subseteq \mathscr{C}_R(0)$ and so, by 1.3(iii), $\mathscr{C}_R(N) = \mathscr{C}_R(0)$. □

Corollary. (Small's theorem) *A right Noetherian ring R has a right Artinian, right quotient ring if and only if $\mathscr{C}_R(0) = \mathscr{C}_R(N)$ where $N = N(R)$.* □

The ring $R = k[x, y]/(x^2, xy)$, with k a field, provides an easy example of a Noetherian ring for which $\mathscr{C}_R(0) \neq \mathscr{C}_R(N)$. Thus R does not have an Artinian quotient ring.

1.5 The remainder of this section concentrates on right Noetherian rings, with the next few results considering Artinian modules over them. These prepare the way for some decomposition theorems.

Lemma. *Let R be right Noetherian with right Artinian, right quotient ring Q. Suppose that M_Q is finitely generated as a right R-module. Then M_R is Artinian and any composition series of M_Q is also one for M_R.*

Proof. Clearly it is sufficient to consider the case when M_Q is simple. Then ann M is a minimal prime of Q and so contains the prime radical, N', of Q. By 1.3(iv), Q/N' is the right quotient ring of the semiprime right Goldie ring R/N. Therefore, by Proposition 3.2.2, $R \cap \text{ann } M$ is a minimal prime of R. But then Proposition 3.2.4 shows that $R/(R \cap \text{ann } M)$ is a right order in $Q/\text{ann } M$. Thus we may suppose, without loss of generality, that R is prime and Q is simple. In that case, M is isomorphic to a minimal right ideal of Q. Therefore $Q \simeq M^k$ for some k and so Q_R is finitely generated. Now if $c \in \mathscr{C}_R(0)$ then for some n one has $c^{-(n+1)} \in R + c^{-1}R + \cdots + c^{-n}R$. It follows that $c^{-1} \in R$. Hence $R = Q$ and M_R is simple, as claimed. □

1.6 For the next result, R, S are, initially, arbitrary rings.

Theorem. *Let $_S M_R$ be a bimodule such that $_S M$ has finite length and M_R is Noetherian. Then M_R has finite length.*

Proof. By induction, one may suppose that $_S M_R$ is a simple bimodule (i.e. has no nonzero proper sub-bimodule), and then, by passing to appropriate factor rings, one may assume that M is both left and right faithful. It follows immediately that R and S are both prime rings. Note that, if $\{a_1, \ldots, a_n\}$ is a generating set for both $_S M$ and M_R, then $\bigcap (\text{r ann } a_i) = 0$ and so the map $r \mapsto (a_1 r, \ldots, a_n r)$ gives a monomorphism $R \to M_R^{(n)}$ and so R is right Noetherian.

Let $N = \{x \in M \mid \exists c \in \mathscr{C}_R(0) \text{ with } xc = 0\}$. Since R is a prime right Goldie ring, it follows from the right Ore condition that N is a sub-bimodule of M, and so is 0 or M. However, if $N = M$, then, taking a common denominator (2.1.16), there exists $c \in \mathscr{C}_R(0)$ with $a_i c = 0$ for $i \in \{1, \ldots, n\}$ and then $Mc = 0$, a contradiction. Therefore $N = 0$.

For any $c \in \mathscr{C}_R(0)$, consider the map $M \to Mc \subseteq M$. This is an S-monomorphism, since $N = 0$; and so, since $_S M$ has finite length, $M = Mc$. Thus M is a right Q-module, for Q the simple Artinian right quotient ring of R. By 1.5, M_R has finite length. □

1.7 Let R be a Noetherian ring. The **Artin radical** of R, written $A(R)$, is the largest left Artinian ideal of R. By 1.6, this is equally the largest right Artinian ideal of R. Clearly $A(R/A(R)) = 0$.

1.8 We can now give a decomposition theorem.

Theorem. *Let R be a Noetherian ring with an Artinian quotient ring Q. Then there is a central idempotent $e \in R$, possibly zero, such that $A(R) = eR$. So $A(R)$ is a direct summand of the ring R.*

Proof. Note first that if $I \triangleleft Q$ with I_R finitely generated then $I \subseteq R$. For if $I = \sum c^{-1} a_i R$ with $a_i \in R$, $c \in \mathscr{C}_R(0)$, then $I = cI = \sum a_i R \subseteq R$. On the other hand, if $I \triangleleft R$ and I_R is Artinian, then $cI = I$ for each $c \in \mathscr{C}_R(0)$ (because $cI \subseteq I$ and $cI \simeq I$, which makes length $cI_R =$ length I). Hence $I = c^{-1} I$, and it follows that I is a left (and also right) ideal of Q.

Secondly, consider the special case when R is semiprime. The remarks above show that $A(R) \triangleleft Q$, which is semisimple Artinian. Therefore, $A(R) = eQ$ for some central idempotent e of Q. Hence e is a central idempotent of R and $A(R) = eR$.

Finally, we consider the general case. Let N, N' be the prime radicals of R, Q respectively. Let e be an idempotent of R such that $eR + N/N = Re + N/N$ is the Artin radical of R/N, with $[e + N]$ central. It will be seen that $A = A(R) = eR = Re$ and that e is central.

The first step is to note that $eN^k + N^{k+1}/N^{k+1}$, viewed as a left R/N-module, is the homomorphic image of a finite direct sum of copies of $eR + N/N$. Hence it has finite length as a left R/N-module and also, by 1.6, as a right R-module. However, $eN^k = eR \cap N^k$ and so $eN^k \cap N^{k+1} = eR \cap N^{k+1} = eN^{k+1}$. Hence

$$eN^k/eN^{k+1} = eN^k/eN^k \cap N^{k+1} \simeq eN^k + N^{k+1}/N^{k+1}.$$

Therefore $(eN^k/eN^{k+1})_R$ has finite length for each k, which implies that eR_R has finite length. Thus $eR \subseteq A$, and, by symmetry, $Re \subseteq A$.

Next consider $M = AN^k/AN^{k+1}$. Since A is an ideal of Q, then $AN^k = AQN^k = AN'^k$. Hence M is a right Q/N'-module. The simple submodules of $M_{Q/N'}$ are all finitely generated over R. The same is true, therefore, of simple submodules of $I = M^*M$ where $M^* = \mathrm{Hom}_{Q/N'}(M, Q/N')$. Since $I \triangleleft Q/N'$, the remarks above show that $I \subseteq R/N$ and so $I(1-e) = 0$. It follows that $1-e$ annihilates all simple submodules of M and so, by induction, $A(1-e) = 0$. It follows that $A \subseteq Re$. Thus $A = eR = Re$. Clearly e is now central. □

To help place this in context consider the following examples.
(i) If
$$R = \begin{bmatrix} \mathbb{Z} & \mathbb{Q} \\ 0 & \mathbb{Q} \end{bmatrix}$$
then R is right, but not left, Noetherian (1.1.9) and the ideal
$$\begin{bmatrix} 0 & \mathbb{Q} \\ 0 & \mathbb{Q} \end{bmatrix}$$
is right, but not left, Artinian. It does not have a central idempotent generator.
(ii) The ring $R = k[x, y]/(x^2, xy)$, as noted in 1.4, is a commutative Noetherian ring which fails to have an Artinian quotient ring. Its Artin radical is generated by $\bar{x}$ and is nilpotent.
(iii) If $R = k[x, y, z]/(xz, yz)$ then R is Noetherian and semiprime, so has a semisimple Artinian quotient ring. In this case $A(R) = 0$.

1.9 This result has some interesting consequences.

Theorem. *A Noetherian ring R is the direct product of a semiprime ring and an Artinian ring if and only if $N = cN = Nc$ for all $c \in \mathscr{C}(N)$.*

Proof. ($\Rightarrow$) We can easily reduce to the case when R is Artinian. Then it is clear since c is a unit.

($\Leftarrow$) Let $c \in \mathscr{C}(N)$. Since the map $N \to N$, given by left multiplication by c, is surjective and N_R is Noetherian, the map is an isomorphism. It follows that $c \in \mathscr{C}'(0)$ and hence, by symmetry, $c \in \mathscr{C}(0)$. Thus 1.3(iii) shows that $\mathscr{C}(0) = \mathscr{C}(N)$, and 1.4 that R has an Artinian quotient ring, Q say. The hypothesis on N ensures that $N = NQ = QN$ and so 1.5 can be applied to show that N_R and $_RN$ have finite length. Thus, if $A(R) = eR$ as prescribed by 1.8, then $N \subseteq eR$.

Note finally that R is the direct product of eR and $(1-e)R$ and that $(1-e)R$ is necessarily semiprime. □

This applies readily to a principal ideal ring.

Corollary. *A principal ideal ring is the direct product of a semiprime ring and an Artinian ring.*

Proof. Suppose $c \in \mathscr{C}(N)$, and consider the right ideal $cR + N$. By assumption, $cR + N = dR$ for $d \in R$ and evidently $d \in \mathscr{C}(N)$. It follows that $N = dN$, and hence $N = cN + N^2$. Nakayama's lemma now shows that $N = cN$. □

One should note that the semiprime ring here is, of course, a principal ideal ring. Therefore 3.4.9 shows that it decomposes as a direct sum of prime rings, each a matrix ring over an integral domain.

1.10 This section ends with a noncommutative version of the 'principal ideal theorem' (see [Kaplansky 70, §3.2] or [Atiyah and Macdonald 69, 11.17].) First some terminology is required. An element a of a ring R is a **normal** element if $aR = Ra$. In particular, of course, any central element is normal.

Note that when a is normal, aR is an ideal of R and $(aR)^n = a^n R$. Furthermore, if R is prime and $a \neq 0$, then a is a regular element.

Lemma. *Let R be a right Noetherian prime ring and $0 \neq a \in R$ a normal element. Then, given any right ideal I of R, there exists $n \geq 0$ such that $I \cap a^n R \subseteq IaR$.*

Proof. Note that $Ia^{-1}R \cap R \subseteq Ia^{-2}R \cap R \subseteq \cdots \subseteq Ia^{-n}R \cap R \ldots$. The a.c.c. shows that $Ia^{-n+1}R \cap R = Ia^{-n}R \cap R$ for some n. Hence, multiplying through by a^n (from the right), $IaR \cap a^n R = IR \cap a^n R$. Hence $I \cap a^n R \subseteq IaR$. □

This property of aR is called the **Artin–Rees property** and is considered further in §2.

1.11 The next result involves the **height** of a prime P, this being the largest length of a chain of prime ideals contained in P, or ∞ if no bound exists.

Theorem. *(Principal ideal theorem) Let R be a right Noetherian ring, a any normal element which is not a unit, and P a prime ideal of R minimal over aR. Then P has height at most 1.*

Proof. Suppose not. Then $P \supset P' \supset P''$ say. Factoring out P'' means we may suppose R is prime and a then is regular. Since P' is a nonzero ideal of R, it contains some regular element, b say.

Now $bR \cap a^n R \subseteq baR$ for some n, by 1.10. Then
$$bR \cap a^{n+1}R = bR \cap a^n R \cap a^{n+1}R \subseteq baR \cap a^{n+1}R = (bR \cap a^n R)a \subseteq (baR)a = ba^2 R.$$
Proceeding in this way we see that $bR \cap a^{2n}R \subseteq ba^n R$.

Now P is minimal over $a^n R$, of course, and a^n is a normal non-unit. Thus, replacing a by a^n, we may assume that $bR \cap a^2 R \subseteq baR$. Then $bR \cap (a^2 R + baR) = baR$. Therefore
$$\frac{a^2 R + bR}{a^2 R + baR} = \frac{a^2 R + baR + bR}{a^2 R + baR} \simeq \frac{bR}{(a^2 R + baR) \cap bR} = \frac{bR}{baR} \simeq \frac{R}{aR}.$$

4.1.12   *Reduced rank and applications*

Therefore $\rho(a^2R + bR/a^2R + baR) = \rho(R/aR)$ with ρ being the reduced rank as modules over $\bar{R} = R/aR$.

The map $r \mapsto a^{-1}ra$ gives an automorphism of R and of $\bar{R}$. It gives a lattice automorphism on the set of right ideals of $\bar{R}$ which, for subfactors, preserves sums, intersections and the property of being torsion with respect to $\mathscr{C}(N)$, where N/aR is the prime radical of $\bar{R}$. Hence it preserves the reduced rank on subfactors. Therefore, if $a^{-1}ba = c$, then

$$\rho(aR + bR/aR) = \rho(aR + cR/aR).$$

Thus

$$\rho\left(\frac{aR+bR}{a^2R+baR}\right) = \rho\left(\frac{aR+bR}{a^2R+acR}\right) = \rho\left(\frac{aR+bR}{aR}\right) + \rho\left(\frac{aR}{a^2R+acR}\right)$$

$$= \rho\left(\frac{aR+cR}{aR}\right) + \rho\left(\frac{R}{aR+cR}\right) = \rho\left(\frac{R}{aR}\right)$$

$$= \rho\left(\frac{a^2R+bR}{a^2R+baR}\right).$$

It follows that

$$\rho\left(\frac{aR+bR}{a^2R+bR}\right) = 0.$$

Therefore there exists $d \in \mathscr{C}_R(N)$ such that $ad \in a^2R + bR$ and so $ad = a^2r + bs$ for some $r, s \in R$. Hence $a(d - ar) = bs \in P'$. However, $a \notin P'$ and, being a normal element, is regular modulo P'. Therefore $d - ar \in P' \subseteq P$ and so $d \in P$. However, P/N is a minimal prime of R/N, so, by 3.2.4(v), $\mathscr{C}(N) \subseteq \mathscr{C}(P)$. This shows that $d \in \mathscr{C}(P)$. Since, of course, $P \cap \mathscr{C}(P) = \emptyset$, this is a contradiction. $\square$

The condition on a cannot be deleted. For example, let $T = \mathbb{Z}[x]$, $R = M_2(T)$ and

$$a = \begin{bmatrix} 2 & 0 \\ 0 & x \end{bmatrix}.$$

Then the only prime P of R containing aR is $M_2(I)$, where $I = (2, x) \triangleleft T$; so height $P = 2$.

1.12 We note a useful extension of 1.11.

Corollary. *Let R be a right Noetherian ring, a any normal element which is not a unit and P a prime ideal of R containing aR. Then height $P \leq$ height$(P/aR) + 1$.*

Proof. Suppose $P = P_0 \supset P_1 \supset \cdots \supset P_m$. If $a \notin P_{m-1}$ we can choose k such that $a \in P_k$, $a \notin P_{k+1}$. Applying 1.11 to R/P_{k+2} shows that P_k is not minimal over

$aR + P_{k+2}$. Therefore there exists P'_{k+1} with $P_k \supset P'_{k+1} \supset P_{k+2}$ and $P'_{k+1} \supseteq aR$. We replace P_{k+1} by P'_{k+1} and then repeat this process. Eventually we arrange that $a \in P_{m-1}$; but then $m \leqslant \text{height}(P/aR) + 1$ as required. □

1.13 This yields a generalized version of 1.11. A sequence $a_1, \ldots, a_n$ of elements of R is called a **normalizing sequence** or, more specially, a **centralizing sequence** if for each $j \in \{0, \ldots, n-1\}$ the image of a_{j+1} in $R/\sum_{i=1}^{j} a_i R$ is a normal (resp. central) element and also $\sum_{i=1}^{n} a_i R \neq R$. The ideal generated by such a sequence is sometimes called **polynormal** or **polycentral**. Such ideals recur in 2.7 and in Chapter 14.

Theorem. (Generalized principal ideal theorem) *Let R be a right Noetherian ring, $a_1, \ldots, a_n$ a normalizing sequence and P a prime ideal containing $A = \sum_{i=1}^{n} a_i R$. Then*

$$\text{height } P \leqslant \text{height}(P/A) + n.$$

Proof. A simple induction argument, based on 1.12, shows this. □

§2 Artin–Rees Property and Localization

2.1 In Chapter 2, §1, localization with respect to a m.c. set was described. Indeed the point of that chapter was to show that if R is a semiprime right Goldie ring then $\mathscr{C}_R(0)$ is a right denominator set; i.e. 0 is a right localizable ideal. On the other hand, Example 2.1.7 showed that even a prime ideal of a Noetherian ring need not be localizable.

In this section and the next we discuss circumstances under which particular ideals of a Noetherian ring are localizable. The emphasis, therefore, will be upon the validity of the right Ore condition.

2.2 In practice, many of the positive results about localization involves ideals having a specific property which has been touched upon already in 1.10 and which is put in perspective by the next result.

Theorem. *The following conditions on an ideal A of a right Noetherian ring R are equivalent:*
 (i) *If $I \triangleleft_r R$ then $I \cap A^n \subseteq IA$ for some n.*
 (ii) *If M_R is finitely generated and $N \triangleleft M$ then $N \cap MA^n \subseteq NA$ for some n.*
 (iii) *If M_R is finitely generated and $N \triangleleft_e M$ with $NA = 0$ then $MA^n = 0$ for some n.*

Proof. (i)⇒(iii) If $m \in M$ then $N \cap mR \triangleleft_e mR$. Hence it will be enough to consider the case when M is cyclic. Therefore we may suppose $M = R/K$ and $N = I/K$ with $I, K \triangleleft_r R$. By hypothesis $NA = 0$, so $IA \subseteq K$. Then (i) gives $I \cap A^n \subseteq IA \subseteq K$; i.e. $MA^n \cap N = 0$. However, $N \triangleleft_e M$; so $MA^n = 0$.

(iii)⇒(ii) Choose a submodule K of M maximal with respect to having $K \cap N = NA$. Letting ¯ denotes images in M/K, note that $\bar{N}A = 0$ and $\bar{N} \triangleleft_e \bar{M}$. By hypothesis $\bar{M}A^n = 0$ for some n. That makes $MA^n \subseteq K$ and then $MA^n \cap N \subseteq K \cap N = NA$.

(ii)⇒(i) This is trivial. □

2.3 The property of the ideal A described in 2.2 is called the **right Artin–Rees property**, or **right AR property** for short. If every ideal A of a ring R has the property, R is called a **right AR ring**. This section concentrates on such ideals and rings.

One example of an ideal with the right AR property was noted in 1.10 and, of course, any nilpotent ideal has the property. Although the next few results will provide further examples it should be pointed out that it is common for ideals to fail to have the right AR property. For instance consider the ideal P of the ring R described in 2.1.7(ii). Since $xy - yx = y$ it is clear that $y \in P^n$ for all n; hence $P^n \cap yR = yR$. On the other hand, $yRP = y^2R + yxR \not\supseteq yR$. Thus P fails to have the right AR property.

2.4 First it should be noted that the property is preserved under Morita equivalence and under homomorphisms.

Proposition. *Let R be a right Noetherian ring and A an ideal with the right AR property.*
 (i) *If $\theta: R \to S$ is a surjective ring homomorphism then $\theta(A)$ has the AR property.*
 (ii) *If $R \stackrel{M}{\sim} S$ and B is the ideal of S corresponding to A, as described in 3.5.9, then B has the right AR property.*

Proof. These are readily checked if (ii) of 2.2 is used as the defining property. □

2.5 An ideal A of a ring R is **invertible** if $AB = BA = R$ for some subset B of some extension ring S of R.

The proof of 1.10 is easily extended to give

Corollary. *Any invertible ideal of a right Noetherian ring R has the right AR property.* □

2.6 Next we extend 1.10 in another way.

Proposition. *Let R be a right Noetherian ring and A an ideal generated by normal elements. Then A has the right AR property.*

Proof. Let M_R be finitely generated and $N \triangleleft_e M$ with $NA = 0$. It is enough, by 2.2, to show that $MA^n = 0$ for some n.

First consider one of the normal elements generating A, x say, and let $\theta: M \to M$ be the abelian group homomorphism defined by $\theta(m) = mx$. The fact that $xR = Rx$ implies that $\ker \theta^k$ and $\operatorname{im} \theta^k$ are submodules of M for each k. Since M_R is Noetherian, $\ker \theta^k = \ker \theta^{k+1}$ for some k. Therefore $\ker \theta^k \cap \operatorname{im} \theta^k = 0$. However, $N \subseteq \ker \theta \subseteq \ker \theta^k$ and $N \triangleleft_e M$; so $\operatorname{im} \theta^k = 0$. Hence $Mx^k = 0$.

Now, suppose $x_1, \ldots, x_t$ are the normal generators of A. The argument above shows that for each i there is a $k(i)$ such that $Mx_i^{k(i)} = 0$. Let $n = \sum k(i)$. Then $A^n \subseteq \sum x_i^{k(i)} R$ and thus $MA^n = 0$. □

2.7 Theorem. *Let R be a right Noetherian ring.*
(i) *Let A be an ideal with a normalizing sequence of generators $x_1, \ldots, x_n$ such that $x_1 A = Ax_1$ and similarly*

$$x_j A + \sum_{i=1}^{j-1} x_i R = Ax_j + \sum_{i=1}^{j-1} x_i R$$

for each $j > 1$. Then A has the right AR property.
(ii) *Any ideal of R with a centralizing sequence of generators has the right AR property.*

Proof. Clearly it is enough to prove (i). By 2.6 the result holds if $n = 1$. We now proceed by induction on n.

Let M_R be finitely generated and $N \triangleleft_e M$ with $NA = 0$. Of course $Nx_1 R = 0$ and so, by 2.6, $Mx_1^t R = 0$ for some t. Now if $Mx_1 R = 0$ then M is an $R/x_1 R$ module, and since $A/x_1 R$ has a normalizing sequence of $n-1$ generators of the appropriate type the induction hypothesis shows that $MA^k = 0$ for some k, as required.

We now introduce the subsidiary inductive hypothesis that the result is valid for modules annihilated by $x_1^{t-1} R$. Consider the exact sequence of abelian groups

$$0 \to K \to M \to Mx_1 \to 0,$$

where $K = \operatorname{ann}_M x_1$. Both K and Mx_1 are submodules of M. The inductive hypothesis on t applied to Mx_1 implies that $Mx_1 A^{s_1} = 0$ for some s_1. However, $x_1 A^{s_1} = A^{s_1} x_1$; so $MA^{s_1} \subseteq K$. Now $Kx_1 R = 0$, and $N \cap K \triangleleft_e K$. Thus $KA^{s_2} = 0$ for some s_2 and hence $MA^{s_1 + s_2} = 0$ as required. □

We will see, in 14.3.4, that the conditions upon the ideal A are satisfied for many ideals in enveloping algebras $U(g)$.

2.8 It is worth noting that in the triangular matrix ring

$$R = \begin{bmatrix} k & k \\ 0 & k \end{bmatrix}$$

over a field k the ideal

$$A = \begin{bmatrix} k & k \\ 0 & 0 \end{bmatrix}$$

has the normalizing sequence of generators e_{12}, e_{11}. However, A does not have the right AR property. For if

$$I = \begin{bmatrix} 0 & k \\ 0 & k \end{bmatrix}$$

then $IA = 0$ yet, since A is idempotent,

$$I \cap A^n = I \cap A = \begin{bmatrix} 0 & k \\ 0 & 0 \end{bmatrix}$$

for each n. This shows that the extra conditions imposed on the normalizing sequence of generators of A in 2.7(i) are not redundant.

Another similar example involving enveloping algebras is provided by the ideal P mentioned in 2.3 and 2.1.7(ii) since y, x form a normalizing sequence of generators.

2.9 Having described some examples of ideals having the right AR property, we next turn to the applications to localization. The first result applies to all such ideals A, but refers to one specific type of multiplicatively closed set, namely $1 - A = \{1 - a | a \in A\}$.

Proposition. *Let R be a right Noetherian ring and A an ideal with the right AR property. Then:*
(i) $1 - A$ *is a right Ore set and so a right denominator set;*
(ii) *writing $\mathcal{S}$ for $1 - A$, we have $A_\mathcal{S} \subseteq J(R_\mathcal{S})$.*

Proof. (i) Given $r \in R, x \in A$ we seek a common right multiple of r and $1 - x$ of the correct form. Set

$$u_i = r(1 - x^i) - (1 - x^i)r$$

for $i = 1, 2, \ldots$ and $U = \sum_{i=1}^{\infty} u_i R$. Note that $u_i \in A^i$ and so $U \subseteq A$. Using the facts that R is right Noetherian and A has the right AR property, one can choose n sufficiently large to have $U = \sum_{i=1}^{n-1} u_i R$ and $U \cap A^n \subseteq UA$. Then

$$u_n = u_1 a_1 + \cdots + u_{n-1} a_{n-1},$$

where $a_i \in A$ for each i. Hence

$$r(1 - x^n) - (1 - x^n)r = (r(1 - x) - (1 - x)r)a_1 + \cdots \\ + (r(1 - x^{n-1}) - (1 - x^{n-1})r)a_{n-1}$$

and so
$$r\left(1 - x^n - \sum_{i=1}^{n-1}(1-x^i)a_i\right) = (1-x)r'$$
for some $r' \in R$. Since $1 - x^n - \sum(1-x^i)a_i \in 1 - A$, then (i) is proved.
(ii) The description of $\mathscr{S}$ makes clear that $A_\mathscr{S} \subseteq J(R_\mathscr{S})$. □

2.10 Next comes a lifting result.

Proposition. *Let R be a right Noetherian ring and A an ideal with the right AR property. Suppose that for each n the multiplicatively closed subset $\mathscr{C}(A/A^n)$ of R/A^n satisfies the right Ore condition. Then $\mathscr{C}_R(A)$ satisfies the right Ore condition.*

Proof. Let $r \in R, c \in \mathscr{C}_R(A)$. For some n,
$$(rR + cR) \cap A^n \subseteq (rR + cR)A = rA + cA.$$
The right Ore condition on $\mathscr{C}(A/A^n)$ gives $r' \in R$, $c' \in \mathscr{C}_R(A)$ such that $rc' - cr' \in A^n$ and so $rc' - cr' \in rA + cA$. Therefore, for some $a_1, a_2 \in A$,
$$rc' - cr' = ra_1 + ca_2.$$
Then $r(c' - a_1) = c(a_2 + r')$ which completes the proof since $c' - a_1 \in \mathscr{C}_R(A)$. □

2.11 The final result here concerning the AR property requires stronger and symmetric hypotheses. First comes a preparatory lemma involving reduced rank.

Lemma. *Let R be a Noetherian AR ring and B any ideal. Then $\rho(B_R) = 0$ if and only if $\rho({}_R B) = 0$.*

Proof. Suppose that $\rho(B_R) = 0$. Since ${}_R B$ is finitely generated it is clear from 1.3(i) that $Ba = 0$ for some $a \in \mathscr{C}(N)$, where $N = N(R)$. Let $A = r\operatorname{ann} B$. Then $B \cap A^n \subseteq BA$ for some n and so $B \cap A^n = 0$. It follows that $A^n B = 0$ and so $a^n B = 0$ since $a \in A$. Therefore, by 1.3(i) again, $\rho({}_R B) = 0$. □

Theorem. *Let R be a Noetherian AR ring and A a semiprime ideal. Then:*
(i) *$\mathscr{C}_R(A)$ satisfies the right and left Ore condition;*
(ii) *$R_{\mathscr{C}(A)}$ exists and $A_{\mathscr{C}(A)} = J(R_{\mathscr{C}(A)})$.*

Proof. (i) It is enough, by 2.10, to show that $\mathscr{C}(A/A^n)$ satisfies the right and left Ore condition in R/A^n for each n. Therefore we may suppose that A is the prime radical of R and that n is the smallest integer with $A^n = 0$. That $\mathscr{C}(A)$ satisfies the Ore condition will now be proved by induction on n. The case $n = 1$ is covered by Goldie's theorem (2.3.6). Suppose that $n > 1$ and, by the inductive hypothesis, that $\mathscr{C}(A/A^{n-1})$ satisfies the Ore condition in R/A^{n-1}. Let $B =$

4.2.13   Artin–Rees property and localization

$\{b \in R \mid cb = 0 \text{ for some } c \in \mathscr{C}(A)\}$, this being the left-hand version of ass $\mathscr{C}(A)$ (see 2.1.3). Of course B is closed under right multiplication and $B \subseteq A$. Let b, c be as above and $r \in R$. The hypothesis on R/A^{n-1} shows that there exists $c' \in \mathscr{C}(A), r' \in R$ such that $c'r - r'c \in A^{n-1}$. Since $b \in A$ it follows that $(c'r - r'c)b = 0$ and so $c'rb = 0$. This shows that B is closed under left multiplication. A similar argument shows that B is closed under addition. Hence $B \triangleleft R$.

Evidently $\rho(_R B) = 0$ and so, by the lemma, $\rho(B_R) = 0$. Now if $c \in \mathscr{C}(A)$ then r ann $c \subseteq B$. Thus $\rho(r\text{ ann }c) = 0$. However, $cR \simeq R/r\text{ ann }c$ and so, by 1.2, $\rho(cR) = \rho(R)$ and then $\rho(R/cR) = 0$. Hence, by 1.3, given any $r \in R$ there exists $c_1 \in \mathscr{C}(A)$ such that $rc_1 \in cR$. So (i) is proved.

(ii) Evidently ass $\mathscr{C}(A) \subseteq A$, so there is no loss in supposing ass $\mathscr{C}(A)$ to be zero. It is then clear that $A_{\mathscr{C}(A)} \subseteq J(R_{\mathscr{C}(A)})$. However, $R_{\mathscr{C}(A)}/A_{\mathscr{C}(A)} \simeq (R/A)_{\mathscr{C}(A)}$ which, by Goldie's theorem, is semisimple. □

2.12 It seems natural to ask whether it would be enough, in Theorem 2.11, to require only that the ideal A have the AR property. Bearing in mind that a nilpotent ideal always has the AR property this leads to the following open question.

Question. *If R is a Noetherian ring, P is a prime ideal and $P^2 = 0$, is P a localizable ideal?*

This question will be discussed further in 5.8. However, we note here that P being localizable is equivalent to the condition that the two sets

$$B_1 = \{r \in R \mid rc = 0 \text{ some } c \in \mathscr{C}(P)\} \quad \text{and} \quad B_2 = \{r \in R \mid cr = 0 \text{ some } c \in \mathscr{C}(P)\}$$

are equal. The necessity of this condition is clear from 2.1.4. Conversely, suppose the sets are equal, to B say. Of course $B \subseteq P$, and the proof of Theorem 2.11 makes clear that $B \triangleleft R$. It follows now from 1.4 that R/B has an Artinian quotient ring and so P is localizable.

2.13 Finally we provide an example demonstrating that the question has a negative answer if R is assumed to be only right Noetherian.

Example. *Let k be a field and*

$$R = \left\{ \begin{bmatrix} f(0) & g(x) \\ 0 & f(x) \end{bmatrix} \middle| f, g \in k[x] \right\} \subseteq M_2(k[x]).$$

Then R is right Noetherian, its nilpotent radical P is prime with $P^2 = 0$ and P is not right localizable.

Proof. If

$$a = \begin{bmatrix} 0 & 1 \\ 0 & 0 \end{bmatrix} \quad \text{and} \quad c = \begin{bmatrix} 0 & 0 \\ 0 & x \end{bmatrix}$$

then $c \in \mathscr{C}(P)$, but there do not exist elements $b \in R$, $d \in \mathscr{C}(P)$ such that $ad = cb$. □

§3 Localization and Prime Ideals

3.1 In this section it will be seen that the failure of localization at individual prime ideals of a Noetherian ring is tied up with an interrelationship between prime ideals. Indeed, one should rather aim to localize at a 'linked' collection of primes, X say; that is, to localize with respect to $\mathscr{C}(X) = \bigcap \{\mathscr{C}(P_i) | P_i \in X\}$. The focus, then, will be upon circumstances under which such a set is a right and left denominator set. A relatively complete answer is obtained, in particular, to the question of which semiprime ideals are localizable.

3.2 Any module M can be embedded, as an essential submodule, in an injective module $E(M)$, the **injective envelope** of M; see [Anderson and Fuller **74**].

Lemma. *Given any ring R and modules T_R, F_R the following conditions are equivalent*:
(i) $\mathrm{Hom}(T_R, E(F_R)) = 0$;
(ii) *for any $t \in T$, $0 \neq f \in F$ there exists $r \in R$ such that $tr = 0$ and $fr \neq 0$*.

Proof. (i)⇒(ii) Suppose (ii) is false; so there exist t, f such that $tr = 0$ implies $fr = 0$. That ensures that there is a nonzero homomorphism $tR \to F$, $tr \mapsto fr$. The diagram

shows that, since $E(F)$ is injective, $\mathrm{Hom}(T, E(F)) \neq 0$.

(ii)⇒(i) Suppose (i) is false and so $0 \neq \alpha \in \mathrm{Hom}(T, E(F))$. Choose $u \in T$ so that $\alpha(u) \neq 0$. Then there exists $s \in R$ such that $0 \neq \alpha(u)s = \alpha(us) \in F$. Set $t = us$ and $f = \alpha(t)$. Then $tr = 0$ implies $fr = 0$. □

Such a module T_R is termed **torsion with respect to** $E(F_R)$.

3.3 A connection with the torsion modules described in 2.1.17 comes next.

Proposition. *Let N be a semiprime ideal of R with R/N right Goldie, and let M be a right R-module. Then M is torsion with respect to $\mathscr{C}(N)$ if and only if M is torsion with respect to $E(R/N_R)$.*

Proof. Suppose M is torsion with respect to $\mathscr{C}(N)$ and that $t \in M$ and $0 \neq f \in R/N$. Then $f = [x + N]$ for some $x \in R \setminus N$. By hypothesis there exists $c \in \mathscr{C}(N)$ such that $tc = 0$; but, since $c \in \mathscr{C}(N)$, $cx \notin N$, i.e. $fc \neq 0$. This verifies condition (ii) of 3.2.

Conversely, suppose M is torsion with respect to $E(R/N_R)$, that $t \in M$, and $0 \neq f \in R/N$. Now $f = [x + N]$ with $x \in R \setminus N$. Applying 3.2(ii) to tx and f gives $r \in R$ such that $txr = 0$ yet $xr \notin N$. It follows immediately that if $D = \{d \in R \mid td = 0\}$ then $D + N/N \triangleleft_e R/N$. Hence $(D + N) \cap \mathscr{C}(N) \neq \varnothing$ by 2.3.5 and so $D \cap \mathscr{C}(N) \neq \varnothing$. Now if $d \in D \cap \mathscr{C}(N)$ then $td = 0$; so M is torsion with respect to $\mathscr{C}(N)$. □

3.4 There is a connection between certain prime ideals of a ring R and submodules of a module M_R. If $M \neq 0$ and $\operatorname{ann} M = \operatorname{ann} M'$ for all nonzero submodules M' of M then M is called a **prime module**. It is immediate that $\operatorname{ann} M$ is a prime ideal. It is called the **affiliated prime** of M_R.

Similarly, if $_S M_R$ is a bimodule then M is a **prime bimodule** if both M_R and $_S M$ are prime. This is equivalent to requiring that $\operatorname{ann} M'_R = \operatorname{ann} M_R$ and $\operatorname{ann} _S M' = \operatorname{ann} _S M$ for all $0 \neq M' \triangleleft _S M_R$.

3.5 We now collect together some basic facts.

Lemma. (i) *If R is satisfies the a.c.c. on ideals then each nonzero module contains a prime submodule.*

(ii) *If R, S satisfy the a.c.c. on ideals then each nonzero bimodule contains a prime sub-bimodule.*

(iii) *Suppose that S is left Noetherian, $_S M_R$ is a bimodule, $_S M$ is finitely generated, M_R is prime with affiliated prime P and R/P is right Goldie. Then M is a torsion-free R/P-module.*

Proof. (i) If $M \supseteq M_1 \supseteq M_2 \supseteq \ldots$ is a descending chain of submodules of M_R then $\operatorname{ann} M_1 \subseteq \operatorname{ann} M_2 \subseteq \ldots$ is an ascending chain of ideals of R.

(ii) This is proved similarly.

(iii) Let N be the torsion submodule of $M_{R/P}$. Evidently N is an (S, R)-sub-bimodule so $_S N$ is finitely generated. Choose $c \in \mathscr{C}(R/P)$ which annihilates each generator of $_S N$. Then $Nc = 0$ and so $\operatorname{ann} N_R \supset P$. Hence $N = 0$. □

3.6 That prime bimodules have some connection with localization is demonstrated by the next result.

Proposition. *Let R be a Noetherian ring which has a right and left denominator set $\mathscr{S}$. Let A, B be ideals of R such that B/A is a prime R-bimodule with affiliated primes P, Q on right and left respectively. Then $\mathscr{C}(P) \supseteq \mathscr{S}$ if and only if $\mathscr{C}(Q) \supseteq \mathscr{S}$.*

Proof. Suppose $\mathscr{C}(P) \supseteq \mathscr{S}$ and let $I = \operatorname{ass} \mathscr{S}$, the kernel of the map $R \to R_{\mathscr{S}}$. Let $b \in I \cap B$; then $bs = 0 \in A$ for some $s \in \mathscr{S} \subseteq \mathscr{C}(P)$. Then $b \in A$ since, by 3.5(iii), B/A

is torsion-free over R/P. Hence $B \cap I = A \cap I$. Furthermore, $IB \subseteq B \cap I \subseteq A$ and so $I \subseteq Q$ and, similarly, $I \subseteq P$. This means, effectively, that we may as well assume that $I = 0$; so then $\mathscr{S} \subseteq \mathscr{C}_R(0)$.

Since $(AR_\mathscr{S} \cap R)/A$ is torsion with respect to $\mathscr{S}$, and hence with respect to $\mathscr{C}(P)$, it is clear that $AR_\mathscr{S} \cap B = A$. By 2.1.16, $AR_\mathscr{S}$ and $BR_\mathscr{S}$ are ideals of $R_\mathscr{S}$ and so

$$(R \cap QR_\mathscr{S})B \subseteq B \cap QBR_\mathscr{S} \subseteq B \cap AR_\mathscr{S} = A.$$

It follows that $R \cap QR_\mathscr{S} = Q$ and hence $\mathscr{S} \subseteq \mathscr{C}(Q)$. Symmetry completes the proof. □

3.7 This result shows that, under these circumstances, one cannot simply localize at P, i.e. with respect to $\mathscr{C}(P)$. Instead one is obliged to look for some subset $\mathscr{S}$ of $\mathscr{C}(P) \cap \mathscr{C}(Q)$. This, of course, applies to any other prime ideal linked, in such fashion, to P or Q.

With this in mind, we say a **second layer link** exists between $P, Q \in \operatorname{Spec} R$, for a Noetherian ring R, if $Q \cap P/QP$ has a homomorphic image which is a prime bimodule with affiliated primes P and Q on right and left respectively. (The idea behind this name will become clearer later; cf. 3.12.) The notation $Q \rightsquigarrow P$ indicates such a link.

This link can be viewed as an edge in a (directed) graph having $\operatorname{Spec} R$ as vertex set. This graph is called the **link graph**. The set of vertices of any connected component of the link graph is called a **clique**.

The remarks above show that the largest subset of $\mathscr{C}(P)$ at which it is feasible to localize is $\bigcap \mathscr{C}(Q)$ as Q ranges through the clique containing P.

3.8 To illustrate this concept we describe some examples of links and cliques.

Example. (i) Let

$$R = \begin{bmatrix} k & k \\ 0 & k \end{bmatrix}$$

with k a field. Then R has only two primes,

$$P = \begin{bmatrix} 0 & k \\ 0 & k \end{bmatrix} \quad \text{and} \quad P' = \begin{bmatrix} k & k \\ 0 & 0 \end{bmatrix}.$$

Note that $P = P^2$, $P' = (P')^2$ and

$$P'P = \begin{bmatrix} 0 & k \\ 0 & 0 \end{bmatrix} = P' \cap P \supset PP' = 0.$$

Furthermore, $P \cap P'/PP'$ is simple on each side with left annihilator P and right annihilator P'. So the link graph is $P \rightsquigarrow P'$. Hence neither prime is localizable and there is just one clique.

4.3.8 *Localization and prime ideals* 109

Note, however, as in 2.1.9, that each prime is localizable on one side. This highlights the symmetry of the conditions imposed in 3.6 and throughout this section.

Example. (ii) Similar calculations show that the only prime ideals of the ring

$$\begin{bmatrix} \mathbb{Z} & \mathbb{Z} \\ 0 & \mathbb{Z} \end{bmatrix} \quad \text{are} \quad P_p = \begin{bmatrix} p\mathbb{Z} & \mathbb{Z} \\ 0 & \mathbb{Z} \end{bmatrix} \quad \text{and} \quad P'_p = \begin{bmatrix} \mathbb{Z} & \mathbb{Z} \\ 0 & p\mathbb{Z} \end{bmatrix}$$

with $p\mathbb{Z}$ any prime ideal of $\mathbb{Z}$. The links here are precisely $P_p \rightsquigarrow P'_p$ for all $p\mathbb{Z}$ and $P_p \rightsquigarrow P_p$ and $P'_p \rightsquigarrow P'_p$ for all nonzero $p\mathbb{Z}$. Once again no prime ideal is (right and left) localizable and each clique has two members.

Example. (iii) As in 1.7.6, let $R = U(g)$ with g the two-dimensional solvable Lie algebra. As noted there $R \simeq k[x][y;\sigma]$, where σ is given by $\sigma(x) = x + 1$; so $xy = y(x + 1)$. Suppose char $k = 0$. It follows easily from 1.8.7(i) that $k[x][y, y^{-1}; \sigma]$ is a simple ring. Thus y belongs to every nonzero prime ideal of R. Of course $yR \triangleleft R$ and $R/yR \simeq k[x]$. Thus yR is the unique height one prime.

Suppose now that k is algebraically closed. Then the nonzero primes of $k[x]$ take the form $(x + \alpha)$ for some $\alpha \in k$. Hence the prime ideals of R apart from 0 and yR are the maximal ideals $yR + (x + \alpha)R = P_\alpha$ say. Note that $yP_\alpha = P_{\alpha-1} y$.

We now describe the links and cliques. It is clear that $\{0\}$ is one clique and that there is no link $0 \rightsquigarrow 0$.

Next consider yR. There is a link $yR \rightsquigarrow yR$ since $yR/(yR)^2 = yR/y^2R \simeq R/yR$. On the other hand, $yR \cap P_\alpha/yRP_\alpha = yR/yP_\alpha = Ry/P_{\alpha-1}y$. Since the left annihilator is $P_{\alpha-1}$, there can thus be no link $yR \rightsquigarrow P_\alpha$. Similarly, there is no link $P_\alpha \rightsquigarrow yR$. So $\{yR\}$ is also a clique.

Finally, consider P_α. Note first that $P_\alpha P_\beta$ contains the element

$$(x + \alpha)y - y(x + \beta) = y(x + \alpha + 1) - y(x + \beta) = y(\alpha + 1 - \beta).$$

So if $\beta \neq \alpha + 1$ then $y \in P_\alpha P_\beta$ and thus

$$P_\alpha \cap P_\beta / P_\alpha P_\beta \simeq (x + \alpha) \cap (x + \beta)/(x + \alpha)(x + \beta)$$

in $k[x]$ and so is 0. On the other hand, if $\beta = \alpha + 1$ then

$$P_\alpha P_\beta = (yR + (x + \alpha)R)(yR + (x + \beta)R)$$
$$= y^2 R + y(x + \beta)R + (x + \alpha)(x + \beta)R$$

and

$$P_\alpha \cap P_\beta = yR + (x + \alpha)(x + \beta)R.$$

Hence $P_\alpha \cap P_\beta / P_\alpha P_\beta \simeq yR/y^2R + y(x + \beta)R$ which is isomorphic to R/P_β as a right module and to R/P_α as a left module. So $P_\alpha \rightsquigarrow P_{\alpha+1}$. Thus the other cliques of R take the form

$$\{\ldots, P_{\alpha-1}, P_\alpha, P_{\alpha+1}, \ldots\}$$

for each $\alpha \in k$. This, together with 3.6, shows that P_0 is not localizable, as was seen in 2.1.7(ii).

Example. (iv) Consider the ring $R = k[y, y^{-1}][x; \sigma]$, where $\sigma(y) = \lambda^{-1} y$ for some $\lambda \in k$, k being a field and λ not being a root of unity. One can see that x is a normal element and, using 1.8.7(ii), that xR is the unique height one prime. Assuming that k is also algebraically closed, the other (maximal) prime ideals are all of the form $P_\alpha = (x, y + \alpha)$. One can check that the cliques are $\{0\}$, $\{xR\}$, and the sets $\{P_{\lambda^n \alpha} | n \in \mathbb{Z}\}$.

3.9 Before proceeding further, some facts about uniform modules are required. Note first that if M_R is uniform and R is right Noetherian then there is a unique prime ideal affiliated to some submodule of M. This prime ideal is called the **associated prime** of M, ass M for short. Of course ass M is the affiliated prime of $\mathrm{ann}_M(\mathrm{ass}\, M)$. For more general M, R see 4.2, 4.4.

Lemma. *Let R be a Noetherian ring and let M_R be finitely generated and uniform with $MP \neq 0$, where $P = \mathrm{ass}\, M$. Then there is a submodule M' of M such that, letting U be the prime module $\mathrm{ann}_{M'} P$:*
(i) $M' \neq U$;
(ii) $\mathrm{ann}\, M' = \mathrm{ann}\, N$ for all $N \triangleleft M'$, $N \not\subseteq U$;
(iii) M'/U is a prime uniform module.

Proof. Let $U' = \mathrm{ann}_M P$, and then choose $M_1 \subseteq M$ to maximize $\mathrm{ann}\, M_1$ whilst keeping $M_1 \not\subseteq U'$. Let $U = M_1 \cap U'$. Let V be any prime uniform submodule of M_1/U and let M' be the inverse image in M_1 of V. The claimed properties are easily verified. □

3.10 Consider a module like M' in Lemma 3.9. Note that U is a finitely generated uniform right R/P-module. If $U_{R/P}$ is torsion-free, then 3.4.3 shows that U is isomorphic to a uniform right ideal of R/P. The next result aims to discern when the module V has the same property with respect to its affiliated prime.

Proposition. *Let R be a Noetherian ring and let $P, Q \in \mathrm{Spec}\, R$. Suppose there is an exact sequence of finitely generated uniform right R-modules $0 \to U \to M \to V \to 0$ in which:*
(i) *for all $N \triangleleft M$ with $N \not\subseteq U$, $\mathrm{ann}\, N = \mathrm{ann}\, M = A$ say;*
(ii) *V is a prime module with affiliated prime Q;*
(iii) *U is a prime module with affiliated prime P, $U = \mathrm{ann}_M P$ and $U_{R/P}$ is torsion-free.*
Then $QP \subseteq A \subseteq P \cap Q$, and
(a) *if $A \subset P \cap Q$ then $V_{R/Q}$ is torsion-free and $Q \rightsquigarrow P$,*
(b) *if $A = P \cap Q$ then $V_{R/Q}$ is not torsion-free and $A = Q \subset P$.*

Proof. It is clear that $QP \subseteq A \subseteq P \cap Q$.

(a) Suppose that $A \subset P \cap Q$. We will show first that $P \cap Q/A$ is a prime bimodule with affiliated primes P and Q; thus $Q \rightsquigarrow P$. To this end, let $I \triangleleft R$ with $A \subset I \subseteq P \cap Q$. It will suffice to show that $\mathrm{ann}\,(I/A)_R = P$ and $\mathrm{ann}_R(I/A) = Q$. The former is a consequence of the fact that $0 \neq MI \subseteq U$. To check the latter, let $T = \mathrm{ann}_R(I/A)$. Then $MTI = 0$ so $\mathrm{ann}\,MT \supset A$ and hence, by condition (i), $MT \subseteq U$. Therefore $T \subseteq Q$. On the other hand, since $QP \subseteq A$ and $I \subseteq P$, one has $QI \subseteq A$ and so $T \supseteq Q$. Thus $Q \rightsquigarrow P$.

We now turn to $V_{R/Q}$ and suppose it is not torsion-free. One can choose $m \in M \setminus U$ and $d \in \mathscr{C}(Q)$ with $md \in U$. Now d is regular on $_R(P \cap Q/A)$, so $d(P \cap Q) + A/A \simeq P \cap Q/A$. Hence $\rho_{R/P}(P \cap Q/d(P \cap Q) + A) = 0$. The kernel of the map $R \to M$, $r \mapsto mr$, contains $d(P \cap Q) + A$. Thus $\rho_{R/P}(m(P \cap Q)) = 0$ and yet $m(P \cap Q) \subseteq U$ which is R/P-torsion-free. Hence $mR(P \cap Q) = 0$ which contradicts (i). Therefore $V_{R/Q}$ must be torsion-free.

(b) Suppose $A = P \cap Q$. Then $MPQ = 0$, yet $MP \neq 0$ since $U \neq M$. However, M is uniform, so $MP \cap U \neq 0$ and $(MP \cap U)Q = 0$. By (iii) this makes $Q \subseteq P$ and so $A = Q$. However, $MP \neq 0$ so $A \neq P$ and thus $Q \subset P$.

Finally, suppose that $V_{R/Q}$ is torsion-free and so, as noted above, is isomorphic to a uniform right ideal of R/Q. Further, if u dim $R/Q = n$ then V^n contains an isomorphic copy of R/Q. Let M' be the inverse image of this copy in M^n. Now the R/Q-homomorphism $M' \to R/Q$ splits since R/Q is free. Therefore $M' \simeq R/Q \oplus (U^n \cap M')$. However, $U \triangleleft_e M$ and so $(U^n \cap M') \triangleleft_e M'$, a contradiction. Thus $V_{R/Q}$ cannot be torsion-free. □

3.11 It is, on the face of it, feasible in case (a) above that $P \subset Q$ or $Q \subset P$; and this is one of the obstructions to localization. Although no examples of such behaviour are known for Noetherian rings, there are easy non-Noetherian examples which illustrate the problem.

For instance, let A be a commutative integral domain and $\theta: A \to A$ a surjective ring homomorphism with kernel $K \neq 0$. (A could be $k[x_1, x_2, \ldots]$ and θ be defined by $\theta(x_i) = x_{i-1}, \theta(x_1) = 0$.) Let R be the subring of $M_2(A)$ consisting of all matrices

$$\begin{bmatrix} a & b \\ 0 & \theta(a) \end{bmatrix},$$

where $a, b \in A$. Let P be those elements of R in which $a = 0$, and Q those in which $a \in K$. It is easily checked that $P \subset Q$ and so $P \cap Q = P$, that $PQ = 0$, and that $(P \cap Q)/PQ$ provides a link $Q \rightsquigarrow P$.

3.12 The localization result at which we aim relies upon case (a) of 3.10 dominating. If a prime ideal P has the property that whenever the situation described in 3.10 or its left-hand version occurs, then case (a) holds, P satisfies the **second layer condition**. A subset of Spec R satisfies the second layer condition if each of its members does; and R satisfies it if $\operatorname{Spec} R$ does.

In the notation of 3.10, note that $M \subseteq E(U)$ and so $V \subseteq E(U)/U$, the 'second layer' of $E(U)$. Also U is isomorphic to a right ideal of R/P and so $E_R(U)^n \simeq E_R(R/P)$, where $n = \operatorname{u\,dim} R/P$. It follows that V is in the 'second layer' of $E_R(R/P)$.

3.13 For cliques, the second layer condition has other formulations.

Proposition. *Let R be a Noetherian ring and let X be a clique in $\operatorname{Spec} R$. Then the following conditions are equivalent:*
(i) *X satisfies the second layer condition.*
(ii) *If $P \in X$ and K is a finitely generated submodule of $E_R(R/P)$ then K has a chain of submodules $K = K_n \supset K_{n-1} \supset \cdots \supset K_0 = 0$ such that for each i, K_i/K_{i-1} is isomorphic to a uniform right ideal of R/P_i for some $P_i \in X$.*
(iii) *For each $P \in X$, the only prime ideal which is the annihilator of a finitely generated submodule of $E_R(R/P)$ is P itself.*

Proof. (i)$\Rightarrow$(ii) As noted above $E_R(R/P) \simeq E_R(W)^n$ for some uniform right ideal W of R/P. It is convenient, for the proof, to aim to demonstrate the conclusion for K a finitely generated submodule of $E(W_1) \oplus \cdots \oplus E(W_t)$ where each W_i is a uniform right ideal of R/Q_i with $Q_i \in X$.

Given such a module K, the short exact sequence

$$0 \to E(W_t) \to E(W_1) \oplus \cdots \oplus E(W_t) \to E(W_1) \oplus \cdots \oplus E(W_{t-1}) \to 0$$

yields a corresponding exact sequence

$$0 \to K \cap E(W_t) \to K \to K' \to 0$$

with $K' \hookrightarrow E(W_1) \oplus \cdots \oplus E(W_{t-1})$. Induction on t reduces the problem to the case when $t = 1$. In that case we write W for W_1, Q for Q_1 and note that, replacing W by $W \cap K$, we may suppose that $W \subseteq K$. Note that $\operatorname{ann}_K Q$ is a uniform torsion-free R/Q-module and so, by 3.4.3, it too is isomorphic to a right ideal of R/Q. Thus we may suppose that $W = \operatorname{ann}_K Q$.

If $W = K$, there is nothing to prove. Otherwise, let V be a uniform submodule of K/W and let M be its inverse image in K. One can then obtain a submodule M' of M as described in 3.9; and it is clear that $U = W \cap M'$. The second layer condition, applied to M', implies that M'/U is isomorphic to a uniform right ideal of R/Q' for some $Q' \in X$. Note that $M'/U \hookrightarrow V$.

Finally, note that K/W is an essential extension of some direct sum of uniform submodules $V_1, \ldots, V_s$ say. The preceding paragraph shows that $K/W \subseteq E(V'_1) \oplus \cdots \oplus E(V'_s)$ with V'_j a uniform right ideal of R/Q_j and $Q_j \in X$. By Noetherian induction, applied to K, we may now deduce that K/W has a composition series of the type required.

(i)$\Rightarrow$(iii) This is a consequence of the argument above. Let K be a finitely generated submodule of $E(R/P)$ with $\operatorname{ann}_R K = A \in \operatorname{Spec} R$, $A \neq P$. Clearly, one of the primes P_i provided by (ii) must equal A and must arise, in the inductive process, as the annihilator of the top piece of an M'. But then ann $M' = A$ contradicting the second layer condition applied to M'.

(ii)$\Rightarrow$(i) Let M, U, V, P, Q be as in 3.10 and suppose the second layer condition fails for M; so ann $M = Q \subset P$ and V is a torsion module over R/Q. By hypothesis, since $M \hookrightarrow E(U)$ there is a chain

$$M = M_n \supset \cdots \supset M_1 \supset M_0 = 0$$

with M_i/M_{i-1} isomorphic to a uniform right ideal of R/P_i and $P_i \in X$. Since M is uniform, $M_1 \cap U \neq 0$. Hence $P = P_1$ and $M_1 \subseteq U = \mathrm{ann}_M P$. If $M_2 \subseteq U$ we may delete M_1 from the chain without loss. So, without loss of generality, suppose $M_2 \not\subseteq U$ and consider $(M_2 \cap U)/M_1$. If it is nonzero then it is isomorphic to a uniform right ideal of R/P_2 and, as a submodule of U/M_1, is annihilated by P. Hence $P \subseteq P_2$, $M_2 P^2 = 0$ and so, since $\mathrm{ann}_{M/U} P = 0$, $M_2 \subseteq U$, contradicting our assumption.

This shows that $(M_2 + U)/U \simeq M_2/M_1$ and hence

$$P_2 = \mathrm{ann}\, M_2/M_1 = \mathrm{ann}\,(M_2 + U)/U = \mathrm{ann}\, V = Q.$$

However, M_2/M_1 is isomorphic to a right ideal of R/P_2 and $(M_2 + U)/U$ is a torsion module over R/Q, a contradiction.

(iii)$\Rightarrow$(i) If, as above, the second layer condition failed for M then ann $M = Q \subset P$, a contradiction. □

3.14 Examples of Noetherian rings satisfying the second layer condition are easily described. In particular, suppose that R has the property that in each prime factor ring every essential one-sided ideal contains a nonzero ideal. Such a ring is termed an FBN ring, and its properties are discussed more fully in Chapter 6, §4. Examples include all commutative Noetherian rings, of course, and also all Noetherian PI rings (see Chapter 13). For an FBN ring R it is clear that $V_{R/Q}$ must be torsion-free. So FBN rings always satisfy the second layer condition.

It is known, too, that if g is a finite dimensional Lie algebra over $\mathbb{C}$ then $U(g)$ has the second layer condition if and only if g is solvable. Also RG has the second layer condition whenever R is commutative Noetherian and G is polycyclic by finite. See [Jategaonkar **86**] for details.

3.15 There are Noetherian rings which fail to satisfy the second layer condition.

Example. *Let R be a prime Noetherian ring with an idempotent prime ideal P which is the unique minimal nonzero prime. (For example take the ring $\mathbb{I}(xA_1)$ described in 1.3.10). Then R does not satisfy the second layer condition.*

Proof. Let $E_R = E(R/P_R)$. There is an embedding $R_R \hookrightarrow P_R$ and thus a surjection

$\text{Hom}(P_R, E_R) \twoheadrightarrow \text{Hom}(R_R, E_R)$. Hence there is a nonzero map $\alpha: P \to E_R$ and, since $P = P^2$, $\alpha(P)P \neq 0$. Hence $\text{ann}_R(E_R) \subset P$. It follows that E_R contains a submodule M of the form described in 3.10. In the notation of 3.10, note that $P \cap Q \supseteq A \supseteq QP$. If $Q \supseteq P$ then $QP \supset P^2 = P$ and so $A = P$ and $MP = 0$, a contradiction. Hence $Q \not\supseteq P$ and so $Q = 0$. Then $A = P \cap Q = 0$ and case (b) of 3.10 must occur. □

The argument can be extended to show that a Noetherian prime ring with a unique minimal nonzero ideal fails to have the second layer condition.

3.16 It is now possible to describe circumstances under which one can localize with respect to a semiprime ideal N. By 3.2.4, we know that $\mathscr{C}(N) = \bigcap \mathscr{C}(P_i)$ as P_i ranges over the set of minimal primes of N. So this, effectively, will deal with the case of a clique X which is a finite set of incomparable prime ideals.

Theorem. *Let R be a Noetherian ring, N a semiprime ideal and X the set of minimal primes of N. Suppose that X is closed under links and satisfies the second layer condition. Then R can be localized at $\mathscr{C}(N)$.*

Proof. We aim to show that $\mathscr{C} = \mathscr{C}(N)$ satisfies the right Ore condition; for this it is enough to demonstrate that for each $c \in \mathscr{C}$ the module R/cR is $\mathscr{C}$-torsion. Suppose not. Then one can choose an image M of R/cR which is not $\mathscr{C}$-torsion but has the property that all its proper factors are $\mathscr{C}$-torsion. Evidently M is uniform. Since $\mathscr{C} = \bigcap \mathscr{C}(P_i)$, it follows that M is not $\mathscr{C}(P)$-torsion for some $P \in X$. Hence by 3.3, there is a nonzero homomorphism $\alpha: M \to E(R/P_R)$. Since $E(R/P_R)$ is $\mathscr{C}(P)$-torsion-free, it follows that α must be an embedding.

Let $U' = \{m \in M \mid mP = 0\}$. It is clear that $U' \neq 0$; in fact we will see that $U' = M$. Suppose otherwise. Then one can choose a submodule M' of M, as described in 3.9, with $U = U' \cap M'$. Let $V = M'/U$. Then U and V are both prime modules, with affiliated primes P, Q respectively, and 3.10 can be applied. By assumption, case (a) occurs; so $V_{R/Q}$ is torsion-free and $Q \leadsto P$. Hence $Q \in X$. However, the construction of M implies that M/U is $\mathscr{C}$-torsion and hence V is $\mathscr{C}(Q)$-torsion. This gives a contradiction and shows that, as claimed, $U' = M$.

Thus $MP = 0$ and so, by 3.4.7, $M \hookrightarrow R/P$. However, M contains a nonzero $\mathscr{C}$-torsion element, namely the image of $[1 + cR]$. This contradiction shows that R/cR must be $\mathscr{C}$-torsion. Symmetry, plus the results of Chapter 2, §1, shows that $\mathscr{C}$ is a right and left denominator set. □

3.17 The proof above can be modified to give a more general result applicable to the case when X is an infinite set of incomparable prime ideals.

Theorem. *Let R be a Noetherian ring and X a set of prime ideals closed under links and satisfying the second layer condition. Suppose further:*

4.3.19   *Localization and prime ideals*   115

(i) given any one-sided ideal I of R with $I \cap \mathscr{C}(P) \neq \emptyset$ for all $P \in X$ then $I \cap \mathscr{C}(X) \neq \emptyset$;
(ii) no two prime ideals in X are comparable.
Then R can be localized with respect to $\mathscr{C}(X)$. □

Condition (i) here is the **intersection condition**. It is not a consequence of the other conditions; for example, let $R = k[x, y]$ for some field k, $I = (x, y)$ and X be the set of all other maximal ideals.

3.18 Naturally the ring $R_{\mathscr{C}(X)}$ obtained in 3.16 or 3.17 is not a local ring in that it will not usually have a unique maximal ideal or a unique simple module. However, it does have several properties reminiscent of semilocal rings.

Theorem. *Let R, X be as in 3.17. The ring $R_X = R_{\mathscr{C}(X)}$ has the following properties:*
(i) R_X/PR_X *is simple Artinian for all $P \in X$;*
(ii) *every simple R_X-module is annihilated by some PR_X with $P \in X$;*
(iii) *any finitely generated essential extension of a simple R_X-module has finite length.*

Proof. If M is a simple right R_X-module then M_R is $\mathscr{C}(X)$-torsion-free and any proper factor of M_R is $\mathscr{C}(X)$-torsion. As in the proof of 3.16, it follows that M_R is a torsion-free R/P-module for some $P \in X$. Condition (ii) is now obvious and so also is (i) once it is noted that if $c \in \mathscr{C}(P)$ then $(cR + P) \cap \mathscr{C}(X) \neq \emptyset$ by condition (i) of 3.17. Thus if U is a uniform right ideal of R/P then $U \otimes R_X$ is simple. Therefore (iii) follows from 3.13(ii). □

3.19 There is a converse result.

Theorem. *Let R be a Noetherian ring, X a set of prime ideals and suppose R_X exists and has the properties listed in 3.18. Then X is as described in 3.17.*

Proof. As before, for simplicity, we will give details only in the case when X is the set of minimal primes of some semiprime ideal N. Suppose $P \in X$ and $Q \rightsquigarrow P$ via the ideal A with $Q \cap P \supset A \supseteq QP$. We check first that

$$Q_X \cap P_X = (Q \cap P)_X \supset A_X \supseteq Q_X P_X = QP_X.$$

This is all clear except, possibly, the strict containment. However, $Q \cap P/A$ is right torsion-free over R/P and so embeds in $(Q \cap P/A)_X$. On the other hand, $(Q \cap P/A)_X$ is also a left $(R/Q)_X$ module; so $R_X \neq Q_X$. Hence Q must be a minimal prime over N. Thus X is closed under links.

Next, suppose that there is an exact sequence

$$0 \to U \to M \to V \to 0$$

as in 3.10. Since R_X is flat, the sequence remains exact after tensoring with R_X. The Ore condition on $\mathscr{C}(X)$ implies that M/U is $\mathscr{C}(X)$-torsion-free. For suppose $mc \in U$ for $c \in \mathscr{C}(X)$. Let $p \in P$. Then there exists $c' \in \mathscr{C}(X)$, $r \in R$ such that $pc' = cr$. Now $\mathscr{C}(X) \subseteq \mathscr{C}(P)$ and so $r \in P$. Hence $mpc' = mcr \in UP = 0$ and so $mp = 0$ and $mP = 0$. That means $m \in U$, as required.

Since M/U is $\mathscr{C}(X)$-torsion-free, $M/U \hookrightarrow M_X/U_X$. However, M_X is an essential extension of U_X and so has finite length over R_X by hypothesis. It follows that V_X must be simple and yet V_X is annihilated by QR_X. Condition (ii) of 3.18 then shows that $Q \in X$. Hence V is R/Q-torsion-free and the second layer condition holds. □

3.20 One further property of R_X can be obtained easily. First we prove a very general fact.

Proposition. *If a ring R has the property that every finitely generated essential extension of a simple module has finite length then $\bigcap J^n = 0$, where J is the Jacobson radical of R.*

Proof. For each $x \in R$, Zorn's lemma provides a right ideal I_x maximal with respect to not containing x. Evidently $\bigcap \{I_x | x \in R\} = 0$. It is evident that $xR + I_x/I_x$ is simple and is an essential submodule of R/I_x. Hence by hypothesis R/I_x has finite length, $n(x)$ say. It follows that $(R/I_x)J^{n(x)} = 0$ and so $J^{n(x)} \subseteq I_x$. Therefore $\bigcap J^n \subseteq \bigcap I_x = 0$. □

Corollary. *If R, X are as described in 3.17 and J is the Jacobson radical of R_X then $\bigcap J^n = 0$* □

§4 Affiliated Primes and Regular Elements

4.1 In 3.4, prime modules and their affiliated primes were introduced. In this section this notion is extended to a more general module M, providing a sequence of affiliated prime ideals. A related idea, the set of associated prime ideals of M, is also described.

These ideas will then be used to show that, in any Noetherian ring, $\mathscr{C}(0) = \bigcap \mathscr{C}(P_i)$ for some finite set of prime ideals. This extends a similar result for semiprime rings, although, unlike that case, the prime ideals may not all be minimal.

4.2 We start by extending from right Noetherian rings to all rings the notion of the **associated prime** of a uniform module M_R. This is defined to be

$$\text{ass } M = \{r \in R | Nr = 0 \text{ for some } 0 \neq N \lhd M_R\}$$
$$= \bigcup \{\text{ann } N | 0 \neq N \lhd M_R\}.$$

4.4.7 *Affiliated primes and regular elements*

That this extends the notion introduced in 3.9 is shown by (iii) of the next result.

4.3 Lemma. *Let R be a ring and M_R a uniform module. Then:*
(i) ass M = ass N for any $0 \neq N \triangleleft M_R$;
(ii) ass $M \in \operatorname{Spec} R$;
(iii) *if R is right Noetherian then* ass M *is the affiliated prime of some submodule of M.*

Proof. Straightforward. □

4.4 More generally, given any module M_R, ass M is defined to be the set of those prime ideals occurring as the associated prime ideals of uniform submodules of M; these are then described as the **associated prime ideals** of M.

Note that if $\operatorname{u\,dim} M = k < \infty$, then ass M contains no more than k prime ideals.

4.5 The connection with affiliated primes is immediate.

Lemma. *If R is right Noetherian and M_R is finitely generated, then $P \in$ ass M if and only if P is the affiliated prime ideal of some prime submodule of M.*

Proof. This is clear, using 4.3(iii). □

4.6 We next extend the notion of affiliated primes to a more general module M_R. If possible, choose a prime submodule M_1 of M with affiliated prime P_1 and with $M_1 = \operatorname{ann}_M P_1$. Repeat this process on M/M_1 etc., obtaining thereby a chain of submodules $0 = M_0 \subset M_1 \subset \cdots$ with M_i/M_{i-1} being prime, $\operatorname{ann} M_i/M_{i-1} = P_i \in \operatorname{Spec} R$ and M_i/M_{i-1} being the annihilator in M/M_{i-1} of P_i. If this chain terminates with $M_n = M$, the chain is called an **affiliated chain** for M_R and the sequence of prime ideals $\{P_i\}$ is called an **affiliated sequence of primes** for M_R.

Note that if M is an (S, R)-bimodule then the M_i will all be sub-bimodules. Also, if M_R and R_R are Noetherian, affiliated chains always exist.

4.7 Of course ass M is uniquely defined by M, but not so the affiliated primes of M_R. They do indeed vary depending upon the choices made in the construction.

For example let A be a commutative Noetherian local domain with maximal ideal $J \neq 0$, and let R be the ring

$$\begin{bmatrix} A & A & A \\ 0 & A & A \\ 0 & 0 & A \end{bmatrix} \bigg/ \begin{bmatrix} 0 & 0 & J \\ 0 & 0 & 0 \\ 0 & 0 & 0 \end{bmatrix} = \begin{bmatrix} A & A & A/J \\ 0 & A & A \\ 0 & 0 & A \end{bmatrix}$$

for short. One can check that

$$0 \subset \begin{bmatrix} 0 & J & 0 \\ 0 & 0 & 0 \\ 0 & 0 & 0 \end{bmatrix} \subset \begin{bmatrix} 0 & J & A/J \\ 0 & 0 & 0 \\ 0 & 0 & 0 \end{bmatrix} \subset \begin{bmatrix} 0 & A & A/J \\ 0 & 0 & 0 \\ 0 & 0 & 0 \end{bmatrix}$$

$$\subset \begin{bmatrix} 0 & A & A/J \\ 0 & 0 & A \\ 0 & 0 & A \end{bmatrix} \subset \begin{bmatrix} 0 & A & A/J \\ 0 & A & A \\ 0 & 0 & A \end{bmatrix} \subset R$$

is an affiliated chain for R_R with the corresponding sequence of affiliated primes

$$\begin{bmatrix} A & A & A/J \\ 0 & 0 & A \\ 0 & 0 & A \end{bmatrix}, \begin{bmatrix} A & A & A/J \\ 0 & A & A \\ 0 & 0 & J \end{bmatrix}, \begin{bmatrix} A & A & A/J \\ 0 & J & A \\ 0 & 0 & A \end{bmatrix},$$

$$\begin{bmatrix} A & A & A/J \\ 0 & A & A \\ 0 & 0 & 0 \end{bmatrix}, \begin{bmatrix} A & A & A/J \\ 0 & 0 & A \\ 0 & 0 & A \end{bmatrix}, \begin{bmatrix} 0 & A & A/J \\ 0 & A & A \\ 0 & 0 & A \end{bmatrix}.$$

On the other hand, replacing the first two nonzero ideals by

$$\begin{bmatrix} 0 & 0 & A/J \\ 0 & 0 & 0 \\ 0 & 0 & 0 \end{bmatrix}$$

gives another affiliated chain for R_R whose affiliated primes do not include the maximal ideal

$$\begin{bmatrix} A & A & A/J \\ 0 & J & A \\ 0 & 0 & A \end{bmatrix}.$$

4.8 We note next some easy properties.

Lemma. *Given an affiliated chain constructed as in 4.6:*
(i) $M_i = \operatorname{ann}_M(P_i \cdots P_2 P_1)$;
(ii) *if M_R is faithful then $\{P_1, \ldots, P_n\}$ includes all minimal primes of R;*
(iii) *if M_R and R_R are Noetherian then the family $\{P_1, \ldots, P_n\}$ contains* ass M.

Proof. (i) By definition, $M_1 = \operatorname{ann}_M P_1$. Suppose that

$$M_{i-1} = \operatorname{ann}_M(P_{i-1} \cdots P_1) \quad \text{and let} \quad M' = \operatorname{ann}_M(P_i \cdots P_i).$$

Certainly $M' \supseteq M_i$; and yet

$$M' P_i \subseteq \operatorname{ann}_M(P_{i-1} \cdots P_1) = M_{i-1}.$$

So $M' = M_i$ and induction completes the argument.

(ii) Evident, since $P_n \cdots P_1 = 0$.

(iii) Suppose $P \in \mathrm{ass}\, M$; so $P = \mathrm{ann}\, N$ for some prime submodule N of M. If $N \cap M_1 \neq 0$ then
$$\mathrm{ann}\, M_1 = \mathrm{ann}\,(N \cap M_1) = \mathrm{ann}\, N = P$$
as required. Otherwise $N \cap M_1 = 0$ and $N \hookrightarrow M/M_1$. Then $P \in \mathrm{ass}\, M/M_1$ and induction completes the proof. □

4.9 The remaining results of this section connect affiliated primes with m.c. sets.

Proposition. *Suppose that R is right Noetherian, $_SM_R$ is a bimodule, M_R is finitely generated and $_SM$ has an affiliated sequence of prime ideals $\{P_i\}$ such that S/P_i is left Goldie for each i. Then*
$$\bigcap \mathscr{C}(P_i) \subseteq \mathscr{C}'_M(0) = \{c \in S \mid cm \neq 0 \text{ for all } 0 \neq m \in M\}.$$

Proof. Let $M_0 \subset M_1 \subset \cdots$ be the appropriate affiliated chain for $_SM$. Suppose that $c \in \bigcap \mathscr{C}(P_i)$ and that $cm = 0$ for some $m \in M$. Choose the least i such that $m \in M_i$. Now M_i/M_{i-1} is torsion-free over S/P_i by the left-hand version of 3.5(iii), yet $cm \in M_{i-1}$ and $c \in \mathscr{C}(P_i)$. It follows that $m \in M_{i-1}$. The conclusion must be that $i = 0$ and so $m = 0$. Hence $c \in \mathscr{C}'_M(0)$. □

4.10 The relationship between affiliated primes and minimal primes is important.

Corollary. *Let R be Noetherian. The following are equivalent:*
(i) *R_R and $_RR$ each have an affiliated sequence of primes with each prime being minimal;*
(ii) *every affiliated sequence of primes of R_R or $_RR$ consists of minimal primes;*
(iii) *R has an Artinian quotient ring.*

Proof. We know that $\mathscr{C}(N) = \bigcap \mathscr{C}(Q_i)$, where N is the prime radical and $\{Q_i\}$ is the set of minimal primes.

(i)$\Rightarrow$(iii) Let $\{P_i\}$ and $\{P'_i\}$ be the given affiliated sequences for R_R and $_RR$. Then $\bigcap \mathscr{C}(P_i) = \bigcap \mathscr{C}(P'_i) = \bigcap \mathscr{C}(Q_i) = \mathscr{C}(N)$, by hypothesis, and so $\mathscr{C}(N) \subseteq \mathscr{C}'(0) \cap \mathscr{C}'(0) = \mathscr{C}(0)$ by 4.9. This, by Corollary 1.4, ensures that R has an Artinian quotient ring.

(iii)$\Rightarrow$(ii) Let $\{P_i\}$ and $\{P'_i\}$ be affiliated sequences for R_R and $_RR$ respectively. It is enough to show that P_i is a minimal prime. Note, by 4.8, that the corresponding submodule of R_R is $M_i = \mathrm{l\,ann}\,(P_i \cdots P_1)$. This is an annihilator ideal; so as in 3.3.1 $Q(R)M_i \cap R = M_i$. Hence
$$P_i Q(R) \subseteq \mathrm{r\,ann}_Q (Q(R)M_i/Q(R)M_{i-1})$$
and so $P_i Q(R) \neq Q(R)$. Since $Q(R)$ is Artinian any nonminimal prime ideal meets $\mathscr{C}(N)$ nontrivially. Therefore P_i must be minimal. □

4.11 Next comes a preparatory result.

Proposition. *Let R be a Noetherian ring, A an associated prime of R_R and $B = \mathrm{l\,ann}\, A$. Let $\{B_i\}$ be an affiliated chain of submodules for $_R B$ with the affiliated sequence of primes $\{P_i\}$. Then $\mathscr{C}'_R(0) = \mathscr{C}$, where $\mathscr{C} = \mathscr{C}'(B) \cap \mathscr{C}(P_1) \cap \cdots \cap \mathscr{C}(P_n)$.*

Proof. Note first that A is the affiliated prime ideal of B_R and so $B_{R/A}$ is torsion-free, by 3.5; and that $\mathscr{C}'_R(0) \subseteq \mathscr{C}'_R(B)$ by 2.1.10 since B is a left annihilator ideal.

Suppose that $c \in \mathscr{C}'_R(0)$. We aim to show that $c \in \mathscr{C}(P_i)$. However, left multiplication by c gives an isomorphism $B_i/cB_i \simeq c^k B_i/c^{k+1} B_i$ for all k. Yet B_i is a finitely generated right R/A-module, so $\rho(B_i)_{R/A} < \infty$ and thus $\rho(B_i/cB_i) = 0$. Therefore B_i/cB_i is a torsion R/A-module.

Now suppose $rc \in P_i$ for some $r \in R$, and so $rcB_i \subseteq B_{i-1}$. However, $\rho(B_i/cB_i)_{R/A} = 0$ and hence for any $b \in B_i$ there exists an essential right ideal E of R/A such that $bE \subseteq cB_i$. Let $H = P_1 P_2 \cdots P_{i-1}$. Then $HrbE \subseteq HrcB_i \subseteq HB_{i-1} = 0$. However, $Hrb \subseteq B$ which is torsion-free over R/A. Therefore $Hrb = 0$. Since b was arbitrary, $HrB_i = 0$. Thus, by 4.8, $rB_i \subseteq \mathrm{ann}_B H = B_{i-1}$. Therefore $r \in P_i$ and so $c \in {'\mathscr{C}}(P_i) = \mathscr{C}(P_i)$. This shows that $\mathscr{C}'_R(0) \subseteq \mathscr{C}$.

Conversely, if $c \in \mathscr{C}$ and $cr = 0$ for some $r \in R$ then $r \in B$ since $c \in \mathscr{C}'(B)$. However, 4.9 shows that $c \in \mathscr{C}'_B(0)$. Therefore $r = 0$ and $\mathscr{C} \subseteq \mathscr{C}'_R(0)$. $\square$

4.12 Corollary. *If R is a Noetherian ring then $\mathscr{C}_R(0) = \bigcap \mathscr{C}(P_i)$ for some finite set of prime ideals $P_1, \ldots, P_m$.*

Proof. Choose $B, P_1, \ldots, P_n$ as in 4.11. By Noetherian induction, we may suppose that
$$\mathscr{C}'_R(B) = \mathscr{C}(P_{n+1}) \cap \cdots \cap \mathscr{C}(P_k)$$
say; and then $\mathscr{C}'(0) = \mathscr{C}(P_1) \cap \cdots \cap \mathscr{C}(P_k)$. By symmetry,
$$'\mathscr{C}(0) = \mathscr{C}(P_{k+1}) \cap \cdots \cap \mathscr{C}(P_m)$$
say; and so $\mathscr{C}(0) = \mathscr{C}(P_1) \cap \cdots \cap \mathscr{C}(P_m)$. $\square$

Despite the many choices made above, it is the case that there is a unique minimal collection of prime ideals such that $\mathscr{C}(0) = \bigcap \mathscr{C}(P_i)$—see [Small and Stafford 82].

§5 Additivity

5.1 Given two Noetherian rings $R \subseteq S$ and a prime ideal P of S, one might expect some connection between the prime ring S/P and the factors of $R/P \cap R$ by its minimal prime ideals. The main result of this section describes a close connection between the uniform dimensions, but it involves more prime ideals of R than indicated above.

5.2 We start with an easy illustrative example. Take $S = M_n(A)$ for some commutative integral domain A, $P = 0$, and $R = \bigoplus_{i=1}^{k} M_{n_i}(A)$, where $\sum n_i = n$. Then R can be embedded in S as blocks down the main diagonal. Evidently R has k minimal primes P_j where $P_j = \bigoplus_{i \neq j} M_{n_i}(A)$. Then u dim $R/P_j = n_j$ and $\sum$u dim $R/P_j = $ u dim S.

There is also an embedding $A \hookrightarrow S$; and in this case u dim $S = (n(\text{u dim } A))$.

By appropriately combining these two types of embedding one can readily obtain, given any positive integers z_i, m_i with $\sum_{i=1}^{k} z_i m_i = n$, a subring R of S having minimal primes $P_1, \ldots, P_k$ such that u dim $R/P_i = m_i$ and so $\sum z_i(\text{u dim } R/P_i) = $ u dim S.

The remainder of this section will indicate how typical this example is.

5.3 First we note an easy special case.

Lemma. *Let $R \subseteq S$ be simple Artinian rings. Then u dim R divides u dim S.*

Proof. If u dim $R = n$ then $R \simeq M^n$ for some minimal right ideal M. Hence $S \simeq R \otimes S \simeq (M \otimes S)^n$ and so u dim $S = n(\text{u dim } M \otimes S)$. □

5.4 Theorem. *Let $R \subseteq S$ be rings such that S is prime right Goldie and R/P is prime right Goldie for each $P \in \text{Spec } R$. Then there is a finite family $X = \{P_1, \ldots, P_n\} \subseteq \text{Spec } R$ which includes all the minimal prime ideals of R and such that*

$$\text{r u dim } S = \sum_{i=1}^{n} z_i(\text{r u dim } R/P_i)$$

with each z_i being an integer ≥ 1.

Proof. Let $Q = Q(S)$ and consider Q as a (Q, R)-bimodule. Since $_Q Q$ is of finite length so too is $_Q Q_R$. Evidently the annihilator in R of each simple composition factor of $_Q Q_R$ is a prime ideal. It follows that there is an affiliated chain

$$0 = M_0 \subset M_1 \subset \cdots \subset M_n = {}_Q Q_R$$

with $P_i = \text{ann}(M_i/M_{i-1})_R \in \text{Spec } R$ and $M_i/M_{i-1} = \text{ann}_{Q/M_{i-1}} P_i$. By 4.8, the sequence $\{P_i\}$ includes all the minimal primes of R.

Now consider $N = M_i/M_{i-1}$. This is a $(Q, R/P_i)$-bimodule and, as in the proofs of 1.5 and 1.6, N is, in fact, a right $Q(R/P_i)$-module of finite length. Hence $Q(R/P_i) \hookrightarrow \text{End}_Q N$. Now u dim $\text{End}_Q N =$ u dim$_Q N$, u dim $Q(R/P_i)$ divides u dim $\text{End}_Q N$, by 5.3, and $\sum_i$u dim$_Q(M_i/M_{i-1}) = $ u dim$_Q Q$ since Q is simple Artinian. The result follows. □

5.5 Next comes an example to show that nonminimal prime ideals can indeed be involved. The construction starts with a commutative integral domain A with a nonzero prime ideal I. We suppose that there is an embedding $\alpha: A/I \hookrightarrow A$. (For instance, A could be $k[x, y]$ and I be yA. Then $A/I \simeq k[x] \hookrightarrow k[x, y]$.)
Next let

$$R = \begin{bmatrix} A & I \\ A & A \end{bmatrix} \quad \text{and note that} \quad P = \begin{bmatrix} I & I \\ A & A \end{bmatrix} \in \operatorname{Spec} R;$$

indeed $R/P \simeq A/I$. Thus there is a ring homomorphism $\theta: R \to A$ with kernel P and with im θ = im α. We consider the embedding of R into $S = M_3(A)$ given by

$$r = \begin{bmatrix} a_1 & a_2 \\ a_3 & a_4 \end{bmatrix} \mapsto \begin{bmatrix} a_1 & a_2 & 0 \\ a_3 & a_4 & 0 \\ 0 & 0 & \theta(r) \end{bmatrix}$$

and identify R with the image. The prime ideals of R making up the family X are then 0 and P. Of course u dim $R = 2$, u dim $R/P = 1$ and

$$\text{u dim } S = \text{u dim } R + \text{u dim } R/P.$$

5.6 We will find circumstances later where the prime ideals $P_i \in X$ are precisely the minimal primes of R. In verifying this, the next result is useful.

Lemma. *Let R, S, X be as in 5.4, but suppose also that S and all prime factors of R are left Goldie. Let $P_i \in X$ and P be a minimal prime ideal of R.*
(i) *There is a nonzero $(R/P, R/P_i)$-bimodule which is both left and right torsion-free and is a subfactor of $_R S_R$.*
(ii) *If, further, RsR_R and $_R RsR$ are Noetherian for each $s \in S$, then the $(R/P, R/P_i)$-bimodule can be chosen to be finitely generated on each side.*

Proof. (i) We continue with the notation of the proof of 5.4 and consider again the bimodule N. Note that since $_Q N$ is semisimple of finite length, then $N_{Q'}$ is also, where $Q' = \text{End}(_Q N)$. View N as an (R, Q')-bimodule. Then the left-hand version of the proof above gives a composition series of (R, Q')-sub-bimodules

$$N = N_m \supset \cdots \supset N_0 = 0$$

and a collection Y' of prime ideals P'_j with $P'_j = \text{ann}(_R N_j/N_{j-1})$. Moreover, Y' includes all minimal primes of R, so P occurs as one of these, say P'_j. Evidently N_j/N_{j-1} has all the desired properties except that it is not a subfactor of $_R S_R$. Hence the bimodule $N_j \cap S/N_{j-1} \cap S$ is as required, provided that it is nonzero.

Now $N = M_i/M_{i-1}$ and $M_i \cap S/M_{i-1} \cap S \neq 0$ by 2.1.16. Also, by Corollary 3.4.6, $\text{End}_S(M_i \cap S/M_{i-1} \cap S)$ is an order in Q'. Let $L = ((M_i \cap S) + M_{i-1})/M_{i-1}$. Then, for each k, $N_k = (N_k \cap L)Q'$; hence $N_j \cap S \supset N_{j-1} \cap S$.

(ii) Consider RbR for any $0 \neq b \in (N_j \cap S)/(N_{j-1} \cap S)$. □

It is, in fact, true that for any $P \in X$ there is a bimodule as described (see [Warfield 83]).

5.7 We will say that R satisfies the **bimodule condition** if there is no nonzero left and right finitely generated torsion-free $(R/P, R/P')$-bimodule with P, P' being distinct comparable primes. There is no known example of a Noetherian ring which fails to have this property; and it is known that the second layer condition implies it [Jategaonkar 86].

5.8 The bimodule condition is closely related to Question 2.12; indeed an affirmative answer to that question is equivalent to all Noetherian rings satisfying the bimodule condition. To see this suppose first that R is a Noetherian ring, $P, P' \in \operatorname{Spec} R$ with $P \subset P'$ and M is an R-bimodule such that $_{R/P}M_{R/P'}$ is torsion-free and finitely generated on each side. The ring

$$S = \left\{ \begin{bmatrix} a & m \\ 0 & a \end{bmatrix} \Big| a \in R/P, m \in M \right\}$$

is Noetherian with prime ideal

$$N = \begin{bmatrix} 0 & M \\ 0 & 0 \end{bmatrix}$$

such that $N^2 = 0$. The two sets B_1, B_2 described in 2.12 are not equal, being respectively N and 0. Therefore N is not a localizable ideal of S.

Conversely, suppose that R, P are as described in Question 2.12 but P is not localizable. Thus $B_1 \neq B_2$. By passing to a factor ring we may suppose $B_1 = 0$ and $B_2 \neq 0$. Then P is a finitely generated torsion-free right R/P-module. Here 3.5 shows that B_2 contains an $(R/P', R/P)$-sub-bimodule, M say, which is finitely generated and torsion-free on each side for some prime ideal P' of R. However, $P' \neq P$ since $B_2 \neq 0$. Hence $P' \supset P$ and R fails to satisfy the bimodule condition.

5.9 Granted the bimodule condition and a finiteness condition, 5.4 can be improved.

Corollary. (Additivity principle) *Let $R \subseteq S$ be rings such that S is prime Goldie, R is Noetherian and, for all $s \in S$, $_R R s R$ and $R s R_R$ are finitely generated. Suppose also that R satisfies the bimodule condition. Then:*
(i) $\operatorname{r} \operatorname{u} \dim S = \sum_{i=1}^{n} z_i (\operatorname{r} \operatorname{u} \dim R/P_i)$, *where $P_1, \ldots, P_n$ are the minimal primes of R and each z_i is an integer ≥ 1;*
(ii) *R is an order in an Artinian ring $Q(R)$ which embeds in $Q(S)$.*

Proof. (i) Incomparability shows that the prime ideals in 5.6 are all minimal.
(ii) It is clear from (i) and 4.10 that R is an order in an Artinian ring $Q(R)$. To see that $Q(R)$ embeds in $Q(S)$, we apply the left-hand version of 4.9 to the

affiliated sequence $P_1, \ldots, P_n$ for $_{Q(S)}Q(S)_R$ described in the proof of 5.4. This shows that $\bigcap \mathscr{C}(P_i) \subseteq \mathscr{C}'_{Q(S)}(0)$; and then symmetry gives $\bigcap \mathscr{C}(P_i) \subseteq \mathscr{C}_{Q(S)}(0)$. Since $\bigcap \mathscr{C}(P_i) = \mathscr{C}_R(0)$, it is now clear that $Q(R) \hookrightarrow Q(S)$. □

§6 Patch Continuity

6.1 If P_1, P_2 are prime ideals of a Noetherian ring R with $P_1 \subset P_2$, some connection between u dim R/P_1 and u dim R/P_2 might be expected. We will, shortly, demonstrate by example that this is not so. Nevertheless over the whole prime spectrum there is some pattern to the uniform dimensions. That is the point of this section.

6.2 First some cautionary examples.

Example. (i) If

$$R = \begin{bmatrix} \mathbb{Z} & 2\mathbb{Z} \\ \mathbb{Z} & \mathbb{Z} \end{bmatrix} \quad \text{and} \quad P = \begin{bmatrix} 2\mathbb{Z} & 2\mathbb{Z} \\ \mathbb{Z} & \mathbb{Z} \end{bmatrix}$$

then R is Noetherian, 0 and P are prime ideals; u dim $R = 2$, u dim $R/P = 1$.

(ii) Let k be a field of characteristic zero and $R = U(g)$ with g the semisimple k-Lie algebra sl$(2, k)$. It is well known (e.g. [Dixmier 77, Chapter 1, §8]) that for each integer $n > 0$ there is a simple R-module of dimension n over k, with endomorphism ring k. If P is the annihilator of this simple module then $R/P \simeq M_n(k)$. So u dim $R/P = n$, whereas u dim $R = 1$ since R is a Noetherian integral domain (see 1.7.5).

(iii) Let $R_1 = U(g)$ as above and $R_2 = R_1 \otimes R_1 \simeq U(g \times g)$. We will show now that given any pair of positive integers n_1, n_2 there is a Noetherian subring T of R_2, equivalent as an order to R_2 and having prime ideals $0 \subset P_1 \subset P_2$ such that u dim $T/P_1 = n_1$, u dim $T/P_2 = n_2$.

First, as in (ii), choose $P \in \mathrm{Spec}\, R_1$ so that $R_1/P \simeq M_{n_1}(k)$. Let $P_1 = P \otimes R_1$; then $R_2/P_1 \simeq M_{n_1}(R_1)$ so u dim $R_2/P_1 = n_1$. Choose an integer t so that $m = n_1 t \geq n_2$ and choose $Q \in \mathrm{Spec}\, R_1$ with $R_1/Q \simeq M_t(k)$. Let Q' be the inverse image of $M_{n_1}(Q)$ in R_2; so $R_2/Q' \simeq M_m(k)$. We consider certain subspaces of $M_m(k)$. Let M be the space of all $(m - n_2) \times n_2$ matrices and let

$$T' = \begin{bmatrix} M_{n_2}(k) & 0 \\ M & M_{m-n_2}(k) \end{bmatrix}, \quad A = \begin{bmatrix} 0 & 0 \\ M & M_{m-n_2}(k) \end{bmatrix}$$

and

$$A' = \begin{bmatrix} M_{n_2}(k) & 0 \\ M & 0 \end{bmatrix}.$$

Then A, A' are prime ideals of T' and, respectively, right or left ideals of $M_m(k)$. Let T be the inverse image of T' in R_2, P_2 that of A and P'_2 that of A'. Note

4.6.5   *Patch continuity*

that $T/P_2 \simeq M_{n_2}(k)$ and so $P_2 \in \operatorname{Spec} T$ and u dim $T/P_2 = n_2$. Furthermore, T/P_1 contains the nonzero ideal Q'/P_1 of R/P_1. Therefore these rings are equivalent orders and u dim $T/P_1 = $ u dim $R_2/P_1 = n_1$. Similarly, the ideal P_1 shows that T and R_2 are equivalent orders.

Finally note that $T = \mathbb{I}_{R_2}(P_2)$ with R_2/P_2 being a finite sum of isomorphic simple modules. An extension of the argument in 1.1.12, which is described fully in 5.5.6, shows that T is right Noetherian. Symmetrically, $T = \mathbb{I}_{R_2}(P'_2)$ and so T is also left Noetherian.

(iv) An inductive procedure, based on (iii), will give, for any sequence $n_1, n_2, \ldots, n_t$, a Noetherian integral domain T contained in $R_t = U(g \times \cdots \times g)$ having a chain of primes $0 \subset P_1 \subset \cdots \subset P_t$ with u dim $T/P_i = n_i$.

6.3 Next we prove an easy positive result.

Proposition. *Suppose that* $Q, P_i \in \operatorname{Spec} R$ *with* $Q = \bigcap P_i$, *and that* u dim $R/P_i \leqslant n$ *for all i. Then* u dim $R/Q \leqslant n$.

Proof. Suppose, without loss of generality, that $Q = 0$. Let $a \in R$ be nilpotent, with index of nilpotency m. The image of a in R/P_i is also nilpotent and so, using 3.2.7, $a^n \in P_i$ for each i. Thus $a^n = 0$ and so $n \geqslant m$. By 3.2.7 again, u dim $R \leqslant n$.   □

6.4 The other positive results regarding uniform dimension and Spec R make use of the following notion. A nonzero module M_R is called **monoform** if for each $K \triangleleft M$ every nonzero homomorphism $K \to M$ is a monomorphism. Evidently a monoform module must be uniform, although the converse is not generally true.

Another easily verified fact is noted next.

Lemma. *If a module* N_R *has a chain of submodules*

$$0 = N_0 \subset N_1 \subset \cdots \subset N_k = N$$

each factor N_{i+1}/N_i *of which can be embedded in a monoform module* M_R *then any refinement of the chain has precisely k nonzero factors embeddable in M.*   □

6.5 Lemma. *If* M_R *is a nonzero Noetherian module then* M *has a monoform submodule.*

Proof. Using Noetherian induction, we may suppose the result holds for any nonzero submodule of each proper factor of M. Let $0 \neq N \triangleleft M$. If N is monoform the result is valid. Otherwise there is a homomorphism $\theta: K \to N$ with $K \subseteq N$ and $0 \neq \ker \theta \neq K$. The inductive hypothesis shows that $K/\ker \theta$

contains a monoform submodule; and an isomorphic copy of this is then contained in N. □

6.6 Proposition. *Let R be a prime right Noetherian ring and M_R be a monoform faithful torsion module. Then, for each n, M^n contains a cyclic essential submodule.*

Proof. For any $k \geq 1$ one can find $x_1, \ldots, x_k \in M$ such that, if $K_i = \operatorname{ann}\{x_1, \ldots, x_i\}$ and $K_0 = R$, then $x_i K_{i-1} \neq 0$ for each i. To see this, suppose that $x_1, \ldots, x_i$ have been already chosen (with x_1 being any nonzero element of M). Now

$$K_i = \operatorname{ann} x_1 \cap \cdots \cap \operatorname{ann} x_i \triangleleft_e R$$

since M_R is torsion. Therefore $MK_i \neq 0$; so choose any x_{i+1} with $x_{i+1} K_i \neq 0$.
Taking the case $k = n$, let $x = (x_1, \ldots, x_n) \in M^n$, and note that the chain

$$0 = xK_n \subset xK_{n-1} \subset \cdots \subset xK_0 = xR$$

has the property described in 6.4. Therefore, xR cannot be embedded in M^{n-1}. It follows that xR intersects nontrivially each of the direct summands in M^n, and so $xR \triangleleft_e M^n$. □

6.7 If M is a right R-module, and P is a prime ideal of R then the **rank of M with respect to P** is defined to be the reduced rank of M/MP over R/P and is written $\rho_P(M)$. It is clear that

$$\rho_P(M) = \operatorname{u dim}(M/MP/\tau(M/MP)),$$

where τ denotes the torsion submodule. In particular $\rho_P(R) = \operatorname{u dim} R/P$; and if R is prime, $\rho_0(M) = \rho(M)$.

6.8 The following properties, similar to those of reduced rank, are readily checked.

Lemma. *Let R be a right Noetherian ring and $P \in \operatorname{Spec} R$. Then:*
 (i) *ρ_P is additive on short exact sequences of R/P-modules;*
 (ii) *if $M = M_1 \oplus M_2$ then $\rho_P(M) = \rho_P(M_1) + \rho_P(M_2)$;*
 (iii) *if $N \triangleleft M$ then*

$$\rho_P(M/N) \leq \rho_P(M) \leq \rho_P(M/N) + \rho_P(N) \leq \rho_P(M) + \rho_P(N);$$

 (iv) *if M has n generators then $\rho_P(M) \leq n\rho_P(R)$;*
 (v) *if $c \in R$ then $\rho_P(R/cR) = 0$ if and only if $c \in \mathscr{C}(P)$.* □

6.9 Theorem. *Let R be a prime right Noetherian ring, M_R a finitely generated torsion module. Then given any $\varepsilon > 0$ there exists $0 \neq I \triangleleft R$ such that $\rho_P(M)/\rho_P(R) < \varepsilon$ for all $P \in \operatorname{Spec} R$ such that $I \nsubseteq P$.*

4.6.10 *Patch continuity* 127

Proof. Proceeding by Noetherian induction we may assume the result holds for all submodules of proper factor modules of M.

Suppose first that M contains an unfaithful submodule N. Let $I_1 = \operatorname{ann} N$. If $P \in \operatorname{Spec} R$ and $I_1 \not\subseteq P$ then N/NP is an unfaithful R/P-module and so is torsion. Thus $\rho_P(N) = 0$. The Noetherian induction provides an ideal I_2 such that $\rho_P(M/N)/\rho_P(R) < \varepsilon$ for all $P \in \operatorname{Spec} R$, $P \not\supseteq I_2$. Letting $I = I_1 \cap I_2$ and applying 6.8(iii) completes the proof in this case.

Next suppose every submodule of M is faithful. Note by 6.5 that M has a monoform submodule, N say, to which 6.6 can be applied. Choose an integer $n > 4/\varepsilon$. Then N^n contains a cyclic essential submodule, C say. It follows easily that $L^n \subseteq C$ for some nonzero submodule L of M.

The induction hypothesis provides $0 \neq I \triangleleft R$ such that

$$\rho_P(M/N)/\rho_P(R) < \varepsilon/2 \quad \text{and} \quad \rho_P(N/L)/\rho_P(R) < 1/n$$

for all $P \in \operatorname{Spec} R$ with $I \not\subseteq P$. For such P

$$\rho_P(N^n/C) \leq \rho_P(N^n/L^n) = n\rho_P(N/L) < \rho_P(R)$$

and $\rho_P(C) \leq \rho_P(R)$, using 6.8. Therefore

$$\rho_P(M) \leq \rho_P(M/N) + \rho_P(N)$$
$$= \rho_P(M/N) + \frac{1}{n}\rho_P(N^n)$$
$$\leq \rho_P(M/N) + \frac{1}{n}(\rho_P(N^n/C) + \rho_P(C))$$

and so

$$\rho_P(M)/\rho_P(R) \leq \varepsilon/2 + \frac{1}{n}(1 + 1) < \varepsilon. \qquad \square$$

6.10 Theorem. *Let R be a prime right Noetherian ring and M_R be finitely generated. For any $\varepsilon > 0$ there exists a nonzero ideal I such that, for all $P \in \operatorname{Spec} R$ with $P \not\supseteq I$, $|\rho_P(M)/\rho_P(R) - \rho(M)/\rho(R)| < \varepsilon$.*

Proof. Let $\rho(M) = k$ and $\rho(R) = n$. Then $\rho(M^n) = nk$ and so $M^n \otimes_R Q(R) \simeq Q(R)^k$. Choose a basis $x_1, \ldots, x_k$ for $M^n \otimes_R Q(R)$ and write each $x_i = m_i \otimes c^{-1}$ for some $m_i \in M^n$ and $c \in \mathscr{C}_R(0)$. Let $N = \sum m_i R$ and note that $(M^n/N) \otimes_R Q(R) = 0$; hence M^n/N is torsion. Application of 6.9 gives $0 \neq I_1 \triangleleft R$ with $\rho_P(M^n/N)/\rho_P(R) < \varepsilon$ for all $P \not\supseteq I_1$. Now $\rho_P(N) \leq k\rho_P(R)$ by 6.8; and so

$$\rho_P(M^n)/\rho_P(R) < \varepsilon + k.$$

Hence $\rho_P(M)/\rho_P(R) < \varepsilon/n + k/n \leq \varepsilon + \rho(M)/\rho(R)$ and so

$$\rho_P(M)/\rho_P(R) - \rho(M)/\rho(R) < \varepsilon.$$

Next, choose an exact sequence $0 \to K \to F \to M \to 0$ with F_R free of rank m say. Since reduced rank is additive, $\rho(K) + \rho(M) = \rho(F) = m\rho(R)$. The argument above, applied this time to K, gives $0 \neq I_2 \triangleleft R$ such that

$$\rho_P(K)/\rho_P(R) - \rho(K)/\rho(R) < \varepsilon$$

for all $P \not\supseteq I_2$. Note, by 6.8, that $\rho_P(K) + \rho_P(M) \geq m\rho_P(R)$. These facts combine to give

$$\rho_P(M)/\rho_P(R) - \rho(M)/\rho(R) \geq (m - \rho_P(K)/\rho_P(R)) - (m - \rho(K)/\rho(R)) > -\varepsilon. \quad \square$$

6.11 One consequence of 6.10 concerns elements regular modulo the different prime ideals. A subset X of Spec R is said to satisfy the **generic regularity** condition (or to be **sparse**) if, for each prime $Q \in \operatorname{Spec} R$ and each $c \in \mathscr{C}(Q)$, there exists $I \triangleleft R$, $I \supset Q$, such that $c \in \mathscr{C}(P)$ for each prime $P \in X$ with $Q \subseteq P, I \not\subseteq P$.

6.12 Proposition. *Let R be right Noetherian, n be a positive integer and X be a subset of Spec R such that u dim $R/P \leq n$ for all $P \in X$. Then X satisfies the generic regularity condition.*

Proof. Let $c \in \mathscr{C}(Q)$ for some $Q \in \operatorname{Spec} R$. Now 6.10 provides an ideal $I \supset Q$ such that, for all $P \in \operatorname{Spec} R$ with $P \supseteq Q$ and $P \not\supseteq I$,

$$|\rho_P(R/cR)/\rho_P(R) - \rho_Q(R/cR)/\rho_Q(R)| < 1/n.$$

However, by 6.8(v), $\rho_Q(R/cR) = 0$; and if we choose $P \in X$ then $\rho_P(R) \leq n$. Hence $\rho_P(R/cR) = 0$ and $c \in \mathscr{C}(P)$. $\quad \square$

6.13 Without the imposition of a bound upon u dim R/P for $P \in X$ the conclusion is false. This is shown by the ring $R = A_1(\mathbb{Z})$ which will be discussed later, in 11.7.8. There it will be shown that, for each prime $p \in \mathbb{Z}$, $pR + x^p R \in \operatorname{Spec} R$. Evidently these primes have zero intersection and yet x is a zero divisor modulo each.

6.14 The preceding results have a topological aspect. One can form a topology on Spec R by defining its closed subsets to be subsets X such that any prime ideal in Spec R which is the intersection of some primes in X belongs to X. This topology is called the **patch topology**. Its closed sets we call **patch-closed** and its open sets are **patch-open**.

For example, 6.3 shows that for any $n \geq 1$ the set

$$X = \{P \in \operatorname{Spec} R \mid \text{u dim } R/P \leq n\}$$

is patch-closed. Other examples arise from the **Zariski topology** in which closed sets have the form

$$V(I) = \{P \in \operatorname{Spec} R \mid P \supseteq I\}$$

for some $I \triangleleft R$ and open sets have the form

$$W(I) = \{P \in \text{Spec } R \mid P \not\supseteq I\}.$$

It is evident that each of these is patch-closed and, since they are complementary, also patch-open. Note also that the set of primes described in the definition of generic regularity is $X \cap V(Q) \cap W(I)$, and that $V(Q) \cap W(I)$ is both patch-closed and patch-open in Spec R.

6.15 We note, but will not use, an alternative description of the patch topology. Suppose we define a new topology by simply taking the sets $V(I)$, $W(I)$, for $I \triangleleft R$, as a sub-basis of the closed sets. In that topology each closed set will be patch-closed since each $V(I)$ and $W(I)$ has that property.

On the other hand, suppose X is a patch-closed set and Y is its complement in Spec R. For each $Q \in Y$, let $I_Q = \bigcap \{P \in X \mid P \supseteq Q\}$. By definition, $I_Q \supset Q$ and so $Q \in V(Q) \cap W(I_Q)$ which, one can check, is contained in Y. Hence Y is a union of such subsets, each of which is open in this new topology. Hence Y is open, and X closed, in the new topology.

This proves that the two topologies coincide.

6.16 We now interpret 6.10.

Corollary. *For any right Noetherian ring R and finitely generated M_R, the map Spec $R \to \mathbb{Q}$ given by $P \mapsto \rho_P(M)/\rho_P(R)$ is continuous with respect to the patch topology.*

Proof. Choose any $Q \in \text{Spec } R$ and any $\varepsilon > 0$. Let $\bar{R} = R/Q$, $\bar{M} = M/MQ$, $\bar{I}$ be the ideal of $\bar{R}$ provided by 6.10, and I its inverse image in R. Then

$$|\rho_P(M)/\rho_P(R) - \rho_Q(M)/\rho_Q(R)| < \varepsilon$$

for all P in the patch-open set $V(Q) \cap W(I)$. □

6.17 There are interesting consequences concerning the uniform dimensions of prime factor rings R/P, R/Q with $P \supseteq Q$ which refine 6.3.

Corollary. *Let R be a right Noetherian ring, $Q \in \text{Spec } R$, and n be any integer > 0.*
  (i) *There exists $I_n \triangleleft R$, $I_n \supset Q$ such that for all $P \in V(Q) \cap W(I_n)$ either $\text{u dim } R/Q$ divides $\text{u dim } R/P$ or $\text{u dim } R/P > n$.*
 (ii) *If $\text{u dim } R/Q$ does not divide n then*

$$\bigcap \{P \in V(Q) \mid \text{u dim } R/P \text{ divides } n\} \supset Q.$$

(iii) *If $Q = \bigcap_i P_i$, where $P_i \in \text{Spec } R$ and $\text{u dim } R/P_i$ divides n for each i, then $\text{u dim } R/Q$ divides n.*

Proof. (i) There is no loss in assuming that $Q = 0$. Applying 6.10 to a uniform right ideal M of R with $\varepsilon = (n(\text{u dim } R))^{-1}$, we obtain

$$\left| \frac{\rho_P(M)\text{u dim } R - \text{u dim } R/P}{\text{u dim } R/P} \right| < \frac{1}{n}.$$

Note that either $|\rho_P(M)\text{u dim } R - \text{u dim } R/P|$ is 0 or else it is at least 1. In the former case u dim R divides u dim R/P, and in the latter u dim $R/P > n$.

(ii) Note that $I_n \subseteq P$ for all $P \in V(Q)$ with u dim R/P dividing n.

(iii) This is simply the contrapositive of (ii). □

6.18 Another consequence, whose proof we omit, concerns cliques.

Corollary. *Let R be a Noetherian ring and $P \in \text{Spec } R$.*
(i) *For any n there are only finitely many prime ideals linked to P with u dim $R/Q \leq n$.*
(ii) *The clique containing P is countable.*

Proof. See [Stafford **P**] or [Jategaonkar **86**, 6.2.15 and 6.2.17]. □

§7 Additional Remarks

7.0 [Jategaonkar **86**] contains further information on most of the topics discussed here and [Chatters and Hajarnavis **80**] covers §§1, 2 more fully.

7.1 (a) Reduced rank was introduced by [Goldie **64**]. It is an important tool in the stability theory described in Chapter 11. Its connection with Ore conditions and Small's theorem appeared in [Chatters *et al.* **79**] and [Warfield **79**].

(b) The importance to 1.4 of the equality of $\mathscr{C}(0)$ and $\mathscr{C}(N)$ was shown by [Talintyre **66**]. Small's theorem appeared in [Small **66, 68**] and 1.4 is a variant, from [Warfield **79**], of his general result. Artinian quotient rings reappear in Chapter 6, Section 8.

(c) 1.6 is derived from [Lenagan **75**]. The Artin radical appeared in [Chatters, Hajarnavis and Norton **77**] and 1.8 comes from [Ginn and Moss **77**]. The original proof of 1.9 is in [Robson **74**].

(d) There are results concerning other types of quotient ring. In [Stafford **82**] it is proved that a Noetherian ring R is a quotient ring if and only if r ann $A(R) \cap$ l ann $A(R) \subseteq J(R)$. In [Chatters and Hajarnavis **80**] there is a discussion of orders in semiprimary rings.

There are also results concerning embeddings of a ring into an Artinian ring; see [Schofield **85**], [Blair and Small **P**] and [Dean and Stafford **P**].

(e) The principal ideal theorems here are due to [Jategaonkar **74, 75**]. The proofs follow [Chatters *et al.* **79**] where a more general 'invertible ideal theorem' is proved. However, that can be deduced from the principal ideal theorem; see [Gray **84**]. The principal ideal theorem is used later, in 13.7.15.

7.2 (a) 2.6, 2.7 are due to [McConnell **68**] with the centralizing case being due to [Nouazé and Gabriel **67**]. As well as applying to enveloping algebras, as shown in Chapter 14, these results apply to group algebras of polycyclic by finite groups. See [Jategaonkar

4.7.6 *Additional remarks*

86] for details. Normal elements appear also in Chapter 7, §3 and in Chapter 10.

(b) 2.9–2.11 come from [P. F. Smith **71**]; see also [P. F. Smith **82**].

7.3 (a) The material in §3 has its origins in [Jategaonkar **73, 74a**]. Subsequent development involved [Müller **76, 76a, 79**], [Brown **81**] and [Brown, Lenagan and Stafford **80**] as well as [Jategaonkar **82**] where the second layer condition appears. This account is based on [Jategaonkar **86**] and [Brown **85**] where further details can be found. One can also consult [Stafford **85c**], [Warfield **86**], [Brown and Warfield **P**], [Bell **84, 87, 87a**] and [Brown and DuCloux **P**].

(b) Another attack on localization at a prime ideal is described in [Goldie **67**]. This involves completions, which also are studied in [McConnell **69, 78, 79**], [Jordan **P**], [Heinicke **81**] and [Small and Stafford **81**].

(c) Granted the second layer condition one can prove that $\bigcap J^n = 0$, that the bimodule conjecture holds and that a primary decomposition theory exists. Details are in [Jategaonkar **86**].

(d) The conditions imposed in 3.17 are not too restrictive. For example they are valid if R is a Noetherian PI ring containing an uncountable field [Müller **79**], [Jategaonkar **86**], or is the enveloping algebra of a solvable complex Lie algebra [Brown **84**] or the group algebra over an uncountable field of a polycyclic by finite group [Brown **86**].

(e) The proof of 3.13 shows that in (ii) it is enough to assume each $P_i \in \operatorname{Spec} R$.

7.4 The material in §4 is drawn from [Stafford **82**] and [Small and Stafford **82**]; see also [Goldie and Krause **84**].

7.5 Additivity was first studied in [Joseph and Small **78**] with later work by [Borho **82**] and [Warfield **83**]. This account is based on Warfield's paper.

7.6 (a) The main results of this section were proved, for a Noetherian ring, by [Stafford **81**]. The proofs here follow [Goodearl **86**] except for the use of monoform modules as suggested by [Jategaonkar **86**].

(b) The map in 6.16 is locally constant if R is an FBN ring [Warfield **80**] but is not so in general [Stafford **P**]; 6.16 is used later, in 11.7.10.

(c) Despite 6.18, it appears that infinite cliques are the rule rather than the exception.

(d) The patch topology was introduced by [Hochster **69**]. The usual definition of patch-closed sets is that described in 6.15.

Chapter 5

SOME DEDEKIND-LIKE RINGS

Dedekind domains can be characterized, amongst commutative integral domains, by various properties; for example, the property that each nonzero ideal is invertible, or is a progenerator, or is projective, or is a generator (see 2.7). Any of these conditions leads to the more familiar arithmetic of ideals and, equally, to the properties of being Noetherian, integrally closed and of dimension 1.

This large selection of attributes suggests many possible generalizations to noncommutative rings. In this chapter some of these are discussed, making use of the general theory of the preceding chapters.

The first section discusses maximal orders. Their relevance here is indicated by the fact that a commutative Noetherian domain is a maximal order if and only if it is integrally closed.

The next section describes those prime Noetherian rings whose nonzero one-sided ideals are all progenerators, or all generators. The classes of rings thus defined are called Dedekind prime rings and Asano prime rings. It is shown that these rings do indeed have an arithmetic of ideals.

In §3, the connection between these rings and the classical maximal orders over a Dedekind domain in a central simple algebra is outlined. Such orders are indeed Dedekind prime rings. It is shown also that there is a tight connection with prime PI rings; and that will be utilized in Chapter 13 to obtain a converse to the preceding result.

The next three sections concern Noetherian hereditary rings, i.e. rings whose one-sided ideals are finitely generated projective. Compared with the commutative case, this leads to a surprisingly rich class of rings. Section 4 reduces the problem to the study of prime rings; but these need not be Dedekind prime rings. The description of such rings depends upon idealizer subrings (as in 1.1.11) and so, of course, has no commutative analogue. The necessary theory of idealizers comes in §5 and then, in §6, a complete description is given, in terms of Dedekind prime rings, of those hereditary Noetherian prime rings which have finitely many idempotent ideals.

Finally, in §7, there is a brief description of modules over a Dedekind prime ring. Analogues are obtained of almost all the results familiar from the commutative case, although there are some interesting differences.

This chapter uses some elementary properties of projective dimension. A summary of these and other homological ideas can be found in Chapter 7, §1.

§1 Maximal Orders

1.1 Let Q be a quotient ring and R a right order in Q, as described in Chapter 3, §1. Then R is called a **maximal** right order if it is maximal within its equivalence class (see 3.1.9).

This section describes some classes of maximal orders and gives some basic facts about them. Further examples will be met in later sections.

1.2 To give some immediate examples we prove

Proposition. *Any simple right Goldie ring R (with 1) is a maximal right order.*

Proof. Suppose $S \supseteq R$, $S \sim R$. By 3.1.10, there are units r_1, r_2 of Q, $r_1, r_2 \in R$ such that $r_1 S r_2 \subseteq R$. However, then

$$R \supseteq R r_1 S r_2 R = (R r_1 R) S (R r_2 R) = RSR = S.$$
□

Corollary. *If k is a field of characteristic zero, then the Weyl algebra $A_n(k)$ is a maximal order.*
□

In fact this result is true if k has finite characteristic as will be shown in Corollary 1.6.

1.3 The special case when R is a commutative integral domain helps put this notion into perspective. Recall that R is **completely integrally closed** in its quotient field Q if, for a and q in Q with $a \neq 0$, $aq^n \in R$ for all n implies $q \in R$. (This implies that R is integrally closed, and is equivalent to that if R is Noetherian; see 3.2, 3.3.)

Proposition. *A commutative integral domain is a maximal order in its quotient field if and only if it is completely integrally closed.*

Proof. (⇒) Let Q be the quotient field of R, and let a, q be as described above. The ring $R' = R[q] \subseteq Q$ satisfies $aR' \subseteq R$. So $R' \sim R$ and thus $R' = R$ and $q \in R$ as required.

(⇒) Say $R' \supseteq R$ with $R' \sim R$, and so $aR' \subseteq R$ for some $0 \neq a \in R$. Pick any $q \in R'$. Then $R[q] \subseteq R'$ and so $aq^n \in R$ for all n. Thus $q \in R$; and hence $R' = R$. □

1.4 The verification that a particular ring R is a maximal right order is most easily effected using R-ideals (see 3.1.13).

Proposition. *If R is a right order in Q then the following conditions are equivalent:*
(i) *R is a maximal right order;*
(ii) *$O_r(I) = O_l(I) = R$ for all fractional R-ideals I;*
(iii) *$O_r(I) = O_l(I) = R$ for all R-ideals I.*

Proof. (i)⇒(ii) If I is a fractional R-ideal then $O_r(I)$ and $O_l(I)$ are equivalent right orders which contain R (by 3.1.12). Therefore $O_r(I) = O_l(I) = R$.
  (ii)⇒(iii) Trivial.
  (iii)⇒(i) Suppose $S \supseteq R$ and $S \sim R$. Using 3.1.10, choose T with $R \subseteq T \subseteq S$, $r_1 S \subseteq T$, $Tr_2 \subseteq R$. The set $I = \{x \in R \mid Tx \subseteq R\}$ is evidently an R-ideal and $T \subseteq O_l(I)$. Therefore $T = R$; and, by symmetry, $S = R$. □

1.5 The next result gives an easy application of 1.4.

Proposition. *Let R be a semiprime right Goldie ring. The following are equivalent:*
(i) *R is a maximal right order;*
(ii) *$M_n(R)$ is a maximal right order;*
(iii) *R is a direct sum of prime right Goldie rings, each of which is a maximal right order.*

Proof. (i)⇔(ii) Note that each $M_n(R)$-ideal has precisely the form $M_n(A)$ with A an R-ideal. Then 1.4 is easily applied.
  (i)⇒(iii) In the notation of Proposition 3.2.4, the ring $R' = \oplus e_i R$ is a direct sum of prime right Goldie rings, is equivalent to R and contains R. So $R = R' = \oplus e_i R$. Since R-ideals all have precisely the form $\oplus A_i$ with each A_i an $e_i R$-ideal, it follows easily that each $e_i R$ is a maximal right order using 1.4.
  (iii)⇒(i) The last part of the proof above is easily reversed. □

1.6 Next, 1.4 will be used to show that enveloping algebras are maximal orders. The technique involves filtered and graded rings as described in §6 of Chapter 1. The terminology and notation of that section will be used.

Theorem. *Let R be a filtered ring with associated graded ring $\operatorname{gr} R$. Suppose that $\operatorname{gr} R$ is a Noetherian integral domain which is a maximal order in its quotient ring. Then R has the same properties.*

Proof. It is shown in 1.6.6 and 1.6.9 that R is a Noetherian integral domain. Let $0 \neq I \triangleleft R$ and $q \in O_r(I)$. So $q = rs^{-1}$ with $r, s \in R$, and so $Ir \subseteq Is$. It will be shown that this implies that $r \in Rs$, this being achieved by induction on the degree of r. Note that, since $\operatorname{gr} R$ is an integral domain, $\bar{a}\bar{c} = \overline{ac}$ for all $a, c \in R$.

Therefore $(\operatorname{gr} I)\bar{r} \subseteq (\operatorname{gr} I)\bar{s}$. However, $\operatorname{gr} I$ is a nonzero ideal of the maximal order $\operatorname{gr} R$. Hence $\bar{r}\bar{s}^{-1} \in \operatorname{gr} R$ and so $\bar{r} \in (\operatorname{gr} R)\bar{s}$. It follows that $r = xs + y$ with $x, y \in R$ and $\deg y < \deg r$. Also $Iy \subseteq Is$ and so, by the induction hypothesis, $y \in Rs$. Therefore $r \in Rs$ as claimed, but that means $rs^{-1} \in R$ and that $O_r(I) = R$. Symmetry completes the proof. □

Corollary. *Let k be any field and $\mathfrak{g}$ a finite dimensional k-Lie algebra.*
(i) *The universal enveloping algebra $U(\mathfrak{g})$ is a maximal order.*
(ii) *If R is an integrally closed commutative Noetherian integral domain and is a k-algebra, then any crossed product $R * U(\mathfrak{g})$ is a maximal order.*
(iii) *The Weyl algebra $A_n(k)$ is a maximal order.*

Proof. (i) This is a special case of (ii).
 (ii) By 1.7.14, $\operatorname{gr}(R * U(\mathfrak{g}))$ is a commutative polynomial ring in finitely many variables over R. It is therefore integrally closed and hence a maximal order.
 (iii) This is also a special case of (ii), by 1.7.11. □

Note that (iii) extends Corollary 1.2 to arbitrary characteristic.

1.7 Next another module theoretic notion is required. Let R be any ring and M_R a right R-module. As in 1.1.5, let $M^* = \operatorname{Hom}(M, R)$, the dual of M. Since M^* is a left R-module, we now define $M^{**} = \operatorname{Hom}(M^*, R)$. There is an obvious homomorphism $M \to M^{**}$ (acting by right multiplication) this being an embedding if and only if M is torsionless. When M is torsionless it is convenient to identity M with the image. The module M_R is **reflexive** if M is torsionless and $M = M^{**}$. In particular, one can quickly check for any M that $M^* = M^{***}$; so $_R M^*$ is always reflexive. Furthermore, projective modules, being direct summands of free modules, are readily checked to be reflexive.

Similar considerations give embeddings

$$\operatorname{End}_R M \subseteq \operatorname{End}_R M^* = \operatorname{End}_R M^{**},$$

for a torsionless module M.

1.8 Consider now the case when R is a right order in a quotient ring Q and I is a fractional right R-ideal. Then $I^* \simeq \{q \in Q \mid qI \subseteq R\}$ and $\operatorname{End} I \simeq O_l(I)$, as was noted in 3.1.15. It is convenient, again, to identify these objects.

There are corresponding considerations regarding fractional left R-ideals. The next result shows a certain symmetry between these notions.

Proposition. *Let R and R' be maximal right orders in a quotient ring Q, and I a fractional (R, R')-ideal. Then $\{q \in Q \mid qI \subseteq R\} = \{q \in Q \mid Iq \subseteq R'\}$.*

Proof. Since $O_r(I) = R$ and $O_l(I) = R'$ one can check that both these sets equal $\{q \in Q \mid IqI \subseteq I\}$. □

These two sets are the duals of I as a right or left R-module. The symmetry shown here means that there is no ambiguity in calling I a **reflexive fractional R-ideal** provided $I = I^{**}$.

1.9 The remainder of this section makes use of reflexive modules. First a result about prime ideals.

Proposition. *Let R be a prime right Goldie ring which is a maximal right order, and let P be a nonzero prime ideal which is reflexive. Then P has height one.*

Proof. Suppose $0 \neq P_1 \subseteq P$ with P_1 prime. Then $P_1^* \supseteq P^*$ and so $R \supseteq P_1^* P_1 \supseteq P^* P_1$. However, $PP^* P_1 \subseteq P_1$; so either $P \subseteq P_1$, as claimed, or else $P^* P_1 \subseteq P_1$. In the latter case $P^* \subseteq O_1(P_1) = R$ by 1.4 and so $P^* = R$; but then $P^{**} = R \neq P$, a contradiction. □

1.10 Next we use these notions to describe the centres of certain maximal orders.

Proposition. *(a) If R is a semiprime, prime or simple ring then $Z(R)$ is, respectively, semiprime, an integral domain or a field.*
  (b) (i) If R is a prime right Goldie ring and a maximal right order then $Z(R)$ is a completely integrally closed integral domain.
  (ii) If, further, R satisfies the a.c.c. for reflexive ideals then $Z(R)$ is a Krull domain.

Proof. (a) Straightforward.
  (b)(i) The proof of 1.3 can be adapted easily.
    (ii) Recall (see [Bourbaki, **61–65**, Chapter 7, §1, Theorem 2]), that one of the characterizations of a Krull domain Z is that it is a completely integrally closed integral domain and satisfies the a.c.c. for reflexive ideals.
  Now let I be a reflexive ideal of $Z = Z(R)$ and let K be the quotient field of Z. Note that $(IR)^{**}$ is a reflexive R-ideal and that $I^* \subseteq (IR)^*$. Hence

$$((IR)^{**} \cap Z)I^* \subseteq (IR)^{**}(IR)^* \cap K \subseteq R \cap K = K.$$

Thus $(IR)^{**} \cap Z \subseteq I^{**} = I$ and so $(IR)^{**} \cap Z = I$. It follows that, since R has a.c.c. on reflexive ideals, so too does Z. □

1.11 Finally, we discuss orders equivalent to a maximal order.

Proposition. *Let R be a maximal right order in a quotient ring Q and let I be a reflexive fractional right R-ideal. Then $O_l(I)$ is a maximal right order in Q.*

Proof. Suppose T is a right order with $T \supseteq O_l(I)$ and $aT \subseteq O_l(I)$ for some unit a of Q. Then

$$I^* aTII^* TI \subseteq I^* aTO_l(I)TI = I^* aTI \subseteq I^* I \subseteq R.$$

It follows that $I^*TI \subseteq O_r(I^*aTI) = R$ and so $TI \subseteq I^{**} = I$. Therefore $T \subseteq O_l(I)$.

If, on the other hand, $Ta \subseteq O_l(I)$, a similar argument, considering I^*TII^*TaI, leads to the same conclusion. These two facts, combined with 3.1.10, show that $O_l(I)$ is indeed a maximal right order. □

Corollary. *Let R be a maximal (right and left) order in a quotient ring Q. Then each order S in Q equivalent to R is contained in a maximal order equivalent to R.*

Proof. By 3.1.14, there is a fractional (S, R)-ideal, I say. Now $S \subseteq O_l(I) \subseteq O_l(I^{**})$ which, by the proposition, is a maximal order. □

The arguments above can be extended to show that if R, S are prime Goldie rings which are context equivalent, and if R is a maximal order then S is contained in an equivalent maximal order.

§2 Asano and Dedekind Prime Rings

In this section we describe some special types of ring defined by properties of their categories of modules related to progenerators. This ensures that each class of rings is closed under Morita equivalence. The rings concerned are rather special since their commutative analogues are merely Dedekind domains.

2.1 To put things in perspective, it is helpful to note some characterizations of semisimple, and simple, Artinian rings.

Proposition. (a) *The following conditions on a ring R are equivalent*:
  (i) *R is semisimple Artinian*;
  (ii) *every right R-module is projective*;
  (iii) *every finitely generated right R-module is projective*.
(b) *The following conditions on a ring R are equivalent*:
  (i) *R is simple Artinian*;
  (ii) *every nonzero right R-module is a generator*;
  (iii) *every nonzero finitely generated right R-module is a progenerator*.

Proof. Exercise (concentrate on simple modules!) □

2.2 The types of ring to be discussed in this chapter will be defined by properties of their right ideals or, equivalently, of the submodules of their (right) progenerators. Such a definition ensures that these classes of rings are closed under Morita equivalence.

One example, of course, is that being a right Noetherian ring is characterized by all submodules of a progenerator being finitely generated. Another example is when every submodule of a progenerator is projective. Such a ring is called **right hereditary**.

2.3 The next few results consider another related class, characterized by submodules of progenerators being generators. We start with an easy lemma, which restricts our attention to prime rings.

Lemma. *If R is a ring such that each nonzero ideal is a (right) generator, then R is a prime ring.*

Proof. Suppose $0 \neq A \triangleleft R$, $x \in R$ and $Ax = 0$. Then $A^*(Ax) = 0$, where $A^* = \mathrm{Hom}\,(A_R, R)$. But, by definition, $A^*A = R$ and so $Rx = 0$ and $x = 0$. Thus R is prime. □

2.4 To get within the scope of this text, we now restrict attention to the case when the prime ring R is (right and left) Goldie. Furthermore, in order to make the theory more complete, conditions henceforth will generally be imposed both on left and right modules.

2.5 Let R be a prime Goldie ring. A fractional R-ideal A is **invertible** if there exists a fractional R-ideal B with $AB = BA = R$. In that case B is usually denoted by A^{-1}.

Lemma. *Let A be a fractional R-ideal, with R a prime Goldie ring. The following are equivalent:*
  (i) A is invertible;
  (ii) A_R and $_RA$ are generators;
  (iii) $O_r(A) = O_l(A) = R$ and $_RA$, A_R are finitely generated projective;
  (iv) $O_r(A) = O_l(A) = R$ and $_RA$, A_R are projective.

Proof. Recall that $A^* = \mathrm{Hom}\,(A_R, R)$ can be identified with $B = \{q \in Q(R) | qA \subseteq R\}$. Similarly, $\mathrm{Hom}\,(_RA, R)$ can be identifed with $C = \{q \in Q | Aq \subseteq R\}$.
  (i)⇒(ii) If A is invertible, it is clear that $A^{-1} \subseteq A^*$; so $A^*A = R$ and so A_R is a generator. Symmetry shows that $_RA$ is also a generator.
  (ii)⇒(i) One has $BA = R = AC$; so $B = C = A^{-1}$.
  (i)⇒(iii) $O_l(A) = O_l(A)R = O_l(A)AA^{-1} = AA^{-1} = R$. But, as above, $A^{-1} \subseteq A^*$ and so $AA^* = O_l(A)$. Identifying $O_l(A)$ with $\mathrm{End}\,(A_R)$, this shows that A_R is finitely generated and projective, by the dual basis lemma (3.5.2).
  (iii)⇒(i) Since $AB = O_l(A) = R = CA$, then $B = C = A^{-1}$.
  (iii)⇔(iv) This follows from Corollary 3.5.2. □

2.6 Proposition. *Let R be a prime Goldie ring. The following conditions are equivalent:*

5.2.7 Asano and Dedekind prime rings

(i) *each nonzero submodule of a (left or right) progenerator is a generator;*
(ii) *R is a maximal order and each ideal is finitely generated projective as a left or right module;*
(iii) *R is a maximal order and each ideal is reflexive;*
(iv) *each nonzero ideal of R is invertible.*

Proof. (i)⇒(ii) Since each nonzero ideal A of R is a right and left generator, 2.5 shows that $O_r(A) = O_l(A) = R$. Therefore, by 1.4, R is a maximal order. The rest is clear from 2.5.

(ii)⇒(iii) As noted in 1.7, projective modules are reflexive.

(iii)⇒(iv) Let $0 \neq A \triangleleft R$, and consider $B = A^*A \triangleleft R$. If $q \in Q(R)$ is such that $qB \subseteq R$ then $qA^* \subseteq A^*$ and so $q \in O_l(A^*) = R$, since R is a maximal order. Thus $B^* = R$, and so $B = B^{**} = R^* = R$. It follows that A is invertible.

(iv)⇒(i) Let M be a nonzero submodule of a progenerator P_R, and let $A = M^*M \triangleleft R$. Now M, being a submodule of a free module, is torsionless, so $A \neq 0$. However, $A^{-1}A = R$ and so $A^{-1}M^*M = R$. This shows that $A^{-1}M^* \subseteq M^*$ and that $M^*M = R$. Thus M is a generator. □

2.7 A prime Goldie ring satisfying the conditions of Proposition 2.6 is called an **Asano prime ring** or an **Asano order**. Evidently this is a Morita invariant property. We now give some examples.

Example. (i) *If R is a commutative integral domain then the following are equivalent:*
(a) *each nonzero ideal is invertible (i.e. R is a Dedekind domain);*
(b) *R is Asano;*
(c) *R is a hereditary Noetherian ring;*
(d) *R is a hereditary ring.*

Proof. That (a)⇔(b)⇒(c)⇒(d) is obvious from 2.6, and (d)⇒(c) comes from Corollary 3.5.2. To see that (c)⇒(a) note, by Lemma 3.5.2, that if $0 \neq A \triangleleft R$ then $AA^* = O_l(A)$; yet $AA^* = A^*A \subseteq R$ and thus $AA^* = R$. □

Example. (ii) Any simple Goldie ring is Asano, since the condition on nonzero ideals is trivially satisfied. Thus $A_n(k)$, with char $k = 0$, is an example, and so too are the $P_{\lambda_1, \ldots, \lambda_n}(k)$ described in 1.8.7. It is worth noting that since gl dim $A_n(k) = n$ (as will be shown in 7.5.8) it is not the case that all one-sided ideals are projective.

Example. (iii) $R = A_n(k)[x]$ with char $k = 0$ is also an example. For, since $A_n(k)[x] \simeq A_n(k) \otimes_k k[x]$ and $A_n(k)$ is simple, the ideals of R take the form $A_n(k) \otimes I$, with $I \triangleleft k[x]$; (see 9.6.9 for a proof). Thus each ideal is principal, both right and left. That, as the next lemma shows, ensures that R is an Asano prime ring.

2.8 Lemma. *If each ideal A of a prime Goldie ring is principal on each side then R is an Asano prime ring. Furthermore, if $A = aR = Rb$, then $A = Ra = bR$.*

Proof. If $0 \neq A = aR = Rb$ then $a, b \in \mathscr{C}_R(0)$ by 3.3.7 since nonzero ideals are essential. Since $(Ra^{-1})A = R = A(b^{-1}R)$ it is clear that A is invertible, and that $Ra^{-1} = b^{-1}R$. It follows that $bR = Ra \triangleleft R$. However, since $aR \triangleleft R$, then $Ra \subseteq aR$ and, similarly, $aR \subseteq Ra$. □

2.9 Next we show that an Asano prime ring has the 'classic' arithmetic of ideals.

Theorem. *If R is an Asano order then each nonzero ideal is a unique (commutative) product of maximal ideals. Indeed the fractional R-ideals form a free abelian group generated by the maximal ideals of R.*

Proof. Let M_1, M_2 be distinct maximal (nonzero) ideals of R and consider $X = M_1^{-1}(M_1 \cap M_2)$. Since $M_1 \cap M_2 \subseteq M_1$, $X \triangleleft R$; and $M_1 X = M_1 \cap M_2 \subseteq M_2$, so $X \subseteq M_2$. Hence
$$M_1 \cap M_2 = M_1 X \subseteq M_1 M_2 \subseteq M_1 \cap M_2.$$
Thus $M_1 M_2 = M_1 \cap M_2 = M_2 M_1$, by symmetry. This shows that multiplication of maximal ideals is commutative. The proof is now routine. □

2.10 The next class of rings is more restrictive, since the aim here is to have the one-sided ideals behaving as in a Dedekind domain.

Theorem. *The following conditions on a ring R are equivalent:*
 (i) *every nonzero submodule of a (right or left) progenerator is a progenerator;*
 (ii) *R is a hereditary Noetherian prime ring and is a maximal order;*
 (iii) *R is a hereditary Noetherian Asano order.*

Proof. This is an easy consequence of 2.6, using the remarks in 2.2. □

Rings satisfying these conditions are called **Dedekind prime rings**, this clearly being a Morita invariant property. If they are integral domains also, they are called **noncommutative Dedekind domains**. When commutative, of course, they are precisely the ordinary Dedekind domains.

2.11 It will be shown in §3 that 'classical' maximal orders over (commutative) Dedekind domains are all Dedekind prime rings. Some further examples are given next.

Example. (i) Let R be any prime principal ideal ring. If $I \triangleleft_r R$, then $I \oplus J \triangleleft_e R$ for complement J; then $I \oplus J = cR$ with $c \in \mathscr{C}_R(0)$. Thus $cR \simeq R$ and I is projective. This, symmetry, and 2.8, shows that R is a Dedekind prime ring.

Example. (ii) It follows from (i) above and 1.3.9 that $B_1(k)$ is a noncommutative Dedekind domain, for k any field. If, further, $\operatorname{char} k = 0$ then 1.3.5 shows that $A_1(k)$ is simple Noetherian and hence Asano, and 7.5.8 will show $A_1(k)$ to be hereditary. So $A_1(k)$ is also a noncommutative Dedekind domain. A similar argument, using 1.8.7 and 7.5.5, shows that the rings $P_\lambda(k)$, for λ not a root of unity, are also noncommutative Dedekind domains.

Example. (iii) If S is any simple Artinian ring then $S[x]$ is a Dedekind prime ring. For $S \simeq M_n(D)$, with D a division ring, and $S[x] \simeq M_n(D[x])$.

Example. (iv) The ring $S = \operatorname{End}(R \oplus A)$, described in 3.5.6, is Morita equivalent to the Dedekind domain R, and so is a Dedekind prime ring. This shows that Dedekind prime rings need not be matrix rings over integral domains.

2.12 Proposition. *A Dedekind prime ring R is Morita equivalent to a noncommutative Dedekind domain.*

Proof. Choose any nonzero uniform right ideal U of R. By 3.3.5, $\operatorname{End} U$ is a right Ore domain; but U_R is a progenerator, so $\operatorname{End} U$ is also a Dedekind prime ring. □

2.13 Theorem. *Let R be a Dedekind prime ring.*
(i) *If a prime Goldie ring S is context equivalent to R and is a maximal order, then S is a Dedekind prime ring and is Morita equivalent to R.*
(ii) *Any maximal order S equivalent to R is a Dedekind prime ring and is Morita equivalent to R.*

Proof. (i) Let

$$\begin{bmatrix} R & V \\ W & S \end{bmatrix}$$

be the prime context concerned (see Chapter 3, §6). It follows from 3.6.6 and 3.4.3 that W_R is a submodule of R^n for some n, and so is a progenerator; moreover, 3.6.6 shows that $S \subseteq \operatorname{End} W$ and $S \sim \operatorname{End} W$. Therefore $S = \operatorname{End} W$, and so R and S are Morita equivalent. Hence S is a Dedekind prime ring.

(ii) By 3.1.14, there is a Morita context

$$\begin{bmatrix} R & B \\ A & S \end{bmatrix}$$

which is readily seen to be prime. Therefore (i) applies. □

2.14 Finally, as mentioned earlier, we consider the arithmetic of one-sided fractional ideals over a Dedekind prime ring R. Of course, if A is a fractional right ideal of R, with $O_r(A) = S$ say, one would hope to have $A^{-1}A = R$, but $AA^{-1} = S$.

Recall that a **Brandt groupoid** G is defined as (the set of morphisms in) a small category whose morphisms are isomorphisms. Thus the elements of G all have two-sided inverses—but with different units on each side; and multiplication $g_1 g_2$ is defined only if the right unit of g_1 is the left unit of g_2.

Theorem. *Let R be a Dedekind prime ring and let G be the set of all fractional right ideals of all Dedekind prime rings equivalent to R. Then G is a Brandt groupoid, with AB being defined only if $O_r(A) = O_l(B)$.*

Proof. Let A be a fractional (R_1, R_2)-ideal and let $B = A^*$, this, by 1.8, being unambiguous and symmetric. It is clear from the proof of 2.13 that A is a progenerator on each side; so $AB = R_1$ and $BA = R_2$ as required. □

2.15 It is the case that modules over a Dedekind prime ring have a structure similar to (but interestingly different to) that over a commutative Dedekind domain. This is outlined in §7.

§3 Classical Orders

3.1 This section concerns a well-known class of orders, namely orders over a Dedekind domain K in a central simple algebra Q. It will be shown that maximal orders over K are Dedekind prime rings; and a converse result appears later, in Chapter 13.

The section starts, however, with a more general setting, allowing K to be any commutative integral domain. Some properties of K-orders are described, with particular emphasis on their relationship with orders in Q, in the sense studied in Chapter 3. This is very closely connected to the theory of prime PI rings; indeed every prime PI ring is a K-order over its centre K. The theory of prime PI rings will be dealt with in Chapter 13.

3.2 We start by discussing some properties of overrings.

Lemma. *Let A, B be rings with A contained in the centre of B and let $b \in B$. Consider the following statements:*
  (i) *$A[b]$ is a Noetherian A-module;*
  (ii) *$A[b]$ is a finitely generated A-module;*
  (iii) *there is a monic polynomial $f(x) \in A[x]$ with $f(b) = 0$;*
  (iv) *$A[b]$ is contained in some finitely generated A-module.*
Then (i) $\Rightarrow$ (ii) $\Leftrightarrow$ (iii) $\Rightarrow$ (iv); and all are equivalent if A is Noetherian.

Proof. This is straightforward. □

5.3.5 Classical orders

There is some related terminology. If (iii) holds, b is said to be **integral** over A; and B is **integral** over A if this is true for all $b \in B$. Similarly, if (iv) holds we will say b is **c-integral** over A and, if this is true for all $b \in B$, then B is **c-integral** over A.

3.3 In the particular case when B is the field of fractions of an integral domain A is **integrally closed** if the elements of B which are integral over A all belong to A. The corresponding property for c-integral elements has already been defined in 1.3, as the next result shows.

Lemma. *Let A be a commutative integral domain with quotient field B. Then:*
(i) *if $b \in B$ is c-integral over A then there exists a finitely generated A-submodule M of B with $A[b] \subseteq M$;*
(ii) *A is completely integrally closed if and only if each element of B which is c-integral over A belongs to A.*

Proof. (i) Since $b \in B$ is c-integral, there is some finitely generated A-module M with $A[b] \subseteq M$. Let N be the torsion submodule of M with respect to $\mathscr{S} = A\setminus\{0\}$, as described in 2.1.17. Since $A[b]$ has zero torsion submodule, then $A[b] \hookrightarrow M/N$. Thus we may suppose M to be torsion-free. In that case $M \hookrightarrow M_{\mathscr{S}} \simeq M \otimes_A B \simeq B^t$ for some t; yet $A[b] \otimes_A B \simeq B$. It follows easily that there is an epimorphism $\alpha: B^t \to B$ with $\ker \alpha \cap A[b] = 0$. Then $A[b] \hookrightarrow M/\ker \alpha \hookrightarrow B$.

(ii) ($\Rightarrow$) Let $b \in B$ be c-integral over A; so $A[b] \subseteq M \subseteq B$ with M_A finitely generated. Using a common denominator, by 2.1.16, then $Ma \subseteq A$ for some $0 \neq a \in A$. Hence $ab^n \in A$ for all n and so $b \in A$.

($\Leftarrow$) If $b \in B$, $0 \neq a \in A$ are such that $ab^n \in A$ for all n, then $A[b] \subseteq Aa^{-1}$ which is a cyclic A-module. Hence $b \in A$ and so A is completely integrally closed. □

3.4 Throughout the remainder of this section we let Q denote a **central simple algebra** over a field Z. This means that Z is the centre of Q and Q is a simple Artinian ring, finite dimensional over Z.

It is evident from 3.2 that Q is integral over Z.

3.5 Suppose now that K is a commutative integral domain, not necessarily Noetherian, whose field of fractions is Z. A subring R of Q will be called a **classical order** in Q over K if $K \subseteq R$, $RZ = Q$ and R_K is finitely generated. For example, if $Q = M_n(Z)$, then $R = M_n(K)$ is a classical order over K. The next result shows that classical orders always exist.

Lemma. *Given any commutative integral domain K with quotient field Z there is a classical order R in Q over K.*

Proof. Let $q_1,\ldots,q_m \in Q$ be a Z basis. Then $q_i q_j = \sum_{k=1}^m a_{ijk} q_k$ with $a_{ijk} \in Z$. By 2.1.16, there exists $0 \neq c \in K$ such that each $a_{ijk} c \in K$. Let $R = K + \sum_{i=1}^m c q_i K$. It is easily verified that R is as claimed. □

Whilst the classical orders (over Dedekind domains) are the main point of this section, it is convenient, as we shall see, to introduce a more general notion. Note first, by 3.2, that a classical order R over a commutative integral domain K is c-integral over K.

3.6 Let K and Z be as in 3.5. A subring R of the central simple algebra Q will be called a *K-order* in Q if $K \subseteq R$, $RZ = Q$ and R is c-integral over K.

This includes, of course, all classical orders.

3.7 We now give some examples of K-orders which are not classical orders.

Example. (i) If K is not Noetherian, with I an ideal which is not finitely generated, then the ring

$$R = \begin{bmatrix} K & I \\ K & K \end{bmatrix}$$

is a K-order in $M_2(Z)$. The centre of R is K, and R is integral over K, by the Cayley–Hamilton theorem; but R_K is not finitely generated.

Example. (ii) Let K be a commutative Noetherian integral domain whose integral closure R in Z is not Noetherian (for an example see [Bourbaki **61**–65, Chapter 5, §1, Exercise 21(b)]). Again R is integral over K but R_K is not finitely generated.

We note that whilst, in Example (i), K is the centre of R and is not Noetherian, in Example (ii), K is not the centre but is Noetherian. Next we give an example of a K-order R which is not integral over its centre K.

Example. (iii) The ingredients for this example are a field L and two automorphisms, σ, τ, of finite order; and G is the subgroup of Aut L generated by σ, τ. The corresponding fixed fields are denoted by L^σ, L^τ and L^G; so $L^G = L^\sigma \cap L^\tau$. The special condition required is that in L^σ there is an element x such that the elements $(\tau\sigma)^n x$ are all distinct.

(One example of such an L, σ, τ is given by taking $L = k(x, y)$ with k a field of characteristic zero, and σ, τ being k-automorphisms of L given by $\sigma(x) = x$, $\sigma(y) = -y$, $\tau(x) = x + y$, $\tau(y) = -y$. Then $(\tau\sigma)^n(x) = x + ny$.)

The order R is a subring of $M_2(L[t])$, namely

$$R = \begin{bmatrix} L^\sigma + tL[t] & tL[t] \\ tL[t] & L^\tau + tL[t] \end{bmatrix}.$$

5.3.9 *Classical orders* 145

The centre K of R is then $(L^\sigma + tL[t]) \cap (L^\tau + tL[t])$; so $K = L^G + tL[t]$. Hence the field of fractions Z of $L[t]$ is also the field of fractions of K and, moreover, $RZ = M_2(Z)$. There is a K-module embedding $L[t] \hookrightarrow tL[t] \subseteq K$, and so $R_K \hookrightarrow K^4$. This shows that R is c-integral over K and completes the demonstration that R is a K-order.

It remains only to note that R is not integral over K. To see this, consider the element

$$r = \begin{bmatrix} x & 0 \\ 0 & 0 \end{bmatrix}.$$

If r were integral over K, then x would be integral over L^G and all the elements $(\tau\sigma)^n(x)$ would be roots of the same monic polynomial. Thus r is not integral over K.

This example has another interesting feature; namely that whilst R is Noetherian, its centre K is not. To prove this, note first that $[L:L^\sigma] < \infty$, $[L:L^\tau] < \infty$ and $[L:L^G] = \infty$ (since $\tau\sigma$ must have infinite order). It follows that $tL[t]/t^2L[t]$ is not a finitely generated K-module, and so K is not Noetherian. On the other hand, $L[t]$ is finitely generated over each of the Noetherian rings $L^\sigma[t]$ and $L^\tau[t]$. Hence, by 1.1.3, $L^\sigma + tL[t]$ and $L^\tau + tL[t]$ are Noetherian. Since $tL[t]$ is then a Noetherian module over each of these rings, 1.1.7 shows that R is Noetherian.

3.8 There are some connections between K-orders and the orders as described in Chapter 3, §1.

Proposition. *Let R be a K-order in Q. Then:*
  (i) *R is an order in Q, with each $q \in Q$ having the form rc^{-1} with $r \in R$, $0 \neq c \in K$;*
  (ii) *each essential right or left ideal of R contains a nonzero element of K;*
  (iii) *if S is an order in Q, equivalent to R, then $c'R \subseteq S$ and $cS \subseteq R$ for some $0 \neq c$, $c' \in K$;*
  (iv) *if K' is the centre of R then R is a K'-order in Q; and if K is completely integrally closed then $K = K'$.*

Proof. (i)(ii) These are clear from 2.1.16.
  (iii) It is easily seen that $aSb \subseteq R$, $a'Rb' \subseteq S$ for some regular elements a, a', b, b' of R (using arguments as in 3.1.14). Each of Ra, bR, $a'R$ and Rb' contains a nonzero element of K, by (ii); and the result then follows.
  (iv) This is clear from the definitions. □

A significant consequence of (iv) is that the condition that K be completely integrally closed is more restrictive than the condition that K be the centre of R.

3.9 These ideas also connect with the theory of prime PI rings. This is expounded in Chapter 13, but all that is required here are the following facts.

(a) A ring R is an order in a central simple algebra if and only if R is a prime PI ring (see 13.6.6).
(b) If R is an order in the central simple algebra Q, K is the centre of R and Z is the centre of Q then Z is the quotient field of K, $RZ = Q$ and $R \subseteq M \subseteq Q$ with M a free K-module of finite rank (see 13.6.5 and 13.6.10).

3.10 These facts, combined with 3.8(i) and the definitions, give

Proposition. *The following conditions on a ring R with centre K are equivalent:*
(i) *R is a prime PI ring;*
(ii) *R is an order in a central simple algebra;*
(iii) *R is a K-order in a central simple algebra.* □

3.11 Corollary. *Let R be a K-order in Q where K is the centre of R.*
(i) *Any order S in Q equivalent to R and containing K is also a K-order in Q.*
(ii) *Any order S in Q equivalent to R is contained in a K-order equivalent to R.*
(iii) *If, further, K is completely integrally closed then any K-order S in Q is equivalent to R.*

Proof. (i) Suppose $s \in S$ and consider $K[s] \subseteq S$. Now, by 3.8(iii), $cS \subseteq R$ for some $0 \neq c \in K$. Therefore $cK[s] \subseteq R \subseteq M$, where $M \simeq K^t$ for some t (using 3.9(b)). That means $K[s] \subseteq c^{-1}M \simeq M$, and so s is c-integral over K. Thus S is a K-order.

(ii) If $S \sim R$ then, by 3.1.14, there is a fractional ideal ${}_S A_R$ and, by 3.1.12, S is contained in, and equivalent to, $O_l(A)$. But also $K \subseteq O_l(A)$ and so, by (i), $O_l(A)$ is a K-order.

(iii) By 3.8, S is an order in Q and K is the centre of S. So, by 3.9(b), $S \subseteq M \subseteq Q$ with M_K finitely generated. Using common denominators, there exists $0 \neq c \in K$ with $cM \subseteq R$ and so $cS \subseteq R$. By symmetry, $c'R \subseteq S$ and so $R \sim S$. □

The extra condition in (iii) above is necessary. This can be seen by considering 3.7(ii). There R is a K-order in Z; but R cannot be equivalent to K or else, by 3.8(iii), R_K is isomorphic to a K-submodule of K, which would make R_K finitely generated.

3.12 One can now exhibit 'natural' K-orders over any integral domain K.

Proposition. *Let K be an integral domain with quotient field Z, and let M be a submodule of K^n. Then $\operatorname{End} M$ is a K-order in $M_t(Z)$, where $t = \operatorname{u\,dim} M_K$.*

Proof. It is easy to see that M embeds in K^t as an essential submodule. By Corollary 3.4.6, $\operatorname{End} M \sim \operatorname{End} K^t \simeq M_t(K)$. Since $K = \operatorname{Centre} M_t(K)$ and $M_t(K)$ is visibly a K-order in $M_t(Z)$, then so too, by 3.11(i), is $\operatorname{End} M$. □

3.13 Next we turn to **maximal K-orders**, these being K-orders in Q which are contained in no larger K-orders in Q.

Lemma. *Each K-order is contained in a maximal K-order.*

Proof. Since the union of a chain of K-orders is again a K-order, Zorn's lemma applies. □

Theorem. *Let R be an order in the central simple algebra Q and let K be the centre of R. Then R is a maximal K-order in Q if and only if R is a maximal order. It follows then that K is completely integrally closed.*

Proof. Suppose R is a maximal order. By 1.10, K is completely integrally closed and, by 3.10, R is a K-order. Hence, by 3.11(iii), R is a maximal K-order.

Conversely, let R be a maximal K-order and suppose $S \supseteq R$ with $S \sim R$. Since $K = $ Centre R, 3.11(i) shows that S is a K-order. Hence $S = R$ and therefore R is a maximal order. □

Corollary. *If R is a K-order with K being completely integrally closed, then any maximal order equivalent to R is a maximal K-order.*

Proof. Use 3.11 and the theorem. □

3.14 Next we consider the case when K is Noetherian.

Proposition. *If R is a K-order in Q with K being the centre of R and K being Noetherian, then R is a classical order over K and R is Noetherian.*

Proof. This is immediate from 3.9(b). □

Examples 3.7(i), (ii) show that neither condition on K can be deleted.

Corollary. *If K is Noetherian and integrally closed then any K-order is a classical order.* □

3.15 To end this section we return to the starting-point of the section, namely the case when K is a Dedekind domain. Since K is then Noetherian and integrally closed, all the preceding results apply although, of course, improvements can be made. The next result elaborates on 3.12.

Proposition. *Let K be a Dedekind domain with quotient field Z and let M be a submodule of K^n. Then $\mathrm{End}\, M$ is a maximal K-order in $M_t(Z)$, for $t = \mathrm{u\, dim}\, M$, and $\mathrm{End}\, M$ is a Dedekind prime ring.*

148 *Some Dedekind-like rings* 5.3.16

Proof. As in the proof of 3.12, End M is a K-order in $M_t(Z)$ and, by 3.8(iv), K is its centre. Since M_K is a progenerator, End M is a Dedekind prime ring and so, by 2.6, is a maximal order. Then by Theorem 3.13, End M is a maximal K-order. □

3.16 The final result of this section extends 3.15 to all maximal K-orders.

Theorem. *If K is a Dedekind domain and R is a maximal K-order in Q, then R is a Dedekind prime ring.*

Proof. Note first that, by 3.8(iv), K is the centre of R and, by 3.14, R is Noetherian and R_K is finitely generated. Moreover, by Theorem 3.13, R is a maximal order.

Next we will show that if $0 \neq P$ is a prime ideal of R then R/P is simple Artinian. To see this note that P is an essential right ideal of R and so, by 3.8(ii), $P \cap K \neq 0$. Since K has dimension 1, then $K/P \cap K$ is an Artinian ring. However, R/P is a finitely generated $K/P \cap K$-module; so R/P is an Artinian, and hence simple, ring. In particular P is maximal.

We now aim to show that P is an invertible ideal. Let $A = P \cap K$. Then $0 \neq AR \triangleleft R$ and $AR \subseteq P$. Since R is Noetherian, 2.2.17 shows that AR contains a product of prime ideals, $P_1 \cdots P_t \subseteq AR$ say, where it may be assumed that t is the minimal possible length. It follows that $P_i \subseteq P$ for some i, and so $P_i = P$ since both are maximal ideals. Thus $BPC \subseteq AR$, where $B = P_1 \cdots P_{i-1}$ and $C = P_{i+1} \cdots P_t$. Therefore $BPCA^{-1} \subseteq R$ and so, using 1.8, $CA^{-1}BP \subseteq R$. Hence $P^* \supseteq CA^{-1}B$. Suppose, for a moment, that $P^* = R$. Then $CA^{-1}B \subseteq R$ and so $CB \subseteq AR$, contradicting the minimality of t. Therefore we deduce that $P^* \supset R$.

If $P^*P = P$, then $P^*P^*P = P$ and hence $(P^*)^2 \subseteq P^*$. However, this makes P^* an order containing R and equivalent to R—which is not allowed. Therefore $P^*P = R$ and, by symmetry, $PP^* = R$. So P is invertible with $P^* = P^{-1}$.

The next step is to show that any nonzero ideal, I say, is invertible. By Noetherian induction we may suppose any larger ideal is invertible. Now $I \subseteq P$ for some maximal ideal P. If $P^{-1}I = I$ then $P^{-1} \subseteq O_l(I)$ and $O_l(I) \supset R$, a contradiction to 1.4. Therefore $P^{-1}I \supset I$, and so $P^{-1}I$ is invertible. Hence I is invertible. Thus R is an Asano order.

By 2.10, it remains only to show that R is hereditary. Note next that since, by 3.8(ii), each essential right ideal A contains a central element, it contains a product of prime ideals. In particular this means that R/A has a finite composition series. It also implies that if A is a maximal right ideal then A contains a prime ideal, P say. Moreover, since R/P is simple Artinian, R/A is a direct summand of R/P. Since P_R is projective it follows that pd $R/A = 1$. This applies to any simple module. A simple induction establishes that pd $M = 1$ for all modules of finite composition length. It follows that pd $R/A = 1$ for A any essential right ideal; and so A_R is projective.

Since every right ideal is a direct summand of an essential right ideal, R is right hereditary. Thus R is a Dedekind prime ring. □

3.17 A converse to this will be established in 13.9.15; namely, if R is an Asano order in a central simple algebra Q then the centre K of R is a Dedekind domain and R is a maximal K-order. Its proof depends rather more heavily on the theory of PI rings.

§4 Hereditary Noetherian Rings

4.1 As noted earlier in 2.7, commutative Dedekind domains are characterized as being hereditary integral domains. Indeed a Noetherian commutative ring is hereditary if and only if it is a finite direct product of fields and Dedekind domains (see [Kaplansky **70**, Theorem 168]). We have also seen, in §2, that Dedekind prime rings are those hereditary Noetherian prime rings which are maximal orders.

In this section we begin an investigation of hereditary Noetherian rings. We will see that such a ring has an Artinian quotient ring, and that it decomposes as a direct product of prime rings and an Artinian ring. The structure of the Artinian portion is described; and the nature of the prime portions is investigated in the remainder of this chapter. It is convenient to defer until later in this chapter the exposition of further examples of hereditary Noetherian rings.

4.2 To start with, only one-sided conditions are required.

Theorem. *A right hereditary right Noetherian ring has a right Artinian right quotient ring.*

Proof. By 4.1.4 it is enough to show, given $c \in \mathscr{C}(N)$, that $c \in \mathscr{C}(0)$. Consider the short exact sequence

$$0 \to \operatorname{r ann} c \to R \to cR \to 0$$

which splits, because cR is projective. Thus $\operatorname{r ann} c$ is a direct summand of R, so has an idempotent generator. However, $\operatorname{r ann} c \subseteq N$; so $\operatorname{r ann} c = 0$. We aim to complete the proof by a symmetric argument—which is possible provided that Rc is projective. The proof of this fact, therefore, is all that remains.

The argument above shows that Rc is projective if and only if $\operatorname{l ann} c$ has an idempotent generator. Let $A = \operatorname{l ann} c = \sum Ra_i$, this summation possibly being infinite, and let $A_n = \sum_{i=1}^{n} Ra_i$. Now $\operatorname{r ann} A_n$ is the kernel of the map $R \to R^{(n)}$ defined by $r \mapsto (a_1 r, a_2 r, \ldots, a_n r)$. Since the image is projective (being a submodule of a free module) the map splits, and so $\operatorname{r ann} A_n$ has an idempotent generator. The descending chain $\operatorname{r ann} A_1 \supseteq \operatorname{r ann} A_2 \cdots$ must terminate—otherwise an infinite direct sum of idempotently generated right ideals would be obtained (using the usual trick that if $e_1R \supset e_2R$ then $e_1R = e_2R \oplus (1-e_2)e_1R$). Thus $r\,\text{ann}\,A = eR$ for some idempotent e and then

$$R(1-e) = l\,\text{ann}\,eR = l\,\text{ann}\,(r\,\text{ann}\,(l\,\text{ann}\,c)) = l\,\text{ann}\,c. \qquad \square$$

4.3 The argument above can be extended to show

Proposition. *Suppose R has no infinite orthogonal set of idempotents.*
(i) *If each principal right ideal is projective then each principal left ideal is projective.*
(ii) *If each finitely generated right ideal is projective then each finitely generated left ideal is projective.* $\qquad \square$

4.4 Proposition. *If R is a right hereditary, right Noetherian ring, and M is a finitely generated projective right or left R-module then $\text{End}\,M$ is a right hereditary right Noetherian ring.*

Proof. By Lemma 3.5.6, or its left-hand version, $\text{End}\,M \simeq eSe$, where $S = M_n(R)$ for some n, and $e = e^2 \in S$. Now $S \overset{M}{\sim} R$ and so S is right Noetherian and right hereditary.

Let $B \lhd_r eSe$; then $B \subseteq BS \lhd_r S$. Moreover, BS is a finitely generated projective right S-module, and so, by the dual basis lemma (3.5.2), $1 \in BS(BS)^*$. Note that $BS(BS)^* = BeS(BS)^*$ and that since $eS(BS)^*B \subseteq eS$ and $Be = B$, then $eS(BS)^*B \subseteq eSe$. This gives a map $eS(BS)^* \to B^*$. Since $1 \in BeS(BS)^*$, then $1 \in BB^*$. Hence, by 3.5.2, B is finitely generated and projective. $\qquad \square$

4.5 We now impose symmetric conditions on the rings being studied. Next we establish a 'co-Artinian' property for such rings.

Proposition. *Let R be a hereditary Noetherian ring, and let I be any essential right ideal. Then $(R/I)_R$ has finite length.*

Proof. Suppose $A \supseteq I$ with $A \lhd_r R$, and consider the dual modules A^* and I^*. Evidently there is a map $_RA^* \to {_RI^*}$, by restriction. In fact, this is a monomorphism. For say $g \in A^*$ and g restricted to I is zero. Choose any $x \in A$. By 2.2.2(iii), the right ideal $B = \{b \in R | xb \in I\}$ is essential. Now $0 = g(xB) = g(x) \cdot B$, so $r\,\text{ann}\,g(x) \lhd_e R$. However, the proof of 4.2 shows that $r\,\text{ann}\,g(x)$ has an idempotent generator. Thus $g(x) = 0$ and then $g = 0$. So indeed $A^* \hookrightarrow I^*$.

Now consider a chain $A_1 \supseteq A_2 \supseteq \cdots \supseteq I$. This yields a chain $A_1^* \subseteq A_2^* \subseteq \cdots \subseteq I^*$. However, $_R(I^*)$ is finitely generated, by the dual basis lemma (3.5.2), and R is left Noetherian. Therefore the ascending chain terminates and, by 3.5.2 again, it follows that the descending chain does so too. $\qquad \square$

4.6 Theorem. *A hereditary Noetherian ring R is a finite direct sum of Artinian hereditary rings and hereditary Noetherian prime rings.*

Proof. Let N be the prime radical of R. Since N is nilpotent (by 2.3.7) then $\operatorname{l ann} N$ is an essential right ideal of R. By 4.2, $\mathscr{C}(N) = \mathscr{C}(0)$; let $c \in \mathscr{C}(0)$. The descending chain $\{c^n R + \operatorname{l ann} N\}$ of right ideals containing $\operatorname{l ann} N$ must stabilize, by 4.5; so $c^n R + \operatorname{l ann} N = c^{n+1} R + \operatorname{l ann} N$ for some n. But then $c^n N = c^{n+1} N$ and so $cN = N$. Symmetry plus 4.1.9 shows that R decomposes as the sum of an Artinian ring and a semiprime ring. Now suppose R is semiprime. Let $R' = \oplus e_i R$ as in 3.2.4 and recall that $aR' \subseteq R$ for some $a \in \mathscr{C}_R(0)$. Since aR'_R is finitely generated projective, so too are R'_R and its direct summands $e_i R$. Therefore the map $\theta : R \to e_i R$ splits; say $R = K \oplus J$ with $K = \ker \theta \subseteq (1 - e_i) Q$ and $J \simeq e_i R$ and so $J = Je_i \subseteq e_i Q$. Now $1 = k + j$ say and multiplying by e_i and $1 - e_i$ gives $k = 1 - e_i$, $j = e_i$ and $R = (1 - e_i) R \oplus e_i R$. An easy induction now shows that $R = R'$. □

4.7 This reduces the study of hereditary Noetherian rings to Artinian rings and prime rings. The Artinian case is a little outside the direction of this book, so we simply describe their structure following [Harada 66]. First consider a triangular array

$$\begin{bmatrix} X_{11} & & & 0 \\ X_{21} & X_{22} & & \\ & & \ddots & \\ X_{n1} & \cdots & & X_{nn} \end{bmatrix}$$

in which each $X_{ii} = D_i$ is a division ring and each X_{ji} is a (D_j, D_i)-bimodule, finite dimensional on each side. Let

$$R = \begin{bmatrix} M_{11} & & & 0 \\ M_{21} & M_{22} & & \\ & & \ddots & \\ M_{n1} & \cdots & & M_{nn} \end{bmatrix}$$

with the M_{ji} being defined recursively by the equation

$$M_{ji} = X_{ji} \oplus \bigoplus_{k=i+1}^{j} M_{jk} \otimes_{D_k} X_{ki}, \qquad j \geq i$$

and matrix multiplication being the obvious one, making use of $\otimes$.

Theorem. (i) R *is a hereditary Artinian ring.*
  (ii) *Each hereditary Artinian ring is Morita equivalent to such a ring R.* □

§5 Idealizer Rings

5.1 Before proceeding with a description of hereditary Noetherian rings it is necessary to study idealizer rings in more depth than in §1 of Chapter 1. That is done in this section, and is used to exhibit some varied examples of hereditary Noetherian prime rings. In the following section it is applied to the general theory of such rings. First, some definitions are required.

5.2 If R is any ring, and $A \triangleleft_r R$ with the property that R/A is an isotypic semisimple module (i.e. R/A is a finite direct sum of isomorphic simple modules), then A is an **isomaximal** right ideal. We intend to study $\mathbb{I}(A)$ for such right ideals. The fact that any maximal right ideal is isomaximal shows that this work extends the results in 1.1.12.

5.3 If R is any ring and $A \triangleleft_r R$ such that $RA = R$, then A is a **generative** right ideal. This implies, of course, that A_R is a generator; but the converse is patently false. It is helpful to consider the special case when A is a maximal right ideal of R; then either $RA = R$ and so A is generative, or else $RA = A$ and A is an ideal. In the latter case $\mathbb{I}(A) = R$, thus only the former case is of interest. A similar dichotomy occurs when A is isomaximal.

Lemma. *If A is an isomaximal right ideal of a ring R then either A is generative or A is an ideal.*

Proof. If $RA \neq R$ then RA annihilates the simple summands of R/RA. Since R/A is isotypic, it follows that RA annihilates R/A. Hence $RA \subseteq A$ and so $A \triangleleft R$. □

Because of this, we will restrict attention to generative isomaximal right ideals.

5.4 In 1.1.11 it was noted that, for any right ideal A of a ring R, End $R/A \simeq \mathbb{I}(A)/A$. Similarly, one can check that, if $B \triangleleft_r R$, then

$$\mathrm{Hom}(R/A, R/B) \simeq \{r \in R \mid rA \subseteq B\}/B.$$

Note that $\mathrm{Hom}(R/A, R/B)$ is a right End R/A-module and hence is a right $\mathbb{I}(A)/A$-module.

5.5 The first step is to analyse the nature of a simple right R-module X when viewed as an $\mathbb{I}(A)$-module. Note that either $X \simeq R/M$ for some maximal right ideal M of R with $M \supseteq A$, or else $\mathrm{Hom}(R/A, X) = 0$.

Proposition. *Let A be a generative isomaximal right ideal of R and let X_R be simple.*
(i) If $\mathrm{Hom}(R/A, X) = 0$ then X is a simple $\mathbb{I}(A)$-module.

(ii) If $X \simeq R/M$ with $M \supseteq A$, then X has a unique composition series of length 2 over $\mathbb{I}(A)$, given by $R \supset \mathbb{I}(A) + M \supset M$. Furthermore, if R/A has length n, then
$$\mathbb{I}(A)/A \simeq (\mathbb{I}(A) + M/M)^n$$
and
$$R/\mathbb{I}(A) \simeq (R/\mathbb{I}(A) + M)^n.$$

Proof. Let M be a maximal right ideal of R with $X \simeq R/M$, and let B/M be a proper $\mathbb{I}(A)$-submodule of R/M. Then
$$(B/M)A = BA + M/M \subseteq B/M \subset R/M$$
and so $BA + M/M$ is a proper R-submodule of R/M. Hence $BA \subseteq M$. Therefore, if $b \in B$, then left multiplication by b maps $R/A \to R/M$. In case (i) this means $bR \subseteq M$; and then $B = M$ and R/M is indeed simple over $\mathbb{I}(A)$.

In case (ii) it means that $M \subseteq B \subseteq \{x \in R \mid xA \subseteq M\} = C$ say. As noted in 5.4, $C/M \simeq \mathrm{Hom}\,(R/A, R/M)$. Also $\mathbb{I}(A)/A \simeq \mathrm{End}\,R/A \simeq M_n(D)$, where $D = \mathrm{End}\,R/M$ is a division ring; and then $\mathrm{Hom}\,(R/A, R/M) \simeq D^n$. Therefore C/M is a simple right $\mathbb{I}(A)/A$-module. Hence either $B = M$ or $B = C$. However, $C \neq M$, since $C/M \simeq \mathrm{Hom}\,(R/A, R/M)$; and $C \neq R$ since $R \not\subseteq M$. Thus R/M does indeed have composition length 2 over $\mathbb{I}(A)$ with the unique series given by $R \supset C \supset M$. However, $C \supseteq \mathbb{I}(A) + M \supset M$; hence $C = \mathbb{I}(A) + M$.

Note next that the simple $\mathbb{I}(A)$-module $\mathbb{I}(A) + M/M$ is annihilated by A, whereas $(R/\mathbb{I}(A) + M)A = R/\mathbb{I}(A) + M$. On the other hand, $\mathbb{I}(A)/A \simeq M_n(D)$, and thus $\mathbb{I}(A)/A$ is isotypic, has length n and is annihilated by A. It follows that $\mathbb{I}(A)/A \simeq (\mathbb{I}(A) + M/M)^n$ and that $R/\mathbb{I}(A) \simeq (R/\mathbb{I}(A) + M)^n$. □

5.6 Corollary. *If A is an isomaximal right ideal of R then R is right Noetherian if and only if $\mathbb{I}(A)$ is right Noetherian.*

Proof. The proof of 1.1.12(iii) can now be used, using the proposition above to replace 1.1.12(ii). □

5.7 The next step concerns projective modules.

Lemma. *Let A be a generative right ideal of R and T a subring of R with $A \subset T \subseteq \mathbb{I}(A)$ and M a right R-module. Then:*
(i) R_T *and* $_TA$ *are finitely generated projective;*
(ii) $R \otimes_T R \simeq R$ *via multiplication;*
(iii) *if M_R is projective then M_T is projective.*

Proof. (i) Left multiplication by any $a \in A$ gives a map in $\mathrm{Hom}\,(R_T, T_T)$. Equally any $r \in R$ gives a map in $\mathrm{Hom}\,(_TA, _TT)$. Since $RA = R \ni 1$ the dual basis lemma shows that R_T, $_TA$ are finitely generated projective.

(ii) By (i) the map $R \otimes_T T \to R \otimes_T R$ is a monomorphism. However, then
$$R \otimes_T R = RA \otimes_T R = R \otimes_T AR = R \otimes_T AT = RA \otimes_T T = R \otimes_T T$$
and so $R \otimes_T R \simeq R$.

(iii) Here M_R is a direct summand of a free R-module which, by (i), is T-projective. □

5.8 Theorem. *If R is a right hereditary ring and A is an isomaximal right ideal then $\mathbb{I}(A)$ is right hereditary.*

Proof. By 5.3, we may suppose A to be generative. Let $T = \mathbb{I}(A)$, and let M be a maximal right ideal of R containing A. By 5.5, R/M has the unique composition series $R \supset T + M \supset M$. Also $R/T \simeq (R/T + M)^n$ and $T/A \simeq (T + M/M)^n$. Since R_T, T_T are projective, by 5.7, then $\text{pd}(R/T)_T \leq 1$ and hence $\text{pd}(R/T + M) \leq 1$.

Now let $B \triangleleft_r T$, and consider BR/BA. For each $b \in B$, there is a map $R/A \to BR/BA$ via $r + A \mapsto bR + A$. Thus BR/BA is the homomorphic image of $(R/M)^I$ for some index set I. Moreover, B/BA is the image of the T-submodule $(T + M/M)^I$ of $(R/M)^I$. It follows that $BR/B \simeq (R/T + M)^J$ for some index set J; therefore $\text{pd}(BR/B)_T \leq 1$.

Since R is right hereditary, $\text{pd}\, BR_R = 0$ and so, by 5.7, $\text{pd}\, BR_T = 0$. From the short exact sequence
$$0 \to B \to BR \to BR/B \to 0$$
one deduces that $\text{pd}\, B_T = 0$, as required. □

5.9 Further results along the lines of 5.7 and 5.8, but for higher global dimensions, can be found in Chapter 7, §5.

5.10 Applied to hereditary Noetherian prime rings 5.8 gives the following result—which preserves the left-hand structure as well as the right.

Theorem. *Let R be a hereditary Noetherian prime ring but not simple Artinian. Let A be a generative isomaximal right ideal. Then $\mathbb{I}(A)$ is a hereditary Noetherian prime ring, A is an idempotent and maximal ideal of $\mathbb{I}(A)$ and $R = O_r(A)$.*

Proof. Here A is idempotent since $A^2 = ARA = AR = A$; and A is maximal since $T/A \simeq \text{End}(R/A) \simeq M_n(D)$, where $T = \mathbb{I}(A)$ and D is a division ring. By 5.8, T is right hereditary and, by 5.6, T is right Noetherian. Since R is not simple Artinian, it follows from 3.3.4(iii) that R has no minimal right ideals. Therefore A must be essential in R. Thus R and T are equivalent orders in a simple Artinian quotient ring, Q say, and thus T is a prime Goldie ring. Furthermore, since $R \subseteq O_r(A)$, then
$$O_r(A) = RO_r(A) = RAO_r(A) = RA = R.$$

5.5.11 *Idealizer rings*

It remains only to obtain the left-handed properties of T. For this it is helpful to consider the ring $S = O_l(A)$. By 3.1.15(iv), $S \simeq \mathrm{End}(A_R)$; yet A_R is a pro-generator by hypothesis. Thus 3.5.10 shows that S is a hereditary Noetherian prime ring also. It is clear that

$$T = \mathbb{I}(A_R) = O_l(A) \cap O_r(A) = S \cap R = \mathbb{I}(_SA).$$

Thus symmetry will complete the proof, provided we check that A is an isomaximal left ideal of S.

To prove this, recall that since A_R is projective it is also reflexive (see 1.7) and so $A = A^{**}$. The same is true for right ideals B containing A, and for left R-modules C with $R \subseteq C \subseteq A^*$. Consequently there is an inverse lattice isomorphism between the submodule lattices of $(R/A)_R$ and $_R(A^*/R)$. It follows that $_R(A^*/R)$ is semisimple isotypic. (To see that it is isotypic, consider a module $M = M_1 \oplus M_2$ with M_1, M_2 being simple. If $M_1 \not\simeq M_2$ then M has exactly two simple submodules; but if $M_1 \simeq M_2$ it has at least three.)

The Morita equivalence between S and R ensures that $A \otimes_R (A^*/R)$ is an isotypic semisimple left S-module. However,

$$A \otimes (A^*/R) \simeq A \otimes A^*/A \otimes R \simeq S/A$$

and that completes the proof. □

5.11 It is now easy to give some examples.

Example. (i) Let K be any commutative Dedekind domain with maximal ideal M and let $R = M_2(K)$. By Morita equivalence R is a Dedekind prime ring. Note that

$$A = \begin{bmatrix} M & M \\ K & K \end{bmatrix}$$

is a maximal right ideal with $RA = R$. Therefore $\mathbb{I}(A)$ is a hereditary Noetherian prime ring. In fact

$$\mathbb{I}(A) = \begin{bmatrix} K & M \\ K & K \end{bmatrix}.$$

Example. (ii) The ring $B_1(k)$ is a principal ideal domain (see 1.3.9) and so is a noncommutative Dedekind domain. Also $A = xB_1$ is a maximal right ideal, and $B_1 A = B_1$. Therefore $\mathbb{I}(xB_1)$ is a hereditary Noetherian domain. One can check, for k of characteristic zero, that $\mathbb{I}(xB_1) = k + xB_1$.

Example. (iii) Similarly, if $R = A_1(k)$, the first Weyl algebra over a field of characteristic zero, then R is a hereditary Noetherian domain, as will be shown in 7.5.8. Again xR is a maximal right ideal and $\mathbb{I}(xR) = k + xR$ is a hereditary Noetherian domain.

§6 Hereditary Noetherian Prime Rings

6.1 One consequence of 5.10 is that one can construct, as idealizer subrings of a hereditary Noetherian prime ring R, other hereditary Noetherian prime rings; but they will automatically contain idempotent maximal ideals. In this section we show that the number of idempotent ideals of R is a measure of its distance from being a Dedekind prime ring. In particular, R has finitely many idempotent ideals precisely when R can be obtained from a Dedekind prime ring by a finite iteration of the process of idealizing at isomaximal right ideals.

6.2 We note first the obvious consequence of 4.5 that if I is any nonzero ideal of a hereditary Noetherian prime ring R then, since $I \triangleleft_e R_R$, the ring R/I must be Artinian. In particular any nonzero prime ideal P is maximal, and R/P is then simple Artinian.

6.3 The significance of idempotent ideals in R is indicated by the next result.

Proposition. *A hereditary Noetherian prime ring R is Dedekind if and only if it has no idempotent ideals other than 0, R.*

Proof. If $0 \neq A \triangleleft R$ then $A^*A \triangleleft R$. Now $AA^* = O_l(A)$ since A_R is finitely generated projective, and hence

$$(A^*A)^2 = A^*O_l(A)A = A^*A.$$

Therefore if R has no nontrivial idempotent ideals then $A^*A = R$. Hence R is Asano and so Dedekind, by 2.10. The converse is trivial. □

6.4 For maximal ideals a dichotomy exists.

Lemma. *A maximal ideal of a hereditary Noetherian prime ring is either idempotent or invertible.*

Proof. For any nonzero ideal A, the argument above shows that A^*A is idempotent. Of course $A \subseteq A^*A$, so if A is maximal and not idempotent then $A^*A = R$. Symmetry then shows that A is invertible. □

6.5 The next key result shows that, associated with each idempotent ideal there is a larger order. For this result we need a stronger equivalence relation on orders. Two orders R, S in a quotient ring Q are called **right equivalent** if $aR \subseteq S$ and $bS \subseteq R$ for some units a, b of Q.

6.6 Theorem. *Let S be a hereditary Noetherian prime ring. There is an inclusion reversing $(1,1)$-correspondence between the nonzero idempotent ideals A of S and* those orders R containing S and right equivalent to S. This correspondence is given by

$$A \mapsto O_r(A), \qquad R \mapsto \{q \in Q \mid qR \subseteq S\}$$

where $Q = Q(S)$.

Proof. An ideal A, as described, contains a regular element, a say. Then $aO_r(A) \subseteq S$ and so $O_r(A)$, S are right equivalent orders. Since $_S A$ is finitely generated and projective, the left-hand version of the dual basis lemma shows that $O_r(A) = BA$, where $B = \{q \in Q \mid Aq \subseteq S\}$. Let

$$A' = \{q \in Q \mid qO_r(A) \subseteq S\}.$$

Evidently $A \subseteq A'$ and $A'O_r(A) \subseteq A$. However,

$$O_r(A)A = BA^2 = BA = O_r(A)$$

and therefore

$$A \subseteq A' \subseteq A'O_r(A) = A'O_r(A)A \subseteq A.$$

Thus $A = A'$.

Next let R be as described, and let $A = \{q \in Q \mid qR \subseteq S\}$. By hypothesis, there is a unit, a say, of Q belonging to A. Thus $0 \neq A \triangleleft S$. Furthermore $R_S \simeq aR_S \subseteq S$ and so R_S is finitely generated projective, and $A = (R_S)^*$. Therefore, by the dual basis lemma, $RA = O_l(R) = R$ and then $A^2 = ARA = AR = A$. Finally, since $RA = R$, it is clear that $O_r(A) = R$. □

6.7 This theorem provides a process which is the reverse of idealizing in the sense that whilst idealizing produces a smaller ring with an idempotent ideal, the step of moving from S to $R = O_r(A)$ uses an idempotent ideal to produce an overring. The next result shows how very precise this connection is.

Theorem. (a) *Let S be a hereditary Noetherian prime ring, A a nonzero idempotent ideal and $R = O_r(A)$. Then R is a hereditary Noetherian prime ring, A is an essential generative right ideal of R and $S \subseteq \mathbb{I}(A_R)$.*

(b) *If, further, A is a maximal ideal of S, then A_R is isomaximal and $S = \mathbb{I}(A)$.*

Proof. (a) Note first that $R \simeq \text{End}(_S A)$ which, by 4.4, is hereditary and Noetherian. Of course, being an order in $Q(S)$ makes R prime. Since $A \triangleleft S$, then $AQ = Q$, by 2.1.16. Hence, by 2.2.12, A is an essential right ideal of R; and the proof of 6.6 shows that A is generative. The fact that $A \triangleleft S$ ensures that $S \subseteq \mathbb{I}(A_R)$.

(b) Since A is maximal, 6.6 shows that there are no orders between S and R; yet, by (a), $S \subseteq \mathbb{I}(A) \subseteq R$. Since A_R is generative, $\mathbb{I}(A) \neq R$; therefore $\mathbb{I}(A) = S$.

Let J/A be the Jacobson radical of the right R-module R/A, i.e. the intersection of its maximal submodules. This is invariant under endomorphisms of R/A, and therefore $\mathbb{I}(A) \subseteq \mathbb{I}(J)$ since $\mathbb{I}(A)/A \simeq \operatorname{End} R/A$. Now $\mathbb{I}(J) \subseteq R$, yet $RJ \supseteq RA = R$ and so $\mathbb{I}(J) \neq R$. Therefore $\mathbb{I}(J) = \mathbb{I}(A) = S$. Note, however, that then $J \triangleleft S$ and $A \subseteq J$. That implies that $A = J$ and so R/A must be semisimple. The fact that $\operatorname{End} R/A \simeq S/A$ which is simple ensures that R/A is isotypic; thus A is isomaximal. □

6.8 The combination of these two processes enables us to characterize certain hereditary Noetherian prime rings.

Theorem. *The following conditions on a ring S are equivalent:*
(i) *S is obtained from a Dedekind prime ring by a finite iteration of the process of forming idealizers of generative isomaximal right ideals;*
(ii) *S is a hereditary Noetherian prime ring with a minimal nonzero idempotent ideal;*
(iii) *S is a hereditary Noetherian prime ring and is equivalent to a Dedekind prime ring.*

Proof. (i)⇒(ii). Say S is obtained from the Dedekind prime ring R as described. It is clear that if $A \triangleleft_e R$ then $\mathbb{I}(A)$ is right equivalent to R. It follows that S is right equivalent to R. By 5.10, S is a hereditary Noetherian prime ring. Since R is a maximal order, there can be no larger order right equivalent to R. The (1, 1)-correspondence provided by 6.6 gives an idempotent ideal, A say, of S corresponding to R and the maximality of R implies the minimality of A.

(ii)⇒(iii) Let A be the minimal idempotent ideal of S, and let R be the order corresponding to A under the correspondence of 6.6. The minimality of A means that R cannot be contained in any larger right equivalent order. By 6.7, R is a hereditary Noetherian prime ring. Applying 6.6 to R now shows that R has no nontrivial idempotent ideals. By 6.3, R is Dedekind.

(iii)⇒(i) Suppose $S \sim R$, with R a Dedekind prime ring. By 3.1.14, there is a fractional (R, S)-ideal B say. Now $R \overset{M}{\sim} O_r(B)$, so $O_r(B)$ is another Dedekind prime ring. However, $S \subseteq O_r(B)$ and, by the proof of 3.1.12, S is right equivalent to $O_r(B)$. This shows that, without loss of generality, we may suppose that $S \subseteq R$ and S, R are right equivalent, say with $bR \subseteq S$ for a unit b of Q.

Note that $bS \subseteq bR \subseteq S$, and that $bS \triangleleft_e S$. Therefore, by 4.5, $(bR/bS)_S$ has finite length. That implies that $(R/S)_S$ has finite length. Choose a maximal chain of orders

$$R = R_0 \supset R_1 \supset R_2 \supset \cdots \supset R_n = S.$$

It is clear from 6.6 and 6.7(a) that each R_i is a hereditary Noetherian prime ring and that R_i corresponds, via 6.6, to a maximal idempotent ideal A_i of R_{i+1}. Hence, by 6.7(b), $R_{i+1} = \mathbb{I}_{R_i}(A_i)$ and A_i is an isomaximal right ideal of R_i. □

6.9 Let S be a hereditary Noetherian prime ring. As noted earlier, S is a PI ring if and only if the quotient ring of S is a central simple algebra. In that case, 13.7.15 together with 6.12 below will show that S does indeed have a minimal idempotent ideal. Thus the theory presented here applies to all such rings.

Furthermore, it will be seen in 13.9.16 that the centre K of S is a Dedekind domain, and hence S is a classical order over K, as described in §3, and is an iterated idealizer from a classical maximal order over K.

6.10 What is not yet fully explained is the collection of idempotent ideals. The next few results do this.

Proposition. *Let S be a hereditary Noetherian prime ring and A be a nontrivial idempotent ideal. Then:*
(i) any maximal ideal containing A is idempotent;
(ii) A is determined uniquely by the maximal ideals which contain it.

Proof. (i) Let M be a maximal ideal containing A. Then $M^n \supseteq A$ for all n and so, since S/A has finite length, $M^n = M^{n+1}$ for some n. If M were invertible that would imply $M = S$. Hence, by 6.4, M must be idempotent.

(ii) Since S/A is Artinian, there are only a finite number of maximal ideals of S containing A, say $M_1, \ldots, M_t$. Then $M_1 \cap \cdots \cap M_t / A$ is the prime radical of S/A and hence is nilpotent. Arguing as in (i), it is clear that $A = (M_1 \cap \cdots \cap M_t)^n$ for $n \gg 0$. Thus $M_1, \ldots, M_t$ determine A. □

Corollary. *If S is a hereditary Noetherian prime ring then S has finitely many idempotent ideals if and only if S has finitely many idempotent maximal ideals.* □

6.11 As shown in 3.6.2, there is a correspondence between certain prime ideals of the rings in a Morita context. This applies, in particular, to the rings R and $S = \mathbb{I}(A)$ where $A \triangleleft_r R$, using the context

$$\begin{bmatrix} R & R \\ A & S \end{bmatrix}.$$

In the special case of interest here, the correspondence becomes very precise.

Theorem. *Let R be a hereditary Noetherian prime ring, A a generative isomaximal right ideal and $S = \mathbb{I}(A)$. There is a set embedding $\gamma: \{P \in \mathrm{Spec}\, R \,|\, P \not\subseteq A\} \to \mathrm{Spec}\, S$ given by $P \mapsto P \cap S$. This preserves idempotence and invertibility. Further:*
(a) If there is no nonzero prime ideal P of R with $P \subseteq A$, then there is only one nonzero prime of S not in the image of γ, namely A, which is idempotent.
(b) If there is a (necessarily unique) nonzero prime ideal $P \subseteq A$ then there are

exactly two nonzero primes of S not in the image of γ, A and, say, A'. Both are idempotent and A' is an isomaximal generative left ideal of R containing P.

Proof. Note first that since $RA = R$ and $AR = A$, then 3.6.2 gives a (1,1)-correspondence α between $\operatorname{Spec} R$ and $\{Q \in \operatorname{Spec} S \mid Q \not\supseteq A\}$ via $P \mapsto Q = \{s \in S \mid RsA \subseteq P\}$. The image of this correspondence is clearly $\operatorname{Spec} S \setminus A$, since A itself is a maximal ideal of S. Note also that since R/A is isotypic, there are at most two prime ideals inside A, namely 0 and $\operatorname{ann}(R/A)_R$. We let γ be the restriction of α to the set $\{P \in \operatorname{Spec} R \mid P \not\subseteq A\}$. Suppose P is in this set; then $P + A = R$ (or else P annihilates a factor of the isotypic module R/A and then $P \subseteq A$). In that case

$$RsA \subseteq P \Rightarrow Rs(A + P) \subseteq P \Rightarrow s \in P$$

and so $Q = P \cap S$ as claimed.

Note next that $AP \subseteq Q$. Therefore

$$P = RP = RAP \subseteq RQ \subseteq RP = P.$$

If $Q^2 = Q$, then $P^2 \supseteq RQ^2 = RQ = P$. On the other hand, if Q is invertible, with $QQ^{-1} = S$, then $PQ^{-1}R = R$ and so P is invertible.

(a) Here the domain of γ is $\operatorname{Spec} R \setminus 0$ and the image of γ is $\operatorname{Spec} S \setminus \{0, A\}$.

(b) This time the nonzero prime ideal $P \subseteq A$ is not included in the range of γ although the correspondence α maps P to $\{s \in P \mid RsA \subseteq P\} = A'$ say. The prime ideals A, A' are the only nonzero primes of S not in the image of γ.

It remains only to establish the properties of A'. However, since $P \subseteq \mathbb{I}(A)$ then, as is easily checked,

$$\mathbb{I}_{R/P}(A/P) \simeq \mathbb{I}(A)/P.$$

As noted in 6.2, R/P is a simple Artinian ring. Clearly A/P is a proper right ideal, and $A'/P = \operatorname{l}\operatorname{ann}_{R/P} A/P$ which is a nonzero left ideal. That implies that R/A' is isotypic, so A' is isomaximal. Moreover, since $0 \neq A'/P \lhd_1 R/P$, it follows that $(A'/P)(R/P) = R/P$. However, $A' \supset P$, and so $A'R = R$. Thus A' is generative, and

$$(A')^2 = A'(RA') = RA' = A'. \qquad \square$$

Corollary. *If R is a hereditary Noetherian prime ring with finitely many idempotent prime ideals and A is a generative isomaximal right ideal, then $\mathbb{I}(A)$ has finitely many idempotent prime ideals.*

Proof. It is clear that $\mathbb{I}(A)$ has at most two extra idempotent prime ideals. $\qquad \square$

6.12 Collecting the facts together establishes

Theorem. *The following conditions on a hereditary Noetherian prime ring S are equivalent*:
 (i) *S has finitely many idempotent ideals*;
 (ii) *S has finitely many idempotent prime ideals*;
 (iii) *S has a minimal nonzero idempotent ideal.*

Proof. (ii)⇒(i) Evident from 6.10 since nonzero prime ideals are maximal.
 (i)⇒(iii) Trivial.
 (iii)⇒(ii) By 6.8, S can be obtained from a Dedekind prime ring, R say, by a finite iteration of the process of forming idealizers of generative isomaximal right ideals. By 6.3, R has no nontrivial idempotent prime ideals. Therefore an induction argument, using Corollary 6.11, shows that S has finitely many idempotent prime ideals. ☐

6.13 Although, as noted in 6.9, this theory covers all hereditary Noetherian orders in central simple algebras, there are examples of hereditary Noetherian prime rings which do indeed have infinitely many idempotent ideals, see [Stafford and Warfield 85]

§7 Modules Over Dedekind Prime Rings

7.1 This section outlines the structure of finitely generated modules over a Dedekind prime ring. This follows a course remarkably similar to the commutative case (see 7.18), especially for the torsion-free modules and those torsion modules which, as in the commutative case, are unfaithful. However, the fact that a Dedekind prime ring can be simple shows that faithful torsion modules also exist. Indeed, it will become clear that the theory of torsion modules is largely concerned with the dichotomy between faithful and unfaithful, with the faithful modules providing some interesting contrasts with the commutative case.

7.2 It is perhaps worth noting first that both faithful and unfaithful simple modules can exist over one noncommutative Dedekind domain.

Example. *Let $R = D[x]$, where D is a division ring which is not algebraic over its centre k. (For example, take D to be the quotient division ring of $A_1(k)$, in characteristic zero.) Then R is a principal ideal domain, so is a Dedekind prime ring. Moreover, R has both faithful and unfaithful simple modules.*

Proof. Evidently R/xR is an unfaithful simple module. Choose $d \in D$ not algebraic over k, and consider $R/(x-d)R$. If this were unfaithful, its annihilator would be a nonzero ideal I contained in $(x-d)R$. Suppose $I = aR$; then a may be taken to be a monic polynomial. Since $fa - af \in I$ for all $f \in D$, it is clear that all the coefficients of a are central. However $x - d$ divides a; so d is algebraic over k, a contradiction. Hence $R/(x - d)R$ is faithful. □

7.3 The first results concern modules which have the property that every nonzero subfactor module is faithful, such modules being called **completely faithful**. Since every module over a simple ring has this property, the relevance of these results is not restricted to Dedekind prime rings.

Lemma. *Let M_R be a module of finite length over a ring R which is not right Artinian, and let $N \triangleleft M$ such that M/N is cyclic.*
(a) *If N is simple and $0 \to N \to M \to M/N \to 0$ is nonsplit then M is cyclic.*
(b) *If N is completely faithful then M is cyclic.*

Proof. (a) Choose any $m \in M$ such that $m + N$ generates M/N. If $mR \not\supseteq N$ then $mR \simeq M/N$ and this can be used to split the sequence. It follows that $mR \supseteq N$; so $mR = M$.

(b) This is proved by induction on the length of N. If $N = 0$, it is trivial. Otherwise, choose a simple submodule S of N. By induction M/S is cyclic, and if the sequence

$$0 \to S \to M \to M/S \to 0$$

is nonsplit, then by (a), M is cyclic.

So, suppose the sequence is split; and then $M \simeq S \oplus M/S$. Suppose $M/S \simeq R/I$, with $0 \neq I \triangleleft_r R$. (Here we used the non-Artinian hypothesis.) Since S is faithful and simple, $\bigcap \{K \mid K$ a maximal right ideal with $R/K \simeq S\} = 0$. Therefore there exists K with $R/K \simeq S$ and $K \not\supseteq I$. But then $K + I = R$ and so

$$R/I \cap K \simeq R/I \oplus R/K \simeq S \oplus M/S \simeq M. \qquad \square$$

Corollary. (i) *Any module of finite length over a simple non-Artinian ring is cyclic.*
(ii) *Any completely faithful module of finite length over a non-Artinian ring is cyclic.* □

7.4 We now consider a finitely generated module M over a hereditary Noetherian prime ring R. Recall from 2.1.17 that M has a torsion submodule $\tau(M)$, and that $M/\tau(M)$ is torsion-free and, by 3.4.2, embeds in $Q(R)^n$ for some n.

Lemma. *If M_R is finitely generated and R is a hereditary Noetherian prime ring, then $M \simeq \tau(M) \oplus M/\tau(M)$. Moreover, $\tau(M)$ has finite length and $M/\tau(M)$ is projective.*

Proof. Let $N = M/\tau(M)$. We may assume that $N \subseteq Q^n$. Since N_R is finitely generated and Q is the left quotient ring, there is a common left denominator

5.7.7   *Modules over Dedekind prime rings*

for those generators, c say. Then $N \simeq cN \subseteq R^n$. Hence N is projective. This shows that the epimorphism $M \to N$ splits; so $M \simeq \tau(M) \oplus M/\tau(M)$.

Note next that any cyclic torsion module is isomorphic to R/I for some essential right ideal and so, by 4.5, has finite length. Since $\tau(M)$ is finitely generated, then $\tau(M)$ has finite length also. □

7.5 The preceding result allows us to consider separately torsion modules and torsion-free modules. We concentrate attention first on the latter.

Lemma. *Let M_R be a finitely generated torsion-free module over a hereditary Noetherian prime ring R. Then $M \simeq \oplus U_i$ for some uniform right ideals U_i of R.*

Proof. We assume $M \neq 0$. If we can show that there is an epimorphism $M \to U$ for some uniform right ideal of R then, since U is projective, $M \simeq M' \oplus U$, and induction on u dim M will complete the proof.

To find U, note first that, by 3.4.2, $M \hookrightarrow M \otimes Q$, where $Q = Q(R)$, and by 3.4.3, u dim $M_R =$ u dim $(M \otimes Q)_Q$. One can choose a submodule N of $M \otimes Q$ with u dim $N =$ u dim $M - 1$, and then $M \otimes Q/N \simeq I$, some minimal right ideal of Q. So $M/M \cap N$ is isomorphic to a finitely generated uniform R-submodule U of I. There is a common left denominator, c say, for the generators of U, and then $U \simeq cU \subseteq R$. □

7.6 Proposition. *Let R be a Dedekind prime ring and let I, J, K be finitely generated projective modules with u dim $I =$ u dim K and $J \neq 0$. Then there exists $L \triangleleft J$ such that $I \oplus J \simeq K \oplus L$.*

Proof. It is easy to reduce to the case when I, J, K are uniform right ideals, using 7.5. Then, since all uniform right ideals are subisomorphic (3.3.4) we may assume that $I \subseteq K$. Let C be a complement to K in R; then $I \oplus C$ and $K \oplus C$ are both essential in R_R. By 4.5, $(K \oplus C)/(I \oplus C)$ has finite length, i.e. K/I has finite length, n say.

The proof now proceeds by induction on n. If $n = 1$, then K/I is simple. However, J is a generator, so there is a nonzero map, and hence an epimorphism, $J \to K/I$. If L is its kernel then, by Schanuel's lemma, 7.1.2, $I \oplus J \simeq K \oplus L$ as required.

Now suppose $n > 1$ and choose K' with $I \subset K' \subset K$ and K'/I simple. By induction $I \oplus J \simeq K' \oplus L'$ for some $L' \triangleleft J$; and also $K' \oplus L' \simeq K \oplus L$ for some $L \triangleleft L'$. □

7.7 Corollary. *Let R be a Dedekind prime ring and $K \triangleleft_r R$.*
(i) *If $K \triangleleft_e R$ then $K \oplus L \simeq R^2$ for some $L \triangleleft_e R$, and K has two generators.*
(ii) *If K is not essential, then $K \oplus L \simeq R$ for some $L \triangleleft_r R$, and K is principal.*

Proof. (i) We apply 7.6 to the case when $I = J = R$ to get $K \oplus L \simeq R^2$, with uniform dimension showing that $L \triangleleft_e R$. The epimorphism $R^2 \to K$ shows that K has two generators.

(ii) Let $\operatorname{u\,dim} K = k < \operatorname{u\,dim} R = n$. Using 7.5, one can write $R = I \oplus J$ with $\operatorname{u\,dim} I = k$ and $J \neq 0$. By 7.6, then $K \oplus L \simeq R$; so K is cyclic. □

The consequence of this corollary, that every right ideal has two generators, will be improved upon in 7.12.

7.8 The description of torsion-free modules is now possible.

Theorem. *Let R be a Dedekind prime ring and M_R be finitely generated torsion-free. Then $M \simeq R^n \oplus I$ for some n and some right ideal I of R.*

Proof. Grouping together the uniform right ideals given by 7.5 we can write $M = I_1 \oplus \cdots \oplus I_n \oplus J$ for some essential right ideals I_i and some nonzero right ideal J. Applying 7.6 repeatedly, we see that $I_1 \oplus I_2 \simeq R \oplus L_1$, $L_1 \oplus I_3 \simeq R \oplus L_2$ etc., with, finally, $L_{n-1} \oplus J \simeq R \oplus I$. Hence $M \simeq R^n \oplus I$. □

We note that the theorem cannot be extended to all hereditary Noetherian prime rings. This can be seen by choosing an idempotent ideal A of R and taking $M = A \oplus A$. Then $MA = M$, yet $RA \neq R$.

7.9 Corollary. *If R is a Dedekind prime ring then*

$$R \simeq \begin{bmatrix} D & \cdots & D & K \\ & \cdots & & \\ D & \cdots & D & K \\ K^{-1} & \cdots & K^{-1} & KK^{-1} \end{bmatrix},$$

where D is a noncommutative Dedekind domain and $0 \neq K \triangleleft_r D$.

Proof. We know, by 2.12, that $R \overset{M}{\sim} D$ for some such D. The right D-module which links R and D may be decomposed in the form $D^n \oplus K$ for some n with $0 \neq K \triangleleft_r D$. Since $R \simeq \operatorname{End}(D^n \oplus K)$ the result is clear. □

7.10 We now turn to torsion modules, starting by considering factor rings.

Theorem. *Let R be a Dedekind prime ring and $0 \neq T \triangleleft R$. Then:*
(i) R/T is an Artinian principal ideal ring;
(ii) if $I \triangleleft_e R$ then $I/IT \simeq R/T$.

Proof. We show first that (i) is a consequence of (ii). To see this, let $I \triangleleft_r R$ with

$I \supseteq T$. Then $I \lhd_e R$ and so, by (ii), I/IT is cyclic over R/T. Hence its homomorphic image I/T is also cyclic. This, together with 4.5, gives (i).

To prove (ii), note first that since $_R T_R$ is a progenerator then 3.5.7 shows that T/IT has the same (finite) length as R/I. Since $I + T/T \simeq I/I \cap T$ and $I + T/I \simeq T/I \cap T$ it follows that length $I/IT =$ length R/T.

Suppose, for a moment, that $T = M$, a maximal ideal. Then by 4.5 R/M is simple Artinian, and $I/IM \simeq R/M$, since their lengths are equal.

Suppose next that $T = M^n$; then R/M^n has Jacobson radical M/M^n and

$$(I/IM^n)/(I/IM^n)(M/M^n) \simeq I/IM$$

which we have seen is cyclic. Hence, by Nakayama's lemma, I/IM^n is cyclic.

Finally, in the general case, $T = M_1^{n_1} \cdots M_k^{n_k}$ say, and $R/T \simeq \oplus R/M_i^{n_i}$. Then $I/IT \simeq \oplus I/IM_i^{n_i}$, each summand of which is cyclic over $R/M_i^{n_i}$. It follows that I/IT is cyclic over R/T. Since they have the same length, then $I/IT \simeq R/T$. □

This result is not true for hereditary Noetherian prime rings. For example if

$$R = \begin{bmatrix} \mathbb{Z} & 2\mathbb{Z} \\ \mathbb{Z} & \mathbb{Z} \end{bmatrix} \quad \text{and} \quad T = \begin{bmatrix} 2\mathbb{Z} & 2\mathbb{Z} \\ 2\mathbb{Z} & 2\mathbb{Z} \end{bmatrix}$$

then

$$R/T \simeq \begin{bmatrix} \mathbb{Z}/2\mathbb{Z} & 0 \\ \mathbb{Z}/2\mathbb{Z} & \mathbb{Z}/2\mathbb{Z} \end{bmatrix}$$

which has the nonprincipal right ideal

$$\begin{bmatrix} \mathbb{Z}/2\mathbb{Z} & 0 \\ \mathbb{Z}/2\mathbb{Z} & 0 \end{bmatrix}.$$

7.11 The next result extends 7.10(i) to factors by right ideals.

Theorem. *Let R be a Dedekind prime ring and $J \subseteq I$ right ideals with* u dim $I =$ u dim J. *Then I/J is cyclic.*

Proof. By adding a complement to both I and J, we may suppose both to be essential in R. The length, n say, of I/J is finite, and the proof is by induction on n. If $n = 1$, it is trivial. If $n > 1$, we choose any simple submodule S of I/J; say $S = J'/J$. By induction I/J' is cyclic. Now Lemma 7.3 shows that I/J is cyclic too unless S is unfaithful and also the sequence

$$0 \to S \to I/J \to I/J' \to 0$$

splits. Thus if I/J is not cyclic then every simple submodule has a complement; and so I/J is a direct sum of unfaithful simple modules. Let $T = \mathrm{ann}\,(I/J)$, and note that $T \neq 0$ since it is the product of a finite number of nonzero primitive ideals. Now I/IT is cyclic, by 7.10, and so its factor I/J is too. □

7.12 It was shown, in 7.7, that non-essential right ideals of R are principal, and that essential right ideals have two generators. The next result shows that the first of the two generators is relatively arbitrary, this being indicated, sometimes, by saying that I has one and a half generators.

Corollary. *Let I be an essential right ideal of a Dedekind prime ring R and let $a \in I \cap \mathscr{C}_R(0)$. Then $I = aR + a'R$ for some $a' \in I \cap \mathscr{C}_R(0)$.*

Proof. Theorem 7.11 shows that $I = aR + a'R$ for some $a' \in I$. The fact that a' can be chosen from $\mathscr{C}_R(0)$ comes from 3.3.6. □

7.13 There is a further consequence of 7.11 which extends 7.6 in the case when $K \triangleleft_e R$. (This restriction is indeed necessary.)

Theorem. *If I, J, K are right ideals of a Dedekind prime ring with $J \triangleleft_e I$ and $K \triangleleft_e R$, then there exists $L \subseteq K$ such that $I/J \simeq K/L$.*

Proof. By adding a complement to both I and J we may assume both to be essential in R; and by multiplying both, on the left, by any element $c \in K \cap \mathscr{C}_R(0)$, we may assume that $J \subseteq I \subseteq K$. Note that $_SK_R$ is a progenerator where $S = O_l(K)$ which therefore is a Dedekind prime ring. We now transfer the problem to S by multiplying by K^{-1}. This gives $JK^{-1} \subseteq IK^{-1} \subseteq KK^{-1} = S$, and we can apply 7.11 to show that IK^{-1}/JK^{-1} is cyclic over S. Thus $IK^{-1}/JK^{-1} \simeq S/H$ say. Translating back to R by multiplying by K gives

$$I/J = IK^{-1}K/JK^{-1}K \simeq SK/HK = K/HK$$

and so we let $L = HK$. □

7.14 In order to complete the description of torsion modules two preliminary results are required. The result at which we aim decomposes such a module as the direct sum of a completely faithful module and an unfaithful module; and the preliminary results each take a step in that direction.

Lemma. *Let R be a Dedekind prime ring and let $K \subseteq I$ be essential right ideals with I/K unfaithful and R/I faithful and simple. Then $R/K \simeq R/I \oplus I/K$.*

Proof. Let $T = \operatorname{ann} I/K$. So $0 \neq T \triangleleft R$ and since R/I is faithful, $T \not\subseteq I$. Therefore $I + T = R$ and, *a fortiori*, $I + (T + K) = R$. We aim to show that $I \cap (T + K) = K$, the result then being clear.

Note first that $R/I \simeq T+K/I \cap (T+K)$ and that $I \cap (T+K) \supseteq K$. So if length $(T+K)/K \leqslant 1$ then $I \cap (T+K) = K$ as required. However, $(T+K)/K \simeq T/K \cap T$ and $IT \subseteq K \cap T$. The fact, seen in the proof of 7.10, that length T/IT = length R/I shows that length $T/K \cap T \leqslant 1$. □

7.15 Lemma. *Let R be a Dedekind prime ring, K an essential right ideal and T a nonzero ideal such that $K \subseteq T$ and T/K is completely faithful. Then $R/K \simeq R/T \oplus T/K$.*

Proof. Since $KT^{-1} \subseteq R$, it will be enough to show that $T + KT^{-1} = R$ and $T \cap KT^{-1} = K$. Now $T/T^2 + K$ is unfaithful, yet is a factor of T/K. Therefore $T = T^2 + K$ and so $R = T + KT^{-1}$. Also, $K \subseteq T \cap KT^{-1}$ and length T/K = length R/KT^{-1} = length $T/T \cap KT^{-1}$; so $T \cap KT^{-1} = K$. □

7.16 The main result about a finitely generated torsion module M can now be obtained. Recall, from 7.4, that M has finite length.

Theorem. *Any finitely generated torsion module M over a Dedekind prime ring R is the direct sum of a completely faithful module and an unfaithful module, both of finite length.*

Proof. The proof starts by showing that M is the extension of a completely faithful module by an unfaithful module. To see this, assume for the moment that every faithful finitely generated torsion module has a faithful simple submodule. Choose N to be maximal amongst the completely faithful submodules of M and suppose M/N is faithful. Then M/N has a faithful simple submodule N'/N say, and N' is completely faithful, a contradiction. Thus we may suppose that M is faithful and aim to show M has a faithful simple submodule. Now M has a faithful cyclic submodule, so we may suppose M cyclic. Furthermore, by induction on length, we may suppose that proper submodules of M are unfaithful. But then 7.14 applies.

That shows that M is the extension of a completely faithful submodule N by an unfaithful module U with $0 \neq \operatorname{ann} U = T$ say. There is a short exact sequence

$$0 \to L \to (R/T)^n \to U \to 0$$

for some n, which leads, see 7.1.12, to the exactness of

$$\operatorname{Hom}_R(L, N) \to \operatorname{Ext}_R^1(U, N) \to (\operatorname{Ext}_R^1(R/T, N))^n.$$

Now $\operatorname{Hom}(L, N)$ must be zero since L is unfaithful and N is completely faithful. Furthermore, 7.14 shows that $\operatorname{Ext}^1(R/T, N) = 0$, since, by Lemma 7.3(b), any extension of N by R/T must be cyclic. Hence $\operatorname{Ext}^1(U, N) = 0$ and so $M \simeq N \oplus U$. □

7.17 This has a familiar-looking consequence.

Corollary. *Every finitely generated torsion module over a Dedekind prime ring is a direct sum of cyclic submodules. Each indecomposable summand is either unfaithful with a unique composition series or else is completely faithful.*

Proof. It is known (see [Jacobson **43**, Theorems 43, 45]) that any finitely generated module over an Artinian principal ideal ring is a direct sum of cyclic modules with unique composition series. This can be applied to the unfaithful part of the module, using 7.10. Then Corollary 7.3(ii) completes the proof. □

7.18 It is salutary to compare these results with the well-known commutative case.

Proposition. *Let R be a commutative Dedekind domain and M be a nonzero finitely generated R-module. Then:*
 (i) *M is the direct sum of a torsion module and a torsion-free module;*
 (ii) *if M is torsion then M is a direct sum of indecomposable cyclic modules each of which has a unique composition series;*
 (iii) *if M is torsion-free then:*
  (a) $M \simeq I_1 \oplus \cdots \oplus I_n$ *for some nonzero ideals I_i of R where $n = \operatorname{u\,dim} M$;*
  (b) $M \simeq R^{n-1} \oplus I$ *with $I = I_1 I_2 \cdots I_n$;*
  (c) *if also $M \simeq R^m \oplus J$ with $0 \neq J \triangleleft R$ then $m = n - 1$ and $J \simeq I$.*

Proof. (i), (ii), (iii)(a) These are clear from 7.4, 7.17 and 7.5 respectively.
 (iii)(b) If $n = 2$ we apply 7.10 to show that $I_1/I_1 I_2 \simeq R/I_2$. Schanuel's lemma then shows that $I_1 \oplus I_2 \simeq R \oplus I_1 I_2$. Induction completes the argument.
 (iii)(c) Uniform dimension forces $m = n - 1$; but to show $I \simeq J$ one needs a determinant argument—see [Cohn **77**, 11.6]. □

We note two points where the commutative and noncommutative theories diverge. Over a noncommutative Dedekind domain, an indecomposable torsion module need not have a unique composition series. Examples of such modules over $A_1(k)$ and $B_1(k)$, with $\operatorname{char} k = 0$, are given in [McConnell and Robson **73**].

For torsion-free modules the analogue of (iii)(c) fails. Indeed, taking $R = A_1(k)$ as an example again, it will be shown in 7.11.6, 7.11.8 or 11.1.4(ii) that $R^2 \simeq R \oplus I$ for every nonzero right ideal I, yet I need not be cyclic. This is connected with K-theory and will be discussed in Chapters 11 and 12.

7.19 Finally, we apply these results to principal ideal rings. The structure of such a ring R is given by 3.4.9 and 4.4.19. This reduces the study to the two cases when R is Artinian or R is prime.

Corollary. *Each finitely generated module M over a principal ideal ring R is a direct sum of cyclic modules.*

Proof. The case when R is Artinian is dealt with by [Jacobson **43**, Theorem 43]. If R is prime then 2.11(i) shows that R is a Dedekind prime ring. Therefore, by 7.4, we need only consider the cases when M is torsion or torsion-free. The former case is covered by 7.17 and the latter by 7.5. □

Since $B_1(k)$ is a principal ideal domain, the example cited in 7.18 also applies here.

§8 Additional Remarks

8.0 There is a well-developed arithmetical ideal theory for orders in central simple algebras described in [Reiner **75**], [Roggenkamp and Huber-Dyson **70**] and [Roggenkamp **70**]; see also [Curtis and Reiner **81**]. The beginnings of the more general theory described here lie in [Chevalley **36**] and [Jacobson **43**, Chapter 6], the latter being an account of work of Asano (which was improved upon in [Asano **49**]).

8.1 (a) The general ideas here come from Asano, but 1.6 is due to [Chamarie **80**] and 1.11 to [Cozzens **76**]. A fuller account of maximal orders is in [Maury and Raynaud **80**].

(b) There is a notion of a noncommutative **Krull ring**; see, for example, [Chamarie **81**, **83**], [Marubayashi **75**] and [Stafford **85b**].

8.2 (a) The results in this section come largely from [Michler **69**] and [Robson **68**] although the approach via progenerators comes from [Robson **71**].

(b) Some further examples of noncommutative Dedekind domains are described in Chapter 7, §11.

8.3 Despite being 'well known' this account has some novel features, with the relationship between prime PI rings and orders over commutative integral domains being carefully delineated. This is made possible by definition 3.6 which is more general than is usual elsewhere; indeed the term K-order is often used for our classical orders.

8.4 (a) 4.2, 4.3 are from [Small **66a**, **67**] and 4.5, 4.6 are from [Chatters **71**, **72**].

(b) More generally, rings of finite global dimension are studied in Chapter 7.

8.5 (a) These results come from [Robson **72**]. There, however, the situation dealt with is more general with A being allowed to be **semimaximal**, i.e. R/A is semisimple. However, it is easy to see that $\mathbb{I}(A)$ can be obtained from R by iterating the process of forming idealizers at isomaximal right ideals, the number of iterations being the number of isomorphism types of simple submodules of R/A. So we are able to avoid the minor complications involved in that more general case.

(b) An alternative process, equivalent to the iteration mentioned in (a), entails idealizing several right ideals simultaneously. This gives a **multiple idealizer**; see [Ely **74**] and [Robson **75a**] for details.

8.6 Some of these results come from [Eisenbud and Robson **70**, **70a**]. The main thrust follows [Robson **72**], but with a substantially simplified proof of 6.7. For further results about the ideal structure of hereditary Noetherian prime rings and about localizations see also [Lenagan **73a**], [Kuzmanovich **72**] and [Goodearl **74a**].

8.7 (a) This section comes from [Eisenbud and Robson **70, 70a**] although the proof of 7.10 is from [Robson **71**]; see also [Goodearl and Warfield **79**] and [Hodges **P**].
   (b) Further details about principal ideal rings can be found in [Jategaonkar **68**].

Part II

DIMENSIONS

Chapter 6
KRULL DIMENSION

This chapter describes one of the dimensions which can be attached to any Noetherian ring or module, namely the Krull dimension. This is a measure of how close the ring or module is to being Artinian. In the case of commutative rings, the usual measure is the maximal length of a chain of prime ideals. However, for noncommutative rings this does not give a good estimate since, for example, there are many simple Noetherian rings which are far from being Artinian.

Instead, for noncommutative rings, the Krull dimension is measured on the lattice of right ideals. Thus, if that lattice is Artinian, the dimension is zero. Then dimension 1 means that any descending chain with non-Artinian factors must terminate.

It proves to be convenient to establish the very basic properties by starting in §1 with a measure, called the deviation, on a partially ordered set (poset, for short). This is applied to modules, via their lattices of submodules, in §2 and to rings in §3. One feature here is that the existence of the Krull dimension, as well as being a consequence of being Noetherian, is itself almost a Noetherian hypothesis.

The relationship with a dimension defined via chains of prime ideals is considered in §4. There is a class of rings, FBN rings, for which these two dimensions coincide; and their properties are also outlined. They share many of the features of commutative rings, and include Noetherian PI rings.

The next two sections estimate and calculate the dimensions of many of the rings introduced in Chapter 1. In §7 it is shown that the Krull dimension is related to stable bounds on numbers of generators of a module. Then, in §8, its relationship with localization is explored. Here, however, only a few features of the Krull dimension are required, and the more general notion of a dimension function is introduced. Finally, in §9, skew polynomials over commutative Noetherian rings are considered in more detail.

§1 Deviation of a Poset

1.1 If a, b belong to a poset A, and $a \geqslant b$, then we define $a/b = \{x \in A \mid a \geqslant x \geqslant b\}$. This is a subposet of A and is called the **factor** of a by b. By a **descending chain** $\{a_n\}$ of elements of A is meant that $a_1 \geqslant a_2 \geqslant \cdots \geqslant a_n \geqslant \cdots$; and the factors a_i/a_{i+1} are called the **factors** of the chain. The poset A is said to satisfy the **d.c.c.** provided that every descending chain in A is eventually constant.

For example, any well-ordered set satisfies the d.c.c. So too does any trivial poset, that being a poset which does not have two distinct, comparable elements.

1.2 We now define the **deviation** of a poset A, dev A for short. If A is trivial then dev $A = -\infty$. If A is nontrivial but satisfied the d.c.c. then dev $A = 0$. For a general ordinal α, we define dev $A = \alpha$ provided:
(i) dev $A \neq \beta < \alpha$;
(ii) in any descending chain of elements of A all but finitely many factors have deviation less than α.

It may help to describe posets of deviation less than α as being 'relatively trivial'. Then dev A is α if (i) A is not 'relatively trivial' but (ii) in any descending chain there are only finitely many factors which are not 'relatively trivial'.

1.3 Example. Consider the linearly ordered set $A = \{a_i \mid i \in \mathbb{N}\}$ in which $a_i > a_j$ if and only if $i < j$. The chain $a_1 > a_2 > \cdots > a_n > \cdots$ shows that dev $A \neq 0$; but any factor a_i/a_j, where $i < j$, has deviation 0. Hence dev $A = 1$.

1.4 It should be noted that a poset need not have a deviation, as we will show in 1.12. In any inequalities, suprema etc., a non-existent deviation will be regarded as greater than any ordinal.

1.5 The next few results relate the deviation of a poset with the deviation of certain subposets.

Lemma. *If B is a subposet of A then* dev $B \leqslant$ dev A.

Proof. This is clear if dev $A = -\infty$ or 0. We suppose it true for posets of deviation less than dev A. The factors of any chain in B are subposets of the factors of the same chain in A; hence all but finitely many have deviation less than dev A. □

1.3 Lemma. (i) *For any poset A,*

$$\sup\{\operatorname{dev}(a/b) \mid a, b \in A\} \leqslant \operatorname{dev} A \leqslant \sup\{\operatorname{dev}(a/b) + 1\}.$$

(ii) *If A has a least element, 0 say, then*

$$\operatorname{dev} A \leqslant \sup\{\operatorname{dev}(a/b) + 1 \mid b \neq 0\}.$$

Proof. (i) The first inequality follows immediately from 1.5. For the second, note that in any chain $\{a_i\}$ of A, each factor will have

$$\text{dev}(a_i/a_{i+1}) < \sup\{\text{dev}(a/b)+1\} = \alpha$$

say, so by definition, dev $A \leq \alpha$.

(ii) In any chain, there will be at most one factor a_i/a_{i+1} with $a_{i+1}=0$ and $a_i \neq 0$. So the result follows by the same argument as in (i). □

1.7 Proposition. *If* dev $A = \alpha$ *and* $\beta < \alpha$ *then there is a chain* $\{a_i | i \in \mathbb{N}\}$ *in A such that* dev $(a_i/a_{i+1}) = \beta$ *for each i.*

Proof. The proof is by induction on α, the case $\alpha = 0$ being clear. Note first that if, in every chain $\{a_i\}$ of A, one had dev $(a_i/a_{i+1}) < \beta$ for all but finitely many i, then by definition dev $A \leq \beta$. This contradiction shows that A contains a chain $\{a_i\}$ having infinitely many factors of deviation β or more. Passing to a subchain, and using 1.5, we may suppose that dev $(a_i/a_{i+1}) \geq \beta$ for all i.

Note next that dev $(a_i/a_{i+1}) < \alpha$ for all but finitely many i. So, passing once more to a subchain, we may suppose dev $(a_i/a_{i+1}) < \alpha$ for each i.

Now if $\alpha = \beta+1$, the chain $\{a_i\}$ is as required. On the other hand, if $\alpha > \beta+1$ the argument above, applied to $\beta+1$ rather than β, certainly gives a factor a/b such that $\beta+1 \leq \text{dev}(a/b) < \alpha$. The induction hypothesis gives a chain $\{a_i\}$ in a/b with dev $(a_i/a_{i+1}) = \beta$ for all i; and, of course, $\{a_i\}$ is also a chain in A. □

1.8 We now turn to some more examples of posets having, or failing to have, a deviation. The first example is, in fact, the link to Noetherian modules as will be seen in §2.

Proposition. *Any poset with a.c.c. has a deviation.*

Proof. By 1.6(i), if each factor a/b in A has a deviation, so too has A. Suppose, to the contrary, that some factor a/b fails to have a deviation. Choose b maximal with respect to this property. Applying 1.6(ii) to the poset a/b shows that dev a/b exists—a contradiction. □

The link shown here between the a.c.c. and deviation can be extended. One can define an ascending version of deviation, called **codeviation**, this being the deviation of the opposite poset. Thus codev $A = 0$ means that A satisfies the a.c.c. An induction, based on 1.8, shows that dev A exists if and only if codev A exists.

1.9 The example described in 1.3 is, the fact, the ordinal ω, but with the opposite ordering to the usual. For any ordinal α, by dev α we will mean the deviation of the opposite poset α^{op}. For example, $(\omega^n)^{\text{op}}$ is just $\mathbb{N}^n$ with the lexicographic ordering $(a_1,\ldots,a_n) > (b_1,\ldots,b_n)$ when, for some $i, a_1 = b_1, a_2 = b_2, \ldots, a_{i-1} = b_{i-1}$ and $a_i < b_i$.

Since α is well ordered, α^{op} satisfied the a.c.c. and so, by 1.8, has a deviation. One can specify this as follows. It is known, cf. [Sierpinski **58**], that any ordinal α has a canonical form

$$\alpha = \omega^{\gamma_1} n_1 + \cdots + \omega^{\gamma_t} n_t,$$

where $n_i \in \mathbb{N}$ and $\gamma_1 > \gamma_2 > \cdots > \gamma_t$. The ordinal γ_1 will be termed the **degree** of α, deg α.

Proposition. *For any ordinal α, dev α = deg α.*

Proof. We proceed by induction on α. Note first, however, that if α_1, α_2 are ordinals, then any chain $\{a_i | i \in \mathbb{N}\}$ in $\alpha_1 + \alpha_2$ has all but finitely many terms in one of the α_i. Hence it follows that

$$\text{dev}(\alpha_1 + \alpha_2) = \sup\{\text{dev}\,\alpha_1, \text{dev}\,\alpha_2\}.$$

Since of course $\deg(\alpha_1 + \alpha_2) = \sup\{\deg \alpha_1, \deg \alpha_2\}$ this means it is enough to prove the proposition for the case $\alpha = \omega^\gamma$, assuming it true for all smaller ordinals.

Note that, in any chain $\{a_i | i \in \mathbb{N}\}$ in ω^γ, all factors are smaller ordinals than ω^γ and so have degrees, and hence deviations, less than γ. Hence dev $\omega^\gamma \leq \gamma$.

To reverse the inequality, consider first the case when γ is a limit ordinal. For each $\delta < \gamma$, there is a factor of ω^γ isomorphic to ω^δ, so having deviation δ. By 1.5, dev $\omega^\gamma \geq \delta$; hence dev $\omega^\gamma \geq \gamma$.

Finally, suppose γ is a nonlimit ordinal, $\gamma = \delta + 1$ say. Then ω^γ has an infinite chain $\{\omega^\delta n | n \in \mathbb{N}\}$ whose factors are isomorphic to ω^δ and so have deviation δ. Hence dev $\omega^\gamma \geq \delta + 1 = \gamma$. □

This shows, of course, that any ordinal can occur as the deviation of a poset.

1.10 Certain ordinals can be used to characterize posets having finite deviations.

Proposition. *Let A be a poset, and n a positive integer. Then:*
(i) dev $A \geq n$ if and only if $(\omega^n)^{op}$ embeds in A;
(ii) dev $A = n$ if and only if $(\omega^n)^{op}$ embeds in A but $(\omega^{n+1})^{op}$ does not.

Proof. (i) By 1.9, dev $(\omega^n) = n$ so if $(\omega^n)^{op} \hookrightarrow A$ then dev $A \geq n$, by 1.5. Conversely, note that the result holds for $n = 1$. We suppose it holds for $n - 1$. By 1.7, A contains a chain $\{a_i\}$ with dev $a_i/a_{i+1} = n - 1$ for each i; so $(\omega^{n-1})^{op} \hookrightarrow a_i/a_{i+1}$ and hence $(\omega^n)^{op} \hookrightarrow A$.

(ii) This is immediate from (i). □

1.11 Proposition 1.10 cannot be extended to posets of infinite deviation. To see this we first note

6.1.14   Deviation of a poset

Lemma. *If A is the disjoint union of posets A_i, with only the partial ordering inherited from these, then*
$$\operatorname{dev} A = \sup \{\operatorname{dev} A_i\}.$$

Proof. Any chain in A is a chain in some A_i, and vice versa. □

Thus if we set $A = \bigcup \{(\omega^n)^{\operatorname{op}} | n \in \mathbb{N}\}$ then $\operatorname{dev} A = \omega$. But if α is an ordinal such that $\alpha^{\operatorname{op}} \hookrightarrow A$ then $\alpha \hookrightarrow \omega^n$ for some n; so $(\omega^\omega)^{\operatorname{op}} \not\hookrightarrow A$.

1.12 Next we describe a poset which fails to have a deviation; namely
$$D = \{m/n \in \mathbb{Q} | 0 \leqslant m/n \leqslant 1, n \text{ a power of } 2\}.$$
This can be constructed by repeated bisection, starting with $\{0,1\}$. Note that for any $0 \neq m/n \in D$, there is an isomorphism between D and $D \cap [m/2n, m/n]$ given by $d \mapsto (d+1)m/2n$. This makes it clear that D has an infinite chain $\{2^{-k} | k \in \mathbb{N}\}$ with factors of the same deviation as D. This contradicts the definition of deviation, unless $\operatorname{dev} D$ does not exist.

1.13 In fact there is a characterization of all posets which fail to have a deviation.

Proposition. *A poset A fails to have a deviation if and only if D embeds in A.*

Proof. One way is clear from the example above together with 1.5. Conversely, suppose $\operatorname{dev} A$ does not exist. Consider, nevertheless, the set of all ordinals which occur as $\operatorname{dev}(a/b)$ for some $a, b \in A$. Being a set, it has a supremum, γ say.

By definition, since $\operatorname{dev} A \not\leqslant \gamma + 1$, then A contains a chain $\{a_i\}$ having infinitely many factors whose deviations are not γ or less. That, of course, means that those factors have no deviation.

Consider one of those factors, a/b say. The argument above, applied to a/b, certainly implies that there is an element c in a/b such that both a/c and c/b fail to have a deviation. Repeated application of this to each factor occurring constructs a copy of D inside a/b. □

1.14 The last few results of this section calculate deviations of related posets.

First recall that if A, B are nonempty posets, then so too is $A \times B$ with the partial ordering given by $(a_1, b_1) \geqslant (a_2, b_2)$ if and only if $a_1 \geqslant a_2$ and $b_1 \geqslant b_2$.

Lemma. $\operatorname{dev}(A \times B) = \sup \{\operatorname{dev} A, \operatorname{dev} B\}$.

Proof. For any $b \in B$, $A \times B \supseteq A \times \{b\} \simeq A$. Hence by 1.5, $\operatorname{dev} A \leqslant \operatorname{dev} A \times B$. Therefore
$$\operatorname{dev} A \times B \geqslant \sup \{\operatorname{dev} A, \operatorname{dev} B\} = \gamma \text{ say.}$$

The reverse inequality is proved by induction on γ. If $\gamma = 0$ it is clearly true. In general, let
$$(a_1, b_1) \geq (a_2, b_2) \geq \cdots$$
be a chain in $A \times B$. For sufficiently large n, $\text{dev}(a_n/a_{n+1}) < \text{dev}\, A$ and $\text{dev}(b_n/b_{n+1}) < \text{dev}\, B$. Hence, by induction,
$$\text{dev}((a_n/a_{n+1}) \times (b_n/b_{n+1})) \leq \sup\{\text{dev}(a_n/a_{n+1}), \text{dev}(b_n/b_{n+1})\} < \gamma.$$
But
$$(a_n/a_{n+1}) \times (b_n/b_{n+1}) \simeq (a_n, b_n)/(a_{n+1}, b_{n+1}).$$
Hence $\text{dev}\, A \times B \leq \gamma$. $\square$

1.15 For some results the presence of a greatest element 1 and a least element 0 is important. If A fails to have these it is sometimes convenient to enlarge A by adjoining such elements, obtaining a new poset A'.

Lemma. *If* $\text{dev}\, A \geq 0$, *then* $\text{dev}\, A' = \text{dev}\, A$ (*but if* $\text{dev}\, A = -\infty$, *then* $\text{dev}\, A' = 0$).

Proof. Clear. $\square$

1.16 For use when considering polynomials of various types, we next consider two constructions based on a poset A. Let $S_c(A)$ denote the set of all eventually constant sequences $\{a_i | i \in \mathbb{N}\}$ of elements of A; that is, for $n \gg 0$, $a_n = a_{n+1} = \cdots = a_\infty$ say. This becomes a poset via the relation $\{a_i\} \geq \{b_i\}$ if and only if $a_i \geq b_i$ for all i.

The subposet of $S_c(A)$ consisting only of the ascending eventually constant sequences is denoted by $S_a(A)$.

Lemma. *For any poset A with 1 and 0,*
$$\text{dev}\, S_a(A) = \text{dev}\, S_c(A) = \text{dev}\, A + 1.$$

Proof. The result is clear if $\text{dev}\, A = -\infty$ (i.e. A is a singleton). Also since $S_a(A) \hookrightarrow S_c(A)$, then $\text{dev}\, S_a(A) \leq \text{dev}\, S_c(A)$. Next we show, by induction on $\text{dev}\, A$, that $\text{dev}\, S_c(A) \leq \text{dev}\, A + 1$.

Consider a decreasing sequence $\{s_n\}$ of elements of S_c, and say
$$s_n = \{a_1, a_2, \ldots\}, \quad s_{n+1} = \{b_1, b_2, \ldots\}.$$
The descending chain of limits of the s_n
$$\cdots \geq a_\infty \geq b_\infty \geq \cdots$$
shows that, for $n \gg 0$, $\text{dev}\, a_\infty/b_\infty < \text{dev}\, A$. Also, given such an n, we can choose $m \gg 0$ so that $a_m = a_\infty$ and $b_m = b_\infty$. But then it is easy to see that
$$s_n/s_{n+1} \simeq (a_1/b_1) \times (a_2/b_2) \times \cdots \times (a_{m-1}/b_{m-1}) \times S_c(a_\infty/b_\infty).$$

Now, by the induction hypothesis,
$$\operatorname{dev} S_c(a_\infty/b_\infty) \leqslant \operatorname{dev}(a_\infty/b_\infty) + 1 \leqslant \operatorname{dev} A + 1$$
Also, for $1 \leqslant i \leqslant m - 1$, one has $\operatorname{dev}(a_i/b_i) \leqslant \operatorname{dev} A < \operatorname{dev} A + 1$. Hence, by 1.14, $\operatorname{dev} s_n/s_{n+1} < \operatorname{dev} A + 1$. Thus $\operatorname{dev} S_c(A) \leqslant \operatorname{dev} A + 1$.

Finally, let $s_n \in S_a(A)$ be the sequence whose first n terms are 0, and all other terms are 1. Evidently $s_n/s_{n+1} \simeq A$ and so $\operatorname{dev} s_n/s_{n+1} = \operatorname{dev} A$. Therefore $\operatorname{dev} S_a(A) \geqslant \operatorname{dev} A + 1$. □

This result is not valid without the hypothesis that $0, 1 \in A$. Indeed one can check that, if A is taken to be the example described in 1.11, then
$$\operatorname{dev} S_a(A) = \operatorname{dev} S_c(A) = \operatorname{dev} A = \omega.$$

1.17 The final result of this section concerns posets linked by a poset map. The main applications will be made in the cases where γ is finite or $\delta = 0$ — in which cases the restriction $\gamma + \delta \leqslant \delta + \gamma$ is automatically satisfied.

Proposition. *Let $f : A \to B$ be a poset map, and let γ, δ be ordinals with $\gamma + \delta \leqslant \delta + \gamma$. Suppose that whenever $\operatorname{dev}(a_1/a_2) \geqslant \delta$ for $a_i \in A$, then $\operatorname{dev}(f(a_1)/f(a_2)) \geqslant \gamma$.*
(i) *If either there is no factor a_1/a_2 in A with $\operatorname{dev}(a_1/a_2) \geqslant \delta$ or there is no factor b_1/b_2 in B with $\operatorname{dev} b_1/b_2 \geqslant \gamma$, then $\operatorname{dev} A \leqslant \delta$.*
(ii) *Otherwise $\gamma + \operatorname{dev} A \leqslant \delta + \operatorname{dev} B$.*

Proof. (i) This is clear.

(ii) In this case $\operatorname{dev} A \geqslant \delta$ and $\operatorname{dev} B \geqslant \gamma$. We proceed by induction on $\operatorname{dev} B$, starting with the case when $\operatorname{dev} B = \gamma$. We claim $\operatorname{dev} A = \delta$, and so the inequality simply states that $\gamma + \delta \leqslant \delta + \gamma$ which, by hypothesis, is so. Suppose, to the contrary, that $\operatorname{dev} A > \delta$. By 1.7, one can choose a chain $\{a_i | i \in \mathbb{N}\} \subseteq A$ with $\operatorname{dev} a_i/a_{i+1} = \delta$ for each i. Then $\{f(a_i)\}$ is a chain in B with $\operatorname{dev} f(a_i)/f(a_{i+1}) \geqslant \gamma$; so $\operatorname{dev} B > \gamma$, a contradiction. So $\operatorname{dev} A = \delta$ as required.

Now suppose that $\operatorname{dev} B > \gamma$. Let β be any ordinal with $\beta < \operatorname{dev} A$ and, using 1.7, let $\{a_i\}$ be a descending chain in A with $\operatorname{dev}(a_i/a_{i+1}) = \beta$ for all i. Note that $\operatorname{dev} f(a_i)/f(a_{i+1}) < \operatorname{dev} B$ for $i \gg 0$. So, by the induction hypothesis, for $i \gg 0$,
$$\gamma + \beta \leqslant \delta + \operatorname{dev} f(a_i)/f(a_{i+1}) < \delta + \operatorname{dev} B.$$
Since this is true for all $\beta < \operatorname{dev} A$, then
$$\gamma + \operatorname{dev} A \leqslant \delta + \operatorname{dev} B. \qquad \square$$

The final deduction is dependent upon having the order of summation of γ and β as given. Examples exist to show the failure of the proposition if the condition on γ and δ is removed.

§2 Krull Dimension of Modules

2.1 The results of §1 apply to a module M_R by viewing as a poset $\mathscr{L}(M)$, the lattice of submodules of M. As well as making such applications, this section also pursues the analogy between satisfying the d.c.c. and having a deviation. In particular, analogues of simple modules, composition series and socles are introduced and studied. Applications to rings are deferred until the next section.

2.2 If M is a right R-module then the **Krull dimension** of M, written $\mathcal{K}(M)$, is defined to be the deviation of $\mathscr{L}(M)$, the lattice of submodules of M. In particular, $\mathcal{K}(R_R)$ is the **right Krull dimension** of R; this is often shortened to $r\mathcal{K}(R)$ or even $\mathcal{K}(R)$.

2.3 A module with Krull dimension is not necessarily Noetherian; for $\mathbb{Z}_{p^\infty}$ is an Artinian, but not Noetherian, $\mathbb{Z}$-module. The converse is true, however, as 1.8 shows.

Lemma. *If M_R is Noetherian then $\mathcal{K}(M)$ exists; and if R is right Noetherian then $\mathcal{K}(R_R)$ exists.* □

This implies that the results below, concerning modules with a Krull dimension, apply to all Noetherian modules. Those which are obviously true for Noetherian modules are indicated by '∗'.

2.4 Lemma. *If N_R is a submodule of M_R then*
$$\mathcal{K}(M) = \sup\{\mathcal{K}(N), \mathcal{K}(M/N)\}.$$

Proof. The natural maps $\mathscr{L}(N) \to \mathscr{L}(M)$ and $\mathscr{L}(M/N) \to \mathscr{L}(M)$ are poset maps preserving proper inclusions. Hence, by 1.17, $\mathcal{K}(M) \geq \sup\{\mathcal{K}(N), \mathcal{K}(M/N)\}$. On the other hand, for a submodule M' of M, the map
$$M' \mapsto (M' \cap N, M' + N/N)$$
gives a poset map
$$\mathscr{L}(M) \to \mathscr{L}(N) \times \mathscr{L}(M/N)$$
which preserves proper inclusions. However, by 1.14,
$$\operatorname{dev}(\mathscr{L}(N) \times \mathscr{L}(M/N)) = \sup\{\operatorname{dev}\mathscr{L}(N), \operatorname{dev}\mathscr{L}(M/N)\}$$
$$= \sup\{\mathcal{K}(N), \mathcal{K}(M/N)\}$$
and so, using 1.17 again, $\mathcal{K}(M) \leq \sup\{\mathcal{K}(N), \mathcal{K}(M/N)\}$. □

We note that there is no analogue of this result for posets—as is easily seen by considering the poset $D \cup \{x\}$ with $0, 1$ adjoined, D being as in 1.13. Then $\operatorname{dev} 1/x = \operatorname{dev} x/0 = 0$, whereas $\operatorname{dev} 1/0$ fails to exist.

2.5 Lemma. *If M_R is finitely generated, then $\mathcal{K}(M) \leqslant \mathcal{K}(R_R)$.*

Proof. It is clear from 2.4 that $\mathcal{K}(R^n) = \mathcal{K}(R)$. The epimorphism $R^n \to M$ shows then that $\mathcal{K}(M) \leqslant \mathcal{K}(R)$ by 2.4. □

Corollary. $\mathcal{K}(R_R) = \sup \{\mathcal{K}(M) | M_R \text{ finitely generated}\}$. □

2.6 The next two results show that having a Krull dimension implies properties of a Noetherian nature.

Lemma*. *A module with Krull dimension has finite uniform dimension.*

Proof. Amongst counter-examples, choose one, M say, of minimal Krull dimension, $\mathcal{K}(M) = \alpha$ say. So $M \supseteq \bigoplus_{i=1}^{\infty} N_i$ for nonzero submodules N_i. Set $M_n = \bigoplus_{j=1}^{\infty} N_{j \cdot 2^n}$. Then the chain $M_0 \supset M_1 \supset \cdots \supset M_n \supset \cdots$ has the property that each composition factor M_n/M_{n+1} has finite uniform dimension. The minimality of α, together with 2.4, shows that $\mathcal{K}(M_n/M_{n+1}) = \alpha$ for all n. But then, by definition, $\mathcal{K}(M) > \alpha$, a contradiction. □

2.7 Given any module M_R it is straightforward to define a polynomial module $M[x]$ over $R[x]$; indeed $M[x]$ is simply $M \otimes_R R[x]$. The usual proof of the Hilbert basis theorem shows easily that M_R is Noetherian if and only if $M[x]_{R[x]}$ is Noetherian.

Proposition. *$M[x]_{R[x]}$ has a Krull dimension if and only if M_R is Noetherian.*

Proof. If M is Noetherian then so is $M[x]$ and hence, by 2.3, $M[x]$ has Krull dimension.

Conversely, suppose that M is not Noetherian, and that $M_0 \subset M_1 \subset \cdots$ is an infinite ascending chain of submodules of M. Set $A = M_1 + M_2 x + M_3 x^2 + \cdots$ and $B = M_0 + M_1 x + M_2 x^2 + \ldots$. Then $A/B \simeq M_1/M_0 \oplus M_2/M_1 \oplus \cdots$ as an $R[x]$-module, with x acting as zero. This infinite direct sum shows, by 2.6, that $M[x]$ fails to have a Krull dimension. □

This implies, of course, that if $\mathcal{K}(R[x])$ exists then R must be right Noetherian.

2.8 The next result is often useful in calculations.

Lemma. *If M has a Krull dimension then $\mathcal{K}(M) \leqslant \sup \{\mathcal{K}(M/E) + 1 | E \text{ an essential submodule of } M\}$.*

Proof. Let α be the supremum. Now, given a chain $M = M_0 \supset M_1 \supset \ldots$, 2.6 shows that, for $n \gg 0$,

$$\operatorname{u dim} M_n = \operatorname{u dim} M_{n+t} \quad \text{for all } t > 0.$$

Choose a submodule L of M maximal with respect to $L \cap M_n = 0$. Now $L \oplus M_{n+t}$ is an essential submodule of M; but

$$M_n/M_{n+t} \simeq (L \oplus M_n)/(L \oplus M_{n+t}) \subseteq M/(L \oplus M_{n+t}).$$

Hence
$$\mathcal{K}(M_n/M_{n+t}) + 1 \leqslant \mathcal{K}(M/L \oplus M_{n+t}) + 1 \leqslant \alpha$$

and so $\mathcal{K}(M_n/M_{n+t}) < \alpha$. Therefore, by definition, $\mathcal{K}(M) \leqslant \alpha$. □

One immediate consequence is the following.

Corollary. *If R is a hereditary Noetherian prime ring then $\mathcal{K}(R) \leqslant 1$.*

Proof. By 5.4.5, $\mathcal{K}(R/E) \leqslant 0$ for all essential right ideals. □

2.9 Next we introduce the analogue of a simple module. A nonzero module M is called α-**critical**, for some ordinal α, provided that $\mathcal{K}(M) = \alpha$ and $\mathcal{K}(M/N) < \alpha$ for each nonzero submodule N. A **critical** module is one which is α-critical for some α.

Of course a 0-critical module is precisely a simple module. As another example, note that if R is a hereditary Noetherian domain but not a division ring then 5.4.5 shows that R_R is 1-critical.

The next few lemmas give some of the properties of critical modules.

2.10 Lemma. *Any nonzero module M with Krull dimension has a critical submodule.*

Proof. Amongst the nonzero submodules of M, choose one, N say, of minimal possible Krull dimension, $\mathcal{K}(N) = \alpha$ say. If N is not α-critical, then it has a nonzero submodule N_1 with $\mathcal{K}(N/N_1) = \alpha$. However, the minimality of α, and 2.4, implies that $\mathcal{K}(N_1) = \alpha$. Applying the same argument to N_1 etc., we obtain a chain $N \supset N_1 \supset N_2 \supset \cdots$ with $\mathcal{K}(N_i/N_{i+1}) = \alpha$ for each i. Since $\mathcal{K}(N) = \alpha$, this chain must terminate. But the process only halts when it reaches an α-critical submodule of N. □

2.11 Lemma. *Any nonzero submodule N of an α-critical module M is α-critical.*

Proof. Let N_1 be a nonzero submodule of N. Then
$$\mathcal{K}(N/N_1) \leqslant \mathcal{K}(M/N_1) < \alpha.$$
Also, since $\mathcal{K}(M) = \alpha = \sup\{\mathcal{K}(M/N), \mathcal{K}(N)\}$ and $\mathcal{K}(M/N) < \alpha$, then $\mathcal{K}(N) = \alpha$. □

2.12 Lemma. *A critical module M is uniform.*

Proof. Suppose $N_1 \oplus N_2 \subseteq M$ with $N_i \neq 0$ and M α-critical. Then, by 2.11, $\mathcal{K}(N_1) = \mathcal{K}(N_2) = \alpha$. But $N_1 \hookrightarrow M/N_2$ and so, by 2.4, $\mathcal{K}(M/N_2) \geq \alpha$, a contradiction. □

2.13 There is a connection with monoform modules (see 4.6.4).

Lemma. *If M is α-critical then M is monoform.*

Proof. Let $0 \neq \theta : N \to M$ for some $N \triangleleft M$. Then $\mathcal{K}(\text{im } \theta) = \alpha$ yet N is α-critical. Therefore $\ker \theta = 0$ □

2.14 Next we turn to the analogue of the socle, that being the sum of the simple submodules. Let M be a module with Krull dimension and suppose that, amongst nonzero submodules of M, the minimal Krull dimension possible is β. By 2.10, M must have a β-critical submodule. The sum of all β-critical submodules of M is termed the **Krull socle** of M, written $S(M)$ or $S_1(M)$. Then, proceeding inductively, we define $S_{\gamma+1}(M)$ to be the inverse image in M of $S(M/S_\gamma(M))$ and $S_\delta(M) = \bigcup_{\gamma < \delta} S_\gamma(M)$ for any limit ordinal δ. The sequence $\{S_\gamma\}$ is called the **Krull socle sequence** of M.

2.15 An immediate consequence of 2.10 is

Lemma. *If M has Krull dimension, then $M = S_\varepsilon(M)$ for some ordinal ε.* □

Of course, if M is Noetherian, $M = S_n(M)$ for some integer n.

2.16 The socle series is an ascending chain, whereas Krull dimension concerns descending chains. The next result links two such chains.

Lemma*. *Suppose that a module M has infinite chains of submodules*
$$M = M_0 \supseteq M_1 \supseteq M_2 \supseteq \cdots \quad \text{and} \quad 0 = A_0 \subseteq A_1 \subseteq A_2 \subseteq \cdots$$
such that, for each $i \geq 0$, $M_i \cap A_{i+1} \not\subseteq M_{i+1} + A_i$. Then M does not have a Krull dimension.

Proof. Let $B = \sum_{i=1}^\infty M_i \cap A_i$. Then $B \subseteq M_{n+1} + A_n$ for each n, and so $M_n \cap A_{n+1} \not\subseteq B$. However,
$$\sum_{i=1}^{n-1} (M_i \cap A_{i+1}) \subseteq A_n$$
and hence
$$\left(\sum_{i=1}^{n-1} (M_i \cap A_{i+1}) \right) \cap (M_n \cap A_{n+1}) \subseteq M_n \cap A_n \subseteq B.$$

Therefore $\sum_{i=1}^{\infty}(M_i \cap A_{i+1})/B$ is an infinite direct sum. Hence, by 2.4 and 2.6, M does not have a Krull dimension. □

2.17 Lemma*. *Let M have Krull dimension and also be the sum of submodules each of which has Krull dimension $\leq \alpha$. Then $\mathcal{K}(M) \leq \alpha$.*

Proof. Every nonzero factor of M contains a nonzero submodule of Krull dimension $\leq \alpha$. Thus we can construct an ascending chain of submodules $\{A_\lambda | \lambda \leq \gamma\}$ with $M = A_\gamma$ and, for each λ, $\mathcal{K}(A_{\lambda+1}/A_\lambda) \leq \alpha$; and $A_\lambda = \bigcup_{\delta < \lambda} A_\delta$ at limit ordinal λ. If $\mathcal{K}(M) \not\leq \alpha$, choose λ to be the least ordinal such that $\mathcal{K}(A_\lambda) \not\leq \alpha$. Without loss, we suppose $M = A_\lambda$. It is evident that λ must be a limit ordinal, for otherwise

$$\mathcal{K}(A_\lambda) = \sup\{\mathcal{K}(A_{\lambda-1}), \mathcal{K}(A_\lambda/A_{\lambda-1})\} \leq \alpha.$$

Since we have assumed $\mathcal{K}(M) \not\leq \alpha$, there must be a chain $M = M_0 \supset M_1 \supset \cdots$ such that $\mathcal{K}(M_i/M_{i+1}) \geq \alpha$ for all i. We aim to delete some M_i and some A_δ from the two chains so as to establish the conditions described in 2.16. This will produce a contradiction, and complete the proof.

Note first that $M \not\subseteq M_1$ and so, for some $\delta < \lambda$, $A_\delta \not\subseteq M_1$. Without loss, suppose $A_\delta = A_1$. Then $A_1 \not\subseteq M_1$ and so, setting $A_0 = 0$, we have $M_0 \cap A_1 \not\subseteq A_0 + M_1$.

Suppose next that, for $0 \leq i \leq n-1$, we have $M_i \cap A_{i+1} \not\subseteq A_i + M_{i+1}$. Now if $M_n \subseteq A_n + M_{n+j}$ for all $j > 0$, then $M_{n+j-1} \subseteq A_n + M_{n+j}$ and so

$$M_{n+j-1} = (M_{n+j-1} \cap A_n) + M_{n+j}.$$

Therefore

$$M_{n+j-1}/M_{n+j} \simeq (M_{n+j-1} \cap A_n)/(M_{n+j} \cap A_n)$$

and so all the composition factors in the chain

$$A_n \cap M_n \supset A_n \cap M_{n+1} \supset A_n \cap M_{n+2} \supset \cdots$$

have Krull dimensions at least α. But $\mathcal{K}(A_n) \leq \alpha$, a contradiction.

We deduce that, for some j, $M_n \not\subseteq A_n + M_{n+j}$ and, without loss, suppose $j = 1$. But $M_n = M_n \cap (\bigcup A_\delta)$ and so there is a $\delta > n$ such that $M_n \cap A_\delta \not\subseteq A_n + M_{n+1}$. Without loss, suppose $\delta = n + 1$. Then $M_n \cap A_{n+1} \not\subseteq A_n + M_{n+1}$, as required. □

2.18 For a module M with Krull dimension we define the α-**torsion submodule** to be

$$\tau_\alpha(M) = \sum\{A | A \subseteq M, \mathcal{K}(A) \leq \alpha\}.$$

The result above has an immediate consequence, noting, for (ii), that M is the sum of its cyclic submodules.

6.2.21 *Krull dimension of modules*

Corollary. *If $\mathcal{K}(M)$ exists then*
(i) $\mathcal{K}(\tau_\alpha(M)) \leq \alpha$, *and*
(ii) $\mathcal{K}(M) \leq \mathcal{K}(R_R)$. □

2.19 The final notion introduced in this section is that of a **critical composition series** for a module M. This is defined to be a chain
$$M = M_0 \supset M_1 \supset \cdots \supset M_n = 0$$
such that each factor M_i/M_{i+1} is critical and, moreover, for each i,
$$\mathcal{K}(M_i/M_{i+1}) \geq \mathcal{K}(M_{i+1}/M_{i+2}).$$
It is immediate from 2.4 that, for each i, $\mathcal{K}(M_i/M_{i+1}) = \mathcal{K}(M_i)$.

For example a composition series of any module of finite length is also a critical composition series. Also, following the comment in 2.9, if R is a hereditary Noetherian domain then $R \supset 0$ is a critical composition series.

2.20 Not every module with Krull dimension has a critical composition series, as $\mathbb{Z}_{p^\infty}$ shows. We now aim to show that a Noetherian module does.

Lemma. *Let M_R be Noetherian and $\beta \leq \mathcal{K}(M)$. Then there exists $M' \subset M$ with M/M' a β-critical module.*

Proof. Choose $M' \subseteq M$ maximal with respect to $\mathcal{K}(M/M') \geq \beta$. Each proper factor of M/M' must have Krull dimension less then β. So, in any descending chain of submodules of M/M', each composition factor, except possibly one, has Krull dimension less then β. Therefore $\mathcal{K}(M/M') \leq \beta$. Clearly then M/M' is β-critical. □

Proposition. *A Noetherian module M has a critical composition series.*

Proof. Set $M = M_0$ and choose $M_1, M_2, \ldots$ as follows. If $\mathcal{K}(M_i) = \alpha_i$, then choose $M_{i+1} \subset M_i$ maximal with respect to $\mathcal{K}(M_i/M_{i+1}) = \alpha_i$. By the lemma, M_i/M_{i+1} is α_i-critical; and
$$\alpha_i = \mathcal{K}(M_i/M_{i+1}) = \mathcal{K}(M_i) \leq \mathcal{K}(M_{i-1}) = \mathcal{K}(M_{i-1}/M_i).$$
Thus it now remains only to check that some $M_n = 0$. Suppose not. The descending sequence of ordinals $\{\alpha_i\}$ must stabilize, say at α_m. Then $\alpha_m = \mathcal{K}(M_m) = \mathcal{K}(M_i/M_{i+1})$ for all $i \geq m$. The chain $M_m \supset M_{m+1} \supset \cdots$ demonstrates that $\mathcal{K}(M_m) > \alpha_m$, a contradiction. □

2.21 As with the usual composition series, there is some uniqueness about the composition factors.

Proposition. *Any two critical composition series of a Noetherian module M have the same length; and their composition factors can be paired so that corresponding factors have an isomorphic nonzero submodule.*

Proof. Let $\mathcal{K}(M) = \alpha$, and let
$$M = M_0 \supset M_1 \supset \cdots \supset M_s = 0 \quad \text{and} \quad M = N_0 \supset N_1 \supset \cdots \supset N_t = 0$$
be the two critical composition series. We proceed by induction on s. If $s = 0$ or 1, the result is immediate from definitions.

Next, we consider the submodules shown in the diagram below.

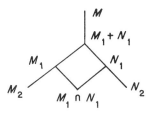

First, suppose that $M_1/M_1 \cap N_1 = 0$; i.e. $M_1 \subseteq N_1$. But M/M_1 is α-critical and so too is M/N_1. Therefore $M_1 = N_1$. Applying the induction hypothesis to the composition series
$$M_1 \supset M_2 \supset \cdots \supset M_s = 0 \quad \text{and} \quad M_1 = N_1 \supset N_2 \supset \cdots \supset N_t = 0$$
the result follows. A symmetric argument deals with the case when $N_1/M_1 \cap N_1 = 0$.

So, we may suppose $M_1/M_1 \cap N_1 \neq 0$ and $N_1/M_1 \cap N_1 \neq 0$. But then, by 2.11, each of these modules is α-critical. We can form a new critical composition series for M_1 starting $M_1 \supset M_1 \cap N_1 \supset \ldots$. The induction hypothesis applied to M_1 shows this must have length $s - 1$. This series also serves to give a composition series for N_1, again of length $s - 1$. The induction hypothesis, applied to N_1, shows that $s - 1 = t - 1$, and so $s = t$. Finally the claimed pairings can be established inductively, bearing in mind that M/M_1 and $M_1/M_1 \cap N_1$ can be paired, respectively, with $N_1/M_1 \cap N_1$ and M/N_1. □

2.22 Finally, as an illustration of the notions introduced in this section, suppose that M is a module whose lattice of nonzero submodules is isomorphic to α^{op} for some ordinal α, with
$$\alpha = \omega^{\gamma_1} n_1 + \omega^{\gamma_2} n_2 + \cdots + \omega^{\gamma_t} n_t$$
as described in 1.9. By 1.9, $\mathcal{K}(M) = \gamma_1$ (and M will be γ_1-critical only if $\alpha = \omega^{\gamma_1}$). The Krull socle $S(M)$ corresponds to the ordinal ω^{γ_t} and $\mathcal{K}(S(M)) = \gamma_t$. Also $\tau_{\gamma_t} M$ corresponds to $\omega^{\gamma_t} n_t$. Furthermore, the unique critical composition series for M will have $\sum_{i=1}^{t} n_i$ terms with, for example, M_1 corresponding to the ordinal
$$\omega^{\gamma_1}(n_1 - 1) + \omega^{\gamma_2} n_2 + \cdots + \omega^{\gamma_t} n_t.$$

§3 Krull Dimension in Rings

3.1 We now turn to Krull dimension in rings. Many results are for later application to Noetherian rings; but, as in §2, there are some results, marked with '*', which specifically concern the non-Noetherian case. These show that having a Krull dimension implies consequences associated with Noetherian rings. For example, 3.7 will show that nil subrings are nilpotent, thus extending 2.3.7.

3.2 Before proceeding further it is well to see an example of a non-Noetherian ring with Krull dimension. Note first that if $B \triangleleft R$ and M is any right R-module then M/MB has the same lattice of submodules over R as over R/B. Hence the notation $\mathcal{K}(M/MB)$ is unambiguous.

We apply this remark to the ring

$$R = \left\{ \begin{bmatrix} a & b \\ 0 & a \end{bmatrix} \middle| a \in \mathbb{Z}, b \in \mathbb{Z}_{p^\infty} \right\}$$

and the ideal

$$B = \left\{ \begin{bmatrix} 0 & b \\ 0 & 0 \end{bmatrix} \middle| b \in \mathbb{Z}_{p^\infty} \right\}.$$

As noted in 2.3, B is not a Noetherian $\mathbb{Z}$-module and yet $\mathscr{L}(B_R) = \mathscr{L}(B_\mathbb{Z})$, since $B^2 = 0$ and $\mathbb{Z} \simeq R/B$. Therefore R is not Noetherian. On the other hand, since $\mathcal{K}(R/B) = \mathcal{K}(\mathbb{Z}) = 1$ (as is clear—or use Corollary 2.8) and $\mathcal{K}(B) = 0$, then, by 2.4, $\mathcal{K}(R_R) = 1$.

3.3 Lemma. *If $A \triangleleft_r R$ and $B \triangleleft R$ and $\mathcal{K}(R/AB)$ exists, then*

$$\mathcal{K}(R/AB) = \sup\{\mathcal{K}(R/A), \mathcal{K}(R/B)\}.$$

Proof. It is clear from 2.4 that

$$\mathcal{K}(R/AB) \geqslant \sup\{\mathcal{K}(R/A), \mathcal{K}(R/B)\}.$$

On the other hand, suppose $\mathcal{K}(R/AB)$ exists. Now A/AB is an R/B-module. Therefore, by 2.18, $\mathcal{K}(A/AB) \leqslant \mathcal{K}(R/B)$. Applying 2.4 to the short exact sequence

$$0 \to A/AB \to R/AB \to R/A \to 0$$

we have

$$\mathcal{K}(R/AB) = \sup\{\mathcal{K}(A/AB), \mathcal{K}(R/A)\}$$
$$\leqslant \sup\{\mathcal{K}(R/B), \mathcal{K}(R/A)\}. \qquad \square$$

3.4 By a **critical right ideal** is meant a right ideal which is a critical module.

Lemma. *Let R be a ring with right Krull dimension and let C be a critical right ideal. Then either $C^2 = 0$ or else there exists $0 \neq c \in C$ such that $\operatorname{r ann} c \cap C = 0$.*

Proof. If $C^2 \neq 0$, choose $0 \neq c \in C$ with $cC \neq 0$. The homomorphism $C \to C$ given by left multiplication by c is a monomorphism, by 2.13. □

3.5 Proposition*. *A semiprime ring R with Krull dimension is a right Goldie ring.*

Proof. By 2.6, R_R has finite uniform dimension. So, by 2.3.6, it is enough to show that the right singular ideal $\zeta(R)$ is zero. If not, $\zeta(R)$ contains a nonzero critical right ideal C, by 2.10; and since R is semiprime, $C^2 \neq 0$. Hence there is $0 \neq c \in C$ with $\operatorname{r ann} c \cap C = 0$. However this contradicts $\operatorname{r ann} c$ being essential. □

The converse is not true. For example consider the ring $R = k[X]$, polynomials in an infinite set X of central indeterminates over a field k. Of course R is not Noetherian although, being a commutative integral domain it is semiprime Goldie. On the other hand, $R \simeq R[x]$ and so, by 2.7, R cannot have Krull dimension.

3.6 Next we consider nil subrings.

Lemma*. *If R is a ring with right Krull dimension then there is no nonzero idempotent ideal $I \subseteq J(R)$.*

Proof. Let I be an idempotent ideal with $I \subseteq J(R)$. Then $MI = 0$ for all simple R-modules M. We aim to show that $MI = 0$ for all modules M with Krull dimension, by induction on $\mathcal{K}(M)$, and then deduce that $RI = 0 = I$.

To this end, consider the Krull socle sequence $\{S_\gamma(M)\}$ for M, as described in 2.14, and assume, for the moment, that $NI = 0$ whenever N_R is a critical module. Evidently then $S_1(M)I = 0$ and so, since I is idempotent, $S_n(M)I = 0$ for all n. But then $S_\omega(M)I = 0$. Transfinite induction completes the proof.

So, it is enough to prove the claim for an α-critical module M assuming, by induction, that it holds for all modules of smaller Krull dimension. However given any nonzero submodule N of M, $\mathcal{K}(M/N) < \alpha$. Therefore $(M/N)I = 0$ and so $MI \subseteq N$. Hence if $MI \neq 0$, then MI is the unique minimal nonzero submodule of M. But then $(MI)I = MI = 0$. □

3.7 Theorem*. *In a ring R with right Krull dimension, nil subrings are nilpotent.*

Proof. The argument used in the proof of 2.3.7 can be applied here, using 3.5, to reduce the problem to proving that $N = N(R)$ is nilpotent.

Using Zorn's lemma, we can choose a chain of ideals of R

$$0 = A_0 \subseteq A_1 \subseteq A_2 \subseteq \cdots \subseteq N$$

in which A_{j+1} is chosen maximal with respect to having $A_{j+1}^2 \subseteq A_j$. Let $I = \bigcup A_j$.

Suppose for a moment that, for each $n, j \in \mathbb{N}$, $I^n \not\subseteq I^{n+1} + A_j$. Then a subsequence $\{B_n\}$ of $\{A_j\}$ can be chosen as follows. If $B_0, \ldots, B_n$ are already chosen, note that $I^n \not\subseteq I^{n+1} + B_n$. However, $\bigcup A_j = I$ and so $I^n \cap A_j \not\subseteq I^{n+1} + B_n$ for some j. Set $B_{n+1} = A_j$. Now 2.16 can be applied to the chains $\{I^n\}$, $\{B_n\}$ giving the contradiction that I does not have Krull dimension.

Therefore, for some n, j, $I^n \subseteq I^{n+1} + A_j$. Hence $(I^n + A_j)/A_j$ is an idempotent ideal contained in N/A_j and so in $J(R/A_j)$. By 3.6, $I^n \subseteq A_j$. It follows that $I^{n-i} \subseteq A_{j+i}$ and so $I = A_k$ for $k = j + n$.

Therefore, letting $\bar{R} = R/A_k$, we see that $\bar{N} = N(\bar{R})$ contains no nonzero ideal A with $A^2 = 0$. However, by 2.10, if $\bar{N} \neq 0$ then $\bar{N}$ contains a critical right ideal, C say. By 0.2.6, $\bar{N}$ is nil; so if $0 \neq c \in C$ then $\operatorname{r\,ann} c \cap C \neq 0$. By 3.4, $C^2 = 0$ and so $(RC)^2 = 0$.

Hence $\bar{N} = 0$, $N = A_k$, and N is nilpotent. □

3.8 Corollary. *Let R be a ring with right Krull dimension. Then:*
(i) *the prime radical N of R is nilpotent;*
(ii) *N is a finite intersection of minimal prime ideals, $P_1, \ldots, P_m$ say;*
(iii) *$\mathcal{K}(R_R) = \sup\{\mathcal{K}(R/P) \mid P \in \operatorname{Spec} R\} = \sup\{\mathcal{K}(R/P_i) \mid i = 1, \ldots, m\} = \mathcal{K}(R/N)$.*

Proof. (i) Clear from 3.7.
(ii) This is a consequence of 3.2.2 since, by 3.5, R/N is semiprime right Goldie.
(iii) $\bigcap_{i=1}^m P_i = N$; so some product of the P_i is zero. Hence, applying 3.3,

$$\mathcal{K}(R) = \sup\{\mathcal{K}(R/P_i) \mid i = 1, \ldots, m\} = \sup\{\mathcal{K}(R/P) \mid P \in \operatorname{Spec} R\}.$$

Likewise $\mathcal{K}(R) = \mathcal{K}(R/N)$. □

3.9 The last few results of this section are concerned with the relationship between the Krull dimension of a ring and that of certain modules.

Lemma. *Let R be a ring with right Krull dimension, and $c \in \mathscr{C}(0)$. Then $\mathcal{K}(R/cR) < \mathcal{K}(R)$.*

Proof. In the chain $\{c^n R \mid n \in \mathbb{N}\}$ the composition factors are all isomorphic to R/cR. Hence, by the definition, $\mathcal{K}(R) > \mathcal{K}(R/cR)$. □

3.10 This has particular application when R is semiprime since then, by 2.3.5, each essential right ideal of R contains an element of $\mathscr{C}(0)$.

Proposition. *Let R be a semiprime ring with right Krull dimension. Then:*
(i) $\mathcal{K}(R) = \sup\{\mathcal{K}(R/E) + 1 \mid E \triangleleft_e R_R\}$;
(ii) $\mathcal{K}(R) = \mathcal{K}(E)$ *for each* $E \triangleleft_e R_R$;
(iii) *critical right ideals are precisely uniform right ideals*;
(iv) $\mathcal{K}(R) = \sup\{\mathcal{K}(C) \mid C \text{ a critical right ideal}\}$.

Proof. (i) Let the supremum be α. By 2.8, $\mathcal{K}(R) \leq \alpha$. On the other hand if $E \triangleleft_e R$ then, as noted above, E contains a regular element, c say. Thus

$$\mathcal{K}(R) \geq \mathcal{K}(R/cR) + 1 \geq \mathcal{K}(R/E) + 1$$

using 3.9 and 2.4.

(ii) This is now clear from 2.4.

(iii) By 2.12, each critical right ideal is uniform. Conversely, suppose that U is a uniform right ideal. By 2.10, U contains a γ-critical right ideal C, for some γ. However, by Corollary 3.3.3, C contains an isomorphic copy of U. Hence, by 2.11, U is γ-critical.

(iv) Given C there is, by 2.2.8, a finite direct sum of uniform right ideals $C_1 \oplus \cdots \oplus C_n$ essential in R with $C = C_1$. By 2.4 and (ii)

$$\mathcal{K}(R) = \mathcal{K}(C_1 \oplus \cdots \oplus C_n) = \sup\{\mathcal{K}(C_i)\}. \qquad \square$$

3.11 For prime rings even stronger results hold.

Proposition. *Let R be a prime ring with right Krull dimension. Then:*
(i) $\mathcal{K}(R) = \mathcal{K}(A)$ *for each nonzero right ideal A*;
(ii) $\mathcal{K}(R/B) < \mathcal{K}(R)$ *for any $0 \neq B \triangleleft R$*;
(iii) *if M_R is finitely generated, then $\mathcal{K}(M) < \mathcal{K}(R)$ if and only if M is a torsion module* (cf. 2.1.17).

Proof. (i) By 3.3.4, all uniform right ideals U are subisomorphic. Hence, by 3.10, $\mathcal{K}(R) = \mathcal{K}(U)$. Since A contains a uniform right ideal, $\mathcal{K}(R) = \mathcal{K}(A)$.

(ii) This follows from 3.10 since $B \triangleleft_e R_R$ by 2.2.1.

(iii) If M is not torsion then there exists $x \in M$ with $\operatorname{r} \operatorname{ann} x$ not essential. Choose $0 \neq A \triangleleft_r R$ with $A \cap \operatorname{r} \operatorname{ann} x = 0$. Then $M \supseteq xA \simeq A$. Hence, by (i) and 2.5, $\mathcal{K}(M) = \mathcal{K}(R)$.

Conversely, if M is torsion then $\mathcal{K}(xR) < \mathcal{K}(R)$ for all $x \in M$, by 3.10(i). Letting x vary through the generators of M shows that $\mathcal{K}(M) < \mathcal{K}(R)$. $\qquad \square$

3.12 The final results allow the ring to be more general, but concentrate on more special ideals and modules.

Lemma. *Let R, S be rings and $_S M_R$ a bimodule such that $_S M$ is finitely generated, M_R is faithful and $\mathcal{K}(M_R)$ exists. Then $\mathcal{K}(M_R) = \mathcal{K}(R_R)$.*

Proof. Suppose $M = \sum_{i=1}^{t} Sm_i$. Then $\bigcap r \operatorname{ann} m_i = \operatorname{ann} M = 0$, and so $R \hookrightarrow M^t$ via the map $r \mapsto (m_1 r, \ldots, m_t r)$. Hence, by 2.4, $\mathcal{K}(R_R) \leq \mathcal{K}(M_R)$, and 2.18 completes the proof. □

Proposition. *Let R be Noetherian and $I \triangleleft R$ such that ${}_R I \triangleleft_e {}_R R$. Then $\mathcal{K}(R_R) = \mathcal{K}(I_R)$.*

Proof. Let $B = \operatorname{r ann} I$; so $I_{R/B}$ is faithful. By the lemma, $\mathcal{K}(I_R) = \mathcal{K}(I_{R/B}) = \mathcal{K}(R/B)$. But B is contained in the left singular ideal of R which, by the left-handed version of 2.3.4, is nilpotent. It follows from 3.3 that $\mathcal{K}(R/B) = \mathcal{K}(R_R)$. □

§4 Prime Ideals and FBN Rings

4.1 The usual definition of the Krull dimension of a commutative Noetherian ring is in terms of the lengths of chains of prime ideals. We will show, in this section, that this gives the same as the module theoretic definition, provided it is finite.

In practice, the arguments used apply equally well to a wider class of rings called FBN rings; so this section also describes them and some of their commutative-like properties.

4.2 The most natural way to define the dimension of a ring R in terms of the poset $\operatorname{Spec} R$ is to take the supremum of the lengths of chains of prime ideals of R, taking ∞ if no supremum exists. However, it is convenient for us to refine this definition, so as to replace ∞ by an ordinal.

4.3 Let A be a poset satisfying the a.c.c. Let A_0 be the subset of all maximal elements of A; and for each ordinal α, let

$$A_\alpha = \{a \in A \mid b \in A, b > a \text{ implies } b \in A_\beta \text{ for some } \beta < \alpha\}.$$

Then $\dim A$ is defined to be the least α such that $A_\alpha = A$. It is an easy exercise to show that there is indeed such an ordinal α.

As an easy illustration, if we take $A = \gamma^{\operatorname{op}}$ for some ordinal γ, it is clear that $\dim A = \gamma$.

4.4 Now let R be any right Noetherian ring. Then $\operatorname{Spec} R$ is a poset satisfying the a.c.c. The **classical Krull dimension** of R is defined to be $\dim(\operatorname{Spec} R)$, or $\dim R$ for short.

Evidently this is the usual definition for a commutative Noetherian ring, provided it is finite.

4.5 We aim to connect $\dim R$ with $\mathcal{K}(R)$, starting with a general result.

Lemma. *If R is a right Noetherian ring then $\dim R \leq \mathcal{K}(R_R)$.*

Proof. We know, by 2.2.16, that R has a finite set of minimal primes, $P_1, \ldots, P_m$ say. Evidently $\dim R = \sup\{\dim R/P_i\}$. Similarly, by 3.8, $\mathcal{K}(R_R) = \sup\{\mathcal{K}(R/P_i)\}$. So we may, without loss, restrict to the case when R is prime; and then a simple induction argument, based on 3.11(ii), completes the proof. □

4.6 In general, these are unequal. For example, if R is a simple Noetherian ring, then $\dim R = 0$ but, provided R is not Artinian, $\mathcal{K}(R) > 0$. So $A_1(k)$ or $P_\lambda(k)$ can be used as counter-examples. Evidently equality is only feasible for a ring with many (two-sided) ideals; and that is the motivation behind the next definition.

4.7 A prime ring is **right bounded** if every essential right ideal contains a nonzero ideal; and a ring R is **right fully bounded** if R/P is right bounded for each prime ideal P of R. Then **bounded**, or **fully bounded**, means that the ring also has the left-handed property. The abbreviation right FBN or FBN is commonly used, respectively, for a right Noetherian right fully bounded or a Noetherian fully bounded ring.

By way of example, note that every commutative ring is of course fully bounded. It is also the case (see 13.6.6) that every PI ring (and hence every classical order) is fully bounded.

4.8 Theorem. *If R is a right FBN ring then $\mathcal{K}(R_R) = \dim R$.*

Proof. This proceeds by induction on $\dim R$. As in 4.5, it is enough to consider the case when R is prime; and we need only show that $\dim R \geq \mathcal{K}(R_R)$.

Let $A_1 \supseteq A_2 \supseteq \cdots$ be a descending chain of right ideals of R. For some N, it is the case that $\mathrm{u\,dim}\, A_n = \mathrm{u\,dim}\, A_{n+1}$ for $n \geq N$, since $\mathrm{u\,dim}\, R_R < \infty$. Choose $B \triangleleft_r R$ complementary to A_N; and then $B \oplus A_n \triangleleft_e R_R$ for all $n \geq N$. By hypothesis, there is a nonzero ideal $I_n \subseteq B \oplus A_n$. Now $(B \oplus A_N)I_n \subseteq B \oplus A_n$ and so $\mathcal{K}(A_N/A_n) \leq \mathcal{K}(R/I_n)$. Hence, from the definition of Krull dimension,

$$\mathcal{K}(R_R) \leq \sup\{\mathcal{K}(R/I)_R + 1 \mid 0 \neq I \triangleleft R\}.$$

However, by the induction hypothesis $\mathcal{K}(R/I) = \dim R/I$ and, since R is prime, 3.11(ii) shows that $\mathcal{K}(R/I) < \mathcal{K}(R)$. Therefore $\mathcal{K}(R_R) \leq \dim R$ as claimed. □

Corollary. *If R is a commutative Noetherian ring then $\mathcal{K}(R) = \dim R$.* □

4.9 This shows that $\mathcal{K}(R)$ coincides with the usual dimension if R is commutative Noetherian and the dimension is finite. To see that $\dim R$ could be infinite, consider the following.

6.4.11 *Prime ideals and FBN rings*

Example. *Let k be a field, $\{x_n \mid n \in \mathbb{N}\}$ a set of indeterminates, and $R = k[x_1, x_2, \ldots]$. Let*

$$a(n) = n(n+1)/2, \quad P_n = (x_{a(n-1)+1}, x_{a(n-1)+2}, \ldots, x_{a(n)}) \quad \text{and} \quad \mathscr{S} = R \setminus \bigcup_{n=1}^{\infty} P_n.$$

Then $R_\mathscr{S}$ is a commutative Noetherian ring with $\dim R_\mathscr{S} = \omega$.

Proof. Note first that P_n is a prime ideal with a generating set of n indeterminates, so P_n contains a chain of prime ideals of length n. This chain is preserved in $R_\mathscr{S}$, so $\dim R_\mathscr{S} \geq n$. Therefore it will be enough to show that if $0 \neq J \triangleleft R_\mathscr{S}$, then $R_\mathscr{S}/J$ is a Noetherian ring of finite dimension.

We may suppose that $J = I_\mathscr{S}$ for $I = J \cap R \triangleleft R$, and there is no loss in supposing I to be principal, $I = (i)$ say. Evidently $i \in k[x_1, x_2, \ldots, x_{a(n)}]$ for some n. Let $T = k[x_{a(n)+1}, x_{a(n)+2}, \ldots]$ and say $0 \neq t \in T$. Then $i + t \notin \bigcup P_n$ and so $t + J$ is a unit in $R_\mathscr{S}/J$. Therefore, if F denotes the quotient field of T, then $R_\mathscr{S}/J$ is a homomorphic image of a localization of $F[x_1, x_2, \ldots, x_{a(n)}]$. Hence $R_\mathscr{S}/J$ is Noetherian and $\dim(R_\mathscr{S}/J) \leq a(n) < \infty$. □

Naturally, $\dim R_\mathscr{S}[y_1, \ldots, y_m] = \omega + m$ if $y_1, \ldots, y_m$ are indeterminates. (This will follow directly from 5.4.) Indeed one can, by modifying these constructions, produce commutative Noetherian domains of arbitrary ordinal dimension; see [Gordon and Robson **73**, 63–67].

4.10 One immediate consequence of 4.8 is of interest.

Corollary. *If R is an FBN ring then $\mathcal{K}(R_R) = \mathcal{K}(_R R)$.* □

The symmetry that this gives leads to several questions. One, unanswered as yet, asks:

Question. *Is it true that $\mathcal{K}(R_R) = \mathcal{K}(_R R)$ for every Noetherian ring?*

4.11 A more general question can be asked concerning bimodules.

Question. *If $_S M_R$ is a bimodule, Noetherian on each side, is $\mathcal{K}(M_R) = \mathcal{K}(_S M)$?*

This has already been partially answered in 4.1.6 which gives a positive answer when $\mathcal{K}(_S M) = 0$, but it remains open in general. We will show, shortly, that the latter question, like the former, has a positive answer for FBN rings. Note that a positive answer to the latter question, for a given Noetherian ring R, implies by 3.12 that the bimodule condition of 4.5.7 also holds for R.

4.12 First we require a result similar to Lemma 3.12.

Proposition. *If R is right FBN and M_R is finitely generated and faithful then $\mathcal{K}(M) = \mathcal{K}(R_R)$.*

Proof. Let $0 = M_0 \subset M_1 \subset \cdots \subset M_n = M$ be an affiliated chain for M, as defined in 4.4.6, and let $P_1, \ldots, P_n$ be the affiliated prime ideals. Then $MP_n P_{n-1} \cdots P_1 = 0$ and so $P_n P_{n-1} \cdots P_1 = 0$. By 3.3, $\mathcal{K}(R) = \sup\{\mathcal{K}(R/P_i)\}$; and, by 2.4, $\mathcal{K}(M) = \sup\{\mathcal{K}(M_{i+1}/M_i)\}$. Therefore it is enough to prove the result in the case when R is prime and $\operatorname{ass} M = \operatorname{ann} M = 0$.

Now suppose $mE = 0$ for some essential right ideal E of R. Then $E \supseteq T$ with $0 \neq T \triangleleft R$ and $mR \cdot T = 0$. This contradicts the assumption that $\operatorname{ass} M = 0$ unless $m = 0$. Therefore M_R is torsion-free and so, by 3.11(iii), $\mathcal{K}(M) = \mathcal{K}(R)$. □

4.13 The answer to Question 4.11 will be obtained, for FBN rings, by considering the lattice of subbimodules $\mathscr{L}(_S M_R)$ of a bimodule $_S M_R$.

Proposition. *Let R be a right FBN ring, S any ring and $_S M_R$ a bimodule such that M_R is finitely generated. Then $\operatorname{dev} \mathscr{L}(_S M_R) = \mathcal{K}(M_R)$.*

Proof. Let $\mu(M)$ denote $\operatorname{dev} \mathscr{L}(_S M_R)$. Since $\mathscr{L}(_S M_R)$ is a sublattice of $\mathscr{L}(M_R)$, it is clear from 1.5 that $\mu(M) \leqslant \mathcal{K}(M)$.

Suppose the opposite inequality is not always true. We suppose M chosen, amongst counter-examples, to minimize $\mu(M)$; say $\mu(M) = \beta$. So, of course, $\mathcal{K}(M) > \beta$. By 2.20, there is an R-submodule N of M with $\mathcal{K}(M/N) = \beta$. Then, by 4.12, if we let $A = \operatorname{ann}(M/N)_R$ and $M_1 = MA$ then

$$\mathcal{K}(M/N) = \mathcal{K}(R/A) = \mathcal{K}(M/M_1) = \beta.$$

Hence, of course, $\mathcal{K}(M_1) > \beta$, using 2.4.

Now $\mu(M/N_1) \leqslant \beta$ and $\mu(M_1) \leqslant \beta$. However, if $\mu(M/M_1) < \beta = \mathcal{K}(M/M_1)$ or if $\mu(M_1) < \beta < \mathcal{K}(M_1)$, the minimality of β is contradicated. Thus $\mu(M/M_1) = \beta$ and M_1 is another counter-example. Iteration gives a chain $\{M_n\}$ in which $\mu(M_n/M_{n+1}) = \beta$ for all n; and this contradicts the hypothesis that $\mu(M) = \beta$. □

Corollary. *(i) Let R, S be FBN rings and let $_S M_R$ be a bimodule finitely generated on each side. Then $\mathcal{K}(M_R) = \mathcal{K}(_S M)$. If, further, M is faithful on each side then $\mathcal{K}(R) = \mathcal{K}(S)$.*

(ii) FBN rings satisfy the bimodule condition. □

4.14 Next we aim to show that $\bigcap J^n = 0$, where J is the Jacobson radical of an FBN ring R. This requires a preliminary result.

6.4.16

Proposition. *Let R be an FBN ring, M_R a finitely generated module and $N \triangleleft_e M$. Then $\mathcal{K}(N) = \mathcal{K}(M)$. In particular, if N is Artinian then so too is M.*

Proof. Let $\mathcal{K}(N) = \alpha$ and choose X with $N \subseteq X \subseteq M$ and maximal with respect to $\mathcal{K}(X) = \alpha$. Let $I = \text{ann } X$; then, by 4.12, $\mathcal{K}(R/I) = \mathcal{K}(X) = \alpha$. If $X \neq M$, choose Y with $X \subset Y \subseteq M$ to maximize $\text{ann}(Y/X)$. If $P = \text{ann}(Y/X)$, then P is prime since it is an associated prime of M/X. Note that $P \supseteq \text{ann } Y$ and that $P \text{ ann } X = PI \subseteq \text{ann } Y$. Thus $P/\text{ann } Y$ is an R/I-module, and so $\mathcal{K}(P/\text{ann } Y) \leqslant \alpha$ both as a right R-module and, by Corollary 4.13, as a left R-module. Choose an affiliated chain for $P/\text{ann } Y$ as a left module with affiliated prime ideals $Q_1, Q_2, \ldots, Q_n$ say. Then 4.12 shows that $\mathcal{K}(R/Q_i) \leqslant \alpha$ for each i. Now $Q_1 Q_2 \cdots Q_n P \subseteq \text{ann } Y$ and so $Y Q_1 Q_2 \cdots Q_n P = 0$. If some $Q_i \subseteq P$ then $\mathcal{K}(R/P) \leqslant \mathcal{K}(R/Q_i) \leqslant \alpha$. Yet Y/X is a right R/P-module and $\mathcal{K}(Y/X) > \alpha$ by the definition of X. This contradiction shows that no $Q_i \subseteq P$. It follows from the choice of Y that $Y Q_1 Q_2 \cdots Q_n \not\subseteq X$ and that $\text{ann}((Y Q_1 Q_2 \cdots Q_n + X)/X) = P$. Therefore, by 4.12, $\mathcal{K}((Y Q_1 Q_2 \cdots Q_n + X)/X) = \mathcal{K}(R/P)$. On the other hand,

$$(Y Q_1 Q_2 \cdots Q_n + X)/X \simeq Y Q_1 Q_2 \cdots Q_n / X \cap Y Q_1 Q_2 \cdots Q_n = W,$$

say. Since $X \triangleleft_e Y$ then
$$X \cap Y Q_1 Q_2 \cdots Q_n \triangleleft_e Y Q_1 Q_2 \cdots Q_n$$

these both being R/P-modules. It follows immediately from 2.2.2(iii) that W is a torsion R/P-module and so, by 3.11, $\mathcal{K}(W) < \mathcal{K}(R/P)$. Since it was shown above that $\mathcal{K}(W) = \mathcal{K}(R/P)$ this gives a contradiction which establishes that $X = M$ and that $\mathcal{K}(M) = \alpha$. □

4.15 Corollary. *If R is an FBN ring then $\bigcap J^n = 0$, where $J = J(R)$.*

Proof. This is immediate from 4.14 and 4.3.20. □

4.16 There is a further consequence of Proposition 4.13 which will be of use later (and which is trivial for commutative rings).

Proposition. *Let R be an FBN ring, $I \triangleleft_r R$, and $T \triangleleft R$. Then $\mathcal{K}(T/IT) \leqslant \mathcal{K}(R/I)$.*

Proof. Let $A = \text{ann}(R/I)$. Then R/I is a faithful R/A-module and so, by 4.12, $\mathcal{K}(R/I) = \mathcal{K}(R/A)$. By Proposition 4.13,

$$\mathcal{K}(T/AT_R) = \text{dev } \mathcal{L}(_{R/A}T/AT_R) = \mathcal{K}(_{R/A}T/AT)$$
$$\leqslant \mathcal{K}(_{R/A}R/A) = \text{dev } \mathcal{L}(_{R/A}R/A_R) = \mathcal{K}(R/A_R)$$

and so $\mathcal{K}(T/IT) \leqslant \mathcal{K}(T/AT) \leqslant \mathcal{K}(R/A) = \mathcal{K}(R/I)$. □

This property of R is called **ideal invariance**. It will be discussed, for more general rings, in 8.13.

§5 Bounds on Krull Dimensions

It was shown in Chapter 1 that many ring constructions preserve the Noetherian property. This section assesses the effect of these constructions on the Krull dimension, on the whole by bounding the possible change. The following section will obtain more precise results, but only under special circumstances.

Recall that by $\mathcal{K}(R)$ is meant $\mathcal{K}(R_R)$.

5.1 First we consider matrices and, more generally, Morita equivalence (see Chapter 3, §5).

Proposition. (i) *If M_R is a progenerator then $\mathcal{K}(M) = \mathcal{K}(R)$.*
(ii) *If $R \overset{M}{\sim} S$, then $\mathcal{K}(R) = \mathcal{K}(S)$.*
(iii) $\mathcal{K}(M_n(R)) = \mathcal{K}(R)$.

Proof. (i) From 2.5 one sees that $\mathcal{K}(M) \leq \mathcal{K}(R)$ since $R^n \twoheadrightarrow M$. However, since M_R is a generator, $M^t \twoheadrightarrow R$ for some t, and so, using 2.4, $\mathcal{K}(R) \leq \mathcal{K}(M^t) = \mathcal{K}(M)$.

(ii) It is clear from 3.5.7 that if $_S M_R$ is the progenerator linking R and S, then $\mathcal{L}(M_R) \simeq \mathcal{L}(S_S)$. This, together with (i), shows that $\mathcal{K}(R) = \mathcal{K}(S)$.

(iii) Clear from (ii). □

5.2 Next we consider the ring $S = \mathbb{I}(A)$ where A is an isomaximal right ideal of R. It was shown in 5.5.6 that S is right Noetherian if and only if R is right Noetherian.

Proposition. *Let R be a right Noetherian ring. Then:*
(i) $\mathcal{K}(R) = \mathcal{K}(\mathbb{I}(A))$;
(ii) *for any finitely generated right R-module M, $\mathcal{K}(M_R) = \mathcal{K}(M_S)$.*

Proof. (i) This follows from (ii) by taking $M = R$, noting that $\mathcal{K}(R/S)_S = 0$, by 5.5.5.

(ii) Let M_R be a finitely generated right R-module. Evidently $\mathcal{K}(M_R) \leq \mathcal{K}(M_S)$, using 1.5. On the other hand, if $\mathcal{K}(M_R) = 0$ then 5.5.5 shows that $\mathcal{K}(M_S) = 0$. This provides the base for the inductive hypothesis that $\mathcal{K}(M_R) = \mathcal{K}(M_S)$ for all modules M with $\mathcal{K}(M_R) < \alpha$ say. We aim, then, to suppose $\mathcal{K}(M_R) = \alpha$ and deduce $\mathcal{K}(M_S) = \alpha$.

Now Proposition 2.20 shows that M_R has a critical composition series. Using 2.4, it is enough to consider the case when M_R is α-critical. Let $0 \neq X \triangleleft M_S$. We will show $\mathcal{K}(M/X)_S < \alpha$, and so $\mathcal{K}(M)_S \leq \alpha$. Suppose first that $XA \neq 0$. Since $XA \triangleleft M_R$, then $\mathcal{K}(M/XA)_R < \alpha$. The inductive hypothesis shows that $\mathcal{K}(M/XA)_S < \alpha$, and since $M/XA \twoheadrightarrow M/X$, then $\mathcal{K}(M/X)_S < \alpha$ as required.

On the other hand, suppose $XA = 0$. The same argument as above shows that $\mathcal{K}(M/XR)_S < \alpha$. Note that $XR = XR/XA$ which is the homomorphic image of $(R/A)^n$ for some n, and that $(R/A)_S$ has finite length. Hence $\mathcal{K}(XR)_S = 0$ and so $\mathcal{K}(M_S) < \alpha$ which is absurd. □

5.3 Before turning, as in Chapter 1, to skew polynomial rings and skew Laurent polynomial rings, a preparatory result about a ring extension $R \subseteq T$ is needed. There are two poset maps of interest, namely $\theta: \mathscr{L}(R_R) \to \mathscr{L}(T_T)$ with $A \mapsto AT$ for $A \triangleleft_r R$, and $\varphi: \mathscr{L}(T_T) \to \mathscr{L}(R_R)$ with $B \mapsto B \cap R$ for $B \triangleleft_r T$.

Lemma. (i) *If θ preserves proper containment then $\mathcal{K}(R) \leqslant \mathcal{K}(T)$.*

(ii) *If either (a) φ preserves proper containment, or (b) $T = R_{\mathscr{S}}$ for some right denominator set $\mathscr{S}$, or (c) T_R is finitely generated then $\mathcal{K}(R) \geqslant \mathcal{K}(T)$.*

Proof. (i), (ii)(a) These follows directly from 1.17, taking $\gamma = \delta = 0$.

(ii)(b) By 2.1.16, φ preserves proper containment.

(ii)(c) From 2.4 and 2.5 it is clear that $\mathcal{K}(T_R) = \mathcal{K}(R_R)$; but 1.5 shows immediately that $\mathcal{K}(T_T) \leqslant \mathcal{K}(T_R)$. □

Corollary. *Let $R \subseteq T$ be a ring extension such that $_RT$ is free. Then $\mathcal{K}(T_T) \geqslant \mathcal{K}(R_R)$. If, further, T_R is finitely generated then $\mathcal{K}(T_T) = \mathcal{K}(R_R)$.* □

It is convenient to note a closely related result which will be of use later.

Proposition. *Let R be a ring of finite Krull dimension, and let $\mathscr{S}$ be a right denominator set in R and suppose that every simple right R-module is $\mathscr{S}$-torsion (cf. 2.1.17). Then $\mathcal{K}(R) > \mathcal{K}(R_{\mathscr{S}})$.*

Proof. If $I, J \triangleleft_r R_{\mathscr{S}}$ with $I \supset J$, then $I \cap R/J \cap R$ is nonzero torsion-free, so cannot have a simple R-submodule. Hence $\mathcal{K}(I \cap R/J \cap R) \geqslant 1$. Applying 1.17, with $\gamma = 1$, $\delta = 0$, gives $1 + \mathcal{K}(R_{\mathscr{S}}) \leqslant \mathcal{K}(R)$. □

5.4 Proposition. *Let R be a right Noetherian ring, σ an automorphism and δ a σ-derivation.*

(i) $\mathcal{K}(R) \leqslant \mathcal{K}(R[x; \sigma, \delta]) \leqslant \mathcal{K}(R) + 1$; *in particular, if $\delta = 0$ then $\mathcal{K}(R[x; \sigma]) = \mathcal{K}(R) + 1$.*

(ii) $\mathcal{K}(R) \leqslant \mathcal{K}(R[x, x^{-1}; \sigma]) \leqslant \mathcal{K}(R) + 1$; *in particular, if $\sigma = 1$ then $\mathcal{K}(R[x, x^{-1}]) = \mathcal{K}(R) + 1$.*

(iii) *If R is right Artinian then $\mathcal{K}(R[x; \sigma, \delta]) = \mathcal{K}(R[x, x^{-1}; \sigma]) = 1$.*

Proof. (i) Let $T = R[x; \sigma, \delta]$. Then Corollary 5.3 applies to give the first inequality. The second inequality is proved along the lines of the proof, in 1.2.9, of the Noetherian property. We associate with a right ideal I of T the ascending chain $\{I_n\}$ of leading right ideals. This chain belongs to $S_a(\mathscr{L}(R_R))$, using the notation of 1.16. Furthermore, the map $\mathscr{L}(T_T) \to S_a(\mathscr{L}(R_R))$ thus obtained is a poset map preserving proper inequality. Hence 1.17 together with 1.16 shows that $\mathcal{K}(T_T) \leq \mathcal{K}(R_R) + 1$.

Suppose now that $\delta = 0$, and consider the descending chain of ideals $\{x^n T\}$. Since $x^n T / x^{n+1} T \simeq R$ it follows that $\mathcal{K}(T) > \mathcal{K}(R)$; so $\mathcal{K}(T) = \mathcal{K}(R) + 1$.

(ii) The first inequality comes directly from Corollary 5.3. The second comes from (i) by applying Lemma 5.3(ii)(b) to the ring extension $R[x; \sigma] \subseteq R[x, x^{-1}; \sigma]$ taking $\mathcal{S} = \{x^n | n \in \mathbb{N}\}$.

(iii) This is an easy exercise. □

This can be applied to the algebra $P_{\lambda_1, \ldots, \lambda_n}(k)$ described in 1.8.7 giving

Corollary. $n \leq \mathcal{K}(P_{\lambda_1, \ldots, \lambda_n}(k)) \leq n + 1$. □

This result is readily extended to give estimates of the Krull dimensions of many more of the rings described in Chapter 1. In each case, more precise results will be obtained in §6.

5.5 First we consider group crossed products.

Proposition. *Let R be right Noetherian, G polycyclic by finite and $R*G$ a crossed product of R by G. Suppose that G has a chain*

$$1 = G_0 \triangleleft G_1 \triangleleft \cdots \triangleleft G_n = G$$

in which h of the factors are infinite cycles, and $n - h$ are finite. Then

$$\mathcal{K}(R) \leq \mathcal{K}(R*G) \leq \mathcal{K}(R) + h.$$

Proof. Since $R*G$ is free over R, Corollary 5.3 gives the first inequality. To prove the second, proceeding by induction on the length of the chain (and recalling by 1.5.9 that $R*G_{i+1} \simeq (R*G_i)*(G_{i+1}/G_i)$), it is enough to consider the case when $n = 1$. If G_1/G_0 is finite then Lemma 5.3(ii)(c) applies; and if G_1/G_0 is an infinite cycle then 5.4(ii) applies since, by 1.5.11, $R*G_1 \simeq R[x, x^{-1}; \sigma]$. □

We note that the number h here is independent of the particular chain chosen and is called the **Hirsch number** of G. See 6.1 for a more precise result.

5.6 Next consider a filtered ring S, as in 1.6.1, with filtration $\{F_n\}$. By 1.6.9, if the associated graded ring gr S is right Noetherian, so too is S.

Lemma. $\mathcal{K}(S) \leq \mathcal{K}(\text{gr } S)$.

Proof. This follows from 1.17 since, by 1.6.7, the map $I \mapsto \text{gr } I$ from $\mathscr{L}(S_s)$ to $\mathscr{L}(\text{gr } S_{\text{gr } S})$ is injective on chains. □

5.7 This is immediately applicable to enveloping algebras and, more generally, to a crossed product of a k-algebra R by $U(g)$, where g is a k-Lie algebra of dimension n over a commutative ring k (cf. 1.7.12).

Corollary. $\mathcal{K}(R) \leqslant \mathcal{K}(R * U(g)) \leqslant \mathcal{K}(R) + \dim g$.

Proof. Since $R * U(g)$ is free over R, Corollary 5.3 gives the first inequality. By 5.6 one knows that $\mathcal{K}(R * U(g)) \leqslant \mathcal{K}(\operatorname{gr}(R * U(g)))$. However, by 1.7.14, $\operatorname{gr}(R * U(g)) \simeq R[x_1, \ldots, x_n]$ which, by 5.4(i), has Krull dimension $\mathcal{K}(R) + n$. □

5.8 This can be applied in particular to $A_n(R) = R \otimes_k A_n(k)$, where R is a right Noetherian k-algebra and $A_n(k)$ is viewed as in 1.7.11(i). It gives

Corollary. $\mathcal{K}(R) + n \leqslant \mathcal{K}(A_n(R)) \leqslant \mathcal{K}(R) + 2n$. □

5.9 Before mentioning a special case of 5.8 a result which, in a sense, extends part of 5.4(i) is required. (A similar result appears in 7.3.16.)

Proposition. *Let I be a proper right ideal of a ring R with $I = \sum_{i=1}^n a_i R$ where:*
(i) $a_i a_j = a_j a_i$ *for each i,j.*
(ii) *If $a_{k+1} r \in \sum_{i=1}^k a_i R$ for some k then $r \in \sum_{i=1}^k a_i R$.*
Then $\mathcal{K}(R) \geqslant \mathcal{K}(R/I)_R + n$.

Proof. Set $I_k = \sum_{i=1}^k a_i R$ with $I_0 = 0$. In the chain
$$R \supset I_{k+1} = I_k + a_{k+1} R \supset I_k + a_{k+1}^2 R \supset \cdots \supset I_k$$
each factor is isomorphic to R/I_{k+1}. Hence $\mathcal{K}(R/I_k) \geqslant \mathcal{K}(R/I_{k+1}) + 1$. Induction on k completes the proof. □

5.10 Corollary. *Let k be a field, D be the quotient division ring of $A_n(k)$, and $S = A_n(D)$. Then $\mathcal{K}(S) = 2n$.*

Proof. Let $\{s_i, t_i | i = 1, \ldots, n\}$ be the generators of $A_n(k) \subseteq D$ and $\{x_i, y_i | i = 1, \ldots, n\}$ then generate $A_n(D)$ over D. The elements $\{t_i - x_i, s_j - y_j | i,j = 1, \ldots, n\}$ commute; call these elements $a_1, \ldots, a_{2n}$. Then $A_n(D)$ is a free right and left D-module with the monomials in the a_i as a basis. Therefore, by 5.9, $\mathcal{K}(A_n(D)) \geqslant 2n$ and 5.8 completes the proof. □

§6 Calculation of Krull Dimension

For each of the results in §5 which involve inequalities, there are cases when either bound is attained. Indeed in 5.4 it has already been noted that the upper bound is attained under certain circumstances. In this section various special cases are dealt with, giving a precise value for the Krull dimension of certain rings.

6.1 First we consider the group ring RG with G polycyclic by finite. Note that R is itself an RG-module with $rg = r$ for each $r \in R$, $g \in G$. We will mean this particular module when we refer to the module R_{RG} henceforth.

Note next that if $H \triangleleft G$ with G/H an infinite cycle generated by $[xH]$ say, then each element of $M = (R_{RH}) \otimes_{RH} RG$ has a unique representation as $\sum_{i \in \mathbb{Z}} r_i \otimes x^i$. It is easy to see that $M(1-x)^n \triangleleft M_{RG}$ and that

$$M(1-x)^n/M(1-x)^{n+1} \simeq M/M(1-x) \simeq R.$$

Thus M_{RG} has a chain of submodules of order type ω^{op} with each factor isomorphic to R.

Proposition. *If the Hirsch number of G is h, then $\mathcal{K}(RG) = \mathcal{K}(R) + h$.*

Proof. Consider first the case when G is poly-infinite cyclic, i.e. it has a chain of length h all of whose factors are infinite cycles. We aim to prove, by induction on h, that there is a poset map $f:(\omega^h)^{\mathrm{op}} \to \mathcal{L}(RG)$ such that for each $\beta < \omega^h$, $f(\beta)/f(\beta+1) \simeq R$. We may suppose that $f':(\omega^{h-1})^{\mathrm{op}} \to \mathcal{L}(RH)$ is as described, where $H \triangleleft G$ with G/H an infinite cycle. As noted above, tensoring over RH with RG transforms each copy of R into an RG-module having a chain of order type ω^{op}, each factor isomorphic to R. Since $\omega \cdot \omega^{h-1} = \omega^h$, the induction claim is established.

Now 1.9 shows that dev $\omega^h = h$. We can apply 1.17, with $\delta = 0$ and $\gamma = \mathcal{K}(R_R)$, to show that $\mathcal{K}(RG) \geq \mathcal{K}(R) + h$. Combined with 5.5 this shows that $\mathcal{K}(RG) = \mathcal{K}(R) + h$.

Finally, suppose that G is an arbitrary polycyclic by finite group. As noted before, G has a chain in which only the top factor is finite, say G/H. The work so far shows that $\mathcal{K}(RH) = \mathcal{K}(R) + h$; then Corollary 5.3 shows that $\mathcal{K}(RG) \geq \mathcal{K}(RH)$. Once again 5.5 completes the argument. □

6.2 Next we consider k-Lie algebras again. A k-Lie algebra $\mathbf{g}$ over a commutative ring k is **solvable** if it has a k-basis $x_1, \ldots, x_n$ such that $kx_1 + \cdots + kx_i \triangleleft kx_1 + \cdots + kx_{i+1}$ for each i. (When k is a field this is equivalent to the usual definition.) The arguments of 6.1 can be adapted to prove

Theorem. *If R is a right Noetherian k-algebra and $\mathbf{g}$ is a solvable k-Lie algebra of dimension n then*

$$\mathcal{K}(R \otimes U(\mathbf{g})) = \mathcal{K}(R) + n. \qquad \square$$

If $\mathbf{g}$ is not solvable then it can happen that $\mathcal{K}(U(\mathbf{g})) < \dim_k \mathbf{g}$; see 8.5.16.

6.6.5 Calculation of Krull dimension

6.3 The next few results are aimed at determining the Krull dimensions of certain skew polynomial and skew Laurent polynomial rings. It will be seen that the lower bounds in 5.4, and hence in 5.5 and 5.7, can be attained.

6.4 The technique to be used involves filtered and graded rings as described in 1.6.4. As in that section, let S be a ring with a filtration $\{F_n\}$ and let $(\operatorname{gr} I)_n$, for $I \triangleleft_r S$, be the nth homogeneous component of $\operatorname{gr} I$, as in 1.6.7. We prefer to write S_0 for F_0 when viewing it as a subring of S.

Proposition. *Let S be a filtered ring with filtration $\{F_n\}$ and let I, J be right ideals of S with $I \subset J$. Then:*
(i) $(\operatorname{gr} J)_n/(\operatorname{gr} I)_n \simeq J \cap (I + F_n)/J \cap (I + F_{n-1})$ *as S_0-modules;*
(ii) *the following conditions are equivalent:*
  (a) $\operatorname{gr} J/\operatorname{gr} I$ *has finite length as an S_0-module;*
  (b) *there is a finite subset X of $\mathbb{N}$ such that $(\operatorname{gr} J)_n/(\operatorname{gr} I)_n = 0$ for $n \notin X$ and $(\operatorname{gr} J)_n/(\operatorname{gr} I)_n$ has finite length, as an S_0-module, for each $n \in X$;*
  (c) J/I *has finite length as an S_0-module.*

Proof. (i) As noted in 1.6.7
$$(\operatorname{gr} I)_n = ((I + F_{n-1}) \cap F_n)/F_{n-1} \simeq I \cap F_n/I \cap F_{n-1}$$
and so
$$\begin{aligned}(\operatorname{gr} J)_n/(\operatorname{gr} I)_n &\simeq (J \cap F_n)/(I \cap F_n) + (J \cap F_{n-1}) \\ &\simeq (J \cap F_n)/(J \cap F_n) \cap ((J \cap F_{n-1}) + I) \\ &\simeq (J \cap F_n) + I/(J \cap F_{n-1}) + I \\ &= J \cap (I + F_n)/J \cap (I + F_{n-1}).\end{aligned}$$

(ii) Here J is the union of the chain of S_0-submodules $\{J \cap (I + F_n) | n \in \mathbb{N}\}$. Therefore (ii) follows from (i). □

6.5 Proposition. *Let S be a filtered ring such that, for each n, F_n/F_{n-1} is a Noetherian right S_0-module (and so S_0 is a right Noetherian ring). Suppose that $I, J \triangleleft_r S$ with $I \subset J$ and J/I not of finite length over S_0. Then the Krull dimension, over $\operatorname{gr} S$, of $\operatorname{gr} J/\operatorname{gr} I$ is at least 1.*

Proof. It follows from 6.4 that either $(\operatorname{gr} J)_n/(\operatorname{gr} I)_n \neq 0$ for $n \in X$, an infinite subset of $\mathbb{N}$ or else, for some n, $(\operatorname{gr} J)_n/(\operatorname{gr} I)_n$ has infinite length over S_0. We consider these two cases separately.

In the first case say $X = \{m_i\}$ with $m_i < m_{i+1}$ for all i. Let
$$L_n = \sum_{i=0}^{m_n} (\operatorname{gr} I)_i \oplus \sum_{i=m_n+1}^{\infty} (\operatorname{gr} J)_i.$$

Then $\operatorname{gr} J \supset L_1 \supset L_2 \supset \cdots \supset \operatorname{gr} I$ and so $\mathcal{K}(\operatorname{gr} J/\operatorname{gr} I) \geq 1$.

In the second case $(\operatorname{gr} J)_n/(\operatorname{gr} I)_n$ cannot be Artinian over S_0, so there is an infinite descending chain of S_0-submodules

$$(\operatorname{gr} J)_n \supset T_1 \supset T_2 \supset \cdots \supset (\operatorname{gr} I)_n.$$

If we let

$$L_m = \sum_{i=0}^{n-1} (\operatorname{gr} I)_i \oplus T_m \oplus \sum_{i=n+1}^{\infty} (\operatorname{gr} J)_i$$

then $\operatorname{gr} J \supset L_1 \supset L_2 \supset \cdots \supset \operatorname{gr} I$ and again $\mathcal{K}(\operatorname{gr} J/\operatorname{gr} I) \geq 1$. □

6.6 We now apply these results to skew polynomial rings.

Theorem. *Let R be a right Noetherian ring of finite Krull dimension and let $S = R[x; \sigma, \delta]$, with σ an automorphism. If no nonzero right S-module has finite length over R then $\mathcal{K}(S) = \mathcal{K}(R)$.*

Proof. In 5.4 it is shown both that $\mathcal{K}(R) \leq \mathcal{K}(S)$ and, since $\operatorname{gr} S = R[x; \sigma]$, that $\mathcal{K}(\operatorname{gr} S) = \mathcal{K}(R) + 1$. However, by 6.5 the map $\mathcal{L}(S) \to \mathcal{L}(\operatorname{gr} S)$ given by $I \mapsto \operatorname{gr} I$ is such that if $J \supset I$ then $\mathcal{K}(\operatorname{gr} J/\operatorname{gr} I) \geq 1$. Hence, by 1.17, $1 + \mathcal{K}(S) \leq \mathcal{K}(\operatorname{gr} S)$. □

6.7 This can be applied in specific cases as follows.

Corollary. *Let R be a commutative Noetherian ring of finite Krull dimension and let $S = R[x; \delta]$.*
(i) *If there are no δ-stable ideals I of R with $\mathcal{K}(R/I) = 0$ then $\mathcal{K}(S) = \mathcal{K}(R)$.*
(ii) *If R is not Artinian and S is simple, then $\mathcal{K}(S) = \mathcal{K}(R)$.*

Proof. (i) Suppose that M_S is a nonzero module of finite length over R, and let $I = \operatorname{ann}_R M$. Then I contains a finite product of maximal ideals of R and so $\mathcal{K}(R/I) = 0$. But I is δ-stable since if $i \in I$ then $0 = Mxi = M(ix - \delta(i)) = M\delta(i)$. This contradicts the hypothesis. Hence 6.6 applies.
(ii) If S is simple then, by 1.8.3, the only proper δ-stable ideal of R is 0. □

This result will be superseded in §9.

6.8 In particular, if k is a field of characteristic zero then the k-algebra $S = k[y_1, y_1^{-1}, \ldots, y_n, y_n^{-1}][x; \delta]$ described in 1.8.6, with the derivation $\delta(y_i) = \lambda_i y_i$ for $\mathbb{Q}$-independent scalars λ_i, is now seen to have $\mathcal{K}(S) = n$. This shows that the lower bound in 5.4(i) and in 5.7 can indeed be attained.

Similarly, it follows that the first Weyl algebra $A_1(k)$ has Krull dimension 1.

6.9 If one could apply 6.7(ii) in the case when R is noncommutative one could easily deduce that $\mathcal{K}(A_n(k)) = n$. Whilst this conclusion is true, as will be shown in 6.15, the deduction is improper as the next example shows.

Example. *Let D be the quotient division ring of $A_1(k)$, with k of characteristic zero. Let $S = A_1(D)$. Then S is simple, $S = R[x; \delta]$, where $R = D[y]$ and $\delta = -\partial/\partial y$, R is Noetherian with $\mathcal{K}(R) = 1$ and $\mathcal{K}(S) = 2$.*

Proof. This is clear from 1.3.8, 5.4 and 5.10. □

6.10 There are results for skew Laurent polynomials similar to those above. If $T = R[x, x^{-1}; \sigma]$ then the role of the associated graded ring will be played by gr S, where $S = R[x; \sigma]$.

Theorem. *Let R be a right Noetherian ring of finite Krull dimension and let $T = R[x, x^{-1}; \sigma]$. If no nonzero right T-module has finite length over R then $\mathcal{K}(T) = \mathcal{K}(R)$.*

Proof. Suppose that $I, J \triangleleft_r T$ with $I \subset J$, and suppose that $J \cap S/I \cap S$ has finite length over R and hence, of course, over S. Then it has a simple S-submodule M say. Now if $mx = 0$ for some $m = [j + I \cap S] \in M$ then $jx \in I$. Hence $j \in I \cap S$ and so $m = 0$. The fact that Mx is an S-submodule of M now shows that $M = Mx$, with x acting bijectively. Hence M is, in fact, a T-module yet, as a submodule of $J \cap S/I \cap S$, has finite length over R. This contradicts the hypothesis.

We conclude, therefore, that $J \cap S/I \cap S$ does not have finite length over R. Then 6.5 shows that $\mathcal{K}(\mathrm{gr}(J \cap S)/\mathrm{gr}(I \cap S)) \geq 1$. We now apply 1.17 to the poset map $\mathscr{L}(T) \to \mathscr{L}(\mathrm{gr}\, S)$ given by $I \mapsto \mathrm{gr}(I \cap S)$ to deduce that

$$1 + \mathcal{K}(T) \leq \mathcal{K}(\mathrm{gr}\, S) = \mathcal{K}(R) + 1.$$

This, together with 5.4, gives the result. □

6.11 Corollary. *Let R be a commutative Noetherian ring of finite Krull dimension and let $T = R[x, x^{-1}; \sigma]$.*
(i) *If there are no σ-stable ideals I of R with $\mathcal{K}(R/I) = 0$ then $\mathcal{K}(T) = \mathcal{K}(R)$.*
(ii) *If R is not Artinian and T is simple then $\mathcal{K}(T) = \mathcal{K}(R)$.*

Proof. Similar to that of 6.7. □

This result will be superseded in §9.

6.12 In particular, if k is a field then the simple k-algebra $P_{\lambda_1,\ldots,\lambda_n}(k)$ described in 1.8.7 must have Krull dimension n. Thus the lower bound in 5.4(ii) can be attained.

6.13 If D is the quotient division ring of $P_\lambda(k)$, with λ not a root of unity, then the algebra $T = P_\lambda(D) = D \otimes_k P_\lambda(k)$ can be viewed as $R[x, x^{-1}; \sigma]$, where $R = D[y, y^{-1}]$ and σ is the identity on D and maps y to $\lambda^{-1}y$. An argument along the same lines as in 6.9 shows that T is simple, $\mathcal{K}(R) = 1$, but $\mathcal{K}(T) = 2$. Thus the hypothesis, in 6.11(ii), that R is commutative is indeed needed.

6.14 It remains now to investigate $\mathcal{K}(A_n(k))$. One case is rather easy.

Proposition. *If k is a field with char $k > 0$ then $\mathcal{K}(A_n(k)) = 2n$.*

Proof. If char $k = p$, it is easily seen that x_i^p and y_i^p are central in $A_n(k)$ for $i = 1, \ldots, n$. Hence the centre contains the polynomial ring $R = k[x_1^p, \ldots, x_n^p, y_1^p, \ldots, y_n^p]$ which has Krull dimension $2n$. Moreover, since $A_n(k)$ is free and finitely generated as an R-module, Corollary 5.3 applies, giving $\mathcal{K}(A_n(k)) = \mathcal{K}(R) = 2n$. $\square$

6.15 When char $k = 0$ we know that $A_n(k)$ is simple. Now if M is a simple right $A_n(k)$-module, and $S = \mathrm{End}\, M$, it follows that $A_n(k) \hookrightarrow \mathrm{End}\,(_S M)$; hence M cannot be finite dimensional over k.

Theorem. *If k is a field of characteristic zero, then $\mathcal{K}(A_n(k)) = n$.*

Proof. We introduce the notation $B_{n,i}(k)$ for the ring $A_n(k) \bigoplus_{k[x_i]} k(x_i)$; and $B_{n,n+i}(k)$ for the ring $A_n(k) \bigotimes_{k[y_i]} k(y_i)$. Set $B = \bigoplus_{j=1}^{2n} B_{n,j}(k)$, this being the direct product of all localizations at one variable, since $B_{n,j} \simeq A_{n-1} \otimes_k B_1$.

Note first that the diagonal map $a \mapsto (a, \ldots, a)$ embeds $A_n(k)$ in B. We aim to show that the map $I \mapsto IB$ from right ideals of $A_n(k)$ to right ideals of B preserves proper containments; and to check this it is enough to show that if I is a maximal right ideal then $IB \neq B$. However, if $IB = B$, then $1 \in IB$ and so $1 \in IB_{n,j}$ for each j. This implies that there are monic polynomials $f_i \in k[x_i] \cap I$ and $g_i \in k[y_i] \cap I$ for each i; and that in turn implies that A/I is finite dimensional over k, a contradiction.

We now proceed by induction on n, with $n = 0$ being trivial. Assuming the result for any field for $n - 1$, we have $\mathcal{K}(A_{n-1}(k(x_n))) = n - 1$. However,

$$B_{n,j}(k) \simeq A_{n-1}(k(x_n))[y_n; \delta]$$

for any j, where $\delta = \partial/\partial x_n$. Hence, by 5.4, $\mathcal{K}(B_{n,j}(k)) \leq n$ and thus $\mathcal{K}(B) \leq n$. Then 5.3(i) shows that $\mathcal{K}(A_n(k)) \leq n$. This, combined with 5.8, completes the proof. $\square$

6.16 This section will end with an application of the above result. First, however, come some results concerned with the extension of the field of scalars of an algebra. This first result is very general.

6.6.18   *Calculation of Krull dimension*   205

Proposition. *Let R be a Noetherian k-algebra and let F be a finitely generated extension field of k of transcendence degree n. Then*
(i) $R \otimes_k F$ *is Noetherian, and*
(ii) $\mathcal{K}(R) \leqslant \mathcal{K}(R \otimes_k F) \leqslant \mathcal{K}(R) + n$.

Proof. Let $x_1, \ldots, x_n$ be a transcendence basis for F over k, and set $S = R \otimes k(x_1, \ldots, x_n)$. This is a localization of $T = R \otimes k[x_1, \ldots, x_n]$, and by 5.4(i), $\mathcal{K}(T) = \mathcal{K}(R) + n$. Hence, using Lemma 5.3(ii), $\mathcal{K}(S) \leqslant \mathcal{K}(R) + n$. However Corollary 5.3 shows that $\mathcal{K}(S) = \mathcal{K}(R \otimes_k F)$.

Finally, note that $R \otimes F$ is free over R. So Corollary 5.3 again applies to give $\mathcal{K}(R \otimes F) \geqslant \mathcal{K}(R)$. $\square$

6.17 There is a special case in which a different lower bound is obtained.

Proposition. *Let R be a Noetherian k-algebra, over a field k, which contains a subfield of transcendence degree n over k; and let F be a finitely generated extension field of k of transcendence degree n. Then $\mathcal{K}(R \otimes F) \geqslant n$.*

Proof. Say $R \supseteq H = k(v_1, \ldots, v_n)$ and $F \supseteq L = k(x_1, \ldots, x_n)$, two rational function fields. Now it is easy to see that in $H \otimes_k L$, each ideal $P_t = \sum_{i=1}^t (x_i - v_i)$ is prime. This chain shows, by 3.11(ii), that $\mathcal{K}(H \otimes L) \geqslant n$. However, R is free over H; so $R \otimes L$ is free over $H \otimes L$. Therefore $\mathcal{K}(R \otimes L) \geqslant \mathcal{K}(H \otimes L)$ by Corollary 5.3. Now $R \otimes F$ is a finitely generated free $R \otimes L$-module, with generators from $1 \otimes F$. Thus Corollary 5.3 applies to give $\mathcal{K}(R \otimes F) = \mathcal{K}(R \otimes L) \geqslant n$. $\square$

6.18 Finally come the promised applications. Let $D_n(k)$ denote the quotient division ring of $A_n(k)$.

Corollary. (i) *If k has characteristic zero, then no subfield of $D_n(k)$ can have transcendence degree greater than n over k.*
(ii) *If char $k \neq 0$, no subfield of $D_n(k)$ has transcendence degree greater than $2n$ over k.*

Proof. (i) If L were a subfield of transcendence degree $> n$, then, by 6.17, $\mathcal{K}(D_n(k) \otimes L) > n$. But $D_n(k) \otimes L$ is evidently a localization of $A_n(k) \otimes L$ and $A_n(k) \otimes L \simeq A_n(L)$. Now $\mathcal{K}(A_n(L)) = n$, by 6.15, and so, by Lemma 5.3(ii), $\mathcal{K}(D_n(k) \otimes L) \leqslant n$. This contradiction establishes the result.
(ii) This is proved likewise. $\square$

We note that the same type of argument is valid for the quotient division ring of the enveloping algebra of any finite dimensional Lie algebra.

6.19 Corollary. (i) *There is no k-algebra embedding $D_n(k) \hookrightarrow D_m(k)$ for $n > m$.*
(ii) *There is no k-algebra embedding $A_n(k) \hookrightarrow A_m(k)$ for $n > m$.* □

§7 Stable Bounds on Generators

7.1 In 5.7.3 it was shown that a module M of finite length over a simple non-Artinian ring is always cyclic. This suggests that the Krull dimension might have some control over numbers of generators. The results of this section confirm this.

7.2 If a finitely generated module M_R has the property, for some n, that whenever $M = \sum_{i=1}^{k+1} a_i R$ with $k \geq n$ then $M = \sum_{i=1}^{k}(a_i + a_{k+1}g_i)R$ for some $g_i \in R$, then we say that n is in the **stable range** of M. Consequences of such a property will be discussed later, in Chapter 11; for now we concentrate on showing some cases when it occurs.

7.3 The first case to be considered is when M is cyclic.

Theorem. *Let R be a right Noetherian ring and M_R a cyclic module with $M = \sum_{i=1}^{s} b_i R + cR$ and $\mathcal{K}(cR) \leq s - 1$. Then there exist $g_i \in R$ such that $M = \sum_{i=1}^{s}(b_i + cg_i)R$.*

Proof. Evidently if $s = 0$ or if $c = 0$, there is nothing to prove. We will assume, as an induction hypothesis, that the result is true for smaller values of s and for all proper factors of M, and that $cR \neq 0$.

By 2.10, cR contains a (nonzero) cyclic critical submodule, crR say. By hypothesis, the result holds for M/crR; so $M = \sum_{i=1}^{s}(b_i + cg_i)R + crR$. This shows that it is enough to consider the case when cR is critical and $\mathcal{K}(cR) \leq s - 1$. We first consider a special case.

(a) Suppose, for some $j \in \{1,\ldots,s\}$ and some $g_j \in R$, that $N \cap cR \neq 0$ where $N = (b_j + cg_j)R$. Then $\mathcal{K}(cR/N \cap cR) \leq s - 2$. So, by induction on s, $M/N = \sum_{i \neq j}(\bar{b}_i + \bar{c}g_i)R$, where $\bar{m} = [m + N]$ for $m \in M$. Thus $M = \sum(b_i + cg_i)R$ as required.

Before proceeding further, we introduce the notation $P = \operatorname{ann} cR$ and $K_j = \operatorname{ann} b_j$, $j = 1,\ldots,s$. The remainder of the proof considers three cases which, together, cover all eventualities.

(b) Suppose $P \not\subseteq K_j$ for some j. Then $b_j P \neq 0$ and, by hypothesis, the result holds in $M/b_j P$. Thus $M/b_j P = \sum_i(\bar{b}_i + \bar{c}g_i)R$. Note that

$$(b_j + cg_j)R \supseteq (b_j + cg_j)P = b_j P$$

and so $M = \sum(b_i + cg_i)R$ as claimed.

(c) Suppose $K_j \not\subseteq P$ for some j. Then $cRK_j \neq 0$ and so $cg_j K_j \neq 0$ for some

6.7.5 *Stable bounds on generators* 207

$g_j \in R$. Therefore
$$(b_j + cg_j)R \supseteq (b_j + cg_j)K_j = cg_jK_j \neq 0$$
and then (a) applies to give the result.

(d) Finally, suppose $P = K_j$ for each j. Note then that $MP = \sum b_i P + cP = 0$, and recall that M is cyclic, $M = mR$ say. There is an epimorphism
$$\theta : R/P \twoheadrightarrow R/\operatorname{ann} m \simeq M \supseteq b_1 R \simeq R/P.$$
Since R is right Noetherian, θ must be a monomorphism. Moreover, then $M \simeq R/P$ and yet $\operatorname{u\,dim} b_1 R = \operatorname{u\,dim} R/P$. It follows that $b_1 R \triangleleft_e M$ and so $b_1 R \cap cR \neq 0$. Thus (a) applies again. □

7.4 This has the following consequences.

Corollary. *Let R be a right Noetherian ring.*
(i) *If M_R is cyclic and $\mathcal{K}(M/MJ(R)_R) = n$ then $n+1$ is in the stable range of M.*
(ii) *If $\mathcal{K}(R/J(R)_R) = n$ then $n+1$ is in the stable range of R_R.* □

7.5 We now turn to the case when M_R is not assumed to be cyclic. Instead some other conditions are imposed.

Proposition. *Let R be a right Noetherian ring and M_R a finitely generated module, with $M = aR + B$ for some submodule B. Suppose that $\mathcal{K}(M) \leq n$ and that $\mathcal{K}(R/\operatorname{ann} B') > n$ for all nonzero subfactor-modules B' of B. Then $\mathcal{K}(M/(a+b)R) < n$ for some $b \in B$.*

Proof. We may assume that $\mathcal{K}(M) = \mathcal{K}(B) = n$, for otherwise the result is true taking $b = 0$. We will suppose, for the moment, that B is a critical and hence uniform module. Now if $aR \cap B \neq 0$, then $\mathcal{K}(M/aR) < \mathcal{K}(B)$, so again the result holds with $b = 0$. If $aR \cap B = 0$, then $M = aR \oplus B$. If $T = \operatorname{ass} B$, then $T = \operatorname{ann} B'$ for some nonzero submodule B' of B. By hypothesis $\mathcal{K}(R/T) > n$. Let $I = \operatorname{ann} a$; so $aR \simeq R/I$ and thus $\mathcal{K}(R/I) \leq n$. It follows that $I \not\subseteq T$, and so, for some $b \in B$, $I \not\subseteq \operatorname{ann} b$. If $J = \operatorname{ann} b$, then $R/J \simeq bR \subseteq B$, and so R/J is n-critical, by 2.11. Hence $\mathcal{K}(R/I + J) < n$. However,
$$(a+b)R \supseteq (a+b)(I+J) = aJ + bI = a(I+J) + b(I+J)$$
and so
$$\mathcal{K}(M/(a+b)R) \leq \sup\{\mathcal{K}(aR/a(I+J)), \mathcal{K}(B/b(I+J))\} < n.$$

That completes the proof when B is critical. In general, consider a critical composition series for B,
$$B = B_0 \supset B_1 \supset \cdots \supset B_k = 0,$$

and proceed by induction on k. Now $M/B_1 = aR + B/B_1$ and so

$$K(M/(a+x)R + B_1) < n$$

for some $x \in B$. But also

$$K((a+x)R + B_1/(a+x+y)R) < n \quad \text{for some } y \in B_1.$$

Setting $b = x + y$ and using 2.4 gives the result. □

7.6 We let $\mathscr{K}_n(R)$ denote the class of all finitely generated modules M_R with $K(M) \leq n$.

Proposition. *Let R be right Noetherian and $n \in \mathbb{N}$. The following conditions are equivalent:*
  (i) *no nonzero factor ring $\bar{R}$ of R has $K(\bar{R}) \leq n$;*
  (ii) *there exists $k \in \mathbb{N}$ such that each $M \in \mathscr{K}_n(R)$ has a set of k generators;*
  (iii) *each $M \in \mathscr{K}_n(R)$ has a set of $n + 1$ generators;*
  (iv) *$n + 1$ is in the stable range of each $M \in \mathscr{K}_n(R)$.*

Proof. Evidently (iv)⇒(iii)⇒(ii). To see that (ii)⇒(i) note that any module of the form $\bar{R}^t$ requires at least t generators. Finally, we need to show that (i)⇒(iv). An easy induction, based on 7.5, shows that if $s \geq n + 1$ and $M = \sum_{i=1}^{s+1} a_i R$, then

$$M = \sum_{i=1}^{s} (a_i + \sum_{j>i} a_j r_{ij})R \quad \text{for some } r_{ij} \in R.$$

It follows that

$$M = \sum_{i=1}^{s} (a_i + a_{s+1} g_i)R \quad \text{for some } g_i \in R. \qquad \square$$

7.7 This provides a generalization of Corollary 5.7.3.

Corollary. *Let R be a right Noetherian ring and M_R a finitely generated completely faithful module with $K(M) < K(R_R)$, and suppose $K(M) = n < \infty$. Then $n + 1$ is in the stable range of M. In particular, M has a set of $n + 1$ generators.* □

7.8 Applied to simple rings it gives the following facts.

Corollary. *Let R be a simple right Noetherian ring.*
  (i) *If M_R is finitely generated with $K(M) < K(R)$ then $K(M) + 1$ is in the stable range of M.*
  (ii) *If $K(R) = n < \infty$ then every right ideal I of R has a set of $n + 1$ generators.*
  (iii) *If R is not right Artinian then every right R-module of finite length is cyclic.*

Proof. (i) This is immediate from 7.7.

(ii) By 3.3.6 there is an element $c \in I$ such that $cR \triangleleft_e I$. Therefore, by 2.8, $K(I/cR) < n$ and so I/cR has a set of n generators.

(iii) Clear. □

This applies, of course, to $A_n(k)$, where k is a field of characteristic zero. However, it is the case that in $A_n(k)$ every right ideal has two generators; see [Stafford 78].

7.9 We end this section by considering polynomials over a simple ring.

Proposition. *Let R be a simple right Noetherian ring with $K(R) = n$. Then each right ideal I of $R[x]$ has a set of $n + 1$ generators.*

Proof. Let $Q = Q(R)$, this being a simple Artinian ring. Hence, by 3.4.11, $Q[x]$ is a principal ideal ring. Now $Q[x] = R[x]_{\mathscr{S}}$ with $\mathscr{S} = \mathscr{C}_R(0)$. Therefore 2.1.16 shows that $I_{\mathscr{S}} = fR[x]_{\mathscr{S}}$ for some $f \in I$ and that I/J is a torsion R-module, where $J = fR[x]$.

Suppose that $I = \sum_{j=1}^k a_j R[x]$. Then, for each j, there exists $s_j \in \mathscr{C}_R(0)$ such that $a_j s_j \in J$. Hence I/J is a homomorphic image of $\bigoplus_{j=1}^k (R[x]/s_j R[x])$. However, $K(\bigoplus_{j=1}^k R/s_j R)_R < n$, since $s_j R \triangleleft_e R$. Therefore $\bigoplus_{j=1}^k (R/s_j R)$ has a set of at most n generators. The same is then true of $\bigoplus_{j=1}^k R[x]/s_j R[x]$ over $R[x]$, and hence of I/J over $R[x]$. □

§8 Quotient Rings and Localization

8.1 Throughout this section R will denote a right Noetherian ring. The methods of Krull dimension lead to some useful criteria for the existence of a quotient ring, or of localizations, of R. The main theme of this section addresses the question: 'If each nonzero right ideal I of R has $K(I) = K(R)$ does $Q(R)$ exist and is it right Artinian?' It will be shown that the answer is positive when R is FBN and also under certain weaker hypotheses. These involve weak ideal invariance and that too is introduced and investigated.

8.2 To indicate the relevance of the question above, consider the prototype of a commutative Noetherian ring R which fails to have an Artinian quotient ring. This is the ring $R = k[X, Y]/(X^2, XY)$, where k is a field. Let x, y denote the images of X, Y in R. Then $N(R) = (x)$; so $\mathscr{C}(N) = R \setminus N$ yet $\mathscr{C}(0) = R \setminus (x, y)$. Thus $\mathscr{C}(0) \neq \mathscr{C}(N)$. Note that $K(R_R) = 1$ and $K(N(R)_R) = 0$.

8.3 The arguments used in this section apply also to other dimensions, such as the Gelfand–Kirillov dimension which will be dealt with in Chapter 8. Therefore it is convenient to introduce the more abstract notion of a dimension function first and to establish results in that generality.

8.4 The possible values of Krull dimension are the ordinals, including 0, together with $-\infty$. For the Gelfand–Kirillov dimension the possible values are either positive real numbers, 0, or $+\infty$. We will eliminate algebras of infinite Gelfand–Kirillov dimension from consideration. That means we must allow $-\infty$, 0, all positive reals, then all ordinals $\geq \omega$. With that as the possible range of values, a **dimension function** δ for a right Noetherian ring R assigns a value $\delta(M)$ to each finitely generated module M and has the following properties:
(i) $\delta(0) = -\infty$.
(ii) If $0 \to M' \to M \to M'' \to 0$ is exact then $\delta(M) \geq \sup\{\delta(M'), \delta(M'')\}$ with equality if the sequence is split.
(iii) If $MP = 0$ for some prime P and M is a torsion module over R/P, then $\delta(M) + 1 \leq \delta(R/P)$.

If, in condition (ii), equality is always true then δ is called an **exact dimension function**.

8.5 It is clear from 2.4 and 3.11 that the Krull dimension is an exact dimension function. It will be seen, in Chapter 8, that the Gelfand–Kirillov dimension is a dimension function, being exact in many cases, including enveloping algebras of Lie algebras and their homomorphic images.

As another easy example, one can use reduced rank to define an exact dimension function by letting $\delta(M) = -\infty$ if $\rho(M) = 0$, and $\delta(M) = 0$ otherwise.

8.6 We note some elementary consequences of the definitions.

Lemma. *Let δ be a dimension function on R and M_R be finitely generated. Then:*
(i) *if $M = \sum_{i=1}^{n} m_i R$ then $\delta(M) = \sup\{\delta(m_i R) | i = 1, \ldots, n\}$;*
(ii) $\delta(M) \leq \delta(R_R)$;
(iii) *if R is semiprime and $E \triangleleft_e R_R$ then $\delta(E) = \delta(R)$;*
(iv) *if R is prime and $0 \neq I \triangleleft_r R$ then $\delta(I) = \delta(R)$.*

Proof. (i) $\delta(m_i R) \leq \delta(M) \leq \delta(\bigoplus_{i=1}^{n} m_i R) = \sup\{\delta(m_i R) | i = 1, \ldots, n\}$.
(ii) Clear from (i).
(iii) By 2.3.5, E contains a regular element c say. Hence $\delta(cR) \leq \delta(E) \leq \delta(R)$. But $cR \simeq R$ and so $\delta(cR) = \delta(R)$.
(iv) Here I contains a uniform right ideal, U say. Using 3.3.4, there is a direct sum of copies of U which is essential in R. Hence, using (iii), $\delta(U) = \delta(R)$; yet $\delta(U) \leq \delta(I) \leq \delta(R)$. □

8.7 Any factor ring $\bar{R}$ of R inherits a dimension function from R by defining

6.8.9 *Quotient rings and localization* 211

$\delta(M_R) = \delta(M_{\bar R})$ for any finitely generated $\bar R$-module M. The dimensions of certain factor rings are related.

Lemma. (i) *If $P \subset Q$ are prime ideals of R then*
$$\delta(R/Q) + 1 \leqslant \delta(R/P).$$

(ii) *If δ is exact, $P_1, \ldots, P_n$ are the minimal prime ideals of R, and N is the prime radical, then*
$$\delta(R) = \delta(R/N) = \sup\{\delta(R/P_i) | i = 1, \ldots, n\}.$$

Proof. (i) This follows from condition (iii) since R/Q is a torsion R/P-module.
(ii) This can be proved using the argument of 3.8. □

8.8 Given a dimension function δ, a module M is **homogeneous** if $\delta(M) = \delta(M')$ for all nonzero submodules M'. More specifically, it will be called γ-**homogeneous** if $\delta(M) = \gamma$. If R_R is homogeneous, R is said to be **right homogeneous**.

Note that any simple module must be homogeneous, so every nonzero finitely generated module will have a homogeneous factor module. Further examples are provided by 8.6(iv), which shows that R/P is right homogeneous for any prime ideal P. Note also that any nonzero submodule of a γ-homogeneous module is again γ-homogeneous.

8.9 It is more convenient, as 8.7 indicates, to restrict attention to exact dimension functions. In fact, associated with any dimension function δ there is an exact dimension function δ^* defined by
$$\delta^*(M) = \sup\{\delta(N) | N \text{ a homogeneous subfactor of } M\}.$$

Lemma. (i) δ^* *is an exact dimension function.*
(ii) $\delta^*(M) \leqslant \delta(M)$ *for all M.*
(iii) $\delta^*(M) = \delta(M)$ *if M is homogeneous.*
(iv) $\delta^*(R/P) = \delta(R/P)$ *for all prime ideals P.*

Proof. (ii), (iii), (iv) These are clear from the definition and 8.6(iv).
(i) If $0 \to M' \to M \to M'' \to 0$ is exact, it is clear that
$$\delta^*(M) \geqslant \sup\{\delta^*(M'), \delta^*(M'')\}.$$

On the other hand, suppose that N/N' is a γ-homogeneous subfactor of M. We show first that either M' or M'' has a nonzero subfactor isomorphic to a submodule of N/N'. Factoring out N' shows that it is sufficient to consider the case when $N' = 0$. If $N \cap M' \neq 0$ then M' has an appropriate subfactor. Otherwise $N \hookrightarrow M/M' \simeq M''$.

It follows therefore that
$$\gamma \leq \sup\{\delta^*(M'), \delta^*(M'')\}$$
and so
$$\delta^*(M) = \sup\{\delta^*(M'), \delta^*(M'')\}.$$
This shows that the first two conditions on an exact dimension function are satisfied, and the third follows immediately from (iv) □

For the remainder of this section we will take δ to be an exact dimension function. The results obtained can be applied to an inexact δ, using δ^*.

8.10 Next we describe a connection with reduced rank, as defined in 4.1.2.

Proposition. (i) If $\delta(M) = \delta(R)$ then $\rho(M) > 0$.
(ii) If R is prime and $\rho(M) > 0$ then $\delta(M) = \delta(R)$.

Proof. (i) Some product of minimal primes of R is zero, say $P_1 P_2 \cdots P_k$. Exactness shows that some factor of the chain
$$M \supseteq MP_1 \supseteq \cdots \supseteq MP_1 P_2 \cdots P_k$$
has the same dimension as M. This reduces the problem to the case when R is prime. It is easy, now, to reduce to the case when M is cyclic, say $M \simeq R/I$ with $I \triangleleft_r R$. Since $\delta(M) = \delta(R)$, axiom (iii) shows that M is not torsion. So I is not essential and thus $\rho(M) \neq 0$.

(ii) As in (i), it is easy to reduce to the case when M is cyclic, say $M \simeq R/I$; and, once again, I cannot be essential. Therefore M has a subfactor module isomorphic to a uniform right ideal U of R. By 8.6(iv), $\delta(U) = \delta(R)$ and so $\delta(M) = \delta(R)$. □

8.11 Corollary. *Let R be semiprime. Suppose $I \triangleleft_r R$, $\delta(R/I) < \delta(R)$, and $b \in R$. Then there exists $c \in I$ with $\delta(R/(b+c)R) < \delta(R)$.*

Proof. By 3.3.6, there is a $c \in I$ such that $(b+c)R \triangleleft_e bR + I$. Therefore $\rho((bR+I)/(b+c)R) = 0$ and so $\delta(bR+I/(b+c)R) < \delta(R)$. Since
$$\delta(R/bR+I) \leq \delta(R/I) < \delta(R),$$
then $\delta(R/(b+c)R) < \delta(R)$. □

8.12 In discussing quotient rings, the set
$$\mathscr{D} = \mathscr{D}(R) = \{c \in R \mid \delta(R/cR) < \delta(R)\}$$
will be used.

Lemma. (i) $\mathscr{C}'(0) \subseteq \mathscr{D}$.
(ii) If R is right homogeneous then $\mathscr{D} \subseteq {'\mathscr{C}}(0)$.

Proof. (i) Let $c \notin \mathscr{D}$. Then $\delta(R/cR) = \delta(R)$ and so, by 8.10(i), $\rho(R/cR) \geq 1$. Now if $c \in \mathscr{C}'(0)$ then $R/cR \simeq c^n R/c^{n+1} R$ for all n, and so $\rho(R/c^n R) \geq n$. Since $\rho(R)$ is finite, a contradiction occurs.

(ii) Suppose $c \in \mathscr{D}$ and $xc = 0$ for some $x \in R$. Then the map $R/cR \to xR$ via $[a + cR] \mapsto xa$ shows that $\delta(xR) \leq \delta(R/cR) < \delta(R)$. Hence $xR = 0$ and $x = 0$. □

8.13 We also require a property of δ akin to that of $\mathcal{K}$ described in 4.16. An ideal T of R is called **right invariant**, with respect to δ, if $\delta(M \otimes_R T) \leq \delta(M)$ for all finitely generated M_R; and T is **right weakly invariant** if $\delta(M \otimes_R T) < \delta(R/T)$ for such M with $\delta(M) < \delta(R/T)$. If every ideal of R has the property, then R is called **right ideal invariant** or **right weakly ideal invariant** respectively.

Note that to check if an ideal T of R is right invariant or right weakly invariant it is enough to consider $M \otimes T$ for cyclic modules M. Thus 4.16 shows that FBN rings are right, and left, ideal invariant with respect to Krull dimension. Other examples will be noted later in Corollary 8.24.

It is not always the case that R must have such a property. For example, let

$$R = \begin{bmatrix} k & k[x] \\ 0 & k[x] \end{bmatrix},$$

where $k[x]$ is a commutative polynomial ring over a field k. Then R is right Noetherian, by 1.1.7, yet the ideal

$$T = \begin{bmatrix} 0 & k[x] \\ 0 & 0 \end{bmatrix}$$

is not right weakly invariant with respect to Krull dimension. For if

$$I = \begin{bmatrix} 0 & k[x] \\ 0 & k[x] \end{bmatrix}$$

then $\mathcal{K}(R/I) = 0$, yet $\mathcal{K}(R/T) = 1 = \mathcal{K}(T/IT)$.

In this example, R is not left Noetherian. However, there are Noetherian rings which are not weakly ideal invariant on either side with respect to Krull dimension; see [Stafford **85**].

8.14 Proposition. *Let $N = N(R)$, the prime radical of R, and suppose N is right weakly invariant. Then:*
(i) *if $\delta(R/cR + N) < \delta(R)$ then $\delta(R/cR) < \delta(R)$;*
(ii) *if $I \triangleleft_r R$ and $\delta(R/I) < \delta(R)$ then there exists $c \in I$ with $\delta(R/cR) < \delta(R)$;*
(iii) *$\mathscr{D}(R)$ satisfies the right Ore condition.*

Proof. (i) Suppose that $\delta(R/cR + N^k) < \delta(R)$ for some k. Since N is weakly invariant and $\delta(R) = \delta(R/N)$ by 8.7, then $\delta(N/cN + N^{k+1}) < \delta(R)$. However,

$$cR + N/cR + N^{k+1} \simeq N/N \cap (cR + N^{k+1})$$

which is a homomorphic image of $N/cN + N^{k+1}$. Hence $\delta(cR + N/cR + N^{k+1}) < \delta(R)$ and so $\delta(R/cR + N^{k+1}) < \delta(R)$. Induction, together with the nilpotency of N, completes the proof.

(ii) This is clear from 8.11 and (i).

(iii) Let $a \in R$, $d \in \mathcal{D}$ and set $I = \{x \in R \mid ax \in dR\}$. The map $R/I \to R/dR$ given by $[r + I] \mapsto [ar + dR]$ is an embedding; so $\delta(R/I) < \delta(R)$. Thus, by (ii), $I \cap \mathcal{D} \neq \emptyset$. Hence $\mathcal{D}$ satisfies the right Ore condition. □

8.15 Theorem. *Suppose that R is right homogeneous with respect to δ and $N(R)$ is right weakly invariant. Then R has a right Artinian right quotient ring and $\mathcal{D}(R) = \mathcal{C}(0)$.*

Proof. By 8.14 $\mathcal{D}$ satisfies the right Ore condition, and by 8.12 $\mathcal{D} \subseteq {}'\mathcal{C}(0)$. Hence, by Proposition 2.1.10(ii), $\mathcal{D} \subseteq \mathcal{C}(0)$. Now 8.12 shows that $\mathcal{D} = \mathcal{C}(0)$. Further, if $c \in \mathcal{C}(N)$ then $\rho(R/cR + N) = 0$ and so 8.10(i) indicates that $\delta(R/cR + N) < \delta(R)$. By 8.14(i), $\delta(R/cR) < \delta(R)$ and so, by definition, $c \in \mathcal{D}(R)$. Hence $\mathcal{C}(N) \subseteq \mathcal{C}(0)$ and the result follows from Corollary 4.1.4. □

8.16 The next few results apply these ideas to FBN rings, starting with a consequence of 8.15.

Corollary. *Let R be an FBN ring which is Krull homogeneous. Then R has an Artinian quotient ring.*

Proof. By 4.16, R is ideal invariant, so 8.15 applies. □

8.17 This can be extended to cover the inhomogeneous case. Recall first that an **irreducible** ring is one in which the intersection of any two nonzero ideals is nonzero (i.e. the ring is uniform as a bimodule).

Proposition. *An irreducible FBN ring is Krull homogeneous.*

Proof. The Noetherian hypothesis shows that there is a right ideal I maximal with respect to $\mathcal{K}(I) < \mathcal{K}(R)$. We will show that $I = 0$, thus proving the theorem. Note first that if $r \in R$ then $\mathcal{K}(rI) \leq \mathcal{K}(I)$, and so $\mathcal{K}(rI + I) = \mathcal{K}(I)$. Hence $I \triangleleft R$. Let $A = \operatorname{r ann} I$; so $A \triangleleft R$ also. By 4.12, $\mathcal{K}(R/A_R) = \mathcal{K}(I_R) < \mathcal{K}(R_R)$ and then Corollary 4.13 shows that $\mathcal{K}({}_R R/A) < \mathcal{K}({}_R R)$. By 8.14, there exists $x \in A$ with $\mathcal{K}(R/Rx) < \mathcal{K}(R)$ (using 4.16 again).

By 2.3.2, there is an n such that $Rx^n \cap \mathrm{l\,ann}\, x^n = 0$. Since $x^n \in A$ and $\mathcal{K}(R/Rx) = \mathcal{K}(R/Rx^n)$ we may as well suppose that $n = 1$.

Let $T = \mathrm{l\,ann}(R/Rx)$; then $\mathcal{K}(R/T) = \mathcal{K}(R/Rx)$ by 4.12. Therefore $\mathcal{K}(R/T) < \mathcal{K}(R)$ and so $T \neq 0$. However, $T \subseteq Rx \subseteq A = \mathrm{r\,ann}\, I$. Hence $I \subseteq \mathrm{l\,ann}\, x$ and then

$$T \cap I \subseteq Rx \cap \mathrm{l\,ann}\, x = 0.$$

Since $T \neq 0$ and R is irreducible, $I = 0$ as required. □

8.18 It is well known, and an easy exercise to show, that any (Noetherian) ring embeds in a (finite) direct product of irreducible rings. Thus 8.16 and 8.17 combine to prove

Corollary. *Any FBN ring can be embedded in an Artinian ring.* □

8.19 The next few results describe connections between right weak invariance of an ideal I and I being a right localizable ideal.

8.20 Proposition. *Let P be a right localizable semiprime ideal with R/P right homogeneous. Then P is right weakly invariant.*

Proof. Suppose $I \triangleleft_r R$ with $\delta(R/I) < \delta(R/P)$. By 8.11 there exists $c \in I$ with $\delta(R/cR + P) < \delta(R/P)$. Then $c \in \mathscr{C}(P)$ by 8.15.

Let $x \in I \cap P$. Since P is right localizable, there exist $a \in R$, $d \in \mathscr{C}(P)$ such that $xd = ca$; but then $a \in P$ and so $xd \in IP$. Thus $x(dR + P) \subseteq IP$. Hence

$$\delta(xR + IP/IP) \leqslant \delta(xR/x(dR + P)) \leqslant \delta(R/dR + P) < \delta(R/P)$$

since $d \in \mathscr{C}(P)$. Therefore $\delta(I \cap P/IP) < \delta(R/P)$. Now

$$P/I \cap P \simeq I + P/I \hookrightarrow R/I.$$

Therefore $\delta(P/I \cap P) < \delta(R/P)$ and so $\delta(P/IP) < \delta(R/P)$. It follows that P is right weakly invariant. □

8.21 The next result is a partial converse.

Proposition. *Let P be a semiprime right weakly invariant ideal and R/P be right homogeneous. Then:*
(i) *$\mathscr{C}(P/P^n)$ is a right Ore set in R/P^n for all n;*
(ii) *if P has the right AR property then P is right localizable.*

Proof. (i) Let $P^{[n]} = \sum \{I \triangleleft_r R \mid I \supseteq P^n,\ \delta(I/P^n) < \delta(R/P)\}$. Then clearly $P^{[1]} = P$ and $P^{[n-1]} \supseteq P^{[n]} \supseteq P^n$ for all n. Moreover, $R/P^{[n]}$ is right homogeneous for each n. Evidently $P/P^{[n]}$ is the prime radical of $R/P^{[n]}$, and so 8.15 applies. Thus, by

4.1.4, $\mathscr{C}(P) = \mathscr{C}(P^{[n]})$ for all n. Furthermore, if $a \in R$ and $c \in \mathscr{C}(P)$, then there exist $a' \in R, c' \in \mathscr{C}(P), q \in P^{[n]}$, such that $ac' - ca' = q$.

Set $I = \{r \in R \mid qr \in P^n\}$; then $R/I \simeq qR + P^n$ and so $\delta(R/I) < \delta(R/P)$. By 8.11 there exists $d \in I$ such that $\delta(R/dR + P) < \delta(R/P)$, and 8.15 shows that $d \in \mathscr{C}(P)$. Thus $ac'd - ca'd \in P^n$ with $c'd \in \mathscr{C}(P)$ as required.

(ii) This now follows from 4.2.10. □

8.22 These combine to yield:

Corollary. (i) *If P is a semiprime right AR ideal and R/P is right homogeneous, then P is right localizable if and only if P is right weakly invariant.*

(ii) *If R is a Noetherian AR ring then every prime ideal of R is right weakly invariant.*

Proof. (i) This follows from 8.20 and 8.21.

(ii) By 4.2.11 each prime ideal of R is localizable. Since, for a prime ideal P, R/P is always right homogeneous, the result follows from (i). □

8.23 The next sequence of results concentrates on the condition, in 8.15, that $N(R)$ is right weakly invariant, establishing its equivalence with other related properties.

8.24 Theorem. *Let α be an ordinal, and suppose that every prime ideal P of R with $\delta(R/P) = \alpha$ is right weakly invariant. Then every ideal T with $\delta(R/T) = \alpha$ is right weakly invariant.*

Proof. By Noetherian induction we will assume the result true for all proper factor rings of R, that $T \triangleleft R$ with $\delta(R/T) = \alpha$ and that every ideal $T' \supset T$ with $\delta(R/T') = \alpha$ is right weakly invariant. Let $I \triangleleft_r R$ with $\delta(R/I) < \alpha$.

If T is prime, there is nothing to prove; so we can suppose that there are ideals A, B properly containing T with $AB \subseteq T$, and B could be chosen to be prime if desired. For the present we do not insist on B being prime.

First we check that AB is right weakly invariant. Note that $\delta(R/AB) = \alpha = \sup\{\delta(R/A), \delta(R/B)\}$. If $\delta(R/A) = \alpha$ then, by hypothesis, A is right weakly invariant, so $\delta(A/IA) < \alpha$. On the other hand, if $\delta(R/A) < \alpha$, then $\delta(I \cap A/IA) < \alpha$ since $I \cap A/IA$ is an R/A-module. Therefore $\delta(A/IA) < \alpha$ in both cases. Repeating the argument for B we see that $\delta(AB/IAB) < \alpha$ as required.

Next we consider the case that $AB \neq 0$. Then T/AB is right weakly invariant in R/AB and so $\delta(T/IT + AB) \leqslant \delta(R/I + AB) < \alpha$. Also

$$\delta(IT + AB/IT) = \delta(AB/AB \cap IT) \leqslant \delta(AB/IAB) < \alpha,$$

and thus $\delta(T/IT) < \alpha$, as required. Symmetry deals with the case when $BA \neq 0$.

Thus we can suppose that $AB = BA = 0$. Then $\delta(R) = \alpha$. Suppose there exists

6.8.26 *Quotient rings and localization* 217

$0 \neq D \triangleleft R$ with $\delta(D) < \alpha$. Then $\delta(R/T + D) = \alpha$ and so $T + D/D$ is a right weakly invariant ideal of R/D. Therefore $\delta(T + D/IT + D) < \alpha$ and so $\delta(T/IT) < \alpha$. Thus we reduce to the case when R is right homogeneous.

Suppose next that $A \cap B \neq T$. In that case, the prime radical N of R strictly contains T, and $\delta(R/N) = \alpha$. Therefore N is right weakly invariant. By 8.14, there exists $c \in I$ with $\delta(R/cR) < \delta(R)$; by 8.15, $c \in \mathscr{C}(0)$. Hence, $c^{n-1}T/c^n T \simeq T/cT$ for $n = 1, 2, \ldots$. Since $\rho(T) < \infty$, $\rho(T/cT) = 0$. Hence, by 8.10, $\delta(T/cT) < \alpha$, and so $\delta(T/IT) < \alpha$ as required.

Thus, finally, we may suppose that $T = A \cap B$; and so $A \not\subseteq B$. We now recall that we can choose that B be prime. Then, by 8.10, $\delta(R/A + B) < \delta(R/B) \leq \alpha$. However, since $T(A + B) = 0$, we have

$$\delta(T/IT) \leq \delta(T) \leq \delta(R/A + B) < \alpha$$

which completes the proof. □

Corollary. *If R is a Noetherian AR ring then R is weakly ideal invariant.* □

This is used in [Brown, Lenagan and Stafford **81**] to show that a polycyclic by finite group algebra over a field of characteristic $p > 0$ or over $\mathbb{Z}$ is weakly ideal invariant.

8.25 As with the Krull dimension, a finitely generated module M is called α-**critical** if $\delta(M) = \alpha$ and $\delta(M/M') < \alpha$ for all submodules $M' \neq 0$. It follows, as before, that M is uniform, that M' is also α-critical, etc.

8.26 Theorem. *Let $\delta(R) = \alpha$. Then the following conditions are equivalent:*
  (i) *$N(R)$ is right weakly invariant;*
  (ii) *each prime ideal P of R with $\delta(R/P) = \alpha$ is right weakly invariant;*
  (iii) *each ideal T of R with $\delta(R/T) = \alpha$ is right weakly invariant;*
  (iv) *each finitely generated α-critical right R-module M is prime (i.e. $\operatorname{ann} M = \operatorname{ass} M$).*

Proof. (i) $\Rightarrow$ (ii) Let P be a prime ideal of R, $\delta(R/P) = \alpha$; and let $I \triangleleft_r R$, $\delta(R/I) < \alpha$. Then $\delta(N/IN) < \alpha$ and so $\delta(IP + N/IP) < \alpha$. But, as is easily checked, each minimal prime of a semiprime ring is right weakly invariant. So P/N is right weakly invariant, and hence $\delta(P/IP + N) < \alpha$. Thus $\delta(P/IP) < \alpha$ and P is right weakly invariant.

(ii) $\Rightarrow$ (iii) This follows from 8.24.

(iii) $\Rightarrow$ (iv) Choose $M_1 \subseteq M$ such that $\operatorname{ass} M = \operatorname{ann} M_1 = P$ say. Then $\delta(R/P) = \alpha$ and P is right weakly invariant. Since $\delta(M/M_1) < \alpha$, $\delta(MP) = \delta(MP/M_1 P) < \alpha$ and so $MP = 0$.

(iv) $\Rightarrow$ (i) Say $I \triangleleft_r R$, $\delta(R/I) < \alpha$. By 8.11, there exists $c \in I$ with $\delta(R/cR + N) < \alpha$. Suppose $\delta(R/cR) = \alpha$. Then R/cR must have an α-critical factor, R/J say, with

$J \supseteq cR$. By (iii), $(R/J)P = 0$ for some prime P. And so $J \supseteq P \supseteq N$ and $\delta(R/J) \leq \delta(R/cR + N) < \alpha$, a contradiction. Hence $\delta(R/cR) < \alpha$ as in 8.14(i). Let $S = \sum \{K \triangleleft_r R \,|\, \delta(K) < \alpha\}$. The proofs of 8.14(ii)(iii) and 8.15 now show that $\mathscr{D}(R/S) \subseteq \mathscr{C}_{R/S}(0)$. Thus $\delta(N + S/cN + S) < \delta(R/S) = \alpha$. Hence

$$\delta(N/IN) \leq \delta(N/cN) \leq \sup \{\delta(N + S/cN + S), \delta(S)\} < \alpha. \qquad \square$$

8.27 Drawing together 8.15 and 8.26 we get

Corollary. *If R is right homogeneous and each minimal prime ideal of R is right weakly invariant, then R has a right Artinian right quotient ring.* $\qquad \square$

8.28 Finally, as an illustration, we drop the convention that δ is exact, and show what 8.27 then says. We use the exact function δ^*, as in 8.9.

Corollary. *Let δ be a dimension function. Suppose that R is right δ-homogeneous and that for each minimal prime P of R and each finitely generated module M_R with $\delta^*(M) < \delta(R/P)$ one has $\delta^*(M \otimes P) < \delta(R/P)$. Then R has a right Artinian right quotient ring.*

Proof. The fact that R is right δ-homogeneous means that $\delta(I) = \delta^*(I)$ for each nonzero right ideal I of R, and so R is δ^*-homogeneous. Also 8.9 shows that $\delta^*(R/P) = \delta(R/P)$; so the conditions, interpreted for δ^*, are those of 8.27. $\qquad \square$

§9 Skew polynomials over commutative rings

9.1 This section concerns skew polynomial rings $R[x; \delta]$ and skew Laurent polynomial rings $R[x, x^{-1}; \sigma]$ over a commutative Noetherian ring R. The obvious symmetry in such rings makes their left and right Krull dimensions the same. We know from Proposition 5.4 that their Krull dimensions are either $\mathcal{K}(R)$ or $\mathcal{K}(R) + 1$, with the latter being the correct value when $\delta = 0$ or $\sigma = 1$ respectively or, of course, if $\mathcal{K}(R) = 0$. Further, it was seen in 6.7 and 6.11 that the former is the correct value if R has no δ-stable, or σ-stable, ideal I with $\mathcal{K}(R/I) = 0$.

In fact there are definitive results describing exactly when each of these possible values occurs; they are the focus of this section. Since, as the comments above indicate, there is a great similarity between the two different types of extension ring, we choose to give them the same name, S, and to give common proofs, wherever possible, for the two cases. In the main, R will be commutative and Noetherian; but some results are proved a little more generally for later use.

9.2 Given a commutative Noetherian ring R it is easy to obtain a chain of ideals

$$R = I_0 \supset I_1 \supset \cdots \supset I_n = 0$$

6.9.7 *Skew polynomials over commutative rings* 219

such that $I_i/I_{i+1} \simeq R/Q_i$ with $Q_i \in \mathrm{Spec}\, R$. Then S has a corresponding chain of right ideals $\{I_i S\}$ with $I_i S/I_{i+1} S \simeq S/Q_i S$. Hence $\mathcal{K}(S) = \sup\{\mathcal{K}(S/PS)\}$ as P runs through the primes (or, indeed, the minimal primes) of R. This explains the emphasis below on prime ideals of R.

9.3 First a general notion is useful. A nonzero module M_R over an arbitrary ring R is **compressible** if for any $0 \neq N \triangleleft M$ there exists a monomorphism $M \hookrightarrow N$.

It is clear from 3.3.4 that any uniform right ideal of a prime right Goldie ring is compressible. In particular R/P is compressible for any prime P of a commutative Noetherian ring R. The next few results will show that S/PS also is compressible.

9.4 Lemma. *A compressible module M_R with Krull dimension is critical.*

Proof. Clear from 2.10 and 2.11. □

9.5 Lemma. *Let R be any ring, $I \triangleleft_r R$ and $S = R[x, x^{-1}; \sigma]$. Then $S/IS \simeq S/\sigma^n(I)S$ for all $n \in \mathbb{Z}$.*

Proof. Map $[x^n + IS] \mapsto [1 + \sigma^n(I)S]$ and note that $\mathrm{ann}_S[x^n + IS] = \sigma^n(I)S$. □

9.6 Proposition. *Let R be a ring and S be $R[x; \delta]$ or $R[x, x^{-1}; \sigma]$. If M_R is compressible then $(M \otimes_R S)_S$ is compressible.*

Proof. Let N be a cyclic submodule of M. Then $M \hookrightarrow N$ and hence $M \otimes_R S \hookrightarrow N \otimes_R S$. Thus there is no loss of generality in supposing M to be cyclic; so we let $M = R/I$ and identify $M \otimes_R S$ with S/IS.

Let $0 \neq L \triangleleft (S/IS)_S$. Choose $0 \neq y \in L$, $y = \sum_{j=0}^n m_j \otimes x^j$, so as to minimize n. Then $m_n \neq 0$ and so $M \hookrightarrow m_n R$. Therefore there exists $r \in R$ with $\mathrm{ann}_R m_n r = I$. Replacing y by $y\sigma^n(r)$ (taking $\sigma = 1$ when $S = R[x; \delta]$) we may suppose $\mathrm{ann}_R m_n = I$. Hence $\mathrm{ann}_R(m_n \otimes x^n) = \sigma^n(I)$ and then the minimality of n makes $\mathrm{ann}_R y = \sigma^n(I)$.

Suppose, for a moment, that $\mathrm{ann}_S y = \sigma^n(I)S$. Then
$$yS \simeq S/\sigma^n(I)S \simeq S/IS \simeq M \otimes_R S$$
and so $M \otimes_R S \hookrightarrow L$ as required.

Thus it remains only to check that $\mathrm{ann}_S y = \sigma^n(I)S$. Suppose that $yz = 0$ for some $z \in S \setminus \sigma^n(I)S$. We may suppose that z has the form $\sum_{i=0}^k r_i x^i$ with $r_k \notin \sigma^n(I)$. However, then the highest coefficient of yz is nonzero, which is a contradiction. □

9.7 We can now draw the conclusion at which we have been aiming.

Corollary. *If R is a commutative Noetherian ring and $P \in \operatorname{Spec} R$ then $(S/PS)_S$ is compressible and critical.* □

9.8 Next we turn to δ-stable and σ-stable prime ideals. In fact a slightly generalized notion is required. A prime ideal P of a ring R will be called σ-**semistable** if $\sigma^n(P) = P$ for some $n \neq 0$. For δ, the characteristic is important; P is called δ-**semistable** if either char $R/P > 0$ or P is δ-stable. Prime ideals which are not semistable are **unstable**.

9.9 When P is a semistable prime ideal, there is a largest stable ideal inside P; this is denoted P^0.

Lemma. (i) *If $P \in \operatorname{Spec} R$ is σ-semistable then $P^0 = \bigcap \sigma^k(P)$.*
(ii) *If $P \in \operatorname{Spec} R$ is δ-semistable and char $R/P = 0$ then $P^0 = P$.*
(iii) *If $P \in \operatorname{Spec} R$ is δ-semistable and char $R/P > 0$ then*

$$P^0 = \{r \in R \mid \delta^n(r) \in P \text{ for all } n \geq 0\}.$$

Proof. Straightforward. □

9.10 The relationship between P and P^0 is important to this section. First we concentrate on the case of finite characteristic. As an example consider $k[y]$ with k a field of characteristic 2, $\delta = d/dy$ and $P = (y)$. Then $P^0 = (y^2)$.

Proposition. *Let R be a commutative Noetherian ring, $P \in \operatorname{Spec} R$, and char $R/P = q > 0$. Then:*
(i) *R/P^0 has characteristic q and P/P^0 is nilpotent;*
(ii) *R/P^0 has an Artinian quotient ring.*

Proof. (i) Let K be the ideal of R generated by qR and the set $\{r^q \mid r \in P\}$. Note that $\delta(K) \subseteq K \subseteq P$ and so $K \subseteq P^0$. Therefore $qR \subseteq P^0$ and char $R/P^0 = q$. Also $r^q \in P^0$ for all $r \in P$, and so P/P^0 is nil and hence nilpotent.

(ii) By passing to R/P^0, we may suppose $P^0 = 0$. Then P is the prime radical of R. We aim to show that $\mathscr{C}(P) = \mathscr{C}(0)$ which, by Theorem 4.1.4, will complete the proof of (ii). First, choose $c \in \mathscr{C}(0)$ and note that $c \notin P$ since P is nilpotent. Therefore $\mathscr{C}(0) \subseteq R \setminus P = \mathscr{C}(P)$.

Next consider the localization R_P of R with respect to $\mathscr{C}(P)$. The map $\alpha: R \to R_P$ extends to a map $R[x;\delta] \to R_P[x;\delta]$; and α is an $R[x;\delta]$-module homomorphism. Hence $\ker \alpha$ is a δ-stable ideal of R contained in P. Therefore $\ker \alpha = 0$ and so $\mathscr{C}(P) \subseteq \mathscr{C}(0)$. □

9.11 It is now easy to describe the connection between semistable prime ideals and stable ideals.

6.9.13 *Skew polynomials over commutative rings* 221

Proposition. *Let R be a commutative Noetherian ring. Then $P \in \operatorname{Spec} R$ is semistable if and only if P is a minimal prime of a stable ideal.*

Proof. (σ) This is straightforward.

(δ)($\Rightarrow$) This is clear from the definition if char $R/P = 0$, and from 9.10 otherwise.

($\Leftarrow$) By passing to the factor of R by the stable ideal, we may suppose that P is a minimal prime of R. It is enough to show that if char $R/P = 0$ then $\delta(P) \subseteq P$. Now the ring R_P is an Artinian local ring with radical $N = P_P$. Suppose $n \in N$ with $\delta(n) \notin N$, and choose the least m with $n^m = 0$. Then $0 = \delta(n^m) = m\delta(n)n^{m-1}$. However, $m\delta(n)$ is a unit of R_P and so $n^{m-1} = 0$. This contradiction shows that $\delta(N) \subseteq N$. Since $N = P_P$ is δ-stable so too is P, its inverse image in R. □

9.12 For either of the two types of extension ring S, semistable primes of R lead to prime ideals in S. As usual it is enough to deal with the case when $P^0 = 0$.

It is convenient to introduce some notation here. Given any right ideal I of S the set of all leading coefficients of elements of I is denoted by I_∞. Evidently $I_\infty = \bigcup I_n$, where the I_n are the leading ideals of I as in 1.2.9 and 1.8.5.

Lemma. *Let R be a commutative Noetherian ring and P a semistable prime with $P^0 = 0$. Then (i) S is prime, and (ii) $K(S) = K(S/PS)$.*

Proof. (i) (δ) If char $R/P = 0$ then $P = 0$ and S is a domain. If char $R/P > 0$ then suppose that A_1, A_2 are ideals of $R[x;\delta]$ with $A_1 A_2 = 0$. Then each $(A_i)_\infty$ is a δ-stable ideal of R and hence so too is $(A_1)_\infty (A_2)_\infty$. Yet $(A_1)_\infty (A_2)_\infty = 0$ and so $(A_i)_\infty \subseteq P$ for some i. Then $(A_i)_\infty = 0$ and $A_i = 0$.

(σ) This is proved similarly.

(ii) Straightforward. □

9.13 If R is a commutative Noetherian ring and $P \in \operatorname{Spec} R$ then the height of P (see 4.1.11) equals that of PR_P in R_P; so $\operatorname{ht} P = K(R_P)$. The main result of this section will show that if $K(R) < \infty$ then $K(S) = K(R) + 1$ if and only if there is a semistable prime P of R with $\operatorname{ht} P = K(R)$. One implication can be established here.

Theorem. *Let R be a commutative Noetherian ring with a semistable prime P such that $\operatorname{ht} P = K(R) < \infty$. Then $K(S) = K(R) + 1$.*

Proof. Consider the localizations R' of R and S' of S with respect to $\mathscr{C}(P^0)$. It will be enough to prove the theorem for these rings since it is clear that $K(R') = K(R)$ and $K(S') \leqslant K(S)$ by Lemma 5.3. So we suppose $R = R'$.

The route followed will be to show that if M_R is simple then $K(M \otimes_R S) > 0$. This will enable us to apply 1.17 to the lattice map $\mathscr{L}(R) \to \mathscr{L}(S)$, taking $\gamma = 1$

and $\delta = 0$, and deduce that $K(S) \geq 1 + K(R)$. We now consider the two cases separately.

(δ) In this case R is local with maximal ideal P and so R/P is the only simple R-module. Note also that $S/P^0S = (R/P^0)[x;\delta]$ which is not Artinian. However, R/P_0 has finite length over R, so its composition factors are isomorphic to R/P. Hence $(R/P \otimes S)_S$ is not Artinian.

(σ) Say $\sigma^n(P) = P$. Then $P, \sigma(P), \ldots, \sigma^{n-1}(P)$ are the maximal ideals of R. Now $(R/P^0)[x, x^{-1}; \sigma]$ is not Artinian; yet $R/P^0 \simeq \oplus R/\sigma^i(P)$ and so, by 9.5, $S/P^0S \simeq ((R/P) \otimes S)^n$. Hence $(R/P) \otimes S$ is not Artinian. □

9.14 In order to prove a converse result, a closer examination of the modules S/PS is needed.

Lemma. *Let I be an ideal of a ring R.*
(i) *If $S = R[x; \delta]$ then each finitely generated R-submodule of S/IS is annihilated by a power of I.*
(ii) *If $S = R[x, x^{-1}; \sigma]$ then each finitely generated R-submodule of S/IS is annihilated by a product of the ideals $\sigma^m(I)$.*

Proof. Note first that $S/IS \simeq (R/I) \otimes_R S$.
(i) Let $\bar{s} \in S/IS$; so $\bar{s} = \sum_{i=0}^n \bar{a}_i \otimes x^i$ with $\bar{a}_n \neq 0$. If $r \in I$ then $\bar{s}r$ has degree $\leq n - 1$. Thus, by induction on degree, $\bar{s}I^{n+1} = 0$.
(ii) If $\bar{s} = \sum_{i=l}^n \bar{a}_i \otimes x^i$ then $\bar{s}\sigma^n(r)$ has shorter length for each $r \in I$ (where $n - l$ is the length of $\bar{s}$). Induction on length gives the result. □

9.15 Lemma. *Let R be a commutative ring, P an unstable prime and $I \triangleleft_r S$ with $I \supset PS$. Then $I \cap R \supset P$.*

Proof. (δ) Let $I_0, I_1, \ldots$ be the leading ideals of I, as in 1.2.9. Since $I \supset PS$, $I_m \supset P$ for some m. If $m > 0$, let $f = a_0 + \cdots + a_m x^m \in I$ with $a_m \notin P$ and choose $p \in P$ such that $\delta(p) \notin P$. Since char $R/P = 0$, $m\delta(p) \notin P$ and so $m\delta(p)a_m \notin P$. However,
$$I \ni pf - fp = \cdots + m\delta(p)a_m x^{m-1}.$$
Thus $I_{m-1} \supset P$ and hence $I_0 \supset P$ as claimed.

(σ) The leading ideals of I are defined as in 1.8.5 to be those of $I \cap R[x; \sigma]$. Suppose f is as above. Since P is unstable either $\sigma^m(P) \not\subseteq P$ or $P \not\subseteq \sigma^m(P)$. In the latter case $\sigma^{-m}(P) \not\subseteq P$ so choose $p \in P$ with $\sigma^{-m}(p) \notin P$ and, using the element $(pf - fp)x^{-1}$, proceed as above. In the former case choose $p \in P$ with $\sigma^m(p) \notin P$ and use the element $(\sigma^m(p)f - f\sigma^m(p))x^{-1}$. □

9.16 Lemma. *Let R be a commutative Noetherian ring and M a finitely generated S-module. Then there is a chain of S-submodules*
$$M = M_0 \supset \cdots \supset M_n = 0$$

6.9.19   *Skew polynomials over commutative rings*   223

such that to each i there corresponds a prime P_i of R and either (a) $M_i/M_{i+1} \simeq S/P_iS$, or else (b) P_i is semistable and M_i/M_{i+1} is a proper homomorphic image of S/P_iS and is also a torsion-free R/P_i^0-module.

Proof. It is enough to show that M has a submodule M_1 as required and then use Noetherian induction. Choose $m \in M$ to maximize $\operatorname{ann}_R m$. So $\operatorname{ann}_R m$ is a prime $P \in \operatorname{ass} M$ and mS is a homomorphic image of S/PS. If $mS \simeq S/PS$ all is well. Also, if P is unstable then 9.15 shows that $mS \simeq S/PS$, using the maximality of P.

Thus we can suppose P to be semistable and mS to be a proper homomorphic image of S/PS. Note that $P^0 S = SP^0$, so mS is an (R/P^0)-module. It remains to show that it is torsion-free. We may, by passing to factor rings, take P^0 to be zero. Suppose $0 \neq m' \in mS$, $c \in \mathscr{C}_R(0)$ and $m'c = 0$. We will obtain contradictions in each case.

(δ) Suppose char $R/P > 0$. Note that by 9.10 $\mathscr{C}_R(0) = \mathscr{C}_R(P)$. Thus $\operatorname{ann} m' \not\subseteq P$. Hence for some $Q \in \operatorname{ass} m'R$, $Q \not\subseteq P$. However, by 9.10 again, P is the unique minimal prime of R. Thus $P \subset Q$, which contradicts the choice of P.

Next suppose char $R/P = 0$. Then $P = P^0 = 0$ and so the fact that $\operatorname{ann}_R m' \neq 0$ contradicts the choice of P.

(σ) The minimal primes $\{P_i\}$ of R are just the ideals $P, \sigma(P), \ldots, \sigma^{n-1}(P)$; and so $\mathscr{C}_R(0) = \bigcap \mathscr{C}_R(P_i)$. Thus $c \notin P_i$ for any i and hence $\operatorname{ass}_R m'R$ is a non-minimal prime, Q say. Therefore $\sigma^i(Q) \supset P$ for some i. However, $\operatorname{ass}_R m'Rx^i = \sigma^i(Q)$, a contradiction. $\square$

9.17 We next aim to discover, in the ideals of leading coefficients I_∞, J_∞ of two right ideals I, J of S, consequences of an isomorphism $S/PS \simeq I/J$. Before doing so it is convenient to introduce some terminology. An element of the form $s = \sum_{k=0}^n r_k x^k$ is a **polynomial element** of S and, if $r_n \neq 0$, its **degree** is n. More generally, if $J \triangleleft_r S$ then $[s + J]$ is a **polynomial element** of S/J if it has a representative s which is a polynomial element of S; the **degree** of $[s + J]$ is the least possible degree of such a representative.

9.18 We also require some simple facts concerning the ideal I_∞ of leading coefficients of a right ideal I of S.

Lemma. *Let R be right Noetherian.*
(i) *If $I \triangleleft_r S$ then $I_\infty = I_n$ for some n.*
(ii) *If $S = R[x; \delta]$ and $I, J \triangleleft_r S$ with $J \subseteq I$ and $I_\infty = J_\infty$ then $(I/J)_R$ is finitely generated.*

Proof. Straightforward. $\square$

9.19 Lemma. *Let R be commutative Noetherian and let $I, J \triangleleft_r S$ with $I/J \simeq S/PS$*

where $P \in \operatorname{Spec} R$. Then there exists $\beta \in I_\infty/J_\infty$ such that $\operatorname{ann}_R \beta \subseteq \sigma^n(P)$ for some n (with $\sigma = 1$ when $S = R[x; \delta]$).

Proof. (δ) Suppose, to the contrary, that $\operatorname{ann} \beta \nsubseteq P$ for all $\beta \in I_\infty/J_\infty$. Then $\operatorname{ann} \beta \cap \mathscr{C}_R(P) \neq \varnothing$ and so, localizing with respect to $\mathscr{C}_R(P)$, $(I_P)_\infty = (J_P)_\infty$. Hence, by 9.18, I_P/J_P is finitely generated over R_P. However, $I_P/J_P \simeq (S/PS)_P \simeq S_P/P_P S_P$ which evidently is not finitely generated over R. This contradiction provides β as required.

(σ) Suppose, to the contrary, that $\operatorname{ann}_R \beta \nsubseteq \sigma^n(P)$ for all $\beta \in I_\infty/J_\infty$ and all n. Let $\varphi: S/PS \to I/J$ be the given isomorphism and $\pi: S \to S/PS$ the canonical surjection. Choose a positive integer u such that $I_u = I_\infty$ and $J_u = J_\infty$; then let M be the set of all polynomial elements of I/J and N be the subset of polynomial elements of degree at most u. Thus M, N are R-submodules of I/J, with N being finitely generated.

Evidently if $[a + J] \in I/J$ then $[ax^k + J] \in M$ for $k \gg 0$. In particular, $\varphi\pi(1)x^k = \varphi\pi(x^k) \in M$ for all $k \geq k_0$ say. Since the sum $\sum \pi(x^k)R \subseteq S/PS$ is direct, so too is the sum $\sum_{k \geq k_0} \varphi\pi(x^k)R \subseteq M$. However, N_R is finitely generated, so there is a $k \geq k_0$ with $\varphi\pi(x^k)R \cap N = 0$.

Let $y \in I \cap R[x; \sigma]$ be such that $[y + J] = \varphi\pi(x^k)$. Since $[y + J]R \cap N = 0$ then degree $[y + J] > u$; and since $\operatorname{ann}_R \pi(x^k) = \sigma^k(P)$ then $\operatorname{ann}_R [y + J] = \sigma^k(P)$. Furthermore, if $r \in R$ is such that $[yr + J] \neq 0$ it follows that degree $[yr + J] > u$ and $\operatorname{ann}_R [yr + J] = \sigma^k(P)$. We choose, amongst such r, one which minimizes degree $[yr + J]$. Suppose that the minimal degree polynomial representing this is $\sum_{i=0}^{v} r_i x^i$. Note that $r_v \in I_v = I_u = I_\infty$. By assumption, $\operatorname{ann}_R [r_v + J_\infty] \nsubseteq \sigma^n(P)$ for any n. Thus there exists $c \in R \setminus \sigma^k(P)$ such that $[r_v + J_\infty]\sigma^{-v}(c) = 0$. It follows that $[yrc + J] \neq 0$ and that degree $[yrc + J] <$ degree $[yr + J]$. This contradiction gives the result. $\square$

It can be shown that there is such a β with $\operatorname{ann}_R \beta = \sigma^n(P)$ for some n.

9.20 The next two results are at the heart of the proof of the converse to 9.13.

Lemma. *Let R be commutative Noetherian of finite Krull dimension and P a semistable prime of R such that $P^0 = 0$ and, for any prime Q with $\operatorname{ht} Q \geq 1$, $K(S/QS) \leq K(R) - \operatorname{ht} Q$. Then $K(S) = K(R)$.*

Proof. Note, by 9.12, that S is prime. Let $K(R) = n$. We know that $K(S)$ is either n or $n + 1$. Hence, by 9.12, it is sufficient to show that $K(S/PS) \neq n + 1$.

Suppose that $K(S/PS) = n + 1$; then there exists $E \triangleleft_r S$ with $E \supset PS$ and $K(S/E) = n$. Since S/PS is critical and hence uniform, $E/PS \triangleleft_e S/PS$ and thus $E \triangleleft_e S$. Since $K(S/E) = n$, 1.7 gives a chain of right ideals

$$S = J_0 \supset J_1 \supset J_2 \supset \cdots \supset E$$

with $\mathcal{K}(J_i/J_{i+1}) = n - 1$ for infinitely many i, say for $i \in \Lambda$. By 9.16 it may be assumed that this chain has been so refined that to each factor J_i/J_{i+1} is associated a prime $P_i \in \operatorname{Spec} R$ and J_i/J_{i+1} has the properties (a) or (b) described in 9.16. If $i \in \Lambda$ then J_i/J_{i+1} is a homomorphic image of $S/P_i S$ and also $\mathcal{K}(J_i/J_{i+1}) = n - 1$. Hence $\mathcal{K}(S/P_i S) \geqslant n - 1$. The hypothesis implies that ht $P_i \leqslant 1$, therefore. Moreover, if ht $P_i = 1$ then $\mathcal{K}(S/P_i S)$ will have to equal $n - 1$ and, since $S/P_i S$ is critical, it follows that $J_i/J_{i+1} \simeq S/P_i S$. So if J_i/J_{i+1} is as described in 9.16(b) and $i \in \Lambda$ then ht $P_i = 0$.

The hypothesis on P ensures, by 9.10, that R has an Artinian quotient ring, $R_\mathscr{C}$ say. Then $S_\mathscr{C} = R_\mathscr{C}[x; \delta]$ or $R_\mathscr{C}[x, x^{-1}; \sigma]$, $\mathcal{K}(S_\mathscr{C}) = 1$ and $S_\mathscr{C}$ is prime. Since $ES_\mathscr{C} \triangleleft_e S_\mathscr{C}$ then $S_\mathscr{C}/ES_\mathscr{C}$ must have finite length. So there can only be finitely many i for which $(J_i/J_{i+1}) \otimes_S S_\mathscr{C} \neq 0$. However, $\mathscr{C} = \mathscr{C}_R(0) = \bigcap \mathscr{C}_R(Q_j)$ as Q_j varies through the minimal primes of R. If J_i/J_{i+1} is as described in 9.16(b), and so by the preceding paragraph is torsion-free over R/P_i^0 with ht $P_i = 0$, then $(J_i/J_{i+1}) \otimes_S S_\mathscr{C} \neq 0$. Therefore, there must be infinitely many $i \in \Lambda$ with $J_i/J_{i+1} \simeq S/P_i S$ and ht $P_i = 1$.

To simplify notation, we write J_i^∞ rather than $(J_i)_\infty$. The infinite chain $\{J_i\}$ gives an infinite chain $\{J_i^\infty\}$ between R and E^∞. Note first that $E^\infty \triangleleft_e R$; for if $0 \neq I \triangleleft R$ then $E \cap IS \neq 0$ and so $0 \neq E^\infty \cap (IS)^\infty = E^\infty \cap I$. Next, suppose i is such that ht $P_i = 1$ and so $J_i/J_{i+1} \simeq S/P_i S$. By 9.19 there exists $\beta_i \in J_i^\infty/J_{i+1}^\infty$ with $\operatorname{ann}_R \beta_i \subseteq \sigma^{n(i)}(P_i)$ for some $n(i)$. Evidently $E^\infty \subseteq \operatorname{ann}_R \beta_i$; but also, since $E \supset PS$, $E^\infty \supset P$. Thus $\sigma^{n(i)}(P_i)$, being a prime of height 1 containing E^∞, must be a minimal prime of E^∞.

Let $\mathscr{D}$ be the complement in R of $\bigcup P'$ with P' ranging over those height 1 primes of R which are minimal primes of E^∞. Then in $R_\mathscr{D}$ all maximal primes have height 1 and so $\mathcal{K}(R_\mathscr{D}) = 1$. However, $E^\infty R_\mathscr{D}$ is essential in $R_\mathscr{D}$ so $\mathcal{K}(R_\mathscr{D}/E^\infty R_\mathscr{D}) = 0$. Thus there can only be finitely many i such that $(J_i^\infty/J_{i+1}^\infty) \otimes_R R_\mathscr{D} \neq 0$. However, we have seen that, for infinitely many i, there exists $\beta_i \in J_i^\infty/J_{i+1}^\infty$ with $\operatorname{ann}_R \beta_i$ contained in one of the primes P' used above; and then $(J_i^\infty/J_{i+1}^\infty) \otimes_R R_\mathscr{D}$ cannot be zero.

This contradiction shows that $\mathcal{K}(S/PS) \neq n + 1$. □

9.21 Lemma. *Let R be commutative Noetherian with $\mathcal{K}(R) < \infty$ and let P be an unstable prime of R such that, for each prime Q with $Q \supset P$, $\mathcal{K}(S/QS) \leqslant \mathcal{K}(R) - \operatorname{ht} Q$. Then $\mathcal{K}(S/PS) \leqslant \mathcal{K}(R) - \operatorname{ht} P$.*

Proof. Let $J \triangleleft_r S$ with $J \supset PS$. By 9.15, $J \cap R \supset P$. Therefore there is a finite chain
$$R = N_0 \supset N_1 \supset \cdots \supset N_t = J \cap R$$
with $N_i/N_{i+1} \simeq R/Q_i$, where $Q_i \in \operatorname{Spec} R$, $Q_i \supset P$. This leads to a chain
$$S = N_0 S \supset N_1 S \supset \cdots \supset N_t S = (J \cap R)S$$
with $N_i S/N_{i+1} S \simeq S/Q_i S$.

By hypothesis,
$$K(S/Q_iS) \leq K(R) - \operatorname{ht} Q < K(R) - \operatorname{ht} P$$
and so
$$K(S/J) \leq K(S/(J \cap R)S) < K(R) - \operatorname{ht} P.$$
Since this holds for all J with $J \supset PS$, then $K(S/PS) \leq K(R) - \operatorname{ht} P$. □

9.22 The promised converse to 9.13 is included in the following result.

Theorem. *Let R be a commutative Noetherian ring with $K(R) < \infty$. Then*
$$K(S) = \sup \{K(R), \operatorname{ht} Q + 1 | Q \text{ a semistable prime}\}.$$

Proof. Granted 9.13, it remains to show that if $K(S) = K(R) + 1$ then R has a semistable prime Q with $\operatorname{ht} Q = K(R)$.

As noted in 9.2
$$K(S) = \sup \{K(S/PS) | P \text{ a minimal prime}\}.$$

Let P be a minimal prime. Then P is semistable by 9.11. Now S/PS is a homomorphic image of S/P^oS. So it is enough to consider the case when R has a minimal prime P with $P^o = 0$.

The proof now proceeds by induction on $n = K(R)$. If $n = 0$ all primes of R are minimal and hence semistable.

Suppose that R has no semistable ideal Q with $\operatorname{ht} Q = K(R) = n$, yet $K(S) = K(R) + 1$. A contradiction will be shown to ensue.

Since $K(S) \neq K(R)$ then $K(S/P_1S) > K(R) = K(R) - \operatorname{ht} P_1$ for some minimal prime P_1 of R. Choose a (not necessarily minimal) prime P_2 maximal with respect to having $K(S/P_2S) > K(R) - \operatorname{ht} P_2$. By 9.21 P_2 cannot be unstable, so P_2 is semistable.

Suppose, for a moment, that P_2 were minimal. Then some conjugate of P_2 by a power of σ must equal P. Since, by 9.5, $S/QS \simeq S/\sigma^k(Q)S$, the hypotheses of 9.20 are satisfied and so $K(S) = K(R)$, a contradiction.

Therefore we may suppose $\operatorname{ht} P_2 \geq 1$. Consider S/P_2^oS. Since $K(R/P_2^o) < K(R)$, the inductive hypothesis implies that
$$K(S/P_2^oS) =$$
$$\sup \{K(R/P_2^o), \operatorname{ht}(Q/P_2^o) + 1 | Q \text{ a semistable prime of } R \text{ with } Q \supseteq P_2^o\}.$$
However, $\operatorname{ht} Q \geq \operatorname{ht}(Q/P_2^o) + \operatorname{ht} P_2$ and $K(R/P_2^o) + \operatorname{ht} P_2 \leq n$. Hence
$$K(S/P_2^oS) \leq \sup \{K(R/P_2^o), \operatorname{ht} Q - \operatorname{ht} P_2 + 1\}$$
$$\leq \sup \{n - \operatorname{ht} P_2, (n-1) - \operatorname{ht} P_2 + 1\}$$
$$= n - \operatorname{ht} P_2$$
which is the contradiction we seek. □

9.23 The result given above does not extend to the case when $K(R) \geq \omega$.

Example. *Let k be a field of characteristic zero, let $R_\mathscr{S}$ be the commutative Noetherian domain of Krull dimension ω described in Example 4.9 and let δ be the k-derivation defined by $\delta(x_{a(n)}) = 1$ for all n and $\delta(x_i) = 0$ if $i \neq a(n)$ for any n. Then no maximal ideal of $R_\mathscr{S}$ is δ-stable, yet $K(R_\mathscr{S}[x; \delta]) = \omega + 1$.*

Proof. It follows from the proof of 4.9 that the maximal ideals of R are precisely the ideals $(P_n)_\mathscr{S}$. These evidently are not δ-stable. Let

$$Q_{nk} = (x_{a(n-1)+1}, \ldots, x_{a(n-1)+k})_\mathscr{S}$$

for $k = 1, 2, \ldots, n-1$. Then

$$0 \subset Q_{n1} \subset \cdots \subset Q_{n,n-1}$$

is a chain of δ-stable prime ideals of $R_\mathscr{S}$. From this one can construct a lattice of S-submodules of S/xS isomorphic to $(\omega^{n-1})^{\text{op}}$. Thus $K(S/xS) \geq \omega$ and $K(S) \geq \omega + 1$. Hence $K(S) = \omega + 1$. □

9.24 In general if $K(R) = \eta + n$, where η is a limit ordinal and n is a natural number, then it is known that it is the prime ideals P with $K(R/P) = \eta$ which control $K(S)$—see [Lenagan 83] for details. However, for simple rings the result for finite Krull dimension carries over.

Proposition. *If R is commutative Noetherian and not Artinian and S is simple then $K(S) = K(R)$.*

Proof. See [Goodearl and Warfield 82] and [Daintree 84]. □

§10 Additional Remarks

10.0 (a) The measurement of a dimension on the lattice of submodules of a right module appeared first in [Gabriel 62], approached categorically, and in [Rentschler and Gabriel 67] approached, as here, using posets. The infinite dimensional case was dealt with by [Krause 70].

(b) A general account appeared in [Gordon and Robson 73] where, in particular, the emphasis was on lattices of submodules rather than posets. Here we have preferred to concentrate on posets, since that helps to unify some concepts and to simplify some of the later proofs.

(c) The Krull dimension is used extensively in Chapter 11.

(d) Another closely related dimension, the Gabriel dimension, is discussed in [Gordon and Robson 74] and [Gordon 74].

10.1 This section basically follows the outline, for finite ordinals, in [Rentschler and Gabriel 67] although there are several extensions and improvements. Other sources were [Lemonnier 72] for 1.13 and [Daintree 84] for the prototype of 1.17.

10.2 The notion of a critical module is implicit in [Gabriel 62] and explicit in [Hart 71]. Its utility was highlighted by [Gordon and Robson 73] which we follow in this section. However 2.6, 2.8 come from [Krause **70**], this carefully distilled proof of 2.17 from [Lenagan **80**] and 2.19–2.22 from [Jategaonkar **74a**] and [Gordon **74**].

10.3 This section too follows [Gordon and Robson **73**]. The first appearance of 3.5 was in [Lemonnier **72a**] and [Gordon and Robson **73a**]. After several partial results, the full version of 3.7 was proved, in different ways, but both relying on 2.17, in [Gordon and Robson **73**] and in [Lenagan **73**].

10.4 (a) The connection between $\mathcal{K}(R)$, dim R and FBN rings comes from [Gabriel **62**] and, for the infinite case, [Krause **72**]. The classical Krull dimension reappears in Chapter 13.

(b) Example 4.9 is due to [Krull **38**] and also appears in [Nagata **62**].

(c) The **Jacobson conjecture** asserts that if R is a Noetherian ring then $\bigcap (J(R))^n = 0$. This is verified for FBN rings by 4.15 which, together with 4.13, 4.14, is due to [Jategaonkar **74a**]. Whilst the conjecture remains open, it is known to be false if R is merely right Noetherian; see [Herstein **65**] or [Jategaonkar **69a**]. See also [Lenagan **77**] and [Goodearl and Schofield **86**] for the case when $\mathcal{K}(R) = 1$.

(d) There are alternative characterizations of FBN rings:

Theorem. ([Cauchon **76**] or see [Goldie **78**].) *If R is right Noetherian then the following are equivalent*:
 (i) *R is right fully bounded.*
 (ii) *The map $M \mapsto \mathrm{ass}\, M$ gives a bijective map from isomorphism types of indecomposable injective modules to $\mathrm{Spec}\, R$.*
 (iii) (Gabriel H-condition) *Given any finitely generated R-module M there exist $m_1, \ldots, m_n \in M$ such that $\mathrm{ann}\, M = \bigcap_{i=1}^n \mathrm{ann}\, m_i$.* □

10.5 The results in this section are 'well known'.

10.6 (a) [P. F. Smith **72**] is the source for 6.1 although the proof here is new; 6.6 and 6.15 come from [Rentschler and Gabriel **67**], 6.7(ii) from [Hart **71**] and 6.10 from [Hodges and McConnell **81**].

(b) [Resco **79**] provided 6.16–6.18 and the proof of 6.19, although the latter result appears also in [Gelfand and Kirillov **66**]. An improved version of 6.17 appears in [Resco **80**]. See also [Goodearl and Lenagan **83**], [Goodearl, Lenagan and Roberts **84**], [Resco, Small and Wadsworth **79**] and [Stafford **83**].

(c) The Krull dimension of other rings and algebras is calculated in later chapters; see 7.8.8, 7.9.4, 7.11.2, 8.5.16, 9.3.17, 10.1.11, and Chapter 15, §3.

10.7 (a) This material is taken from [Stafford **76, 81a**] with 7.9 coming from [Robson **75**]. However, 7.8(iii) is due to [Eisenbud and Robson **70**] and is true without the hypothesis that R is right Noetherian; see Corollary 5.7.3.

(b) In [Stafford **80**] it is shown that if R is a right Noetherian ring with a bound on the number of generators required for right ideals then dim $R \leq 1$; see also [Stafford **83a**].

10.8 (a) This account is based upon the original draft of [Jategaonkar **86**] and upon [Borho **82a**], both of which introduced the notion of a dimension function and extended to it earlier results about Krull dimension. Ideal invariance comes from [Stafford **77**].

(b) The Krull dimension version of 8.16–8.18 appeared in [Gordon **75**], of 8.24 in [Brown, Lenagan and Stafford **80**] and of 8.27 in [Krause, Lenagan and Stafford **78**].

6.10.9 *Additional remarks*

(c) The approach outlined in this section may well be superseded by the approach to localization at semiprime ideals described in Chapter 4.

10.9 (a) This is a unified account of results for $R[x;\delta]$ from [Goodearl and Warfield **82**] and for $R[x, x^{-1}; \sigma]$ from [Hodges **84**]. However, 9.15 is due to [Hart **71**].

(b) Results about several indeterminates can be found in [Goodearl and Lenagan, **83a, 84**]. We note one here.

Proposition. *Let R be right Noetherian and*
$$R = R_0 \subset R_1 \subset \cdots \subset R_n = S$$
where, for each i, either $R_{i+1} = R_i[x; \delta_i]$ or $R_{i+1} = R_i[x, x^{-1}; \sigma_i]$. If $0 \neq M_S$ is finitely generated as an R-module then $\mathcal{K}(M \otimes_R S)_S \geq \mathcal{K}(M_S) + n$.

Proof. First suppose $n = 1$. By 7.5.2 and 12.5.3 there is a short exact sequence of S-modules
$$0 \to M \otimes_R S \to M \otimes_R S \to M \to 0.$$
Thus $\mathcal{K}(M \otimes_R S) > \mathcal{K}(M_S)$ as required.

We now proceed by induction on n. Let $T = R_1$; then M_T is finitely generated and so, by hypothesis, $\mathcal{K}(M \otimes_T S) \geq \mathcal{K}(M_S) + n - 1$. The short exact sequence
$$0 \to M \otimes_R T \to M \otimes_R T \to M \to 0$$
as above, when tensored over T with S, yields the short exact sequence
$$0 \to M \otimes_R S \to M \otimes_R S \to M \otimes_T S \to 0.$$
Hence $\mathcal{K}(M \otimes_R S) \geq \mathcal{K}(M \otimes_T S) + 1 \geq \mathcal{K}(M_S) + n$. □

(c) There is another more general result concerning $\mathcal{K}(M \otimes_R S)$ obtained by adapting the proof of 5.4.

Proposition. *Let R be right Noetherian and $S = R[x;\delta]$ or $R[x, x^{-1}; \sigma]$. If M is a finitely generated right R-module then*
$$\mathcal{K}(M_R) \leq \mathcal{K}(M \otimes_R S)_S \leq \mathcal{K}(M_R) + 1$$
□

Chapter 7
GLOBAL DIMENSION

The aim of this chapter is to describe some basic results about global dimensions of related rings and thereby to calculate, or at least estimate, the global dimensions of the various Noetherian rings introduced in Chapter 1.

We have made a deliberate attempt to make these results easily accessible by using as little as possible of the machinery of homological algebra; what is required is described briefly in an introductory section. There is a price to pay for the limited use of the machinery; for occasionally a less than definitive result is proved, with a reference elsewhere for the definitive case.

The chapter proceeds as follows. Some basis results about subrings, factor rings and localization take up §§2, 3 and 4. Then §5 applies these results to give estimates of the global dimensions of most of the rings described in Chapter 1 except for enveloping algebras. They are dealt with in §6 by using filtered and graded techniques. There is also a brief discussion of regular rings.

The last few sections then look more closely at certain types of rings: firstly the fixed ring R^G for G a finite group, and then skew polynomials over R with R, successively, being Noetherian, commutative Noetherian and finally a Dedekind domain. The chapter ends by showing that when k is a field with char $k = 0$ then $M_n(A_1(k))$ is a principal ideal ring for all $n \geqslant 2$ even though $A_1(k)$ is not a principal ideal ring.

§1 Preliminaries

1.1 In this section we will outline those standard parts of homological algebra which will be used later in the chapter. On the whole these can be found in full detail in several texts [Rotman **79**], [Stenstrom **75**], [Anderson and Fuller **74**], [Kaplansky **72**] although the results special to Noetherian rings are not quite so easily available.

1.2 We recall first the long version of Schanuel's lemma. This says that if

$$0 \to K \to P_{n-1} \to \cdots \to P_1 \to P_0 \to M \to 0$$
$$0 \to K' \to P'_{n-1} \to \cdots \to P'_1 \to P'_0 \to M' \to 0$$

7.1.6 *Preliminaries*

are exact sequences of R-modules in which each P_i, P_i' is projective and $M \simeq M'$ then

$$K \oplus P'_{n-1} \oplus P_{n-2} \oplus P'_{n-3} \oplus \cdots \simeq K' \oplus P_{n-1} \oplus P'_{n-2} \oplus P_{n-3} \oplus \cdots,$$

where the last terms are P_0 and P_0' as appropriate. (If $n = 1$, this is **Schanuel's lemma**).

One defines the **projective dimension** of a module M_R, written $\operatorname{pd} M_R$, to be the shortest length n of a projective resolution

$$0 \to P_n \to P_{n-1} \to \cdots \to P_0 \to M \to 0$$

or ∞ if no finite projective resolution exists. The long Schanuel lemma shows that any projective resolution of M can be terminated at this length.

1.3 There is a corresponding notion of the injective dimension, $\operatorname{id} M_R$, this being the shortest length of an injective resolution

$$0 \to M \to I_0 \to I_1 \to \cdots \to I_n \to 0$$

and an injective version of Schanuel's lemma establishes a similar claim.

1.4 Recall that a module M_R is **flat** when $M \otimes_R -$ is an exact functor and that to prove M_R flat it is enough to check that $0 \to M \otimes I \to M \otimes R$ is exact for each $I \triangleleft_1 R$. The **flat** (or **weak**) dimension $\operatorname{fd} M_R$ of a module M_R is defined as the shortest length of a **flat resolution** of M_R,

$$0 \to F_n \to \cdots \to F_1 \to F_0 \to M \to 0.$$

Although there is no flat version of Schanuel's lemma it is still the case that flat resolutions of M can all be terminated at the same length, this being proved by using 1.3 together with the facts that $\operatorname{Hom}_{\mathbb{Z}}(-, \mathbb{Q}/\mathbb{Z})$ is an exact functor and that M_R is flat if and only if the left R-module $\operatorname{Hom}_{\mathbb{Z}}(M, \mathbb{Q}/\mathbb{Z})$ is injective.

In practice, flat dimension will be useful to us because it behaves better than projective dimension under localization and products.

1.5 Since a projective module is flat, then clearly $\operatorname{fd} M_R \leq \operatorname{pd} M_R$. Less obviously, $\operatorname{fd}(\varinjlim M_i) \leq \sup\{\operatorname{fd} M_i\}$ and in particular any direct limit of flat modules is flat. Furthermore, any flat module is a direct limit of projective modules; indeed, a finitely presented flat module is projective. It follows that if M_R is finitely generated and R is right Noetherian then $\operatorname{pd} M_R = \operatorname{fd} M_R$.

1.6 Given a short exact sequence of right R-modules

$$0 \to A \to B \to C \to 0$$

it is the case that if two have finite projective, injective or flat dimension then so does the third. In fact $\operatorname{pd} B = \sup\{\operatorname{pd} A, \operatorname{pd} C\}$ unless $\operatorname{pd} B < \operatorname{pd} C = 1 + \operatorname{pd} A$.

The same relationship holds for flat dimension; and it also holds for injective dimension except that the equality occurs unless $\operatorname{id} B < \operatorname{id} A = 1 + \operatorname{id} C$.

Note that if $0 \to A_n \to A_{n-1} \to \cdots \to A_0 \to M \to 0$ and $\operatorname{pd} A_i \leq k$ for all i, then $\operatorname{pd} M \leq n + k$. Likewise for $\operatorname{fd} M$ and similarly (in reverse) for $\operatorname{id} M$.

1.7 Given a family of right R-modules $\{M_i | i \in I\}$ one has $\operatorname{pd}(\oplus M_i) = \sup\{\operatorname{pd} M_i\}$, $\operatorname{fd}(\oplus M_i) = \sup\{\operatorname{fd} M_i\}$ and $\operatorname{id}(\prod M_i) = \sup\{\operatorname{id} M_i\}$.

In the special case of a left Noetherian ring R it is also true that $\operatorname{fd}(\prod M_i) = \sup\{\operatorname{fd} M_i\}$; and for a right Noetherian ring $\operatorname{id}(\oplus M_i) = \sup\{\operatorname{id} M_i\}$. Indeed the property that a direct sum of injective modules is injective characterizes right Noetherian rings.

1.8 The following numbers are all equal:
 (i) $\sup\{\operatorname{pd} M | M \text{ any right } R\text{-module}\}$;
 (ii) $\sup\{\operatorname{pd} M | M \text{ any cyclic right } R\text{-module}\}$;
 (iii) $\sup\{\operatorname{id} M | M \text{ any right } R\text{-module}\}$;
 (iv) $\sup\{\operatorname{id} M | M \text{ any cyclic right } R\text{-module}\}$.

The common number is called the **right global dimension** of R, written $\operatorname{r\,gld} R$.

Of course $\operatorname{r\,gld} R = 0$ means precisely that R is a semisimple Artinian ring; and $\operatorname{r\,gld} R = 1$ means that R is right hereditary.

We note two further useful consequences, the former coming from (ii) and 1.6, and the latter from 1.6:
(a) $\operatorname{r\,gld} R = \sup\{\operatorname{pd} I | I \triangleleft_r R\} + 1$ unless R is semisimple;
(b) if $0 \to A \to B \to C \to 0$ is an exact sequence of R-modules and $\operatorname{pd} A = \operatorname{r\,gld} R$, then $\operatorname{pd} B = \operatorname{r\,gld} R$.

1.9 There are corresponding results for flat dimension. In particular

$$\sup\{\operatorname{fd} M | M \text{ any right } R\text{-module}\}$$

and

$$\sup\{\operatorname{fd} M | M \text{ any cyclic right } R\text{-module}\}$$

are equal. The number concerned is called the **weak global dimension** of R, $\operatorname{w\,gld} R$. Note, from 1.5 that $\operatorname{w\,gld} R \leq \operatorname{r\,gld} R$ with equality if R is right Noetherian.

1.10 Some facts about the Tor groups will be needed occasionally, although not too much is lost if those places are omitted.
(i) If L is a left R-module then $\operatorname{fd}_R L \leq n$ if and only if $\operatorname{Tor}^R_{n+1}(M, L) = 0$ for all M_R.
(ii) If $0 \to A \to B \to C \to 0$ is an exact sequence of right R-modules and L is a left R-module then there is an exact sequence of abelian groups

$$\cdots \to \operatorname{Tor}^R_2(C, L) \to \operatorname{Tor}^R_1(A, L) \to \operatorname{Tor}^R_1(B, L) \to \operatorname{Tor}^R_1(C, L)$$
$$\to A \otimes L \to B \otimes L \to C \otimes L \to 0.$$

7.1.13 *Preliminaries*

In particular, if B_R is projective then $\operatorname{Tor}_p^R(A,L) \simeq \operatorname{Tor}_{p+1}^R(C,L)$ for all $p > 0$.
(iii) If A_R, $_RB_S$, $_SC$ are given with $_RB$ and B_S flat then $\operatorname{Tor}_n^S(A \otimes_R B, C) \simeq \operatorname{Tor}_n^R(A, B \otimes_S C)$.

1.11 One consequence of 1.10(i) is that weak global dimension is symmetric, being equal to the corresponding left-hand version. Therefore w gld $R \leqslant$ l gld R also.

It is not generally true that r gld $R =$ l gld R. However, 1.9 together with the symmetry of w gld R shows that when R is Noetherian r gld $R =$ l gld $R =$ gld R for short.

1.12 We note some facts about Ext groups, although they will not be used in this chapter:
(i) pd $M_R \leqslant n$ if and only if $\operatorname{Ext}_R^{n+1}(M,N) = 0$ for all N_R;
(ii) $\operatorname{Ext}_R^1(M,N) = 0$ if and only if every extension

$$0 \to N \to X \to M \to 0$$

of N by M is split;
(iii) if $0 \to A \to B \to C \to 0$ is exact then the sequence

$$0 \to \operatorname{Hom}(C,N) \to \operatorname{Hom}(B,N) \to \operatorname{Hom}(A,N)$$
$$\to \operatorname{Ext}^1(C,N) \to \operatorname{Ext}^1(B,N) \to \operatorname{Ext}^1(A,N)$$
$$\to \operatorname{Ext}^2(C,N) \to \cdots$$

is exact for any N_R.

1.13 For Noetherian rings there is an improved version of 1.8. Since this is less standard a proof is included, using the following result.

Proposition. *Let R be a Noetherian ring of global dimension $n < \infty$ and let A_R, $_RB$ be such that $\operatorname{Tor}_n^R(A,B) \neq 0$. Then $\operatorname{Tor}_n^R(C,B) \neq 0$ for some simple subfactor module C of A.*

Proof. Since $\operatorname{Tor}_n^R(C,B)$ commutes with direct limits we may suppose that A is cyclic. Choose a submodule A' of A maximal with respect to $\operatorname{Tor}_n^R(A/A',B) \neq 0$. Replacing A by A/A' we may assume that if $0 \neq D \triangleleft A$ then $\operatorname{Tor}_n^R(A/D,B) = 0$.

Now let $C = \bigcap_j A_j$, the intersection of all nonzero submodules A_j of A. Evidently either C is simple or $C = 0$. Suppose first that $C = 0$. Then the diagonal map $A \to \prod_j (A/A_j)$ is an embedding. Now $\operatorname{Tor}_n^R(\prod (A/A_j), B) \simeq \prod (\operatorname{Tor}_n^R(A/A_j, B) = 0$ and so, using the long exact sequence of 1.10(ii), $\operatorname{Tor}_n^R(A,B) = 0$.

This contradiction shows that C must be simple. The short exact sequence

$$0 \to C \to A \to A/C \to 0$$

leads to the exact sequence

$$0 \to \operatorname{Tor}_n^R(C, B) \to \operatorname{Tor}_n^R(A, B) \to \operatorname{Tor}_n^R(A/C, B) = 0$$

and so $\operatorname{Tor}_n^R(C, B) \neq 0$ as required. □

1.14 Corollary. *If R is a Noetherian ring with $\operatorname{gld} R < \infty$ then $\operatorname{gld} R = \sup \{\operatorname{pd} C_R | C_R \text{ simple}\}$.* □

1.15 Finally we record some results from [Kaplansky 70] and [Kunz 85] concerning a commutative Noetherian ring R of finite global dimension d. Further comments are in 15.2.8.
(i) R is semiprime; indeed R is a finite direct sum of integral domains and $\operatorname{gld} R = \dim R$.
(ii) If P is a prime ideal then $\operatorname{gld} R_P \leq d$ and

$$\operatorname{gld} R_P = \operatorname{height} P = \dim R_P.$$

(iii) If M is a finitely generated R-module then $\operatorname{pd} M_R = \sup \{\operatorname{pd}(M_P)_{R_P}\}$, over the maximal ideals P of R.
(iv) If R is local with maximal ideal P then $\operatorname{pd} M_R = d$ if and only if $\operatorname{ann} m = P$ for some $m \in M$, and $\operatorname{pd} M_R = 0$ if and only if M_R is free.

§2 Change of Rings

2.1 This section and the next two are concerned with the relationship between r gld R and r gld S when there is a ring homomorphism $\theta: R \to S$. In this generality there is little to say. Therefore certain special cases are considered one by one. This section concentrates on the case when θ is a monomorphism. The next two sections then deal with θ being an epimorphism or a localization. Thereby, as will be seen later in the chapter, the global dimensions of most of the rings described in Chapter 1 can be estimated.

2.2 First consider the general case where $\theta: R \to S$ is any ring homomorphism. Then any right S-module M may be viewed as a right R-module. The first lemma is, in fact, an obvious consequence of the following proposition. However its proof is very elementary.

Lemma. *Let $\theta: R \to S$ and M_S be given.*
(i) If M_S is projective and S_R is projective then M_R is projective.
(ii) If M_S is flat and S_R is flat then M_R is flat.

Proof. (i) M_S is a direct summand of a free S-module.
(ii) $M \otimes_R -$ is naturally isomorphic to the composite of the exact functors $M \otimes_S -$ and $S \otimes_R -$. So by definition M_R is flat. □

7.2.4   *Change of rings*   235

Proposition. *Let $\theta: R \to S$ and M_S be given. Then:*
(i) $\operatorname{pd} M_R \leq \operatorname{pd} M_S + \operatorname{pd} S_R$;
(ii) $\operatorname{fd} M_R \leq \operatorname{fd} M_S + \operatorname{fd} S_R$.

Proof. (i) We proceed by induction on n where $n = \operatorname{pd} M_S$. If $n = 0$ then M_S is a direct summand of a free S-module F say. Hence

$$\operatorname{pd} M_R \leq \operatorname{pd} F_R = \operatorname{pd} S_R = \operatorname{pd} M_S + \operatorname{pd} S_R.$$

If $n > 0$, there is a short exact sequence of right S-modules

$$0 \to K \to F \to M \to 0$$

with F_S free and $\operatorname{pd} K_S = n - 1$. By the inductive hypothesis $\operatorname{pd} K_R \leq n - 1 + \operatorname{pd} S_R$; and, of course, $\operatorname{pd} F_R = \operatorname{pd} S_R$. Therefore, by 1.6, $\operatorname{pd} M_R \leq n + \operatorname{pd} S_R$.

(ii) We proceed by induction on n where $n = \operatorname{fd} M_S$. If $n = 0$ then, by 1.5, M_S is a direct limit of projective modules. But if P_S is projective then $\operatorname{fd} P_R \leq \operatorname{fd} S_R$. So, by 1.5, $\operatorname{fd} M_R \leq \operatorname{fd} S_R$, as required. The proof when $n > 0$ is almost identical to that given in (i). $\square$

2.3 A module $_RM$ is **faithfully flat** if it is flat and also, for any N_R, $N \otimes_R M = 0$ implies $N = 0$. For example, any nonzero free module is faithfully flat.
There is a useful criterion for faithful flatness.

Proposition. *Let $_RM$ be flat. Then $_RM$ is faithfully flat if and only if $IM \neq M$ for all proper right ideals I of R.*

Proof. Let $I \triangleleft_r R$, $I \neq R$. Then

$$0 \to I \otimes_R M \to R \otimes_R M \to (R/I) \otimes_R M \to 0$$

is exact. Moreover, $R \otimes_R M \simeq M$ and, under this isomorphism, $I \otimes M \to IM$.

Now suppose $_RM$ is faithfully flat. Then $(R/I) \otimes M \neq 0$ and so $IM \neq M$.

For the converse, suppose $0 \neq N_R$ and choose a nonzero cyclic submodule N' of N. Then $N' \simeq R/I$ for some I. By hypothesis $IM \neq M$, so the exact sequence above implies that $N' \otimes M \neq 0$. However, $N' \otimes M$ embeds in $N \otimes M$ since $_RM$ is flat, and thus $N \otimes M \neq 0$. $\square$

2.4 In fact faithful flatness appeared, essentially, in 6.6.15, which concerned $A_n(k)$ with k a field with $\operatorname{char} k = 0$. An extension ring $B = \oplus B_{nj}(k)$ was constructed there with each $B_{nj}(k)$ being a localization of $A_n(k)$ and hence, by 2.1.16, being flat over $A_n(k)$. Thus B is flat over $A_n(k)$ and, as the proof of 6.6.15 shows, B is faithfully flat. That proof used this faithful flatness to connect the Krull dimensions of these two rings. A similar connection will now be obtained for global dimension.

2.5 First a preliminary fact.

Lemma. *Let $R \subseteq S$ be rings with $_R S$ being faithfully flat and let M be a right R-module. The map $M \to M \otimes_R S$ via $m \mapsto m \otimes 1$ is an embedding.*

Proof. Let K be the kernel of the map. Then $0 \to K \to M$ is exact and so $0 \to K \otimes S \to M \otimes S$ is also exact. Given $k \in K$ and $s \in S$ then, in $M \otimes S$, $k \otimes s = (k \otimes 1)s = 0s = 0$. This shows that $K \otimes S = 0$ and, since $_R S$ is faithfully flat, that $K = 0$. □

2.6 Theorem. *Let R, S be rings with $R \subseteq S$, $_R S$ faithfully flat and $\operatorname{r\,gld} R < \infty$. If either*
(i) S_R is projective, or
(ii) R is right Noetherian and S_R is flat,
then $\operatorname{r\,gld} R \leq \operatorname{r\,gld} S$.

Proof. (i) Choose M_R with $\operatorname{pd} M_R = \operatorname{r\,gld} R = n$, say. By 2.5, there is a short exact sequence of R-modules
$$0 \to M \to M \otimes_R S \to C \to 0.$$
Now $\operatorname{pd} C \leq n$, of course; so by 1.6 $\operatorname{pd}(M \otimes_R S)_R = n$. However, Proposition 2.2 shows that $\operatorname{pd}(M \otimes S)_R \leq \operatorname{pd}(M \otimes S)_S$ and so $\operatorname{r\,gld} S \geq n$.

(ii) The same argument, but replacing projective dimension by flat dimension, shows that $\operatorname{w\,gld} R \leq \operatorname{w\,gld} S$; and then 1.9 completes the proof. □

2.7 There are several applications of this in §5 and later. One problem in applying it arises if the known information is mainly about S. For if it is not known that $\operatorname{r\,gld} R < \infty$, the theorem no longer shows that $\operatorname{r\,gld} R \leq \operatorname{r\,gld} S$. That this is a genuine obstacle is shown by the next example.

Example. *Let $R = k[y]/(y^2)$ with k a field of characteristic 2, let δ be the derivation d/dy (i.e. $\delta(y) = 1$) and let $S = R[x; \delta]$. Then $\operatorname{gld} R = \infty$ and $\operatorname{r\,gld} S = 1$; yet S is free (and hence faithfully flat) over R on each side and R is Noetherian.*

Proof. The nonsplit short exact sequence
$$0 \to yR \to R \to R/yR \to 0$$
together with the fact that $yR \simeq R/yR$ shows that $\operatorname{gld} R = \infty$. Since $_R S$ is free, the sequence
$$0 \to yR \otimes_R S \to R \otimes_R S \to (R/yR) \otimes_R S \to 0$$
is exact. Now $R \otimes_R S \simeq S$ and so $R/yR \otimes S \simeq S/yS$ and $yR \otimes S \simeq yS$; thus $S/yS \simeq yS$. Note next that xy is idempotent, say $xy = e$, and so xyS is projective.

7.3.4 Factor rings 237

But $xyS \simeq yS$ and so $S \simeq eS \oplus eS$. It follows that $S \simeq M_2(eSe) = M_2(k[t])$ with $t = ex^2$. Since $k[t]$ is a principal ideal domain it is hereditary; so r gld S = gld $(k[t]) = 1$. □

2.8 The next result does not suffer the disadvantage of requiring r gld $R < \infty$.

Theorem. *Let R, S be rings with $R \subseteq S$ such that R is an R-bimodule direct summand of S. Then:*
(i) r gld $R \leqslant$ r gld S + pd S_R;
(ii) w gld $R \leqslant$ w gld S + fd S_R.

Proof. (i) Let $_R S_R = R \oplus I$ and let M_R be any right R-module. Then $M \otimes_R S \simeq M \oplus (M \otimes_R I)$ and so pd $M_R \leqslant$ pd $(M \otimes S)_R$ by 1.7. Proposition 2.2 now shows that

$$\text{pd } M_R \leqslant \text{pd}(M \otimes S)_R \leqslant \text{pd}(M \otimes S)_S + \text{pd } S_R \leqslant \text{r gld } S + \text{pd } S_R.$$

(ii) This is proved similarly. □

§3 Factor Rings

3.1 Next we study the relationship between r gld S and r gld S/I for an ideal I of a ring S. As in Section 2, the results concern special cases only. There are results when $I = xS = Sx$ with x regular, when $I^n = I^{n+1}$ for some n and when $I \subseteq J(S)$. These, in conjunction with the results of §§2 and 4, will enable us to estimate and determine global dimensions of many examples, this being achieved in later sections.

Whilst the earlier results of this section continue to rely only on elementary homological techniques, some of the later ones make use of some simple properties of Tor groups.

3.2 First we note a useful fact.

Lemma. *If $I \triangleleft S$ and M_S is projective then M/MI is a projective S/I-module.*

Proof. This is clear for free S-modules, and hence for their direct summands. □

3.3 The first collection of results deals with the case when $I = xS$ with x a normal regular non-unit. So $I = Sx$ also, and $I \simeq S$ as a right or left S-module. Indeed x induces an inner automorphism σ of S via $sx = x\sigma(s)$ and this, in turn, gives an automorphism, again labelled σ, of S/I.

3.4 Let R be any ring with an automorphism σ and let M_R be a module. One can construct a new R-module, denoted by M^σ, as follows. It has the same underlying abelian group but with elements labelled m^σ rather than m; and multiplication is given by $m^\sigma r = (m\sigma^{-1}(r))^\sigma$.

Lemma. pd $M_R^\sigma =$ pd M_R and fd $M_R^\sigma =$ fd M_R.

Proof. Note first that $R_R^\sigma \simeq R_R$ via the obvious map. Hence any projective or flat resolution of M_R can be transformed into one of M^σ and vice versa. □

3.5 Theorem. *Let S be a ring and $I = xS$ with x a regular normal nonunit.*
(i) *If $M_{S/I}$ is nonzero and pd $M_{S/I} = n < \infty$ then pd $M_S = n + 1$.*
(ii) *If r gld $S/I < \infty$ then r gld $S \geq$ r gld $S/I + 1$.*

Proof. (i) Since $Mx = 0$, M cannot be a submodule of a free S-module. Hence pd $M_S \neq 0$. Note also that $I = xS \simeq S$; it follows that pd $(S/I)_S = 1$.

The proof now proceeds by induction on n. When $n = 0$, $M_{S/I}$ is a direct summand of a free S/I-module. Hence pd $M_S = 1$.

Now suppose $n > 0$. There is a short exact sequence of right S/I-modules $0 \to K \to F \to M \to 0$ with F free and, by 1.6, pd $K_{S/I} = n - 1$. By the induction hypothesis pd $K_S = n$. Hence, by 1.6 again, pd $M_S = n + 1$ except possibly in the case when $n = 1$; for then pd $K_S = $ pd $F_S = 1$ and 1.6 only shows that pd $M_S \leq 2$.

We now concentrate on this case. As noted above, pd $M_S \neq 0$. Suppose that pd $M_S = 1$. Then there is a short exact sequence of S-modules $0 \to Q \to P \to M \to 0$ with pd $P_S = $ pd $Q_S = 0$. Since $Mx = 0$ then $Px \subseteq Q$ and there are induced exact sequence of S/I modules

$$0 \to Q/Px \to P/Px \to M \to 0$$

and

$$0 \to Px/Qx \to Q/Qx \to Q/Px \to 0.$$

By 3.2, both P/Px and Q/Qx are projective S/I-modules; and by assumption pd $M_{S/I} = 1$; thus pd $(Q/Px)_{S/I} = 0$. The latter sequence therefore must split, and so pd $(Px/Qx)_{S/I} = 0$.

However, using the notation of 3.4, it is clear that $0 \to Q^\sigma \to P^\sigma \to M^\sigma \to 0$ is exact. Now $P^\sigma \simeq Px$ and $Q^\sigma \simeq Qx$ and so pd $M_{S/I}^\sigma = 0$. Hence by 3.4, pd $M_{S/I} = 0$, which contradicts the hypothesis.

It follows that pd $M_S = 2$, as required.
(ii) This is clear. □

3.6 Next we consider modules M_S on which x is regular; i.e. $mx = 0 \Rightarrow m = 0$.

Proposition. (a) *Let S be a ring, $I = xS$ with x a regular normal non-unit, and suppose that x is regular on a module M_S. Then pd $(M/MI)_{S/I} \leq$ pd M_S.*

(b) *If in addition S is right Noetherian, M_S is finitely generated, and $x \in J(S)$, then* $\operatorname{pd}(M/MI)_{S/I} = \operatorname{pd} M_S$.

Proof. (a) If $\operatorname{pd} M_S = \infty$ there is nothing to prove. We suppose $\operatorname{pd} M_S = n$ and proceed by induction on n. The case $n = 0$ is clear from 3.2, so suppose that $n > 0$. There is a short exact sequence of S-modules $0 \to K \to F \to M \to 0$ with F free and $\operatorname{pd} K_S = n - 1$. Since K is a submodule of F, x is regular on K. Therefore by the induction hypothesis $\operatorname{pd}(K/KI)_{S/I} \leqslant n - 1$.

There is an induced exact sequence of S/I-modules

$$0 \to K + FI/FI \to F/FI \to M/MI \to 0$$

and $K + FI/FI \simeq K/K \cap FI$. Since x is regular on M, $K \cap Fx = Kx$; i.e. $K \cap FI = KI$. Thus

$$0 \to K/KI \to F/FI \to M/MI \to 0$$

is exact and so $\operatorname{pd}(M/MI)_{S/I} \leqslant n$.

(b) If $\operatorname{pd}(M/MI)_{S/I} = \infty$ then (a) shows that $\operatorname{pd} M_S = \infty$. We suppose $\operatorname{pd}(M/MI)_{S/I} = n$ and proceed by induction on n. We start with the special case when M/MI is free over S/I, and aim to show then that M_S is free. Let $\bar{m}_1, \ldots, \bar{m}_k \in M/MI$ be a free basis for $(M/MI)_{S/I}$. By Nakayama's lemma, $M = \sum m_i S$. Suppose $\sum m_i s_i = 0$ with some $s_i \neq 0$, say s_1. Evidently, however, each $s_i \in I = Sx$; so $s_i = s_i' x$ for some s_i' and then $\sum m_i s_i' = 0$. This process can be repeated, leading to the chain $s_1 S \subset s_1' S \subset s_1'' S \subset \cdots$; that this is strictly ascending is clear since if, for example, $s_1 S = s_1' S$ then $s_1 SJ = s_1 S$ in contradiction to Nakayama's lemma. However, S is right Noetherian, and so the conclusion is that M_S is indeed free.

Now consider the case $n = 0$, when M/MI is projective over S/I. If $0 \to K \to F \to M \to 0$ is an exact sequence of S-modules with F_S free of finite rank, then, as in (a), there is an induced exact sequence of S/I-modules

$$0 \to K/KI \to F/FI \to M/MI \to 0.$$

This must split, and then $M/MI \oplus K/KI$ is free over S/I. It follows from the above that $M \oplus K$ is free over S and so M_S is projective.

Finally, suppose $n > 0$, let $0 \to K \to F \to M \to 0$ and

$$0 \to K/KI \to F/FI \to M/MI \to 0$$

be as above. Then $\operatorname{pd}(K/KI)_{S/I} = n - 1$. Moreover, K_S is finitely generated and x is regular on K. By the induction hypothesis, $\operatorname{pd} K_S = n - 1$ and so $\operatorname{pd} M_S = n$. □

3.7 Theorem. *Let S be a right Noetherian ring and x a regular normal element belonging to $J(S)$. If $\operatorname{r gld} S/I < \infty$ then $\operatorname{r gld} S = \operatorname{r gld} S/I + 1$.*

Proof. In 3.5 it is shown that r gld $S \geq$ r gld $S/I + 1$. Conversely, let M_S be finitely generated. We aim to show that pd $M_S \leq n + 1$ where $n =$ r gld S/I. If pd $M_S \leq 1$ there is nothing to prove. Otherwise, form an exact sequence of S-modules $0 \to K \to F \to M \to 0$ with F_S free of finite rank. Then 3.6(b) can be applied to K to give pd $K_S =$ pd $(K/KI)_{S/I} \leq n$. Hence pd $M \leq n + 1$. □

3.8 That the restriction r gld $S/I < \infty$ is required in 3.5 and 3.7 is easily shown. For example let $S = \mathbb{Z}_{(2)}$, the integers localized at the ideal (2). Since $\mathbb{Z}_{(2)}$ is a principal ideal domain, gld $\mathbb{Z}_{(2)}$ must be 1, yet gld $\mathbb{Z}_{(2)}/4\mathbb{Z}_{(2)} = \infty$ of course, since $\mathbb{Z}_{(2)}/4\mathbb{Z}_{(2)} \simeq \mathbb{Z}/(4)$.

3.9 We now turn to some more general, but less precise, results, starting with some straightforward facts.

Lemma. *Let S be a ring, I an ideal and M_S a module.*
(i) *If M_S is projective then* pd $MI_S \leq$ pd I_S.
(ii) *If $_SI$ is flat and M_S is a submodule of a free module then* pd $MI_S \leq$ pd $M_S +$ pd I_S.

Proof. (i) If $M \oplus K$ is free then $MI \oplus KI \simeq I^X$ for some index set X; so 1.7 shows that pd $MI \leq$ pd I.

(ii) We may, of course, suppose pd $M_S < \infty$. If M_S is projective then (i) applies. Otherwise, form an exact sequence of S-modules

$$0 \to K \to F \to M \to 0$$

with F_S free. Then pd $K_S =$ pd $M_S - 1$ so, by induction, pd $KI \leq$ pd $K +$ pd I. Since $_SI$ is flat the sequence

$$0 \to K \otimes I \to F \otimes I \to M \otimes I \to 0$$

is exact. Of course $F \otimes I \simeq FI$ since F_S is free; and, restricted to $K \otimes I$, this shows that $K \otimes I \simeq KI$. The same argument applies to M, since M is a submodule of a free module; thus $M \otimes I \simeq MI$. Therefore the sequence

$$0 \to KI \to FI \to MI \to 0$$

is exact. Now pd $FI =$ pd I and pd $KI \leq$ pd $K +$ pd I. Therefore, by 1.6,

$$\text{pd } MI \leq \text{pd } K + \text{pd } I + 1 = \text{pd } M + \text{pd } I.$$
□

3.10 Theorem. *Let S be a ring, I an ideal with I_S projective and $I^m = I^{m+1}$ for some m. Then:*
(i) *if M is a right S/I-module then* pd $M_{S/I} \leq$ pd $M_S + 2m - 2$;
(ii) r gld $S/I \leq$ r gld $S + 2m - 2$.

Proof. It will be enough to prove (i), and we may suppose $\operatorname{pd} M_S = n < \infty$. If $n = 0$ then 3.2 shows that $\operatorname{pd} M_{S/I} = 0$. The proof now proceeds by induction on n. We form an exact sequence of S-modules

$$0 \to K \to F \to M \to 0$$

with F_S free and $K \subseteq F$. Note that since $MI = 0$ then $FI \subseteq K$ and so

$$F \supseteq K \supseteq FI \supseteq KI \supseteq FI^2 \supseteq \cdots.$$

Since $F/K \simeq M$ the sequence of S/I-modules

$$0 \to K/FI \to F/FI \to M \to 0$$

is exact. Proceeding in the same way one obtains the long exact sequence

$$0 \to KI^{m-1}/KI^m \to \cdots \to KI/KI^2 \to FI/FI^2 \to K/KI \to F/FI \to M \to 0,$$

where the left-hand zero is explained by the fact that $FI^m = FI^{m+1}$.

Suppose now that $\operatorname{pd} M_S = 1$. Then K_S is projective and so, by 3.9 and an induction on i, FI^i and KI^i are projective over S for each i. Then 3.2 shows that FI^i/FI^{i+1} and KI^i/KI^{i+1} are projective over S/I for each i. Hence the long exact sequence above is a projective resolution of $M_{S/I}$ and therefore $\operatorname{pd} M_{S/I} \leq 2m - 1$ as claimed.

Finally, suppose $\operatorname{pd} M_S = n > 1$. Note that $\operatorname{pd}(F/FI)_S = \operatorname{pd} S/I_S \leq 1$. Hence 1.6, applied to the short exact sequence above, shows that $\operatorname{pd}(K/FI)_S = n - 1$. By induction

$$\operatorname{pd}(K/FI)_{S/I} \leq n - 1 + 2m - 2.$$

However, $\operatorname{pd}(F/FI)_{S/I} = 0$, of course, and so

$$\operatorname{pd} M_{S/I} \leq \operatorname{pd}(K/FI)_{S/I} + 1 \leq \operatorname{pd} M_S + 2m - 2. \qquad \square$$

3.11 The next few results of this section rely upon the following proposition. The necessary facts about Tor are outlined in 1.10.

Proposition. *Let S be a ring, I an ideal and M_S a module such that $\operatorname{Tor}_p^S(M, S/I) = 0$ for all $p > 0$. Then $\operatorname{pd}(M/MI)_{S/I} \leq \operatorname{pd} M_S$.*

Proof. If M_S is projective then so is $(M/MI)_{S/I}$ by 3.2. Suppose $\operatorname{pd} M_S = n > 0$; we proceed by induction on n. Form an exact sequence of S-modules $0 \to K \to F \to M \to 0$ with F_S free and so $\operatorname{pd} K_S = n - 1$. Now, for all $p > 0$

$$\operatorname{Tor}_p^S(K, S/I) = \operatorname{Tor}_{p+1}^S(M, S/I) = 0$$

and therefore, by induction,

$$n - 1 = \operatorname{pd} K_S \geq \operatorname{pd}(K/KI)_{S/I}.$$

Note next that
$$0 \to K \otimes S/I \to F \otimes S/I \to M \otimes S/I \to 0$$
is exact since $\operatorname{Tor}_1^S(M, S/I) = 0$, and that $K \otimes S/I \simeq K/KI$ etc. Therefore
$$0 \to K/KI \to F/FI \to M/MI \to 0$$
is exact. Since $(F/FI)_{S/I}$ is free, then
$$\operatorname{pd}(M/MI)_{S/I} \leqslant \operatorname{pd}(K/KI)_{S/I} + 1 \leqslant n. \qquad \square$$

3.12 We can now obtain a result very similar to 3.10.

Theorem. *Let S be a ring, I an ideal with $_SI$ flat and $I^m = I^{m+1}$ for some m. Then:*
(i) *if M is a right S/I-module then* $\operatorname{pd} M_{S/I} \leqslant \operatorname{pd} M_S + (m-1)\operatorname{pd} I_S + 2m - 2$;
(ii) $\operatorname{r\,gld} S/I \leqslant \operatorname{r\,gld} S + (m-1)\operatorname{pd} I_S + 2m - 2$.

Proof. First suppose that K_S is any submodule of a free S-module, F say, and consider the exact sequence $0 \to K \to F \to F/K \to 0$. Since $\operatorname{fd} S/I \leqslant 1$ then $\operatorname{Tor}_{p+1}^S(F/K, S/I) = 0$ for all $p > 0$. Hence $\operatorname{Tor}_p^S(K, S/I) = 0$ and so 3.11 can be applied to give $\operatorname{pd}(K/KI)_{S/I} \leqslant \operatorname{pd} K_S$.

Next consider the long exact sequence used in the proof of 3.10. Applying the above argument to KI^{i-1} rather than K and then using 3.9(ii) one obtains
$$\operatorname{pd}(KI^{i-1}/KI^i)_{S/I} \leqslant \operatorname{pd}(KI^{i-1})_S \leqslant \operatorname{pd} K_S + (i-1)\operatorname{pd} I_S;$$
and a similar inequality holds for FI^{i-1}/FI^i. Thus one can apply 1.6 to obtain
$$\operatorname{pd} M_{S/I} \leqslant \operatorname{pd} K_S + (m-1)\operatorname{pd} I_S + 2m - 1.$$
Now if $\operatorname{pd} M_S \neq 0$, then $\operatorname{pd} K_S = \operatorname{pd} M_S - 1$ and the result follows. Of course, if $\operatorname{pd} M_S = 0$ then 3.2 applies. $\qquad \square$

3.13 One example of an ideal I with $I^m = I^{m+1}$ for some m is any nilpotent ideal. In this case another inequality, in the opposite direction, can be obtained. This requires an improved version of 3.11. It is related to 3.6(b), whose proof it follows closely.

Proposition. *Let S be a ring, I a nilpotent ideal and M_S a module such that $\operatorname{Tor}_p^S(M, S/I) = 0$ for all $p > 0$. Then $\operatorname{pd} M_S = \operatorname{pd}(M/MI)_{S/I}$.*

Proof. If $\operatorname{pd}(M/MI)_{S/I} = \infty$ then 3.11 gives the result. We suppose $\operatorname{pd}(M/MI)_{S/I} = n$ and start an inductive proof with the special case when M/MI is free over S/I. Let $\{\bar{m}_i | i \in X\}$ be a free basis for M/MI, let m_i be an inverse image of $\bar{m}_i$ in M for each i and let $E = \sum m_i S$. Then $E + MI = M$ and so

$(M/E)I = M/E$. The fact that I is nilpotent implies that $M/E = 0$ and $M = E$. Let F_S be free on a basis $\{e_i | i \in X\}$, and define $\alpha: F \to M$ by $\alpha(e_i) = m_i$ for each i. This gives an exact sequence $0 \to K \to F \xrightarrow{\alpha} M \to 0$. Since $\operatorname{Tor}_1^S(M, S/I) = 0$ the sequence

$$0 \to K \otimes S/I \to F \otimes S/I \xrightarrow{\alpha \otimes 1} M \otimes S/I \to 0$$

is exact. However, $\alpha \otimes 1$ is an isomorphism of free S/I-modules, so $K \otimes S/I = K/KI = 0$. Thus $K = KI$ and $K = 0$. This shows that M_S is free. The same argument as in 3.6(b) now completes the proof. $\square$

Note that the nilpotency of I was used only in deducing from the equations $(M/E)I = M/E$ and $KI = K$ that $M/E = 0 = K$. Thus a minor modification of the proof above using Nakayama's lemma gives the next result.

Corollary. *Let S be a right Noetherian ring, I an ideal with $I \subseteq J(S)$ and M_S a finitely generated module such that $\operatorname{Tor}_p^S(M, S/I) = 0$ for all $p > 0$. Then* $\operatorname{pd} M_S = \operatorname{pd} M/MI_{S/I}$. $\square$

3.14 Theorem. *Let S be a ring and I an ideal. If*
(i) *I is nilpotent, or*
(ii) *S is right Noetherian and $I \subseteq J(S)$*
then

$$\operatorname{r\,gld} S \leq \operatorname{r\,gld} S/I + \operatorname{fd}_S(S/I).$$

Proof. (i) Let M be an arbitrary right S-module, and suppose $\operatorname{fd}_S(S/I) = n < \infty$. Form an exact sequence

$$0 \to X_n \to F_{n-1} \to \cdots \to F_0 \to M \to 0$$

in which each F_i is a free S-module. Now for all $p > 0$

$$\operatorname{Tor}_p^S(X_n, S/I) = \operatorname{Tor}_{n+p}^S(M, S/I)$$

which is zero, since $\operatorname{fd}_S(S/I) = n$. Hence X_n satisfies the conditions of Proposition 3.13 so that

$$\operatorname{pd}(X_n)_S = \operatorname{pd}(X_n/X_n I)_{S/I} \leq \operatorname{r\,gld} S/I.$$

Hence $\operatorname{pd} M_S \leq \operatorname{r\,gld} S/I + n$.

(ii) If M and all the F_i are chosen to be finitely generated then so too is X_n. Then Corollary 3.13 can be used instead of Proposition 3.13. $\square$

3.15 For nilpotent ideals there is another more elementary upper bound on $\operatorname{r\,gld} S$, obtainable by setting the ideals in the next result equal to each other.

Proposition. *Let S be a ring and $I_1, I_2, \ldots, I_m$ be ideals such that $I_1 I_2 \cdots I_m = 0$. Then*

$$\operatorname{r\,gld} S \leqslant \sup_i \{\operatorname{r\,gld}(S/I_i) + \operatorname{pd}(S/I_i)_S\}.$$

Proof. Given M_S, consider the submodules $M_i = MI_1 I_2 \cdots I_i$ and the factors M_{i-1}/M_i. If

$$0 \to A_n \to A_{n-1} \to \cdots \to A_0 \to M_{i-1}/M_i \to 0$$

is an S/I_i-projective resolution of M_{i-1}/M_i, then 1.6 applies to show that

$$\operatorname{pd}(M_{i-1}/M_i)_S \leqslant \operatorname{r\,gld}(S/I_i) + \operatorname{pd}(S/I_i)_S.$$

An easy induction down the chain $M = M_0 \supseteq M_1 \supseteq \cdots \supseteq M_m = 0$ shows, by 1.6, that

$$\operatorname{pd} M_S \leqslant \sup_i \{\operatorname{r\,gld}(S/I_i) + \operatorname{pd}(S/I_i)_S\}. \qquad \square$$

3.16 Finally, we state a result concerning the projective dimension of a factor module R/I for a rather special right ideal I. The proof, which uses the Koszul resolution, is omitted; see [Resco **79**, 2.8] and [Cartan and Eilenberg **56**, p. 150, Theorem 4.2] for the details, and 6.5.9 for a related result.

Theorem. *Let I be a proper right ideal of a ring R with $I = \sum_{i=1}^n a_i R$ where the generators satisfy:*
  (i) $a_i a_j = a_j a_i$ *for each i, j;*
  (ii) *if $a_{k+1} r \in \sum_{i=1}^k a_i R$ for some k, then $r \in \sum_{i=1}^k a_i R$.*
Then $\operatorname{pd}(R/I)_R = n$ if and only if $\sum_{i=1}^n Ra_i \neq R$. $\qquad \square$

§4 Localization

4.1 This section is concerned with the global dimensions of a ring R and a localization Q of R with respect to some right denominator set $\mathscr{S}$. In general, $\operatorname{r\,gld} Q \leqslant \operatorname{r\,gld} R$. For Noetherian rings, it will be shown that the possibility of equality here is determined by the simple R-modules.

4.2 Proposition. *Let $Q = R_{\mathscr{S}}$ for some right denominator set $\mathscr{S}$, let M be a right Q-module, and let N be a right R-module. Then:*
  (i) $Q \otimes_R Q \simeq Q$ *as a Q-bimodule, and $M \otimes_R Q \simeq M$ as a right Q-module;*
  (ii) $\operatorname{pd}(N \otimes Q)_Q \leqslant \operatorname{pd} N_R$, *and* $\operatorname{fd}(N \otimes Q)_Q \leqslant \operatorname{fd} N_R$;
  (iii) *if $\mathscr{S}$ is also a left denominator set, then $\operatorname{fd} M_Q = \operatorname{fd} M_R$.*

7.5.1   *Estimates of global dimension*   245

Proof. (i) Let μ be the multiplication map $Q \otimes Q \to Q$ given by $a \otimes b \mapsto ab$. This is clearly a Q-bimodule surjection and is injective when restricted to either $Q \otimes 1$ or $1 \otimes Q$. Note, for any $z \in Q \otimes Q$ that $zt \in Q \otimes 1$ for some $t \in \mathcal{S}$. If $z \in \ker \mu$ then $zt \in \ker \mu$ and so $zt = 0$. Hence $z = 0$ and μ is an isomorphism. Also
$$M_Q \simeq M \otimes_Q Q \simeq M \otimes_Q (Q \otimes_R Q) \simeq M \otimes_R Q.$$

(ii) Applying $- \otimes_R Q$ to a projective, or flat, resolution of N_R gives a similar resolution of $(N \otimes_R Q)_Q$.

(iii) Note that Q_R is flat, by 2.1.16. Hence, by Lemma 2.2, any flat right Q-module is flat over R. Thus a flat resolution of M_Q is equally a flat resolution of M_R and so fd $M_R \leq$ fd M_Q. However, $M_Q \simeq M \otimes_R Q$, by (i), and so fd $M_Q \leq$ fd M_R, by (ii). □

4.3 Corollary. *If $\mathcal{S}$ is a right denominator set in a ring R then* r gld $R_\mathcal{S} \leq$ r gld R *and* w gld $R_\mathcal{S} \leq$ w gld R.

Proof. Let M be a right Q-module. Then by 4.2(i), $M \simeq M \otimes_R Q$, and so pd $M_Q \leq$ pd M_R and fd $M_Q \leq$ fd M_R, by 4.2(ii). □

4.4 Theorem. *Let R be Noetherian with* gld $R = n < \infty$ *and $\mathcal{S}$ be a right and left denominator set. Then* gld $R_\mathcal{S} =$ gld R *if and only if there is a simple module M_R with* pd $M = n$ *and* $M_\mathcal{S} \neq 0$.

Proof. First suppose M_R is as described and let $Q = R_\mathcal{S}$. Evidently $M \hookrightarrow M_\mathcal{S} = M \otimes Q$ and so, by the flat version of 1.8(b), fd $(M_\mathcal{S})_R = n$. Hence, by 4.2(iii), fd $(M_\mathcal{S})_Q = n$ and therefore w gld $Q \geq n$. It follows that gld $Q = n$.

Conversely, suppose gld $Q = n$ and let $_Q N$ be a left module with pd$_Q N = n$. Then, by 4.2(iii), fd$_R N = n$ and so Tor$_n^R (M, N) \neq 0$ for some right R-module M which, by 1.13, may be taken to be simple. However, by 1.10(iii),
$$\operatorname{Tor}_n^R(M, N) = \operatorname{Tor}_n^R(M, Q \otimes_R N) \simeq \operatorname{Tor}_n^R(M \otimes_R Q, N).$$
Therefore $M \otimes Q \neq 0$. Of course, fd $M_R \geq n$ and thus pd $M_R = n$. □

The analogy with Proposition 6.5.3 is worthy of note.

§5 Estimates of Global Dimension

We now turn to the examples of Chapter 1 and will see that the results of the previous two sections combine to give bounds on their global dimensions. More precise results will then be obtained in later sections.

5.1 Matrix rings $M_n(R)$ are readily dealt with since, as noted in 3.5.10, global dimension is preserved under Morita equivalence. For triangular rings some estimates are also easily obtained.

Proposition. *Let A, B be rings, M an (A, B)-bimodule and*

$$S = \begin{bmatrix} A & M \\ 0 & B \end{bmatrix}.$$

Then

$$\sup\{\operatorname{r\,gld} A, \operatorname{r\,gld} B\} \leq \operatorname{r\,gld} S \leq \sup\{\operatorname{r\,gld} B, \operatorname{r\,gld} A + \operatorname{pd} M_B + 1\}.$$

Proof. Let

$$I = \begin{bmatrix} 0 & M \\ 0 & B \end{bmatrix} \quad \text{and} \quad J = \begin{bmatrix} A & M \\ 0 & 0 \end{bmatrix}.$$

Evidently $I = I^2$, $S/I \simeq A$, and $_S I$ is projective. Hence 3.12 shows that $\operatorname{r\,gld} A \leq \operatorname{r\,gld} S$. Similarly, using J and 3.10, $\operatorname{r\,gld} B \leq \operatorname{r\,gld} S$.

Since $IJ = 0$, 3.15 gives

$$\operatorname{r\,gld} S \leq \sup\{\operatorname{r\,gld} A + \operatorname{pd} A_S, \operatorname{r\,gld} B + \operatorname{pd} B_S\}.$$

Note that $\operatorname{pd} B_S = 0$ and so $\operatorname{pd} M_S \leq \operatorname{pd} M_B$, and that $\operatorname{pd} J_S = 0$. Hence the short exact sequence

$$0 \to M \to J \to A \to 0$$

shows that $\operatorname{pd} A_S \leq 1 + \operatorname{pd} M_S \leq 1 + \operatorname{pd} M_B$. □

5.2 Before obtaining the basic results concerning global dimensions of skew polynomials and Laurent polynomials a preparatory result is required. Recall from 3.4 that if M is a right R-module and σ is an automorphism of R then there is an R-module M^σ.

Proposition. *Let $S = R[x; \sigma, \delta]$ or $S = R[x, x^{-1}; \sigma]$, where σ is an automorphism, and let M be a right S-module.*
(i) *There is an exact sequence of S-modules*

$$0 \to M^\sigma \otimes_R S \xrightarrow{\beta} M \otimes_R S \xrightarrow{\alpha} M \to 0$$

with $\alpha(m \otimes s) = ms$ and $\beta(m^\sigma \otimes s) = mx \otimes s - m \otimes xs$.
(ii) $\operatorname{pd} M_S \leq \operatorname{pd} M_R + 1$; $\operatorname{fd} M_S \leq \operatorname{fd} M_R + 1$.

Proof. (i) We use the same argument for each case. Each element of $M \otimes S$ has a unique expression as $\sum m_i \otimes x^i$. Define $\psi : M^\sigma \times S \to M \otimes_R S$ by

$$\psi(m^\sigma, s) = mx \otimes s - m \otimes xs.$$

7.5.3 Estimates of global dimension

One can verify that $\psi(m^\sigma r, s) = \psi(m^\sigma, rs)$ for any $r \in R$ and so, by universality, ψ factors through $M^\sigma \otimes_R S$. Thus β is an S-homomorphism.

It is clear that α is surjective and that $\operatorname{im} \beta \subseteq \ker \alpha$. Also, if $0 \neq z \in M^\sigma \otimes_R S$ with $z = \sum m_i^\sigma \otimes x^i$ and n is the largest index with $m_n^\sigma \neq 0$, then the coefficient of x^{n+1} in $\beta(z)$ is m_n. Thus $\ker \beta = 0$.

It remains to check that $\operatorname{im} \beta = \ker \alpha$. If that is not so, then there will be nonzero elements in $\ker \alpha \backslash \operatorname{im} \beta$ of the form $\sum_{i=0}^n m_i \otimes x^i$. Amongst these choose one, z say, with least possible value of n. Then

$$z = (m_n \otimes x - m_n x \otimes 1)x^{n-1} + (m_n x \otimes x^{n-1} + \sum_{i=0}^{n-1} m_i \otimes x^i).$$

The first term here belongs to $\operatorname{im} \beta$ and so to $\ker \alpha$. Hence the second term belongs to $\ker \alpha \backslash \operatorname{im} \beta$. However, this contradicts the minimality of n; so $\ker \alpha = \operatorname{im} \beta$.

(ii) By 3.4, $\operatorname{pd} M_R^\sigma = \operatorname{pd} M_R$; and since $_R S$ is free, the proof of 4.2(ii) shows that $\operatorname{pd}(M \otimes_R S)_S \leq \operatorname{pd} M_R$. Applying 1.6 to the exact sequence of (i) now gives the result for projective dimension. An analogous argument deals with flat dimension. □

5.3 Theorem. *Let R be any ring, and let σ be an automorphism. Then:*
(i) $\operatorname{r gld} R \leq \operatorname{r gld} R[x; \sigma, \delta] \leq \operatorname{r gld} R + 1$ *if* $\operatorname{r gld} R < \infty$;
(ii) $\operatorname{r gld} R \leq \operatorname{r gld} R[x, x^{-1}; \sigma] \leq \operatorname{r gld} R + 1$;
(iii) $\operatorname{r gld} R[x, \sigma] = \operatorname{r gld} R + 1$;
(iv) $\operatorname{r gld} R[x, x^{-1}] = \operatorname{r gld} R + 1$;
(v) $\operatorname{r gld} R[[x; \sigma]] = \operatorname{r gld} R + 1$ *if R is right Noetherian*;
(vi) *if R is semisimple Artinian then* $\operatorname{r gld} R[x; \sigma, \delta] = \operatorname{r gld} R[x, x^{-1}; \sigma] = 1$.

Proof. (i) It is clear from 5.2(ii) that $\operatorname{r gld} R[x; \sigma, \delta] \leq \operatorname{r gld} R + 1$; and the other inequality comes from 2.6(i).

(ii) Again 5.2(ii) gives the second inequality, and 2.8(i) can be used to give the first since R is an R-bimodule direct summand of $R[x, x^{-1}; \sigma]$.

(iii) Using 2.8(i) rather than 2.6(i) shows that the inequalities of (i) hold here, even if $\operatorname{r gld} R = \infty$. Also if $\operatorname{r gld} R < \infty$ then, since $R \simeq R[x; \sigma]/xR[x; \sigma]$, 3.5 can be applied, showing that $\operatorname{r gld} R[x; \sigma] \geq \operatorname{r gld} R + 1$.

(iv) This is proved similarly, but using the regular normal nonunit $1 - x$ in applying 3.5.

(v) If $\operatorname{r gld} R = \infty$ then 2.8(ii), together with 1.9, shows that

$$\operatorname{r gld} R = \operatorname{w gld} R \leq \operatorname{w gld} R[[x; \sigma]] = \operatorname{r gld} R[[x; \sigma]]$$

since $R[[x; \sigma]]$ is a flat left R-module by 1.7. When $\operatorname{r gld} R < \infty$ then 3.7 applies.

(vi) This is clear from (i), (ii) since these two rings are not Artinian. □

The restriction to finite global dimension in (i) is shown to be necessary by 2.7.

5.4 The global dimension of the ring $P_{\lambda_1,\ldots,\lambda_n}(k)$ described in 1.8.7 can now be estimated.

Corollary. *For any choice of* $\lambda_1,\ldots,\lambda_n$
$$n \leq \operatorname{r\,gld}(P_{\lambda_1,\ldots,\lambda_n}(k)) \leq n+1.$$

Proof. This is clear from 5.3(ii) and (iv). □

Further details will be given in 10.3.

5.5 Later sections will describe circumstances under which one or other of the inequalities in 5.3(i), (ii) become equalities. In particular the case when R is a commutative Noetherian ring will be analysed. However, one very special case is readily dealt with.

Theorem. *Let R be a commutative Noetherian ring with* $\operatorname{gld} R = d < \infty$, *let S be either $R[x;\delta]$ or $R[x,x^{-1};\sigma]$ and suppose S is a simple ring. Then* $\operatorname{r\,gld} S = \sup\{1,d\}$.

Proof. If $d=0$ then 5.3(vi) applies. Suppose $d \geq 1$ and let M_S be any simple S-module. By 5.3(i), (ii), $\operatorname{pd} M_S \leq d+1$ and, by 1.14, it will suffice to show $\operatorname{pd} M_S \leq d$.

Recall first that the conditions on R ensure that R is semiprime. So if $\mathscr{C} = \mathscr{C}_R(0)$, then $R_\mathscr{C}$ is semisimple Artinian and $S_\mathscr{C}$, by 5.3(vi), has global dimension 1.

Consider the sequence of S-modules
$$0 \to M \xrightarrow{\theta} M \otimes_S S_\mathscr{C} \to K \to 0$$
where $K = \operatorname{coker} \theta$ and $\theta(m) = m \otimes 1$. Note, by 4.2(iii), that
$$\operatorname{fd}(M \otimes S_\mathscr{C})_S = \operatorname{fd}(M \otimes S_\mathscr{C})_{S_\mathscr{C}} \leq 1.$$
Suppose now that θ is injective. It follows from the short exact sequence that if $\operatorname{fd} M_S = d+1$ then $\operatorname{fd} K_S = d+2$ which is impossible. So $\operatorname{fd} M_S \leq d$ and so $\operatorname{pd} M_S \leq d$.

Suppose next that θ is not injective and so, since M is simple, $\theta = 0$. Then $M \simeq S/I$, where $I \triangleleft_r S$ and $IS_\mathscr{C} = S_\mathscr{C}$. By 6.9.16, there corresponds to S/I a prime ideal P of R with $P = \operatorname{ann}_R m$ for some $m \in S/I$. Since $IS_\mathscr{C} = S_\mathscr{C}$ it follows that $P \cap \mathscr{C} \neq \emptyset$. Thus P cannot be a minimal prime of R. However, by 1.8.3, R has no nonzero stable ideals. Hence, by 6.9.11, P cannot be semistable. Thus 6.9.16(a) applies and
$$M \simeq S/I \simeq S/PS \simeq R/P \otimes_R S.$$
Hence $\operatorname{pd} M_S \leq \operatorname{pd}(R/P)_R \leq d$. □

5.6 Next we turn to group rings. The first part of the next result is a version of Maschke's theorem.

Theorem. *Let R be a ring, G a finite group with $|G|$ a unit in R, and $S = R*G$, a crossed product of R by G. Let M be any right S-module.*
(i) *If $K \triangleleft M_S$ and K is a direct summand of M as an R-module then K is a direct summand over S.*
(ii) $\operatorname{pd} M_R = \operatorname{pd} M_S$.
(iii) $\operatorname{r\,gld} R = \operatorname{r\,gld} S$.

Proof. (i) Let $\pi: M \to K$ be the R-module splitting homomorphism; so $\pi(k) = k$ for all $k \in K$. Define $\lambda: M \to K$ by

$$\lambda(m) = |G|^{-1} \sum_{g \in G} \pi(m\bar{g})\bar{g}^{-1}.$$

Then λ is an S-module splitting homomorphism.
(ii) First note that if M_R is projective and

$$0 \to K \to F \to M \to 0$$

is a short exact sequence of S-modules with F free, then the sequence splits over R and hence over S. Thus M_S is projective. Furthermore, S_R is free.

It now follows that an S-projective resolution of any module M_S is also an R-projective resolution which terminates when a kernel is, equally, R-projective or S-projective. Thus $\operatorname{pd} M_R = \operatorname{pd} M_S$.
(iii) Clear from (ii) and 2.8(i). □

Corollary. *Let $R*G$ be a crossed product of a ring R and group G. Then:*
(i) $\operatorname{r\,gld} R \leq \operatorname{r\,gld} R*G$;
(ii) *if G is poly-infinite cyclic with Hirsch number h then $\operatorname{r\,gld} R*G \leq \operatorname{r\,gld} R + h$;*
(iii) *if R is a $\mathbb{Q}$-algebra and G is polycyclic by finite with Hirsch number h then*

$$\operatorname{r\,gld} R*G \leq \operatorname{r\,gld} R + h.$$

Proof. (i) This follows from 2.8(i).
(ii) Proposition 1.5.11 shows that $R*G$ can be viewed as a skew Laurent extension, iterated h times. Then 5.3(ii) applies.
(iii) This is proved by a straightforward induction using (ii) and Theorem 5.6(iii). □

The hypothesis that $|G|^{-1} \in R$ is necessary above since one standard example of a ring of infinite global dimension is kG, where k is a field of characteristic 2 and G is cyclic of order 2.

5.7 A crossed product $R*U(\mathfrak{g})$ is defined in 1.7.12 for any commutative ring k, a k-algebra R with $k \subseteq R$ and a k-Lie algebra g. In 6.6.2, g being solvable is defined, but a stronger condition is now required. We call g **completely solvable** when g has a k-basis $\{x_1, \ldots, x_n\}$ such that for each $i < n$, $kx_1 + \cdots + kx_i$ is an ideal of g. (Lie's theorem, 14.5.3, asserts that if k is an algebraically closed field of characteristic zero and g is solvable then g is completely solvable.) The crucial point here for us is that if g is completely solvable then x_1 is a regular normal nonunit of $R \otimes U(g)$, with factor isomorphic to $R \otimes U(g/kx_1)$.

Theorem. (a) *If* r gld $R < \infty$ *then* r gld $R \leqslant$ r gld $R * U(g)$, *and if g is solvable then* r gld $R * U(g) \leqslant$ r gld $R + \dim_k g$.

(b) r gld $R \leqslant$ r gld $R \otimes U(g)$, *and if g is completely solvable then* r gld $R \otimes U(g) =$ r gld $R + \dim_k g$.

Proof. (a) Here $R * U(g)$ is free over R, so 2.6 shows the first inequality. When g is solvable, $R * U(g)$ can be regarded as a skew polynomial extension, iterated $\dim_k g$ times. So 5.3(i) applies.

(b) Here R is an R-bimodule direct summand, so 2.8 gives the first inequality. The remark preceding the theorem shows that 3.5 can be applied $\dim_k g$ times to give

$$\text{r gld } R \otimes U(g) \geqslant \text{r gld } R + \dim_k g. \qquad \square$$

We note that the second inequality of (a) is valid even without g being solvable. This will be proved in 6.19.

5.8 Next we consider the Weyl algebras.

Theorem. (i) *Let R be a ring with* r gld $R < \infty$. *Then*

$$\text{r gld } R + n \leqslant \text{r gld } A_n(R) \leqslant \text{r gld } R + 2n.$$

(ii) *If k is a field with* char $k = p > 0$ *then* r gld $A_n(k) = 2n$.
(iii) *If k is a field with* char $k = 0$ *then* r gld $A_n(k) = n$.

Proof. (i) Note first that $A_n(R)$ is a free left and right module over the subring $R[y_1, \ldots, y_n]$. Now r gld $R[y_1, \ldots, y_n] =$ r gld $R + n$ by 5.3(iii). So r gld $A_n(R) \geqslant$ r gld $R + n$ by 2.6. On the other hand, $A_n(R)$ can be viewed as an iterated skew polynomial ring over R, with $2n$ iterations. So 5.3(i) gives r gld $A_n(R) \leqslant$ r gld $R + 2n$.

(ii) As noted in 6.6.14, $A_n(k)$ is free over the central subring $k[x_1^p, \ldots, x_n^p, y_1^p, \ldots, y_n^p]$ which, of course, has global dimension $2n$. Hence by 2.6, r gld $A_n(k) \geqslant 2n$.

(iii) Let B be as described in 6.6.15. Then B is faithfully flat over $A_n(k)$, as was noted in 2.4, and so by 2.6 r gld $A_n(k) \leqslant$ r gld B. Now $B = \bigoplus_{j=1}^{2n} B_{n,j}(k)$

with $B_{n,j} \simeq A_{n-1}(K)[y_n; \delta]$, where $K \simeq k(x_n)$. By the appropriate induction hypothesis r gld $A_{n-1}(K) = n - 1$ and so r gld $B_{n,j} \leq n$ by 5.3(i). Hence r gld $B \leq n$, as required. □

5.9 The fact that k is a field in 5.8(iii) is important. To see this, let D denote the quotient division ring of $A_n(k)$, where k is a field of characteristic zero.

Proposition. r gld $A_n(D) = 2n$.

Proof. Let $a_1, \ldots, a_{2n}$ be as described in the proof of 6.5.10 and let $I = \sum_{i=1}^{2n} a_i A_n(D)$. Then, by 3.16, pd $A_n(D)/I = 2n$. □

In fact, by using quotient division rings D of other Weyl algebras one can arrange, still in characteristic zero, that r gld $A_n(D) = i = K(A_n(D))$ for any i between n and $2n$; see [McConnell 74], [Goodearl, Hodges and Lenagan 84].

Further comments on r gld $A_n(R)$, for other rings R, can be found in Chapter 9.

5.10 It seems worth noting one related result.

Proposition. *Let D be a division ring which contains a central subfield k, and which also contains a commutative subfield of transcendence degree n over k. Then* r gld $D \otimes_k k(x_1, \ldots, x_n) = n$.

Proof. This is proved along the same lines as 6.6.17, using the technique of 5.9. □

5.11 Finally, we turn to a different type of example, namely the idealizer rings. We note that the arguments here will, to an extent, rely upon the latter part of §3. First, however, an extension of 5.5.7 is required.

Proposition. *Let R be a ring, A a generative right ideal and T a subring with $A \subseteq T \subseteq \mathbb{I}(A)$.*
(i) *If M is a right R-module, then* pd $M_R =$ pd M_T.
(ii) *If N is a left R-module then* pd $_R N \leq$ pd $_T N$.

Proof. (i) By 5.5.7, if M_R is projective so too is M_T. Conversely, suppose that M_T is projective and suppose an exact diagram of R-modules

$$\begin{array}{c} M \\ {}^{\alpha}\swarrow \downarrow \\ V \to W \to 0 \end{array}$$

is given. Since M_T is projective, a T-homomorphism α exists making this commute, viewed as a T-module diagram. Hence the R-module diagram

$$\begin{array}{c} M \otimes_T R \\ \alpha \otimes 1 \swarrow \quad \downarrow \\ V \otimes_T R \to W \otimes_T R \to 0 \end{array}$$

also commutes. By 5.5.7, $R \otimes_T R \simeq R$ and so

$$M \otimes_T R \simeq M \otimes_R R \otimes_T R \simeq M \otimes_R R \simeq M;$$

and similarly for $V \otimes R$ and $W \otimes R$. It follows that α is in fact an R-module homomorphism, so M_R is projective.

This means that any minimal R-projective resolution of M is equally a minimal T-projective resolution. Thus $\mathrm{pd}\, M_R = \mathrm{pd}\, M_T$.

(ii) If $0 \to P_n \to \cdots \to P_0 \to N \to 0$ is a T-projective resolution of N, then the fact (5.5.7) that R_T is projective means that

$$0 \to R \otimes_T P_n \to \cdots \to R \otimes_T P_0 \to R \otimes_T N \to 0$$

is exact. This is thus an R-resolution of N since, as above, $R \otimes_T N \simeq N$. □

5.12 Following the same line of proof as that of 5.5.8, but making use of the above result, one can prove

Corollary. *If A is an isomaximal right ideal of a ring R then*

$$\mathrm{r\, gld}\, \mathbb{I}(A) = \sup\{1, \mathrm{r\, gld}\, R\}. \qquad \square$$

5.13 One can also estimate the left global dimension; but that can be achieved in more general circumstances.

Theorem. *Let R be a ring, A a generative right ideal and let T be a ring with $A \subset T \subseteq \mathbb{I}(A)$. Then*

$$\sup\{\mathrm{r\, gld}\, R, \mathrm{r\, gld}\, T/A\} \leqslant \mathrm{r\, gld}\, T \leqslant \mathrm{r\, gld}\, T/A + \mathrm{r\, gld}\, R + 1,$$

and

$$\sup\{\mathrm{l\, gld}\, R, \mathrm{l\, gld}\, T/A\} \leqslant \mathrm{l\, gld}\, T \leqslant \mathrm{l\, gld}\, T/A + \mathrm{l\, gld}\, R + 1.$$

Proof. First consider $\mathrm{r\, gld}\, T$. It is clear from 5.11(i) that $\mathrm{r\, gld}\, R \leqslant \mathrm{r\, gld}\, T$. Note that $A^2 = A$ and, by 5.5.7, that $_T A$ is projective. Thus 3.12 can be applied to give $\mathrm{r\, gld}\, T/A \leqslant \mathrm{r\, gld}\, T$. Next let $V \triangleleft_r T$ and consider the exact sequence

$$0 \to VA \to V \to V/VA \to 0.$$

By Proposition 2.2, and recalling from 5.5.7 that R_T is projective,

$$\mathrm{pd}\, VA_T \leqslant \mathrm{pd}\, VA_R + \mathrm{pd}\, R_T = \mathrm{pd}\, VA_R$$

and

$$\mathrm{pd}\, (V/VA)_T \leqslant \mathrm{pd}\, (V/VA)_{T/A} + \mathrm{pd}\, (T/A)_T.$$

7.5.14 Estimates of global dimension

Hence, by 1.6,

$$\operatorname{pd} V_T \leqslant \sup\{\operatorname{pd} VA_R, \operatorname{pd}(V/VA)_{T/A} + \operatorname{pd}(T/A)_T\}.$$

Now $VA \triangleleft_r R$ and $\operatorname{pd}(T/A)_T \leqslant 1 + \operatorname{pd} A_T = 1 + \operatorname{pd} A_R$. Therefore, provided R is not semisimple,

$$\operatorname{pd} V_T \leqslant \sup\{\operatorname{r\,gld} R - 1, \operatorname{r\,gld} T/A + \operatorname{r\,gld} R\}.$$

Hence

$$\operatorname{r\,gld} T \leqslant \operatorname{r\,gld} T/A + \operatorname{r\,gld} R + 1.$$

Of course, if R is semisimple then $A = eR$ for some idempotent e. Hence $A = eT$ and $T/A \simeq (1-e)T$ which is projective. Therefore the inequalities above show that $\operatorname{pd} V_T \leqslant \operatorname{r\,gld} T/A$ and so $\operatorname{r\,gld} T \leqslant 1 + \operatorname{r\,gld} T/A$.

We now turn to $\operatorname{l\,gld} T$. Since $_T A$ is projective and $A^2 = A$, 3.10 shows that $\operatorname{l\,gld} T/A \leqslant \operatorname{l\,gld} T$. On the other hand, 5.11(ii) shows that $\operatorname{l\,gld} R \leqslant \operatorname{l\,gld} T$.

Now let $V \triangleleft_l T$ and consider the exact sequence

$$0 \to V \to RV \to RV/V \to 0,$$

noting that RV/V is a left T/A-module. By Proposition 2.2

$$\operatorname{pd}_T(RV/V) \leqslant \operatorname{pd}_{T/A}(RV/V) + \operatorname{pd}_T(T/A)$$

and

$$\operatorname{pd}_T(RV) \leqslant \operatorname{pd}_R(RV) + \operatorname{pd}_T R.$$

Hence, if R is not semisimple, then

$$\operatorname{l\,gld} T \leqslant \sup\{\operatorname{l\,gld} R + \operatorname{pd}_T R, \operatorname{l\,gld} T/A + \operatorname{pd}_T(T/A)\}.$$

However, taking $V = T$ in the exact sequence and using 1.6 and 2.2 gives

$$\operatorname{pd}_T R \leqslant \sup\{\operatorname{pd}_T(R/T), \operatorname{pd}_T T\}$$
$$= \operatorname{pd}_T R/T \leqslant \operatorname{pd}_{T/A}(R/T) + \operatorname{pd}_T(T/A) \leqslant \operatorname{pd}_{T/A}(R/T) + 1.$$

Hence

$$\operatorname{l\,gld} T \leqslant \operatorname{l\,gld} R + \operatorname{l\,gld} T/A + 1.$$

Finally, if R is semisimple then $A = eR$, as above, and then

$$T \simeq \begin{bmatrix} eRe & eR(1-e) \\ 0 & (1-e)T(1-e) \end{bmatrix}.$$

Noting that eRe is semisimple and applying the left-hand version of 5.1, we see that

$$\operatorname{l\,gld} T \leqslant \operatorname{l\,gld}(1-e)T(1-e) + 1 = \operatorname{l\,gld} T/A + 1. \qquad \square$$

5.14 Finally, a specific illustration.

Corollary. *Let R be the Weyl algebra $A_n(k)$ with k a field of characteristic zero.*

(So, by 5.8, r gld R = l gld R = n). Let A be any nonzero right ideal and $T = k + A$. Then
$$n \leq \{\text{r gld } T, \text{l gld } T\} \leq n+1. \qquad \square$$

§6 Filtered and Graded Modules

6.1 One major class of examples described in Chapter 1, but not covered by the preceding section except in special circumstances, is the crossed product $S = R * U(g)$, where R is a k-algebra over a commutative ring k with r gld $R = d$ and g is a finite dimensional k-Lie algebra. We recall, from Chapter 1, §7, that this means g is a free k-module, of dimension n say, and S is a filtered ring with gr $S \simeq R[x_1, \ldots, x_n]$, these being central indeterminates. Thus r gld gr $S = d + n$.

A consequence of the main result of this section is that r gld $S \leq d + n$ also. The main result shows that if M is a right S-module and gr M is a certain module over gr S obtained from M then suitable projective resolutions of gr M yield projective resolutions of M. This relies upon a study of filtered and graded modules which takes up this section. The details are elementary but a little lengthy.

6.2 Let $T = \bigoplus_{n \in \mathbb{N}} T_n$ be a graded ring as described in 1.6.3. A **grading** of a right T-module M is an abelian group decomposition $M = \bigoplus_{n \in \mathbb{N}} M_n$ such that $M_i T_j \subseteq M_{i+j}$ for all $i, j \in \mathbb{N}$. A module M may have many different gradings. For example if $s \in \mathbb{N}$ then setting $M'_0 = M'_1 = \cdots = M'_{s-1} = 0$ and $M'_{n+s} = M_n$ gives another grading $M = \bigoplus M'_n$, this being called a **translate** of the original one.

6.3 A **graded module** is a module together with a fixed grading. The nonzero elements of the subgroup M_n are called **homogeneous** of degree n, and if $m = \sum_{n \in \mathbb{N}} m_n$ with $m_n \in M_n$ then m_n is the nth **homogeneous component** of m. A **graded submodule** N of M is a submodule with a grading $N = \bigoplus_{n \in \mathbb{N}} N_n$ such that $N_n \subseteq M_n$. A **graded homomorphism** θ is a T-homomorphism between two graded T-modules M, M' such that $\theta(M_n) \subseteq M'_n$ for all n.

6.4 Let MR denote the category of all right R-modules and R-module homomorphisms over any ring R. Similarly, let $M_{\text{gr}} T$ be the category of graded right T-modules and graded homomorphisms over a graded ring T.

A graded module M_T can be viewed as an object of $M_{\text{gr}} T$ or, via the forgetful functor, as an object of MT, and it will be important to know that some of its categorical properties are independent of the choice of category. For example an object M of $M_{\text{gr}} T$ being finitely generated should mean that M has a finite set of homogeneous generators. Evidently this is true if and only if M_T is finitely generated.

6.5 First consider freeness. A graded module F is **free-graded** on the basis

7.6.7 Filtered and graded modules

$\{e_j|j\in J\}$ if F is free as a right T-module on the basis $\{e_j\}$ and also each e_j is homogeneous, say of degree $n(j)$. It follows that F is isomorphic to a direct sum of translates of T and that $F_n = \sum_j e_j T_{n-n(j)}$. One can check that F and $\{e_j\}$ have the universal property that any map $\theta:\{e_j\}\to M$ with $M\in M_{gr}T$ and $\theta(e_j)\in M_{n(j)}$ can be extended to a graded T-module homomorphism $\theta:F\to M$. By appropriate choice of F it can be arranged that θ is a surjection.

6.6 Now we consider projectivity.

Proposition. *Let T be a graded ring and P a graded right T-module. The following are equivalent:*
(i) *P is a direct summand, in $M_{gr}T$, of a free-graded module;*
(ii) *given any exact diagram*

$$\begin{array}{c} P \\ {}^{\alpha}\swarrow \downarrow \\ M \to N \to 0 \end{array}$$

in $M_{gr}T$ there exists a unique graded homomorphism α making the diagram commute;
(iii) *P_T is projective.*

Proof. We discuss only the proof (iii)$\Rightarrow$(i), the other arguments being similar or trivial. Let $\{p_j|j\in J\}$ be homogeneous generators of P with degree $p_j = n(j)$. Let F be the free-graded module with basis $\{e_j|j\in J\}$ and degree $e_j = n(j)$. There is then a graded homomorphism $\pi:F\to P$ which is surjective and so is split in MT. The splitting homomorphism, θ say, need not, of course, be a graded homomorphism. However, let $\theta':P\to F$ be defined to be the graded homomorphism such that if $p\in P_n$ then $\theta'(p)$ is the nth homogeneous component of $\theta(p)$. Since π is a graded homomorphism, one sees that $m = \pi\theta(m) = \pi\theta'(m)$ and that θ' splits π in $M_{gr}T$. □

A module P as above is called a **projective-graded** module. The result shows that this is the same as being both projective and graded.

6.7 For the applications, the graded ring T will arise as the associated graded ring gr S of a filtered ring S. Suppose, then, that S is a filtered ring as described in 1.6.1, having the filtration $S = \bigcup F_n$. A **filtration** of a right S-module M is a family of subgroups M_n, $n\in\mathbb{N}$, such that
(i) for each i, j, $M_i F_j \subseteq M_{i+j}$,
(ii) for each $i < j$, $M_i \subseteq M_j$, and
(iii) $\bigcup M_n = M$.
If $m\in M_n\setminus M_{n-1}$, m is said to have **degree** n.

As with gradings, the filtration of a module is not unique. Indeed, given any

F_0-submodule M_0 of M which generates M_S, one gets a **standard** filtration by setting $M_n = M_0 F_n$.

6.8 A module with a filtration is called a **filtered module**. A **filtered homomorphism** θ is an S-homomorphism between filtered modules M, M' such that $\theta(M_j) \subseteq M'_j$ for all j. The collection of all such modules and homomorphisms makes up a category $M_{\text{fl}}S$; and there is a forgetful functor linking $M_{\text{fl}}S$ and MS.

6.9 Given S as above, let gr S be its associated graded ring, as described in 1.6.4. The same process yields a graded module gr M over gr S from any filtered module M_S. Explicitly, $\text{gr } M = \bigoplus_{n \in \mathbb{N}} (\text{gr } M)_n$ where $(\text{gr } M)_n = M_n/M_{n-1}$; and multiplication is based, via the distributive laws, on the rule that if $0 \neq \bar{m} \in (\text{gr } M)_i$ and $0 \neq \bar{s} \in (\text{gr } S)_j$ then $\bar{m}\bar{s}$ is the image of ms in $(\text{gr } M)_{i+j}$. Similarly, a filtered homomorphism $\theta: M \to M'$ yields a graded homomorphism $\text{gr } \theta: \text{gr } M \to \text{gr } M'$ by defining $(\text{gr }\theta)(\bar{m})$ to be the image of $\theta(m)$ in $(\text{gr } M')_i$.

Proposition. *If S is a filtered ring then* gr *is a functor from* $M_{\text{fl}}S$ *to* $M_{\text{gr}}(\text{gr } S)$. $\square$

6.10 Given any module M_S over the filtered ring S and any filtration of M, one obtains the graded module gr M over gr S. The point of this section is to connect the homological properties of M and gr M. More precisely, given a projective resolution for gr M of length n, we aim to discover a similar one for M_S. This involves results linking MS, $M_{\text{fl}}S$ and $M_{\text{gr}}(\text{gr } S)$.

6.11 First comes finite generation.

Lemma. *Let S be a filtered ring and M an S-module.*
(i) *If M is filtered and* gr M *is finitely generated over* gr S *then M_S is finitely generated.*
(ii) *If M_S is finitely generated then M has a standard filtration such that* gr M *is finitely generated over* gr S.

Proof. (i) Choose a finite set of homogeneous generators of gr M and an inverse image for each in M. They generate M.

(ii) Choose a finitely generated F_0-submodule M_0 of M with $M_0 S = M$. The standard filtration using M_0 makes gr M finitely generated. $\square$

It is easy to see that if an arbitrary filtration is chosen in (ii) then gr M need not be finitely generated (cf. 1.6.9).

6.12 We now aim at exact sequences. Recall first that if $\theta: M \to N$ is a filtered homomorphism then $\theta(M_j) \subseteq \theta(M) \cap N_j$. If it happens that $\theta(M_j) = \theta(M) \cap N_j$ for each j then θ is called **strict**.

7.6.15 *Filtered and graded modules*

For example if $\alpha: M \to N$ is any homomorphism and M is given the induced filtration $M_j = \alpha^{-1}(\alpha(M) \cap N_j)$ then α is a strict filtered homomorphism. Similarly, if α is a surjection and if N is given the induced filtration $N_j = \alpha(M_j)$, then α is strict.

6.13 Proposition. *Let S be a filtered ring and let $K \xrightarrow{\theta} M \xrightarrow{\phi} N$ be an exact sequence of filtered modules and filtered homomorphisms. Then $\operatorname{gr} K \xrightarrow{\operatorname{gr}\theta} \operatorname{gr} M \xrightarrow{\operatorname{gr}\phi} \operatorname{gr} N$ is exact if and only if θ and ϕ are strict.*

Proof. First suppose the graded sequence is exact. We aim to show that ϕ, θ are strict. Suppose $n \in N_j \cap \operatorname{im} \phi$ and $n \notin N_{j-1}$. Then $n = \phi(m)$ for some $m \in M_{j+r}$ for some $r \geq 0$. We need only show that m can be chosen so that $r = 0$. If $r \neq 0$, then $\operatorname{gr}\phi(\bar{m}) = 0$ and so, by exactness, $\bar{m} = (\operatorname{gr}\theta)(\bar{k})$ where $k \in K_{j+r}$. Let $m' = m - \theta(k)$. Then $n = \phi(m')$ and $m' \in M_{j+r-1}$. This together with induction on r shows that ϕ is strict. A similar argument deals with θ.

Next suppose that θ, ϕ are strict. To show exactness at $\operatorname{gr} M$ it is enough to check that if $m \in M_j \setminus M_{j-1}$ with $(\operatorname{gr}\phi)(\bar{m}) = 0$ in $(\operatorname{gr} N)_j$ then $\bar{m} \in \operatorname{im}(\operatorname{gr}\theta)$. However, $\phi(m) \in N_{j-1} \cap \operatorname{im}\phi$; so $\phi(m) = \phi(m')$ for some $m' \in M_{j-1}$. Then $m - m' \in M_j \cap \operatorname{im}\theta$. Hence $m - m' = \theta(k)$ for some $k \in K_j$. Therefore $\operatorname{gr}\theta(\bar{k}) = \overline{(m - m')} = \bar{m}$ as required. □

6.14 Corollary. *Let S be a filtered ring and $\phi: M \to N$ a filtered homomorphism. Then $\operatorname{gr}\phi$ is injective (surjective) if and only if ϕ is injective (resp. surjective) and ϕ is strict.*

Proof. ($\Rightarrow$) Suppose ϕ is not injective. Choose $0 \neq m \in M_j$ with $\phi(m) = 0$ and j as small as possible. Then $0 \neq \bar{m} \in (\operatorname{gr} M)_j$ and yet $(\operatorname{gr}\phi)(\bar{m}) = 0$, being the image of $\phi(m)$ in N_j/N_{j-1}. Thus if $\operatorname{gr}\phi$ is injective so too is ϕ. Applying 6.13 to the exact sequence $0 \to M \to N$ shows that ϕ is strict. A similar argument deals with surjectivity.

($\Leftarrow$) This is clear from 6.12 and 6.13. □

6.15 A filtered module $M = \bigcup M_n$ over a filtered ring $S = \bigcup F_n$ is **free-filtered** on the basis $\{e_j | j \in J\}$ if M_S is free on $\{e_j\}$ with degree $e_j = n(j)$ say, and $M_n = \bigoplus_j e_j F_{n-n(j)}$.

Proposition. *Let S be a filtered ring.*
(i) *If M_S is free-filtered then $\operatorname{gr} M$ is free-graded over $\operatorname{gr} S$.*
(ii) *If M' is free-graded over $\operatorname{gr} S$ then $M' \simeq \operatorname{gr} M$ for some free-filtered module M_S.*
(iii) *If M_S is free-filtered, N is filtered and $\alpha: \operatorname{gr} M \to \operatorname{gr} N$ is a graded surjection, then $\alpha = \operatorname{gr}\beta$ for some strict filtered surjection $\beta: M \to N$.*

Proof. (i), (ii) These are straightforward verifications.

(iii) Let $\{e_j | j \in J\}$ be the free basis of M with degree $e_j = n(j)$. For each j, choose $x_j \in N$ with degree $x_j = n(j)$ so that the image of x_j in $(\operatorname{gr} N)_{n(j)}$ is $\alpha(\bar{e}_j)$. Then let β be defined by $\beta(e_j) = x_j$. By construction, β is a filtered homomorphism and $\operatorname{gr} \beta = \alpha$ since they coincide on generators. Therefore β is a strict filtered surjection by 6.14. □

6.16 The last piece of information required concerns projectivity.

Proposition. *Let S be a filtered ring and P a filtered S-module such that $\operatorname{gr} P$ is a projective $\operatorname{gr} S$-module. Then P is a projective S-module.*

Proof. Let α be a graded surjection $\operatorname{gr} F \to \operatorname{gr} P$, where F is a free-filtered S-module and so $\operatorname{gr} F$ is a free-graded $\operatorname{gr} S$-module. By 6.15, $\alpha = \operatorname{gr} \beta$ for some strict filtered surjection $\beta : F \to P$. Let $K = \ker \beta$, with the induced filtration from F. There is an exact sequence $0 \to K \xrightarrow{\gamma} F \xrightarrow{\beta} P \to 0$, and hence, by 6.12 and 6.13, the sequence $0 \to \operatorname{gr} K \to \operatorname{gr} F \to \operatorname{gr} P \to 0$ is exact. By 6.6, this splits in $M_{\operatorname{gr}}(\operatorname{gr} S)$. The graded homomorphism $\operatorname{gr} F \to \operatorname{gr} K$ which splits this can, as above, be assumed to take the form $\operatorname{gr} \theta$ for some strict filtered surjection $\theta : F \to K$. Note that $\operatorname{gr}(\theta \gamma) = 1_{\operatorname{gr} K}$ and so, by 6.14, $\theta \gamma$ is an automorphism of K. Thus $F \simeq K \oplus P$ and so P is projective. □

6.17 Next comes the main result.

Theorem. *Let S be a filtered ring and M_S a filtered module. Let*

$$0 \to K' \to F'_n \to \cdots \to F'_0 \to \operatorname{gr} M \to 0 \qquad (*)$$

be an exact sequence in $M_{\operatorname{gr}}(\operatorname{gr} S)$ with the F'_i being free-graded.
(i) *There is a corresponding exact sequence*

$$0 \to K \to F_n \to \cdots \to F_0 \to M \to 0 \qquad (**)$$

in $M_{\operatorname{fl}}(S)$ with the F_i being free-filtered, the maps being strict and with $(*) \simeq \operatorname{gr}(**)$.
(ii) *If K' is projective over $\operatorname{gr} S$ then K is projective over S.*
(iii) *If the $\operatorname{gr} S$ modules in $(*)$ are finitely generated then the S-modules in $(**)$ are also finitely generated.*

Proof. (i) By 6.15 the map $F'_0 \to \operatorname{gr} M$ in $(*)$ has the form $\operatorname{gr} \beta$ for some strict filtered surjection $\beta : F_0 \to M$, where $F'_0 = \operatorname{gr} F_0$ and F_0 is free-filtered. Let $K_0 = \ker \beta$ with the induced filtration from F_0. Then

$$0 \to \operatorname{gr} K_0 \to \operatorname{gr} F_0 \to \operatorname{gr} M \to 0$$

is exact. The map $F'_1 \to F'_0$ in $(*)$ thus factors through $\operatorname{gr} K_0$. Hence the construction can be repeated, starting with $\operatorname{gr} K_0$ rather than $\operatorname{gr} M$.

(ii)(iii) These are immediate from 6.16 and 6.11 respectively. □

6.18 Corollary. *Let S be a filtered ring. Then:*
(i) $\operatorname{r} \operatorname{gld} S \leqslant \operatorname{r} \operatorname{gld} \operatorname{gr} S$;
(ii) *if M_S is a filtered module then* $\operatorname{pd} M_S \leqslant \operatorname{pd} (\operatorname{gr} M)_{\operatorname{gr} S}$.

Proof. (i) This will follow immediately from (ii) since every S-module has a filtration.

(ii) It is clear from 6.5 that one can construct a resolution (∗) as in 6.17. If $\operatorname{pd}(\operatorname{gr} M) = n + 1$ then K' and hence K are projective. □

6.19 This applies easily to enveloping algebras.

Corollary. *Let k be a commutative ring, R be a k-algebra with $\operatorname{r} \operatorname{gld} R = d$ and g a k-Lie algebra (cf. 1.7.8) free of rank n over k. Then $\operatorname{r} \operatorname{gld}(R \ast U(g)) \leqslant d + n$.*

Proof. By 1.7.14, $\operatorname{gr}(R \ast U(g)) \simeq R[x_1, \ldots, x_n]$. Hence, by 5.3,

$$\operatorname{r} \operatorname{gld}(R \ast U(g)) \leqslant n + d.$$
□

6.20 The inequalities above can be strict. For example, 5.8 shows that if k is a field of characteristic zero then $\operatorname{r} \operatorname{gld} A_n(k) = n$, whereas $\operatorname{r} \operatorname{gld}(\operatorname{gr} A_n(k)) = 2n$.

§7 Regular Rings

7.1 For several later applications it is not necessary to know the precise value of the global dimension of the ring. Nor, indeed, is it required that it be finite. What is needed is that every finitely generated right module has finite projective dimension—or, equivalently, that every cyclic module has finite projective dimension. A ring with this property is called **right regular**.

7.2 This section contains some rather straightforward extensions to right regular rings of earlier results. First, however, it seems sensible to consider an example.

Example. *Let R be the commutative Noetherian integral domain of Krull dimension ω described in 6.4.9. Then $\operatorname{gld} R = \infty$ but R is regular.*

Proof. Any nonzero element of R is contained in only finitely many maximal ideals. So if $0 \neq I \triangleleft R$ then

$$\operatorname{pd}(R/I)_R = \sup \{\operatorname{pd}(R_P/I_P)_{R_P}\},$$

where the supremum is over the finitely many maximal ideals P containing I.

Each R_P is a localization of a polynomial ring in finitely many variables over a field; so $\operatorname{pd}(R/I)_R < \infty$ and R is regular. Since the global dimension of the rings R_P is unbounded as P ranges over all maximal ideals of R, it follows from 4.3 that $\operatorname{gld} R = \infty$. □

7.3 We now extend 4.3.

Proposition. *If Q is a right localization of a right regular ring R then Q is right regular.*

Proof. Let M_Q be finitely generated, say $M = \sum_{i=1}^n m_i Q$, and let $N = \sum m_i R$. Now $N \otimes Q \hookrightarrow M \otimes Q \simeq M$ and so $N \otimes Q \simeq NQ = M$. Hence, by 4.2,

$$\operatorname{pd} M_Q = \operatorname{pd} N \otimes Q_Q \leqslant \operatorname{pd} N_R < \infty$$

as required. □

7.4 The filtered and graded techniques of §6 apply easily. If R is a filtered ring and M_R is finitely generated then, by 6.11, M has a filtration so that $\operatorname{gr} M$ is finitely generated over $\operatorname{gr} R$. By 6.18, $\operatorname{pd} M_R \leqslant \operatorname{pd}(\operatorname{gr} M)_{\operatorname{gr} R}$. This effectively proves the next result.

Proposition. *Let R be a filtered ring such that $\operatorname{gr} R$ is a right regular right Noetherian ring. Then R is right regular and right Noetherian.* □

7.5 It remains to deal with the various constructions.

Theorem. *Let R be a right Noetherian right regular ring and σ an automorphism. Then each of the following rings is right Noetherian and right regular:*
  (i) $R[x; \sigma]$;
  (ii) $R[x; \sigma, \delta]$;
  (iii) $R[x, x^{-1}; \sigma]$;
  (iv) $R * G$ *with G poly-infinite cyclic;*
  (v) $R * U(\mathfrak{g})$, *a crossed product over a commutative ring k, with R a k-algebra and $\mathfrak{g}$ a k-Lie algebra of finite dimension.*

Proof. First we show that all are easy consequences of (i). Suppose, for now, that (i) is true.
  (ii) The filtration of $R[x; \sigma, \delta]$ by degree gives an associated graded ring isomorphic to $R[x; \sigma]$. Now apply 7.4 and (i).
  (iii) This follows from (i) and 7.3.
  (iv) By 1.5.11, $R * G$ can be obtained from R by iterated extensions as in (iii). So induction on (iii) yields (iv).
  (v) By 1.7.14, $\operatorname{gr}(R * U(\mathfrak{g}))$ is a polynomial ring over R in a finite set of central

indeterminates. Induction on (i) shows that $\mathrm{gr}(R*U(g))$ is right Noetherian and right regular. Now apply 7.4.

(i) It remains only to prove this. Consider any finitely generated right S-module M, where $S = R[x; \sigma]$. By 5.2(ii), $\mathrm{pd}\, M_S \leq \mathrm{pd}\, M_R + 1$; so it is enough to show that $\mathrm{pd}\, M_R < \infty$. Since M_R need not be finitely generated this is not immediately obvious.

Choose any finitely generated R-submodule M_0 of M such that $M_0 S = M$ and let $M_n = \sum_{i=0}^{n} M_0 x^i$. Evidently M_n, and hence M_n/M_{n-1}, is finitely generated over R. Moreover, since $M_{n+1} = M_n x + M_n$, right multiplication by x gives a sequence of abelian group surjections

$$M_0 \to M_1/M_0 \to M_2/M_1 \to \cdots.$$

Let K_n be the kernel of the resulting map $M_0 \to M_n/M_{n-1}$. Then K_n is a submodule of M_0 and $\{K_i\}$ is an increasing chain. Hence $K_{n+t} = K_n$ for some n and all t. Consequently,

$$M_{n+t+1}/M_{n+t} \simeq M_n/M_{n-1}$$

as abelian groups. Using 3.4 and its notation, it is clear that

$$M_{n+t+1}/M_{n+t} \simeq (M_n/M_{n-1})^\alpha$$

as R-modules, where $\alpha = \sigma^{t+1}$, and that

$$\mathrm{pd}\,(M_{n+t+1}/M_{n+t}) = \mathrm{pd}\,(M_n/M_{n-1}).$$

Hence, by 1.6,

$$\mathrm{pd}\, M_{n+t} \leq \sup\{\mathrm{pd}\, M_0, \mathrm{pd}\, M_1/M_0, \ldots, \mathrm{pd}\, M_n/M_{n-1}\} = w,$$

say, and so $(\mathrm{pd}\, \bigoplus_{n=0}^{\infty} M_n) = w$.

However, there is an exact sequence

$$0 \to \bigoplus M_n \xrightarrow{\iota} \bigoplus M_n \xrightarrow{\pi} M \to 0,$$

where $\iota: (m_n) \mapsto (m_n')$ with $m_n' = m_n - m_{n-1}$ and $\pi: (m_n) \mapsto \sum m_n$. Therefore $\mathrm{pd}\, M_R \leq w + 1 < \infty$. □

§8 Fixed Rings

8.1 This section concerns a skew group ring $S = R\#G$, as described in 1.5.4, in the case when G is finite, and the **fixed ring**

$$R^G = \{r \in R \mid r^g = r \text{ for all } g \in G\}.$$

This notation will be kept fixed throughout. As well as investigating their global dimensions this section also shows when they are simple and provides thereby some interesting examples.

8.2 Earlier results give information about S. Thus 1.1.3 shows that if R is right Noetherian then so is S and 6.5.3 shows that $\mathcal{K}(S) = \mathcal{K}(R)$. Similarly, 5.6 shows that if $|G|$ is a unit in R then r gld S = r gld R.

The next few results establish a Morita context linking S with R^G and so providing a channel for the transfer of some properties from S to R^G. In the special case when $|G|^{-1} \in R$ and S is simple this context gives a Morita equivalence.

8.3 Recall, from 1.5.6, that R has a right S-module structure, indicated by the notation R_S, in which $r_1(gr_2) = r_1^g r_2$. There is also a left module structure $_S R$ given by $(r_1 g^{-1}) r_2 = r_1 (r_2)^g$. By 3.5.1, there is a Morita context

$$\begin{bmatrix} S & (R_S)^* \\ R_S & \mathrm{End}(R_S) \end{bmatrix}.$$

The next few results concern the components of this context.

8.4 Let $f = \sum_{g \in G} g \in S$.

Lemma. (i) $gf = f = fg$ for all $g \in G$.
(ii) $f^2 = |G| f$.
(iii) $fS = fR$ and $Sf = Rf$.

Proof. Straightforward. □

8.5 Proposition. (i) $R_S \simeq (fS)_S \simeq (Sf)^*$, and $_S R \simeq {_S}(Sf) \simeq (fS)^*$.
(ii) $\mathrm{End}(R_S) \simeq R^G \simeq \mathrm{End}(_S R)$, where R^G acts by left multiplication on R_S and by right multiplication on $_S R$.

Proof. (i) If $x \in fS$ then $x = fr_1$ for some $r_1 \in R$. Now $x(gr_2) = fr_1 gr_2 = fgr_1^g r_2 = fr_1^g r_2$. Hence $R_S \simeq fS$. Note next the embedding $R_S \to (Sf)^*$ which sends $r \in R$ to the right multiplication map $sf \mapsto sfr$. To see this is an isomorphism, let $\alpha \in (Sf)^*$. This is determined by $\alpha(f) = \sum gr_g$ say. For each $h \in G$,

$$\alpha(f) = \alpha(hf) = h\alpha(f) = \sum hgr_g$$

and therefore $r_{hg} = r_g$. Hence $\alpha(f) = fr$ for some $r \in R$. Symmetry completes the argument.
(ii) A similar verification gives this. □

8.6 It follows that the Morita context of 8.3 may be rewritten

$$\begin{bmatrix} S & Sf \\ fS & R^G \end{bmatrix}.$$

7.8.9 *Fixed rings*

with multiplications
$$Sf \times fS \to S \quad \text{via } (s_1 f, f s_2) \mapsto s_1 f s_2$$
and
$$fS \times Sf \to R^G \quad \text{via } (f r_1, r_2 f) \mapsto \sum_{g \in G} (r_1 r_2)^g.$$

The element $\sum_g r^g$ is called the **trace** of $r \in R$. It is clear from 3.5.4 that this context yields a Morita equivalence between S and R^G provided that $S = SfS$ and the trace map is surjective. The former condition is, of course, true when S is simple and this will be of use later. We concentrate next on a case in which the latter condition holds.

8.7 Suppose that $|G|$ is a unit in R and let $e = |G|^{-1} f$. Then $e^2 = e$. Hence $R_S \simeq fS = eS$ which is a projective module. Also, $R^G \simeq \operatorname{End} eS \simeq eSe$ and the trace map is a surjection.

Corollary. *If $|G|^{-1} \in R$ then the Morita context may be rewritten*
$$\begin{bmatrix} S & Se \\ eS & eSe \end{bmatrix}$$
using the ordinary multiplication within S. □

8.8 In these circumstances some of the properties of R^G can be deduced via S from those of R.

Theorem. *Suppose that $|G|^{-1} \in R$.*
(i) If R is right Noetherian then so too is R^G and $\mathcal{K}(R^G) \leq \mathcal{K}(R)$.
(ii) If R is right hereditary then so too is R^G.

Proof. (i) From 1.1.3 and 6.5.3 one sees that S is right Noetherian and $\mathcal{K}(S_S) = \mathcal{K}(R_R)$. The map sending a right ideal I of eSe to IS embeds the lattice of right ideals of eSe in the lattice of submodules of eS_S. Therefore, by 6.5.3, $\mathcal{K}(eSe) \leq \mathcal{K}(eS_S) \leq \mathcal{K}(S_S)$.

(ii) By 5.6, S is right hereditary. So it is enough to prove the next result. □

8.9 The next result extends 5.4.4

Proposition. *If a right hereditary ring T contains a nonzero idempotent e then eTe is right hereditary.*

Proof. Let $I \triangleleft_r eTe$, let F be a free T-module on a basis $\{b_i | i \in I\}$ and let F' be the image of the T-homomorphism $\sigma \colon F \to F$ given by $\sigma(\sum b_i t_i) = \sum b_i e t_i$. Define

$\theta\colon F \to T$ by $\theta(\sum b_i t_i) = \sum i t_i$. Now im $\theta = IT$ which is a projective right T-module. Therefore $F' \simeq IT \oplus K$ for some right T-module K. Hence $F'e \simeq ITe \oplus Ke$ this being an eTe-isomorphism. However $ITe = I$ and $F'e$ is a free eTe-module on the basis $\{b_i | i \in I\}$. Therefore I is projective. □

8.10 One might expect a similar result for higher global dimensions. Of course 5.6 shows that r gld S = r gld R. However, 8.8(ii) and 8.9 do not extend to higher global dimensions.

Example. Let k be a field with char $k \ne 2$. Let R be the commutative polynomial ring $k[x,y]$ and $G = \{1, g\}$ with g being the k-algebra automorphism sending x to $-x$, y to $-y$; so $|G|^{-1} \in R$. One can check that $R^G = k[x^2, xy, y^2]$, that $P = (x^2, xy, y^2)$ is a maximal ideal and that P_P requires three generators over $(R^G)_P$. However, by 8.8(i), $\mathcal{K}(R^G) \le 2$ and so $\mathcal{K}((R^G)_P) \le 2$. Therefore 1.15(ii) shows that gld $(R^G)_P = \infty$ and that gld $R^G = \infty$.

8.11 Nevertheless, there are circumstances when there is a result of this kind.

Proposition. (i) If $|G|^{-1} \in R$ and $SeS = S$ then r gld R^G = r gld R.
(ii) If $|G|^{-1} \in R$ and S is simple then r gld R^G = r gld R.

Proof. (i) In this case $R^G \overset{M}{\sim} S$.
(ii) This follows from (i). □

8.12 There are circumstances under which simplicity of S can be guaranteed. An **outer** automorphism of a ring R is an automorphism which is not inner. If each $g \in G$, other than the identity element, induces an outer automorphism of R, G is called a **group of outer automorphism**.

Proposition. *Let R be a simple ring and G a group of outer outomorphisms. Then S is simple.*

Proof. (cf. the proof of 1.8.5) Let I be a nonzero ideal of S and let $0 \ne x = \sum_g gr_g \in I$ have the smallest possible support. We may suppose, without loss of generality, that $r_1 \ne 0$ and then, since R is simple, that $r_1 = 1$. For any $r \in R$, $rx - xr \in I$ has smaller support than x, so $rx - xr = 0$. In particular, if $r_g \ne 0$ for some $g \ne 1$ then $r^g r_g = r_g r$. Hence $Rr_g = r_g R$, which implies that r_g is a unit and then that g is an inner automorphism of R. This contradiction shows that $x = 1$ and $I = S$. □

8.13 It will be useful later to be able to specify the uniform dimension of S in such circumstances.

7.8.14   Fixed rings

Proposition. *Let R be a right Noetherian integral domain, S be prime, and $|G| = n$. Then R^G is a right Ore domain, $\operatorname{u\,dim} S_S = n$, and, letting $Q(-)$ denote quotient rings, $Q(S) \simeq M_n(Q(R^G))$.*

Proof. Consider the context

$$\begin{bmatrix} S & Sf \\ fS & R^G \end{bmatrix}.$$

Its alternative description in 8.3 makes clear that this is a prime context, as in 3.6.5. Moreover, S is right Noetherian, R^G is an integral domain, and $fS \triangleleft_r S$. It follows from 3.6.6 that R^G is a right Ore domain and that $\operatorname{u\,dim} fS_S = 1$. Since S is prime right Goldie, 3.3.4 shows that S contains a finite direct sum of uniform right ideals, $U_1 \oplus \cdots \oplus U_k = E$ say, with $k = \operatorname{u\,dim} S_S$, with $U_i \simeq fS$ for each i, and $E \triangleleft_e S_S$. Now E contains a regular element of S and so $\operatorname{u\,dim} E_R = \operatorname{u\,dim} S_R = |G|$. It follows that $|G| = k(\operatorname{u\,dim}(fS_R)) = k$ since $fS = fR \simeq R$. Thus $k = \operatorname{u\,dim} S_S = |G| = n$. Note next that $Q(R^G) \overset{M}{\sim} Q(S)$, by 3.6.9. However, $Q(R^G)$ is a division ring, and $\operatorname{u\,dim} Q(S) = \operatorname{u\,dim} S = n$. Therefore $Q(S) \simeq M_n(Q(R^G))$. □

8.14 The preceding results are now applied to yield some examples.

Example. (i) We show that the ring $T = \mathbb{R}[X, Y]/(X^2 + Y^2 - 1)$ is a Dedekind domain. To see this note that T is a subring of $R = \mathbb{C}[X, Y]/(X^2 + Y^2 - 1)$ and that $R = \mathbb{C} \otimes_\mathbb{R} T$. Let x, y denote the images of X, Y in R and set $z = x + iy$. Since $(x + iy)(x - iy) = 1$, $R = \mathbb{C}[z, z^{-1}]$ which is a principal ideal domain.

Let τ denote complex conjugation on R and set $G = \{1, \tau\}$. Then $R^G = T$, so T is a hereditary Noetherian domain and so is a Dedekind domain, by 5.2.7.

Further discussion of T will be given in 12.1.6 where, in particular, it will be seen that T is not a principal ideal domain, and in 15.3.13.

Example. (ii) Let k be a field of characteristic zero and $R = A_1(k)$. Suppose, for some integer $n > 1$, that k contains a primitive nth root of 1, ε say. Then σ defined by $\sigma(x) = \varepsilon x$, $\sigma(y) = \varepsilon^{-1} y$ is an automorphism of R, and must be outer since the units of R are the nonzero elements of k. Thus the cyclic group G generated by σ is outer and so $A_1(k) \# G$ is simple Noetherian hereditary of uniform dimension n, and R^G is a simple Noetherian hereditary integral domain.

Example. (iii) Let $R = P_\lambda(k)$, where k is a field and $\lambda \in k$ is not a root of unity. The units of R belong to $x^i y^j k$ for $i, j \in \mathbb{Z}$ and so the image of x under an inner automorphism must be $\lambda^h x$ for some $h \in \mathbb{Z}$. In particular, if $\varepsilon \in k$ is a primitive nth root of unity and if σ is defined by $\sigma(x) = \varepsilon x$, $\sigma(y) = \varepsilon^{-1} y$ then the group G generated by σ is outer. Hence $R \# G$ is a simple Noetherian hereditary ring and, if $|G|^{-1} \in k$, so too is R^G.

Similarly, if $\tau(x) = x^{-1}$ and $\tau(y) = y^{-1}$ and $G = \{1, \tau\}$ then $R\#G$ is simple, Noetherian and hereditary and, if char $k \neq 2$, so too is R^G.

§9 Skew Polynomial Rings

9.1 Throughout this section R is a ring, σ an automorphism, and $S = R[x; \sigma, \delta]$. It is clear from 5.3 and 5.5 that if r gld $R = n$ then r gld S is either n or $n+1$ and that each is a possibility. The aim of this section is to describe when each occurs. The main results come when R is Noetherian. Then r gld $S = n+1$ if and only if there is a finitely generated right R-module of projective dimension n which also has the structure of a right S-module. For simplicity's sake one of these implications will be proved only when R is semiprime.

Similar results for the ring $T = R[x, x^{-1}; \sigma]$ will also be obtained.

9.2 The first few results concern a localization of S.

Lemma. *Let $S = R[x; \sigma, \delta]$, $\mathscr{D}$ be the set of monic polynomials in S, M be a right S-module and $m \in M$. Then:*
(i) *$\mathscr{D}$ is a m.c. set and $\mathscr{D} \subseteq \mathscr{C}_S(0)$;*
(ii) *mS is finitely generated over R if and only if $md = 0$ for some $d \in \mathscr{D}$.*

Proof. (i) Straightforward.

(ii) ($\Rightarrow$) Suppose mS is finitely generated over R. Then it is generated by $\{m, mx, \ldots, mx^t\}$ for some t, in which case $mx^{t+1} = \sum_{i=0}^{t} mx^i r_i$. Thus $md = 0$, where $d = x^{t+1} - \sum x^i r_i$.

($\Leftarrow$) If $md = 0$ then mS is a homomorphic image of S/dS which, since d is monic, is finitely generated over R. □

9.3 Proposition. *Suppose that R is right Noetherian and S, $\mathscr{D}$ and M are as above.*
(i) *$\mathscr{D}$ is a right Ore set and so $S_\mathscr{D}$ exists.*
(ii) *The following conditions on M are equivalent:*
   (a) *$M \otimes_S S_\mathscr{D} = 0$;*
   (b) *M is $\mathscr{D}$-torsion;*
   (c) *each finitely generated S-submodule of M is finitely generated over R.*
(iii) *$S_\mathscr{D}$ is a faithfully flat left R-module.*

Proof. (i) Given $f \in S$ and $d \in \mathscr{D}$, set $J = \{s \in S \mid fs \in dS\}$. There is an S-module embedding $S/J \to S/dS$ via $[s + J] \mapsto [fs + dS]$. By 9.2(ii), S/dS is finitely generated over R and so S/J is too. The same result then shows that $J \cap \mathscr{D} \neq \emptyset$. Hence $\mathscr{D}$ is right Ore.

(ii) (a)$\Leftrightarrow$(b). This is clear from 2.1.17.

(b)$\Leftrightarrow$(c). This follows from 9.2(ii) since $\mathscr{D}$ is right Ore.

(iii) $S_\mathscr{D}$ is a flat left S-module, by 2.1.16, and S is a flat left R-module, being free. Hence $S_\mathscr{G}$ is a flat left R-module. Also, if N_R is a nonzero finitely generated R-module then $N \otimes_R S$ is finitely generated over S but not over R, since $\{\sum_{i=0}^{n} N \otimes x^i | n \in \mathbb{N}\}$ provides an infinite ascending chain of R-submodules. Hence, by (ii), $N \otimes_R S \otimes_S S_\mathscr{G} \neq 0$; i.e. $N \otimes_R S_\mathscr{G} \neq 0$. □

9.4 The aim of the next few results is to determine $\operatorname{r gld} S_\mathscr{G}$. First, however, the Krull dimension is dealt with.

Theorem. *Let R, S be as above with R right Noetherian. Then $\mathcal{K}(S_\mathscr{G}) = \mathcal{K}(R)$.*

Proof. By 6.5.3, since $S_\mathscr{G}$ is faithfully flat, $\mathcal{K}(S_\mathscr{G}) \geq \mathcal{K}(R)$. Conversely, consider the map $\mu: \mathscr{L}(S_\mathscr{G}) \mapsto \mathscr{L}(R)$ given by $L \mapsto (L \cap S)_\infty$, this being the right ideal of leading coefficients of elements of $L \cap S$, when written in the form $\sum r_i x^i$. It is clear, using 6.9.18, that μ preserves proper containment and so, by 6.1.17, $\mathcal{K}(S_\mathscr{G}) \leq \mathcal{K}(R)$. □

9.5 One can define the **leading coefficient** m' of an element m of S^n by viewing m as a polynomial in x over R^n with coefficients on the left. The set M_∞ of all leading coefficients of elements in a submodule M of S^n is an R-submodule of R^n. It will be seen that there is a close relationship between M_∞ and $M_\mathscr{G}$.

9.6 Lemma. *Let R be right Noetherian, let M be an S-submodule of S^n and let $m_1, \ldots, m_t \in M$. If $M_\infty = \sum_{i=1}^{t} m'_i R$ then $M_\mathscr{G} = \sum_{i=1}^{t} m_i S_\mathscr{G} \subseteq S_\mathscr{G}^n$.*

Proof. If u is the maximal degree of the elements $m_1, \ldots, m_t$ then it is clear that
$$M \subseteq \sum_{i=1}^{t} m_i S + \sum_{j<u} R^n x^j.$$
Hence $M / \sum m_i S$ is finitely generated over R and so, by 9.3, $(M/\sum m_i S) \otimes S_\mathscr{G} = 0$. □

9.7 Continuing with the notation of 9.6, note that multiplication of any m_i by a power of x will not affect the situation. So it can be arranged that all the m_i have the same degree, u say.

There is an S-homomorphism $\phi: S^t \to M$ given by $\phi(s_1, \ldots, s_t) = \sum_{i=1}^{t} m_i s_i$ and, similarly, an R-homomorphism $\phi': R^t \to M_\infty$ given by $\phi'(r_1, \ldots, r_t) = \sum_{i=1}^{t} m'_i r_i$. Let $K = \ker \phi$ and $L = \ker \phi'$. The twist caused by the powers of x means that L may not equal K_∞.

Lemma. $K_\infty = \sigma^u(L)$.

Proof. Let $(s_1, \ldots, s_t) \in K$ and let v be the maximum degree of the s_i. The coefficient of x^{u+v} in $\sum m_i s_i$ is $\sum \{m'_i \sigma^{-u}(s'_i) | \deg s_i = v\}$. It follows that $K_\infty \subseteq \sigma^u(L)$.

To obtain the converse, we use the R-submodule
$$A = \{a \in M \mid \deg a < u\} \triangleleft M.$$
Given any $a_1 \in A$ there is a process leading to another element $a_2 \in A$. First choose $j(1)$ so that $\deg(a_1 x^{j(1)}) = u$; then note that the leading coefficient of $a_1 x^{j(1)}$ belongs to $M_\infty = \sum m'_i R$. So one can choose $r_{11}, \ldots, r_{1t} \in R$ such that $\sum m_i r_{1i}$ has the same leading term. Then $a_2 = a_1 x^{j(1)} - \sum m_i r_{1i}$ is as claimed.

Now, suppose $(r_1, \ldots, r_t) \in L$ and set $a_1 = \sum m_i \sigma^u(r_i)$. Clearly $a_1 \in A$, so the process above can be applied repeatedly to give $a_2, a_3, \ldots$. However, A_R is finitely generated and thus $a_{w+1} = \sum_{j=1}^w a_j b_j$ for some w and some $b_j \in R$. Rewriting this equation in terms of the m_i yields an equation $\sum m_i s_i = 0$ for some $s_i \in S$. Evidently $(s_1, \ldots, s_t) \in K$. Moreover, the highest power of x in $(s_1, \ldots, s_t)$ comes from a_1 and so its leading coefficient is $\sigma^u(r_1, \ldots, r_t)$. Thus $\sigma^u(L) \subseteq K_\infty$. □

9.8 The next result connects the projective dimensions of M_∞ and $M_\mathcal{D}$.

Proposition. *Let R be right Noetherian and M be an S-submodule of S^n. Then $\text{pd}\,(M_\mathcal{D})_{S_\mathcal{D}} \leq \text{pd}\,(M_\infty)_R$.*

Proof. Evidently we may suppose $\text{pd}\, M_\infty < \infty$. The proof starts with the case when M_∞ is free, say on generators $m'_1, \ldots, m'_t$, where $m_1, \ldots, m_t \in M$. By considering the highest degree terms in any relation $\sum m_i s_i = 0$, $s_i \in S$, it is clear that $\{m_i\}$ is independent over S and so over $S_\mathcal{D}$. By 9.6, $M_\mathcal{D} = \sum m_i S_\mathcal{D}$; so $M_\mathcal{D}$ is free.

Next suppose M_∞ is projective, with $M_\infty \oplus N$ free of finite rank. Note that $M_\infty \oplus N = (M \oplus (N \otimes_R S))_\infty$, where $M \oplus (N \otimes S)$ is regarded as a submodule of the free S-module $S^n \oplus ((M_\infty \oplus N) \otimes S)$. It follows that $M_\mathcal{D} \oplus (N \otimes S_\mathcal{D})$ is free over $S_\mathcal{D}$ and hence $M_\mathcal{D}$ is projective.

Finally, induction is applied to $\text{pd}\, M_\infty = d \geq 1$. There is a short exact sequence
$$0 \to K \to S^t \xrightarrow{\phi} M \to 0.$$
By 9.7 there is an induced sequence
$$0 \to L \to R^t \xrightarrow{\phi'} M_\infty \to 0$$
with $K_\infty = \sigma^u(L)$. Now $\text{pd}\, L = d - 1 = \text{pd}\, K_\infty$ and so, by induction, $\text{pd}\, K_\mathcal{D} \leq d - 1$. The exact sequence
$$0 \to K_\mathcal{D} \to S^t_\mathcal{D} \to M_\mathcal{D} \to 0$$
shows that $\text{pd}\, M_\mathcal{D} \leq d$. □

9.9 We can now complete our study of $S_\mathcal{D}$.

7.9.11

Theorem. *Let R be Noetherian with* gld $R = n < \infty$ *and let* $S = R[x;\sigma,\delta]$. *Then* gld $S_\mathcal{D} = n$.

Proof. By 9.3 and symmetry, $S_\mathcal{D}$ is faithfully flat on each side over R. Hence, by 2.6, gld $S_\mathcal{D} \geq n$.

Conversely, consider first the case when $n = 0$; i.e. when R is semisimple Artinian. Let eR be a minimal right ideal of R and let $0 \neq m \in eS$. Now $m'R = eR$ and so mS contains an element of the form

$$e(r_0 + r_1 x + \cdots + r_{t-1} x^{t-1} + x^t) = ed, \quad d \in \mathcal{D}.$$

It follows that $mS_\mathcal{D} = eS_\mathcal{D}$ and hence $eS_\mathcal{D}$ is a minimal right ideal of $S_\mathcal{D}$. Hence $S_\mathcal{D}$ is a direct sum of minimal right ideals and so gld $S_\mathcal{D} = 0$.

Next suppose $n > 0$ and let $I \triangleleft_r S_\mathcal{D}$. Then $I = J_\mathcal{D}$ for some $J \triangleleft_r S$. By 9.8 pd $I \leq$ pd $J_\infty <$ gld R and so gld $S_\mathcal{D} \leq$ gld R. □

9.10 We now apply this to S itself.

Theorem. *Let R be Noetherian with* gld $R = n < \infty$ *and let* $S = R[x;\sigma,\delta]$. *Suppose* gld $S = n+1$ *and* M_S *satisfies* pd $M_S = n+1$. *Then* M_S *has a submodule* N_S *such that* pd $N_S = n+1$, pd $N_R = n$ *and* N_R *is finitely generated.*

Proof. M_S is the direct limit of its finitely generated submodules. By 1.5 one of these, K say, must have fd $K_S = n+1$. Hence, by 1.5, pd $K_S = n+1$. Now let N be the kernel of the map $K \to K_\mathcal{D}$. We will check that N is as described. Note first that there is a short exact sequence of S-modules

$$0 \to K/N \to K_\mathcal{D} \to H \to 0.$$

Since gld $S_\mathcal{D} = n$, by 9.9, then fd $(K_\mathcal{D})_{S_\mathcal{D}} \leq n$ and so fd $(K_\mathcal{D})_S \leq n$, by 2.2(ii). The flat version of 1.8(b) shows that fd $K/N \neq n+1$. Since fd $K = n+1$ it follows that fd $N = n+1$ and so pd $N_S = n+1$. It is clear from 9.2 that N_R is finitely generated. Finally, 5.2 shows that pd $N_S \leq$ pd $N_R + 1$; hence pd $N_R = n$. □

9.11 Consider now the case when $\delta = 0$ and $T = R[x, x^{-1}; \sigma]$. Then T has two subrings $S = R[x;\sigma]$ and $S' = R[x^{-1};\sigma]$. Let $\mathcal{D}$ be as before and $\mathcal{D}'$ be the set of monic polynomials in x^{-1} of S'.

Lemma. *Let R be right Noetherian, M be a right T-module, and* $m \in M$. *Then* mT *is finitely generated over R if and only if* $md = md' = 0$ *for some* $d \in \mathcal{D}$, $d' \in \mathcal{D}'$.

Proof. ($\Rightarrow$) If mT_R is finitely generated then so too are mS_R and mS'_R. Now apply 9.2.

($\Leftarrow$) Clear. □

Theorem. *Let R be Noetherian with $\gld R = n < \infty$. Let $T = R[x, x^{-1}; \sigma]$ and suppose $\gld T = n+1$ and $\pd M_T = n+1$. Then M_T has a submodule N_T such that $\pd N_T = n+1$, $\pd N_R = n$ and N_R is finitely generated.*

Proof. Evidently $S_\mathscr{D} = S'_{\mathscr{D}'} = T_\mathscr{D} = T_{\mathscr{D}'}$. The same argument as in 9.10 can be used, but with S replaced by T and N being the kernel of the map $K \to K_\mathscr{D} \oplus K_{\mathscr{D}'}$. □

Note that, by 1.14, it is possible to choose M in Theorems 9.10 and 9.11 to be simple, in which case $M = N$, of course.

9.12 The route to a converse is through the study of S-modules finitely generated over R. Recall that the commutation relation in $S = R[x; \sigma, \delta]$ is $rx = x\sigma(r) + \delta(r)$. Suppose that M_R is a right R-module and hence a $(\mathbb{Z}, R)$-bimodule. It is easy to verify:

Lemma. *M_R can be given a right S-module structure if and only if there exists $\alpha \in \Hom(_\mathbb{Z}M, {_\mathbb{Z}}M)$ such that*

$$r\alpha = \alpha\sigma(r) + \delta(r).$$
□

9.13 This process can be carried further.

Lemma. *Suppose that M_S is a right S-module, that F_R is free and that $0 \to K \to F \xrightarrow{\pi} M \to 0$ is an exact sequence of R-modules. Then F and K can be given S-module structures making it an exact sequence of S-modules.*

Proof. Let $F = R^I$, $\{e_i\}$ be the standard basis and $m_i = \pi(e_i)$. Note that $m_i x = \sum_{j \in I} m_j a_{ji}$ for some $a_{ji} \in R$. Then one can define multiplication of F by x (i.e. the map α of 9.12) by setting

$$e_i r_i x = \sum_{j \in I} e_j a_{ji} \sigma(r_i) + e_i \delta(r_i)$$

or, in the usual notation, $vx = A\sigma(v) + \delta(v)$ for $v \in F$. It is clear that this makes π into an S-homomorphism with K, as the kernel, inheriting an S-structure from F. □

9.14 We now restrict attention to the case when R is semiprime.

Proposition. *Let R be semiprime right Noetherian.*
(i) *S is semiprime right Noetherian.*
(ii) *If J is an essential right ideal of S then J contains a polynomial whose leading coefficient lies in $\mathscr{C}_R(0)$.*

Proof. (i) Let P be a minimal prime of R. Then $\sigma^k(P)$ is also a minimal prime and $P^0 = \bigcap \sigma^k(P)$ is a σ-stable ideal. It follows easily that, if $S' = R[x;\sigma]$, then $P^0 S'$ is a prime ideal of S' and hence S' is semiprime. Now $S' = \mathrm{gr}\, S$ using the filtration by degree and so, by a similar argument to the proof of 1.6.6, S is semiprime. Of course, by 1.2.9, S is right Noetherian.

(ii) We will write coefficients to the left of powers of x (although the result is clearly independent of this choice). Choose a direct sum of uniform right ideals $I_1 \oplus \cdots \oplus I_k = I \triangleleft_e R_R$. Since $J \triangleleft_e S_S$ there is a nonzero element $a_i \in J \cap I_i S$ and it can be arranged, by multiplying by appropriate powers of x, that the degrees of the a_i are equal, to n say. Let b_i be the leading coefficient of a_i. Then $0 \neq b_i R \subseteq I_i$ so $\sum b_i R \triangleleft_e R_R$. Hence there is a regular element of the form $\sum b_i r_i$, and this is the leading coefficient of $\sum a_i \sigma^n(r_i)$. □

9.15 In dealing with projective dimensions, a preparatory result is required.

Lemma. *Let R be a semiprime Noetherian ring, $I \triangleleft_r R$ and $a \in R$. Let Q be the quotient ring of S.*
(i) *If $q(I + (x - a)S) \subseteq S$ for some $q \in Q$ then there exists $s \in S$ such that $(q - s)I = 0$.*
(ii) *If $\mathrm{pd}\,(I + (x - a)S)_S = 0$ then $\mathrm{pd}\,(R/I)_R = 0$.*

Proof. (i) Since $q(x - a)S \subseteq S$ then $q \in S(x - a)^{-1} = R(x - a)^{-1} + S$ and so $q = r(x - a)^{-1} + s$ for some $r \in R$, $s \in S$. Since Q is the left quotient ring of S, $Eq \subseteq S$ for some essential left ideal E. By the left-hand version of 9.14, $yq \in S$ for some $y \in E$ with leading coefficient in $\mathscr{C}_R(0)$. It follows that $yr = t(x - a)$ for some $t \in S$ and so $y^{-1}t = r(x - a)^{-1}$. Now

$$r(x - a)^{-1}I = (q - s)I \subseteq S,$$

so $y^{-1}tI \subseteq S$ and $tI \subseteq yS$. However, the nonzero elements of yS have degree at least that of y, since the leading coefficient of y is regular. If $tI \neq 0$ this implies that $\deg t \geq \deg y$, which contradicts the equation $yr = t(x - a)$. Hence $tI = 0$ and

$$(q - s)I = r(x - a)^{-1}I = y^{-1}tI = 0.$$

(ii) Let $J = I + (x - a)S$ and suppose J_S is projective. There is an R-epimorphism $\pi: J \to I \simeq J/(x - a)S$ which gives I an S-module structure. Hence there is an S-epimorphism $\alpha: S^n \to I$ for some n. Projectivity of J provides β making the diagram

commute. Now $J \triangleleft_e S_S$ so $\beta = (q_1, \ldots, q_n)$ for some $q_i \in Q$. By (i), there exists $\gamma = (s_1, \ldots, s_n)$ such that $(q_1 - s_1, \ldots, q_n - s_n)I = 0$. Viewing γ as a map from R to

S^n gives us a commutative diagram

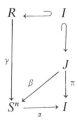

Evidently $\alpha\gamma$ restricted to I is the identity map. Hence $R \simeq I \oplus R/I$ and R/I is projective. □

9.16 Theorem. *Let R be semiprime Noetherian and $S = R[x; \sigma, \delta]$. If M_S is finitely generated over R then* $\operatorname{pd} M_S = \operatorname{pd} M_R + 1$.

Proof. We already know that $\operatorname{pd} M_R \leqslant \operatorname{pd} M_S \leqslant \operatorname{pd} M_R + 1$ by 2.2 and 5.2. The case when $\operatorname{pd} M_R = \infty$ is thus dealt with.

The proof now proceeds by induction on $\operatorname{pd} M_R = n$, say. First suppose $n = 0$. Let $0 \neq m \in M$. Then $mS \simeq S/J$ and, by 9.2, J contains a monic polynomial which is regular. Hence mS is a $\mathscr{D}$-torsion module and so M is not projective. It follows, as required, that $\operatorname{pd} M_S = 1$.

Next let $n = 1$ and suppose, contrary to the claim, that $\operatorname{pd} M_S = 1$. Note that M_R is finitely generated, say by k elements. It is easily verified that M^k is a cyclic right $M_k(R)$-module and also a right $M_k(S)$-module. The Morita equivalence of R with $M_k(R)$ and S with $M_k(S)$ maps M to M^k. Finally, note that $M_k(S) \simeq M_k(R)[x; \sigma, \delta]$, where σ, δ act on the entries of each matrix. This shows that there is no loss of generality in supposing M_R to be cyclic.

In that case $M \simeq R/I$, say, and $[1 + I]x = [a + I]$ for some $a \in R$. Thus $M_S \simeq S/J$ with $J = I + (x - a)S$. Now if $\operatorname{pd} M_S = 1$ then $\operatorname{pd} J_S = 0$ and so $\operatorname{pd}(R/I)_R = 0$ by 9.15(ii). This contradicts the fact that $\operatorname{pd} M_R = 1$. We conclude that $\operatorname{pd} M_S$ must be 2.

Finally, let $n \geqslant 2$ and assume the result for $n - 1$. There is a short exact sequence of R-modules
$$0 \to K \to F \to M \to 0$$
with F_R free of finite rank. By 9.13 this may be regarded as an exact sequence of S-modules. Since K_S is finitely generated over R and $\operatorname{pd} K_R = n - 1$ then $\operatorname{pd} K_S = n$. Also $\operatorname{pd} F_S = 1$ since $\operatorname{pd} F_R = 0$. Hence $\operatorname{pd} M_S = n + 1$. □

9.17 A similar result holds for Laurent polynomials.

Corollary. *Let R be semiprime Noetherian and $T = R[x, x^{-1}; \sigma]$. If M_T is finitely generated over R then* $\operatorname{pd} M_T = \operatorname{pd} M_R + 1$.

Proof. Given M_T we may view it as a module over the subring $S = R[x;\sigma]$. By 9.16, pd M_S = pd M_R + 1. However,

$$\text{pd } M_S = \text{fd } M_S = \text{fd } M_T = \text{pd } M_T$$

by 1.5 and 4.2. □

9.18 Combining these results with Theorems 9.10 and 9.11 gives

Corollary. *Let R be a Noetherian semiprime ring with* gld $R = n < \infty$ *and let U be one of the rings $S = R[x;\sigma,\delta]$ or $T = R[x,x^{-1};\sigma]$. Then* gld $U = n + 1$ *if and only if there is a simple U-module M which is finitely generated over R and with* pd $M_R = n$. □

§10 Skew Polynomials Over Commutative Rings

10.1 Throughout this section we let R denote a commutative Noetherian ring with gld $R = d < \infty$ and $S = R[x;\delta]$ or $R[x, x^{-1};\sigma]$. We know already that gld $S = d$ or $d + 1$. Moreover, since R is semiprime, by 1.15, the results of Section 9 can be applied. However, there is another criterion, analogous to that described, for the Krull dimension, in Chapter 6, §9, which involves semistable prime ideals. That is the content of this section. Of course if gld $R = 0$ then gld $S = 1$. So we also will assume $d > 0$.

10.2 The first result applies standard commutative theory to semistable ideals.

Proposition. *Let P be a semistable prime ideal with* ht $P < d$ *and let $H = P^\circ$. Then* pd $R/H < d$.

Proof. By 1.15, pd R/H is the supremum, over the maximal ideals I of R, of pd $(R_I/H_I)_{R_I}$. So to show pd $R/H < d$ it is enough to consider $I \supseteq H$ with gld $R_I = d$. In that case I_I has height d and so I_I is not a minimal prime of H_I.
If H and hence H_I is semiprime, it follows immediately that $I_I \cap \mathscr{C}_{R_I}(H_I) \neq \varnothing$.
If H is not semiprime then 6.9.9 and 6.9.10 show that P is the unique minimal prime of H and that $I \cap \mathscr{C}(P) = I \cap \mathscr{C}(H) \neq \varnothing$. Again it follows that $I_I \cap \mathscr{C}_{R_I}(H_I) \neq \varnothing$. Hence, by 1.15,

$$\text{pd}_{R_I}(R_I/H_I) < \text{gld } R_I = d.$$

□

10.3 Theorem. *With the notation as above,* gld $S = d + 1$ *if and only if there is a semistable prime of height d.*

Proof. ($\Rightarrow$) Suppose that there is no such semistable prime and let M_S be finitely generated. We will show that pd $M_S \leq d$ and hence gld $S \neq d + 1$. Lemma 6.9.16

shows that M has a chain of submodules of a special kind. An easy induction argument reduces the problem to the case where the chain has length 1. Two possibilities then remain. If $M \simeq S/PS$ then

$$\text{pd } M_S \leqslant \text{pd } (R/P)_R \leqslant d.$$

Otherwise M is a proper homomorphic image of S/PS and is torsion-free over R/P^0, where P is a semistable prime ideal necessarily of height $< d$. By 6.9.9 and 6.9.10, R/P^0 has a quotient ring, Q say. Since M is torsion-free, M embeds in $M \otimes_{R/P^0} Q$ so there is an exact sequence

$$0 \to M \to M \otimes Q \to C \to 0.$$

Let R^0 denote R localized at $\mathscr{C}(P^0)$. Since ht $P < d$, gld $R^0 < d$. However, $M \otimes Q$ is an R^0-module and so

$$d > \text{fd } (M \otimes Q)_{R_0} = \text{fd } (M \otimes Q)_R$$

using 1.5 and 4.2. Of course fd $C_R \leqslant d$ and so fd $M_R < d$ by 1.6. Hence fd $M_S \leqslant d$ by 5.2 and so pd $M_S \leqslant d$ as required.

($\Leftarrow$) Suppose P is a semistable prime of height d. Let R^0 and S^0 denote R and S localized at $\mathscr{C}(P^0)$. Since ht $P = d$ then gld $R^0 = d$. Furthermore, gld $S \geqslant$ gld S^0; so it is sufficient to show that gld $S^0 = d + 1$. Thus we may suppose, without loss of generality, that $R = R^0$ and $S = S^0$.

If $S = R[x; \delta]$ then R is local with maximal ideal P, so pd $R/P = d$. Also R/P^0 has finite length with simple factors isomorphic to R/P. It follows easily that pd $R/P^0 = d$.

On the other hand, if $S = R[x, x^{-1}; \sigma]$ then $\sigma^n(P) = P$ for some n and $R/P^0 \simeq \bigoplus_{i=1}^n R/\sigma^i(P)$. Hence pd $R/P^0 = \text{pd } R/P = d$.

In either case R/P^0 is an S-module, finitely generated over R with pd $(R/P^0)_R = d$. Hence, by 9.16 and 9.17, pd $(R/P^0)_S = d + 1$. Therefore gld $S = d + 1$. □

§11 Simple Dedekind Domains

11.1 Let R be a commutative Dedekind domain and let $S = R[x; \delta]$ or $R[x, x^{-1}; \sigma]$. We will deduce from earlier results that S is simple precisely when S is a noncommutative Dedekind domain. In that case there is a close relationship between the torsion-free S-modules and R-modules which yields strengthenings of the module theoretic results of Chapter 5, §7. This in turn gives interesting information about some of the simple integral domains introduced in Chapter 1.

11.2 First the basic characterizations.

Theorem. *Let R, S be as above. Then the following conditions on S are equivalent:*
(i) *S is simple;*
(ii) *S is hereditary;*

(iii) $K(S) = 1$;
(iv) S is a noncommutative Dedekind domain.

Proof. (i)$\Rightarrow$(ii) This is a special case of 5.5.
 (ii)$\Rightarrow$(iii) This is clear from 5.4.5 and 6.3.10.
 (iii)$\Rightarrow$(i) Suppose first that R has a proper nonzero stable ideal, I say. Now $K(R/I) = 0$ and so $K(R/I[x;\delta]) = 1$ and $K(R/I[x, x^{-1};\sigma]) = 1$. Note that these rings are isomorphic to S/IS. Since $K(S) = 1$ and $IS \triangleleft_e S$ it follows from 6.3.10 that $K(S/IS) = 0$, a contradiction. Hence R cannot have a proper nonzero stable ideal. This implies by 1.8.5 that $R[x, x^{-1};\sigma]$ is simple since commutative rings cannot have inner automorphisms. Similarly, 1.8.4 shows that $R[x;\delta]$ is simple if R is a $\mathbb{Q}$-algebra. There remain the possibilities either that char $R = 0$ and R is not a $\mathbb{Q}$-algebra, or that char $R = p > 0$. In the former case there exists $0 \neq n \in \mathbb{Z}$ such that $nR \neq R$ and then nR is a proper nonzero stable ideal; and in the latter case so is $r^p R$ for any nonzero nonunit r. Therefore, these are not possibilities and S is indeed simple.
 (iv)$\Rightarrow$(ii) This is part of the definition of Dedekind prime rings.
 (i) and (ii)$\Rightarrow$(iv) See 5.2.10 and 5.2.7(ii). $\square$

11.3 Several of the simple rings described in Chapter 1, §8 are covered by this.

Corollary. *Let k be a field. Then the following rings are simple Dedekind domains:*
 (i) $P_\lambda(k) = k[y, y^{-1}][x, x^{-1};\sigma]$, where $\sigma(y) = \lambda y, 0 \neq \lambda \in k, \lambda$ not a root of unity.
 (ii) $A_1(k) = k[x][y; d/dx]$ when char $k = 0$.
 (iii) $A'_1(k) = k[y, y^{-1}][x; y d/dy]$ when char $k = 0$. $\square$

11.4 Next we turn to the module theory over S when S is a simple Dedekind domain as above.

Lemma. *Let I, J be nonzero right ideals of S with $I \supseteq J$.*
 (i) *There is a chain $J = J_0 \subset J_1 \subset \cdots \subset J_n = I$ such that, for each i, $J_i/J_{i-1} \simeq S/P_i S$ for some nonzero prime P_i of R.*
 (ii) $I \oplus AS \simeq J \oplus S$, *where $A = P_1 P_2 \cdots P_n$.*

Proof. (i) Since S is simple R has no nonzero proper semistable prime ideals. The result is then a consequence of 6.9.16.
 (ii) Since $J_1/J_0 \simeq S/P_1 S$, Schanuel's lemma shows that $J_0 \oplus S \simeq J_1 \oplus P_1 S$. Next note that
$$J_2/J_1 \simeq S/P_2 S \simeq P_1 S/P_1 P_2 S$$
since $P_1/P_1 P_2 \simeq R/P_2$ (e.g. by 5.7.10). Hence
$$J_0 \oplus S \simeq J_1 \oplus P_1 S \simeq J_2 \oplus P_1 P_2 S.$$

An easy induction completes the proof. $\square$

11.5 We now elaborate on 5.7.6.

Theorem. *Let R be a Dedekind domain, S be simple and I, J, K be any nonzero right ideals of S. Then $I \oplus J \simeq K \oplus AS$ for some $A \triangleleft R$.*

Proof. By 5.7.6, $I \oplus J \simeq S \oplus L$ for some $0 \neq L \triangleleft_r S$. Let $0 \neq k \in K$; then $kL \subseteq K$ and, by 11.4,
$$K \oplus AS \simeq kL \oplus S$$
for some $A \triangleleft R$. However, $kL \simeq L$ and so $I \oplus J \simeq K \oplus AS$. □

11.6 Corollary. *Let M_S be a finitely generated torsion-free module. Then either $M \simeq I$ for some $I \triangleleft_r S$ or else $M \simeq S^n \oplus AS$ for some $A \triangleleft R$. In particular, if R is a principal ideal domain and M is not uniform then M_S is free.*

Proof. By 5.7.8, $M \simeq S^n \oplus I$ for some $I \triangleleft_r S$. If $n \neq 0$ then $M \simeq S^{n-1} \oplus S \oplus I \simeq S^{n-1} \oplus S \oplus AS$ by 11.5. And if A is principal, then $M \simeq S^{n+1}$. □

11.7 This has an interesting consequence which applies to the rings $P_\lambda(k)$, $A_1(k)$ and $A_1'(k)$ mentioned in 11.3 since in each case the ring R is a principal ideal domain. It should be compared with 3.4.10 and 11.3.8.

Corollary. *If R is a principal ideal domain and S is simple then $M_n(S)$ is a prime principal ideal ring for each $n \geq 2$.*

Proof. Under the Morita equivalence between $M_n(S)$ and S an essential right ideal E of $M_n(S)$ corresponds to an essential submodule M of S^n. Since $M \simeq S^n$ then $E \simeq M_n(S)$ and so E is principal. Every right ideal is a direct summand of an essential right ideal and so again is principal. □

11.8 As shown in 3.4.10, if a ring T is a principal ideal ring then so too is $M_n(T)$. So it is worth showing, for k a field with char $k = 0$, that $A_1(k)$ is not a principal ideal ring.

Example. *The right ideal I of $A_1(k)$ generated by x^2 and $1 + xy$ is not principal.*

Proof. Note that $(1 + xy)x = x^2 y$ and so, writing S for $A_1(k)$, it is clear that $I = x^2 S + (1 + xy)k[y]$. Evidently $I \cap k[y] = 0$ and so if $I = aS$ then the degree, in x, of a must be at least 1. However $1 + xy = ab$ for some $b \in S$. Therefore a has degree 1, say $a = a_0 + xa_1$, and b has degree 0. Then $ab = a_0 b + xa_1 b$ and therefore $b \in k$. That shows that $I = (1 + xy)S$. However, the degree in y shows that $x^2 \notin (1 + xy)S$, a contradiction. □

See also 11.2.13.

§12 Additional Remarks

12.0 (a) The study of global dimension goes back to [Hilbert **1898**] with [Cartan and Eilenberg **56**] marking out the area decisively.
 (b) Some other homological aspects are dealt in Chapters 11 and 12.
 (c) The structure of Noetherian rings of finite global dimension is not discussed here, apart from the case of hereditary rings in Chapter 5. There are positive results, but only in special cases; [Brown, Hajarnavis and MacEachern **82, 83**], [Brown and Hajarnavis **84**], [Robson **85**], [Kirkman and Kuzmanovich **P**], [Hajarnavis **P**], [Stafford **Pa**].

12.1 Corollary 1.14 is due to [Bhatwadekar **76**] and [Goodearl **75**]. In [Rainwater **P**] it is shown that the hypothesis that gld $R < \infty$ can be removed if R is fully bounded.

12.2 (a) 2.6 is due to [McConnell **77**] inspired by [Roos **72**]. A similar result for finitistic global dimension appears in [Cortzen **82**].
 (b) 2.7 is due to [Goodearl **74**].

12.3 The first portion of §3, up to 3.7, is adapted from [Kaplansky **72**]. The remainder, except for 3.16, is a combination of [Small **68a**] and [Fields **70**].

12.4 The proof of 4.4 was provided by K. Goodearl.

12.5 Much of 5.3 appears in [Fields **69**] and both 5.5 and 5.9 are from [Hart **71**]. The proof of 5.6 was supplied by M. Lorenz. [Rinehart **62**] gave 5.8(i), (ii) and, when $n = 1$, (iii). The general proof of (iii) appeared in [Roos **72**]. 5.11 comes from [Robson **72**] and [Goodearl **73**] and 5.13 from [Robson and Small **P**]. A strengthening of 5.10 appears in [Resco **80**]; see also [Hart **72**]. For $R*G$ see also [Aljadeff and Rosset **86**].

12.6 (a) Filtered and graded techniques were used in [Serre **58**]. We use them again in Chapter 12.
 (b) 6.16–6.19 come from [Roy **65**]; see also [Feldman **82**], [Stafford **82b**].

12.7 7.5(i) is due to [Swan **68**] for the case $\sigma = 1$ and to [Farrell and Hsiang **70**] in general.

12.8 (a) The ingredients of this Morita context are all described more fully in [Montgomery **80**]. The context itself reappears in Chapter 10.
 (b) 8.8(ii) is in [Bergman **71**] and 8.9 in [Webber **70**].

12.9 (a) The material up to 9.9 comes from [Resco, Small and Stafford **82**] although the proofs have been simplified following [Schofield **86**].
 (b) [Rinehart and Rosenberg **76**] and [Rosenberg and Stafford **76**] give 9.10 and a more general result than 9.16. However, 9.16 appeared in [Goodearl **75**] and 9.17 in [Hodges **84**]. A similar result for Krull dimension is noted in 6.10.9(b).

12.10 These results, with different proofs, appear in [Goodearl and Warfield **82**]; see also [Goodearl **78**], [Bjork **72**].

12.11 (a) This section is based on [Webber **70**] which built upon [Rinehart **62**] where 11.8 appeared first.
 (b) Further related results appear in Chapter 11.

Chapter 8

GELFAND—KIRILLOV DIMENSION

This chapter introduces and describes the Gelfand–Kirillov dimension. Since this dimension applies only to algebras over a field, k say, the chapter concentrates exclusively on these.

The Gelfand–Kirillov dimension, or GK dimension for short, is a measure of the rate of growth of the algebra in terms of any generating set. For a group algebra kG it measures the rate of growth of the group; and for a commutative affine domain R it measures the transcendence degree over k of the quotient field of R and so equals the Krull dimension of R.

The first section defines and establishes some basic properties of the GK dimension and the next section calculates the GK dimensions of many of the rings introduced in Chapter 1. Then, in §3, come the main details of the module–theoretic properties of this dimension. As noted earlier, in Chapter 6, §8, for a Noetherian algebra it is a dimension function and whilst not necessarily exact it is ideal invariant.

These facts highlight one point which the reader should bear in mind, namely that the comparison with the Krull dimension sometimes favours one, sometimes the other. On the whole, this comparison is left to the reader.

Section 4 discusses almost commutative algebras, otherwise recognized as homomorphic images of enveloping algebras of finite dimensional Lie algebras. For them the GK dimension is particularly well behaved. The explanation of this is that they, like commutative affine algebras, have a polynomial, the Hilbert polynomial, which describes their growth rate. Its degree is the GK dimension and its leading coefficient gives information about chains of submodules.

Next, in Section 5, two examples are given of the use of the GK dimension. One gives information about modules over $A_n(k)$, the Weyl algebra; and the other helps in the calculation of $\mathcal{K}(U(g))$, where g is the semisimple Lie algebra $\mathrm{sl}(2,k)$. Finally, §6 extends the results of §4 from almost commutative to somewhat commutative algebras.

Further details concerning the GK dimension, and, in particular, details of many more examples, can be found in [Krause and Lenagan **85**]. We make specific reference to it occasionally.

§1 Definition and Examples

1.1 This section introduces the definitions associated with the GK dimension and illustrates them with some examples.

1.2 We start with an example which serves as a prototype for what follows. Let $R = k[x_1, \ldots, x_s]$, the commutative polynomial ring over a field k. Let V be the k-subspace of R generated by $\{x_1, \ldots, x_s\}$ and let V^n be the k-subspace of R generated by the n-fold products of elements in V. Evidently the monomials of degree n, i.e. the words

$$\{x_1^{i(1)} \cdots x_s^{i(s)} \mid i(1) + \cdots + i(s) = n\}$$

form a basis for V^n.

Corresponding to the generating set $\{x_1, \ldots, x_n\}$ there is a standard filtration, $\{R_n\}$ say, as described in 1.6.2. It is clear that $R_n = \bigoplus_{i=0}^n V^i$, where $R_0 = V^0 = k$.

1.3 Our interest here is in the dimensions of these different k-spaces.

Lemma. $\dim V^n = \binom{n+s-1}{s-1}$, $\dim R_n = \binom{n+s}{s}$.

Proof. Each monomial in R_n can be written as $1^{i(0)} x_1^{i(1)} \cdots x_s^{i(s)}$ where $\sum i(j) = n$. This corresponds, in an obvious fashion, to the monomial $x^{i(0)} y x^{i(1)} y \cdots y x^{i(s)}$ in the free algebra $k\langle x, y \rangle$. The number of such monomials is determined by the number of ways of choosing the s positions, out of $n+s$, for the letters y. Hence

$$\dim R_n = \binom{n+s}{s}.$$

A similar argument deals with $\dim V^n$. □

1.4 One can write

$$\binom{n+s}{s} = p(n)$$

for some $p \in \mathbb{Q}[x]$, the polynomial p having degree s and leading coefficient $1/s!$ In what follows the degree s will appear as the GK dimension of R. The polynomial itself appears in a later section as the Hilbert polynomial.

1.5 Before turning to more general k-algebras, it is convenient to discuss a number $\gamma(f)$ associated with any function from $\mathbb{N}$ to $\mathbb{R}'$, where $\mathbb{R}' = \{r \in \mathbb{R} \mid r \geq 1\}$. The function f has **polynomially bounded growth**, or **polynomial growth** for short, if, for some $v \in \mathbb{R}$, $f(n) \leq n^v$ for $n \gg 0$; and then

$$\gamma(f) = \inf \{v \mid f(n) \leq n^v \text{ for } n \gg 0\}.$$

If f is not polynomially bounded then $\gamma(f) = \infty$.

The example we have in mind is $f(n) = \dim R_n$ in which case it is clear that $\gamma(f) = s$.

1.6 There is an alternative description of $\gamma(f)$.

Lemma. $\gamma(f) = \limsup (\log f(n)/\log n)$.

Proof. The three statements:
(a) $f(n) \leqslant n^v$ for $n \gg 0$;
(b) $\log f(n)/\log n \leqslant v$ for $n \gg 0$; and
(c) $\limsup (\log f(n)/\log n) \leqslant v$;
are equivalent. □

The former description has the advantage of avoiding any difficulties caused by the distinction between $\limsup (\log f(n)/\log n)$ and $\lim (\log f(n)/\log n)$.

1.7 The properties of γ which will be needed later are summarized next.

Lemma. *Let* $f, g: \mathbb{N} \to \mathbb{R}'$.
(i) $\gamma(f+g) = \sup \{\gamma(f), \gamma(g)\}$.
(ii) $\gamma(fg) \leqslant \gamma(f) + \gamma(g)$.
(iii) *If* $f(n) = p(n)$ *for* $n \gg 0$, *where* $p \in \mathbb{R}[x]$, *then* $\gamma(f) = \deg p$.
(iv) *If* $\gamma(f) = \lim (\log f(n)/\log n)$ *then* $\gamma(fg) = \gamma(f) + \gamma(g)$.
(v) *If* $g(n) \leqslant f(an+b)$ *for* $n \gg 0$, *where* $a, b \in \mathbb{N}$, *then* $\gamma(g) \leqslant \gamma(f)$.

Proof. (i)(ii)(iii) Straightforward.

(iv) In this case if $\gamma(f) = \alpha$ and $\gamma(g) = \beta$ then, for any $\varepsilon > 0$, $f(n) \geqslant n^{\alpha-\varepsilon}$ if $n \gg 0$ and $g(n) \geqslant n^{\beta-\varepsilon}$ for all n in some infinite subset I of $\mathbb{N}$. Evidently $f(n)g(n) \geqslant n^{\alpha+\beta-2\varepsilon}$ for all $n \gg 0$, $n \in I$, so $\gamma(fg) \geqslant \alpha + \beta$.

(v) If $\alpha = \gamma(f)$ and $\varepsilon > 0$ then $f(n) < n^{\alpha+\varepsilon}$ for $n \gg 0$. Hence

$$g(n) \leqslant f(an+b) < (an+b)^{\alpha+\varepsilon} = n^{\alpha+\varepsilon}\left(a + \frac{b}{n}\right)^{\alpha+\varepsilon} < n^{\alpha+2\varepsilon}$$

for $n \gg 0$, and thus $\gamma(g) \leqslant \alpha = \gamma(f)$. □

In the light of (iii) above, it seems natural to call $\gamma(f)$ the **degree** of f.

1.8 We note, although it will not be used here, that $f: \mathbb{N} \to \mathbb{R}$ has **exponential growth** if for some $v \in \mathbb{R}$ with $v > 1$, $f(n) \geqslant v^n$ for $n \gg 0$.

1.9 After these preliminaries we return to k-algebras and introduce some terminology. Let R be a k-algebra with a filtration $\{R_n\}$ and M a right R-module

8.1.12   *Definition and examples*   281

with a filtration $\{M_n\}$. The filtration of R is **standard** if $R_1^n = R_n$ for each n and is called **finite dimensional** if $R_0 = k$ and $\dim_k R_n < \infty$ for each n. The filtration of M is **standard** if $M_n = M_0 R_n$ for each n and is **finite dimensional** if $\dim_k M_n < \infty$ for each n.

For example, suppose that R is an **affine** k-algebra, that being one which is finitely generated as a k-algebra by a finite set of elements or, equivalently, by a finite dimensional k-subspace V. Then V is called a **generating subspace** of R, and R has a standard finite dimensional filtration $\{R_n\}$ with $R_0 = V^0 = k$ and $R_n = \sum_{i=0}^n V^i$. If M_R is finitely generated it has a finite dimensional subspace M_0 such that $M_0 R = M$. This is called a **generating subspace** for M, and M has a standard finite dimensional filtration $\{M_n\}$ with $M_n = M_0 R_n$.

Henceforth, until §6, affine k-algebras and their finitely generated modules will be understood to have standard finite dimensional filtrations as described here.

1.10 Suppose that R is affine and M_R is finitely generated and, after the choice of V and M_0, the filtrations $\{R_n\}$ and $\{M_n\}$ are obtained. If we let $f(n) = \dim R_n$ and $g(n) = \dim M_n$ then f and g measure the rate of growth of R and M with respect to that choice. The next result demonstrates some independence from that choice.

Lemma. *The degrees $\gamma(f)$ and $\gamma(g)$ are independent of the choice of V and M_0.*

Proof. Let V', M_0' be some other choice, $\{R_n'\}$, $\{M_n'\}$ the resulting filtrations and f', g' the resulting functions. Since $\bigcup R_n = R$, $V' \subseteq R_a$ for some $a \in \mathbb{N}$; similarly, $M_0' \subseteq M_b$ for $b \in \mathbb{N}$ and therefore $M_n' \subseteq M_{an+b}$. Hence $g'(n) \leq g(an+b)$ and then 1.7(v) shows that $\gamma(g') \leq \gamma(g)$. Symmetry shows that $\gamma(g') = \gamma(g)$. The case $M = R$, $M_0 = k$ then gives $\gamma(f') = \gamma(f)$. □

1.11 Continuing with the same notation, we can now define the **Gelfand–Kirillov dimension**, or **GK dimension** for short; namely $GK(R) = \gamma(\dim R_n)$, $GK(M) = \gamma(\dim M_n)$ for any choice of generating subspaces.

It is perhaps worth noting that if $1 \in V$ then $V^n = R_n$ and hence $GK(R) = \gamma(\dim V^n)$. However, this is not true when $1 \notin V$ as 1.3 and 1.4 demonstrate.

1.12 There is, as mentioned earlier, a corresponding notion for groups. If G is a group with a finite generating set X then the rate of growth of G is measured by the cardinality $f(n)$ of the subset consisting of all words, in elements of X and their inverses, of length at most n. Of course, if X is chosen to be closed under inverses then $f(n) = \dim(kG)_n$, where kX is the chosen generating subspace. Thus the degree of the rate of growth of G is the GK dimension of kG.

1.13 The next result notes some immediate consequences of the definitions. The first was used implicitly to finish the proof of 1.10.

Lemma. *Let R be affine and M_R finitely generated.*
(i) $\mathrm{GK}(R) = \mathrm{GK}(R_R) = \mathrm{GK}(_R R)$.
(ii) *If R' is an affine subalgebra of R then $\mathrm{GK}(R') \leq \mathrm{GK}(R)$; and if M' is a finitely generated R'-submodule of M then $\mathrm{GK}(M'_{R'}) \leq \mathrm{GK}(M_R)$.*

Proof. (i) Take $M_0 = k$.
(ii) Using 1.10, we can choose the generating subspaces so that $V' \subseteq V$ and $M'_0 \subseteq M_0$. Then 1.7(v) applies. □

1.14 Another immediate consequence concerns gr R, the associated graded ring of an affine algebra R with respect to a standard finite dimensional filtration $\{R_n\}$. If we choose the generating subspace R_1/R_0 for gr R it is clear that gr R has the same growth function f as R. This, together with a similar argument for M, proves

Proposition. *If R is affine and M_R is finitely generated then $\mathrm{GK}(\mathrm{gr}\, R) = \mathrm{GK}(R)$ and $\mathrm{GK}(\mathrm{gr}\, M) = \mathrm{GK}(M)$.* □

1.15 It is now easy to calculate the GK dimension of several algebras making use of the fact, given by 1.7(iii), that if $\dim R_n = p(n)$ for $n \gg 0$ for some $p(x) \in \mathbb{R}[x]$ then $\mathrm{GK}(R) = \deg p(x)$.

Proposition. *Let k be any field.*
(i) *If $R = k[x_1, \ldots, x_s]$, the commutative polynomial ring, then $\mathrm{GK}(R) = s$.*
(ii) *If $R = A_m(k)$, the m-th Weyl algebra, then $\mathrm{GK}(R) = 2m$.*
(iii) *If $\mathbf{g}$ is a finite dimensional Lie algebra then $\mathrm{GK}(U(\mathbf{g})) = \dim \mathbf{g}$.*
(iv) *If R is the free associative algebra $k\langle x_1, \ldots, x_s\rangle$ with $s \geq 2$ then $\mathrm{GK}(R) = \infty$.*

Proof. (i) Clear from 1.4.
(ii) (iii) The associated graded rings, with respect to the obvious generating subspaces, are commutative polynomial rings in $2m$ and $\dim \mathbf{g}$ indeterminates respectively.
(iv) This time $f(n) = 2^{n+1} - 1$, so R has exponential growth. □

1.16 The definition of the GK dimension is readily extended to arbitrary k-algebras S and modules N_S. One defines

$$\mathrm{GK}(S) = \sup\{\mathrm{GK}(R) \mid R \text{ an affine } k\text{-subalgebra of } S\},$$

and

$$\mathrm{GK}(N) = \sup\{\mathrm{GK}(M_R) \mid R \text{ an affine } k\text{-subalgebra of } S \text{ and } M \text{ a finitely generated } R\text{-submodule of } N\}.$$

8.1.18 *Definition and examples*

It is clear from 1.13 that this is in accord with the definitions for affine algebras and finitely generated modules.

As an immediate example, if S is the commutative polynomial ring over k in an infinite set of indeterminates then $GK(S) = \infty$.

1.17 We turn next to a brief review of the possible values of $GK(R)$, starting with the smallest values. A module N_R is **locally finite dimensional** if every finitely generated submodule has finite dimension; for example, take $R = k[\partial/\partial y]$ acting on $N = k[y]$.

Proposition. *Let R be an affine k-algebra and N_R a right R-module. Then:*
(i) $GK(N) = 0$ if and only if N is locally finite dimensional;
(ii) if $GK(N) > 0$ then $GK(N) \geq 1$.

Proof. Let M be a finitely generated submodule of N with standard filtration $\{M_n\}$. It is clear from the definition that $GK(M) = 0$ if and only if M is finite dimensional over k, i.e. if and only if $M = M_n$ for some n.

On the other hand, if $M \neq M_n$ for any n then $M_{n+1} \supset M_n$ for each n. Hence $\dim M_n \geq n$ and $GK(M) \geq 1$. □

1.18 It follows, of course, that $GK(R)$ cannot lie strictly between 0 and 1. This, together with the examples above, suggests that $GK(R)$ is perhaps always an integer or ∞. That this is true when R is commutative and in some more general circumstances will be shown in Sections 4 and 6. However, in general this is false.

Proposition. *Given any $r \in \mathbb{R}$ with $r \geq 2$ there is an affine algebra R with $GK(R) = r$.*

Proof. First consider the factor algebra $k\langle X, Y\rangle/I^3$, where I is the ideal generated by Y. Letting x, y denote the images of X, Y, it is clear that this algebra has a k-basis consisting of all words of the form x^h, $x^h y x^i$, and $x^h y x^i y x^j$. The number of words of length n of each of these three types is, respectively, 1, n, and $n(n-1)/2$.

Let $0 < a < 1$. We now choose to factor out an ideal generated by certain of the monomials of the third type; namely, for each n, all words of length n of the form $x^h y x^i y x^j$ in which $i < n - n^a$. It is easy to see that the factor ring R, in which we retain the notation x, y, has a basis consisting of all words of the first two types together with the other words of the third type.

Let $V = kx + ky$. We seek to estimate $\dim V^n$. For each $i \geq n - n^a$ the number of words of the third type in V^n is $h + j + 1$, i.e. is $n - i - 1$. Since $i \geq n - n^a$ then $n - i - 1 \leq n^a - 1$ and so $\dim V^n \leq 1 + n + n^a(n^a - 1)$.

Next consider those i with $n - n^a \leq i \leq n - \tfrac{1}{2}n^a$. For each such i,

$$n - i - 1 \geq n - (n - \tfrac{1}{2}n^a) - 1 = \tfrac{1}{2}n^a - 1$$

and so $\dim V^n \geq 1 + n + \tfrac{1}{2}n^a(\tfrac{1}{2}n^a - 1)$.

It follows that if $\frac{1}{2} \leq a < 1$ then

$$bn^{2a} < \dim V^n < cn^{2a}$$

for some positive real numbers b, c, for $n \gg 0$. Therefore

$$b \int_0^n x^{2a}\, dx < \dim R_n < c \int_1^{n+1} x^{2a} dx$$

and hence $\operatorname{GK}(R) = 1 + 2a$.

This gives an algebra of GK dimension r for any r with $2 \leq r < 3$. By taking polynomials in an appropriate number of indeterminates one can obtain an algebra of GK dimension r for any $r \geq 2$. □

The restriction here to the case $r \geq 2$ is necessary since a result of [Bergman **Ud**] (or see [Krause and Lenagan **85**, 2.5]) shows that there are no algebras R with $1 < \operatorname{GK}(R) < 2$.

1.19 One might suspect that when R is affine and $\operatorname{GK}(R) < \infty$ then R must be Noetherian. In fact one can check easily that the rings described in 1.18 are not Noetherian; and the next example shows that even if $\operatorname{GK}(R) = 1$ it need not be Noetherian.

Example. *Let R be the factor of the free algebra $k\langle X, Y\rangle$ by the ideal generated by X^2, XYX and YXY. Then $\operatorname{GK}(R) = 1$ and yet R contains an infinite direct sum of ideals.*

Proof. Let x, y denote the images of X, Y in R and let $V = xk + yk$. It is easy to see, for $n \geq 3$, that V^n has the k-basis $\{xy^{n-2}x, xy^{n-1}, y^{n-1}x, y^n\}$ and that $\dim R_n = 4(n-1)$. Hence $\operatorname{GK}(R) = 1$. However $A_i = xy^i xk \triangleleft R$ and the sum $\sum A_i$ is direct. □

1.20 Despite this, there is a positive result which depends upon the next fact.

Proposition. *Let R be an algebra which is an integral domain and which has no subalgebra isomorphic to the free algebra $k\langle X, Y\rangle$. Then R is a right and left Ore domain.*

Proof. Let x, y be nonzero elements of R. By the hypothesis there is a relation between x and y. Choose one of least total degree and write it in the form $a + xf + yg = 0$ with $a \in k$ and f, g in the subalgebra generated by x and y. Clearly at least one of f and g is nonzero, g say. Then

$$x(a + fx) = (a + xf)x = y(-gx) \neq 0$$

giving the right Ore condition and, by symmetry, the left Ore condition too.
□

8.2.4 Dimensions of related algebras

1.21 Corollary. *Let R be a k-algebra which is an integral domain and has $GK(R) < \infty$. Then R is a right and left Ore domain.*

Proof. Apply 1.13 and 1.15(iv). □

§2 Dimensions of Related Algebras

2.1 This section describes how the GK dimension of algebras behaves under many of the constructions described in Chapter 1.

2.2 We start with a rather general result.

Proposition. *If R is a k-algebra and S is either a subalgebra or a homomorphic image then $GK(S) \leq GK(R)$.*

Proof. The case of a subalgebra is dealt with by the definition, so suppose S is a homomorphic image of R. Let S' be an affine subalgebra of S, with filtration $\{S_n'\}$. One can choose a generating subspace for an affine subalgebra R' of R such that R_1' maps onto S_1', and then R_n' maps onto S_n'. Hence $\dim R_n' \geq \dim S_n'$ and $GK(R) \geq GK(S)$. □

2.3 The GK dimension is particularly well suited to tensor products.

Proposition. *Let R, R' be k-algebras. Then*
$$\sup\{GK(R), GK(R')\} \leq GK(R \otimes_k R') \leq GK(R) + GK(R').$$

Proof. Since $R \simeq R \otimes 1$ and $R' \simeq 1 \otimes R'$ the first inequality is immediate from the definition. Next, choose some finite dimensional subspace, W say, of $R \otimes R'$ and let T be the affine subalgebra generated by W. Now $W \subseteq U \otimes U'$ for some finite dimensional subspaces of R and R'. Let S, S' be the affine subalgebras they generate. Then $T_n \subseteq S_n \otimes S_n'$ and so $\dim T_n \leq (\dim S_n)(\dim S_n')$ which, by 1.7(ii), gives the second inequality. □

2.4 The case $R = R' = k[x]$ indicates that the first inequality is often strict. The next result, whose hypothesis should be compared with the alternative description of $\gamma(f)$ in 1.6, shows that the second is often an equality.

Lemma. *Let R' contain an affine subalgebra S' with a standard finite dimensional filtration $\{S_n'\}$ such that*
$$\lim_{n \to \infty} (\log \dim S_n' / \log n) = GK(S') = GK(R').$$
Then $GK(R \otimes_k R') = GK(R) + GK(R')$.

Proof. Let S be any affine subalgebra of R, let $T = S \otimes S'$ and $T_1 = (S_1 \otimes 1) + (1 \otimes S'_1)$. Then $T_{2n} \supseteq S_n \otimes S'_n$ and so $\dim T_{2n} \geq (\dim S_n)(\dim S'_n)$. Therefore, by 1.7, $\gamma(T) \geq \gamma(S) + \gamma(S')$. Hence $\gamma(R \otimes R') \geq \gamma(R) + \gamma(R')$. □

2.5 It is not too difficult to devise an example of an affine algebra S such that

$$\mathrm{GK}\,(S) \neq \lim_{n \to \infty} (\log \dim S_n / \log n).$$

We now proceed to sketch such a construction which is a modification of that described in 1.18.

The process described there involved factoring out, for each n, all words of the form $x^h y x^i y x^j$ of degree n with $i < n - n^a$. For any other monotonic increasing function $g(x)$, one could factor out all such words with $i < g(n)$ with similar effect, obtaining an algebra, S say.

Suppose, then, that $\tfrac{1}{2} < a < b < 1$ and $g(x)$ is as sketched in Figure 8.1.

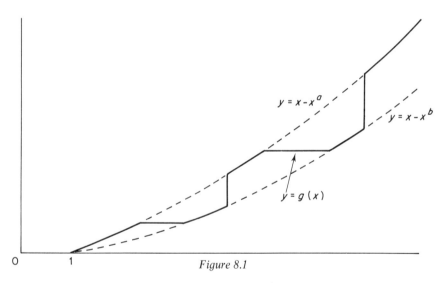

Figure 8.1

If $g(x)$ follows the two curves $x - x^a$ and $x - x^b$ for appropriately lengthening periods it can be arranged that

$$\limsup \{\log \dim S_n / \log n\} = 1 + 2b$$

and

$$\liminf \{\log \dim S_n / \log n\} = 1 + 2a.$$

2.6 It is very plausible that if two algebras S, S' are constructed along the lines indicated in 2.5 but with the oscillations of the dimensions of their filtrations arranged to be opposite, then one could have $\mathrm{GK}\,(S \otimes S') < \mathrm{GK}\,(S) + \mathrm{GK}\,(S')$. This can be done; see [Warfield **84**] or [Krause and Lenagan **85**, 3.14] for details.

8.2.9 Dimensions of related algebras

2.7 There are some immediate applications of 2.4 which are listed next.

Proposition. *For any k-algebra R*:
 (i) $GK(M_n(R)) = GK(R)$;
 (ii) $GK(RG) = GK(R)$ *for any finite group G*;
 (iii) $GK(R[x]) = GK(R) + 1$;
 (iv) $GK(R \otimes U(g)) = GK(R) + \dim g$ *for any finite dimensional Lie algebra g*;
 (v) $GK\, A_n(R) = GK(R) + 2n$.

Proof. (i) Note that $GK(M_n(k)) = 0$ since $M_n(k)$ is finite dimensional. Hence 2.4 applies with $R' = M_n(k)$.
 (ii) A similar argument applies with $R' = kG$.
 (iii)(iv)(v) These follow from 2.4 and the proof of 1.15 taking R' to be $k[x]$, $U(g)$ and $A_n(k)$ respectively. □

The rest of the section extends these results to more general algebras.

2.8 First we turn to endomorphism rings starting with a very general observation.

Lemma. *Let R be any ring and M_R a faithful module generated by a finite set of generators, n in number. Then there is a subring S of $M_n(R)$ and a surjection $S \to \mathrm{End}\, M$; and if R is a k-algebra so too is S.*

Proof. There is a surjection $\pi: R^n \to M$. We identify $\mathrm{End}(R^n)$ with $M_n(R)$ and consider the subring
$$S = \{\alpha \in M_n(R) \mid \alpha(\ker \pi) \subseteq \ker \pi\}.$$
It is readily checked that each such α induces a map $\alpha': M \to M$ and that the map $\alpha \mapsto \alpha'$ is a surjective homomorphism. □

2.9 This leads to some results which extend 2.7(i), (ii).

Proposition. *Let R be a k-algebra.*
 (i) *If M_R is finitely generated then $GK(\mathrm{End}\, M) \leq GK(R/\mathrm{ann}\, M) \leq GK(R)$.*
 (ii) *If $R \subseteq R'$ with R' a k-algebra finitely generated as a right, or left, R-module then $GK(R') = GK(R)$.*
 (iii) *If $R' \overset{M}{\sim} R$ then $GK(R') = GK(R)$.*
 (iv) $GK(R * G) = GK(R)$ *for any finite group G.*

Proof. (i) It is enough to consider the case when $\mathrm{ann}\, M = 0$. Then, with S as described in 2.8,
$$GK(\mathrm{End}\, M) \leq GK(S) \leq GK(M_n(R)) = GK(R).$$

(ii) View R' as an (R', R)-bimodule; so R' embeds in $\operatorname{End}(R'_R)$. Hence, by (i), $\operatorname{GK}(R') \leq \operatorname{GK}(R)$.

(iii) In this case $R' \simeq \operatorname{End}(P_R)$ with P_R a progenerator; so (i) applies.

(iv) Clear from (ii). □

2.10 The extension of 2.7(iii), (iv) to more general cases is non-trivial.

Proposition. *Let R be a k-algebra, g a finite dimensional k-Lie algebra and S a crossed product, $S = R * U(g)$. Then $\operatorname{GK}(S) \geq \operatorname{GK}(R) + \dim g$. If R is affine then $\operatorname{GK}(S) = \operatorname{GK}(R) + \dim g$.*

Proof. Suppose $\dim g = d$. Fix a basis for g and let X_m be the set of all standard monomials of length m (see 1.7.5). We need the fact, used in 1.15, that

$$\dim \sum_{i=0}^{n} (X_i k) = \binom{n+d}{d}$$

which has degree d as a polynomial in n.

First, let R' be any affine subalgebra of R and S' be the affine subalgebra of S with generating subspace $S'_1 = R'_1 + g = R'_1 + X_1 k$. Evidently

$$S'_{2n} \supseteq R'_n + X_1 R'_n + \cdots + X_n R'_n$$

and so

$$\dim S'_{2n} \geq (\dim R'_n) \binom{n+d}{d}.$$

Therefore, by 1.7, $\operatorname{GK}(S') \geq \operatorname{GK}(R') + d$ and thus $\operatorname{GK}(S) \geq \operatorname{GK}(R) + d$.

Next suppose that R is affine with standard finite dimensional filtration $\{R_n\}$. From the definition of crossed products it is clear one can choose m so that both

$$R_1 X_1 \subseteq X_1 R_1 + R_m \quad \text{and} \quad X_1^2 k \subseteq X_2 k + X_1 k + R_m.$$

One can then show, for each n, that

$$R_n X_1 \subseteq X_1 R_n + R_{m+n}$$

and

$$X_n X_1 k \subseteq X_1^{n+1} k \subseteq X_{n+1} k + \sum_{j=0}^{n} X_{n-j} R_{mj}.$$

Let $S_1 = R_m + g = R_m + X_1 k$. A straightforward induction on n, using these relations, shows that

$$S_n \subseteq \sum_{i=0}^{n} X_i R_{mn}.$$

Therefore

$$\dim S_n \leq (\dim R_{mn}) \binom{n+d}{d}$$

and so, by 1.7, $\operatorname{GK}(S) \leq \operatorname{GK}(R) + d$. □

More light will be shed on this result in §6.

2.11 Corollary. *If R is an k-algebra and δ is a k-derivation then $\mathrm{GK}(R[x,\delta]) \geq \mathrm{GK}(R) + 1$ with equality if R is affine.* □

2.12 If R is not affine, the inequalities above may be strict. For example, if $R = k[x_1, x_2, \ldots]$, the commutative algebra on generators $\{x_i | i \in \mathbb{N}\}$ with the relations $x_i x_j = 0$ for all i, j, then R is locally finite dimensional, so $\mathrm{GK}(R) = 0$. However, if δ is defined by $\delta(x_i) = x_{i+1}$ and $S = R[x; \delta]$ then $\mathrm{GK}(S) \geq 2$. To see this let T be the affine subalgebra with generating subspace $W = x_1 k + xk$. Then it is easy to check that W^n has basis

$$\{x^n, x^{n-1}x_1, \ldots, xx_{n-1}, x_n\}$$

and so $\dim T_n = n(n+1)/2$.

In fact it can be arranged that $\mathrm{GK}(R) = 0$ and $\mathrm{GK}(R[x;\delta]) = \infty$ by appropriately modifying this example. See [Krause and Lenagan 85, 3.9] for details.

2.13 The basic result on localization is also rather restricted.

Proposition. *Let R be an algebra and $\mathscr{S}$ be a multiplicatively closed set of central regular elements of R. Then*

$$\mathrm{GK}(R_\mathscr{S}) = \mathrm{GK}(R).$$

Proof. Let T be an affine subalgebra of $R_\mathscr{S}$ with filtration $\{T_n\}$. Since T_1 is finite dimensional, there exists $s \in \mathscr{S}$ such that $sT_1 \subseteq R$. Let U be the affine subalgebra of R with $U_1 = sT_1 + k$. Then $T_n \simeq s^n T_n \subseteq U_n$ and so $\dim T_n \leq \dim U_n$. Hence $\mathrm{GK}(R_\mathscr{S}) \leq \mathrm{GK}(R)$. Of course R is a subalgebra of $R_\mathscr{S}$, so $\mathrm{GK}(R) \leq \mathrm{GK}(R_\mathscr{S})$. □

The restriction here to central elements is not completely necessary (see 15.4.2); but it cannot simply be removed. For example, whilst $\mathrm{GK}(A_1(k)) = 2$ it is known that $\mathrm{GK}((A_1)_\mathscr{S}) = 3$ if $\mathscr{S}$ is the multiplicatively closed subset generated by $\{y^n, x^n | n \in \mathbb{N}\}$ and that $\mathrm{GK}(D_1(k)) = \infty$, where $D_1(k)$ is the quotient division ring of $A_1(k)$. See [Makar–Limanov 83] or [Krause and Lenagan 85, 8.18].

2.14 One consequence of 2.13 connects the GK dimension, Krull dimension, and transcendence degrees for commutative affine k-algebras.

Theorem. *(i) Let R be a commutative affine k-algebra. Then R contains a polynomial subalgebra $S = k[x_1, \ldots, x_t]$ such that R is a finitely generated S-module, and $\mathrm{GK}(R) = \mathcal{K}(R) = t$.*

(ii) If R is also an integral domain then t is the transcendence degree of the

quotient field Q of R. Moreover, if $0 \neq f \in R$ and R_f is the localization of R at the powers of f then $\mathrm{GK}(R_f) = \mathcal{K}(R_f) = t$.

Proof. (i) The Noether normalization theorem [Zariski and Samuel **60**, p. 200, Theorem 25] provides S as described. Evidently $\mathrm{GK}(S) = \mathcal{K}(S) = t$ and, by 2.9 and 6.5.3, $\mathrm{GK}(R) = \mathrm{GK}(S) = t$ and $\mathcal{K}(R) \leq \mathcal{K}(S) = t$. However, if $P \triangleleft R$ is chosen maximal with respect to $P \cap S = 0$ then P is prime, $S \hookrightarrow R/P$ and R/P is a finite module over S. It will be enough now to prove (ii); for then $\mathcal{K}(R) \geq \mathcal{K}(R/P) = t$.

(ii) Let $Q' = k(x_1, \ldots, x_t)$. Then Q is finite dimensional over Q' and so, by 2.9, $\mathrm{GK}(Q) = \mathrm{GK}(Q')$. Hence, by 2.13, $\mathrm{GK}(R) = \mathrm{GK}(S) = t$. However by [Zariski and Samuel **60**, p. 193, Theorem 20], $t = \dim R$ which, by 6.4.8, equals $\mathcal{K}(R)$. Finally, note that R_f is affine and has the same quotient ring as R. □

2.15 There is another obvious consequence of 2.13.

Corollary. *For any k-algebra R,*

$$\mathrm{GK}(R[x, x^{-1}]) = \mathrm{GK}(R) + 1.$$

Proof. Apply 2.13 to $R[x]$ with $\mathcal{S} = \{x^n\}$ and use 2.7(iii). □

2.16 The restriction, in 2.10 and 2.15, to skew polynomials and Laurent polynomials not involving an automorphism σ is necessary as the next example shows.

Example. *Let $R = k[y, y^{-1}, z, z^{-1}]$, $S = R[x; \sigma]$ and $T = R[x, x^{-1}; \sigma]$, where $\sigma(y) = yz$, $\sigma(z) = z$. Then $\mathrm{GK}(R) = 2$ and $\mathrm{GK}(S) = \mathrm{GK}(T) = 4$.*

Proof. It follows from 2.15 that $\mathrm{GK}(R) = 2$. The fact that S is a subalgebra of T ensures that $\mathrm{GK}(S) \leq \mathrm{GK}(T)$. So it will be enough to show that $\mathrm{GK}(T) \leq 4$ and $\mathrm{GK}(S) \geq 4$.

For the former, consider the filtration $\{T_n\}$ whose generating subspace has $\{x, x^{-1}, y, y^{-1}\}$ as basis. The monomials $x^a y^b z^c$ with $a, b, c \in \mathbb{Z}$ form a k-basis for T, and the number of such monomials in T_n is easily bounded above. For if $x^a y^b z^c \in T_n$ then $|a| \leq n$, $|b| \leq n$ and, since $y^r x^s = x^s y^r z^{rs}$ and each z arises only in such a way, $|c| \leq n^2$. Thus $\mathrm{GK}(T) \leq 4$.

For the latter, let S' be the subalgebra generated by x, y, and $\{S'_n\}$ be the filtration whose generating subspace has basis $\{x, y\}$. If $0 \leq c < n^2$ then $c = np + q$ with $0 \leq p, q \leq n-1$ and

$$x^n y^n z^c = x^{n-p-1} y^q x y^{n-q} x^p \in S'_{2n}.$$

Hence $x^a y^b z^c \in S'_{4n}$ for all a, b, c with $n \leq a, b \leq 2n$, $0 \leq c \leq n^2$ and so $\dim S'_{4n} \geq n^4$. Therefore, by 1.7(v), $\mathrm{GK}(S') \geq 4$ and so $\mathrm{GK}(S) \geq 4$. □

8.3.1 Module theory

2.17 The preceding example T is, of course, the group algebra kG where G is the group generated by x, y, z with relations $zy = yz$, $zx = xz$, and $y^{-1}x^{-1}yx = z$. The representation

$$x \mapsto \begin{bmatrix} 1 & 0 & 0 \\ 0 & 1 & 1 \\ 0 & 0 & 1 \end{bmatrix}, \quad y \mapsto \begin{bmatrix} 1 & 1 & 0 \\ 0 & 1 & 0 \\ 0 & 0 & 1 \end{bmatrix}, \quad z \mapsto \begin{bmatrix} 1 & 0 & 1 \\ 0 & 1 & 0 \\ 0 & 0 & 1 \end{bmatrix}$$

gives an isomorphism of G with the group of 3×3 matrices

$$\left\{ \begin{bmatrix} 1 & a & b \\ 0 & 1 & c \\ 0 & 0 & 1 \end{bmatrix} \,\middle|\, a, b, c \in \mathbb{Z} \right\}$$

i.e. the **discrete Heisenberg group**.

2.18 As noted in 1.12, the GK dimension of kG is determined by the rate of growth of G. There are results known about this which give the following facts.

Theorem. (i) *Let G be a finitely generated group. Then $\mathrm{GK}(kG) < \infty$ if and only if there is a nilpotent subgroup N of finite index in G.*

(ii) *Let N be a finitely generated nilpotent group, with lower central series $N = N_1 \supset N_2 \supset \cdots \supset N_{m+1} = 1$. Let*

$$d_i = \operatorname{rank} N_i/N_{i+1} = \dim((N_i/N_{i+1}) \otimes_{\mathbb{Z}} \mathbb{Q})_{\mathbb{Q}}.$$

Then $\mathrm{GK}(kN) = \sum_{i=1}^{m} i d_i$.

Proof. (i) See [Gromov **81**] or, for a proof when G is solvable, [Krause and Lenagan **85**, 11.2].

(ii) See [Krause and Lenagan **85**, 11.11]. □

Of course (ii) can be applied to the discrete Heisenberg group. In that case, if G' is the derived group, then G' is generated by z. Hence $G/G' \simeq \mathbb{Z}^2$, $G' \simeq \mathbb{Z}$ and $\mathrm{GK}(kG) = 1 \cdot 2 + 2 \cdot 1 = 4$, as seen in 2.16.

Note also that since there are polycyclic groups which are not nilpotent by finite, the GK dimension can jump from finite to infinite under a skew Laurent extension. Indeed if R is the commutative algebra $k[y, y^{-1}, z, z^{-1}]$ and $S = R[x, x^{-1}; \sigma]$ with $\sigma(z) = y$ and $\sigma(y) = zy^2$ then $\mathrm{GK}(R) = 2$ and $\mathrm{GK}(S) = \infty$.

§3 Module Theory

3.1 This section outlines the basic module–theoretic properties of the GK dimension. In particular it shows that, over a right Noetherian algebra R, it is a dimension function, as mentioned in Chapter 6, §8 and that R is right and left ideal invariant.

3.2 The first result lists the more obvious properties.

Proposition. *Let R be any k-algebra and M_R any module.*
(i) *If $M = \sum L_i$ for some submodules L_i then $\mathrm{GK}(M) = \sup\{\mathrm{GK}(L_i)\}$.*
(ii) *If $0 \to M' \to M \to M'' \to 0$ is a short exact sequence of R-modules then $\mathrm{GK}(M) \geq \sup\{\mathrm{GK}(M'), \mathrm{GK}(M'')\}$. If the sequence is split, equality holds.*
(iii) $\mathrm{GK}(M) \leq \mathrm{GK}(R)$.
(iv) *If $L_1, \ldots, L_t$ are submodules of M such that $\bigcap L_i = 0$ then $\mathrm{GK}(M) = \sup\{\mathrm{GK}(M/L_i)\}$.*
(v) *If $MI = 0$ with $I \triangleleft R$ then $\mathrm{GK}(M_R) = \mathrm{GK}(M_{R/I})$.*

Proof. (i) Let R' be an affine subalgebra of R and M' a finitely generated R'-submodule of M, with generating subspace M'_0. Now $M'_0 \subset \sum_i L_{i0}$, where L_{i0} is a finite dimensional subspace of L_i and the summation is over some finite set, say $i = 1, \ldots, t$. It is clear that

$$\dim(M'_0 R'_n) \leq \sum_{i=1}^{t} \dim(L_{i0} R'_n)$$

and hence

$$\mathrm{GK}(M') \leq \sup\{\mathrm{GK}(L_{i0} R')\} \leq \sup\{\mathrm{GK}(L_i)\}$$

using 1.7(i). Therefore $\mathrm{GK}(M) \leq \sup\{\mathrm{GK}(L_i)\}$. However, 1.16 shows that $\mathrm{GK}(M) \geq \sup\{\mathrm{GK}(L_i)\}$.

(ii) That $\mathrm{GK}(M) \geq \mathrm{GK}(M')$ is clear from 1.16. Next let R' be an affine subalgebra of R, and N'' a finitely generated R'-submodule of M'' with generating subspace N''_0. There is a finite dimensional subspace N_0 of M mapping onto N''_0; let $N = N_0 R'$. The resulting filtrations have $\dim N_s \geq \dim N''_s$; so $\mathrm{GK}(M) \geq \mathrm{GK}(M'')$ as claimed.

(iii)(iv) These are clear from (i) and (ii), using the canonical embedding $M \hookrightarrow \bigoplus_i M/L_i$.

(v) Straightforward. □

3.3 This has the following easy consequence.

Corollary. *Let R be any k-algebra.*
(i) *If $R = S \oplus T$ for algebras S, T then $\mathrm{GK}(R) = \sup\{\mathrm{GK}(S), \mathrm{GK}(T)\}$.*
(ii) *If $I_1, \ldots, I_t$ are ideals of R with $\bigcap I_j = 0$ then $\mathrm{GK}(R) = \sup\{\mathrm{GK}(R/I_j)\}$.* □

3.4 Before proceeding further, an example is given to show that equality does not always occur in 3.2(ii)

Example. *Let $S = k\langle X, Y \rangle$, the free algebra, and A be the ideal generated by Y. Let $R = S/A^2$ and $B = A/A^2$. In the short exact sequence of R-modules*

$$0 \to B \to R \to R/B \to 0$$

$\mathrm{GK}(R) = 2$ *and* $\mathrm{GK}(B) = \mathrm{GK}(R/B) = 1$.

8.3.6 *Module theory*

Proof. Let x, y denote the images of X, Y in R, and let V be the k-subspace generated by $\{x, y\}$. The monomials $\{x^n, x^i y x^j | i + j = n - 1\}$ are a basis for V^n. Thus $\dim V^n = n + 1$, $\dim R_n = (n + 1)(n + 2)/2$, and $\mathrm{GK}(R) = 2$. Note next that $\mathrm{GK}(B_R) = \mathrm{GK}(B_{R/B}) \leq \mathrm{GK}(R/B)$. However, $R/B \simeq k[x]$ and so $\mathrm{GK}(R/B) = 1$. Since B_R is not locally finite dimensional, $\mathrm{GK}(B) = 1$. □

In this example B_R is not finitely generated. However, there are similar short exact sequences in which all three modules are finitely generated; see [Bergman **81**] or [Krause and Lenagan **85**, 5.2].

3.5 There is one useful special case where more is known about short exact sequences.

Proposition. *Let R be a k-algebra and*

$$0 \to M' \xrightarrow{\alpha} M \xrightarrow{\beta} M'' \to 0$$

be an exact sequence of finitely generated R-modules with $M' \simeq M$. Then $\mathrm{GK}(M'') \leq \mathrm{GK}(M) - 1$.

Proof. We may suppose that $M' \subseteq M$ via α and that there is an isomorphism $\phi: M \to M'$. Let M_0'' be any finite dimensional generating subspace of M_R'' and let S be an affine subalgebra of R, with standard filtration $\{S_n\}$.

Choose a finite dimensional generating subspace M_0 of M_R so that $\beta(M_0) = M_0''$. Since $\phi(M_0)$ is finite dimensional then $\phi(M_0) \subseteq M_0 T_1$ for some finite dimensional subspace T_1 of R such that $T_1 \supseteq S_1$. Let T be the affine subalgebra generated by T_1; then $\phi^i(M_0) \subseteq M_0 T_i$.

Next, for each n, choose a subspace D_n of $M_0 S_n$ complementary to $M' \cap M_0 S_n$ and such that $\beta(D_n) = M_0'' S_n$. Note that $\dim D_n = \dim M_0'' S_n$, and, using induction on the number of summands, that the sum $\sum_{i=0}^{n} \phi^i(D_n)$ is direct. But

$$\sum_{i=0}^{n} \phi^i(D_n) \subseteq \sum \phi^i(M_0 S_n) = \sum \phi^i(M_0) S_n \subseteq \sum M_0 T_i T_n \subseteq M_0 T_{2n}.$$

Therefore $\dim M_0 T_{2n} \geq (n + 1) \dim D_n = (n + 1) \dim M_0'' S_n$ and hence, by 1.7, $\mathrm{GK}(M) \geq \mathrm{GK}(M'') + 1$. □

3.6 One of the immediate consequences is the dimension function result mentioned earlier.

Corollary. *Let R be a right Noetherian k-algebra and suppose $\mathrm{GK}(R) < \infty$.*
(i) *If R is prime and E is an essential right ideal then $\mathrm{GK}(R/E) \leq \mathrm{GK}(R) - 1$.*
(ii) *If M_R is finitely generated, $MP = 0$ for some $P \in \mathrm{Spec}\, R$ and $M_{R/P}$ is torsion then $\mathrm{GK}(M) \leq \mathrm{GK}(R/P) - 1$.*
(iii) *The GK dimension is a dimension function for R.*

(iv) If $P_0 \subset P_1 \subset \cdots \subset P_m$ is a chain of distinct prime ideals of R then $\mathrm{GK}(R) \geq \mathrm{GK}(R/P_0) \geq \mathrm{GK}(R/P_m) + m$.

Proof. (i) By 2.3.5, E contains a regular element, c say. Then $cR \simeq R$ and so $\mathrm{GK}(R/cR) \leq \mathrm{GK}(R) - 1$ by 3.5. However, by 3.2(ii), $\mathrm{GK}(R/E) \leq \mathrm{GK}(R/cR)$.

(ii) Each $m \in M$ is annihilated by an essential right ideal of R/P; so by (i) $\mathrm{GK}(mR) \leq \mathrm{GK}(R/P) - 1$. This, by 3.2(i), is enough.

(iii) This is now clear from the definition in 6.8.4.

(iv) This follows from 6.8.7(i). □

We note that (i), (ii) and (iv) do not require that R be right Noetherian but rather that each prime factor ring of R be right Goldie.

3.7 There are circumstances under which the GK dimension is an exact dimension function, and these involve the hypothesis that the associated graded ring $\mathrm{gr}\, R$, with respect to a standard finite dimensional filtration $\{R_n\}$, be right Noetherian. That, of course, implies that the original algebra R must be right Noetherian.

Before discussing that, it seems sensible to give an example that shows that R being right Noetherian does not imply that $\mathrm{gr}\, R$ is too.

3.8 For the example, and for later, the following lemma is useful.

Lemma. *If* $M = \bigoplus_{n=0}^{\infty} M_n$ *is a graded module over a graded ring* $B = \bigoplus_{n=0}^{\infty} B_n$ *and* M_B *is finitely generated then there exists an m such that*

$$M_{m+n} = \sum_{j=0}^{m} M_j B_{m+n-j}$$

for all n. If, further, $B_n = B_1^n$ for all n then $M_{m+n} = M_m B_n$.

Proof. Choose m large enough for $\sum_{j=0}^{m} M_j$ to contain a generating set for M_B. Then

$$M_{m+n} \subseteq \sum_{j=0}^{m} M_j B_{m+n-j} \subseteq M_{m+n}.$$

In the latter case,

$$M_j B_{m+n-j} = M_j B_{m-j} B_n \subseteq M_m B_n$$

and the result follows. □

3.9 Example. Let $R = k[u,v][x;\delta]$, where $k[u,v]$ is a commutative polynomial ring and $\delta = uv^2 \partial/\partial u$. Let $\{R_n\}$ be the standard filtration whose generating subspace has basis $\{u,v,x\}$ and let $B = \mathrm{gr}\, R$. Then R is Noetherian and B is neither right nor left Noetherian.

Proof. Here R has a basis, over k, consisting of the words $w = u^a v^b x^c$. First consider what restrictions are imposed upon b if $w \in R_n$. Then w is in the support of a product of at most n letters, a of which are u, at least c are x and so at most $n - (a + c)$ are v. The form of δ shows that

$$b \leq n - (a + c) + 2c \leq n + (a + c) - 2a \leq 2n - 2a.$$

In particular, $uv^{2n-1} \notin R_n$.

Next consider uv^{2n}. It is clear that $uv^{2n} \in uR \cap R_{n+1}$ but we ask now whether $uv^{2n} \in (uR \cap R_n)(ku + kv + kx)$. If this were so, uv^{2n} would be in the support of an expression uwu, uwv or uwx, after rewriting it in terms of the basis, where w also belongs to that basis, $w = u^a v^b x^c$ say. However, the words arising from uwu all have degree 2 or more in u; those from uwx have degree at least 1 in x; and $uwv = u^{a+1} v^{b+1} x^c$. So one would have $w = v^{2n-1}$ and $uw = uv^{2n-1} \in R_n$, a contradiction.

Finally, let $B = \operatorname{gr} R$ and $I = \operatorname{gr}(uR)$. The arguments above demonstrate that $I_{n+1} \neq I_n B_1$ for any $n \geq 1$ and so, by 3.8, I_B is not finitely generated. Hence B is not right Noetherian and, similarly, not left Noetherian. □

3.10 If, in this algebra R, the generating subspace is chosen to have the basis $\{u, v, v^2, x\}$ then it can be shown that $\operatorname{gr} R$ is Noetherian. Also it will be shown in §6 that R has a finite dimensional filtration for which $\operatorname{gr} R$ is commutative affine. It is not known whether there is a right Noetherian affine algebra R for which no finite dimensional filtration leads to a right Noetherian associated graded ring.

3.11 After this digression we return to the GK dimension.

Proposition. *Let R be an affine algebra with standard finite dimensional filtration $\{R_n\}$ and suppose that $\operatorname{gr} R$ is right Noetherian. Then the GK dimension is exact on short exact sequences of finitely generated right R-modules.*

Proof. Let $0 \to L \xrightarrow{\imath} M \xrightarrow{\pi} N \to 0$ be a short exact sequence of finitely generated right R-modules. We need to show that $\operatorname{GK}(M) = \sup\{\operatorname{GK}(L), \operatorname{GK}(N)\}$. Clearly, it will be enough to show that there are generating subspaces L_0, M_0, N_0 such that for each n the sequence $0 \to L_n \to M_n \to N_n \to 0$ is exact. To ease notation, let L be a submodule of M, and $\imath$ be containment.

First let M'_0 be any generating subspace of M and $\{M'_n\}$ the resulting filtration; so $M'_n = M'_0 R_n$. By 7.6.13, if L and N are given the filtrations defined by $L'_n = M'_n \cap L$ and $N'_n = \pi(M'_n)$ then there is an exact sequence of graded modules over $\operatorname{gr} R$

$$0 \to \operatorname{gr} L \to \operatorname{gr} M \to \operatorname{gr} N \to 0.$$

It is clear that $\operatorname{gr} M$ is finitely generated, by M'_0.

Since $\operatorname{gr} R$ is right Noetherian, $\operatorname{gr} L$ also is finitely generated. By 3.8 there exists m such that $L'_{n+m} = L'_m R_n$ for all n.

Now set $L_0 = L'_m$, $M_0 = M'_m$ and $N_0 = N'_m$ and let $\{L_n\}$, $\{M_n\}$, $\{N_n\}$ be the resulting standard filtrations. Note that

$$M_n \cap L = M'_{m+n} \cap L = L'_{m+n} = L'_m R_n = L_n.$$

It follows immediately that the sequence

$$0 \to L_n \to M_n \to N_n \to 0$$

is exact. □

3.12 This, together with 6.8.7 gives

Corollary. *Let R be affine with standard finite dimensional filtration $\{R_n\}$ such that $\operatorname{gr} R$ is right Noetherian. Let $P_1, \ldots, P_n$ be the minimal prime ideals of R and N the prime radical. Then*

$$\operatorname{GK}(R) = \operatorname{GK}(R/N) = \sup\{\operatorname{GK}(R/P_i) | i = 1, \ldots, n\}.$$
□

3.13 In fact 3.3(ii) shows that the latter equality is valid provided only that the minimal primes of R are finite in number. Also, the example described in 3.4 demonstrates that the former equality can fail under weaker hypotheses on R. Nevertheless, there is a positive result in this direction.

Proposition. *Let R be a k-algebra.*
(i) *If I, J are ideals with $IJ = 0$ then $\operatorname{GK}(R) \leq \operatorname{GK}(R/I) + \operatorname{GK}(R/J)$.*
(ii) *If N is an ideal with $N^m = 0$ then $\operatorname{GK}(R) \leq m \operatorname{GK}(R/N)$.*

Proof. An easy induction based on (i) will give (ii). So we consider (i). It is known (see 7.3(b)) that there is an $(R/I, R/J)$-bimodule M such that R embeds as a subalgebra of the ring

$$S = \begin{bmatrix} R/I & M \\ 0 & R/J \end{bmatrix}.$$

Any finite dimensional subspace of S is contained in one of the form $V = X \oplus Y \oplus Z$, where $X \subseteq R/I$, $Y \subseteq R/J$ and $Z \subseteq M$. Let

$$T = \begin{bmatrix} A & C \\ 0 & B \end{bmatrix}$$

be the affine algebra generated by V. It is easy to see that

$$\dim T_n = \dim A_n + \dim B_n + \dim C_{n-1}$$

using the generating subspace X for A, Y for B, and Z for C, but viewing C as a right $A^{op} \otimes B$-module and giving $A^{op} \otimes B$ the usual generating subspace $X^{op} \otimes 1 + 1 \otimes Y$. Hence, by 2.3 and 3.2(iii),

$$\text{GK}(T) \leq \sup\{\text{GK}(A), \text{GK}(B), \text{GK}(C_{A^{op} \otimes B})\}$$
$$\leq \text{GK}(A) + \text{GK}(B) \leq \text{GK}(R/I) + \text{GK}(R/J). \qquad \square$$

The algebra $R = k\langle x, y\rangle/N^m$, where $N = (y)$ shows that the bound above may be attained. In this case $\text{GK}(R/N) = \text{GK}(k[x]) = 1$ and yet the argument in 1.3, applied to the set of words $\{x^{i(1)}yx^{i(2)}y\cdots yx^{i(m)}\}$ shows that $\text{GK}(R) \geq m$.

3.14 The results at which we aim next make use of bimodules. When discussing bimodules over k-algebras it is understood that the left and right actions of k coincide.

Proposition. *Let R, S be k-algebras and $_SM_R$ be a bimodule such that $_SM$ is finitely generated.*
  (i) *Suppose that R' is an affine subalgebra of R with standard finite dimensional filtration $\{R'_n\}$, and M_0 is a finite dimensional generating subspace of $_SM$. Then there is an affine subalgebra T of S with standard finite dimensional filtration $\{T_n\}$ such that $M_0 R'_n \subseteq T_n M_0$.*
  (ii) $\text{GK}(M_R) \leq \text{GK}(_SM)$.
  (iii) *If N is a right S-module then $\text{GK}(N \otimes_S M_R) \leq \text{GK}(N_S)$.*

Proof. (i) It is clear that $M_0 R'_1 \subseteq W M_0$ for some finite dimensional subspace W. Let T be the affine subalgebra of S generated by W and let $\{T_n\}$ be the resulting filtration. Then it follows that $M_0 R'_n \subseteq T_n M_0$.

(ii) We may suppose that R is affine and that X is some finite dimensional subspace of M. Of course $X \subseteq M_0$ for some finite dimensional subspace with $SM_0 = M$. By (i), $M_0 R_n \subseteq T_n M_0$ and therefore $\text{GK}(M_R) \leq \text{GK}(_T T M_0) \leq \text{GK}(_S M)$.

(iii) Again, we may suppose R to be affine and choose some finite dimensional subspace X of $N \otimes M$. Then $X \subseteq N_0 \otimes M_0$, where N_0, M_0 are finite dimensional subspaces of N, M respectively and $SM_0 = S$. By (i), $M_0 R_n \subseteq T_n M_0$ and therefore

$$(N_0 \otimes M_0)R_n \subseteq N_0 \otimes T_n M_0 = N_0 T_n \otimes M_0.$$

Hence
$$\dim X R_n \leq (\dim N_0 T_n)(\dim M_0)$$
and so
$$\text{GK}(N \otimes M_R) \leq \text{GK}(N_0 T_T) \leq \text{GK}(N_S). \qquad \square$$

3.15 One can obtain an alternative proof of 2.9(ii) from 3.14(ii). For suppose S is an algebra containing R such that S_R is finitely generated. Of course $_SS$ is finitely generated, so 3.14(ii) shows that $\text{GK}(_SS) = \text{GK}(S_R)$. Now $\text{GK}(S_R) \leq \text{GK}(R)$ by 3.2(iii) and yet, by definition, $\text{GK}(R) \leq \text{GK}(S)$.

3.16 The next consequence concerns right ideal invariance (see 6.8.13).

Corollary. *If R is a left Noetherian k-algebra of finite GK dimension, M_R is a right R-module and $I \triangleleft R$ then $\mathrm{GK}((M \otimes_R I)_R) \leqslant \mathrm{GK}(M_R)$.*

Proof. Clear. □

One can, therefore, using 3.11 and 3.16, apply results such as 6.8.15 to any affine algebra whose graded ring is right and left Noetherian.

3.17 The section ends by discussing a property of some affine algebras R of finite GK dimension. R is **right finitely partitive** if, given any finitely generated M_R, there is an integer $n > 0$ such that, for every chain

$$M = M_0 \supset M_1 \supset \cdots \supset M_m$$

with $\mathrm{GK}(M_i/M_{i+1}) = \mathrm{GK}(M)$, one has $m \leqslant n$.

In Sections 4 and 6, we will describe classes of algebras which are finitely partitive. However, as yet no examples are known which fail to have this property.

3.18 One could make a similar definition for any dimension function. This was not done for the Krull dimension in Chapter 6 since a right Noetherian ring automatically has the analogous property, by 6.2.21. In fact, as the results below show, this property can serve as a bridge between these two dimensions.

Proposition. *Let R be a right finitely partitive affine algebra with $\mathrm{GK}(R) < \infty$.*
(i) $\mathrm{GK}(R) \geqslant \mathcal{K}(R)$ and $\mathrm{GK}(N) \geqslant \mathcal{K}(N)$ for all finitely generated modules N_R.
(ii) Let $a, b \in \mathbb{N}$ and suppose, for all finitely generated modules M_R with $\mathcal{K}(M) = a$, that $\mathrm{GK}(M) \geqslant a + b$. Then $\mathrm{GK}(R) \geqslant \mathcal{K}(R) + b$, and if N_R is finitely generated and $\mathcal{K}(N) \geqslant a$ then $\mathrm{GK}(N) \geqslant \mathcal{K}(N) + b$.

Proof. (i) This follows from (ii) by taking $a = b = 0$.

(ii) An easy induction on $\mathcal{K}(N) - a$ establishes the result for any N_R of finite Krull dimension. However, the bound thus obtained shows that every finitely generated right R-module has finite Krull dimension. The rest is clear. □

3.19 We present one consequence here; another comes in 5.16.

Corollary. *Let R be a simple right Noetherian affine k-algebra with $0 < \mathrm{GK}(R) < \infty$ and suppose that R is right finitely partitive. Then*

$$\mathcal{K}(R_R) \leqslant \mathrm{GK}(R) - 1.$$

Proof. The hypotheses imply that each simple right R-module is infinite dimensional and so has GK dimension at least 1. Thus 3.18 applies with $a = 0$ and $b = 1$. □

3.20 Finally, for use in Chapter 15, we consider how the GK dimension behaves on passing to the associated graded module, a special case of which was dealt with in 1.14. Here, however, it is not assumed that the filtration is either finite dimensional or standard.

Lemma. *Let R be a filtered k-algebra and M_R a filtered R-module. Then $\mathrm{GK}\,(\mathrm{gr}\,M_{\mathrm{gr}\,R}) \leq \mathrm{GK}\,(M_R)$ and so $\mathrm{GK}\,(\mathrm{gr}\,R) \leq \mathrm{GK}\,(R)$.*

Proof. Let W be any finite dimensional subspace of $\mathrm{gr}\,R$ with $1 \in W$ and F be any finite dimensional subspace of $\mathrm{gr}\,M$. There exists a finite dimensional subspace V of R, with $1 \in V$, and a finite dimensional subspace E of M such that $W \subseteq \mathrm{gr}\,V$ and $F \subseteq \mathrm{gr}\,E$. Now for each n,
$$FW^n \subseteq \mathrm{gr}\,E(\mathrm{gr}\,V)^n \subseteq \mathrm{gr}\,E\,\mathrm{gr}\,(V^n) \subseteq \mathrm{gr}\,(EV^n).$$
Thus
$$\dim FW^n \leq \dim(\mathrm{gr}\,(EV^n)) = \dim EV^n$$
and so $\mathrm{GK}\,(\mathrm{gr}\,M_{\mathrm{gr}\,R}) \leq \mathrm{GK}\,(M_R)$. □

A related result, where equality holds, appears in 6.5.

§4 Almost Commutative Algebras and Hilbert Polynomials

4.1 In this section some well-known properties of commutative affine algebras will be established. In fact they will be obtained for the much larger class of almost commutative algebras R defined below.

The major step is to demonstrate that given any finitely generated M_R there is a polynomial $p(x) \in \mathbb{Q}[x]$ such that $\dim M_n = p(n)$ for $n \gg 0$. From this follow the consequences that the GK dimension is an exact dimension function for R, that $\mathrm{GK}\,(M)$ is an integer and that there is a connection between $\mathrm{GK}\,(M)$ and $\mathcal{K}(M)$.

These results will be extended to more general algebras in §6.

4.2 An **almost commutative algebra** R is an affine algebra which has some standard finite dimensional filtration with respect to which $\mathrm{gr}\,R$ is commutative. It should be noted that this condition is stronger than merely requiring that R be affine and have some associated graded ring which is commutative; see 6.10 for an explicit example.

We start by considering where almost commutative algebras come from. Of course, if R is almost commutative via a standard finite dimensional filtration

$\{R_n\}$ and gr R is the associated graded ring then gr R is affine too, generated by R_1/R_0. Thus gr R is a homomorphic image of $k[x_1,\ldots,x_n]$ for some n and so, using 1.6.9, both gr R and R are Noetherian.

4.3 In fact, R itself is the homomorphic image of an algebra met earlier.

Proposition. *An algebra R is almost commutative if and only if R is a homomorphic image of $U(g)$ for some finite dimensional k-Lie algebra g.*

Proof. ($\Rightarrow$) Since gr R is commutative it follows that $[a,b] \in R_1$ for each $a,b \in R_1$; and it is easy to check that, under this bracket product, R_1 is a Lie algebra. By the universal property of enveloping algebras (1.7.2), there is a homomorphism $U(R_1) \to R$ which is surjective since its image contains R_1.

($\Leftarrow$) Conversely, say there is a surjection $\pi: S = U(g) \to R$ with $I = \ker \pi$. If S is given the filtration with generating subspace g, R that with generating subspace $\pi(g)$ and I the filtration $I_n = S_n \cap I$, then 7.6.13 shows that gr $R \simeq$ gr S/gr I. Of course gr S is commutative and so gr R is commutative as well. □

One particular example, of course, is the Weyl algebra $A_n(k)$, as noted in the proof of 1.15(ii).

4.4 Before beginning the theory of almost commutative algebras we record some elementary facts which will be used.

Proposition. *Let $f: \mathbb{N} \to \mathbb{Q}$.*
(a) The following conditions, for $n \gg 0$, are equivalent:
  (i) $f(n) = p(n)$ for some $p \in \mathbb{Q}[x]$ with $\deg p = d$.

  (ii) $f(n) = a_d \binom{n}{d} + \cdots + a_1 \binom{n}{1} + a_0$ with $a_i \in \mathbb{Q}, a_d \neq 0$.

  (iii) $f(n+1) - f(n) = a_d \binom{n}{d-1} + \cdots + a_2 \binom{n}{1} + a_1$ with $a_i \in \mathbb{Q}, a_d \neq 0$.

(b) Granted these conditions:
  (i) p and the a_i are uniquely determined by f;
  (ii) if $f(n) \in \mathbb{Z}$ for $n \gg 0$ then each $a_i \in \mathbb{Z}$;
  (iii) if $f(n) \in \mathbb{N}$ for $n \gg 0$ then $a_d \in \mathbb{N}$.

Proof. (a) (i)$\Leftrightarrow$(ii) Note that

$$n^d - \binom{n}{d} d!$$

is a polynomial in n of degree less than d and then use induction on d.
(ii)⇒(iii) The identity
$$\binom{n+1}{j} - \binom{n}{j} = \binom{n}{j-1}$$
makes this clear.
(iii)⇒(ii) Choose m so that (iii) is valid for all $n \gg m$. Define
$$g(n) = a_d\binom{n}{d} + \cdots + a_1\binom{n}{1} + a_0,$$
where a_0 is chosen to make $g(m) = f(m)$. The fact that
$$g(n+1) - g(n) = f(n+1) - f(n)$$
now ensures that $g(n) = f(n)$ for $n \geq m$.
(b) (i) Clear.
(ii) The fact that $f(n+1) - f(n)$ is a polynomial of degree $d-1$ leads to an easy induction argument.
(iii) Clear from (ii). □

4.5 The next result provides the required polynomials.

Theorem. *Let R be an almost commutative algebra via the filtration $\{R_n\}$, and M be a finitely generated R-module with standard finite dimensional filtration $\{M_n\}$. Then there is a polynomial $p \in \mathbb{Q}[x]$ with $\deg p \leq \dim(R_1/R_0)$ such that $\dim M_n = p(n)$ for $n \gg 0$.*

Proof. Let $B = \operatorname{gr} R$. Then $\operatorname{gr} M$ is a finitely generated B-module, with generating subspace M_0 and with a filtration whose nth term is
$$M_0 \oplus M_1/M_0 \oplus \cdots \oplus M_n/M_{n-1}.$$
However, this also is the nth term of the filtration obtained by viewing B as the homomorphic image of a commutative polynomial algebra $k[x_1,\ldots,x_s]$, where B_1 is the image of the generating subspace $\sum x_i k$. Since
$$\dim M_n = \dim(M_0 \oplus M_1/M_0 \oplus \cdots \oplus M_n/M_{n-1})$$
this shows that it is enough to prove the result in the case when $R = k[x_1,\ldots,x_s]$ with generating subspace $\sum_{i=1}^s x_i k$. So $\dim R_1/R_0 = s$ and the proof proceeds by induction on s.

The case $s=0$ is trivial, so suppose $s>0$. Define $\theta: M \to M$ by $m \mapsto mx_s$, and let $K = \ker\theta$, $C = \operatorname{coker}\theta$ with the induced filtrations $\{K_n\}$ and $\{C_n\}$. As in 3.11 one can make these filtrations be standard over R and hence also over $k[x_1,\ldots,x_{s-1}]$. Moreover, by 7.6.13 the exact sequence
$$0 \to K \to M \to M \to C \to 0$$

induces the exact sequence
$$0 \to K_n/K_{n-1} \to M_n/M_{n-1} \to M_{n+1}/M_n \to C_{n+1}/C_n \to 0.$$
Let $g(n) = \dim M_n/M_{n-1}$; then
$$g(n+1) - g(n) = \dim C_{n+1}/C_n - \dim K_n/K_{n-1}.$$
If $n \gg 0$ then, by the induction hypothesis, $\dim K_n$ and $\dim C_n$ are polynomials in $\mathbb{Q}[x]$ of degree at most $s-1$. It follows by 4.4 that $\dim(K_n/K_{n-1})$ and $\dim(C_{n+1}/C_n)$ are polynomials of degree at most $s-2$. Hence so too is $g(n+1) - g(n)$. But then 4.4 shows that $g(n)$ is a polynomial of degree at most $s-1$, and another application of 4.4 shows that $\dim M_n$ is a polynomial of degree at most s. □

4.6 The polynomial p given by 4.5 is easily seen to depend upon the choices of generating subspace for R and M. For example, if $R = M = k[x]$ and $M_0 = k$ and if R_1 is chosen, in turn, to be $k + xk$ or $k + xk + x^2k$ then $p(n) = n+1$ or $2n+1$ respectively.

The polynomial p is called the **Hilbert polynomial** for M, relative to that choice, and may be written as p_M. Of course, $\deg p_M = \mathrm{GK}(M)$. This shows that $\deg p_M$ is independent of the choices made and gives the next result.

Corollary. *If R is almost commutative and M is an R-module then $\mathrm{GK}(M)$ is a non-negative integer.* □

4.7 By 4.4,
$$p(n) = \sum_{i=0}^{d} a_i \binom{n}{i} \quad \text{with } a_d \in \mathbb{N}.$$
This number a_d is called the **Bernstein number** or **multiplicity** of M, denoted by $e(M)$. The example mentioned in 4.6 demonstrates that $e(M)$ depends upon the choice of R_1. However, if R_1, and hence $\{R_n\}$, is fixed, then $e(M)$ is independent of the choice of M_0.

To see this, suppose that M'_0 is another choice, and that $q(n) = \dim M'_n$ for $n \gg 0$. Of course, $M'_0 \subseteq M_m$ for some m; hence $M'_n \subseteq M_{m+n}$ and so $q(n) \leq p(n+m)$. Conversely, $p(n) \leq q(n+m')$ for some m', from which it is evident that the leading coefficients of p and q must coincide.

4.8 These concepts are specially useful in dealing with exact sequences and chains of submodules.

Theorem. *Let R be an almost commutative algebra via the fixed standard finite dimensional filtration $\{R_n\}$ and let $0 \to L \to M \to N \to 0$ be a short exact sequence of finitely generated right R-modules. Then:*

(i) *Generating subspaces can be so chosen that* $p_M = p_L + p_N$.
(ii) $\mathrm{GK}(M) = \sup\{\mathrm{GK}(L), \mathrm{GK}(N)\}$.
(iii) *Precisely one of the following statements is true*:
  (a) $\mathrm{GK}(L) < \mathrm{GK}(N) = \mathrm{GK}(M)$ *and* $e(M) = e(N)$;
  (b) $\mathrm{GK}(N) < \mathrm{GK}(L) = \mathrm{GK}(M)$ *and* $e(M) = e(L)$;
  (c) $\mathrm{GK}(L) = \mathrm{GK}(N) = \mathrm{GK}(M)$ *and* $e(M) = e(L) + e(N)$.

Proof. (i) The proof of 3.11 shows that there is a choice of L_0, M_0, N_0 such that the sequence $0 \to L_n \to M_n \to N_n \to 0$ is exact for all n.
(ii), (iii) These follow immediately. □

4.9 Corollary. *Let R be an almost commutative algebra. Then*:
(i) *GK dimension over R is an exact dimension function*;
(ii) *R is (right and left) finitely partitive*;
(iii) *if M_R is finitely generated then* $\mathcal{K}(M) \leq \mathrm{GK}(M)$.

Proof. (i) Clear from 4.8(ii).
(ii) If $P \triangleleft M$ and $\mathrm{GK}(P) = \mathrm{GK}(M) = \mathrm{GK}(M/P)$ then $e(P) < e(M)$. Hence, there cannot be a chain of submodules $M = P_0 \supset P_1 \supset \cdots \supset P_m$ with $m > e(M)$ and $\mathrm{GK}(P_i/P_{i+1}) = \mathrm{GK}(M)$ for each i.
(iii) Apply 3.18. □

4.10 The Bernstein number can be used when subalgebras are being considered.

Proposition. *Let R, S be almost commutative algebras via the fixed standard finite dimensional filtrations $\{R_n\}$ and $\{S_n\}$ with $R \subseteq S$, and let M_S, N_R be finitely generated modules with $N_R \triangleleft M_R$. Then*:
(i) $R_1 \subseteq S_t$ *for some t*;
(ii) $\mathrm{GK}(N_R) \leq \mathrm{GK}(M_S)$;
(iii) *if* $\mathrm{GK}(N_R) = \mathrm{GK}(M_S) = d$, *say, then* $e(N_R) \leq t^d e(M_S)$.

Proof. (i), (ii) Clear.
(iii) Choose generating subspaces N_0, M_0 for N_R, M_S so that $N_0 \subset M_0$. Then $N_n \subseteq M_{nt}$ and so $p_N(n) \leq p_M(nt)$ for $n \gg 0$. Since the degrees of p_N and p_M are equal it is clear from 4.4 that $e(N_R) \leq t^d e(M_S)$. □

§5 Applications

5.1 GK dimension is one of the tools used in the study of semisimple Lie algebras, but its applications there are beyond the scope of this book and the reader is referred to [Jantzen 83]. We can, however, indicate its usefulness here by describing two applications linked to enveloping algebras. The first concerns the Weyl algebra $A_n(k)$ and shows that certain modules have finite length. The second concerns $U(\mathbf{sl}(2,\mathbb{C}))$. It is still an open problem to determine $\mathcal{K}(U(g))$ for g semisimple; but, using the GK dimension, we will show that $\mathcal{K}(U(\mathbf{sl}(2,\mathbb{C})))=2$.

Further applications to PI algebras appear later, in Chapter 13.

5.2 First we record a simple consequence of the definitions.

Lemma. *If S is a subalgebra of R and $I \triangleleft_r R$ such that $I \cap S = 0$ then $\mathrm{GK}\,(R/I)_R \geqslant \mathrm{GK}\,(S)$.*

Proof. The embedding $S \hookrightarrow R/I$ makes this clear. □

5.3 The next few results concern modules over the nth Weyl algebra $A_n(k)$.

Proposition. *If $R = A_n(k)$ and $z \in R \setminus k$ then $\mathrm{GK}\,(R/zR) = 2n - 1$.*

Proof. Let $S_1,\ldots,S_{2n}$ be the different subalgebras of R generated by all but one of the $2n$ generators. Since $\bigcap S_i = k$ then $z \notin S_i$ for some i. It follows that $zR \cap S_i = 0$ and so, by 5.2, $\mathrm{GK}\,(R/zR) \geqslant \mathrm{GK}\,(S_i)$. However, it is clear from 1.15 that $\mathrm{GK}\,(S_i) = 2n - 1$ and from 3.5 that $\mathrm{GK}\,(R/zR) \leqslant 2n - 1$. □

5.4 One might, of course, hope that such a result was true quite generally. However, if G is the discrete Heisenberg group then 2.16 shows that $\mathrm{GK}\,(kG) = 4$. In the notation of 2.16, the element $z - 1$ is a central non-unit and yet $\mathrm{GK}\,(kG/(z-1)) = 2$ since

$$kG/(z-1) \simeq k[x, x^{-1}, y, y^{-1}].$$

5.5 Proposition. *Let $R = A_n(k)$ with $\operatorname{char} k = 0$ and let M_R be non-zero. Then $\mathrm{GK}\,(M) \geqslant n$.*

Proof. The proof proceeds by induction on n, the case $n = 0$ being trivial.

Suppose the result is valid for $n - 1$, that M_R is non-zero, and that $\mathrm{GK}\,(M) < n$ and so $\mathrm{GK}\,(M) \leqslant n - 1$ by 4.6. A contradiction will ensue.

Let $S = A_{n-1}(k)$, viewed as a subalgebra of R and let N be any nonzero finitely generated S-submodule of M. Then $\mathrm{GK}\,(N_S) \leqslant \mathrm{GK}\,(M_R) \leqslant n - 1$ and yet, by the inductive hypothesis, $\mathrm{GK}\,(N_S) \geqslant n - 1$. So $\mathrm{GK}\,(N_S) = \mathrm{GK}\,(M_R) = n - 1$.

Choosing the generating subspaces of S and R to be those spanned by the generators $\{x_i, y_i\}$ it is automatic that $S_1 \subseteq R_1$. Hence, by 4.10, $e(N) \leqslant e(M)$. It follows immediately from 4.9(ii) that N_S has finite length, the length being no more than $e(M)$. Since this is valid for all such N, it must be that M_S has length at most $e(M)$. It will be seen in 9.5.5 that, under such circumstances, $\mathrm{End}\,(M_S)$ must be algebraic over k. However, since x_n and y_n centralize S and M_R is, perforce, faithful, there is an embedding $A_1(k) \hookrightarrow \operatorname{End} M_S$; and $A_1(k)$ is not algebraic by, for example, 1.3.3. □

5.6 Corollary. *For any finitely generated module M over $A_n(k)$, with $\operatorname{char} k = 0$,*

$$\operatorname{GK}(M) \geqslant \mathcal{K}(M) + n.$$

Proof. Use 3.18 and 4.9(ii) together with 5.5. □

Since $\operatorname{GK}(A_n(k)) = 2n$ this gives another proof that $\mathcal{K}(A_n(k)) = n$. Since this is not true if $\operatorname{char} k > 0$, the restriction to characteristic zero here is necessary.

5.7 Corollary. *If $\operatorname{char} k = 0$, $R = A_n(k)$ and M_R is a finitely generated module with $\operatorname{GK}(M) = n$ then M_R has finite length bounded above by $e(M)$.*

Proof. Apply 4.9 and 5.5. □

5.8 In general a module M over an algebra R is **holonomic** if $\operatorname{GK}(M) = \frac{1}{2}\operatorname{GK}(R/\operatorname{ann} M)$. In particular if $\operatorname{char} k = 0$ and $R = A_n(k)$ then M_R is holonomic if $\operatorname{GK}(M) = n$, as above. It is clear, by 5.5 and 5.3 that each simple module is holonomic when $n = 1$. However, this is not true for larger values of n—see [Stafford 85] for details.

5.9 The result at which we aim needs an extension of 5.7. This shows that growth conditions can imply that a module not assumed to be finitely generated is, in fact, of finite length. Bearing in mind 5.5, the growth is 'as small as possible'.

Lemma. *Let $\operatorname{char} k = 0$ and $R = A_n(k)$. Suppose that M_R has a filtration $\{M_t\}$ which is not necessarily standard, and that there exist $a, b \in \mathbb{N}$ such that, for all $t \gg 0$,*

$$\dim M_t \leqslant a \binom{t}{n} + b \binom{t}{n-1}.$$

Then M has finite length, at most a.

Proof. Let N be any finitely generated R-submodule of M, with standard filtration $\{N_t\}$. Then $N_0 \subseteq M_p$ for some p and therefore $N_t \subseteq M_{p+t}$ for all t. It follows that for some $c \in \mathbb{N}$

$$\dim N_t \leqslant a \binom{t}{n} + c \binom{t}{n-1}$$

and so $e(N) \leqslant a$ and, by 5.7, length $N \leqslant a$. Hence length $M \leqslant a$. □

5.10 We now come to the first of the promised applications of the GK dimension. It concerns the Weyl algebra $R = A_n(k)$, with $\operatorname{char} k = 0$, and the module

$M = R/\sum x_i R$. Of course, $M \simeq k[y_1, \ldots, y_n] = k[\mathbf{y}]$, and M is a simple R-module on which y_i acts by multiplication and x_i as $-\partial/\partial y_i$. If $0 \neq f \in k[\mathbf{y}]$ then the localization of $k[\mathbf{y}]$ with respect to $\{f^m | m \in \mathbb{N}\}$ is again an R-module, denoted by M_f.

Theorem. *With the notation above, $\operatorname{GK}(M_f) = n$ and M_f has finite length over $A_n(k)$. Indeed, length $M_f \leqslant (d+1)^n$, where $d = \deg f$.*

Proof. Let $L = M_f$. For $t \geqslant 0$ we define
$$L_t = \{gf^{-t} | g \in k[\mathbf{y}], \deg g \leqslant t(d+1)\}$$
and check first that $\{L_t\}$ is a filtration of L compatible with the usual standard filtration $\{R_t\}$ of R.

It is clear that $L_t \subseteq L_{t+1}$ and that $L_t y_j \subseteq L_{t+1}$ for each j. Moreover, if $gf^{-t} \in L_t$ then
$$\frac{\partial}{\partial y_j}(gf^{-t}) = \frac{\partial g}{\partial y_j} f^{-t} - \left(tg \frac{\partial f}{\partial y_j}\right) f^{-(t+1)} \in L_{t+1}.$$

Hence $L_t R_1 \subseteq L_{t+1}$ and so $L_t R_m \subseteq L_{t+m}$. Finally, it remains to check that $\bigcup L_t = L$. Let $gf^{-s} \in L$ and let $c = \deg g$. Then $gf^{-s} = (gf^c)f^{-(s+c)}$ and
$$\deg(gf^c) = c(d+1) \leqslant (s+c)(d+1).$$

Hence $gf^{-s} \in L_{s+c}$.

In order to apply 5.9, we consider $\dim L_t$. If $k[\mathbf{y}]$ is given the usual filtration by degree, it is clear that $\dim L_t = \dim(k[\mathbf{y}])_{t(d+1)}$. Therefore
$$\dim L_t = \binom{t(d+1)+n}{n}$$
by 1.3. This can be rewritten as a polynomial in t of degree n with leading coefficient $(d+1)^n/n!$ and hence, using 4.4,
$$\dim L_t \leqslant (d+1)^n \binom{t}{n} + b \binom{t}{n-1}$$
for some $b \in \mathbb{N}$. The theorem follows from 5.9. $\square$

5.11 The theorem above plays a major role in a proof, due to Bernstein, that certain functions on the complex half-plane can be extended to meromorphic functions on the whole plane. An account of this can be read in [Krause and Lenagan **85**, 8.12].

5.12 The second application, to determine the Krull dimension of $U(\mathfrak{sl}(2, \mathbb{C}))$, requires the following facts which can be found in [Humphreys **80**] or [Dixmier **77**]:

8.5.15   *Applications*   307

(i) if k is a field of characteristic zero then $g = \text{sl}(2,k)$ is the k-Lie algebra with basis $\{e,f,h\}$ and products $[h,e] = 2e$, $[h,f] = -2f$ and $[e,f] = h$;
(ii) g is semisimple, so all finite dimensional modules are semisimple modules;
(iii) for each $n \geq 0$, g has a unique simple module of dimension $n+1$ with a basis $\{v_0, \ldots, v_n\}$ such that

$$v_j e = j(n+1-j)v_{j-1},$$
$$v_j f = v_{j+1} \quad \text{if } j \neq n, \qquad v_n f = 0,$$
$$v_j h = (n-2j)v_j;$$

(iv) the element $q = 4ef + h^2 - 2h = 4fe + h^2 + 2h$ is central in $U(g)$ and $q - n(n+2)$ annihilates the simple module of dimension $n+1$.

5.13 In the remainder of this section, let $g = \text{sl}(2,k)$.

Lemma. *If M is a finite dimensional simple $U(g)$-module and N is a cyclic $U(g)$-module of finite length with composition factors all isomorphic to M, then length $N \leq \dim M$.*

Proof. If $P = \text{ann}_{U(g)} M$ then $NP = 0$ and so N is a homomorphic image of $U(g)/P$. However, $U(g)/P$ embeds in $\text{End}_k M$ and so has dimension at most $(\dim M)^2$. □

5.14 The first steps concern certain subalgebras.

Lemma. (i) *The subalgebras $k[e,q]$ and $k[f,q]$ of $U(g)$ are isomorphic to the commutative polynomial ring in two indeterminates and have GK dimension 2.*
(ii) *The subalgebra $k[e,h]$ is a skew polynomial ring $k[h][e;\sigma]$ with $\sigma(h) = h+2$ and has Krull dimension 2.*
(iii) $\mathcal{K}(U(g)) \geq 2$.

Proof. (i) Evidently each is a homomorphic image of the commutative polynomial ring in two indeterminates. Suppose that $0 \neq p(x,y) \in k[x,y]$ satisfies $p(e,q) = 0$ and p has minimal degree in x. Comparison of $p(e,q)$ with $[h,p(e,q)]$ leads to an immediate contradiction since $[h,e^m] = 2me^m$. Now apply 1.15. A similar argument deals with $k[f,q]$.

(ii) It is clear from 1.7.5 that $k[e,h]$ is a skew polynomial ring as described, and then 6.5.4 applies.

(iii) Using 1.7.5 again, one sees that $U(g)$ is a free module over $k[e,h]$. Therefore, by 6.5.3 and (ii), $\mathcal{K}(U(g)) \geq 2$. □

5.15 The main step comes next.

Proposition. *If M is a finitely generated $U(g)$-module with $\mathcal{K}(M) = 1$ then $\text{GK}(M) \geq 2$.*

Proof. By 6.2.20, M has a 1-critical factor module. It will suffice to consider the case when M is itself 1-critical.

By 4.6, GK (M) is an integer. It is evident from 1.17 that GK $(M) \neq 0$. Therefore, we will assume that GK $(M) = 1$ and aim to derive a contradiction. By 4.9 there is a bound, given by a Bernstein number, on the length of a chain of submodules of M which has factors of GK dimension 1. Therefore we may assume that M is cyclic and all proper factors of M have GK dimension zero and so are finite dimensional. They are, therefore, semisimple. By 5.13 there is a fixed bound on the number of copies of any particular simple module in a cyclic module of finite length. Since M is not of finite length, it must be that infinitely many distinct simple modules occur as factors of M. Consequently, by 5.12(iii), there is no bound on the dimensions of simple factors of M.

We write R for $U(g)$; then $M \simeq R/I$ for some right ideal I. If $I \cap k[e, q] = 0$ then, by 5.2 and 5.14, GK $(R/I) \geqslant$ GK $(k[e, q]) = 2$, a contradiction. Similarly, if $I \cap k[f, q] = 0$ a contradiction ensues. Therefore there are nonzero elements

$$p_1(e, q) \in I \cap k[e, q] \quad \text{and} \quad p_2(f, q) \in I \cap k[f, q].$$

We view p_1, p_2 as polynomials in e, f respectively with coefficients in $k[q]$, and of degree n_1, n_2 respectively. Since their leading coefficients can have only a finite number of roots in k, one can choose $m \in \mathbb{N}$ so that if $n \geqslant m$ with $n \in \mathbb{N}$ then $r_1 = p_1(e, n(n+2))$ and $r_2 = p_2(f, n(n+2))$ still have degree n_1 and n_2 respectively.

As proved above, there is a maximal right ideal J of R, $J \supset I$, such that $\dim (R/J) > \sup \{m, n_1 + n_2\}$. Put $n + 1 = \dim (R/J)$. By 5.12, $q - n(n+2)$ annihilates R/J; so $q - n(n+2) \in J$. However, $p_1(e, q) \in J$ and $p_2(f, q) \in J$ and therefore $r_1, r_2 \in J$.

We now view R/J in the way described in 5.12(iii). Then some element v must have annihilator J. Say $v = v_s c_s + \cdots + v_t c_t$, where $0 \leqslant s \leqslant t \leqslant n$, $c_s \neq 0$, and $c_t \neq 0$. Since $r_1, r_2 \in J$ we see that $v r_1 = v r_2 = 0$. By considering the action on v_t of the smallest power of e occurring in r_1 it becomes clear that $n_1 > t$. Similarly, $n_2 + s > n$ and hence $n_1 + n_2 > n + t - s \geqslant n$. However, $n \geqslant n_1 + n_2$, by choice, so a contradiction has been obtained. □

5.16 Theorem. *If* char $k = 0$ *then* $\mathcal{K}(U(\mathrm{sl}\,(2, k))) = 2$.

Proof. By 1.15, GK $(U(g)) = 3$. By 4.9, we see that $U(g)$ is finitely partitive and so 3.18 can be applied, using 5.15, to prove that $\mathcal{K}(U(g)) \leqslant 2$. Now use 5.14. □

§6 Somewhat Commutative Algebras

6.1 There are two main topics in this section. The main one involves the relaxation of the restriction, imposed so far, that filtrations be standard. Applied to the definition of almost commutative algebras, this will give the larger class of somewhat commutative algebras. It will be shown that the theory described in §4 extends to this class. This will have applications in Chapter 15.

The other topic concerns almost centralizing extensions which are defined below and which include crossed products by enveloping algebras. Although the GK dimension is not well behaved under almost normalizing extensions, we will see that it is under almost centralizing extensions. It will be shown that the class of somewhat commutative algebras is closed under almost centralizing extensions.

6.2 It is convenient to start with some general facts about filtrations. Given a filtered ring R, one can view gr R not only as a graded ring, with homogeneous components $(\text{gr } R)_n$, but also as a filtered ring with filtration subspaces

$$R_{[n]} = (\text{gr } R)_0 \oplus \cdots \oplus (\text{gr } R)_n.$$

We adopt the convention that $R_{[n]}$ and R_n are zero for negative n.

Lemma. *Let R be a filtered ring with gr R being a finitely generated ring extension of R_0. There exists $t \in \mathbb{N}$ such that gr R is generated, as a ring, by $R_{[t]}$. Also, for each $n > 0$,*

(a) (i) $R_{[n]} = \sum_{i=1}^{t} R_{[i]} R_{[n-i]} = \sum \{ R_{[i(1)]} \cdots R_{[i(m)]} \mid 1 \leqslant i(j) \leqslant t, \sum i(j) = n \}$,

(ii) $R_{[n]} \subseteq (R_{[t]})^n \subseteq R_{[nt]}$,

(b) (i) $R_n = \sum_{i=1}^{t} R_i R_{n-i} = \sum \{ R_{i(1)} \cdots R_{i(m)} \mid 1 \leqslant i(j) \leqslant t, \sum i(j) = n \}$,

(ii) $R_n \subseteq (R_t)^n \subseteq R_{nt}$.

Proof. The existence of t is clear.

(a)(i) Straightforward.

(ii) $(R_{[t]})^n \subseteq R_{[nt]}$ by definition. The other inclusion will be proved by induction on n, the case $n \leqslant t$ being obvious. For $n > t$,

$$(R_{[t]})^n = R_{[t]}(R_{[t]})^{n-1} \supseteq R_{[t]} R_{[n-1]}$$
$$\supseteq R_{[1]} R_{[n-1]} + R_{[2]} R_{[n-2]} + \cdots + R_{[t]} R_{[n-t]} = R_{[n]}$$

using the induction hypothesis and (i).

(b) We obtain (i) from (a)(i), and (ii) follows as above. □

6.3 Similar facts apply to a filtered R-module M. In the same way, we let

$$M_{[n]} = (\text{gr } M)_0 \oplus \cdots \oplus (\text{gr } M)_n$$

and adopt the convention that $M_{[n]}$ and M_n are zero for negative n. We note that a filtration of M is called a **good filtration** if gr M is a finitely generated gr R-module. We recall, from 7.6.11, that then M_R must be finitely generated and that finitely generated modules always have good filtrations.

6.4 Lemma. *Let R, t be as in 6.2 and let M_R be finitely generated. Let $\{M_n\}$ be a good filtration of M, and let $s \in \mathbb{N}$ be chosen so that $M_{[s]}$ generates gr M as a gr R-module. Then for each n*

$$M_{[n]} \subseteq M_{[s]} R_{[n]} \subseteq M_{[s]} (R_{[t]})^n \subseteq M_{[s]} R_{[nt]} \subseteq M_{[s+nt]}$$

and

$$M_n \subseteq M_s R_n \subseteq M_s (R_t)^n \subseteq M_s R_{nt} \subseteq M_{s+nt}.$$

Proof. The argument in 6.2 is easily extended. □

6.5 If the filtrations involved are finite dimensional, the GK dimensions are left unchanged by these gradings.

Proposition. *Let R have a finite dimensional filtration $\{R_n\}$ such that gr R is affine and let M_R have a good finite dimensional filtration $\{M_n\}$. Then*

$$\mathrm{GK}\,(R) = \gamma(\dim R_n) = \gamma(\dim R_{[n]}) = \mathrm{GK}\,(\mathrm{gr}\,R)$$

and

$$\mathrm{GK}\,(M) = \gamma(\dim M_n) = \gamma(\dim M_{[n]}) = \mathrm{GK}\,(\mathrm{gr}\,M).$$

Proof. Using 1.7(v) and 6.2, we see that $\mathrm{GK}\,(R) = \gamma(\dim R_n)$ and that $\mathrm{GK}\,(\mathrm{gr}\,R) = \gamma(\dim R_{[n]})$. However, $\dim R_n = \dim R_{[n]}$. A similar argument, using 6.4, completes the proof. □

Using this one can easily remove the condition, in 3.11 and 3.12, that the filtration $\{R_n\}$ is standard.

6.6 We now consider two rings $R \subseteq S$. We call S an **almost centralizing extension** of R if S is generated as a ring by R together with a finite set $\{x_1, \ldots, x_p\} \subseteq S \setminus R$ satisfying
 (i) $rx_i - x_i r \in R$, and
 (ii) $x_i x_j - x_j x_i \in \sum_{f=1}^{p} x_f R + R$
for each i, j and each $r \in R$.

The definitions in 1.6.10 and 1.7.12 make clear that this is a special type of almost normalizing extension and that, if R is a k-algebra and g a finite dimensional k-Lie algebra, then any crossed product $R * U(g)$ is an almost centralizing extension of R (see also 6.21).

6.7 The next result shows that if R has a filtration as in 6.5 then so too has S. (It is worth noting that when $S = R * U(g)$ the standard filtration

$$S_0 = R, S_1 = R \oplus \sum \{x_i R \mid x_i \in g\}$$

8.6.7 Somewhat commutative algebras

will not be appropriate since S_0 will not usually be finite dimensional.)

Proposition. *Let R, S be k-algebras with R having a finite dimensional filtration $\{R_n\}$ such that gr R is affine and with S being an almost centralizing extension of R generated by $x_1, \ldots, x_p$. Let t be as specified in 6.2 and let x^i denote the standard monomial $x_1^{i(1)} \cdots x_p^{i(p)}$.*

(i) *There exists $h > t$ such that*
$$rx_i - x_i r \in R_{h-1} \quad \text{and} \quad x_i x_j - x_j x_i \in R_{h-1} + \sum_f x_f R_{h-1}$$
for all i, j and all $r \in R_t$.

(ii) *S has a finite dimensional filtration given by $S_n = R_n$ for $0 \leq n \leq h-1$,*
$$S_h = R_h + \sum x_f k \quad \text{and} \quad S_n = \sum \{S_i S_j | 0 < i < n, i+j = n\} \quad \text{for } n > h.$$

(iii) *The inclusion $R \hookrightarrow S$ induces a graded ring homomorphism $\vartheta : \operatorname{gr} R \to \operatorname{gr} S$.*

(iv) *gr S is affine, being generated over im ϑ by the central elements $\bar{x}_1, \ldots, \bar{x}_p$.*

(v) $\operatorname{GK}(R) \leq \operatorname{GK}(S) \leq \operatorname{GK}(R) + p.$

(vi) *If $R \cap \sum \{x^i R | i \neq 0\} = 0$ then $\vartheta : \operatorname{gr} R \to \operatorname{gr} S$ is an inclusion.*

(vii) *gr S is the polynomial ring over gr R in the central indeterminates $\bar{x}_1, \ldots, \bar{x}_p$ if and only if the set $\{x^i\}$ is a free basis for S_R. In this case $\operatorname{GK}(S) = \operatorname{GK}(R) + p$.*

Proof. (i), (ii) Clear.

(iii) By construction, the inclusion $R \hookrightarrow S$ is a filtered homomorphism.

(iv) It is clear that gr S is generated by $S_{[h]}$; so gr S is affine. Note that ϑ restricts to an isomorphism $R_{[h-1]} \to S_{[h-1]}$. Furthermore, since $R_h = \sum_{i=1}^{t} R_i R_{h-i}$, one sees that $(\operatorname{gr} S)_h$ is contained in the algebra generated by $S_{[h-1]} + \sum_f k \bar{x}_f$. Hence gr S is generated by the image of $R_{[h-1]}$ together with $\bar{x}_1, \ldots, \bar{x}_p$ and so by im ϑ together with $\bar{x}_1, \ldots, \bar{x}_p$.

It follows from 6.2 that gr S is also generated by the image of $R_{[t]}$ together with $\bar{x}_1, \ldots, \bar{x}_p$. The relations in (i) ensure that each $\bar{x}_i$ commutes with these generators, and thus is central.

(v) Of course $\operatorname{GK}(R) \leq \operatorname{GK}(S)$. Now (iv) shows that $\operatorname{GK}(\operatorname{gr} S) \leq \operatorname{GK}(\operatorname{gr} R) + p$; and 6.5 demonstrates that $\operatorname{GK}(\operatorname{gr} R) = \operatorname{GK}(R)$ and $\operatorname{GK}(\operatorname{gr} S) = \operatorname{GK}(S)$.

(vi) Here ϑ is an embedding if and only if $S_{n-1} \cap R_n = R_{n-1}$ for each n. We suppose that ϑ is not an embedding and choose some n such that $S_{n-1} \cap R_n \neq R_{n-1}$. It is evident from the description of S_{n-1} in (ii) that $n \geq h$. Choose $r \in (S_{n-1} \cap R_n) \setminus R_{n-1}$. Combining 6.2(b)(i) with the relations in (i), (ii) one can write r in the form
$$r = \sum x^i r_i$$

with each $x^i r_i$ in S_{n-1} and each $r_i \in R_{n-1}$. Then
$$0 \neq r - r_0 \in R \cap \sum \{x^i R | i \neq \mathbf{0}\},$$
a contradiction.

(vii) This is straightforward. □

Of course, all these facts are applicable to $R * U(g)$ when R is affine; cf. 2.10.

6.8 We note in particular that 6.7(v) is false for almost normalizing extensions as Examples 2.16 and 2.18 show.

6.9 We now turn to the main topic in this section. We define a k-algebra R to be **somewhat commutative** if it has a finite dimensional filtration such that gr R is a commutative affine k-algebra. By 6.2 it is clear that R itself is affine.

By definition, any almost commutative algebra is somewhat commutative. When combined with the preceding theory, this provides a large fund of examples.

Proposition. *A k-algebra which is an almost centralizing extension of a somewhat commutative k-algebra is somewhat commutative.* □

6.10 In particular, if R is a commutative affine k-algebra then $S = R * U(g)$ is somewhat commutative. However, it is not, in general, almost commutative. For example, let $R = k[y]$ and $S = R[x; -y^2 \partial/\partial y]$. Then $S \simeq R * U(g)$ with $\dim_k g = 1$ yet, as will be proved in 14.3.9, S is not almost commutative.

6.11 We can now outline how the results of §4 will be extended to a finitely generated module M over a somewhat commutative k-algebra R. Using the given finite dimensional filtration $\{R_n\}$ of R and some good filtration of M we consider gr R and gr M. Since gr R is commutative affine, the results of §4 are applicable. They provide, given some choice of standard finite dimensional filtrations of gr R and gr M, a Hilbert polynomial p which, in turn, gives GK (gr M) and e(gr M). We already know, from 6.5, that GK (gr M) = GK (M); so this, at least, is independent of the choices made.

The remaining results will demonstrate that e(gr M) is independent of the choices of filtration of M and gr M although, from §4, it is clearly dependent upon the choices of filtration of R and gr R.

Once this is established, versions of 4.8 and 4.9 will easily be deduced.

6.12 Two filtrations $\{M_n\}$, $\{M'_n\}$ are **equivalent**, or of **bounded difference**, if there exist $a, b \in \mathbb{N}$ such that for all n
$$M_{n-a} \subseteq M'_n \subseteq M_{n+b}.$$

6.13 We will be interested specifically in good filtrations.

Proposition. *Let R be a filtered ring such that $\operatorname{gr} R$ is a finitely generated ring extension of R_0 and let M_R be finitely generated. Then any two good filtrations of M are equivalent.*

Proof. Let $\{M_n\}$, $\{M'_n\}$ be the two good filtrations. Let s be as specified by 6.4 and choose $a \geq s$ so that $M_s \subseteq M'_a$. Then, for all n,
$$M_n \subseteq M_s R_n \subseteq M'_a R_n \subseteq M'_{a+n};$$
and symmetry completes the argument. □

6.14 Some obvious examples of equivalent filtrations are provided by defining a **translate** of $\{M_n\}$ to be a filtration $\{M'_n\}$ with $M'_n = M_{n-c}$ for some $c \in \mathbb{N}$. This gives a filtration equivalent to $\{M_n\}$; and $\operatorname{gr} M'$ is a translate of $\operatorname{gr} M$ (see 7.6.2). In particular $\operatorname{gr} M' \simeq \operatorname{gr} M$ as $\operatorname{gr} R$-modules.

6.15 We will see below that the use of translates reduces the study of equivalent filtrations to the case when $a = 0$, using the notation of 6.12. The next result will then enable us to reduce further to the situation where $b = 1$.

Lemma. *Let R be a filtered ring and let M_R have equivalent filtrations $\{M_n\}$, $\{M'_n\}$ such that, for some $b \geq 0$,*
$$M_n \subseteq M'_n \subseteq M_{n+b}$$
for all n. Define $\{M''_n\}$ by $M''_n = M'_n \cap M_{n+1}$. Then $\{M''_n\}$ is a filtration, $M_n \subseteq M''_n \subseteq M_{n+1}$, and $M''_n \subseteq M'_n \subseteq M''_{n+b-1}$ for all n.

Proof. Straightforward. □

6.16 In the case when $b = 1$, certain graded modules arise.

Lemma. *Let R be a filtered ring and let M_R have equivalent filtrations $\{M_n\}$ and $\{M'_n\}$ such that $M_n \subseteq M'_n \subseteq M_{n+1}$ and $M_0 = M'_0 = 0$. There are graded $\operatorname{gr} R$-modules*
$$Y = \bigoplus_{n=0}^{\infty} M_{n+1}/M'_n \quad \text{and} \quad Z' = \bigoplus_{n=0}^{\infty} M'_{n+1}/M_{n+1} \subseteq Z = \bigoplus_{n=0}^{\infty} M'_n/M_n$$
with $Z \simeq Z'$ as a $\operatorname{gr} R$-module, and short exact sequences of graded modules
$$0 \to Z \to \operatorname{gr} M \to Y \to 0$$
$$0 \to Y \to \operatorname{gr} M' \to Z' \to 0.$$

Proof. It is evident that Y, Z', Z are graded modules. Also, for each n, there are exact sequences
$$0 \to M'_n/M_n \to M_{n+1}/M_n \to M_{n+1}/M'_n \to 0$$
and
$$0 \to M_{n+1}/M'_n \to M'_{n+1}/M'_n \to M'_{n+1}/M_{n+1} \to 0$$
which yield the cited sequences of graded modules. □

6.17 Next we consider the prime radical of the annihilator ideal.

Proposition. *Let R be a filtered ring and $\{M_n\}$, $\{M'_n\}$ be equivalent filtrations of M_R. Then $N(\mathrm{ann}_{\mathrm{gr}\,R}(\mathrm{gr}\,M)) = N(\mathrm{ann}_{\mathrm{gr}\,R}(\mathrm{gr}\,M'))$.*

Proof. Let A, A' denote these two annihilator ideals. If $\{M'_n\}$ is a translate of $\{M_n\}$ then $\mathrm{gr}\,M \simeq \mathrm{gr}\,M'$ and so $A = A'$ and $N(A) = N(A')$. This enables one to reduce to the case described in 6.15 and hence to that described in 6.16. The short exact sequences make it clear that $A^2 \subseteq A'$ and $(A')^2 \subseteq A$; so $N(A') = N(A)$. □

6.18 We can now apply these ideas, as promised in 6.11.

Theorem. *(a) Let R have a filtration such that $\mathrm{gr}\,R$ is commutative affine and let M_R have good filtrations $\{M_n\}$, $\{M'_n\}$. Then $\mathrm{GK}\,(\mathrm{gr}\,M) = \mathrm{GK}\,(\mathrm{gr}\,M')$.*
(b) Choose standard finite dimensional filtrations of $\mathrm{gr}\,R$, $\mathrm{gr}\,M$ and $\mathrm{gr}\,M'$.
  (i) The leading terms of the Hilbert polynomials obtained for $\mathrm{gr}\,M$ and $\mathrm{gr}\,M'$ are equal.
  (ii) $e(\mathrm{gr}\,M) = e(\mathrm{gr}\,M')$.

Proof. (a) Let $A = \mathrm{ann}\,\mathrm{gr}\,M$. Evidently $\mathrm{GK}\,(\mathrm{gr}\,M) = \mathrm{GK}\,((\mathrm{gr}\,R)/A)$ if $\mathrm{gr}\,M$ is cyclic; and the exactness provided by 4.9(i) extends this to the finitely generated case. Also
$$\mathrm{GK}\,((\mathrm{gr}\,R)/A) = \mathrm{GK}\,((\mathrm{gr}\,R)/N(A))$$
by 3.12 applied to $\mathrm{gr}\,R$. Hence, by 6.13 and 6.17, $\mathrm{GK}\,(\mathrm{gr}\,M) = \mathrm{GK}\,(\mathrm{gr}\,M')$.

(b) Arguing as in the proof of 6.17, one reduces to the case described in 6.16. If 4.8(iii) is applied to the two short exact sequences given by 6.16, it demonstrates that $e(\mathrm{gr}\,M) = e(\mathrm{gr}\,M')$. This proves (ii); and (i) follows from (a). □

6.19 Combining all these results gives the required facts for somewhat commutative algebras.

Theorem. *Let R be a somewhat commutative k-algebra and M_R a finitely generated module with a good filtration. Choose standard finite dimensional filtrations of $\mathrm{gr}\,R$ and $\mathrm{gr}\,M$, and let p be the resulting Hilbert polynomial of $\mathrm{gr}\,M$. Then:*

(i) $\deg p = \mathrm{GK}(M) = \mathrm{GK}(\mathrm{gr}\, M) \in \mathbb{N}$. *This integer is independent of the specific choices of filtrations of R, M, $\mathrm{gr}\, R$ and $\mathrm{gr}\, M$.*
(ii) *For a fixed choice of the filtrations of R and $\mathrm{gr}\, R$, the leading term of p is independent of the choice of filtrations of M and $\mathrm{gr}\, M$.*

Proof. (i) Combine 6.5 and 6.18.
(ii) Clear from 6.18. □

6.20 In these circumstances it now makes sense to define $e(M)$ to be the leading coefficient of the Hilbert polynomial of $\mathrm{gr}\, M$, where we fix the filtrations of R and $\mathrm{gr}\, R$.

Corollary. *If R is a somewhat commutative algebra then the statements in Theorem 4.8(ii), (iii) and Corollary 4.9 are all valid.*

Proof. Given a short exact sequence

$$0 \to L \to M \to N \to 0$$

of finitely generated right R-modules we can use 7.6.13, as in the proof of 3.11, to obtain finite dimensional filtrations such that

$$0 \to \mathrm{gr}\, L \to \mathrm{gr}\, M \to \mathrm{gr}\, N \to 0$$

is exact and $\mathrm{gr}\, M$ is finitely generated. Hence $\mathrm{gr}\, L$ and $\mathrm{gr}\, N$ also are finitely generated. By 4.8(i), given a standard finite dimensional filtration of $\mathrm{gr}\, R$, there are standard finite dimensional filtrations of $\mathrm{gr}\, M$, $\mathrm{gr}\, L$ and $\mathrm{gr}\, N$ such that

$$p_{\mathrm{gr}\, M} = p_{\mathrm{gr}\, L} + p_{\mathrm{gr}\, N}.$$

One can now argue as in 4.8 and 4.9. □

6.21 Finally, we return briefly to the algebra $S = R * U(g)$ with R a commutative affine k-algebra. The standard, but not finite dimensional, filtration mentioned in 6.7 makes $\mathrm{gr}\, S$ a commutative affine k-algebra. This suggests another weakening of the definition of almost commutative algebras which would also include this example. Our final result, however, demonstrates that this class is already included in the class of somewhat commutative algebras.

Proposition. *The following conditions upon a k-algebra S are equivalent:*
(i) *S is an almost centralizing extension of a commutative affine k-algebra R;*
(ii) *S has a standard filtration with respect to which $\mathrm{gr}\, S$ is a commutative affine k-algebra.*

Proof. (i)$\Rightarrow$(ii) Take $S_0 = R$ and $S_1 = R + \sum_i x_i R$. Then $\mathrm{gr}\, S$ is generated over R by the central elements $\bar{x}_1, \ldots, \bar{x}_p$.

(ii)⇒(i) Set $R = S_0$. Note that S_1 must be finitely generated as an R-module, say by $x_1, \ldots, x_p$. The commutativity of gr S ensures that these elements satisfy the conditions of 6.6. □

§7 Additional Remarks

7.0 (a) [Gelfand and Kirillov **66, 66a**] contained a conjecture about quotient division rings of enveloping algebras, namely that, for g algebraic,

$$Q(U(g)) \simeq Q(A_n[z_1, \ldots, z_m])$$

where $z_1, \ldots, z_m$ are central indeterminates. The GK dimension arose from the aim to show that the integers m, n here were uniquely determined by the division ring. The dimension was then axiomatized and its basic properties analysed in [Borho and Kraft **76**].

(b) This account shares some features with that in [Krause and Lenagan **85**], in part because there were interchanges of ideas during the writing of this chapter and that book. We have exploited [Krause and Lenagan **85**] here as a ready source of additional examples and results.

(c) The GK dimension reappears frequently in later chapters, especially 9, 13, 14 and 15. It is also used and discussed in [Jantzen **83**] and [Joseph **80**]. Results related to the Gelfand–Kirillov conjecture appear in Chapter 14, see 14.7.6 for example.

7.1 (a) In fact the alternative description of $\gamma(f)$ given in 1.6 is the more common definition elsewhere.

(b) Much of the material in this section comes from [Borho and Kraft **76**] although the proof of 1.18 is new and 1.20 is due to [Jategaonkar **69**] and [Kosevoi **70**]. In relation to 1.20, see [Irving and Small **83**].

(c) It is shown in [Small, Stafford and Warfield **85**] that an affine k-algebra R with GK $(R) = 1$ is rather special. It must be a PI ring, $N(R)$ is nilpotent and $R/N(R)$ is a finite module over its centre which is Noetherian.

(d) Other rates of growth, such as exponential and subexponential, are discussed in [M. K. Smith **77**] and [Krause and Lenagan **85**].

7.2 (a) 2.3 appears in [Borho and Kraft **76**] except that the right-hand inequality is due to [Warfield **84**]. So too is 2.5. Walters, unpublished, proved 2.10, [Lorenz **82**] contains 2.11 and 2.12, and 2.13 comes from [Borho and Kraft **76**].

(b) 2.18(i) combines results of [Milnor **68, 68a**], [Wolf **68**] and [Gromov **81**]; and 2.18(ii) is due to [Bass **72**].

7.3 (a) 3.5 appears in [Borho **82**], 3.11 in [Tauvel **82**], 3.14 in [Lenagan **81**] and [Borho **82**], and 3.18 is in [S. P. Smith **81**]. 3.9 and 3.13 are unpublished results of J. C. McConnell and L. W. Small respectively. Further results related to 3.12, 3.13 appear in [Bergman P], [Lorenz and Small **82**], 3.11 and 3.12 are improved in 6.5.

(b) In 15.1.8 there is a description of the module of differentials and the corresponding universal derivation. In a similar vein, given ideals I, J of a k-algebra R, one can construct an $(R/I, R/J)$-bimodule M together with a universal k-derivation $d: R \to M$. This is described in [Lewin **74**] where it is proved that the kernel of the map

$$R \to \begin{bmatrix} R/I & M \\ 0 & R/J \end{bmatrix}, \quad r \mapsto \begin{bmatrix} [r+I] & d(r) \\ 0 & [r+J] \end{bmatrix}$$

is IJ, as required in 3.13.

8.7.6 *Additional remarks* 317

(c) One could, alternatively, define R to be right finitely partitive if, given any finitely generated module M_R, any chain

$$M = M_0 \supset M_1 \supset \cdots \supset M_n \supset \cdots$$

with GK (M_i/M_{i+1}) = GK (M) must terminate. This weaker condition is clearly enough for 3.18 and 3.19. However, the stronger condition is satisfied by the examples of interest to us.

7.4 (a) Almost commutative algebras were introduced by [Duflo **73**] where 4.3 appears. They are studied and used in Chapters 9 and 14.

(b) The remainder of the section derives from [Bernstein **71, 72**].

(c) 4.5 can be modified to cover the case when $\{M_n\}$ is a good filtration.

7.5 5.10 is due to [Bernstein **72**] and 5.16 to [S. P. Smith **81**]. For further results see [Levasseur **82, 86**].

7.6 (a) The material in this section is new although it is influenced by [Bernstein **71**] which, in particular, contains 6.17. The ideal N(ann gr M) is the **characteristic ideal** of M.

(b) Let R be a filtered k-algebra such that gr R is commutative affine; and M be a finitely generated R-module with a good filtration. In the light of 6.5, it is natural to ask if GK (M) = GK (gr M). This is true if $R = A_n(k)$ with the standard filtration $R_0 = k[y_1,\ldots,y_n]$, $R_1 = R_0 + \sum x_i R_0$ (see [Bernstein **71**, Theorem 3.1] or [Bjork **79**, Chapter 3, A.2.4]). It is also true for S as in 6.7(vii) this having unpublished proofs due to McConnell and to Stafford. In general it is unknown.

(c) There are circumstances other than those of 4.9 and 6.20 in which the GK dimension is known to be an exact dimension function, namely, when the algebra is a Noetherian PI ring; see [Krause and Lenagan **85**, 10.20].

(d) Some related results appear in [Van den Essen **86**].

Part III

EXTENSIONS

Chapter 9
THE NULLSTELLENSATZ

One formulation of the Hilbert Nullstellensatz states that if R is a commutative affine algebra over a field k and I is a maximal ideal then R/I is finite dimensional over k. One could interpret this as a result about the endomorphism ring of each simple module and it is in that guise that it appears in this chapter.

The results presented here concern algebras over either a field k or, more generally, a commutative ring K — and **that notation will remain fixed throughout the chapter**. The first two sections present and discuss some properties which, together, will comprise 'the Nullstellensatz', firstly over k, then over K.

The question of which algebras have the properties is not approached directly but rather via another property, generic flatness. This is defined and studied in §3, and its connections with the Nullstellensatz are made clear. Since generic flatness is preserved under many types of extension, this enables one to show in §4 that many of the algebras of Chapter 1 do satisfy the Nullstellensatz and, as §5 shows, have quite strong properties on endomorphism rings.

The final section provides a Nullstellensatz in a rather different setting, namely polynomial rings over division rings.

§1 Algebras Over a Field

1.1 This section introduces two properties, of a ring R, and of an algebra R over a field k, which are suggested by different versions of the commutative Nullstellensatz. It then shows how these properties are linked, and that if R is countably generated and k is uncountable then both hold. This result covers several of the rings described in Chapter 1, and later sections will substantially extend it.

1.2 The first property to be defined is ring-theoretic. A ring R has the **radical property** if the Jacobson radical of each factor ring of R is nil.

This property is closely related to the standard notion of a **Jacobson ring**, this being a ring in which every prime ideal is an intersection of primitive ideals or, equivalently, every prime factor ring has zero Jacobson radical.

Lemma. (i) *Each Jacobson ring has the radical property.*
(ii) *A ring whose prime factor rings are all right Goldie has the radical property if and only if it is a Jacobson ring.*

Proof. (i) This follows from the fact that the prime radical of a ring is nil.
(ii) By 2.3.2, prime right Goldie rings have no nonzero nil ideals. □

Of course, all prime factor rings are right Goldie if R is either commutative or right Noetherian or a PI ring (see 13.6.5).

1.3 The next result shows that this property is preserved under finite module extensions of right Noetherian rings.

Proposition. *Let R be a right Noetherian subring of S with S_R finitely generated.*
(i) $J(S) \cap R \subseteq J(R)$.
(ii) *If R is a Jacobson ring then so too is S.*

Proof. (i) Let $y \in J(S) \cap R$ and set $1 - y = z$; so $z^{-1} \in S$. Since S_R is Noetherian, there exists n such that $z^{-(n+1)} \in \sum_{i=0}^{n} z^{-i} R$. Therefore

$$z^{-(n+1)} = r_0 + z^{-1} r_1 + \cdots + z^{-n} r_n$$

for some $r_i \in R$, and $z^{-1} = z^n r_0 + \cdots + r_n \in R$. Hence $J(S) \cap R \subseteq J(R)$.
(ii) To show S is Jacobson it is enough to consider the case when S is prime and to prove then that $J(S) = 0$. Note first, by (i) and the hypothesis, that $J(S) \cap R$ is nilpotent. Therefore

$$\mathcal{K}(R_R) = \mathcal{K}(R/(J(S) \cap R)_R) \leq \mathcal{K}(S/J(S)_R) \leq \mathcal{K}(S_R) = \mathcal{K}(R_R)$$

using 6.3.3 and 6.2.5. In particular $\mathcal{K}(S/J(S)_R) = \mathcal{K}(S_R)$. However, S is a right Noetherian prime ring, using 1.1.3, and so if $J(S) \neq 0$ then $J(S)$ contains a regular element c. The argument of 6.3.9 demonstrates that $\mathcal{K}(S/cS_R) < \mathcal{K}(S_R)$ which provides the contradiction that $\mathcal{K}(S/J(S)_R) < \mathcal{K}(S_R)$. □

1.4 Next suppose that R is an algebra over k. Then R has the **endomorphism property over** k if, for each simple module M_R, $\operatorname{End} M$ is algebraic over k. If, in addition, R has the radical property then R **satisfies the Nullstellensatz over** k.

1.5 Before demonstrating a link between these properties an elementary result is required.

Lemma. *Let R be a ring, $R[x]$ the polynomial ring in a central indeterminate and $a \in R$. Then a is nilpotent if and only if $(1 - ax)R[x] = R[x]$.*

Proof. If $a^{n+1} = 0$ then $(1 - ax)(1 + ax + \cdots + a^n x^n) = 1$. Conversely, if
$$(1 - ax)(1 + a_1 x + \cdots + a_n x^n) = 1$$
then $a_i = a^i$ and $a^{n+1} = 0$. □

1.6 Proposition. *Let R be a k-algebra such that $R[x]$ has the endomorphism property over k. Then R satisfies the Nullstellensatz over k.*

Proof. Note first that the endomorphism property passes to factor rings. Therefore R too has the endomorphism property; and also any factor ring of R satisfies the hypothesis. So it is enough to show that $J(R)$ is nil.

Let $a \in J(R)$ and suppose a is not nilpotent. By 1.5, $(1 - ax)R[x] \neq R[x]$, so there is a maximal right ideal I containing $(1 - ax)R[x]$. Let $M = R[x]/I$ and $m_0 = [1 + I] \in M$; and let $d \in \text{End } M$ be the map $m \mapsto mx$ for $m \in M$.

Note that $0 = m_0(1 - ax) = m_0 - dm_0 a$ and so $d \neq 0$. The subfield $k(d^{-1})$ of End M is algebraic, and hence finite dimensional, over k. Therefore $d = g(d^{-1})$ for some polynomial g over k. Now $d^{-1}m_0 = m_0 a$, so

$$dm_0 = g(d^{-1})m_0 = m_0 g(a) \quad \text{and} \quad 0 = m_0 - dm_0 a = m_0(1 - g(a)a).$$

However, this contradicts the fact that, since $a \in J(R)$, $1 - g(a)a$ is a unit. □

1.7 In one special case the endomorphism property is automatic.

Proposition. *If k is an uncountable field and R is a countably generated k-algebra then R has the endomorphism property over k.*

Proof. Let D be the division ring End M. Since R is countably generated, M has a countable k-basis. Given $0 \neq m \in M$ there is a k-vector space embedding $D \to M$ given by $d \mapsto dm$. This shows that D has a countable k-basis.

Suppose that $\alpha \in D$ is not algebraic over k; so $k(\alpha)$, the subfield of D generated by α, is isomorphic to $k(x)$, the field of rational functions. However, if $\{c_i | i \in I\}$ is any uncountable subset of k, the set $\{(x - c_i)^{-1} | i \in I\}$ is linearly independent over k since, if $\sum_{i=1}^{n} a_i(x - c_i)^{-1} = 0$ for some $n \in \mathbb{N}$ and some $a_i \in k$, then

$$a_1(x - c_2)(x - c_3) \cdots (x - c_n) + \cdots + a_n(x - c_1) \cdots (x - c_{n-1}) = 0.$$

Hence D is algebraic over k. □

1.8 Corollary. *Let k be an uncountable field.*
(i) *If R is a countably generated k-algebra then R satisfies the Nullstellensatz over k.*
(ii) *The following algebras all satisfy the Nullstellensatz over k and are Noetherian Jacobson rings:*
  (a) $A_n(k)$, *the Weyl algebra*;

(b) $U(g)$, where g is a finite dimensional Lie algebra;
(c) kG, where G is a polycyclic by finite group.

Proof. (i) Since $R[x]$ is also countably generated, 1.6 combines with 1.7 to yield this.
(ii) This is clear from (i) and 1.2. □

1.9 That 1.8(ii) is valid over any field k, and for more general algebras, is the point of the next three sections. However, we will now show that the restriction, in 1.7 and 1.8(i), that k be uncountable is necessary. The construction is based on the following example.

Example. *Let $F = k\langle x, y\rangle$ be the free algebra over any field k and let $S = k + xF$. Then S is a free k-algebra on the generators $\{xy^i | i = 0, 1, \ldots\}$.*

Proof. Let $a_i = xy^i$; it is clear that $\{a_i\}$ generates S and that the sum $\sum Fa_i$ is direct. Suppose there are nontrivial relations on the a_i. Amongst these, choose one involving the shortest possible words in the a_i and, subject to that, having minimal support. Since $\sum Fa_i$ is direct, each word in the relation must belong to the same summand, Fa_i; but then one can cancel the last letter a_i contradicting the minimality. □

1.10 Proposition. *Let k be a countable field, let F, S be as above and let T be the matrix ring*

$$\begin{bmatrix} S & xF \\ F & F \end{bmatrix}.$$

Then T has a homomorphic image which is an affine prime k-algebra whose Jacobson radical is not nil.

Proof. It is readily checked that T is affine, being generated over k by $\{e_{11}, e_{12}x, e_{21}, e_{22}, e_{22}y\}$ so any homomorphic image will be affine.

Consider now the commutative local domain $L = k[t]_{(t)}$ for some indeterminate t. This is countably generated over k, so is a homomorphic image of S; thus $L \simeq S/P$ for some $P \in \operatorname{Spec} S$. Evidently $P \not\supseteq xF$; so 3.6.2 and the proof of 3.6.5 show that T has a prime ideal Q such that $e(T/Q)e \simeq S/P \simeq L$, where $e = [e_{11} + Q]$. Let $R = T/Q$ and M_R be simple; then Me is either zero or else a simple eRe-module. In each case $MeJ(eRe) = 0$. Hence $0 \neq J(eRe) \subseteq eJ(R)e$ and so $J(R) \neq 0$. Finally, if I were a nonzero nil ideal of R then eIe would be a nonzero nil ideal of eRe; so R has no nonzero nil ideal. □

1.11 In fact this algebra R has no nonzero nil right ideals. That is clear from the next result.

Lemma. *If a prime ring R has a nonzero nil right ideal I and a nonzero idempotent e then eRe has a nonzero nil right ideal.*

Proof. Note that $0 \neq RI \triangleleft R$ so $0 \neq eRIe \triangleleft eRe$. Choose $a \in R$, $i \in I$ such that $0 \neq eaie$, and let $b \in eRe$. Since I is nil, $(iba)^n = 0$ for some n. But then

$$(eaib)^{n+1} = ea(iba)^n ib = 0.$$

Hence $eaieRe$ is a nonzero nil right ideal of eRe. □

1.2 Next we note one application of the endomorphism property. We recall from 6.5.8 and 7.5.8, that if R is a right Noetherian ring and $\dim R < \infty$, where $\dim R$ denotes right Krull dimension or right global dimension, then

$$\dim R + n \leqslant \dim A_n(R) \leqslant \dim R + 2n.$$

Moreover $R = k$ and $R = Q(A_n(k))$ show that these extremes can occur even in characteristic zero.

Proposition. *Let k be a field of characteristic zero and R be a Noetherian k-algebra with $\dim R < \infty$ such that $R[t]$, the polynomial ring in a central indeterminate, has the endomorphism property over k. Then $\dim A_1(R) = \dim R + 1$.*

Proof. Let $H = R[t]$. First we consider $H \otimes_{k[t]} k(t)$. Now if M is a simple H-module then, by hypothesis, t acts algebraically on M. So $Mf = 0$ for some $0 \neq f \in k[t]$ and then $M \otimes_{k[t]} k(t) = 0$. Hence, by Proposition 6.5.3 and 7.4.4,

$$\dim H \otimes_{k[t]} k(t) < \dim H = \dim R + 1.$$

Next, as in 6.6.15, we use two localizations of $A_1(R)$, namely

$$B_1 = A_1(R) \otimes_{k[y]} k(y) = (R[y] \otimes_{k[y]} k(y))[x; \delta]$$

and

$$B_2 = A_1(R) \otimes_{k[x]} k(x) = (R[x] \otimes_{k[x]} k(x))[y; \delta].$$

It is clear from the above that $\dim (B_1 \oplus B_2) \leqslant \dim R + 1$. As noted in 7.2.4 the diagonal embedding $A_1(R) \hookrightarrow B_1 \oplus B_2$ is faithfully flat, so $\dim A_1(R) \leqslant \dim (B_1 \oplus B_2)$. □

1.13 Similar results hold for $A'_1(R)$ and also, if k is any field and $\lambda \in k$ is nonzero and not a root of unity, for $P_\lambda(R)$. Further comment will be made in 4.23.

1.14 An extension of 1.12 will be useful later.

Proposition. *Let k be a field of characteristic 0 and H be a Noetherian k-algebra which satisfies the endomorphism condition. Suppose that $S = H[x; \delta]$ and that $\delta(y) = 1$ for some $y \in Z(H)$. Then $K(S) = K(H)$ and if $\mathrm{gld}\, H < \infty$ then $\mathrm{gld}\, S = \mathrm{gld}\, H$.*

Proof. Since $A_1(k)$ is simple, the obvious map $A_1(k) \to S$ is injective and so y is an indeterminate. We consider two localizations of S, with respect to $\mathcal{T} = k[y]\setminus\{0\}$, and with respect to $\mathcal{D}$, the set of monic polynomials in x, as described in 7.9.2. As in 1.12 one sees that $\dim H_{\mathcal{T}} < \dim H$; and it follows from 7.9.4 and 7.9.9 that $\dim S_{\mathcal{D}} = \dim H$.

Now consider the diagonal embedding $S \to S_{\mathcal{T}} \oplus S_{\mathcal{D}}$; it will be enough, by 6.5.3 and 7.2.6, to show this is faithfully flat. So let $I \triangleleft_1 S$ such that $S_{\mathcal{T}}I = S_{\mathcal{T}}$ and $S_{\mathcal{D}}I = S_{\mathcal{D}}$. Then there exists $0 \neq f \in k[y] \cap I$ and also $I \cap \mathcal{D}$ contains a monic polynomial, of degree m say. Let

$$I_j = I + H + Hx + \cdots + Hx^{j-1} \quad \text{for } j = 0, \ldots, m;$$

these are H-submodules of S with $I_0 = I$, $I_m = S$. It is clear that $I_j x + I_j = I_{j+1}$ and induction on j shows that $fI_{j+1} \subseteq I_j$. Therefore, $f^m S \subseteq I$ and thus the annihilator of S/I in the subalgebra $A_1(k)$ is nonzero. However, $A_1(k)$ is simple; so $S/I = 0$ as required. □

§2 Algebras Over Rings

2.1 This section extends the ideas of §1 to algebras over a commutative ring K.

2.2 First consider a K-algebra R; that means, of course, that there is a ring homomorphism, θ say, from K to the centre of R. Evidently R is an algebra over $\operatorname{im}\theta$ and often one can reduce to the case where $K = \operatorname{im}\theta \subseteq R$. However, even if θ is not an embedding, if X is a subset of R the notation $X \cap K$ will be used to denote $\theta^{-1}(X \cap \operatorname{im}\theta)$.

2.3 If $P \in \operatorname{Spec} R$ and $p = P \cap K$ then $p \in \operatorname{Spec} K$. If, whenever P is primitive then p is maximal, we say R has the **primitive property over** K; and if, further, each R/P has the endomorphism property over the field K/p then R has the **endomorphism property over** K. Finally, if R also has the radical property then R **satisfies the Nullstellensatz over** K.

2.4 Apart from the primitive property these are natural extensions of the earlier definitions. The primitive property itself is, of course, held by every K-algebra when K is a field.

2.5 As 1.6 indicated, the behaviour of $R[x]$ influences R. This phenomenon, which recurs throughout the chapter, is studied next, starting with a rather special result.

Lemma. *If R is an algebra over a field k and $R[x]$ has the primitive property over $k[x]$, then R has the endomorphism property over k.*

Proof. Let M_R be simple with $M = mR$ and let $d \in \text{End } M_R$. We make M into an $R[x]$-module by letting $mx = dm$. The primitive property ensures that $(\text{ann } M_{R[x]}) \cap k[x]$ is maximal and so contains a nonzero element $f(x)$ say. However, $f(d)m = mf(x) = 0$ and so $f(d) = 0$ as required. □

2.6 The main result of this section echoes 1.6.

Theorem. *Let R be a K-algebra.*
(i) *If $R[x]$ has the primitive property over K then R has the primitive property over K.*
(ii) *If $R[x]$ has the endomorphism property over K then R satisfies the Nullstellensatz over K.*

Proof. (i) Let P be a primitive ideal of R. Then $P + xR[x]$ is a primitive ideal of $R[x]$ and so $(P + xR[x]) \cap K$ is a maximal ideal. However, $(P + xR[x]) \cap K = P \cap K$.

(ii) The same sort of argument shows immediately that R has the endomorphism property over K, so it is enough to establish the radical property in R. Since any factor ring of R satisfies the same hypothesis, we need only show that $J(R)$ is nil.

Let $a \in J(R)$ and suppose $(1 - ax)R[x] \neq R[x]$. Let I be a maximal right ideal of $R[x]$ containing $(1 - ax)R[x]$; and let $M = R[x]/I$, $P = \text{ann } M$ and $p = P \cap K$. So p is a maximal ideal and, by 1.6, the image of a in $R[x]/pR[x]$ is nilpotent. Then 1.5 shows that

$$R[x] \supset I \supseteq (1 - ax)R[x] + pR[x] = R[x].$$

This contradiction means that $(1 - ax)R[x] = R[x]$ and so, by 1.5 again, a is nilpotent. □

§3 Generic Flatness

3.1 The key to our study of the Nullstellensatz in noncommutative algebras over a commutative ring is the notion of generic flatness. This, together with some stronger variants, is described in this section and its relevance to the Nullstellensatz demonstrated. The following section will provide many examples of algebras which have all these properties.

3.2 The definition of generic flatness starts with the case when K is an integral domain, R is a K-algebra and M_R is a module. If $0 \neq f \in K$, then K_f denotes the localization of K at the powers of f, $R_f = R \otimes_K K_f$, and

$$M_f = M \otimes_R R_f = M \otimes_K K_f.$$

We say R is **generically flat over** K if for each finitely generated M_R there exists $0 \neq f \in K$ such that M_f is free over K_f.

In general, with K not assumed to be an integral domain, if $p \in \operatorname{Spec} K$ and R/pR is generically flat over K/p, then R is **generically flat at** p. However, if $p = 0$ we will continue to say 'generically flat over K'. By convention, if $pR = R$ then R is generically flat at p.

Although 'generically free' might seem a more natural name, this terminology was established for geometric reasons.

3.3 Of course every k-algebra is generically flat over k. As noted earlier, §4 will provide many examples over any K. However, not all algebras are generically flat. For example $k(x)$ is not generically flat over $k[x]$ since it is clear (or use 8.2.14) that, for any $0 \neq f \in k[x]$, $k(x) \neq k[x]_f$ and so $k(x)$ is not free over $k[x]_f$.

3.4 We now start to study generic flatness.

Lemma. *If R is a K-algebra which is generically flat over an integral domain K then any factor ring S of R has the same property.*

Proof. Any finitely generated right S-module is also a finitely generated right R-module. □

3.5 The next result shows that in checking generic flatness it is enough to consider cyclic modules.

Lemma. *Let R be a K-algebra over an integral domain K. If, for each cyclic module N_R there exists $0 \neq f \in K$ such that N_f is free over K_f, then R is generically flat over K.*

Proof. Suppose $M_R = \sum_{i=1}^n m_i R$. The proof proceeds by induction on n, the case $n = 1$ being given. We can, therefore, assume the existence of $0 \neq v, w \in K$ such that $(m_1 R)_v$ is free over K_v and $(M/m_1 R)_w$ is free over K_w. Let $f = vw$; then both $(m_1 R)_f$ and $(M/m_1 R)_f$ are free over K_f. Hence the short exact sequence

$$0 \to (m_1 R)_f \to M_f \to (M/m_1 R)_f \to 0$$

splits and M_f is free over K_f. □

3.6 In applying generic flatness, some bimodule techniques are useful. The point here is that if R is a K-algebra and M_R is faithful over K then $K \subseteq \operatorname{End} M$. Furthermore, if E is a commutative ring such that $K \subseteq E \subseteq \operatorname{End} M$, then M is an (E, R)-bimodule and hence a right $R \otimes_K E$-module. Also, $R \otimes_K E$ is an E-algebra and M is faithful over E.

3.7 Lemma. *Let K be an integral domain, R a K-algebra, M a finitely generated right R-module faithful over K and E an integral domain with $K \subseteq E \subseteq \text{End } M_R$. If $0 \neq f \in K$ then $K_f \subseteq E_f \subseteq \text{End}(M_f)_{R_f}$.*

Proof. The remarks in 3.6 mean we may suppose $E = K$. It is clear that M_f is a right R_f-module and it remains to show that K_f acts faithfully upon it. Suppose, then, that $kM_f = 0$ for some $k \in K$. Therefore, if $\{m_1, \ldots, m_n\}$ forms a generating set for M_R, there must exist $s(i)$ such that $kf^{s(i)}m_i = 0$ for each i. It follows that $kf^s M = 0$ for $s = \sup\{s(i)\}$. Thus $kf^s = 0$ and $k = 0$. □

3.8 Generic flatness provides free modules—when the next result applies.

Lemma. *If ${}_S M_R$ is a bimodule such that ${}_S M$ is free then*
(i) the map $I \mapsto IM$ embeds $\mathscr{L}(S_S)$ into $\mathscr{L}(M_R)$, and
(ii) $K(S_S) \leq K(M_R)$. □

3.9 Recall that a commutative integral domain K is a **G-domain** if, for some $0 \neq f \in K$, the ring K_f is a field. More generally a prime ideal of a commutative ring is a **G-ideal** if the factor by it is a G-domain.

Evidently any maximal ideal of a commutative ring K is a G-ideal; conversely, if K is a Jacobson ring, then each G-ideal is maximal. Thus $k[x]$ is not a G-domain.

3.10 We now turn to consequences of generic flatness.

Theorem. *Let K be a commutative ring, and let R be a K-algebra which is generically flat at all $p \in \text{Spec } K$.*
(i) If P is a primitive ideal of R then $P \cap K$ is a G-ideal of K.
(ii) If K is a Jacobson ring then R has the primitive property over K.

Proof. (i) By passing to the factor rings $K/P \cap K$ and $R/(P \cap K)R$ it may be assumed that $P \cap K = 0$. Let M_R be a simple module with $\text{ann } M = P$. Then M_K is faithful. By hypothesis, M_f is free over K_f for some $0 \neq f \in K$. By 3.7, K_f acts faithfully on M_f which is simple over R_f. Hence, by 3.8, K_f must be a field.
(ii) Clear from (i). □

3.11 Next we turn to the endomorphism property.

Theorem. *Let R be a k-algebra such that $R[y]$ is generically flat over $k[y]$ and let M_R be a module of finite length. Then $\text{End } M_R$ is algebraic over k.*

Proof. Let $0 \neq d \in \text{End}(M_R)$ and suppose d is not algebraic over k. So $k[y] \hookrightarrow \text{End}(M_R)$ via $y \mapsto d$. By hypothesis, there exists $0 \neq f \in k[y]$ such that

M_f is free over $k[y]_f$. Therefore, by 3.7 and 3.8, $k[y]_f$ must be Artinian, and so be a field. This, however, is false (see 3.3); so d is algebraic over k. □

3.12 Before turning to the Nullstellensatz, terminology is introduced which reflects the influence on R of properties of polynomial rings.

Suppose that R is a K-algebra over a commutative ring K and that $p \in \operatorname{Spec} K$. Then R is (n, m)-**generically flat at** p if $(R/pR)[x_1, \ldots, x_n, y_1, \ldots, y_m]$ is generically flat over $(K/p)[y_1, \ldots, y_m]$, these being polynomial rings in central indeterminates. When $p = 0$ we continue to say that R is (n, m)-**generically flat over** K. Thus, for example, the condition on R in 3.11 is that R is $(0, 1)$-generically flat over k.

3.13 These strands can now be combined to give a criterion for satisfying the Nullstellensatz, based solely upon generic flatness.

Theorem. *Let K be a commutative Jacobson ring and let R be a K-algebra which is $(1, 0)$- and $(1, 1)$-generically flat at all primes of K. Then R satisfies the Nullstellensatz over K.*

Proof. Let M be a simple $R[x]$-module, P be the primitive ideal $\operatorname{ann} M$ and $p = P \cap K$. By 3.10, $R[x]$ has the primitive property over K and so p is maximal. By hypothesis, $(R/pR)[x, y]$ is generically flat over $(K/p)[y]$. Therefore, by 3.11, $\operatorname{End} M$ is algebraic over K/p. Thus $R[x]$ has the endomorphism property over K and so 2.6 applies. □

3.14 This section ends with more results about endomorphism rings. First we note a useful lemma.

Lemma. *Let K, E be commutative integral domains such that $K \subseteq E \subseteq Q$, where Q is the quotient field of K. If there is a nonzero E-module M which is free over K then $K = E$.*

Proof. It is easy to check that M_E is torsion-free and that if m is a basis element for M_K then $mE = mK$. □

3.15 Proposition. *Let R be a k-algebra, M_R be finitely generated and D be a division ring with $k \subseteq D \subseteq \operatorname{End} M$. If R is $(0, 1)$-generically flat over k then D is algebraic over k.*

Proof. Suppose $0 \neq d \in D$ with d not algebraic over k. Then $k(y) \hookrightarrow \operatorname{End} M$ via $y \mapsto d$. We know that M_f is free over $k[y]_f$ for some $0 \neq f \in k[y]$, and 3.7 shows that $k[y]_f \subseteq k(y)_f \subseteq \operatorname{End} M_f$. Hence, by 3.14, $k[y]_f = k(y)_f = k(y)$ which, as noted in 3.3, is not true. Therefore D is algebraic over k. □

3.16 If, for some n and all $m \in \mathbb{N}$, R is (n,m)-generically flat at $p \in \operatorname{Spec} K$ then R is $(n, \mathbb{N})$-**generically flat at** p. In the same way one also defines $(\mathbb{N}, m)$- and $(\mathbb{N}, \mathbb{N})$-**generic flatness at** p.

3.17 Corollary. *Let R be a k-algebra which is $(0, \mathbb{N})$-generically flat over k, let M_R be finitely generated, and let A be any affine commutative k-subalgebra of $\operatorname{End} M$. Then $\mathcal{K}(A) \leqslant \mathcal{K}(M_R)$.*

Proof. We know from 8.2.14 that A contains a polynomial subalgebra $K = k[y_1, \ldots, y_m]$, where $\mathcal{K}(A) = m$. Let $R' = R \otimes_k K = R[y_1, \ldots, y_m]$; so R' is a K-algebra and, as noted in 3.6, M is a finitely generated right R'-module. By hypothesis there exists $0 \neq f \in K$ such that M_f is free over K_f. Therefore 3.8 shows that $\mathcal{K}(K_f) \leqslant \mathcal{K}((M_f)_{R'})$. However, it is clear that $\mathcal{K}((M_f)_{R'}) \leqslant \mathcal{K}(M_R)$ and, by 8.2.14, $\mathcal{K}(K_f) = m = \mathcal{K}(A)$. $\square$

§4 Constructible Algebras

4.1 The aim of this section is to show that many of the K-algebras described in Chapter 1 are $(\mathbb{N}, \mathbb{N})$-generically flat at all primes of K. Consequently, when K is a Jacobson ring they all satisfy the Nullstellensatz.

4.2 In demonstrating generic flatness, 3.5 shows that one can concentrate on cyclic modules. Our proofs depend upon the investigation of cyclic modules over a particular type of extension ring S of a given ring R; namely one generated, over R, by an element y such that $yR = Ry$. (This does not assume that y is an indeterminate.)

As in 1.2.9 one can define the leading right ideal I_n of a right ideal I of S by

$$I_n = \{r \in R \mid ry^n + s \in I \quad \text{for some } s \in \sum_{i=0}^{n-1} y^i R\}.$$

4.3 These are used in investigating the cyclic module S/I.

Lemma. *With the notation above:*
(i) $I_0 \subseteq I_1 \subseteq \cdots \subseteq I_n \subseteq \cdots$;
(ii) *the module $M = (S/I)_R$ has a chain of submodules*

$$0 = M_{-1} \subseteq M_0 \subseteq \cdots \subseteq M_n \subseteq \cdots \subseteq M$$

with $M = \bigcup M_n$, and with $M_n/M_{n-1} \simeq R/I_n$ as a module over any subring of R whose elements commute with y.

Proof. Both parts are easily checked, taking $M_n = (I + \sum_{i=0}^n Ry^i)/I$, and noting that $[ry^n + I] \in M_{n-1}$ if and only if $r \in I_n$. $\square$

4.4 Lemma. *Let T be any ring, let M_T be a module having a chain of submodules*

$$0 = M_{-1} \subseteq M_0 \subseteq \cdots \subseteq M_n \subseteq \cdots \subseteq M$$

with $M = \bigcup M_n$, and let $N = \bigoplus_{i=0}^{\infty} M_i/M_{i-1}$. If N_T is free then $M \simeq N$ and so M_T is free.

Proof. It is clear that M_i/M_{i-1} is projective and so $M_i = M_{i-1} \oplus N_i$ with $N_i \simeq M_i/M_{i-1}$. Hence $M \simeq \bigoplus N_i \simeq N$. □

4.5 Before proceeding to demonstrate that various K-algebras R are $(\mathbb{N},\mathbb{N})$-generically flat at all prime ideals p of K, we note that this is often less daunting than at first appears. For one thing, by passing to R/pR and K/p it is enough to deal with the zero ideal of an integral domain K. Frequently results will be stated only in this form.

Secondly, if the hypotheses on R and K remain valid under polynomial extension of R, then generic flatness implies $(\mathbb{N},0)$-generic flatness. Similarly, if the hypotheses remain valid under simultaneous polynomial extension of R and K, then generic flatness implies $(0,\mathbb{N})$-generic flatness.

4.6 There is one result which is special to Noetherian rings.

Proposition. *Let K be an integral domain and R be a right Noetherian K-algebra which is generically flat over K. Then R is $(\mathbb{N},0)$-generically flat over K.*

Proof. Clearly, it is enough to show that $S = R[x]$ is generically flat over K. Let $I \triangleleft_r S$ and note that the chain $\{I_i\}$ of leading right ideals given by 4.3 must terminate, say with $I_t = I_{t+1} = \ldots$. One can choose $0 \neq f \in K$ so that each module $(R/I_i)_f$ is free over K_f for $i = 0, \ldots, t$ and hence for all i. Therefore $\bigoplus_{i=0}^{\infty} (R/I_i)_f$ is free, and so, by 4.3 and 4.4, $(S/I)_f$ must be free. □

4.7 This has an easy corollary which will, however, be superseded by 4.14 and 4.20.

Corollary. *Any commutative affine algebra R over a Noetherian ring K is $(\mathbb{N},\mathbb{N})$-generically flat at all primes of K.*

Proof. We may suppose K to be an integral domain. Of course K is generically flat over K so, by 4.6, K is $(\mathbb{N},0)$-generically flat over K. Since R is a homomorphic image of a polynomial algebra over K, R must be generically flat over K. The remarks in 4.5 then show that R is $(\mathbb{N},\mathbb{N})$-generically flat over K. □

4.8 The next few results concern related pairs of algebras.

9.4.11 *Constructible algebras* 333

Lemma. *Suppose that R, S are algebras over an integral domain K, that $R \subseteq S$ with S_R being finitely generated and that R is (n, m)-generically flat over K. Then so too is S.*

Proof. The remarks in 4.5 show it is enough to consider the case when R is generically flat. Since a finitely generated S-module is also finitely generated over R, this is immediate. □

4.9 Note that a **filtered K-algebra** S is a K-algebra with a filtration $\{S_i\}$ whose components S_i are K-submodules of S. If $K \subseteq S$ this is merely saying that $K \subseteq S_0$.

Proposition. *Let S be a filtered K-algebra over an integral domain K and suppose that $\mathrm{gr}\, S$ is (n, m)-generically flat over K. Then so too is S.*

Proof. As usual, we can take $n = m = 0$. Let M_S be finitely generated. By 7.6.11, M can be filtered so that $\mathrm{gr}\, M$ is finitely generated over $\mathrm{gr}\, S$. Therefore, for some $0 \neq f \in K$, $(\mathrm{gr}\, M)_f$ is free over K_f. However, $(\mathrm{gr}\, M)_f = \mathrm{gr}(M_f)$ if M_f is given the filtration $\{(M_i)_f\}$. It follows from 4.4 that $M_f \simeq \mathrm{gr}(M_f)$ as a K_f-module, and so is free. □

4.10 Next we turn towards almost normalizing extensions (see 1.6.10). It is at this point that $(\mathbb{N}, 0)$-generic flatness, rather than generic flatness is important.

Proposition. *Let R, S be K-algebras over an integral domain K with $R \subseteq S$ and S generated over R by an element y such that $Ry = yR$, and suppose that R is $(\mathbb{N}, m)$-generically flat over K. Then so too is S.*

Proof. It is enough to deal with the case $m = 0$; and to prove that, it will suffice to prove that S is generically flat.

Let $I \triangleleft_r S$, $M = S/I$, and $\{I_i\}$ be the leading right ideals of I. By 3.5 we need only show that M_f is free over K_f for some f. Let $N = R[x]/\sum_{i=0}^{\infty} I_i x^i$. This is a cyclic $R[x]$-module. By hypothesis, N_f is free over K_f for some $0 \neq f \in K$. Hence by 4.3 and 4.4, M_f is free over K_f. □

4.11 Corollary. *Let R, S be K-algebras over an integral domain K such that S is an almost normalizing extension of R. If R is $(\mathbb{N}, m)$-generically flat over K then so too is S.*

Proof. In 1.6.14 it was seen that S can be filtered so that $\mathrm{gr}\, S$ is obtainable from R by a finite number of extensions of the form described in 4.10. Hence $\mathrm{gr}\, S$ is $(\mathbb{N}, m)$-generically flat over K and then 4.9 applies. □

4.12 These results can all be combined. We will say that a K-algebra S is **constructible** from a subalgebra R if it can be obtained from R by a finite number of extensions, each being either an almost normalizing extension or a finite module extension. If $R = K$, then S is called a **constructible K-algebra**.

For example, it is clear from Chapter 1 that if g is a K-Lie algebra of finite rank then a K-algebra $R*U(g)$ is constructible from R; and if G is a polycyclic by finite group then a K-algebra $R*G$ is constructible from R.

4.13 Corollary. *Suppose K is an integral domain, and S is constructible from the K-algebra R. If R is $(\mathbb{N}, m)$-generically flat over K then so too is S.* □

4.14 If 4.7, applied to K itself, is combined with 4.13 one obtains:

Corollary. *Any constructible K-algebra over a Noetherian ring K is $(\mathbb{N}, \mathbb{N})$-generically flat at all primes of K.* □

The next few results will show, using methods also required in §6, that the Noetherian hypothesis here is redundant.

4.15 First some notation and terminology is required. We consider the set $\mathscr{W}$ of all words in commuting indeterminates $x_1, \ldots, x_t$; so each $w \in \mathscr{W}$ has the form $w = x_1^{n(1)} \cdots x_t^{n(t)}$ and then has **total degree** $n(1) + \cdots + n(t)$. The set $\mathscr{W}$ is ordered by using first the total degree and then, subject to that, a lexicographic ordering. (For example $x_1 < x_2 < x_1^2 x_3 < x_1 x_3^2$.) This well orders $\mathscr{W}$.

4.16 Now let R be a ring, $S = R[x_1, \ldots, x_t]$, the polynomial ring in central indeterminates, and $I \triangleleft_r S$. We define the **leading right ideals** $I(w)$ of I, for $w \in \mathscr{W}$, by setting

$$I(w) = \{r \in R \mid rw + s \in I \quad \text{for some } s \in \sum \{Rv \mid v \in \mathscr{W}, v < w\}\}.$$

4.17 The same argument as in 4.3 gives the next result.

Lemma. *S/I has an ascending chain $\{M(w) \mid w \in \mathscr{W}\}$ of R-modules, ordered by $\mathscr{W}$, whose union is S/I and with composition factors isomorphic to $R/I(w)$, for $w \in \mathscr{W}$ in the given order.* □

4.18 If $\mathscr{W}$ is as above and $u, v, w \in \mathscr{W}$ are such that $uv = w$, we write $u|w$. Suppose now that $\mathscr{S}$ is some subset of $\mathscr{W}$. If $w \in \mathscr{S}$ and there are no words $u \in \mathscr{S}$, other than w, such that $u|w$ then w is a **minimal element** of $\mathscr{S}$.

4.19 Lemma. *(i) Any infinite sequence $\mathscr{T}$ of words from $\mathscr{W}$ has an infinite subsequence in which each word divides the next.*
(ii) Any subset $\mathscr{S}$ of $\mathscr{W}$ has only finitely many minimal elements.

9.4.23 Constructible algebras

Proof. (i) Clearly $\mathscr{T}$ has a subsequence $\mathscr{T}_1$ of words whose first index (degree in x_1) is nondecreasing. Iteration produces a subsequence $\mathscr{T}_t$ in which each index is nondecreasing—as required.

(ii) Order the set of minimal elements of $\mathscr{S}$ using the ordering inherited from $\mathscr{W}$. If the sequence $\mathscr{T}$ thus obtained were infinite it would have an infinite subsequence as described in (i). This contradicts their minimality. □

4.20 Theorem. *Any constructible K-algebra over a commutative ring K is $(\mathbb{N}, \mathbb{N})$-generically flat at all primes of K.*

Proof. The earlier arguments show that we may suppose K to be an integral domain and need only demonstrate that K is $(n, 0)$-generically flat.

Let $S = K[x_1, \ldots, x_n]$, $I \triangleleft S$ and $I(w)$ be the leading ideals of I. Let

$$\mathscr{S} = \{w \in \mathscr{W} \mid I(w) \neq 0\}$$

and let $w_1, \ldots, w_t$ be the minimal elements of $\mathscr{S}$. Note that if $w_i | w$ then $I(w_i) \subseteq I(w)$; so if we choose $0 \neq f \in I(w_1) \cap \cdots \cap I(w_t)$ then

$$(K/I(w))_f = \begin{cases} 0 & \text{if } w \in \mathscr{S}, \\ K_f & \text{if } w \notin \mathscr{S}. \end{cases}$$

It follows from 4.17 and 4.4 that since $\oplus \{(K/I(w))_f \mid w \in \mathscr{W}\}$ is free over K_f then so too is $(S/I)_f$. □

4.21 We end this section by returning to the Nullstellensatz.

Theorem. *If K is a commutative Jacobson ring and R is a constructible K-algebra then (i) R satisfies the Nullstellensatz over K, and (ii) R is a Jacobson ring.*

Proof. (i) This is immediate from 4.20 and 3.13.

(ii) Let P be a prime ideal of R. By passing to R/P and $K/P \cap K$ we may suppose that R is prime and K is an integral domain. By (i) we know that $J(R)$ is nil; we need to show it is nilpotent.

Let F be the quotient field of K and $Q = R \otimes_K F$. Then Q is a constructible F-algebra and so is a Noetherian prime ring. By 2.3.7, this has no nonzero nil ideals. However, $J(R)$ generates a nil ideal of Q; so $J(R) = 0$. □

4.22 Corollary. *Let R be a constructible K-algebra over a commutative Jacobson ring K. Let $\mathfrak{g}$ be a K-Lie algebra of finite rank and G be a polycyclic by finite group. Then the K-algebras $A_n(R)$, $R*U(\mathfrak{g})$ and $R*G$ all satisfy the Nullstellensatz.* □

4.23 As in 1.12 we let dim denote Krull or global dimension.

Corollary. *Let k be a field of characteristic zero and R be a constructible k-algebra with $\dim R < \infty$. Then $\dim A_n(R) = \dim R + n$.* □

§5 Finite Dimensional Endomorphism Rings

5.1 This section will show that if M is a simple module over a constructible k-algebra then $\operatorname{End} M$ is finite dimensional over k. The methods involved are rather different to those of earlier sections. They rely on one property of constructible algebras, namely that they remain Noetherian under field extensions.

5.2 The section starts with some results about fields and division rings.

Proposition. *Let D be a division ring with centre Z and let C be a maximal subfield of D. If $\dim C_Z < \infty$ then $\dim D_C < \infty$ and $\dim D_Z < \infty$.*

Proof. Note first that if ${}_R M$ is a simple module over a left Artinian ring R and $S = R/\operatorname{ann} M$, then S is simple Artinian. Hence $S \simeq M_n(E)$ for the division ring $E = \operatorname{End}({}_S M) = \operatorname{End}({}_R M)$ and $M \simeq E^n$.

We will apply this with $R = D \otimes_Z C$ and ${}_R M = D$ via $(d \otimes c)(d') = dd'c$. Since C is finite dimensional over Z, ${}_D R$ is finite dimensional and so R is left Artinian. Evidently ${}_R M$ is simple and $E = \operatorname{End}{}_R M \subseteq \operatorname{End}_D M \simeq D$ with E consisting of right multiplications by those elements of D which commute with C. By maximality, $E = C$ and so D_C is finite dimensional.

Finally $\dim D_Z = (\dim D_C)(\dim C_Z)$. □

5.3 Proposition. *Let C be an extension field of a field k. Then $C \otimes_k C$ is a Noetherian ring if and only if C is a finitely generated field extension of k.*

Proof. ($\Leftarrow$) If $C = k(\alpha_1, \ldots, \alpha_n)$ then $C \otimes_k C$ is a localization of its subring $C \otimes_k k[\alpha_1, \ldots, \alpha_n]$ which clearly is Noetherian.

($\Rightarrow$) Suppose, to the contrary, that C is not a finitely generated field extension of k. Then there is an infinite chain of subfields of C.

$$k \subset k(1) \subset k(2) \subset \cdots \subset C.$$

Consider the map

$$\theta_i : C \otimes_{k(i)} C \to C \otimes_{k(i+1)} C.$$

If $x \in k(i+1) \setminus k(i)$ then

$$0 \neq 1 \otimes x - x \otimes 1 \in \ker \theta_i.$$

Hence if I_i denotes the kernel of the map $C \otimes_k C \to C \otimes_{k(i)} C$ then $\{I_i\}$ is a strictly ascending chain of ideals of $C \otimes_k C$. □

5.4 The applications of these results to constructible k-algebras comes via the next result.

Proposition. *Let R be a k-algebra, M_R a finitely generated module and C be a subfield of $\mathrm{End}\,(M_R)$ containing k. If $R \otimes_k C$ is right Noetherian then C is a finitely generated field extension.*

Proof. The proof centres on $M \otimes_k C$. Note first that this is a finitely generated right $(R \otimes_k C)$-module via $(m \otimes c)(r \otimes c') = mr \otimes cc'$ and is therefore Noetherian. Also, it is a left $(C \otimes_k C)$-module via $(c_1 \otimes c_2)(m \otimes c) = c_1 m \otimes c_2 c$; and since M is free over C then $M \otimes_k C$ is free over $C \otimes_k C$. By 3.8 the map $I \mapsto I(M \otimes_k C)$ gives an embedding of the lattice of ideals of $C \otimes C$ into the lattice of $R \otimes C$-submodules of $M \otimes C$. Thus $C \otimes C$ is Noetherian and 5.3 applies. □

5.5 We now combine this result with results of earlier sections. Note first that if R is a constructible k-algebra and $k \subseteq C$, a field, then $R \otimes_k C$ is a constructible C-algebra and so is Noetherian. Note also that if C is both algebraic over k and a finitely generated field extension of k, then C must be finite dimensional over k.

Theorem. *Let R be a constructible k-algebra and M, N be finitely generated right R-modules.*
  (i) *If D is a division ring with $k \subseteq D \subseteq \mathrm{End}\,M$ then D is finite dimensional over k.*
  (ii) *If M_R is simple then $\mathrm{End}\,M$ is finite dimensional over k.*
  (iii) *If M, N have finite length then $\mathrm{Hom}\,(M, N)$ is finite dimensional over k.*

Proof. (i) Clear from the remarks above together with 3.15 and 4.14.
  (ii) Clear from (i).
  (iii) Let M', N' be simple submodules of M, N respectively. One can obtain the diagram below which has exact rows and columns.

$$\begin{array}{ccccc}
& 0 & & 0 & & 0 \\
& \downarrow & & \downarrow & & \downarrow \\
0 \to \mathrm{Hom}\,(M/M', N') & \to & \mathrm{Hom}\,(M, N') & \to & \mathrm{Hom}\,(M', N') \\
& \downarrow & & \downarrow & & \downarrow \\
0 \to \mathrm{Hom}\,(M/M', N) & \to & \mathrm{Hom}\,(M, N) & \to & \mathrm{Hom}\,(M', N) \\
& \downarrow & & \downarrow & & \downarrow \\
0 \to \mathrm{Hom}\,(M/M', N/N') & \to & \mathrm{Hom}\,(M, N/N') & \to & \mathrm{Hom}\,(M', N/N')
\end{array}$$

An easy induction on length completes the proof. □

This also applies to $A_n(R)$, $R * U(\mathfrak{g})$ and $R * G$, with G polycyclic by finite, and so, in particular, to $A_n(k)$, $U(\mathfrak{g})$ and kG.

§6 Polynomials Over Division Rings

6.1 The emphasis now changes from algebras over commutative rings to polynomials in central indeterminates over a division ring D. The main result is that any simple module over the ring $D[x_1,\ldots,x_n]$ is finite dimensional over D. This has some consequences concerning primitive rings.

6.2 The arguments centre on the division ring D, its centre k and the polynomial rings $R = D[x]$ and $S = D[x_1,\ldots,x_n]$. **This notation will now remain fixed.** The aim is to use the techniques of 4.15–4.19 to relate properties of S to properties of R. Therefore we start with some results concerning R.

6.3 Lemma. (i) *R is a principal right and left ideal domain.*
(ii) *Each ideal A of R is generated by a monic polynomial $\alpha \in k[x]$; and A is maximal if and only if α is irreducible in $k[x]$.*

Proof. (i) See 1.2.9.
(ii) If $A \neq 0$, then A contains a polynomial α which has minimal possible degree and is monic. If $0 \neq d \in D$ then $\alpha - d^{-1}\alpha d$ must be zero; so $\alpha d = d\alpha$ and $\alpha \in k[x]$. The rest is clear. □

6.4 The next result shows that any nonzero element of R has a regular image in almost all simple factor rings. (Recall, by 2.3.4, that right and left regularity coincide here.)

Proposition. *Suppose that $0 \neq p \in R$, and that $A_1,\ldots,A_n$ are maximal ideals of R for some n with $n > \deg p$. Then, for some j, the image of p in R/A_j is regular.*

Proof. We proceed by induction on $\deg p$. If $\deg p = 0$ it is obvious. We will assume it holds for polynomials of degree less than $\deg p$ and that it fails for p, and then derive a contradiction.

For each i, let α_i be a central generator for A_i. By assumption, there exists $g_i \in R \setminus A_i$ such that $g_i p \in A_i$. Now $pR + A_n = qR$ for some $q \in R$, and then $p = qr$, $\alpha_n = qs$ for some $r, s \in R$. Note that $qR \neq R$ or else

$$g_n \in g_n qR \subseteq g_n pR + A_n = A_n.$$

It follows that $\deg q \geq 1$ and so $\deg r < n - 1$. By the induction hypothesis, r must have a regular image in R/A_j for some $j \in 1,\ldots,n-1$. However, $g_j qr = g_j p \in A_j$; therefore $g_j q \in A_j$. Finally, note that since α_j and α_n are irreducible elements of $k[x]$, $1 = \alpha_j \beta + \alpha_n \gamma$ for some $\beta, \gamma \in k[x]$. Then

$$g_j = g_j(\alpha_j\beta + \alpha_n\gamma) = g_j\alpha_j\beta + g_j qs\gamma \in A_j$$

which is the contradiction sought. □

6.5 Corollary. *Let $p_1,\ldots,p_m$ be a finite set of nonzero elements of R. Then there is a maximal ideal A such that the image of each p_i in R/A is regular.*

Proof. There are infinitely many maximal ideals in $k[x]$. □

6.6 We now turn to the relationship between S and R, where R is identified with the subring $D[x_n]$ of S. Thus elements of S are viewed as taking the form $s = \sum r_w w$ with $r_w \in R$ and $w \in \mathcal{W}$, the set of all words in $x_1,\ldots,x_{n-1}$ ordered as in 4.15. The largest w with $r_w \neq 0$ is the **degree** of s, and r_w is then the **leading coefficient**.

6.7 The first result concerns the leading right ideals as defined in 4.16.

Lemma. *If $I \triangleleft_r S$ then there are only finitely many distinct right ideals of R of the form $I(w)$ for some $w \in \mathcal{W}$.*

Proof. Let $\mathcal{T}$ be a sequence of words from $\mathcal{W}$ chosen so that $\{I(t)|t\in\mathcal{T}\}$ is a set of distinct right ideals. By 4.19, if $\mathcal{T}$ were infinite it would have an infinite subsequence $\mathcal{T}'$ in which each word divides the next. However, then the corresponding leading right ideals would form a strictly ascending chain in R. □

6.8 The main result at which we are aiming is (ii) of the next theorem.

Theorem. *If I is any maximal right ideal of S then*:
 (i) $I \cap R \neq 0$;
 (ii) S/I is finite dimensional over D;
 (iii) $I \cap R$ is an isomaximal right ideal of R.

Proof. (i) By 6.7 there are only finitely many distinct leading right ideals of I, generated by $p_1,\ldots,p_m$ say. Then 6.5 provides a maximal ideal A of R modulo which each p_i is regular. Say $A = \alpha R$ with $\alpha \in k[x_n]$. If $\alpha \in I$ the result is obvious. So we assume that $\alpha \notin I$ in which case $\alpha S + I = S$ and so $1 - \alpha q \in I$ for some $q \in S$. Amongst all such q we choose one of minimal degree, w say, and with leading coefficient q_w. Note that $q \neq 0$. If $w = 1$ then $0 \neq 1 - \alpha q \in I \cap R$. If $w \neq 1$ then $\alpha q_w \in I(w) = pR$ for some $p \in \{p_1,\ldots,p_m\}$. Thus $\alpha q_w = pr$ for some $r \in R$. However, p is regular, modulo αR; so $r = \alpha r'$ and $q_w = pr'$ for some $r' \in R$.

Now we choose $s \in I$ of degree w and with leading coefficient p. Let $q' = q - sr'$. Then $\deg q' < \deg q$ and

$$1 - \alpha q' = 1 - \alpha(q - sr') = (1 - \alpha q) + s(\alpha r') \in I$$

contradicting the choice of q. Hence $I \cap R \neq 0$.

(ii) By (i) I contains a monic polynomial of $D[x_i]$ for each i.
(iii) Let $M = S/I$, let N be a maximal R-submodule of M and let

$$N_w = \{m \in M \mid mw \in N\} \quad \text{for} \quad w \in \mathcal{W}.$$

Clearly, N_w is an R-submodule of M and $\bigcap_{w \in \mathcal{W}} N_w$ is an S-submodule of M, and so is zero. On the other hand, right multiplication by w gives an R-monomorphism $M/N_w \to M/N$. Hence either $M/N_w = 0$ or else $M/N_w \simeq M/N$. It follows from (ii) that $(S/I)_R$ is isomorphic to a finite direct sum of copies of M/N; and, of course, $R/I \cap R \hookrightarrow (S/I)_R$. □

6.9 We next turn to some applications concerned with primitive rings. First, however, come two preparatory lemmas.

Lemma. *Let U be a simple ring with centre k and V any k-algebra. Then:*
(i) *the map $A \mapsto A \otimes U$ provides a $(1,1)$-correspondence between ideals of V and ideals of $V \otimes_k U$;*
(ii) *if V is prime then $V \otimes U$ is prime;*
(iii) *$Z(V) \simeq Z(V \otimes U)$ via $z \mapsto z \otimes 1$.*

Proof. (i) Let $B \triangleleft V \otimes U$ and $A = \{v \in V \mid v \otimes 1 \in B\}$. Of course $A \triangleleft V$ and $A \otimes U \subseteq B$. By considering $V \otimes U / A \otimes U$ it will be enough to prove that if $B \neq 0$ then $A \neq 0$.

We choose $0 \neq b = \sum_{i=1}^{n} v_i \otimes u_i \in B$ to minimize n. Then $u_1 \neq 0$ and hence $Uu_1U = U$. Therefore we may assume that $u_1 = 1$. But then, for any $u \in U$, $(1 \otimes u)b - b(1 \otimes u)$ has fewer than n terms and so is zero. Hence each $u_i \in k$ and then $0 \neq \sum v_i u_i \in A$.

(ii) Clear from (i).

(iii) Fix a k-basis $\{v_i\}$ for V and pick $z \in Z(V \otimes U)$. Then $z = \sum v_i \otimes u_i$, where the u_i are uniquely determined by z. Commuting with any $1 \otimes u$, with $u \in U$, shows that each $u_i \in Z(U) = k$ and so $z = v \otimes 1$ for some $v \in V$. Evidently $v \in Z(V)$ and the rest is clear. □

6.10 Lemma. *Let T be a prime ring and M be a faithful module of finite length. Then T is primitive.*

Proof. If $P_1, \ldots, P_n$ are the primitive annihilators of the composition factors of M_T, then $P_1 P_2 \cdots P_n = 0$ and so some $P_i = 0$. □

6.11 We now return to the ring $S = D[x_1, \ldots, x_n]$.

Theorem. *S is primitive if and only if, for some m, $M_m(D)$ contains, as a k-subalgebra, a copy of the field $k(x_1, \ldots, x_n)$.*

Proof. ($\Rightarrow$) Let M_S be a simple faithful module, say $M \simeq S/I$ with $I \triangleleft_r S$, and let $m = \dim M_D$. Since $I \cap k[x_1,\ldots,x_n] = 0$ then

$$k[x_1,\ldots,x_n] \hookrightarrow \operatorname{End} M_S \hookrightarrow \operatorname{End} M_D = M_m(D).$$

($\Leftarrow$) Suppose $M_m(D)$ contains $k(x_1,\ldots,x_n)$. Let $M = D^n$; this is a right D-module of finite length and $V = k[x_1,\ldots,x_n] \hookrightarrow \operatorname{End}(D^n)_D$. Now $S \simeq V \otimes_k D$ and M is a right S-module via $m(v \otimes d) = vmd$. Let $B = \operatorname{ann} M_S$. By 6.9, $B = A \otimes D$ and thus $AMD = AM = 0$. Since $A \subseteq \operatorname{End} M$ this makes $A = 0$; so M_S is faithful. Hence, by 6.10, S is primitive. $\square$

6.12 Corollary. *Let k be a field of characteristic zero and D_n be the quotient division ring of $A_n(k)$. Then the polynomial ring $D_n[t_1,\ldots,t_r]$ is primitive if $r \leqslant n$ and not primitive if $r > n$.*

Proof. The proof of 6.6.18 is easily modified to show that no subfield of $M_m(D_n)$ can have transcendence degree greater than n over k. Of course $M_m(D_n)$ does have the subfield $k(x_1,\ldots,x_n)$. So the result is clear from 6.11. $\square$

§7 Additional Remarks

7.0 (a) These results form a clear development from the well-known commutative theory. This account is influenced by [Hall **59**], [Quillen **69**], [Duflo **73**], [Irving **79**] and [McConnell **82**].

(b) The approach here is new. It leans heavily on generic flatness, following [Quillen **69**], and its connection with the Jacobson property, following [Duflo **73**]; but the definitions in 3.12 and 4.12 and many of the proofs are new.

(c) There is not, as yet, a standard definition of the 'Nullstellensatz' in the literature; 1.4 and 2.3 represent our preference.

7.1 (a) 1.3(ii) is an unpublished result of L. W. Small; 1.6 comes from [Duflo **73**], 1.7 from [Amitsur **56**], and 1.12 from [McConnell **84**]. Example 1.10 is due to L. W. Small following the earlier example of [Beidar **81**].

(b) The focus here is upon the Nullstellensatz rather than simply the Jacobson property. There are noteworthy results in that direction, however. If R is a Noetherian Jacobson ring then $R[x;\sigma]$ and $R[x;\delta]$ are Jacobson rings (see [Goldie and Michler **74**] and [Jordan **75**] respectively). There are examples showing that the Noetherian hypothesis is needed, in [Pearson and Stephenson **77**] and [Ferrero and Kishimoto **85**] respectively. However, [Watters **75**] shows that when x is a central indeterminate the Noetherian hypothesis can be removed.

(c) More general questions about the description of $J(R \# U(g))$ are discussed in [Bergen, Montgomery and Passman **P**].

7.2 This approach is heavily influenced by [Duflo **73**].

7.3 (a) This section springs from the proof, in [Quillen **69**], of a result closely related to 3.11, namely:

Lemma. (Quillen's lemma) *Let R be a filtered k-algebra such that $\operatorname{gr} R$ is a commutative affine k-algebra. If M is a simple R-module then $\operatorname{End} M$ is algebraic over k.*

Proof. Here $\operatorname{gr} R$ is $(\mathbb{N}, \mathbb{N})$-generically flat over k by 4.7. Hence R is, by 4.9. Hence, by 3.13, R satisfies the Nullstellensatz over k. □

(b) The proof of 3.17 comes from [Joseph **77**].
(c) G-domains were studied by [Goldman **51**]. Noncommutative generalizations are discussed in 13.9.12 and 14.5.6.

7.4 (a) [McConnell **82, 84**] proved 4.21 and 4.23 respectively. The proofs here are new.
(b) Some related results for group rings appear in [Brown **82**] and [Snider **82**].

7.5 (a) A version of 5.5 for enveloping algebras was proved by P. Gabriel; see [Dixmier **77**, 2.6.9]. The results here are developed from an unpublished theorem of L. W. Small.
(b) One sees from 5.5 that if M is a simple $A_1(k)$-module then $\operatorname{End} M$ is a finite dimensional extension of k. [Farkas and Snider **81**], extending [Quebbemann **79**], shows that all division rings finite dimensional over k arise in this way. In [Robson and Small **86**] subalgebras of $A_1(k)$ are considered.

7.6 (a) This follows [Amitsur and Small **78**] although 6.12 is, in part, due to [Resco **79**]; see also [Resco, Stafford and Warfield **86**].
(b) Another application related to 6.6.17 and 7.5.10 appears in [Resco **80**].

Chapter 10
PRIME IDEALS IN EXTENSION RINGS

The importance of prime and primitive ideals has been made clear in earlier chapters. In the preceding chapter the investigation of simple modules provided some information about primitive rings. This chapter describes the prime ideals of certain extension rings S in terms of those of the original ring R.

One situation where one might expect to be able to provide such a description is when S_R and $_RS$ are both finitely generated. However, the existence of an example, due to [Bergman **83**], of such rings R, S with S being a division ring and R a free algebra on an infinite set of generators shows that to be an unreasonable expectation.

Much of the chapter concerns the more special case when S is finitely generated, as an R-module, by elements which normalize R (i.e. satisfy $sR = Rs$); S is then called a finite normalizing extension of R. The results obtained are those described, in the commutative case, as 'lying over', 'going up', 'incomparability', etc. These are proved, in part, in §2 after a preliminary section which, amongst other things, links the chain conditions satisfied by the two rings. The definitive results, in §4, require the use of a quotient ring of a prime ring R even when R is not Noetherian. Therefore, in §3, more general quotient rings are introduced, and normal and central closure is described.

The final two sections apply the results to two cases: first the crossed product $R*G$ and the fixed ring R^G, where G is a finite group acting on R; and then the skew Laurent polynomial ring $R[x, x^{-1}; \sigma]$.

One point to notice is that, although in general the rings are not assumed to satisfy chain conditions, the techniques are Noetherian in nature. In particular the observation in 2.2.14 that a prime ring R is uniform as an (R, R)-bimodule plays a crucial role.

§1 Finite Extensions and Chain Conditions

1.1 This section describes several types of finite normalizing extension S of R and proceeds to demonstrate that S is right Noetherian if and only if R is. The techniques involved here are module theoretic, and will be used again in later sections.

1.2 First, however, consider the more general situation where R, S are rings with $R \subseteq S$ and S_R is finitely generated. Such an extension is called a **finite extension**. We noted earlier, in 1.1.3, that if R is right Noetherian then so is S. The example cited in the introduction shows that the converse, in this generality, is false. This example also demonstrates that although, by 6.5.3, $K(S) \leq K(R)$ nevertheless equality need not occur unless extra conditions are imposed.

1.3 The extra conditions we have in mind concern the generators of S_R. An element $a \in S$ will be said to **normalize** R if $aR = Ra$, to **centralize** R if $ar = ra$ for each $r \in R$, and to be **automorphic** on R if there is some automorphism σ of R such that $ra = ar^\sigma$ for each $r \in R$. Evidently a centralizing element is automorphic, taking $\sigma = 1$, and an automorphic element is normalizing. Also, any regular normalizing element is automorphic.

If S_R has a finite set of generators $\{a_i | i = 1, \ldots, t\}$ each of which normalizes, or centralizes, or is automorphic, then S is called a **finite normalizing, centralizing,** or **automorphic** extension of R.

1.4 Several examples of such extensions have already been encountered.

Example. (i) The ring $S = M_n(R)$ is a finite centralizing extension, with generators the matrix units $\{e_{ij} | i, j \in 1, \ldots, n\}$. However, our results will not add to our knowledge of this example.

Example. (ii) If G is a finite group then RG is a finite centralizing extension and $R*G$ is a finite automorphic extension with the elements of G serving as generators. These will be discussed further in §5.

Example. (iii) Let D be an integral domain, $S = D[x; \sigma]$, and $R = D[x^t; \sigma^t]$ for some fixed positive integer t. Then S is a finite automorphic extension of R, using $\{x^i | i = 0, \ldots, t-1\}$ as generators. Since the units of S all belong to D it is clear that S is not a crossed product $R*G$ for any G. This example occurs again in §6.

Example. (iv) Finally, an example of a finite normalizing extension which is not automorphic. Let k be a field and A the commutative polynomial algebra over k in indeterminates $\{x_i | i \in \mathbb{N}\}$. Let σ be the k-endomorphism of A defined by $\sigma(x_0) = 0$, $\sigma(x_i) = x_{i-1}$ for $i > 0$. Let $S = M_2(A)$ and

$$R = \left\{ \begin{bmatrix} a & 0 \\ 0 & \sigma(a) \end{bmatrix} \middle| a \in A \right\}.$$

10.1.9 *Finite extensions and chain conditions* 345

It is easy to check that S is generated over R by the matrix units, and that they normalize R but are not automorphic. Later results will show that there is no finite automorphic generating set of S over R (see 2.6, 2.7).

1.5 For the rest of this section let $S = \sum_{i=1}^{t} a_i R$ with each a_i normalizing R and let M be a right S-module. Of course M is also an R-module, and we now study these two structures on M. This will be used to connect the chain conditions on R and on S.

1.6 If N is an R-submodule of M then Na_i^{-1} denotes the set $\{m \in M \mid ma_i \in N\}$; this notation does not imply that a_i is a unit. Note that since a_i is normalizing Na_i^{-1} is also an R-submodule of M.

Lemma. *The map $M/Na_i^{-1} \to M/N$ given by $m + Na_i^{-1} \mapsto ma_i + N$ is a group monomorphism. It induces a lattice embedding $\mathscr{L}(M/Na_i^{-1})_R \to \mathscr{L}(M/N)_R$. If a_i centralizes R the map is an R-homomorphism.*

Proof. This is easily verified. □

1.7 Given $N \triangleleft M_R$, as in 1.6, there is a largest S-submodule of M contained in N. This is called $b(N)$, the **bound** of N.

Lemma. *(i) If $N \triangleleft M_R$ then $b(N) = \bigcap_{i=1}^{t} Na_i^{-1}$.*
(ii) If, further, $N \triangleleft_e M_R$ then $Na_i^{-1} \triangleleft_e M_R$, $b(N) \triangleleft_e M_R$, and $b(N) \triangleleft_e M_S$.

Proof. (i) Straightforward.
(ii) Let $A \triangleleft M_R$. If $Aa_i = 0$ then $A \subseteq Na_i^{-1}$. Otherwise $Aa_i \cap N \neq 0$ and so $A \cap Na_i^{-1} \neq 0$ again. The rest is clear. □

1.8 Lemma. *M contains an R-submodule N maximal with respect to having $b(N) = 0$.*

Proof. Let $\{N_k \mid k \in I\}$ be a chain of R-submodules of M such that $b(N_k) = 0$ for all $k \in I$. If $b(\bigcup N_k) \neq 0$ then $\bigcup N_k \supseteq xS$ for some $0 \neq x \in \bigcup N_k$. Hence $xa_i \in \bigcup N_k$ and so $xa_i \in N_{k(i)}$ for some $k(i) \in I$. Choosing $k = \sup\{k(i) \mid i = 1, \ldots, t\}$ we see that $xa_i \in N_k$ for each i and so $x \in b(N_k)$, a contradiction. Therefore $b(\bigcup N_k) = 0$. Thus Zorn's lemma can be applied to give N as required. □

1.9 Uniform dimension will play a major role in what follows. The basic result concerning this comes next.

Proposition. *(i) If $\operatorname{u\,dim} M_S = d$ and N is an R-submodule of M maximal with respect to having $b(N) = 0$ then*

$$\operatorname{u\,dim}(M/N)_R \leq d \leq \operatorname{u\,dim} M_R \leq dt.$$

(ii) If M_S is simple then M_R is semisimple of length at most t. Furthermore, if S is a finite centralizing extension then M_R is isotypic.

(iii) If M_S has finite length then so too does M_R.

Proof. (i) Evidently $d = \operatorname{u\,dim} M_S \leq \operatorname{u\,dim} M_R$. Let $A_1, \ldots, A_k$ be R-submodules of M strictly containing N whose sum, modulo N, is direct. Of course $b(A_j) \neq 0$ for each j. If $k > d$ then, for some j,

$$0 \neq b(A_j) \cap \sum_{i \neq j} b(A_i) \subseteq A_j \cap \sum_{i \neq j} A_i \subseteq N$$

which contradicts the fact that $b(N) = 0$. Hence $\operatorname{u\,dim}(M/N)_R \leq d$.

Finally, by 1.6, $\operatorname{u\,dim}(M/Na_i^{-1})_R \leq d$ and, by 1.7(i), $\bigcap Na_i^{-1} = 0$. Thus

$$M_R \hookrightarrow \bigoplus (M/Na_i^{-1})$$

and so $\operatorname{u\,dim} M_R \leq td$.

(ii) In this case it is clear that N is a maximal R-submodule of M. Hence, by 1.6, M/Na_i^{-1} is either simple or zero, and so M_R is semisimple with length $M \leq t$. If a_i centralizes R and $M/Na_i^{-1} \neq 0$ then, by 1.6 again, $M/Na_i^{-1} \simeq M/N$.

(iii) This is clear from (ii). □

1.10 Corollary. M_S *is Noetherian if and only if* M_R *is Noetherian. Moreover, then* $\mathcal{K}(M_S) = \mathcal{K}(M_R)$.

Proof. We know, by 6.1.5, that if M_R is Noetherian then so too is M_S and $\mathcal{K}(M_S) \leq \mathcal{K}(M_R)$. Therefore it is sufficient, using 6.2.20, to consider the case when M_S is α-critical for some α and to show that M_R is Noetherian with $\mathcal{K}(M_R) \leq \alpha$. Moreover, by induction, the result can be assumed true for proper factors of M_S.

Using 1.8, choose $N \triangleleft M_R$ maximal with respect to having $b(N) = 0$. If $H \supset N$ then $b(H) \neq 0$ and so $(M/b(H))_R$ is Noetherian with $\mathcal{K}(M/b(H))_R < \alpha$. Hence $(M/N)_R$ is Noetherian and $\mathcal{K}(M/N)_R \leq \alpha$. By 1.6 the same is true of M/Na_i^{-1}. The embedding of M_R in $\bigoplus M/Na_i^{-1}$ now shows that M_R is Noetherian and $\mathcal{K}(M_R) \leq \alpha$. □

1.11 We can now deduce results linking the properties of R and S themselves.

Corollary. (i) S *is right Artinian if and only if* R *is right Artinian.*

(ii) S *is right Noetherian if and only if* R *is right Noetherian, and then* $\mathcal{K}(S) = \mathcal{K}(R)$.

(iii) *If* S *is a right Goldie ring then so too is* R.

Proof. (i), (ii) Consideration of S_R makes this clear from 1.9 and 1.10.

(iii) 1.9 shows that $\operatorname{u\,dim} R < \infty$ and, being a subring of S, R must satisfy the a.c.c on right annihilators. □

1.12 The converse to 1.11(iii) is false. To see this, choose a commutative integral domain A with a homomorphic image B which is not a Goldie ring; say $\vartheta: A \to B$ is the epimorphism. Let
$$R = \{(a, \vartheta(a)) | a \in A\} \subseteq S = A \oplus B.$$
Of course $R \simeq A$, yet S is not Goldie. The elements $(1,0)$ and $(0,1)$ generate S_R and centralize R.

1.13 Nevertheless there are circumstances when a converse is valid. This is so, for example, if S is prime [Lanski 80]. Another case, useful later, is noted next.

Lemma. *If R is semiprime right Goldie and S_R is torsionless, then S is right Goldie.*

Proof. Since S_R is finitely generated, 3.4.3 shows that $\operatorname{u\,dim} S_R < \infty$ and so $\operatorname{u\,dim} S_S < \infty$. By 3.4.5, $\operatorname{End}(S_R)$ is right Goldie. The embedding of S in $\operatorname{End} S_R$ means that S must satisfy the a.c.c. on right annihilators. □

This covers the case when $S = R * G$ for a finite group G, since then S_R is free.

1.14 As noted earlier in 2.2.13, results about modules are readily extended to bimodules. For a finite normalizing extension, special considerations apply. There are ring homomorphisms
$$R \otimes R^{\mathrm{op}} \xrightarrow{\alpha} S \otimes R^{\mathrm{op}} \xrightarrow{\beta} S \otimes S^{\mathrm{op}}$$
induced by the embedding of R in S; and $S \otimes S^{\mathrm{op}}$ is a finite normalizing extension of $\operatorname{im} \beta$, with generators $\{1 \otimes a_i | i \in \{1, \ldots, t\}\}$ and also of $\operatorname{im} \beta\alpha$, with generators $\{a_j \otimes a_i | i,j \in \{1, \ldots, t\}\}$.

If M is an (S, S)-bimodule then M is equally a right $S \otimes S^{\mathrm{op}}$-module. So any (R, S)-sub-bimodule Y is simply a submodule of $M_{\operatorname{im} \beta}$, and an (R, R)-sub-bimodule X is a submodule of $M_{\operatorname{im} \beta\alpha}$. Then $b(Y) = \bigcap a_i^{-1} Y$ and $b(X) = \bigcap a_i^{-1} X a_j^{-1}$ are simply the largest (S, S)-sub-bimodules of M contained, respectively, in Y and X.

By this means, bimodule versions of the module-theoretic results above are made available and will be used without further ado.

§2 Prime Ideals

2.1 Throughout this section S denotes a finite normalizing extension $\sum_{i=1}^{t} a_i R$ of a ring R. The bimodule theory described in 1.14 will be applied to demonstrate 'lying over', 'going up' and 'incomparability', the last being proved only when the rings are right Noetherian. The non-Noetherian case will be dealt with in §4.

2.2 First come two simple observations.

Lemma. (i) *If $a \in S$ normalizes R then there is a ring isomorphism $R/\operatorname{l ann}_R a \to R/\operatorname{r ann}_R a$ given by $r \mapsto r'$, where $ra = ar'$.*
 (ii) *If $I \triangleleft S$ then S/I is a finite normalizing extension of $R/I \cap R$ with generators $\{a_i + I \mid i = 1, \ldots, t\}$.* □

2.3 When S is commutative and I is prime, the embedding $R/I \cap R \hookrightarrow S/I$ shows that $R/I \cap R$ is an integral domain and $I \cap R$ is prime. In the noncommutative case this is no longer true. For example, take $S = M_2(A)$ for some integral domain A, let

$$R = \begin{bmatrix} A & 0 \\ 0 & A \end{bmatrix},$$

and choose the automorphic generators 1 and

$$\begin{bmatrix} 0 & 1 \\ 1 & 0 \end{bmatrix}$$

for S_R.

2.4 Nevertheless the next result provides a description of $I \cap R$ for any S.

Theorem. (Cutting down) *Let S be a finite normalizing extension of R and $I \in \operatorname{Spec} S$. Then $I \cap R$ is semiprime and has at most t minimal primes. The corresponding prime factor rings are all isomorphic.*
 If S is a finite centralizing extension then $I \cap R$ is prime.

Proof. By 2.2(ii) we can reduce to the case when $I = 0$ and so S is prime. That case is dealt with by 2.5, which also provides extra information needed later. □

2.5 Suppose, therefore, that S is prime. By 1.8 (and 1.14), there exists ${}_R Y_S \in \mathcal{L}({}_R S_S)$ maximal with respect to $b(Y) = 0$, i.e. with respect to not containing any nonzero ideal of S. Let $P = Y \cap R$ and $P_i = (a_i^{-1} Y) \cap R$.

Proposition. *Let S be prime. With the notation as specified:*
 (i) *if $r \in R$, $s \in S$ and $rRs \subseteq Y$ then $r \in P$ or $s \in Y$;*
 (ii) *$P \in \operatorname{Spec} R$;*
 (iii) *if $a_i \in Y$ then $a_i^{-1} Y = S$ and $P_i = R$;*
 (iv) *if $a_i \notin Y$ then*
   (a) *$\operatorname{l ann}_R(a_i) \subseteq P$ and $\operatorname{r ann}_R(a_i) \subseteq P_i$,*
   (b) *the map defined by 2.2(i) induces an isomorphism $R/P \to R/P_i$,*
   (c) *$P_i \in \operatorname{Spec} R$ and $a_i P_i = P a_i$, and*
   (d) *$a_i^{-1} Y$ is maximal in $\mathcal{L}({}_R S_S)$ with respect to having zero bound;*

(v) $\bigcap P_i = 0$, R is semiprime and the family $\{P_i | i = 1, \ldots, t\}$ includes all the minimal primes of R;
(vi) if R has w minimal primes then $w \leq u \dim_R S_S \leq t$.

Proof. (i) Suppose $r \notin P$ and $s \notin Y$. Then $A_1 = b(RrS + Y)$ and $A_2 = b(RsS + Y)$ are nonzero ideals of the prime ring S. Therefore

$$0 \neq A_1 A_2 \subseteq (RrS + Y)A_2 \subseteq RrA_2 + Y \subseteq Rr(RsY + Y) + Y \subseteq Y$$

contradicting the hypothesis that $b(Y) = 0$.
(ii) Clear from (i).
(iii) Obvious.
(iv)(a) Evidently $(l \operatorname{ann}_R (a_i))Ra_i = 0$ and so, by (i), $l \operatorname{ann}_R (a_i) \subseteq P$. By definition $r \operatorname{ann}_R (a_i) \subseteq P_i$.
(b), (c) Suppose $ra_i = a_i r' = s$. By (i) we see that $r \in P \Leftrightarrow s \in Y \Leftrightarrow r' \in P_i$. The rest is now clear.
(d) Note that

$$b(a_i S + Y)b(a_i^{-1} Y) \subseteq (a_i S + Y)b(a_i^{-1} Y) \subseteq a_i(a_i^{-1} Y) + Y \subseteq Y.$$

However, $b(a_i S + Y) \neq 0$ and $b(Y) = 0$; hence $b(a_i^{-1} Y) = 0$, since S is prime.
Next, suppose that $_R W_S \supset a_i^{-1} Y$. Then $a_i W \nsubseteq Y$ and so $b(a_i W + Y) \neq 0$; say $b(a_i W + Y) = A$. Then, for each $x \in A$, $a_i x \in A \subseteq a_i W + Y$; so $a_i x = a_i w + y$ with $w \in W$, $y \in Y$. But then $x - w \in a_i^{-1} Y$ and so $A \subseteq W + a_i^{-1} Y = W$ and $b(W) \neq 0$.
(v) $0 = b(Y) = \bigcap_{i=1}^t a_i^{-1} Y \supseteq \bigcap_{i=1}^t P_i$.
(vi) Suppose the a_i are renumbered so that $\bigcap_{i=1}^u a_i^{-1} Y = 0$, for some $u \leq t$, with this intersection being irredundant. As in (v), $\bigcap_{i=1}^u P_i = 0$ so $w \leq u$. However, since $_S S_S$ is uniform, the bimodule version of 1.9 shows that $_R(S/a_i^{-1} Y)_S$ is uniform for $i = 1, \ldots, u$. Hence $u \dim _R S_S = u \leq t$. $\square$

Note that the proof of 2.4 is now complete.

2.6 The family $\{P_1, \ldots, P_t\}$ described in 2.5 will often include repetitions as, indeed, the commutative case demonstrates. Moreover, the primes which belong to this family need not all be minimal. For example, if R, S are the rings described in 1.4(iv), with the four matrix units as generators, and if $_R Y_S$ is chosen to be $e_{11} S$, the ideals $\{P_i\}$ can be calculated to be $\{R, R, 0, K\}$, where

$$K = \begin{bmatrix} \ker \vartheta & 0 \\ 0 & 0 \end{bmatrix}.$$

2.7 Nevertheless, under stronger hypotheses, more can be said.

Theorem. (i) *If S is a prime finite centralizing extension of R then R is prime.*
(ii) *If S is a prime finite automorphic extension of R then each of the ideals P,*

$P_1, \ldots, P_t$ provided by 2.5 is a minimal prime or equal to R. Furthermore, the minimal primes of R form a single orbit under the action of the automorphisms corresponding to the generators of S.

Proof. (i) The map associated, by 2.2(i), with a_i must be the identity. Hence, by 2.5, either $P_i = P$ or $P_i = R$.

(ii) The isomorphisms described in 2.5(iv) are induced by the given automorphisms σ_i. Any such automorphism must, of course, permute the minimal primes of R. So if $P_i \neq R$ then P_i is minimal. Suppose, now, that $P_1, \ldots, P_h$ is an orbit and $P_{h+1}, \ldots, P_w$ are the remaining minimal primes of R. It is easy to see that the ideals of S generated by $P_1 \cap \cdots \cap P_h$ and $P_{h+1} \cap \cdots \cap P_w$ must have product zero. The rest is clear. □

2.8 Having established the nature of $I \cap R$, for $I \in \operatorname{Spec} S$, we next consider which prime ideals Q of R are involved. If Q is a minimal prime of $I \cap R$ we will describe I as **lying over** Q. Of course, for commutative rings, or if S is a finite centralizing extension of R, then I lies over Q precisely when $I \cap R = Q$.

2.9 Theorem. (Lying over) *If $Q \in \operatorname{Spec} R$ and S is a finite normalizing extension of R then there is a prime ideal I of S lying over Q.*

Proof. By Zorn's lemma, there is an ideal I of S maximal with respect to having $I \cap R \subseteq Q$; and I is easily checked to be prime. It remains to show that Q is a minimal prime of $I \cap R$. By passing to $R/I \cap R \subseteq S/I$ it is enough to deal with the case when $I = 0$, i.e. the case when every nonzero ideal J of S satisfies $J \cap R \nsubseteq Q$.

If Q is not a minimal prime of R then Q is essential in ${}_R R_R$ by 2.2.14. Hence if ${}_R K_R$ is a complement to R in ${}_R S_R$ then $Q \oplus K$ is essential in ${}_R S_R$. By 1.7, applied to (R, R)-bimodules, $b(Q \oplus K)$ is an ideal of S essential as an (R, R)-sub-bimodule. Thus

$$0 \neq b(Q \oplus K) \cap R \subseteq (Q \oplus K) \cap R = Q.$$

This contradiction yields the result. □

2.10 Next, we note some immediate consequences, the first of which involves the prime radicals $N(R)$ and $N(S)$.

Corollary. *Let S be a finite normalizing extension of R.*
(i) $N(R) = N(S) \cap R$.
(ii) (Going up). *If $Q_1, Q_2 \in \operatorname{Spec} R$ and $I_1 \in \operatorname{Spec} S$ with $Q_1 \subset Q_2$ and I_1 lying over Q_1, then there exists $I_2 \in \operatorname{Spec} S$ with $I_1 \subset I_2$ and I_2 lying over Q_2.*
(iii) *If $I \in \operatorname{Spec} S$ lies over $Q \in \operatorname{Spec} R$ and I is maximal then so is Q.*
(iv) *If $I \in \operatorname{Spec} S$ lies over $Q \in \operatorname{Spec} R$ and I is primitive then so is Q.*
(v) $J(R) \subseteq J(S) \cap R$.

10.2.13 *Prime ideals* 351

Proof. (i) Clear.
(ii) Apply 2.9 to S/I_1 and $R/I_1 \cap R$.
(iii) Clear from (ii).
(iv) Let M_S be a simple module with $\mathrm{ann}_S M = I$. By 1.9, M_R is semisimple of finite length; let $Q_1, \ldots, Q_k$ be the primitive ideals of R annihilating its simple factors. Then, for each i,

$$Q_1 Q_2 \cdots Q_k \subseteq \mathrm{ann}_R M = I \cap R \subseteq Q_i,$$

and so each minimal prime of $I \cap R$ equals some Q_i.
(v) This is clear from (iv). □

2.11 If 2.10(v) is combined with 9.1.3 one sees that $J(S) \cap R = J(R)$ when R is right Noetherian. In fact this is true for any ring R, and also the implications in 2.10(iii), (iv) can be reversed. This will be proved in §4, making use of two further properties of finite normalizing extensions. We describe these properties next and show how they apply.

If $I \in \mathrm{Spec}\, S$ has the property that $(J \cap R)/(I \cap R)$ is essential as an ideal of $R/I \cap R$ for each $J \triangleleft S$, $J \supset I$, then S satisfies **essentiality at** I. If this holds for all $I \in \mathrm{Spec}\, S$ then S satisfies **essentiality**.

2.12 First we establish this property in the Noetherian case.

Proposition. *If S is a finite normalizing extension of R and S is right Noetherian then essentiality is satisfied.*

Proof. It is enough to consider the case when $0 = I \subset J$ and S is prime. Suppose $\mathrm{u\,dim}\, S_S = d$. By 1.9, $\mathrm{u\,dim}\, S_R \leqslant td < \infty$. By 2.3.5, J contains a regular element c of S and then $S \simeq cS \subseteq J$. Hence $\mathrm{u\,dim}\, J_R = \mathrm{u\,dim}\, S_R$ and so $J \triangleleft_e S_R$ and $J \cap R \triangleleft_e R_R$. □

Of course all that was used here was that S/I is prime right Goldie for each $I \in \mathrm{Spec}\, S$. By 6.3.5 this is so if S has a right Krull dimension; and, as will be shown in 13.6.6, it also holds if S is a PI ring.

2.13 Essentiality leads to some interesting consequences regarding $\mathrm{Spec}\, R$ and $\mathrm{Spec}\, S$, the first being known as **incomparability**.

Proposition. *Let S be a finite normalizing extension of R satisfying essentiality.*
(i) *If $I_1, I_2 \in \mathrm{Spec}\, S$ with $I_1 \cap R = I_2 \cap R$ then $I_1 \not\subseteq I_2$ and $I_2 \not\subseteq I_1$.*
(ii) *If $I_1, I_2 \in \mathrm{Spec}\, S$ with $I_1 \subset I_2$ then $I_1 \cap R \subset I_2 \cap R$.*
(iii) *If $I \in \mathrm{Spec}\, S$ lies over $Q \in \mathrm{Spec}\, R$ then I is maximal if and only if Q is maximal.*

Proof. (i) (ii) It is easy to see that (i) and (ii) are equivalent; and (ii) is immediate.
(iii) By 2.10(iii), we need deal only with the case when Q is maximal. As usual,

we may suppose that $I = 0$ and so Q is a minimal prime of R. It is clear, from 2.4, that all the minimal primes of R are maximal. However, if $0 \neq J \triangleleft S$ then $J \cap R$ is essential, so is not contained in any minimal prime, by 2.2.14. Hence $J \cap R = R$ and $J = S$. □

2.14 The second property, **properness**, requires that if H is a proper right ideal of R then HS is a proper right ideal of S. Once again there are circumstances under which this is clear.

Lemma. *If S is a finite normalizing extension of R and $_RS$ is free then properness holds.*

Proof. Trivial. □

2.15 Granted this property, the Jacobson radicals are easily linked.

Proposition. *If S is a finite normalizing extension of R and properness holds then $J(R) = J(S) \cap R$.*

Proof. By 2.10(v), $J(R) \subseteq J(S) \cap R$. On the other hand, let H be a maximal right ideal of R. Since $HS \neq S$ there is a maximal right ideal K of S with $HS \subseteq K$ and $K \cap R = H$. Hence $J(R) \supseteq J(S) \cap R$. □

2.16 Granted both properties we can improve upon 2.10(iv).

Proposition. *Let S be a finite normalizing extension of R and suppose that essentiality and properness hold. If $I \in \operatorname{Spec} S$ lies over $Q \in \operatorname{Spec} R$ then I is primitive if and only if Q is primitive.*

Proof. Using 2.10(iv) it is enough to deal with the case when $I = 0$ and Q is primitive. We choose a maximal right ideal H of R with $\operatorname{ann}_R R/H = Q$ and let K be as in the preceding proof. If $(S/K)_S$ is not faithful then essentiality shows that $(\operatorname{ann}_S S/K) \cap R \triangleleft_{eR} R_R$. Hence $Q \triangleleft_{eR} R_R$ which contradicts 2.2.14. Therefore $(S/K)_S$ is faithful and I is primitive. □

§3 and §4 (optional on a first reading) demonstrate that these two properties always hold, so the conclusions of 2.13, 2.15 and 2.16 are always true.

§3 Quotient Rings and Closure

3.1 The proof of incomparability in finite normalizing extensions makes use of a more general quotient ring than that described in Chapter 2, namely one concerned more with ideals rather than regular elements. The point here is that the nonzero ideals of a prime ring R satisfy a type of Ore condition caused by $_RR_R$ being uniform.

There is also a need for closure operations in this quotient ring, namely the central closure and the normal closure. The construction and detailed properties of these comprise the second part of this section. They involve a more general notion of graded rings which is also described.

3.2 The construction of the right localization $R_{\mathcal{S}}$ for a right denominator set $\mathcal{S}$ of a ring R was described in 2.1.12. It used maps from right ideals of R, containing elements of $\mathcal{S}$, to the ring R; the verification that $R_{\mathcal{S}}$ was a ring depended upon the properties of that collection of right ideals.

What we do now is simply to axiomatize the properties required for that verification.

3.3 A collection $\mathcal{F}$ of right ideals of a ring R is called a **right localization set** if for any $I_1, I_2 \in \mathcal{F}$ and any $\alpha \in \mathrm{Hom}(I_2, R)$, there exist $I_3, I_4 \in \mathcal{F}$ such that:
(i) $I_3 \subseteq I_1 \cap I_2$; and
(ii) $I_4 \subseteq I_2$ and $\alpha(I_4) \subseteq I_1$.

3.4 Examples. (i) The set $\mathcal{F}$ described in 2.1.12 is, of course, a right localization set.

(ii) The example of particular interest to us here occurs when R is a prime ring and $\mathcal{E}$ is the set of nonzero ideals of R. The requirements above are fulfilled, with $I_3 = I_1 \cap I_2$ and $I_4 = I_2 I_1$. More generally, if R is semiprime and $\mathcal{E}$ is the set of ideals essential in $_RR_R$, then 2.2.14 makes clear that $\mathcal{E}$ is a right localization set.

(iii) Finally, we note an example of importance in other contexts. A right ideal D of a ring R is called **dense** if, for each $a \in R$, the right ideal $\{r \in R \mid ar \in D\}$ has zero left annihilator. It can be checked that the set $\mathcal{D}$ of all dense right ideals of R is a right localization set. For details see [Lambek 66] or [Stenstrom 75].

3.5 Given such a set $\mathcal{F}$, the **localization** $R_{\mathcal{F}}$ of R with respect to $\mathcal{F}$ is defined to be the set

$$\bigcup \{\mathrm{Hom}(I, R) \mid I \in \mathcal{F}\}$$

modulo the equivalence relation given by $\alpha_1 \sim \alpha_2$ if $\alpha_1 : I_1 \to R$, $\alpha_2 : I_2 \to R$ and $\alpha_1 = \alpha_2$ when restricted to some $I_3 \in \mathcal{F}$, $I_3 \subseteq I_1 \cap I_2$. The operations in $R_{\mathcal{F}}$ are defined as follows: $[\alpha_1] + [\alpha_2] = [\beta]$, where β is the sum of α_1, α_2 when restricted to $I_3 \in \mathcal{F}$, with $I_3 \subseteq I_1 \cap I_2$ and $[\alpha_1][\alpha_2] = [\gamma]$ with γ being the composite of α_1 with α_2 restricted to $I_4 \in \mathcal{F}$, where $I_4 \subseteq I_2$ and $\alpha_2(I_4) \subseteq I_1$.

As in 2.1.12, one can verify that $R_{\mathcal{F}}$ is a ring. There is a canonical map $\varphi : R \to R_{\mathcal{F}}$, sending $a \in R$ to the map $\alpha : R \to R$ given by $\alpha(x) = ax$. The kernel of

φ is $\{a \in R \mid aI = 0 \text{ for some } I \in \mathscr{F}\}$, this being called the $\mathscr{F}$**-singular ideal** of R. In particular, φ is an embedding if and only if $1\operatorname{ann}_R I = 0$ for all $I \in \mathscr{F}$. It is clear that this is so for both $\mathscr{F} = \mathscr{E}$ and $\mathscr{F} = \mathscr{D}$ with these sets as described in 3.4. The ring $R_\mathscr{E}$ we term the **Martindale right quotient ring** of the prime, or semiprime, ring R; and the ring $R_\mathscr{D}$ is the **maximal right quotient ring**.

In particular, if R is prime right Goldie then $\mathscr{D} = \mathscr{F} \supseteq \mathscr{E}$ and $R_\mathscr{E} \subseteq R_\mathscr{D} = R_\mathscr{F} = R_\mathscr{S}$, where $\mathscr{S} = \mathscr{C}_R(0)$. Thus if R is simple but not Artinian then $R = R_\mathscr{E} \neq R_\mathscr{S}$.

3.6 As in 2.1.17, an exactly similar process applies to a right module M_R. Thus

$$M_\mathscr{F} = \bigcup \{\operatorname{Hom}(I, M) \mid I \in \mathscr{F}\}/\sim,$$

where $\gamma_1: I_1 \to M$, $\gamma_2: I_2 \to M$ are equivalent if their restrictions to $I_3 \in \mathscr{F}$, with $I_3 \subseteq I_1 \cap I_2$, are equal. Addition, and multiplication by elements of $R_\mathscr{F}$, are defined in a way analogous to that above in 3.5, making $M_\mathscr{F}$ into a right $R_\mathscr{F}$-module. There is a canonical map $\varphi: M \to M_\mathscr{F}$ having kernel

$$\{m \in M \mid mI = 0 \quad \text{for some } I \in \mathscr{F}\},$$

this being called the $\mathscr{F}$**-singular submodule** of M.

3.7 We note some useful properties of $R_\mathscr{F}$ in the case when $\varphi: R \to R_\mathscr{F}$ is an embedding. As usual, R is identified with its image in $R_\mathscr{F}$.

Lemma. *Let S be any ring with $R \subseteq S \subseteq R_\mathscr{F}$.*
 (i) *If $0 \neq A \triangleleft_r S$ then $A \cap R \neq 0$.*
 (ii) *If $\operatorname{u\,dim} R_R \leq n$ then $\operatorname{u\,dim} S_S \leq n$.*
 (iii) *If $sI = 0$ for $I \in \mathscr{F}$ and $s \in S$ then $s = 0$.*
 (iv) *If R is prime or semiprime then so too is S.*

Proof. Straightforward. □

3.8 Throughout the remainder of this section our attention is concentrated on a prime ring R and its Martindale right quotient ring $R_\mathscr{E}$. We let N denote the **normalizer** of R in $R_\mathscr{E}$; i.e. the set of elements in $R_\mathscr{E}$ which normalize R. Similarly, the elements in $R_\mathscr{E}$ which centralize R form the **centralizer** C of R.

For example if $R = M_2(\mathbb{Z})$ then $C = \mathbb{Q}$ and N is the set of all rational multiples of units of $M_2(\mathbb{Z})$.

3.9 Lemma. *If $0 \neq n \in N$ then n has an inverse in N. If $nI \subseteq R$ with $I \in \mathscr{E}$ then $In \subseteq R$.*

Proof. We note first that n is a regular element of $R_\mathscr{E}$. To see this, let $0 \neq q \in R_\mathscr{E}$. By the construction of $R_\mathscr{E}$, there are nonzero ideals I, I' of R with $0 \neq nI \triangleleft_r R$,

10.3.12   *Quotient rings and closure*   355

$0 \neq qI' \triangleleft_r R$. Note that
$$qnR = qRnR \supseteq qI'nI \neq 0$$
and that
$$RnqR = nRqR \supseteq nIqI' \neq 0.$$
Hence $nq \neq 0$ and $qn \neq 0$ and n is regular.

It follows that the map represented by n, namely $I \to nI \subseteq R$, given by $x \mapsto nx$, is an isomorphism. Since $0 \neq nI \triangleleft R$ the map has an inverse $n^{-1} \in R_\mathscr{E}$. Evidently $n^{-1} \in N$ and
$$In = n^{-1}nIn \subseteq n^{-1}Rn = R. \qquad \square$$

3.10 It follows from 3.9 that each nonzero $n \in N$ induces an inner automorphism σ say of $R_\mathscr{E}$ which restricts to an automorphism σ of R such that $rn = nr^\sigma$. Such a σ is called an **X-inner automorphism** of R.

For any $\sigma \in \mathrm{Aut}\, R$ one can define
$$N_\sigma = \{n \in N \mid rn = nr^\sigma \text{ for all } r \in R\}$$
and then
$$N = \bigcup \{N_\sigma \mid \sigma \in \mathrm{Aut}\, R\}.$$

3.11 Some facts are immediate.

Lemma. (i) $N_\sigma \neq 0$ if and only if σ is an X-inner automorphism.
(ii) $N_1 = C$ and is a field.
(iii) If σ is X-inner then N_σ is a one-dimensional C-vector space.
(iv) If σ, τ are X-inner then $N_\sigma N_\tau = N_{\sigma\tau}$. $\qquad \square$

3.12 The next result provides, within R, a way of detecting elements of N_σ and hence of N.

Proposition. (i) *Let $n \in N_\sigma$ and choose $0 \neq a \in R$ such that $an \in R$, $an = b$ say. Then $arb = br^\sigma a^\sigma$ for all $r \in R$.*
(ii) *Conversely, suppose that $\sigma \in \mathrm{Aut}\, R$, $0 \neq a, b \in R$ and $arb = br^\sigma a^\sigma$ for all $r \in R$. Then $b = an$ for some $n \in N_\sigma$.*

Proof. (i) $arb = aran = anr^\sigma a^\sigma = br^\sigma a^\sigma$.
(ii) Let $I = Ra^\sigma R$ and define a map $I \to R$ via $\sum r_i^\sigma a^\sigma s_i \mapsto \sum r_i b s_i$. To see that this is well defined, suppose that $\sum r_i^\sigma a^\sigma s_i = 0$. Then, for each $y \in R$,
$$0 = by^\sigma \sum r_i^\sigma a^\sigma s_i = \sum b(yr_i)^\sigma a^\sigma s_i$$
$$= \sum a(yr_i)bs_i = ay\sum r_i bs_i$$
and so $aR\sum r_i bs_i = 0$. Hence $\sum r_i bs_i = 0$ as required.

Let $n \in R_{\mathscr{E}}$ be the element corresponding to this map. Then $nr^\sigma x = rnx$ for all $x \in I$, and so $nr^\sigma = rn$ for all $r \in R$. This shows that $n \in N_\sigma$; and its definition ensures that $an = na^\sigma = b$. □

3.13 We turn next to closure. Let G be any subgroup of Aut R and $N_G = \bigcup \{N_\sigma | \sigma \in G\}$. Evidently, if G is trivial then $N_G = C$; and if $G = \text{Aut } R$ then $N_G = N$. It is clear from the above that $N_G R$ is a ring; it is called the **G-closure** of R in $R_{\mathscr{E}}$. If $R = N_G R$ then R is **G-closed**.

In particular, NR is the **normal closure** and CR the **central closure**; and if $R = NR$ or $R = CR$ it is, respectively, **normally closed** or **centrally closed**.

3.14 One natural question is answered by the next result.

Lemma. *The G-closure of R is G-closed.*

Proof. Let $\sigma \in G$ and suppose n is an element of the Martindale right quotient ring of $N_G R$ inducing σ on $N_G R$. Of course, n induces σ on R; so it will be enough to show that $n \in R_{\mathscr{E}}$.

By definition, there exists $0 \neq I \triangleleft N_G R$ such that $nI \subseteq N_G R$. It is not difficult to check that $0 \neq I \cap R \cap n^{-1} R = I' \triangleleft R$ and that $nI' \subseteq R$; so $n \in R_{\mathscr{E}}$. □

3.15 This implies, of course, that the central closure of R is centrally closed. It does not, however, imply that the normal closure of R is normally closed, for the natural embedding Aut $R \hookrightarrow$ Aut NR may be proper. For example, let k be a field and R be the k-algebra $k[t, x, y]$ with the relations that t is central and $txy = yx$. In this case $N = C = k(t)$, and x^{-1} normalizes NR although $x^{-1} \notin NR$. The details of this example, which is due to G. Bergman, are given in [Cohen and Montgomery 79].

3.16 Thus far, the nature of N_G has not been discussed. It is clear from 3.11 that it is closed under multiplication. However, N_G is not necessarily closed under addition. For example, if $R = M_2(k)$ for some field k, then

$$\begin{bmatrix} 1 & 0 \\ 0 & 1 \end{bmatrix} \text{ and } \begin{bmatrix} 0 & 1 \\ 1 & 0 \end{bmatrix} \in N \text{ yet } \begin{bmatrix} 1 & 1 \\ 1 & 1 \end{bmatrix} \notin N.$$

3.17 The ring generated by N_G is called the **algebra of G in $R_{\mathscr{E}}$** and is denoted here by R_G. (It is often denoted $B(G)$.) Evidently $R_G = \sum_{\sigma \in G} N_\sigma$. In order to clarify its structure we need to extend the notion of a graded ring.

3.18 Given a semigroup, or most commonly a group, G, a ring T is said to be **G-graded** if T has a collection of subgroups $\{T_g | g \in G\}$ such that $T = \oplus T_g$ and $T_g T_h \subseteq T_{gh}$ for $g, h \in G$.

10.4.2   Incomparability

Thus the graded rings defined earlier in 1.6.3 are $\mathbb{N}$-graded rings, although they can also be regarded as $\mathbb{Z}$-graded with $T_m = 0$ for all $m < 0$. Another useful example of a $\mathbb{Z}$-graded ring is the skew Laurent polynomial ring $T = R[x, x^{-1}; \sigma]$, taking $T_n = x^n R$ for each $n \in \mathbb{Z}$.

As before, a **homogeneous element** is one belonging to some T_n and a **graded right ideal** is a right ideal generated by its homogeneous elements.

3.19 Any homomorphic image S of a G-graded ring T retains some features of T. In particular, if S_g is the image of T_g then $S_g S_h \subseteq S_{gh}$ and $S = \Sigma S_g$, although this sum need no longer be direct. Conversely, if S is a ring having subgroups S_g with these two properties, it is easy to see that the ring $T = \oplus S_g$, the external direct sum with the multiplication extended from the products $S_g S_h$, is G-graded and has S as a homomorphic image.

In such circumstances, we define a **homogeneous element** of S to be any member of an S_g and a **graded right ideal** to be one generated by its homogeneous elements. Furthermore, if S has no nontrivial graded ideals it will be called **graded-simple**; if each nonzero homogeneous element is a unit it will be called a **graded-division ring**; and so on.

3.20 With this terminology, we are now able to describe the structure of the algebra of G in R_g.

Proposition. (i) R_G *is a homomorphic image of a G-graded ring, with N_G being the set of homogeneous elements.*

(ii) R_G *is a graded-division ring.*

Proof. Clear, using 3.9.   □

§4   Incomparability

4.1 Throughout this section $S = \sum_{i=1}^{t} a_i R$ with each a_i normalizing R. As shown in 2.13, 2.15, 2.16, many further links between Spec S and Spec R can be obtained once essentiality and properness are established, and that is the aim of this section.

As usual, the work will concentrate mainly on the case when S is prime. It will make use of the Martindale quotient ring of S to reduce to the case of a finite automorphic extension with both S and R being prime. That special case, which it is convenient to deal with first, involves use of a normal closure of R in its Martindale quotient ring.

4.2 We start, therefore, with the case when S is a finite automorphic extension of R, both S and R are prime and N_G and N are as in 3.13. **This will remain**

fixed until 4.10. As will become clear, the route followed involves the structure of the set of elements of S which normalize R.

4.3 Lemma. (i) *If* $0 \neq D, D' \triangleleft R$ *then there exists* $0 \neq B \triangleleft R$ *such that* $SBS \subseteq DS \cap SD'$.
(ii) *If* $0 \neq a \in S$ *normalizes* R, *then* a *induces a unique automorphism* σ *of* R *via* $ra = a\sigma(r)$.

Proof. (i) If σ_i is the automorphism of R associated with a_i then $Da_i = a_i\sigma_i(D)$ and $\sigma_i^{-1}(D')a_i = a_iD'$. The ideal $B = \bigcap_{i=1}^{t}(\sigma_i(D) \cap \sigma_i^{-1}(D'))$ is as required.
(ii) It is clear from (i) that $\operatorname{r ann}_R a = \operatorname{l ann}_R a = 0$ since each is an ideal of R and S is prime. □

4.4 Since R and S are prime, each has a Martindale right quotient ring, $R_\mathscr{E}$, $S_\mathscr{E}$ respectively.

Proposition. *The embedding of* R *in* S *extends to an embedding of* NR *in* $S_\mathscr{E}$.

Proof. A map $\alpha: NR \to S_\mathscr{E}$ is defined as follows. If $x = \sum n_i r_i \in NR$, there exists $0 \neq D \triangleleft R$ with $n_i D \subseteq R$ for each i and, using 4.3, $0 \neq E \triangleleft R$ with $SES \subseteq DS$. Then $\alpha(x)$ is defined to be the restriction to SES of the map $\alpha(x): DS \to S$ given by

$$\sum_j d_j a_j \mapsto \sum_{i,j}(n_i(r_i d_j))a_j.$$

To see this is well defined, let $D' = \bigcap_i n_i D n_i^{-1}$ and let B be as in 4.3. For any $d' \in D'$,

$$d'(\alpha(x)(\sum_j d_j a_j)) = \sum_{i,j}(d'n_i r_i)(d_j a_j) = (d'x)(\sum d_j a_j).$$

If either $x = 0$ or $\sum d_j a_j = 0$ it follows that $SBS(\alpha(x)(\sum d_j a_j)) = 0$ and so $\alpha(x)(\sum d_j a_j) = 0$. One can now check easily that $\alpha(x)$ is a well-defined right S-homomorphism and that α is a well-defined ring embedding. □

We note that this result fails if S is not assumed to be automorphic. For example, for the rings described in 1.4(iv), if $0 \neq a \in \ker \sigma$ then

$$\begin{bmatrix} a & 0 \\ 0 & 0 \end{bmatrix}$$

is a unit in NR and a zero divisor in S.

4.5 The next result will be used to effect a reduction to the case when R is closed.

Lemma. *Let* G *be a subgroup of* $\operatorname{Aut} R$ *containing each automorphism* σ_i *associated with the generating set* $\{a_i\}$. *Then* $N_G S = SN_G$ *and is a prime ring generated as*

10.4.7   *Incomparability*   359

an $N_G R$-module by the automorphic elements $\{a_i\}$. Further, for each $n \in N_G$ there exist $n_i, n_i' \in N_G$ such that $na_i = a_i n_i$ and $a_i n = n_i' a_i$.

Proof. It is easily checked, using 3.9, that each σ_i extends to an automorphism of $N_G R$ which maps N_G to N_G. Hence the elements n_i, n_i' exist as described, and also $N_G S$ is a ring generated over $N_G R$ by the automorphic elements a_i. By 4.4, $N_G S \subseteq S_\mathscr{E}$ and so, by 3.7, $N_G S$ is prime. Similarly, $N_G R$ is prime and, by 3.14, $N_G R$ is G-closed. $\square$

4.6 We now consider the case when G is as in 4.5 and R is G-closed. By 4.3, if $m \in S$ normalizes R then m induces a unique automorphism of R. For each $\sigma \in G$ we let $M_\sigma = \{m \in S \mid m \text{ induces } \sigma \text{ on } R\}$ and set $M_G = \bigcup \{M_\sigma \mid \sigma \in G\}$. The subring of S generated by M_G is denoted by S_G and called the **algebra of G in S**. As with R_G this is a homomorphic image of a G-graded ring. The next few results aim towards showing that S_G is graded-simple Artinian (cf. 3.20).

4.7 First a basis is obtained.

Lemma. *Let G be as above, R be G-closed, and Γ be a subset of M_G. Then there exist $m_1, \ldots, m_h \in \Gamma$, with $h \leq t$, such that $\Gamma R = \bigoplus_{i=1}^{h} m_i R$. Moreover, $\Gamma \subseteq \sum_{i=1}^{h} m_i N_G$.*

Proof. Choose a maximal R-independent subset of $\{a_1, \ldots, a_t\}$, say $\{a_1, \ldots, a_k\}$ after renumbering. For each $i \in \{1, \ldots, t\}$ let $D_i = \{r \in R \mid a_i r \in \bigoplus_{j=1}^{k} a_j R\}$ and let $D = \bigcap D_i$. Then $0 \neq SD \subseteq \bigoplus_{j=1}^{k} a_j R$ and hence, using 4.3(i), $\operatorname{u dim}_R S_R = k \leq t$. Therefore, $\operatorname{u dim}_R (\Gamma R)_R \leq k$ and so there exists a maximal R-independent subset, say $\{m_1, \ldots, m_h\}$, of Γ with $h \leq k$.

Now suppose $0 \neq m_0 \in \Gamma$; we need only show that $m_0 \in \sum_{i=1}^{h} m_i N_G$. However, $m_0 R \cap \bigoplus_{i=1}^{h} m_i R \neq 0$ and so $\sum_{i=0}^{h} m_i r_i = 0$ for some $r_i \in R$ with $r_0 \neq 0$. Each m_i induces an automorphism, τ_i say, of R. Hence, for each $r \in R$,

$$0 = \tau_0^{-1}(r_0)r(\sum m_i r_i) - (\sum m_i r_i)\tau_0(r)r_0 \in \bigoplus_{i=1}^{h} m_i R$$

and therefore $\tau_i \tau_0^{-1}(r_0)\tau_i(r)r_i = r_i \tau_0(r)r_0$ for each $i \in \{0, \ldots, h\}$. By 3.12, there exists $n_i \in N_{\tau_0 \tau_i^{-1}}$ such that $r_i = n_i r_0$, with $n_0 = 1$. Note that

$$r \sum_{i=0}^{h} m_i n_i = \sum_{i=0}^{h} m_i n_i \tau_0(r) \quad \text{and so} \quad \sum_{i=0}^{h} m_i n_i \in M_{\tau_0}.$$

However,

$$0 = \sum_{i=0}^{h} m_i r_i = \sum_{i=0}^{h} m_i n_i r_0$$

and therefore, by 4.3(ii),

$$\sum_{i=0}^{h} m_i n_i = 0 \quad \text{and} \quad m_0 \in \sum_{i=1}^{h} m_i N_G. \quad \square$$

4.8 Proposition. *Let G be as above and R be G-closed.*
(i) If X is a graded right ideal of S_G then $XS = XR = RX$ and $(XS \cap M_G)S_G = X$.
(ii) If X, Y are graded right ideals of S_G with $X \subseteq Y$ and $\mathrm{u\,dim}_R XS_R = \mathrm{u\,dim}_R YS_R$ then $X = Y$.
(iii) S_G is a graded-simple, graded-Artinian ring.

Proof. (i) The only step requiring an explanation is to show that $(XS \cap M_G)S_G \subseteq X$; i.e. that $XR \cap M_G \subseteq X \cap M_G$. Suppose that $m_1, \ldots, m_k$ is a free R-basis for XR, as provided by 4.7, and say $\sum m_i r_i \in M_\sigma$. It is easy to see that each $r_i \in N_G$ and that $m_i r_i \in M_\sigma$. Hence $\sum m_i r_i \in X \cap M_G$.

(ii) If $y \in Y \cap M_G$ then $yR \cap XR \neq 0$ and so, by 4.7, $y \in \sum m_i N_G \subseteq X$.

(iii) By 4.7, $\mathrm{u\,dim}_R S_R \leq t$ and so, by (ii), S_G has graded-composition length at most t and so is graded-Artinian. Note next that if X, Y are nonzero graded-ideals of S_G then, by (i), $RXY = XSY \neq 0$ and so S_G is graded-prime. The usual proof that each ideal of a semiprime Artinian ring is generated by a central idempotent (see, for example, [Herstein **68**, §1.4.2]) is readily adapted to S_G, making use of the fact that the only automorphism induced by an idempotent element is the identity map. Hence S_G is graded-simple. □

4.9 We can now prove the result we seek in this special case.

Theorem. *If R, S are prime rings and S is a finite automorphic extension of R, then essentiality at 0 and properness are both satisfied.*

Proof. As before, let G be the subgroup of $\mathrm{Aut}\,R$ generated by the automorphisms σ_i associated with the a_i, $i = 1, \ldots, t$. By 3.14 and 4.5, $N_G R$ is G-closed and $N_G S$ is a prime automorphic extension of $N_G R$.

First note, by 4.7, that if K is a proper right ideal of R then $K N_G S \neq N_G S$. Therefore, by 4.5, $KS \neq S$ and properness is established.

Next, let $0 \neq I \triangleleft S$. We are required to show that $I \cap R \triangleleft_{eR} R_R$; i.e. that $I \cap R \neq 0$. However, $0 \neq N_G I \triangleleft N_G S$ and if $0 \neq x \in N_G I \cap N_G R$ then one can choose $0 \neq A \triangleleft R$ such that $0 \neq Ax \subseteq I \cap R$ as required. This shows that we may suppose that R is G-closed.

In that case, 4.7 shows that S is freely generated, over R, by a subset of $\{a_1, \ldots, a_t\}$, say $\{a_1, \ldots, a_h\}$. Choose $0 \neq x \in I$ of minimal support on $\{a_1, \ldots, a_h\}$, say $x = \sum_{i=1}^{h} a_i r_i$ with $r_i \in R$ and, rearranging if necessary, $r_1 \neq 0$. Let σ_i be the automorphism linked with a_i.

For each $r \in R$, the element $\sigma_1^{-1}(r_1 r)x - xrr_1 \in I$ and also has smaller support than has x. Hence, for each i, $\sigma_i \sigma_1^{-1}(r_1 r) r_i = r_i r r_1$. Therefore, by 3.12, there exists $n_i \in N_{\sigma_1 \sigma_i^{-1}}$ such that $r_i = \sigma_i \sigma_1^{-1}(r_1) n_i = n_i r_1$, and so $0 \neq x = (\sum a_i n_i) r_1$. Note that $a_i n_i \in M_{\sigma_1}$; so if $m = \sum a_i n_i$ then $0 \neq m \in M_{\sigma_1}$ and $x = m r_1$. However, $I \supseteq S m r_1 S = S m R r_1 RS$ and by 4.3, $R r_1 RS \supseteq SBS$ for some $0 \neq B \triangleleft R$. Therefore $I \supseteq S m SB$. By 4.8(iii), $1 \in S_G m S_G \subseteq S m S$ and thus $0 \neq B \subseteq I \cap R$. □

4.10 We now turn towards the general case again. As usual, the main work takes place when S is prime and so, by 2.5, R is semiprime of finite rank. We recall that, in 2.5, $_R Y_S \subseteq S$ was chosen maximal with respect to having $b(Y) = 0$. After renumbering the a_i, it was arranged, in the proof of 2.5, that $\bigcap_{i=1}^{u} a_i^{-1} Y = 0$ for some $u \leq t$ with this intersection being irredundant. Each $P_i = a_i^{-1} Y \cap R$ is prime, and all minimal primes of R occur in this form. We rearrange the a_i once again so as to ensure that $P_1, \ldots, P_v$ is a set of representatives of the primes occurring, and $P_1, \ldots, P_w$ is the set of minimal primes; so $w \leq v \leq u \leq t$.

We introduce the notation

$$U_j = \bigcap \{a_i^{-1} Y \mid i \in \{1, \ldots, u\}, i \neq j\}, \text{ for } j \leq u,$$

and

$$Z_i = \bigoplus \{U_j \mid j \in \{1, \ldots, u\}, P_i = P_j\}, \text{ for } i \leq v,$$

and let $E = \bigoplus_{i=1}^{v} Z_i = \bigoplus_{j=1}^{u} U_j$ and $B = b(E)$. For each $i \leq v$, we define $f_i : B \to S$ to be the homomorphism given below, using projection and injection on direct sums:

$$f_i : B \hookrightarrow E = \bigoplus Z_i \to Z_i \subseteq E \subseteq S.$$

Thus $f_i \in S_\mathscr{E}$ since, as shown below, $B \neq 0$.

4.11 We now note several facts about these objects.

Lemma. *Let S be a prime finite normalizing extension of R. With the notation as above:*
(i) $E \triangleleft_{eR} S_S$ *and* $0 \neq B \triangleleft S$;
(ii) *the elements $f_i \in S_\mathscr{E}$ are nonzero orthogonal idempotents with $\sum_{i=1}^{v} f_i = 1$, each f_i centralizing R and $\mathrm{ann}_R f_i = P_i$;*
(iii) *if $x \in R$ and $q \in S_\mathscr{E}$ then $xRf_i q = 0$ implies $x \in P_i$ or $f_i q = 0$; and $qf_i Rx = 0$ implies $x \in P_i$ or $qf_i = 0$.*

Proof. (i) As noted in the proof of 2.5(vi), $S/a_j^{-1} Y$ is a uniform (R, S)-bimodule and $\mathrm{u\,dim}\,_R S_S = u$. However $0 \neq U_j \hookrightarrow S/a_j^{-1} Y$; so U_j is a uniform (R, S)-bimodule and $\mathrm{u\,dim}\,_R E_S = u$. Thus $E \triangleleft_{eR} S_S$. It follows from 1.7(ii) that B is nonzero, and hence $f_i \in S_\mathscr{E}$.

(ii) These facts are easily verified.

(iii) First suppose $xRf_i q = 0$, and choose $0 \neq A \triangleleft S$ such that $qA \subseteq S$. Then $qAB \subseteq B$ and so $f_i qAB \subseteq Z_i$. Thus it is enough to deal with the case when $f_i q \in Z_i$ or, even better, when $f_i q \in U_j$, where $P_j = \mathrm{r\,ann}\,U_j = P_i$. However, $U_j \subseteq S/a_j^{-1} Y$ and 2.5(i) can be applied to deduce that either $x \in P_i$ or $f_i q = 0$.

Next suppose $f_i qRx = 0$, and suppose $x \notin P_i$; so, by (ii) $f_i x = xf_i \neq 0$. It follows from the above that $RxZ_i \triangleleft_{eR} (Z_i)_S$ and hence

$$(\bigoplus \{Z_j \mid j \in \{1, \ldots, w\}, j \neq i\}) \bigoplus RxZ_i \triangleleft_{eR} S_S.$$

Being essential in $_RS_S$ implies that it contains a nonzero ideal, A' say, of S. However, $qf_iZ_j = 0$, for $i \neq j$, and $qf_iRxZ_i = 0$; thus $qf_iA' = 0$ and so $qf_i = 0$. □

4.12 Our aim is to reduce from R and S to f_iR and f_iSf_i. Since f_i centralizes R, it is clear from 4.11(ii) that $f_iR \simeq R/P_i$. However, it is not clear even that f_iSf_i is a ring. The next result prepares the ground for such a result, at least when P_i is a minimal prime.

Lemma. *Let S be a prime ring and a finite normalizing extension of R. With the same notation as before, let P_i be a minimal prime of R, and $a \in S$ be such that $aR = Ra$ and $f_jaf_i \neq 0$ for some j. Then, for all $s \in S$, $f_jaf_is = f_jas$.*

Proof. Let D be the complement to P_i in $_RR_R$. Then $D \neq 0$ and $f_id = df_i = d$ for each $d \in D$. Since $f_ja \neq 0$ and $D \not\subseteq P_i$ it is clear from 4.11(iii) that $f_jaf_iD \neq 0$. We choose $D' \triangleleft _RR_R$ with $D'a = aD$. Then

$$0 \neq f_jaf_iD = f_jaDf_i = f_jD'af_i$$

and so $f_jD' \neq 0$. Pick $d' \in D'$ and then $d \in D$ so that $ad = d'a$. For each $s \in S$

$$d'f_jaf_is = f_jd'af_is = f_ja(df_i)s = f_jads = f_jd'as = d'f_jas.$$

It follows that $D'f_j(af_is - as) = 0$ and so, by 4.11(iii), $f_jaf_is = f_jas$. □

4.13 Theorem. *Let S be a prime ring and a finite normalizing extension of R, let $0 \neq I \triangleleft S$, let P_i be a minimal prime ideal of R and let f_i be the corresponding idempotent described above. Then f_iSf_i is a prime ring and is a finite automorphic extension of the prime ring f_iR, and f_iIf_i is a nonzero ideal of f_iSf_i.*

Proof. That f_iR is prime and $f_iR \subseteq f_iSf_i$ is clear. It is also clear from 4.12 that f_iSf_i is a ring and that $\{f_ia_jf_i | j = 1, \ldots, t\}$ is a set of normalizing generators over f_iR. Moreover, the ideals of f_iSf_i all have the form f_iAf_i with $A \triangleleft S$, and $f_iAf_if_iA'f_i = f_iAA'f_i$; so S is prime. However, 4.11(iii) shows that if $f_ia_jf_i \neq 0$ then

$$l \, \mathrm{ann}_R \, f_ia_jf_i = r \, \mathrm{ann}_R \, f_ia_jf_i = P_i$$

and therefore, by 2.2, $f_ia_jf_i$ induces a (unique) automorphism of f_iR. So the extension is automorphic.

Finally, let D be as in the preceding proof; so $Df_i = f_iD = D$. Then, since S is prime,

$$0 \neq DID = Df_iIf_iD$$

and so $f_iIf_i \neq 0$. □

4.14 We now draw the conclusion we have been seeking in this section.

Theorem. *If S is any finite normalizing extension of a ring R then essentiality and properness are both satisfied.*

Proof. First we consider properness. Let K be a maximal right ideal of R and $P = \operatorname{ann} K$. There is, by 2.9, a prime ideal I of S lying over P. After factoring out I and $I \cap R$, we may suppose I to be zero and P to be a minimal prime ideal of R, say $P = P_1$. Therefore $P = \operatorname{ann}_R f_1$ and so $R/K \simeq f_1 R/f_1 K$. Hence $f_1 K \neq f_1 R$. We can now apply 4.9 and 4.13 to deduce that $f_1 K f_1 S f_1 \neq f_1 S f_1$. However, $f_1 K f_1 S f_1 = f_1 K S f_1$; so $KS \neq S$. This is enough to prove properness.

Next comes essentiality. Again we may suppose S to be prime and $0 \neq I \triangleleft S$. It is required to show that $I \cap R \triangleleft_e {}_R R_R$. By 4.13, $0 \neq f_i I f_i \triangleleft f_i S f_i$ for each i corresponding to a minimal prime and, by 4.9, $f_i I f_i \cap f_i R \neq 0$. With D being the (nonzero) complement to P_i in ${}_R R_R$, as in 4.12, note that

$$0 \neq D(f_i I f_i \cap f_i R) D \subseteq I \cap D = (I \cap R) \cap D.$$

Thus $I \cap R$ has nonzero intersection with the complement of each minimal prime ideal P_i and so $I \cap R \triangleleft_e {}_R R_R$. □

4.15 Corollary. *Let S be a finite normalizing extension of R.*
(i) *(Incomparability) If $I_1, I_2 \in \operatorname{Spec} S$ with $I_1 \subset I_2$ then $I_1 \cap R \subset I_2 \cap R$.*
(ii) *If $I \in \operatorname{Spec} S$ lies over $Q \in \operatorname{Spec} R$ then I is maximal if and only if Q is maximal, and I is primitive if and only if Q is primitive.*
(iii) $J(R) = J(S) \cap R.$

Proof. Clear from 4.14, 2.13, 2.15 and 2.16. □

§5 Crossed Products and Fixed Rings

5.1 As noted in 1.4(ii), a crossed product $R*G$, with G a finite group, is a finite automorphic extension of R. The results of earlier sections can therefore be applied to $R*G$, and they can also be improved upon. This is the first aim of this section. Then the results are applied to the fixed ring R^G, if G is a finite group of automorphisms of R, making use of the Morita context, described in 7.8.6, which links R^G with $R\#G$.

5.2 Throughout this section R is a ring and G is a finite group. Furthermore, until 5.13, S denotes a crossed product $R*G$, as defined in 1.5.8.

There is, therefore, a set map $G \to S$ with $g \mapsto \bar{g}$, the latter being a unit in S. By assumption $\bar{g}R = R\bar{g}$; so $\bar{g}$ induces an automorphism of R, given by $r \mapsto r^g$, where $r\bar{g} = \bar{g}r^g$.

Note that the collection of R-modules $\{\bar{g}R | g \in G\}$ under multiplication, forms a group isomorphic to G.

5.3 The links between chain conditions in R and S are clear from 1.11 and 1.13.

Proposition. *If one of R and S is right Artinian or right Noetherian of right Krull dimension α, then so too is the other.* □

5.4 As in 1.8.2, an ideal A of R is G-**stable** if $A = A^g$ for all $g \in G$. Evidently, if A is G-stable then AS is an ideal of S, denoted by $A*G$, and $S/AS \simeq (R/A)*G$.

If A is G-stable and has the property that, whenever $BC \subseteq A$ for G-stable ideals B and C, then $B \subseteq A$ or $C \subseteq A$, then A is a G-**prime ideal**.

5.5 Such ideals are readily constructed.

Lemma. *If $D \triangleleft R$ then $\bigcap D^g$ is a G-stable ideal of R; and if $D \in \operatorname{Spec} R$ then $\bigcap D^g$ is a G-prime ideal.*

Proof. The fact that $\bar{g}R\bar{h}R = \overline{gh}R$ makes plain that $D^{gh} = (D^g)^h$. The proof is then straightforward. □

5.6 There is also a direct link with ideals of S.

Lemma. *(i) (Cutting down) If $I \in \operatorname{Spec}(R*G)$ then $I \cap R$ is a G-prime ideal of R.*
(ii) (Lying over) If A is a G-prime ideal of R then $A = I \cap R$ for some $I \in \operatorname{Spec}(R*G)$.

Proof. (i) Straightforward.
(ii) Choose $I \supseteq A*G$ maximal with respect to having $I \cap R = A$; then $I \in \operatorname{Spec}(R*G)$. □

5.7 The combination of 2.7 with 5.5 and 5.6 gives another description of G-prime ideals.

Corollary. *An ideal A of R is G-prime if and only if $A = \bigcap P^g$, where P is any of the minimal primes of A.* □

Of course the number of minimal primes of A is therefore at most $|G|$. Indeed, if $H = \{g \in G | P^g = P\}$ then the number of minimal primes of A is $|G|/|H|$.

5.8 Shortly, conditions will be established under which S is semiprime. First, however, we note a result converse to the remarks above.

Proposition. *Suppose that R is a G-prime ring and $S = R*G$ is semiprime. Then S has at most $|G|$ minimal primes. They are distinguished as those prime ideals having zero intersection with R.*

Proof. If $w = \mathrm{u\,dim}\,{_R}R_R$, this being at most $|G|$ from the remarks above, then $\mathrm{u\,dim}\,{_R}S_R = w|G|$. In particular if $0 \neq I \triangleleft S$ then $0 < \mathrm{u\,dim}\,{_S}I_S \leq \mathrm{u\,dim}\,{_S}S_S < \infty$. Suppose that $I \in \mathrm{Spec}\,S$ and so, by 5.6 and 5.7, $I \cap R = \bigcap P^g$ for some $P \in \mathrm{Spec}\,R$.

If P is not a minimal prime of R then neither is P^g for each g. Hence $P^g \triangleleft_{eR} R_R$ and so $I \cap R \triangleleft_{eR} R_R$. However, S is free over R and so $\sum \bar{g}(I \cap R) \triangleleft_{eR} S_R$ and $I \triangleleft_{eS} S_S$. Therefore, by 2.2.15, I cannot be a minimal prime of S.

On the other hand, if P is a minimal prime then $\bigcap P^g = 0 = I \cap R$ and 4.15 shows I to be minimal.

Since $\mathrm{u\,dim}\,{_S}S_S < \infty$, S has only finitely many minimal primes, say $I_1, \ldots, I_n$. Let K_j be the complement to I_j in S. By 2.2.15, $K_j \neq 0$ and, of course, K_j embeds in S/I_j as an ideal. However, S/I_j is a finite normalizing extension of R and so $K_j \cap R \triangleleft_{eR} R_R$ by 4.14. Hence $\mathrm{u\,dim}\,{_R}(K_j)_R \geq \mathrm{u\,dim}\,{_R}R_R = w$. Since the sum $\sum_{j=1}^n K_j \subseteq S$ is direct, it follows that $\mathrm{u\,dim}\,{_R}S_R \geq nw$. It was noted above that $\mathrm{u\,dim}\,{_R}S_R = |G|w$; so $n \leq |G|$. □

5.9 An R-module M is $|G|$-**torsion-free** if $m|G| = 0$ implies $m = 0$ for $m \in M$. Suppose that R itself is $|G|$-torsion-free. We will see that the hypothesis, in 5.8, that S is semiprime is then a consequence of R being G-prime. The demonstration of this depends upon the next result which could also be used to simplify the proof of 5.8, in this case by avoiding reference to 4.14 and 4.15.

5.10 Proposition. *Let $S = R*G$, M_S be $|G|$-torsion-free and $N \triangleleft_e M_S$. Then $N \triangleleft_e M_R$.*

Proof. Choose $L \triangleleft M_R$ so that $N \oplus L \triangleleft_e M_R$. Let $B = b(N \oplus L)$, the bound of $N \oplus L$ as in 1.7. By 1.7, $B \triangleleft_e M_R$ and $B \triangleleft_e M_S$. Moreover, since $B \supseteq N$, by definition, it follows that $B = N \oplus K$ for some $K \triangleleft B_R$.

Let $\pi: N \oplus K \to N$ be the natural projection and $\vartheta: B \to N$ be defined by $x \mapsto \sum \pi(x\bar{g})\bar{g}^{-1}$ for $x \in B$. One can check that ϑ is an S-module homomorphism and that, for $n \in N$, $\vartheta(n) = n|G|$. Thus $N \cap \ker \vartheta = 0$ and so $\ker \vartheta = 0$. However, for $x \in B$,

$$\vartheta(x|G| - \vartheta(x)) = \vartheta(x)|G| - \vartheta(x)|G| = 0$$

and so $\vartheta(x) = x|G| \in N$. Therefore $B|G| \subseteq N$. However, $B|G| \triangleleft_e B_R \triangleleft_e M_R$ and so $N \triangleleft_e M_R$ and $K = L = 0$. □

This result is not true for arbitrary finite normalizing extensions. For example, let S be the 2×2 upper triangular matrix ring over R, take $M = S$ and $N = Se_{22}S$.

Then $N \triangleleft_e M$ yet $N \cap R = 0$. The necessity of the assumption that M is $|G|$-torsion-free is made clear by considering the next result in the case when $R = k$, a field, and $S = kG$, viz. Maschke's theorem; for that too depends upon $|G|$-torsion-freeness.

5.11 Theorem. *If R is semiprime, and R is $|G|$-torsion-free then $R*G$ is semiprime.*

Proof. If N is a nilpotent ideal of S then, by 2.2.1, $\operatorname{l ann}_S N \triangleleft_e S_S$. However, S is $|G|$-torsion-free so, by 5.10, $\operatorname{l ann}_S N \triangleleft_e S_R$ and so $\operatorname{l ann}_R N \triangleleft_e R_R$. Since R is semiprime it follows that $\operatorname{r ann}_R (\operatorname{l ann}_R N)$ must be zero. The fact that S is free over R now ensures that $\operatorname{r ann}_S (\operatorname{l ann}_R N) = 0$ and so $N = 0$. □

5.12 We note two consequences regarding $\operatorname{Spec} R$ and $\operatorname{Spec} R*G$ in the case when $|G|$ is a unit of R, this ensuring, of course, that all factor rings of R are $|G|$-torsion-free.

Corollary. *Suppose that $|G|^{-1} \in R$.*
(i) *If A is a G-prime ideal of R then $A*G$ is semiprime and has at most $|G|$ minimal primes. Moreover, if $I \in \operatorname{Spec} R*G$ then I is a minimal prime of $A*G$ if and only if $I \cap R = A$.*
(ii) *(Going down) If $I \in \operatorname{Spec} R*G$ and A, A' are G-prime ideals of R with $A' \subset A$ and $I \cap R = A$, then there exists $I' \in \operatorname{Spec} R*G$ with $I' \subset I$ and $I' \cap R = A'$.*

Proof. (i) Pass to R/A and $R*G/A*G$ which takes the form $(R/A)*G$. By 5.11, this ring is semiprime and, by 5.8, has at most $|G|$ minimal primes, each distinguished by having zero intersection with R/A.

(ii) By (i), $A'*G$ is a finite intersection of prime ideals. Since $A'*G \subseteq I$, one of these primes, I' say, is contained in I. By (i) again, $I' \cap R = A'$. □

5.13 The second half of this section shows how the preceding results can be used in the study of prime ideals of fixed rings. As before R is a ring and G a finite group. However, it is now assumed that there is a group homomorphism $G \to \operatorname{Aut} R$ and that $S = R\#G$, the skew group ring defined in 1.5.4. This is, of course, also a crossed product so that earlier results apply. The strategy employed in the study of the relationship between primes of R and of the fixed ring R^G is to use $R\#G$ as a connecting link.

5.14 First we consider the special case when $|G|$ is a unit in R. We recall that there is, by 7.8.7, a Morita context

$$\begin{bmatrix} S & Se \\ eS & eSe \end{bmatrix}$$

with $e = e^2 \in S = R\#G$ and $R^G \simeq eSe$. The elementary facts about prime ideals in Morita contexts described in 3.6.3(iv) can thus be applied.

Proposition. *If $|G|^{-1} \in R$ then there is an order preserving $(1,1)$-correspondence between the set $\{I \in \operatorname{Spec} R\#G | e \notin I\}$ and $\operatorname{Spec} R^G$.* □

5.15 This is easily combined with the preceding results.

Theorem. *Suppose $|G|^{-1} \in R$.*
 (i) (Cutting down) *If $P \in \operatorname{Spec} R$ then $P \cap R^G$ is a semiprime ideal having at most $|G|$ minimal primes.*
 (ii) (Lying over) *If $Q \in \operatorname{Spec} R^G$ then there exists $P \in \operatorname{Spec} R$ such that Q is a minimal prime of $P \cap R^G$.*
 (iii) (Going up and going down) *If P, Q are as in (ii) and $Q', Q'' \in \operatorname{Spec} R^G$ with $Q' \subset Q \subset Q''$, then there exist $P', P'' \in \operatorname{Spec} R$ with $P' \subset P \subset P''$ such that Q' is minimal over $P' \cap R$ and Q'' is minimal over $P'' \cap R$.*

Proof. (i) Note that $A = \bigcap P^g$ is a G-prime ideal by 5.7 and, by 5.12, there are at most $|G|$ prime ideals I_i of $S = R\#G$ minimal over $A\#G$ with $I_i \cap R = A$ and $\bigcap I_i = A\#G$. Evidently $\bigcap eI_ie = eAe$ and, by 5.14, each ideal eI_ie is either a prime ideal of eRe or else, if $e \in I_i$, is eRe itself.

On the other hand, $P \cap R^G = P^g \cap R^G$ for each $g \in G$ and so $P \cap R^G = A \cap R^G$. The isomorphism $R^G \simeq eR^Ge$ gives $ePe = eAe = \bigcap eI_ie$ and the result follows.

 (ii) Let $I \in \operatorname{Spec} S$ correspond, via 5.14, with $Q \in \operatorname{Spec} R^G$ and suppose $I \cap R = \bigcap P^g$. Then the argument above demonstrates that Q is a minimal prime of $P \cap R^G$.

 (iii) These now follow easily using 5.12. □

5.16 Finally, we aim towards Goldie conditions in the more general case when R is $|G|$-torsion-free, although our first remarks do not rely upon that hypothesis. Rather they concern the trace map $\tau: R \to R^G$ which was defined in 7.8.6 by $\tau(r) = \sum \{r^g | g \in G\}$. If $A \triangleleft_r R$ then $\tau(A) = \{\tau(a) | a \in A\}$. In particular, $\tau(R)$ is called the **trace ideal**.

The next result indicates the value of the hypothesis of being $|G|$-torsion-free.

Lemma. (i) *τ is an (R^G, R^G)-bimodule homomorphism from R to R^G.*
 (ii) *If $A \triangleleft_r R^G$ then $|G|A \subseteq \tau(AR) \subseteq A$.*

Proof. Straightforward. □

5.17 Recall that f denotes the element $\sum\{g | g \in G\} \in S = R\#G$. Thus $fg = f$, $f^2 = |G|f$ and $frf = f\tau(r)$ for any $r \in R$. Moreover, $R \simeq fR = fS$, and this gives

R a right S-module structure. The submodules of R_S are then simply the G-stable right ideals of R.

Lemma. (i) *If $E \triangleleft_r R$ then $(RfE)^{n+1} = Rf(\tau(E))^n E$ for all $n \geq 0$.*

(ii) *If E is a G-stable right ideal of R then fE is a right ideal, and RfE an ideal, of $R\#G$. Moreover, $fEf = f\tau(E) \subseteq f(E \cap R^G)$.*

Proof. Straightforward. □

5.18 We now impose some hypotheses upon R.

Proposition. *Let R be semiprime and $|G|$-torsion-free, let $E \triangleleft_r R$ and $A \triangleleft_r R^G$.*
(i) *If E is a nonzero G-stable right ideal then $0 \neq \tau(E) \subseteq E \cap R^G$ and $\operatorname{u\,dim} E \leq |G| \operatorname{u\,dim} E^G$.*
(ii) *If $E \triangleleft_e R$ then $E^G \triangleleft_e R^G$.*
(iii) $\operatorname{u\,dim} A \leq \operatorname{u\,dim} AR$.
(iv) *If $A \triangleleft_e R^G$ then $AR \triangleleft_e R$.*

Proof. (i) Since $R_S \simeq (fR)_S$ then $0 \neq fE \triangleleft_r S = R\#G$. However, S is semiprime, by 5.11, so $fEf \neq 0$. Hence by 5.17, $0 \neq \tau(E) \subseteq E \cap R^g$. It is clear now that $\operatorname{u\,dim} E_S = \operatorname{u\,dim} (fE)_S \leq \operatorname{u\,dim} E^G$ and so, by 1.9, $\operatorname{u\,dim} E_R \leq |G| \operatorname{u\,dim} E^G$.

(ii) Let $E' = \bigcap \{E^g \mid g \in G\}$; then $E' \triangleleft_e R_R$ and E' is G-stable. Let $0 \neq X \triangleleft_r R^G$. Then $E' \cap XR \neq 0$ and so, by (i), $0 \neq \tau(E' \cap XR) \subseteq E^G \cap X$.

(iii) If $A_1 \oplus \cdots \oplus A_k$ is a direct sum of right ideals of R^G, then (i) shows that the sum $\sum A_i R$ is direct.

(iv) It is clear using (i) that $AR \triangleleft_e R_S$ and so, by 5.10, $AR \triangleleft_e R_R$. □

5.19 Theorem. *Let R be semiprime and $|G|$-torsion-free. Then:*
(i) R^G *is semiprime;*
(ii) $(\zeta(R))^G = \zeta(R^G)$;
(iii) $\operatorname{u\,dim} R^G \leq \operatorname{u\,dim} R \leq |G|(\operatorname{u\,dim} R^G)$;
(iv) R^G *is semisimple Artinian if and only if R is semisimple Artinian;*
(v) R^G *is right Goldie if and only if R is right Goldie and then $Q(R^G) = (Q(R))^G$.*

Proof. (i) Let $0 \neq A \triangleleft_r R^G$ and $E = AR$; so $\tau(E) \subseteq A$. However, $0 \neq RfE$ which is an ideal of S and so, by 5.17, $0 \neq (\tau(E))^n$ for all n. Thus A is not nilpotent.

(ii), (iii) Clear from 5.18.

(iv) If R^G is semisimple Artinian and $E \triangleleft_e R$ then 5.18(ii) shows that $E^G \triangleleft_e R^G$ and so $E^G = R^G$. Thus $1 \in E$ and so R is semisimple Artinian.

Conversely, if R is semisimple Artinian it is clear that $|G|^{-1} \in R$. If $A \triangleleft_e R^G$ it follows that $AR = R$ and so $|G| \in \tau(A) \subseteq A$. Hence R^G is semisimple Artinian.

(v) It is clear from 5.18 that if R is right Goldie then so too is R^G. The converse is also clear once it is noted that $\zeta(R)$ is G-stable.

10.6.4   *Primes in polynomial rings*   369

Now suppose c is a regular element of R^G. Then $cR^G \triangleleft_e R^G$ and so $cR \triangleleft_e R$ by 5.18(iv). Therefore $c^{-1} \in Q(R)^G$ and so $Q(R^G) \subseteq Q(R)^G$.

On the other hand, if $x \in \mathscr{C}_R(0)$ then $xR \supseteq cR$ for some regular $c \in R^G$, using 5.18(ii). Therefore $x^{-1} = rc^{-1}$ for some $r \in R$ and hence each $q \in Q(R)$ has the form $q = rc^{-1}$ with $r \in R$, $c \in R^G$. It is clear that $q \in Q(R)^G$ only if $r \in R^G$. Thus $Q(R)^G \subseteq Q(R^G)$. □

§6 Primes in Polynomial Rings

6.1 This section concerns the relationship between the prime ideals of three rings; namely, a ring R with an automorphism σ, the ring $S = R[x; \sigma]$ and the ring $T = R[x, x^{-1}; \sigma]$. **This notation will remain fixed throughout the section.**

It will be seen that the investigation divides into two cases depending upon whether some power of σ is, in a sense, X-inner. When it is, the results of the preceding sections become applicable.

6.2 The prime ideals of S fall naturally into two classes, those containing x and those not doing so. The former correspond directly with the prime ideals of R and require no further attention. The latter class is thus the object of our attention. It will be shown that it is related closely to Spec T.

6.3 Note that σ extends to automorphisms of both S and T. Thus the notions of σ-stable ideals and σ-prime ideals, as in 5.4, make sense in all three rings. We let σ-**Spec** denote the set of σ-prime ideals of a ring. The next two results summarise some properties of σ-stable ideals and σ-prime ideals.

Lemma. (i) *If $A \triangleleft T$ then A is σ-stable, $A = (A \cap S)T$ and $A \cap S$ is a σ-stable ideal of S.*

(ii) *If $B \triangleleft S$ and B is σ-stable then $BT \triangleleft T$ and $B \cap R$ is a σ-stable ideal of R.*

(iii) *If C is a σ-stable ideal of R then CS and CT are σ-stable ideals of S and T respectively. Furthermore, $S/CS \simeq (R/C)[x; \sigma]$ and $T/CT \simeq (R/C)[x, x^{-1}; \sigma]$.*

Proof. Straightforward. □

6.4 Lemma. (i) *σ-Spec $T =$ Spec T; and if $A \in$ Spec T then $A \cap S \in \sigma$-Spec S and $x \notin A \cap S$.*

(ii) *If $B \in$ Spec S and $x \notin B$ then $B \in \sigma$-Spec S.*

(iii) *If $0 \neq B \in \sigma$-Spec S with $x \notin B$, then $BT \in$ Spec T, $B = BT \cap S \nsubseteq xS$ and $B \cap R \in \sigma$-Spec R.*

(iv) *If $C \in \sigma$-Spec R then $CS \in \sigma$-Spec S with $x \notin CS$, and $CT \in$ Spec T.*

(v) *If $A \in$ Spec T then $A \cap R \in \sigma$-Spec R and $(A \cap R)T \in$ Spec T.*

Proof. (i) Clearly, $x \notin A \cap S$ and 6.3(i) makes clear that σ-Spec $T =$ Spec T and that $A \cap S$ is σ-stable. If I, J are σ-stable ideals of S with $IJ \subseteq A \cap S$, then IT, JT

are ideals of T with $(IT)(JT) = (IJ)T \subseteq A$ and thus $IT \subseteq A$ or $JT \subseteq A$. Hence $I \subseteq A \cap S$ or $J \subseteq A \cap S$.

(ii) It is enough to show that B is σ-stable. Now $B \supseteq Bx = x\sigma(B) = xS\sigma(B)$, and then $\sigma(B) \subseteq B$ since $x \notin B$. However, $\sigma(B) \in \operatorname{Spec} S$, $x \notin \sigma(B)$ and yet $\sigma(B) \supseteq BSx$; so $B \subseteq \sigma(B)$.

(iii) Note first that $B \nsubseteq xS$ or else $B = xSB'$ for some σ-ideal $B' \supset B$. To see that $BT \cap S = B$ consider $c \in BT$, say $c = \sum_{j=0}^{n} b_j x^{-j}$ with $b_j \in B$, $n \in \mathbb{N}$. Then $cx^n \in B$ and indeed $\sigma^m(c) \in B$ for all $m \in \mathbb{Z}$ since $B = \sigma(B)$. Therefore, if $c \in BT \cap S$ then $(\sum_{m \in \mathbb{Z}} S\sigma^m(c)S)Sx^nS \subseteq B$. However, $x \notin B$ and so $x^n \notin B$; and thus $ScS \subseteq B$ and $c \in B$.

The rest is straightforward.

(iv) It is enough to consider the case when $C = 0$. Let $I, J \triangleleft T$ with $IJ = 0$ and let I', J' denote the ideals of leading coefficients, in R, of these ideals. Evidently I', J' are σ-stable ideals of R and $I'J' = 0$. Thus I or J is zero and T is prime. Hence, by (i), S is σ-prime.

(v) Clear from above. □

6.5 It is possible for R and S to be σ-prime without being prime. For example, let k be a field and $k_i \simeq k$ for $i \in \mathbb{Z}$. Let $R = \prod_{i \in \mathbb{Z}} k_i$ and σ be the shift map given by $k_i \xrightarrow{\sim} k_{i+1}$ for each i. It is easy to check that R is σ-prime since, if $0 \neq a, b \in R$, then $a\sigma^m(b) \neq 0$ for some $m \in \mathbb{Z}$. Nevertheless, the ideal $I = Sxk_0S$ satisfies $I^2 = 0$. By 6.4(iv), S is σ-prime and T is prime.

In fact S shows that a σ-prime ring need not even be semiprime, in contrast to the Noetherian case.

Lemma. *If R is right Noetherian and σ-prime then R is semiprime.*

Proof. Here $N(R)$ is σ-stable and nilpotent. Hence $N(R)T$ is a nilpotent ideal of T; yet T is prime. □

6.6 The next main result can be viewed as a weak form of incomparability. It asserts that a chain of prime ideals of T, each of which has the same intersection with R, can involve no more than two distinct primes. Before proving this in general we first consider the case when R is Noetherian—when a much simpler proof is available.

Theorem. *Let R be Noetherian and let $P_0 \subset P_1 \subseteq P_2$ be prime ideals of T such that $P_0 \cap R = P_2 \cap R$. Then $P_1 = P_2$.*

Proof. It is clear from 6.4(v) that we may suppose that $P_0 = 0$ and so $P_2 \cap R = 0$. This means that R is σ-prime and hence R must be semiprime. By Goldie's theorem, R has a quotient ring, Q say, which is semisimple Artinian. The automorphism σ extends uniquely to Q so $T \subseteq Q[x, x^{-1}; \sigma] = T'$, say.

10.6.9 *Primes in polynomial rings* 371

It is clear from 2.1.16 that $P_1 T'$ and $P_2 T'$ are prime ideals of T' and that $P_i T' \cap T = P_i$, since $P_i \cap R = 0$. However, 6.5.4 shows that T' has Krull dimension 1; so $P_1 T' = P_2 T'$ and $P_1 = P_2$. □

6.7 We now aim at the general result, starting with some notation. For $t \in T$, $\hat{t}$ denotes its highest degree term. The **length** of t is the difference in degree between its highest and lowest degree terms; and t is **homogeneous** if length $t = 0$.

6.8 Lemma. *Let $0 \neq I \triangleleft T$ and let $a, b \in I$ be nonzero with a having minimal possible length; say length $a = m$, length $b = n$.*
(i) *For any homogeneous element $t_0 \in T$*
$$\text{length } (\hat{a} t_0 b - a t_0 \hat{b}) < \text{length } b.$$
(ii) *For any $t \in T$, $\hat{a} t a = a t \hat{a}$.*
(iii) *Given any homogeneous elements $t_1, \ldots, t_{n-m} \in T$ there exists $c \in T$ such that, for all $t \in T$,*
$$\hat{a} t \hat{a} t_{n-m} \cdots \hat{a} t_1 b = a t c.$$

Proof. (i) Clear.

(ii) When t is homogeneous this comes from (i); and the general case then follows.

(iii) If $n = m$ the argument in (ii) also shows that $\hat{a} t b = a t \hat{b}$ and we may set $c = \hat{b}$. In general we proceed by induction on n. Let $b' = \hat{a} t_1 b - a t_1 \hat{b}$ and $p = $ length b'. By (i), $p < n$. Hence, by induction, there exists $c' \in T$ such that
$$\hat{a} t \hat{a} t_{p-m+1} \cdots \hat{a} t_2 b' = a t c'$$
for all $t \in T$. Replacing t by $t \hat{a} t_{n-m} \cdots \hat{a} t_{p-m+2}$ and using (ii), we obtain the equation
$$\hat{a} t \hat{a} t_{n-m} \cdots \hat{a} t_1 b = a t c,$$
where
$$c = \hat{a} t_{n-m} \cdots \hat{a} t_{p-m+2}(c' + \hat{a} t_{p-m+1} \cdots \hat{a} t_1 \hat{b}).$$
□

6.9 Theorem. (i) *If $P_0 \subset P_1 \subseteq P_2$ are prime ideals of T with $P_0 \cap R = P_2 \cap R$, then $P_1 = P_2$.*
(ii) *If $P_0 \subset P_1 \subseteq P_2$ are σ-prime ideals of S with $P_0 \cap R = P_2 \cap R$, then $P_1 = P_2$.*

Proof. (i) By passing to $R/P_0 \cap R$ and $T/(P_0 \cap R)T$ it may be supposed that T is prime and $P_2 \cap R = 0$. Let $a \in P_2$ and $b \in P_1$ be of minimal length, m and n respectively, amongst nonzero elements of those ideals. Note that $m \geq 1$ because $P_2 \cap R = 0$. Since T is prime, there exists $t \in T$ such that $\hat{a} t b \neq 0$. Hence there is a homogeneous element t_1 such that $\hat{a} t_1 b \neq 0$. By induction we can obtain homogeneous elements $t_2, \ldots, t_{n-m}$ such that the element $d = \hat{a} t_{n-m} \cdots \hat{a} t_1 b$ is nonzero.

By 6.8(iii), there exists $c \in T$ such that $\hat{a}td = atc$ for each $t \in T$. One can choose four homogeneous values for t so that, in turn, the product of the highest or lowest degree terms in $\hat{a}td$ or atc is nonzero. It follows readily that

$$\text{length } \hat{a} + \text{length } d = \text{length } a + \text{length } c$$

and so

$$\text{length } c = \text{length } d - \text{length } a$$
$$\leqslant \text{length } b - \text{length } a = n - m < n.$$

Thus $c \notin P_1$. However, $aTc \subseteq Tb \subseteq P_1$ and so $a \in P_1$.

Finally, we aim to show, by induction on length, that each element $e \in P_2$ belongs to P_1. Note that, for any homogeneous element $t_0 \in T$,

$$\text{length}(\hat{a}t_0 e - at_0 \hat{e}) < \text{length } e$$

and, moreover, $\hat{a}t_0 e - at_0 \hat{e} \in P_2$. By induction, $\hat{a}t_0 e - at_0 \hat{e} \in P_1$ and so $\hat{a}t_0 e \in P_1$. Hence $\hat{a}Te \subseteq P_1$, yet $\hat{a} \notin P_1$ since $P_1 \cap R = 0$. Thus $e \in P_1$ and $P_2 = P_1$.

(ii) Once again it may be supposed that T is prime and $P_2 \cap R = 0$. If $x \in P_2$ then $P_2 = (x)$, since $P_2 \cap R = 0$. By 6.4 (iii), it follows that $x \in P_1$ and so $P_1 = P_2$.

On the other hand, if $x \notin P_2$, and so $x \notin P_1$, then 6.4(iii) shows that $0, P_1 T$ and $P_2 T$ belong to Spec T, that $0 \subset P_1 T \subseteq P_2 T$ and that $0 = P_1 T \cap R = P_2 T \cap R$. Hence (i) shows that $P_1 T = P_2 T$ and so, by 6.4(iii), $P_1 = P_2$. □

The remainder of this section aims to determine precisely when there exists a chain of two prime ideals of T with equal intersection with R. This requires some preliminary work.

6.10 First comes an elementary result.

Lemma. *If $B \triangleleft S$ and $A = \text{l ann } B$, then $\sigma(A) \subseteq A$. Furthermore, if B is σ-stable then so also is A.*

Proof. Here $x\sigma(A)B = AxB \subseteq AB = 0$ and so $\sigma(A)B = 0$ and $\sigma(A) \subseteq A$. If B is σ-stable then $AB = 0$ if and only if $\sigma(A)\sigma(B) = 0$, i.e. if and only if $\sigma(A)B = 0$. □

6.11 Recall from 6.9.8, 6.9.9 that if $P \in \text{Spec } R$ and $\sigma^n(P) = P$ for some n then P is σ-semistable and $P^0 = \bigcap \sigma^k(P)$ is the largest σ-stable ideal contained in P. Such an ideal P^0 is called σ-**cyclic**.

Lemma. *A σ-cyclic ideal is σ-prime.*

Proof. Straightforward. □

6.12 For right Noetherian rings a converse is valid, although it will not be used here.

Proposition. *If R is right Noetherian and $A \in \sigma$-Spec R then A is σ-cyclic and $AS \in$ Spec S. Also $AT \in$ Spec T.*

Proof. It is shown in 6.5 that A must be semiprime. Suppose $P_1, \ldots, P_n$ are the minimal primes of A. So $P_1 \cap \cdots \cap P_n = A = \sigma(A) = \sigma(P_1) \cap \cdots \cap \sigma(P_n)$. Thus σ permutes the set $\{P_1, \ldots, P_n\}$. If $P_1, \ldots, P_r$ is one cycle of this permutation and $C = P_1 \cap \cdots \cap P_r, D = P_{r+1} \cap \cdots \cap P_n$ then both C and D are σ-stable and $CD \subseteq A$. Hence $A = C$ and is σ-cyclic.

To show that $AS \in$ Spec S it is enough to assume $A = 0$ and prove S prime. Let B, C be ideals of S with $BC = 0$. We may suppose $B = \mathrm{l\,ann}_S C$ and $C = \mathrm{r\,ann}_S B$. By 6.10, $\sigma(B) \subseteq B$ and so there is an ascending chain $B \subseteq \sigma^{-1}(B) \subseteq \cdots \subseteq \sigma^{-n}B \subseteq \ldots$. The Noetherian hypothesis implies that $B = \sigma(B)$. Hence B and, similarly, C are σ-stable, and so too are their ideals of leading coefficients, B', C' say. But $B'C' = 0$ so one is zero. Hence S is prime.

Finally, 6.4 shows that AT is prime. □

6.13 Returning to an arbitrary ring R, we now study the situation when some power of σ, say σ^m, is inner. Thus $\sigma^m(r) = uru^{-1}$ for some unit u of R. In this case there is a connection with earlier sections.

Lemma. *Let σ^m be inner, being given by a unit u as above. Then S is a finite normalizing extension of the ring $S' = R[ux^m]$, this being a polynomial ring in the central indeterminate ux^m. Also T is a crossed product $T' * G$, where $T' = R[ux^m, (ux^m)^{-1}]$, the Laurent polynomial ring in the central indeterminate ux^m, and G is cyclic of order m.*

Proof. (i) This is easily checked.

(ii) If g generates G we map g^i to x^i for $i = 0, \ldots, n-1$. The rest is straightforward. □

6.14 A result similar to 6.12 is valid in the case.

Proposition. *If some power of σ is inner and $A \in \sigma$-Spec R then A is σ-cyclic and $AS \in$ Spec S.*

Proof. Note first that if $B \triangleleft S$ with $\sigma(B) \subseteq B$ then $B = \sigma^m(B) \subseteq \cdots \subseteq \sigma(B) \subseteq B$ and so B is a σ-ideal. The same proof as in 6.12 now shows that $AS \in$ Spec S.

Before proving that A is σ-cyclic note that it is enough to consider the case when $A = 0$. Then S is prime and so S', as described in 6.13, must be semiprime of finite rank, using 2.4. Let $Q_1, \ldots, Q_n$ be the minimal primes of S'. It is evident that $ux^m S'$ is an essential ideal of S', so $ux^m \notin Q_i$ for $i = 1, \ldots, n$. Therefore, by 6.4, $Q_i \cap R \in$ Spec R. Now $\bigcap (Q_i \cap R) = 0$ so R is semiprime of finite rank. It follows, as in the proof of 6.12, that 0 is σ-cyclic. □

6.15 Before proceeding further a modified version of normal closure is required. Let R be a σ-prime ring and let $\mathcal{F}$ be the collection of nonzero σ-stable ideals of R. Clearly, this is a right localization set, so a right localization $R_{\mathcal{F}}$ exists. Let $Q = R_{\mathcal{F}}$, let $N_i = \{n \in Q \mid rn = n\sigma^i(r) \text{ for all } r \in R\}$, and let $N = \bigcup_{i \in \mathbb{Z}} N_i$.

It is clear that $\sigma(N_i) = N_i$ and $\sigma(N) = N$, and that σ extends to Q and to the ring NR. Thus one can form rings $NS = NR[x;\sigma]$ and $NT = NR[x, x^{-1};\sigma]$.

If σ^m is inner on NR for some m then σ^m is called **X-inner**, as in 3.10. The strategy now is to study this circumstance by reducing to the closure NR and applying 6.14.

6.16 First the reduction.

Proposition. *Let R be σ-prime and $\mathcal{F}, N$ be as above.*
(i) There is a $(1,1)$-correspondence

$$\{P \in \sigma\text{-Spec } S \mid P \cap R = 0\} \leftrightarrow \{I \in \sigma\text{-Spec } NS \mid I \cap NR = 0\}$$

via $P \mapsto P^e = \{q \in NS \mid qX \subseteq P \text{ for some } X \in \mathcal{F}\}$ and $I \mapsto I \cap S$.

(ii) The analogous correspondence between primes of T and of NT also holds.

Proof. (i) First we check that $I = (I \cap S)^e$ and that $P = P^e \cap S$. For the former, suppose $i \in I$. Of course $iX \subseteq S$ for some $X \in \mathcal{F}$; so $iX \subseteq I \cap S$ and $i \in (I \cap S)^e$. Conversely, if $q \in (I \cap S)^e$ then $qX \subseteq I \cap S$ for some $X \in \mathcal{F}$. Note that $qNSX = qXNS$, since X is σ-stable, and thus $qNSX \subseteq I$. However, I is σ-prime and NSX is σ-stable so, using 6.10, $q \in I$. Thus $I = (I \cap S)^e$.

For the latter, suppose $s \in P^e \cap S$. Then $sX \subseteq P$ for some $X \in \mathcal{F}$ and so $sSX = sXS \subseteq P$. Thus $s \in P$ and $P^e \cap S \subseteq P$. The reverse containment is trivial.

Next we check that $I \cap S$ and P^e are σ-prime. Suppose A, B are σ-stable ideals of S with $AB \subseteq I \cap S$. Then NA, NB are σ-stable ideals of NS and $NANB = NAB \subseteq I$. Therefore $A \subseteq I \cap S$ or $B \subseteq I \cap S$, as required.

Suppose next that $A \triangleleft NS$ with $A \cap S \subseteq P$. For any $a \in A$ there exists $X \in \mathcal{F}$ with $aX \subseteq S$. Hence $aX \subseteq A \cap S \subseteq P$ and so $a \in P^e$. Therefore $A \subseteq P^e$. It follows easily that P^e is σ-prime.

Finally, note that if $I \cap NR = 0$ then $(I \cap S) \cap R = 0$; and conversely, if $0 \neq i \in I \cap NR$ then $0 \neq iX \subseteq I \cap R$ for some $X \in \mathcal{F}$, and so if $I \cap NR \neq 0$ then $(I \cap S) \cap R \neq 0$.

(ii) This is proved similarly. □

6.17 As promised, the final theorem combines these results to determine when two comparable primes of T can have equal intersection with R.

Theorem. *The following statements are equivalent.*
(i) Some power of σ is X-inner on some σ-prime factor ring of R.

(ii) *There exist* $P_0, P_1 \in \sigma\text{-Spec } S$ *with* $P_0 \subset P_1$, $P_0 \cap R = P_1 \cap R$ *and* $x \notin P_1$.
(iii) *There exist* $I_0, I_1 \in \text{Spec } T$ *with* $I_0 \subset I_1$ *and* $I_0 \cap R = I_1 \cap R$.

Proof. (i)⇒(ii) It is enough to consider the case when R is σ-prime and σ^m is X-inner on R. Then $0 \in \sigma\text{-Spec } S$, so we need only construct a nonzero σ-prime P_1 of S with $x \notin P_1$ and $P_1 \cap R = 0$.

Suppose, for a moment, that there is a nonzero σ-prime ideal P_1' of NS with $x \notin P_1$ and $P_1' \cap NR = 0$. Then 6.16 shows that $P_1 = P_1' \cap S$ is as required. Now σ^m is inner on NR, so it is enough to deal with the case when σ^m is inner on R.

In that case, let $S' = R[ux^m]$, as in 6.13, and let J be the ideal of S' generated by $1 + ux^m$. Then $J \cap R = 0$ and $S'/J \simeq R$. Thus, by 6.14, J is σ-cyclic; so $J = \bigcap_{k=1}^m \sigma^k(H)$ for some $H \in \text{Spec } S'$. By 2.9 there is a prime ideal, K say, of S lying over H; i.e. H is a minimal prime of $K \cap S'$. Moreover, 2.7(ii) shows that $K \cap S' = \bigcap_{k=1}^m \sigma^k(H) = J$. The same is true, of course, for $\sigma^k(K)$ with $1 \leqslant k \leqslant m$. Thus if $P_1 = \bigcap_{k=1}^m \sigma^k(K)$ then P_1 is σ-prime, by 6.11, and $P_1 \cap R = J \cap R = 0$. Since $ux^m \notin J$ it is clear that $x \notin P_1$.

(ii)⇒(iii) Let $I_0 = P_0 T$, $I_1 = P_1 T$ and apply 6.4.

(iii)⇒(i) We may suppose that $I_0 \cap R = I_1 \cap R = 0$, that R is σ-prime and that $I_1 \neq 0$. Suppose that the minimal length of a nonzero element of I_1 is m; so there exist elements of the form $p_0 + xp_1 + \cdots + x^m p_m$ with $p_0 \neq 0$ and $p_m \neq 0$. The minimality ensures that such elements are uniquely specified by their first coefficient p_0. The set of all such p_0, together with 0, forms a σ-ideal, B say, of R. The map $B \to R$ given by $p_0 \mapsto p_m$ gives an element q in the quotient ring $Q = R_\mathcal{F}$. However, for all $a \in R$,

$$a(p_0 + xp_1 + \cdots + x^m p_m) = ap_0 + x\sigma(a)p_1 + \cdots + x^m \sigma^m(a)p_m$$

and so $qap_0 = \sigma^m(a)p_m = \sigma^m(a)qp_0$ for all $p_0 \in B$. Therefore $qa = \sigma^m(a)q$ and so σ^m is X-inner. □

§7 Additional Remarks

7.0 (a) This material has its origins in the study of finite central extensions such as orders over Dedekind domains and in finite group rings and crossed products and in the versions of the Krull relations between their prime ideals. Normalizing extensions form a natural common generalization. The module-theoretic approach adopted here is developed from ideas of [Bergman **U, Ua, Ub**].

(b) In the light of the commutative theory one might have expected an approach involving integrality. There is a type of integrality available here, as [Lorenz **81**] shows, but it does not appear to yield the results described here. It is used, however, by [Stewart **83**]. See also 13.11.8(b).

7.1 (a) This follows [Bit-David and Robson **80**] and [Bit-David **80**]. However, 1.11(i), (ii) are due to [Formanek and Jategaonkar **74**] and 1.11(iii) to [Lanski **80**].

(b) [Lemonnier **78**] improves upon 1.10 by showing, without the Noetherian hypothesis, that $K(M_S) = K(M_R)$.

(c) Some related types of extension are considered in [Robson and Small **81**], [Heinicke and Robson **84**], [Lemonnier **84**] and [Whelan **86**].

7.2 (a) This follows [Heinicke and Robson **81**] and [Lorenz **81**] except that 2.9 comes from [Bit-David and Robson **80**].

(b) A rather different approach to some of these results can be found in [Passman **81**].

7.3 (a) The Martindale quotient ring and the central closure appeared first in [Martindale **69**], X-inner automorphisms in [Kharchenko **75**], and the normal closure in [Montgomery **79**].

(b) The conditions in 3.3 are slightly weaker than usual in order to encompass the Martindale quotient ring. As mentioned in 2.4.1(b) there are many types of quotient ring.

7.4 This section follows [Heinicke and Robson **81**]. The automorphic case was also dealt with in [Lorenz **81**]. Further generalizations appear in [Heinicke and Robson **84**].

7.5 (a) This is only a brief account of some aspects of the area covered by [Montgomery **80**]; see also [Passman **86**], [Montgomery **P**].

(b) 5.8 appears in [Lorenz and Passman **79**], 5.10 in [Passman **83**], 5.11 in [Fisher and Montgomery **78**] and 5.15 in [Montgomery **81**]. 5.19 combines results of [Bergman and Isaacs **73**], [Kharchenko **74**], [Fisher and Osterburg **78**], [Cohen **75**], [Levitzki **35**] and [Cohen and Montgomery **75**].

7.6 (a) This again is only a brief account of an area. In particular, more general results about $R*G$ for certain infinite groups can be found in [Chin **87**], [Chin and Quinn **P**], and [Passman **Pa**].

(b) This account builds upon [Goldie and Michler **74**], [Pearson, Stephenson and Watters **81**], [Pearson and Stephenson **77**] and [Jordan **78**], with 6.9 coming from [Bergman **Uc**].

(c) Results concerning $R[x;\delta]$ and, more generally, $R*U(g)$ are discussed in Chapter 14 §2.

(d) Prime ideals in $R[x;\sigma,\delta]$ for R commutative are discussed by [Irving **79a**, **79b**].

Chapter 11
STABILITY

It is a common phenomenon that the structure of modules over a ring R becomes more regular as their size increases relative to R_R, with this process continuing up to a certain point at which stability is achieved. This chapter concerns various aspects of this phenomenon.

The first aspect involves stably free modules, a module A_R being stably free of rank t provided that $A \oplus R^s \simeq R^{s+t}$ for some s. It is clear from 7.11.6 that this need not entail A being free, and §2 provides many further examples. The point at which this property stabilizes is the rank beyond which all stably free modules are free. In §1 it is shown that this is determined by the behaviour of the general linear group $Gl(n, R)$. In §3 it is shown that there is a connection, via the elementary group $E(n, R)$, with the stable range as introduced in 6.7.2. The values of these ranges for certain rings are described in §5.

The other aspects of stability discussed here, concern the process of adding copies of R to more general modules A, B. On the one hand, one asks whether, for 'large' A, an isomorphism $A \oplus R \simeq B \oplus R$ implies that $A \simeq B$; and on the other hand, whether every 'large' A takes the form $A = C \oplus R$ for some C. The former question is investigated in §4, the answers obtained once again involving $Gl(n, R)$ and the stable range.

So far the results do not rely on the rings being Noetherian. However, for commutative Noetherian rings there is a well-known and complete answer to both questions involving the local behaviour of modules at prime ideals. Rather remarkably [Stafford 81] showed that these answers can be extended to the noncommutative case. In §6 the local information is described, this involving reduced rank and Krull dimension. Then in §7 the answers are obtained and compared with those for commutative rings.

§1 Stably Free Modules

1.1 As noted in the introduction, a module P_R is **stably free of rank** t if $P \oplus R^s \simeq R^{s+t}$ for some s. In this section this property is shown to have connections with free resolutions and with invertible matrices, and an upper bound on the rank of nonfree stably free modules is noted.

1.2 The rank of a stably free module is readily seen to be well defined provided that R has the **invariant basis property**, namely that $R^m \simeq R^n$ only if $m = n$. Clearly, this is so for any ring which has a factor ring with finite uniform dimension. In particular it holds for right Noetherian rings and for commutative rings. **Throughout this chapter** it will be assumed that R has the invariant basis property.

1.3 If R_R has finite uniform dimension and P_R is stably free it is easily checked that

$$\operatorname{rank} P = \operatorname{u\,dim} P_R / \operatorname{u\,dim} R_R.$$

In particular, rank $P = \operatorname{u\,dim} P$ if R is a right Ore domain.

1.4 Evidently, every finitely generated free module is stably free, and every stably free module is finitely generated projective. However, as the examples below indicate, these three concepts are distinct.

Example. (i) Let R be a commutative Dedekind domain and P any nonzero ideal; so P is finitely generated projective. If P were stably free, it would follow from 5.7.18(iii) that $P \simeq R$. Hence, if P is chosen not to be principal, then P is not stably free.

Example. (ii) Let k be a field with char $k = 0$ and let R be the first Weyl algebra $A_1(k)$. By 7.11.8, R has a nonprincipal right ideal, P say and, since R is a hereditary Noetherian ring, P is finitely generated projective. Now if P were free then uniform dimension shows that it would have to be free of rank 1, and hence principal. It follows that P is not free. However, P is stably free with $P \oplus R \simeq R^2$ since, by 7.11.6, every projective module of uniform dimension 2 or more is free.

1.5 These examples serve to illustrate some further points. Firstly, it is clear from the above that $A_1(k)$ has the property that all finitely generated projective modules are stably free; it will be shown, in the next chapter, that many rings share this property.

Secondly, note that over $A_1(k)$ all stably free modules of uniform dimension at least 2 are free; and over a commutative Dedekind domain every stably free module P is free, this following as in 1.4(i).

The remainder of this section concerns stably free modules and conditions under which they are free.

1.6 A **finite free resolution**, FFR for short, of a module is a finite resolution by finitely generated free modules. There is an immediate connection with stably free modules.

11.1.9 *Stably free modules* 379

Proposition. *A finitely generated projective R-module P is stably free if and only if P has a FFR. Furthermore, if*
$$0 \to F_n \to F_{n-1} \to \cdots \to F_0 \to P \to 0$$
is a FFR then
$$\operatorname{rank} P = \sum_{k=0}^{n} (-1)^k \operatorname{rank} F_k.$$

Proof. Straightforward. □

1.7 Another characterization of stably free modules comes from linear transformations.

Proposition. *The following conditions on a module P_R are equivalent:*
 (i) *P is stably free with $P \oplus R^m \simeq R^n$;*
 (ii) *$P \simeq \operatorname{coker} \alpha$ for some split monomorphism $\alpha: R^m \to R^n$;*
 (iii) *$P \simeq \ker \beta$ for some split epimorphism $\beta: R^n \to R^m$.*

Proof. Straightforward. □

1.8 Maps α, β as in 1.7 are given by left multiplication by an $n \times m$ matrix A and an $m \times n$ matrix B respectively.

Lemma. (i) *α is a split monomorphism if and only if A is left invertible.*
 (ii) *β is a split epimorphism if and only if B is right invertible.*

Proof. Clear. □

A left invertible $n \times 1$ matrix is called a **unimodular column**, the set of these being denoted $U_c(n, R)$. Similarly, $U_r(n, R)$ is the set of $1 \times n$ **unimodular rows**.

It will be convenient at times to write columns in the form $(r_1, \ldots, r_n)$ without special mention, and to let e_i denote the unimodular column, or row, which has 1 as the ith coordinate and zeros elsewhere.

1.9 We note a simple fact concerning columns. This, like several later results, has an obvious analogue concerning rows.

Lemma. *If $r = (r_1, \ldots, r_n) \in R^n$, then the following are equivalent:*
 (i) *r is a unimodular column;*
 (ii) *$\sum_{i=1}^{n} R r_i = R$;*
 (iii) *the map $R_R \to R_R^n$ given by $a \mapsto ra$ is a split monomorphism;*
 (iv) *the map ${}_R R^n \to {}_R R$ given by $x \mapsto x \cdot r = \sum x_i r_i$ is a split epimorphism.* □

This highlights the involvement of both R_R and ${}_R R$ in this theory.

1.10 We now start to investigate when stably free modules are necessarily free.

Proposition. *Let P_R be stably free and let α and A be as in 1.7 and 1.8. The following conditions are equivalent:*
  (i) *P is free;*
  (ii) *α can be lifted to an isomorphism $\hat{\alpha}: R^m \oplus R^{n-m} \to R^n$ such that $\hat{\alpha}\iota = \alpha$, where $\iota: R^m \to R^m \oplus R^{n-m}$ is injection into the first summand;*
  (iii) *A can be completed to an invertible $n \times n$ matrix by the addition of $n - m$ further columns.*

Proof. (i)⇔(ii). If $\hat{\alpha}$ exists then $P = \operatorname{coker} \alpha \simeq \operatorname{coker} \iota = R^{n-m}$. Conversely, if $P \simeq R^{n-m}$ then one can construct $\hat{\alpha}$ via

$$\hat{\alpha}: R^m \oplus R^{n-m} \xrightarrow{\sim} R^m \oplus P \xrightarrow{\sim} R^n.$$

(ii)⇔(iii) This is clear. □

There is, of course, an analogous result for β and B.

1.11 We concentrate now on unimodular elements. Recall that the general linear group $Gl(n, R)$ is the group of all invertible $n \times n$ matrices over R. It acts on $U_c(n, R)$ by left multiplication and on $U_r(n, R)$ by right multiplication.

1.12 Proposition. *Let $\mathscr{S}_{n-1}$ denote the collection of all modules P_R such that $P \oplus R \simeq R^n$. The following are equivalent:*
  (i) *each $P \in \mathscr{S}_{n-1}$ is free;*
  (ii) *$Gl(n, R)$ acts transitively on $U_c(n, R)$;*
  (iii) *each $\mathbf{a} \in U_c(n, R)$ can be extended to a matrix in $Gl(n, R)$.*

Proof. (i)⇒(iii) This is a special case of 1.10.
  (iii)⇒(ii) Given $\mathbf{a}, \mathbf{b} \in U_c(n, R)$ we need to find a matrix $C \in Gl(n, R)$ such that $C\mathbf{a} = \mathbf{b}$. Suppose that $\mathbf{a}$ and $\mathbf{b}$ are the first columns of the matrices $A, B \in Gl(n, R)$ and $C = BA^{-1}$; then $C\mathbf{a} = \mathbf{b}$.
  (ii)⇒(i) Let $P \oplus R \simeq R^n$ and $\mathbf{a}$ be the image of $(0, 1)$ in R^n. Then $\mathbf{a} \in U_c(n, R)$ and so $C\mathbf{a} = e_1$ for some $C \in Gl(n, R)$. Then $P \simeq R^n/e_1 R \simeq R^{n-1}$. □

In such circumstances we say that $n - 1$ is in the **general linear range** of R.

1.13 Symmetry of the general linear range is demonstrated by the next result.

Lemma. *$Gl(n, R)$ acts transitively on $U_c(n, R)$ if and only if $Gl(n, R)$ acts transitively on $U_r(n, R)$.*

Proof. It is enough to prove one implication. Let $\mathbf{b} \in U_r(n, R)$; then $\mathbf{b} \cdot \mathbf{a} = 1$ for some $\mathbf{a} \in U_c(n, R)$. Suppose that $C\mathbf{a} = e_1$ for some $C \in Gl(n, R)$ and yet

11.1.16 *Stably free modules* 381

$(bC^{-1})(Ca) = 1$. Evidently the first entry of bC^{-1} is 1. Hence $bC^{-1}D = e_1$ for some D in $Gl(n, R)$. It follows that $Gl(n, R)$ acts transitively on $U_r(n, R)$. □

1.14 The least integer $t \geq 1$ such that s is in the general linear range of R for all $s \geq t$ is the **general linear rank** of R, glr R. The next result, when combined with the symmetry provided by 1.13, gives several characterizations of glr R.

Theorem. *The following properties of a ring R and integer $t \geq 1$ are equivalent:*
(i) glr $R = t$;
(ii) t is minimal such that each stably free module P with rank $P \geq t$ is free;
(iii) t is minimal such that each $a \in U_c(n, R)$ with $n \geq t + 1$ can be extended to a matrix in $Gl(n, R)$.

Proof. (i)⇔(ii) Granted (ii) it is clear from 1.12 that glr $R \leq t$. On the other hand, if $P \oplus R^m \simeq R^n$ with $n - m =$ rank $P \geq t =$ glr R then $(P \oplus R^{m-1}) \oplus R \simeq R^n$. Hence, by 1.12, $P \oplus R^{m-1}$ is free. Induction on m shows that P is free.
(i)⇔(iii) Clear from 1.12. □

1.15 Note that glr R is not necessarily the least integer in the general linear range. For example it is shown, in Theorem 9.3 of [Swan 77] and the following remarks, that for each m with $m \equiv 2 \mod 4$ there is a commutative Noetherian ring R such that:
(i) all projective R modules are stably free; and
(ii) there are stably free nonfree projectives of rank m but none of any other rank.
It is clear from 1.12 that if $t \neq m$ then t is in the general linear range of R. On the other hand, by 1.14, m is not in the general linear range. Thus glr $R = m + 1$.

1.16 In §§3 and 5, some bounds on glr R will be obtained. This section ends with a result special to commutative rings.

Proposition. *Let R be a commutative ring. Then:*
(i) *every stably free module of rank 1 is free;*
(ii) *1 is in the general linear range of R;*
(iii) *glr $R \neq 2$.*

Proof. Evidently it is sufficient to prove (i). Let A be the matrix corresponding to a split epimorphism $R^n \to R^{n-1}$ with kernel P; so A is a right invertible $(n-1) \times n$ matrix. Let $b_1, \ldots, b_n$ be the maximal minors of A, i.e. the determinants of the $(n-1) \times (n-1)$ submatrices.

Suppose that some maximal ideal M of R contains $b_1, \ldots, b_n$, and let $\bar{}$ denote images modulo M. Then $\bar{R}$ is a field and so $\bar{P} \simeq \bar{R}$. Therefore $\bar{A}$ can be completed to a matrix in $Gl(n, \bar{R})$. However, expanding along the last row of this matrix makes clear that its determinant is zero.

This contradiction shows that $\sum b_i r_i = 1$ for some $r_i \in R$. Therefore A can be completed to an $n \times n$ matrix of determinant 1 by adjoining the last row

$$((-1)^{n-1}r_1, (-1)^{n-2}r_2, \ldots, r_n).$$

Hence, by 1.10, P is free. □

That this result is special to commutative rings is easily seen. Let $R = A_1(k)$ with char $k = 0$. It is clear from 1.4(ii) and 1.14 that glr $R = 2$ and that 1 is not in the general linear range.

1.17 In §3, it will be shown that if n is in the stable range of a ring R then glr $R \leqslant n$ and, if R is right Noetherian, then glr $R \leqslant K(R_R) + 1$.

§2 Stably Free Nonfree Modules

2.1 This section shows how to construct some stably free nonfree modules. It starts with a brief survey of some examples over commutative rings. Then it turns to the special case of modules of rank 1, where, by 1.16, noncommutativity is a necessity. Unlike the commutative case, the techniques required are not topological.

2.2 Recall first that no examples can be found for polynomial rings over a field.

Theorem. (Quillen–Suslin) *For any field k and any n, each projective module over $k[X_1, \ldots, X_n]$ is free.*

Proof. See [Lam 78, Chapter 5, 2.7]. □

2.3 The next result provides an example of minimal possible rank over commutative rings.

Proposition. *The ring $R = \mathbb{R}[X, Y, Z]/(X^2 + Y^2 + Z^2 - 1)$ has a stably free nonfree module P with $P \oplus R \simeq R^3$ and so rank $P = 2$.*

Proof. Let x, y, z denote the images in R of X, Y, X. Note that (x, y, z) is a unimodular column which defines a split monomorphism $R \to R^3$ with cokernel P say. So $P \oplus R \simeq R^3$. Suppose that P is free. Then, by 1.10, there is an invertible matrix $A \in Gl(3, R)$ of the form

$$A = \begin{bmatrix} x & a_1 & b_1 \\ y & a_2 & b_2 \\ z & a_3 & b_3 \end{bmatrix}$$

where, without loss, we may suppose $\det A = 1$. Thus if $\mathbf{r} = (x, y, z)$, $\mathbf{a} = (a_1, a_2, a_3)$ and $\mathbf{b} = (b_1, b_2, b_3)$ then $\mathbf{b} \cdot (\mathbf{r} \wedge \mathbf{a}) = 1$. At each point on $\mathbb{S}^2$, the 2-sphere, $\mathbf{r}$ gives the unit outward normal to $\mathbb{S}^2$. Thus $r \wedge a$ gives a continuous tangent vector field on $\mathbb{S}^2$ which vanishes nowhere, contradicting a well-known result; see, for example, [Milnor **78**]. Therefore P is not free. □

2.4 Further examples over commutative Noetherian rings are provided by a special case of a result of [Raynaud **68**].

Theorem. *Let k be a field of characteristic zero, $k[X_1, \ldots, X_n, Y_1, \ldots, Y_n]$ be the commutative polynomial ring with $n \geqslant 2$ and*

$$R = k[X_1, \ldots, X_n, Y_1, \ldots, Y_n] \Big/ \left(\sum_{i=1}^n X_i Y_i - 1 \right)$$

with x_i, y_i denoting the images in R of X_i, Y_i. Then the cokernel of the unimodular column $(x_1, \ldots, x_n)$ is stably free and nonfree. □

2.5 In sharp contrast to the commutative case as described in 1.16, it will be shown in the remainder of this section that there are many examples of stably free nonfree modules of rank 1 over noncommutative Noetherian rings, even including polynomial rings over division rings. Thus the behaviour of $A_1(k)$ is not untypical.

2.6 Lemma. *Let a, b be regular elements of a ring R. Then*

$$aR \cap bR \simeq \{x \in R | ax \in bR\}.$$

If, further, $aR + bR = R$ then $aR \cap bR$ is stably free.

Proof. The first claim is obvious, and the second is clear from the short exact sequence

$$0 \to aR \cap bR \to aR \oplus bR \to aR + bR \to 0.$$ □

2.7 The constructions will rely upon the next result.

Theorem. *Let R be a right Noetherian integral domain and $S = R[x; \sigma, \delta]$ with σ an automorphism. Suppose r, $r' \in R$ are such that $S = rS + (x + r')S$ and let $K = \{s \in S | rs \in (x + r')S\}$. Then:*
 (i) *K is a stably free right ideal;*
 (ii) *K is free if and only if K is cyclic;*
 (iii) *if K is free then $K = (x + d)S$ for some $d \in R$;*
 (iv) *if r is a non-unit then K is not free.*

Proof. (i) This is clear from 2.6.
 (ii) Note that $\operatorname{u dim} K_S = 1$.

(iii) Note first that
$$rx = x\sigma(r) + \delta(r) = (x+r')\sigma(r) + (\delta(r) - r'\sigma(r)).$$
Since R is a right Ore domain, there exist $c_1, c_2 \in R$, with $c_1 \neq 0$, such that $(\delta(r) - r'\sigma(r))c_1 = rc_2$. Then $r(xc_1 - c_2) = (x+r')\sigma(r)c_1$ and so $xc_1 - c_2 \in K$. So if $K = kS$ then $\deg k \leq 1$. On the other hand, $rk \in (x+r')S$, so $\deg k = 1$, say $k = xd_1 + d_2$. However, the set of monic polynomials of S is a right Ore set by 7.9.3. Therefore, K contains a monic polynomial. Hence d_1 must be a unit and so $K = (x+d)S$ with $d = d_2 d_1^{-1}$.

(iv) Suppose that K is free; so $K = (x+d)S$ as above. Since $S = rS + (x+r')S$, then $1 = rf + (x+r')g$ for some $f, g \in S$. However, it is evident that $S = (x+d)S + R$; so $f = (x+d)s + r''$ for some $s \in S$, $r'' \in R$. Therefore
$$1 = r(x+d)s + rr'' + (x+r')g.$$
By definition of K, $r(x+d) = (x+r')s'$ for some $s' \in S$, yet $r(x+d) = x\sigma(r) + \delta(r) + rd$. Thus $s' = \sigma(r)$ and $r(x+d) = (x+r')\sigma(r)$. Hence
$$1 = (x+r')(\sigma(r)s + g) + rr''$$
and so $\sigma(r)s + g = 0$ and $rr'' = 1$. This contradicts the hypothesis on r and shows that K cannot be free. □

2.8 Essentially the same argument can be applied to skew Laurent polynomials.

Theorem. *Let R be a right Noetherian integral domain, $T = R[x, x^{-1}; \sigma]$ and $S = R[x; \sigma]$. Suppose $r, r' \in R$ are such that $S = rS + (x+r')S$, where r is a non-unit. Then $rT \cap (x+r')T$ is a stably free nonfree right ideal.* □

2.9 We now begin to draw some conclusions from 2.7 and 2.8. The first is a striking demonstration of the effect of noncommutativity.

Corollary. *Let D be a division ring and let $n \geq 2$. Then $S = D[x_1, x_2, \ldots, x_n]$, the polynomial ring in central indeterminates, has a stably free nonfree right ideal if and only if D is not a field.*

Proof. ($\Rightarrow$) This follows from 1.16 or, of course, from 2.2.

($\Leftarrow$) Choose elements $a, b \in D$ such that $ab \neq ba$. Note that
$$ab - ba = (x_n + a)(x_{n-1} + b) - (x_{n-1} + b)(x_n + a)$$
and so $(x_n + a)S + (x_{n-1} + b)S = S$. Let $R = D[x_1, \ldots, x_{n-1}]$, $x = x_n$, $r' = a$ and $r = x_{n-1} + b$. Then 2.7 shows that $(x_n + a)S \cap (x_{n-1} + b)S$ is a stably free nonfree right ideal. □

2.10 Corollary. *Let R be a right Noetherian integral domain, σ be an automorphism*

11.2.14 *Stably free nonfree modules*

and $S = R[x; \sigma, \delta]$. *Suppose that* $\sum_{i=0}^{\infty} \delta^i(r)R = R$ *for some non-unit* $r \in R$. *Then* $rS \cap xS$ *is stably free but not free.*

Proof. Note that $\delta^{i+1}(r) = \delta^i(r)x - x\sigma(\delta^i(r))$ for each i. Since $\sum \delta^i(r)R = R$, it follows that $1 \in rS + xS$ and so $S = rS + xS$. Then 2.7 applies. $\square$

2.11 Corollary. *Let* $S = A_n(K)$ *be the n-th Weyl algebra over a right Noetherian integral domain* K. *Then* S *has a stably free nonfree right ideal.*

Proof. View S as $R[x; -\partial/\partial y]$, where $R = A_{n-1}(K)[y]$. Then 2.10 applies with $r = y$. $\square$

2.12 Corollary. *Let* $S = R[x; \sigma, \delta]$, *where* σ *is an automorphism of the ring* R. *Then* S *is a pri-domain if and only if* R *is a division ring.*

Proof. In 1.2.9 it is shown that if R is a division ring then S is a pri-domain. Conversely, suppose S is a pri-domain. Since the map $I \mapsto IS$ embeds $\mathscr{L}(R_R)$ into $\mathscr{L}(S_S)$ it is clear that R is a right Noetherian integral domain. Choose any nonzero element $r \in R$ and consider the right ideal $rS + xS$. By hypothesis, this is principal; and, since it contains r, its generator is a scalar, a say. However then $x = a(cx + d)$ with $c, d \in R$ and so $ac = 1$. Thus $rS + xS = S$. Now $rS \cap xS$ is principal, and so free. Therefore, by 2.7, r is a unit and R is a division ring. $\square$

2.13 The connection between stably free nonfree modules and noncommutativity is highlighted in the final result.

Corollary. *Let* R *be a commutative Noetherian domain, let* $S = R[x; \delta]$ *or* $R[x, x^{-1}; \sigma]$, *and suppose* S *is simple. Then the following are equivalent:*
 (i) R *is a field*;
 (ii) S *is a pri-domain*;
 (iii) S *has no stably free nonfree right ideal.*

Proof. (i) $\Rightarrow$ (ii) $\Rightarrow$ (iii) Clear.
 (iii) $\Rightarrow$ (i) Suppose R is not a field and $S = R[x; \delta]$. Since S is simple, δ leaves invariant no proper nonzero ideal of R, by 1.8.3. However, if $0 \neq r \in R$ then $I = \sum_{i=0}^{\infty} \delta^i(r)R$ is an invariant ideal and so $I = R$. Hence if r is picked to be a non-unit then 2.10 applies. The case when $S = R[x, x^{-1}; \sigma]$ is dealt with in a similar fashion. $\square$

2.14 These methods can be used to obtain stably free nonfree right ideals in enveloping algebras of any finite dimensional non-abelian Lie algebra, [Stafford **85a**].

2.15 Finally, we note without proof some related results.
(i) If k is a field with char $k = 0$ then 2 is in the stable range of $A_n(k)$. Hence the only stably free nonfree modules over this ring are of rank 1, by 1.17; see [Stafford **78**].
(ii) If D is a division ring with an infinite centre then any finitely generated projective module over $D[x_1, x_2, \ldots, x_n]$ is either free or isomorphic to a right ideal; see [Stafford **80a**].

§3 Stable and Elementary Ranks

3.1 This section starts by discussing the stable range of a ring, as introduced in 6.7.2. In particular, it is shown that if t is in the stable range of R then stably free modules of rank t or more are all free. An application of this to the theory of principal ideal rings is noted. The notion of elementary rank is then introduced and a unified description is given of the three ranks: stable, elementary and general linear.

3.2 We recall from 6.7.2 that to say t is in the stable range of R_R means that whenever $\sum_{i=1}^{n+1} a_i R = R$ with $n \geq t$ then $\sum_{i=1}^{n} (a_i + a_{n+1} r_i) R = R$ for some $r_i \in R$. The condition that $\sum_{i=1}^{n+1} a_i R = R$ says that $(a_1, \ldots, a_{n+1}) \in U_r(n+1, R)$, of course. The additional condition, that

$$(a_1 + a_{n+1} r_1, \ldots, a_n + a_{n+1} r_n) \in U_r(n, R)$$

for some $r_i \in R$, makes $(a_1, \ldots, a_{n+1})$ a **stable** unimodular row.

For example, over $\mathbb{Z}$, $(5, 2)$ is a stable unimodular row while $(2, 5)$ is not.

3.3 Lemma. *Given $t \geq 1$, the following are equivalent:*
(i) *t is in the stable range of R_R;*
(ii) *each $\boldsymbol{a} \in U_r(t+1, R)$ is stable.*

Proof. It is enough to prove that (ii)$\Rightarrow$(i). Suppose that

$$\boldsymbol{x} = (x_1, \ldots, x_{n+1}) \in U_r(n+1, R) \quad \text{with } n \geq t;$$

so $\sum_{i=1}^{n+1} x_i y_i = 1$ for some $y_i \in R$. Let $z = \sum_{i=t+1}^{n+1} x_i y_i$ and note that

$$(x_1, \ldots, x_t, z) \in U_r(t+1, R).$$

By hypothesis there exist $r_1, \ldots, r_t \in R$ such that

$$(x_1 + z r_1, \ldots, x_t + z r_t) \in U_r(t, R).$$

Then

$$(x_1 + x_{n+1} y_{n+1} r_1, \ldots, x_t + x_{n+1} y_{n+1} r_t, x_{t+1}, \ldots, x_n) \in U_r(n, R)$$

as required. $\square$

11.3.4 Stable and elementary ranks

It is an interesting contrast with the general linear range that the range of values of t for which (ii) holds has no gaps.

3.4 The least positive integer t in the stable range of R_R is called the **stable rank** of R_R, sr R_R for short. Terminology and notation are simplified by noting the symmetry of this invariant.

Proposition. *The stable ranks of R_R and of $_R R$ are equal.*

Proof. Suppose sr $_R R = t$ and that $y \in U_r(t+1, R)$. It is enough to show that y is stable.

Since y is unimodular $y \cdot x = 1$ for some $x \in R^{t+1}$. Note that

$$(x_1, \ldots, x_t, y_{t+1} x_{t+1}) \in U_c(t+1, R)$$

and so, by hypothesis, is stable. Therefore there exists $r \in R^t$ such that

$$x' = (x_1 + r_1 y_{t+1} x_{t+1}, \ldots, x_t + r_t y_{t+1} x_{t+1}) \in U_c(t, R).$$

Therefore there exists $z \in R^t$ with $z \cdot x' = -x_{t+1}$. Consider the $(t+2) \times (t+2)$ matrix

$$\alpha = \begin{bmatrix} 1 & 0 & 0 \\ 0 & I_t & 0 \\ 0 & z & 1 \end{bmatrix} \begin{bmatrix} 1 & 0 & 0 \\ -r & I_t & 0 \\ 0 & 0 & 1 \end{bmatrix} \begin{bmatrix} 1 & 0 & 0 \\ 0 & I_t & r y_{t+1} \\ 0 & 0 & 1 \end{bmatrix} \begin{bmatrix} 1 & y \\ 0 & I_{t+1} \end{bmatrix} \begin{bmatrix} 1 & 0 \\ -x & I_{t+1} \end{bmatrix}.$$

Clearly $\alpha \in Gl(t+2, R)$; and one can check that α has the form

$$\alpha = \begin{bmatrix} 0 & y_1 & y_2 & \cdots & y_t & y_{t+1} \\ & & & & & 0 \\ & & * & & & \vdots \\ & & & & & 0 \\ 0 & -w_1 & -w_2 & \cdots & -w_t & 1 \end{bmatrix}$$

where $*$ denotes unspecified entries and where $w = (w_1, \ldots, w_t)$ is thereby defined. Then the matrix

$$\beta = \alpha \begin{bmatrix} 1 & 0 & 0 \\ 0 & I_t & 0 \\ 0 & w & 1 \end{bmatrix}$$

$$= \begin{bmatrix} 0 & y_1 + y_{t+1} w_1 & \cdots & y_t + y_{t+1} w_t & y_{t+1} \\ & & & & 0 \\ & & * & & \vdots \\ & & & & 0 \\ 0 & 0 & \cdots & 0 & 1 \end{bmatrix} \in Gl(t+2, R).$$

Evidently the matrix obtained by deleting the final row and column of β is invertible, so its top row is unimodular. Therefore y is stable. □

Henceforth we write sr R for the stable rank of R, either on the left or right.

3.5 The connection between stably free modules and the stable rank comes through the **elementary group** $E(n, R)$. This is the subgroup of $Gl(n, R)$ generated by all elementary transvection matrices, these being matrices of the form $I + e_{ij}r$, where e_{ij} is one of the matrix units, $i \neq j$, and $r \in R$. For example the matrices α, β in the preceding proof belong to $E(t + 2, R)$.

It is easily verified that any triangular matrix of the form

$$\begin{bmatrix} 1 & & * \\ & 1 & \\ & & \ddots \\ 0 & & 1 \end{bmatrix} \quad \text{or} \quad \begin{bmatrix} 1 & & 0 \\ & 1 & \\ & & \ddots \\ * & & 1 \end{bmatrix}$$

belongs to $E(n, R)$.

3.6 Lemma. *Let $x \in R^{n+1}$ be a stable unimodular column. Then*
(i) $\sigma x = e_{n+1}$ *for some $\sigma \in E(n + 1, R)$, and*
(ii) *x can be completed to a matrix in $Gl(n + 1, R)$ by the addition of n further columns.*

Proof. (i) Let $x = (x_1, \ldots, x_{n+1})$. By hypothesis there exist $r_1, \ldots, r_n \in R$ such that

$$x' = (x'_1, \ldots, x'_n) \in U_c(n, R),$$

where $x'_i = x_i + r_i x_{n+1}$. Therefore $\sum_{i=1}^{n} s_i x'_i = 1 - x_{n+1}$ for some $s_i \in R$. Let $\alpha, \beta, \gamma \in E(n + 1, R)$ be the matrices sending $(a_1, \ldots, a_{n+1}) \in R^{n+1}$ to, respectively,

$$(a_1 + r_1 a_{n+1}, \ldots, a_n + r_n a_{n+1}, a_{n+1}),$$
$$(a_1, \ldots, a_n, a_{n+1} + s_1 a_1 + s_2 a_2 + \cdots + s_n a_n)$$

and

$$(a_1 - x'_1 a_{n+1}, \ldots, a_n - x'_n a_{n+1}, a_{n+1}).$$

Then one can check that $\gamma \beta \alpha x = e_{n+1}$.

(ii) It is clear from 1.9 that x defines a split monomorphism whose cokernel P is stably free. However, $P \simeq R^{n+1}/e_{n+1}R$. So P is free and 1.10 applies. □

3.7 Theorem. (i) *glr $R \leq$ sr R and so any stably free module P with rank $P \geq$ sr R is free.*
(ii) *If R is right Noetherian then glr $R \leq K(R/J(R)_R) + 1$ and so any stably free module P with rank $P \geq K(R/J(R)_R) + 1$ is free.*

Proof. Combine 3.6, 1.14 and 6.7.4. □

3.8 Next we note an interesting application of 3.7. It was demonstrated in 7.11.7 and 7.11.8 that if char $k = 0$ then, although $A_1(k)$ is not a pri- or pli-ring, $M_n(A_1(k))$ is, for any $n \geqslant 2$. In fact this behaviour is typical. We recall from 3.4.9 that a prime pri-ring is a matrix ring over a right Noetherian domain.

Theorem. *Let R be a prime principal ideal ring with $R \simeq M_n(A)$ for some n and some integral domain A. Then $M_t(A)$ is a prime principal ideal ring for all $t \geqslant 2$.*

Proof. Note first, by 5.2.11(i) and 6.2.8, that $K(R) = 1$. Morita equivalence shows that both A and $M_t(A)$ are Noetherian prime rings with Krull dimension 1. Consequently, 2 is the stable range of A_A. By 3.4.10, $M_s(R) = M_{sn}(A)$ is a principal ideal ring, so we may suppose $n > t$. Next, note that if each essential right ideal E of $M_t(A)$ is principal then $M_t(A)$ is a pri-ring.

However, under the Morita equivalence between $M_t(A)$ and A, E corresponds to an essential submodule, H say, of the free module A^t. Evidently $H \oplus A^{n-t} \triangleleft_e A^n$. The Morita equivalence between A and $M_n(A)$ makes $H \oplus A^{n-t}$ correspond to an essential right ideal I of R; and $I \simeq R$, since by 3.3.7 its generator must be a regular element. Hence $H \oplus A^{n-t} \simeq A^n$ and so H is stably free of rank t. Since $t \geqslant 2$, 3.7 shows that $H \simeq A^t$. Hence $E \simeq M_t(A)$ as required. □

3.9 Next we introduce a third related rank. By analogy with 1.12, we say that t is in the **elementary range** of R if $E(t + 1, R)$ acts transitively on the set of unimodular rows of length $t + 1$. The proof of 1.13 is easily modified to show that this is a symmetric property of R.

The elementary range, like the general linear range, can have 'gaps'. For instance, let k be a field with char $k = 0$ and let R be the free associative k algebra in $2n$ variables $x_1, \ldots, x_n, y_1, \ldots, y_n$ modulo the relation $\sum_{i=1}^n x_i y_i = 1$. Then 1, 2, ..., $n - 2$ are in the elementary range of R yet $n - 1$ is not.

To see this one notes first, from [Cohn **85**, 2.11.2], that R satisfies a certain algorithm, the $n - 1$ term weak algorithm, which ensures, almost immediately, that $E(t, R)$ acts transitively on $U_r(t, R)$ for each $t \leqslant n - 1$. On the other hand, $x = (x_1, \ldots, x_n)$ is the universal unimodular row. So if $x\beta = e_1$ for some $\beta \in E(n, R)$ it would follow that $E(n, S)$ acts transitively on $U_r(n, S)$ for all k-algebras S; yet that is not so, by 1.12 and 2.4. So $n - 1$ is not in the elementary range.

3.10 The **elementary rank** of R, er R, is the smallest integer $t \geqslant 1$ such that n is in the elementary range of R for all $n \geqslant t$. Evidently this too is left–right symmetric.

It is clear from 3.6 that er R has some significance. This becomes more apparent in connection with the K-group K_1; see [Bass **64**].

3.11 One connection between the elementary rank and the ranks defined earlier is clear from the definitions together with 3.6.

Proposition. $1 \leqslant \text{glr } R \leqslant \text{er } R \leqslant \text{sr } R$. □

3.12 In fact there is a unified approach to these ranks which clarifies their relationship. Before describing it some terminology is required.

Let H be a sequence of groups $\{H(n,R) | n = 2, 3, \ldots\}$ with $H(n,R)$ a subgroup of $Gl(n,R)$ for each n. For example one could take $H(n,R) = Gl(n,R)$ in which case we write $H = Gl$. Similarly, $H = E$ if $H(n,R) = E(n,R)$. Finally, we let $B_r = \{B_r(n,R)\}$ and $B_c = \{B_c(n,R)\}$, where $B_r(n,R)$ and $B_c(n,R)$ are the subgroups of $E(n,R)$ consisting of all matrices of the respective forms

$$\begin{bmatrix} 1 & & & \\ & 1 & & \\ & & \ddots & \\ b_1 & b_2 & \cdots & b_{n-1} & 1 \end{bmatrix} \quad \text{or} \quad \begin{bmatrix} 1 & & & b_1 \\ & 1 & & b_2 \\ & & \ddots & \vdots \\ & & & b_{n-1} \\ & & & 1 \end{bmatrix}$$

with $b_i \in R$ and with the unspecified entries being 0.

3.13 Given such a sequence H, we say that H-**rank** $R = t$ if t is the least integer such that for all $n \geq t$, given a split epimorphism α and the canonical inclusion γ, there exists $\beta \in H(n+1, R)$ such that the composed map

$$R^n \xrightarrow{\gamma} R^n \oplus R \xrightarrow{\beta} R^{n+1} \xrightarrow{\alpha} R$$

is a split epimorphism.

3.14 Proposition. *For any ring R:*
(i) $\operatorname{sr} R = B_r\text{-rank } R$;
(ii) $\operatorname{er} R = E\text{-rank } R$;
(iii) $\operatorname{glr} R = Gl\text{-rank } R$.

Proof. (i) The fact that if

$$\begin{bmatrix} 1 & & & \\ & 1 & & \\ & & \ddots & \\ b_1 & b_2 & \cdots & 1 \end{bmatrix} \in B_r(n+1, R)$$

and $\alpha = (x_1, \ldots, x_{n+1})$ then

$$\alpha\beta = (x_1 + x_{n+1}b_1, \ldots, x_n + x_{n+1}b_n, x_{n+1})$$

makes this clear.

(ii) Suppose $n \geq \operatorname{er} R$ and that α, γ are given. Now α is unimodular, so there exists $\beta \in E(n+1, R)$ such that $\alpha\beta = e_1$. This β is as required by 3.13; so $\operatorname{er} R \geq E\text{-rank } R$.

Conversely, suppose that $n \geq E\text{-rank } R$ and let $\alpha \in U_r(n+1, R)$. By hypothesis there exists $\beta \in E(n+1, R)$ such that $\alpha\beta\gamma : R^n \to R$ is a split epimorphism. Thus if

$\alpha\beta = x = (x_1, \ldots, x_{n+1})$ then $(x_1, \ldots, x_n)$ is unimodular. It follows from 3.6(i) that $x\sigma = e_{n+1}$ for some $\sigma \in E(n+1, R)$. This shows that $E(n+1, R)$ acts transitively on $U_r(n+1, R)$ and so E-rank $R \geqslant$ er R.

(iii) This is proved in the same way as (ii). □

3.15 Symmetry, or a modified version of 3.14, shows that the E-rank and Gl-rank of R could equally well have been defined in terms of a split monomorphism α, the canonical projection $\gamma: R^n \oplus R \to R^n$ and the existence of β making $\gamma\beta\alpha$ a split monomorphism since the same values, er R and glr R, would result. The same holds for sr R provided that this time B_c rather than B_r is used.

§4 Cancellation of Modules

4.1 The fact that if t is in the stable range of R then stably free modules of rank t or more are free can be viewed as a cancellation result. It asserts that if $P \oplus R \simeq R^n \oplus R$ where $n \geqslant t$ then $P \simeq R^n$. This section is aimed towards results in which a module M_R, rather than R_R, is to be cancelled.

4.2 We start with a property analogous to that described in 3.13. Let M, N be right R-modules and H be some subgroup of $\text{Aut}(N \oplus M)$. If, given any split epimorphism $\alpha: N \oplus M \to M$ and the canonical inclusion $\gamma: N \to N \oplus M$, there is a $\beta \in H$ such that $\alpha\beta\gamma$ is a split epimorphism, we say that N is H-**permeated** by M. Of course, this implies that N has a direct summand isomorphic to M.

The two cases of interest to us here are when $H = \text{Aut}(N \oplus M) = Gl$, for short, and where

$$H = B = \left\{ \begin{bmatrix} 1 & 0 \\ \beta' & 1 \end{bmatrix} \middle| \beta' \in \text{Hom}(N, M) \right\}.$$

4.3 The same argument as in 3.14 gives the following result.

Lemma. (a) *Let M be a right R-module. Then*:
  (i) M^n *is Gl-permeated by M if and only if n is in the general linear range of* $\text{End } M_R$.
  (ii) M^n *is B-permeated by M if and only if n is in the stable range of* $\text{End } M_R$.
(b)(i) R^n *is Gl-permeated by R if and only if n is in the general linear range of R.*
  (ii) R^n *is B-permeated by R if and only if n is in the stable range of R.* □

4.4 The next result extends 3.7(i).

Theorem. *Let N, M, L be right R-modules such that N is Gl-permeated by M and $N \oplus M \simeq L \oplus M$. Then $N \simeq L$.*

Proof. Let $\sigma: N \oplus M \to L \oplus M$ be an isomorphism, $\pi: L \oplus M \to M$ be the canonical epimorphism and $\alpha = \pi\sigma: N \oplus M \to M$, this being a split epimorphism. By hypothesis, there is an automorphism β of $N \oplus M$ such that $\pi\sigma\beta\gamma: N \to M$ is a split epimorphism, where γ is the canonical inclusion of N in $N \oplus M$. Let $\tau: M \to N$ be a splitting of $\pi\sigma\beta\gamma$, let $\tau(M) = M'$ and let W be a complement to M' in N. The isomorphism $\sigma\beta$ induces an isomorphism from $(N \oplus M)/\gamma(M')$ to $(L \oplus M)/\sigma\beta\gamma(M')$. However,

$$(N \oplus M)/\gamma(M') \simeq W \oplus M \simeq W \oplus M' \simeq N$$

and $(L \oplus M)/\sigma\beta\gamma(M') \simeq L$ since $L = \ker \pi$. □

4.5 Corollary. (i) *If n is in the general linear range of* End M *and* $M^n \oplus M \simeq L \oplus M$ *then* $M^n \simeq L$.
(ii) *If* $n \geqslant \mathrm{glr}(\mathrm{End}\, M)$ *and* $M^n \oplus M^s \simeq L \oplus M^s$ *then* $M^n \simeq L$. □

4.6 As 4.3 demonstrated, B-permeation is related to the stable range. Before describing further examples and cancellation results, a preliminary result is required.

Lemma. *If n is in the stable range of a finitely generated module M_R then n is in the stable range of any factor module M/K.*

Proof. Suppose $[m_1 + K], \ldots, [m_{t+1} + K]$ generate M/K for some $t \geqslant n$, and let $M = \sum_{i=1}^{s} x_i R$. Then $x_i = \sum_{j=1}^{t+1} m_j r_{ij} + k_i$ for some $k_i \in K$, $r_{ij} \in R$ and so M is generated by $m_1, \ldots, m_{t+1}, k_1, \ldots, k_s$. Since $t \geqslant n$ it follows that

$$M = \sum_{i=1}^{t}\left(m_i + m_{t+1}a_i + \sum_{j=1}^{s} k_j b_{ij}\right) R$$

for some $a_i, b_{ij} \in R$. Therefore $M/K = \sum_{i=1}^{t} [m_i + m_{t+1}a_i + K]R$. □

4.7 Proposition. *If X, M are right R-modules, $S = \mathrm{End}\, M$ and $N = X \oplus M^n$ with $n \geqslant \mathrm{sr}\, S$ then N is B-permeated by M.*

Proof. Let $\alpha: X \oplus M^n \oplus M = N \oplus M \to M$ be a split epimorphism, say $\alpha = (\alpha', s_1, \ldots, s_{n+1})$, where $\alpha' \in \mathrm{Hom}(X, M)$ and $s_i \in S$. There is a splitting $\sigma: M \to X \oplus M^{n+1}$, say $\sigma = (\sigma', t_1, \ldots, t_{n+1})$, where $\sigma' \in \mathrm{Hom}(M, X)$ and $t_i \in S$, such that $1 = \alpha'\sigma' + \sum_{i=1}^{n+1} s_i t_i$. The application of 4.6 to S_S implies that $S = \alpha'\sigma' S + \sum_{i=1}^{n} s'_i S$, where $s'_i = s_i + s_{n+1} u_i$ for some $u_i \in S$. Then the map

$$\beta' = (0, u_1, \ldots, u_n): X \oplus M^n = N \to M$$

gives the automorphism β required by the definition in 4.2. □

4.8 Corollary. (i) *Suppose that M, X, Y are right R-modules, $n \geq \mathrm{sr}\,(\mathrm{End}\,M_R)$, and $X \oplus M^{n+1} \simeq Y \oplus M$. Then $X \oplus M^n \simeq Y$.*

(ii) *Suppose that X, Y are right R-modules, $n \geq \mathrm{sr}\,R$, and $X \oplus R^{n+1} \simeq Y \oplus R$. Then $X \oplus R^n \simeq Y$.*

Proof. (i) Combine 4.7 with 4.4.

(ii) This is a special case of (i). □

4.9 One can improve upon 4.8(i) when 1 is in the stable range.

Theorem. *Let M, X, Y be right R-modules, $1 = \mathrm{sr}\,(\mathrm{End}\,M_R)$ and $X \oplus M \simeq Y \oplus M$. Then $X \simeq Y$.*

Proof. Let $S = \mathrm{End}\,M_R$. Given inverse isomorphisms $X \oplus M \underset{\beta}{\overset{\alpha}{\rightleftarrows}} Y \oplus M$ one can view each as a 2×2 matrix; namely

$$\alpha = \begin{bmatrix} \alpha_{YX} & \alpha_{YM} \\ \alpha_{MX} & \alpha_{MM} \end{bmatrix}, \quad \beta = \begin{bmatrix} \beta_{XY} & \beta_{XM} \\ \beta_{MY} & \beta_{MM} \end{bmatrix},$$

where $\beta_{XY} \in \mathrm{Hom}\,(Y, X)$ etc. Since $\beta\alpha = 1$ then $\beta_{MY}\alpha_{YM} + \beta_{MM}\alpha_{MM} = 1_S$. Hence $S = S\alpha_{MM} + S\beta_{MY}\alpha_{YM}$ and so, by hypothesis, $\alpha_{MM} + s\beta_{MY}\alpha_{YM}$ is a unit, u say, of S for some choice of $s \in S$.

Define $\beta': Y \oplus M \to X \oplus M$ by

$$\beta' = \begin{bmatrix} \beta_{XY} & \beta_{XM} \\ s\beta_{MY} & 1_S \end{bmatrix}.$$

Then

$$\beta'\alpha = \begin{bmatrix} 1_{YY} & 0 \\ * & u \end{bmatrix}$$

and so β' is an isomorphism. So too, therefore, is the map $\gamma: Y \oplus M \to X \oplus M$ given by

$$\gamma = \begin{bmatrix} 1_{XX} & -\beta_{XM} \\ 0 & 1_S \end{bmatrix} \beta' \begin{bmatrix} 1_{YY} & 0 \\ -s\beta_{MY} & 1_S \end{bmatrix} = \begin{bmatrix} -\beta_{XM}s\beta_{MY} + \beta_{XY} & 0 \\ 0 & 1_S \end{bmatrix}.$$

Hence $-\beta_{XM}s\beta_{MY} + \beta_{XY}$ is an isomorphism from Y to X. □

§5 Ranks of Certain Rings

5.1 The earlier sections have shown how the different ranks of a ring influence the module structure. This section concerns the ranks themselves. For some examples the ranks will be calculated or described. In particular, polynomial extensions and matrix extensions will be considered.

5.2 It seems clear from 3.11 and 4.9 that rings of stable rank 1 are of special interest.

Proposition. sr $S = 1$ *if either*:
(i) *S is right Artinian; or*
(ii) *S/J(S) is right Artinian; or*
(iii) $S = R_X$, *the localization of a Noetherian ring R with respect to a set X of prime ideals as described in* 4.3.17.

Proof. (i)(ii) Since $K(S/J(S)_S) = 0$, then 6.7.4 applies.

(iii) If X is a finite set, this follows directly from (ii) since $S/J(S)$ is semisimple Artinian. If X is an infinite set, the proof is nontrivial and can be found in [Warfield 86]. □

5.3 Next we calculate the three ranks for some simple examples.

Proposition. *Let* $R = \mathbb{Z}$ *or* $R = D[x; \sigma, \delta]$ *with D a division ring. Then* glr $R =$ er $R = 1$ *and* sr $R = 2$.

Proof. First consider er R. Given any $x \in U_c(t, R)$ with $t \geqslant 2$, the Euclidean algorithm gives $\alpha \in E(t, R)$ such that $y = \alpha x$ has a unit as one entry. One can arrange that the unit is not the final entry of y, essentially by using the fact that

$$\begin{bmatrix} 0 & -1 \\ 1 & 0 \end{bmatrix} = \begin{bmatrix} 1 & -1 \\ 0 & 1 \end{bmatrix} \begin{bmatrix} 1 & 0 \\ 1 & 1 \end{bmatrix} \begin{bmatrix} 1 & -1 \\ 0 & 1 \end{bmatrix} \in E(2, R).$$

Then y is a stable unimodular column and so, by 3.6, $\beta y = e_t$ for some $\beta \in E(t, R)$. This shows that er $R = 1$ and so, by 3.11, glr $R = 1$.

Finally, consider sr R. Since $K(R) = 1$, we know from 6.7.4 that sr $R \leqslant 2$. On the other hand, $(2, 5) \in \mathbb{Z}^2$ is unimodular, but not stable. Similarly $(x, 1 - x^2) \in (D[x; \sigma, \delta])^2$ is unimodular but not stable. Hence sr $R = 2$. □

5.4 Proposition. *If* $R = \mathbb{R}[x, y, z]/(x^2 + y^2 + z^2 - 1)$ *then* glr $R =$ er $R =$ sr $R = 3$.

Proof. In 2.3 it is shown that R has a nonfree stably free module of rank 2. Therefore, by 1.14, glr $R \geqslant 3$. On the other hand $K(R) = 2$ so sr $R \leqslant 3$ by 6.7.4. Now apply 3.11. □

5.5 Next we survey what is known of these ranks for commutative polynomial rings over a field k. Combining Theorem 2.2 with Theorem 1.14 deals with the general linear rank.

11.5.9 *Ranks of certain rings*

Corollary. $\mathrm{glr}(k[x_1,\ldots,x_n]) = 1$ *for any field* k *and any* n. □

Thus if $\bar{R}$ is a homomorphic image of a ring R, it may happen that $\mathrm{glr}\,\bar{R} > \mathrm{glr}\,R$. This contrasts with stable rank where $\mathrm{sr}\,\bar{R} \leqslant \mathrm{sr}\,R$ by 4.6.

5.6 The elementary rank, like the general linear rank, can be relatively small.

Theorem. *Let A be a commutative Noetherian ring and let $R = A[x_1, x_1^{-1},\ldots,x_t, x_t^{-1}, x_{t+1},\ldots,x_n]$. Then* $\mathrm{er}\,R \leqslant \sup\{2, \mathcal{K}(A)+1\}$.

Proof. See [Suslin 77, Theorem 7.2]. □

5.7 This can be made precise for polynomials over a field.

Corollary. *For any field k,* $\mathrm{er}(k[x_1,\ldots,x_n])$ *is 1 if $n=1$, and is 2 if $n>1$.*

Proof. The case when $n=1$ is covered by 5.3. When $n>1$ it is enough, by 5.6, to show $\mathrm{er}\,R \neq 1$. Consider, then, the case $n=2$, and suppose $\mathrm{er}\,R = 1$. Note that

$$A = \begin{bmatrix} 1+x_1x_2 & x_1^2 \\ -x_2^2 & 1-x_1x_2 \end{bmatrix} \in Gl(2,R)$$

and so $(1+x_1x_2, x_1^2) \in U_r(2,R)$. There must exist $E_1 \in E(2,R)$ with $(1+x_1x_2, x_1^2)E_1 = (1,0)$ and then

$$AE_1 = \begin{bmatrix} 1 & 0 \\ f & g \end{bmatrix}$$

say. However, since $\det AE_1 = 1$, $g = 1$ and so $AE_1E_2 = I$ for some $E_2 \in E(2,R)$. It follows that $A \in E(2,R)$ which contradicts [Cohn 66, Proposition 7.3].

One can now deduce, for $n \geqslant 2$, that $\mathrm{er}\,R \neq 1$ by using the injection and projection

$$k[x_1,x_2] \to k[x_1,x_2,\ldots,x_n] \to k[x_1,x_2]. \qquad \square$$

5.8 Now we turn to the stable rank of polynomial rings. First we quote a general result for affine algebras from 6.7.4 and [Suslin 81, Theorems 12, 13].

Theorem. *Let R be a commutative affine k-algebra over a field k.*
(i) $\mathrm{sr}(R) \leqslant \mathcal{K}(R) + 1$.
(ii) *If k has infinite transcendence degree over its prime subfield then* $\mathrm{sr}\,R = \mathcal{K}(R) + 1$.
(iii) *If k is an algebraic extension of a finite field then* $\mathrm{sr}\,R \leqslant \sup\{2, \mathcal{K}(R)\}$. □

5.9 Once again, for polynomial rings more precise bounds can be given. Let k be a field, k_0 its prime subfield and t the transcendence degree of k over k_0. The **Kronecker dimension** of k is defined to be t if $\mathrm{char}\,k > 0$, and $t+1$ if $\mathrm{char}\,k = 0$.

Theorem. *Let d be the Kronecker dimension of a field k.*
(i) *If $n \leq d$ then $\operatorname{sr} k[x_1,\ldots,x_n] = n+1$.*
(ii) *If $n > d$ then $\lceil(n+d+1)/2\rceil \leq \operatorname{sr} k[x_1,\ldots,x_n]$.*

Proof. See [Suslin **81**, Theorem 11]. □

5.10 Corollary. (i) $\operatorname{sr} \mathbb{R}[x_1,\ldots,x_n] = n+1$.
(ii) *If k is an algebraic extension of a finite field then $\operatorname{sr} k[x_1] = 2$, $\operatorname{sr} k[x_1,x_2] = 2$, $\operatorname{sr} k[x_1,x_2,x_3]$ is 2 or 3 and $\operatorname{sr} k[x_1,x_2,x_3,x_4]$ is 3 or 4.*

Proof. These follow from 5.3, 5.8 and 5.9. However, we give a direct proof of (i) for the case $n = 2$ and note that the general case can be established in a similar fashion (see [Gabel **75**] for details).
So let $S = \mathbb{R}[x, y]$. Since $K(S) = 2$, we know that $\operatorname{sr} S \leq 3$. Consider the unimodular column $(x, y, x^2 + y^2 - 1) \in S^3$ and let $R = \mathbb{R}[X, Y, Z]/(X^2 + Y^2 + Z^2 - 1)$ as in 2.3. There is an embedding $S \to R$ with x the image in R of X and y that of Y. Under this $(x, y, x^2 + y^2 - 1) \mapsto (x, y, z^2)$.
Suppose now that $(x, y, x^2 + y^2 - 1)$ were a stable unimodular column in S^3. Then (x, y, z^2) would be a stable unimodular column in R^3 and so too would (x, y, z). However 2.3 showed that (x, y, z) could not be extended to an invertible 3×3 matrix over R and so, by 3.6, (x, y, z) is not stable. It follows that $(x, y, x^2 + y^2 - 1)$ is not stable in S^3 and so $\operatorname{sr} S = 3$. □

5.11 The results above make clear that the behaviour of stable rank and elementary rank under ring extensions is rather unpredictable. We make no attempt, therefore, to discuss the ranks of the various rings described in Chapter 1 except for one special case, namely matrix rings, where fairly complete results can be obtained. The arguments used for the three ranks differ according to how easily the Morita equivalence can be exploited.

5.12 First we deal with the general linear rank where the Morita equivalence is readily used.

Theorem. $\operatorname{glr} M_k(R) = t$ *if and only if for all $n \geq t$ every stably free R-module of rank nk is free.*

Proof. Let $S = M_k(R)$ and $_SP_R$ be the progenerator which gives the Morita equivalence between R and S; so $P \simeq Se_{11} \simeq R^k$.
($\Rightarrow$) Suppose M_R is stably free of rank nk. Then $(M \otimes P^*)_S$ is stably free of rank n, so is free of rank n. However, $M \simeq M \otimes P^* \otimes P$, so M_R is free of rank nk.
($\Leftarrow$) Suppose M_S is stably free of rank n. Then $(M \otimes P)_R$ is stably free of rank nk and so M_S is free of rank n. □

5.13 Corollary. glr $M_k(R) \leq \lceil \text{glr } R/k \rceil$. □

5.14 The inequality in this corollary can be strict. To verify this, consider the ring R described in 1.15. This has nonfree projectives only in rank m, and these are stably free. It is clear from 5.12 that if $k|m$ then $M_k(R)$ has stably free nonfree modules of rank m/k, whereas if $k \nmid m$ then all stably free modules over $M_k(R)$ will be free. In the former case glr $M_k(R) = (m/k) + 1 = \lceil \text{glr } R/k \rceil$; and in the latter case glr $M_k(R) = 1$.

5.15 Next we turn to elementary rank.

Theorem. er $M_k(R) \leq \lceil \text{er } R/k \rceil$.

Proof. Let $S = M_k(R)$. By 3.15, to show er $S \leq t$ we need to show that if $n \geq t$ and α, γ are given then there exists $\beta \in E(n+1, S)$ such that, in the diagram

$$S \xrightarrow{\alpha} S^n \oplus S \xrightarrow{\beta} S^n \oplus S \xrightarrow{\gamma} S^n$$

$\gamma\beta\alpha$, like α, is a split monomorphism.

Translated by the Morita equivalence to R that diagram becomes

$$R^k \xrightarrow{\alpha} R^{nk} \oplus R^k \xrightarrow{\beta} R^{nk} \oplus R^k \xrightarrow{\gamma} R^{nk}.$$

We will, in fact, show in this case that there exists $\beta \in E(nk+k, R)$ such that $\beta\alpha e_i = e_i$ for $i = 1, \ldots, k$. This is proved by induction.

Note first that there is a β_1 such that $\beta_1 \alpha e_1 = e_1$, by hypothesis. Factor out $e_1 R$ on both sides. One now has

$$R^{k-1} \xrightarrow{\beta_1 \alpha} R^{nk} \oplus R^{k-1}.$$

By an appropriate induction hypothesis, there exists $\beta_2' \in E(nk+k-1, R)$ such that $\beta_2'\alpha(e_i) = e_i$ for all i. Let β_2 be the matrix

$$\begin{bmatrix} 1 & 0 \\ 0 & \beta_2' \end{bmatrix} \in E(nk+k, R).$$

Then $\beta_2\beta_1\alpha e_1 = e_1$ and $\beta_2\beta_1\alpha e_i = e_i + e_1 c_i$ for some $c_i \in R$. Let β_3 be defined by $e_1 \mapsto e_1$, $e_i \mapsto e_i - e_1 c_i$ for $i = 2, \ldots, k$, $e_j \mapsto e_j$ for $j > k$. Then $\beta_3\beta_2\beta_1\alpha e_i = e_i$ for $i = 1, \ldots, k$ as required.

Let $\beta = \beta_3\beta_2\beta_1$; then $\beta \in E(nk+k, R)$ and $\gamma\beta\alpha$ is a split monomorphism. However, $E(nk+k, R)$ can be identified with $E(n+1, S)$, making use of such simple relations as $I + e_{12}r = ABA^{-1}B^{-1}$, where $A = I + e_{1\,k+1}$ and $B = I + e_{k+1\,2}r$. Hence the corresponding map $\beta \in E(n+1, S)$ is as required. □

5.16 Our approach to stable rank is similar to that above for elementary rank, but it is expressed in matrix terms rather than module terms. We therefore extend the terminology of stable unimodular columns to matrices. A left invertible $(n+k) \times k$ matrix α is **stable** if for some $B \in B_c(n+k, R)$ the $(n+k-1) \times k$ matrix obtained by deleting the final row of $B\alpha$ is left invertible. If every $(n+k) \times k$ left invertible matrix is stable we write $(n, k) \in$ stable range R. Thus $(n, 1) \in$ stable range R means that $n \geq \operatorname{sr} R$.

5.17 Proposition. *Let k be any integer with $k \geq 1$. Then $n \geq \operatorname{sr} R$ if and only if $(n, k) \in$ stable range R.*

Proof. ($\Rightarrow$) Let α be a left invertible $(n+k) \times k$ matrix with $k > 1$. The first column of α is left unimodular. Therefore, for some $B_1 \in B_c(n+k, R)$,

$$B_1 \alpha = \begin{bmatrix} a & \alpha_1 \\ g & \end{bmatrix},$$

where $a \in U_c(n+k-1, R)$, g is 1×1 and α_1 is $(n+k) \times (k-1)$. Since $k > 1$, a is a stable unimodular column, so $Ca = e_1$ for some $C \in E(n+k-1, R)$, by 3.6. If

$$C_1 = \begin{bmatrix} C & 0 \\ 0 & 1 \end{bmatrix}$$

then

$$C_1 B_1 \alpha = \begin{bmatrix} 1 & x \\ 0 & \alpha_2 \\ g & \end{bmatrix}$$

where x is $1 \times (k-1)$ and α_2 is $(n+k-1) \times (k-1)$. Now let

$$F = \begin{bmatrix} 1 & -x \\ 0 & I_{k-1} \end{bmatrix} \in Gl(k, R)$$

and note that

$$C_1 B_1 \alpha F = \begin{bmatrix} 1 & 0 \\ 0 & \alpha_3 \\ g & \end{bmatrix}.$$

This matrix is left invertible, so the $(n+k-1) \times (k-1)$ matrix α_3 is left invertible too. By induction on k, for some $B_2 \in B_c(n+k-1, R)$

$$B_2 \alpha_3 = \begin{bmatrix} \alpha_4 \\ h \end{bmatrix}$$

with α_4 being a left invertible $(n+k-2) \times (k-1)$ matrix. If

$$B_3 = \begin{bmatrix} 1 & 0 \\ 0 & B_2 \end{bmatrix} \in B_c(n+k, R)$$

then
$$B_3C_1B_1\alpha F = \begin{bmatrix} 1 & 0 \\ 0 & \alpha_4 \\ g & h \end{bmatrix} = \begin{bmatrix} \alpha_5 \\ f \end{bmatrix}$$

say with α_5 a left invertible $(n+k-1) \times k$ matrix. Finally, note that

$$C_1^{-1}B_3C_1B_1\alpha = \begin{bmatrix} C^{-1}\alpha_5 F^{-1} \\ m \end{bmatrix}.$$

This matrix has the form $B\alpha$ with $B \in B_c(n+k, R)$ since $C_1^{-1}B_3C_1 \in B_c(n+k, R)$; and $C^{-1}\alpha_5 F^{-1}$ is a left invertible $(n+k-1) \times k$ matrix.

($\Leftarrow$) Suppose $a \in U_c(n+1, R)$. Consider the matrix

$$\alpha = \begin{bmatrix} I_{k-1} & 0 \\ 0 & a \end{bmatrix},$$

noting that it is a left invertible $(n+k) \times k$ matrix and so is stable. Therefore

$$B\alpha = \begin{bmatrix} \alpha_1 \\ b \end{bmatrix},$$

where $B \in B_c(n+k, R)$ and α_1 is left invertible. Now, for some $B_1 \in B_c(n+1, R)$

$$B = \begin{bmatrix} I_{k-1} & \alpha_2 \\ 0 & B_1 \end{bmatrix}$$

and it follows easily that the first n entries of $B_1 a$ form a unimodular column; i.e. a is stable. □

There is, of course, an analogous result for $k \times (n+k)$ matrices.

5.18 Theorem. sr $M_k(R) = \lceil (\text{sr } R - 1)/k \rceil + 1$.

Proof. Suppose sr $R = d$, $S = M_k(R)$ and $\alpha \in U_c(n+1, S)$. One can view α as a left invertible $(n+1)k \times k$ matrix over R and, if $(n+1)k \geq d + k$, then 5.17 shows that this matrix must be stable. In that case by adding multiples of the final row to earlier rows, and then deleting the final row, one can obtain an $((n+1)k - 1) \times k$ matrix which is left invertible. If $((n+1)k - (k-1)) \geq d + k$, this process can be repeated k times in all to obtain an $nk \times k$ left invertible matrix. It follows, viewing α again as an element of S^{n+1}, that α is stable. Hence sr $S \leq \lceil (d-1)/k \rceil + 1$.

Conversely, suppose $n = $ sr S and $a \in U_c(nk - k + 2, R)$. Consider the $(n+1)k \times k$ matrix

$$\alpha = \begin{bmatrix} I_{k-1} & 0 \\ 0 & a \\ 0 & 0 \end{bmatrix}$$

in which the final $k-1$ rows are zero. Then α is left invertible, so is a stable unimodular element of S^{n+1}. It follows easily that a is stable. Therefore $nk - k + 2 - 1 \geqslant \operatorname{sr} R = d$ and so $n \geqslant \lceil (d-1)/k \rceil + 1$. □

5.19 The difference in behaviour between the stable rank and the other two ranks in passing from R to $S = M_k(R)$ seems a little surprising. It is particularly apparent when $k \geqslant \operatorname{sr} R > 1$. For then $\operatorname{glr} S = \operatorname{er} S = 1$, whereas $\operatorname{sr} S = 2$.

5.20 Note, however, that the major use we have made of the stable rank of a ring, rather than the other two ranks, is in cancellation results like 4.8. Suppose we define yet another rank, the **cancellation rank** of R, $\operatorname{cr} R$, to be the least integer t such that if $n \geqslant t$ and $X \oplus R^{n+1} \simeq Y \oplus R$ then $X \oplus R^n \simeq Y$. Thus 4.8(ii) asserts that $\operatorname{sr} R \geqslant \operatorname{cr} R$.

5.21 We will show now that the cancellation rank behaves in a more regular fashion for matrix rings.

Proposition. $\operatorname{cr} M_k(R) \leqslant \lceil \operatorname{cr} R/k \rceil$.

Proof. Let $S = M_k(R)$ and let ${}_R P_S$ be the progenerator $e_{11}S$. Suppose, for some n, X_S and Y_S, that $X \oplus S^{n+1} \simeq Y \oplus S$. Then $X \oplus P^{kn+k} \simeq Y \oplus P^k$. Since $\operatorname{End}(P_S) = R$, 4.8(i) shows that one can cancel k copies of P. Hence $X \oplus S^n \simeq Y$. □

In particular this means that if $k \geqslant \operatorname{sr} R$ then

$$X \oplus S^2 \simeq Y \oplus S \Rightarrow X \oplus S \simeq Y$$

which is better than that which follows from 5.18 and 4.8(ii).

We note also that the example and argument of 5.14 can be used to show that $\operatorname{cr} S$ may not equal $\lceil \operatorname{cr} R/k \rceil$.

5.22 Finally, we note that 5.17 can be used to extend the results of Chapter 6, §7.

Proposition. (i) $\operatorname{sr}(R^k)_R = k - 1 + \operatorname{sr} R$.
(ii) *If M_R is a module having k generators, then* $\operatorname{sr} M \leqslant k - 1 + \operatorname{sr} R$.

Proof. (i) Let $\alpha : R^t \to R^k$ be an epimorphism. By 5.17, the matrix of α is stable if and only if $t \geqslant k + \operatorname{sr} R$.
(ii) M is a homomorphic image of R^k. So (i) together with 4.6 gives the result. □

§6 Local Information

6.1 Throughout this section R will denote a right Noetherian ring of finite Krull dimension and M a finitely generated right R-module.

11.6.3 Local information

One can view 5.22 as providing an upper bound on sr M in terms of 'global' information, namely the minimal number of generators of M and the stable rank of R. As mentioned earlier, §7 will give more precise results based on 'local' information.

The general aim of this section is to describe and study the local information about M at a prime $P \in \operatorname{Spec} R$. This information is provided by two numbers. The first to be dealt with is the basic dimension, this being of particular relevance when M/MP is a torsion R/P-module. The second is the local number of generators. This is of importance only when M/MP is not torsion.

6.2 The section starts with the basic dimension. This is a variant on the Krull dimension which concentrates on faithful critical subfactor modules and, in a way, reflects their special status as shown earlier by 6.7.7.

First suppose that R is prime. The **basic dimension** of M at 0, $B(M,0)$, is defined as the least integer t such that M has a critical composition series

$$M = M_n \supset M_{n-1} \supset \cdots \supset M_0 = 0$$

in which, for each i, if M_i/M_{i-1} is faithful then $\mathcal{K}(M_i/M_{i-1}) \leqslant t$; in the case where M is unfaithful, $B(M,0) = -1$.

For the general ring R, if P is a prime ideal, then $B(M,P)$ denotes $B(M/MP, 0)$ over the prime ring R/P. Any chain between M and MP which exhibits the value of $B(M,P)$ will be called a **basic composition series** of M at P.

Finally, for a set X of prime ideals, $B(M,X) = \sup\{B(M,P) | P \in X\}$, with the convention that $B(M, \varnothing) = -1$.

6.3 The next few results outline some of the properties of the basic dimension.

Proposition. *Let R be a prime ring and N a submodule of M. Then:*
 (i) $B(M,0) = \sup\{B(N,0), B(M/N,0)\}$;
 (ii) $-1 \leqslant B(M,0) \leqslant \mathcal{K}(M) \leqslant \mathcal{K}(R_R)$;
 (iii) $B(M,0) = -1$ if and only if $\operatorname{ann} M \neq 0$;
 (iv) $B(M,0) = \mathcal{K}(R_R)$ if and only if M is not a torsion module;
 (v) *if R is commutative or FBN then $B(M,0)$ equals -1 if M is torsion and equals $\mathcal{K}(R)$ otherwise.*

Proof. (i), (ii), (iii) Straightforward.

 (iv) Recall, from 6.3.11, that M is torsion if and only if $\mathcal{K}(M) < \mathcal{K}(R)$ in which case, by (ii), $B(M,0) < \mathcal{K}(R)$. On the other hand, if $B(M,0) < \mathcal{K}(R)$ then each basic composition factor M_i/M_{i-1} of M is either critical with $\mathcal{K}(M_i/M_{i-1}) \leqslant B(M,0) < \mathcal{K}(R)$ or else is unfaithful. In either case M_i/M_{i-1} is torsion; hence M is torsion.

 (v) In this case M is torsion if and only if $\operatorname{ann} M \neq 0$. $\square$

6.4 Corollary. (a) *If* $P \in \operatorname{Spec} R$ *then* $-1 \leq B(M, P) \leq \mathcal{K}(M/MP) \leq \mathcal{K}(R/P)$, *and* $B(M, P) = \mathcal{K}(R/P)$ *if and only if* M/MP *is not torsion over* R/P.
(b) *If* $X \subseteq \operatorname{Spec} R$ *and* $N \triangleleft M$ *then*
  (i) $B(M, X) \leq \mathcal{K}(R_R)$, *and*
  (ii) $B(M/N, X) \leq B(M, X) \leq \sup\{B(N, X), B(M/N, X)\}$. □

6.5 The next result indicates the control exercised by primitive ideals.

Proposition. *If* $B(M, P) = -1$ *for all primitive ideals* P *of* R *then* $M = 0$.

Proof. Suppose $M \neq 0$. Let N be a maximal submodule of M and let $P = \operatorname{ann} M/N$. Then P is primitive and, since $MP \subseteq N$, $\operatorname{ann} M/MP = P$. Therefore, by 6.3(iii), $B(M, P) \neq -1$, a contradiction. □

6.6 We recall from 4.6.14 that a subset X of $\operatorname{Spec} R$ is patch-closed if every prime ideal which is an intersection of prime ideals in X belongs to X. Of course, $\operatorname{Spec} R$ has this property. Another example of importance here is J-**Spec** R, this being the set of all prime ideals which are intersections of primitive ideals; such primes are termed J-**primes**.

6.7 Given such a set X, the set $\Lambda = \{P \in X \mid B(M, P) = B(M, X)\}$ plays an important part in what follows. First we note that Λ may well be an infinite set.

For example, let S be any ring such that $\operatorname{Spec} S$ is infinite. Let $R = S[x]$, $M = R/(x-1)R$ and $X = \{P \in \operatorname{Spec} R \mid x \in P\}$. Then X is patch-closed and infinite. For all $P \in X$, $M = MP$ and so $B(M, P) = B(M, X) = -1$; thus $\Lambda = X$.

6.8 Nevertheless, there are conditions under which Λ must be finite. Before demonstrating this some preliminary results are needed.

Lemma. *Let* N, L *be submodules of* M *with* $L \subset N$ *and* N/L *critical and suppose* $A \triangleleft R$ *with* $A \nsubseteq \operatorname{ann} N/L$. *Then* $\mathcal{K}(N + MA/L + MA) < \mathcal{K}(N/L)$.

Proof. If the canonical epimorphism

$$N/L \to (N + MA)/(L + MA) \to N/N \cap (L + MA)$$

were an isomorphism one would have

$$L = N \cap (L + MA) \supseteq NA$$

in which case $A \subseteq \operatorname{ann} N/L$. □

6.9 Given any subset X of $\operatorname{Spec} R$ we will let X_t denote the set $\{P \in X \mid \mathcal{K}(R/P) \geq t\}$.

11.6.11 *Local information* 403

Proposition. *If X is a patch-closed subset of* Spec R *and* $t \in \mathbb{N}$ *then the set*
$$Y = \{P \in X_t \setminus X_{t+1} \mid B(M, P) \geq \max\{0, B(M, X_{t+1})\}\}$$
is finite.

Proof. Suppose, to the contrary, that Y is infinite. Amongst infinite intersections of distinct primes in Y, choose Q to be maximal; say $Q = \bigcap P_j$. Note that Q is prime, $Q \in X$ and $K(R/Q) > t$; so $Q \in X_{t+1}$. Moreover, if P is one of the P_j then $M \neq MP$ and so $M \neq MQ$. Let
$$M = M_s \supset M_{s-1} \supset \cdots \supset M_0 = MQ$$
be a basic composition series for M at Q, let $A_i = \operatorname{ann}(M_i/M_{i-1})$ and $N_i = M_i + MP$. Evidently $A_i \supseteq Q$.

If $A_i = Q$ then 6.8 shows that $K(N_i/N_{i-1}) < K(M_i/M_{i-1})$; and if $A_i \nsubseteq P$ then $\operatorname{ann}(N_i/N_{i-1}) \supset P$. In the former case, this shows that N_i/N_{i-1} makes a smaller contribution to $B(M, P)$ than M_i/M_{i-1} makes to $B(M, Q)$; and in the latter, it makes no contribution at all.

However, since $Q \in X_{t+1}$,
$$B(M, Q) \leq B(M, X_{t+1}) \leq B(M, P).$$

It follows that the set $W = \{i \mid A_i \supset Q\}$ is non-empty and that there exists $i \in W$ with $A_i \subseteq P$.

Since there are infinitely many choices of P, for some $i \in W$, A_i is contained in infinitely many of the prime ideals P_j. This contradicts the maximality of Q and establishes the result. $\square$

6.10 Corollary. (i) *If X is a patch-closed subset of* Spec R *and* $B(M, X) \neq -1$ *then the set* $\Lambda = \{P \in X \mid B(M, P) = B(M, X)\}$ *is finite.*
(ii) *If $X = J$-Spec R and $M \neq 0$ then $B(M, X) \neq -1$ and so Λ is finite.*

Proof. (i) If Λ were infinite then for some $t \leq K(R)$ the set Y in 6.9 would be infinite.
(ii) Clear from 6.5. $\square$

6.11 The next result, which echoes results in Chapter 6, §7, aims towards stable generation.

Proposition. *Let R be prime and $0 \neq A \triangleleft R$. Let M be torsion and $x, y \in M$ be such that $B(M/xR + yR, 0) < B(M, 0)$. Then there exists $a \in A$ such that $B(M/(x + ya)R, 0) < B(M, 0)$.*

Proof. By 6.3(i), $B(yA, 0) \leq B(M, 0)$. Let
$$yA = N_s \supset \cdots \supset N_0 = 0$$

be a basic composition series of yA at 0 and consider the statement: $\mathscr{S}(i)$: There exists $a_i \in A$ with $B(M/N_i + (x + ya_i)R, 0) < B(M, 0)$.

Note that $B(yR + xR/yA + xR, 0) = -1$ since $\operatorname{ann}(yR + xR/yA + xR) \supseteq A$. Therefore, by 6.3(i),

$$B(M/yA + xR, 0) = B(M/yR + xR, 0) < B(M, 0)$$

and so $\mathscr{S}(s)$ holds with $a_s = 0$. Note also that the result at which we aim is $\mathscr{S}(0)$.

Suppose now that $\mathscr{S}(i)$ holds for some $i \in \{1, \ldots, s\}$ and write $z = x + ya_i$. It is enough to deduce $\mathscr{S}(i - 1)$. If $\operatorname{ann} N_i/N_{i-1} \neq 0$ then $\operatorname{ann}(N_i + zR/N_{i-1} + zR) \neq 0$ and so $\mathscr{S}(i - 1)$ holds with $a_{i-1} = a_i$. On the other hand, suppose that $\operatorname{ann} N_i/N_{i-1} = 0$ in which case N_i/N_{i-1} is critical. If $I = \operatorname{ann} z$ then $I \neq 0$, since M is torsion, and therefore $N_i I \nsubseteq N_{i-1}$. Thus $nf \notin N_{i-1}$ for some $n \in N_i$, $f \in I$, and yet $nf \in (N_{i-1} + (z + n)R) \cap N_i$. Hence

$$\mathcal{K}(N_i + (z + n)R/N_{i-1} + (z + n)R) < \mathcal{K}(N_i/N_{i-1}) \leq B(M, 0).$$

Of course, $N_i + zR = N_i + (z + n)R$ and $n = ya'$ for some $a' \in A$. Therefore $\mathscr{S}(i - 1)$ holds with $a_{i-1} = a_i + a'$. □

6.12 Corollary. *Let R be prime and $0 \neq A \triangleleft R$. Let M be a nonzero torsion module with $M = \sum_{i=1}^{s} m_i R$ and $B(M, 0) \geq 0$. Then there exist $a_2, \ldots, a_s \in A$ such that $B(M/cR, 0) < B(M, 0)$, where $c = m_1 + \sum_{i=2}^{s} m_i a_i$.*

Proof. Let $M' = \sum_{i=2}^{s} m_i R$. If $M' = 0$ we can take $a_i = 0$. Otherwise, by induction on s, we may suppose that $B(M'/yR, 0) < B(M', 0)$, where $y = m_2 + \sum_{i=3}^{s} m_i b_i R$ and $b_i \in A$. The epimorphism $M'/yR \to M/m_1 R + yR$ shows that

$$B(M/m_1 R + yR, 0) \leq B(M'/yR, 0) < B(M', 0) \leq B(M, 0)$$

and then 6.11 gives the result. □

6.13 We now turn to the second piece of local information about M at $P \in \operatorname{Spec} R$; we let $Q = Q(R/P)$, the right quotient ring of R/P.

If R is commutative, the local number of generators of M at P is defined to be the minimal number of generators of M_P over R_P and is written as $g(M, P)$. It is easy, using Nakayama's lemma, to see that this equals the minimal number of generators of $M \otimes_R Q$ over Q and so is just the dimension of $M \otimes Q$ over the field Q.

For a noncommutative ring R we have seen that localization at P might not be feasible. However, it still makes sense to define the **local number of generators of M at P**, written $g(M, P)$, to be the minimal number of generators of $M \otimes_R Q$ over Q. Note that $M \otimes_R Q \simeq M/MP \otimes_{R/P} Q$.

6.14 It is easy to relate $g(M, P)$ with $\rho_P(M)$, the rank of M with respect to P, as defined in 4.6.7.

11.6.16

Lemma. *If $P \in \operatorname{Spec} R$ and $Q = Q(R/P)$, then the following numbers are all equal to $g(M, P)$:*
  (i) the minimal number of generators of $M/MP \otimes_{R/P} Q$ over Q;
  (ii) $\lceil \operatorname{length}(M/MP \otimes Q)_Q / \operatorname{length} Q \rceil$;
  (iii) $\lceil \rho_P(M)/\rho_P(R) \rceil$.

Proof. Straightforward. □

6.15 We note next two ways in which the behaviour of $g(M, P)$ over noncommutative rings is not as orderly as over commutative rings. The first point is that one can find pairs of prime ideals $P_1 \subset P_2$ with $g(M, P_1) \neq 0$ and $g(M, P_2) = 0$. To see this let R be the integral domain $k + xA_1(k)$ described in 1.3.10. This has a nonzero prime ideal $P_2 = xA_1(k)$ with $P_2^2 = P_2$. If $M = P_2$ and $P_1 = 0$ then it is clear that $g(M, P_2) = 0$ and $g(M, P_1) = 1$.

The second point is that one can have $g(M, P) = 0$ for all primes P without M being zero. For example, let R be any simple Noetherian domain and M any proper homomorphic image of R_R. In this case 0 is the only prime ideal and, since M is torsion, $g(M, 0) = 0$.

6.16 Our study of $g(M, P)$ or, equivalently of $\rho_P(M)$, starts with a rather general stability result.

Proposition. *Let F, G be right R-modules with F being projective and let $\varphi: F \to M$, $\gamma: G \to M$ be given homomorphisms. Let $P \in \operatorname{Spec} R$ and $A \triangleleft R$ with $A \not\subseteq P$. Then there exists $\beta: F \to GA$ such that*

$$\rho_P(M/(\varphi + \gamma\beta)(F)) = \max\{\rho_P(M/\varphi(F) + \gamma(G)), \rho_P(M) - \rho_P(F)\}.$$

Proof. It is clear that, for any choice of β, the left-hand side is greater than or equal to the right-hand side. So it is enough to establish the reverse inequality. There is no loss of generality in supposing that $P = 0$ (so $\rho_P = \rho$), that the torsion submodule of M is zero, and that $\gamma(G) \neq 0$. Furthermore, an easy induction argument allows us to assume that $\gamma(G)$ is uniform.

If $\ker \varphi = 0$ then F embeds in M and so $\rho(\varphi(F)) = \rho(F)$. Therefore, since reduced rank is additive, $\rho(M/\varphi(F) + \gamma(G)) \leq \rho(M) - \rho(F)$ and $\beta = 0$ is satisfactory.

Next suppose that $\ker \varphi \neq 0$ and choose $k \in \ker \varphi \setminus \{0\}$. Since F is projective there is a map $\alpha: F \to R$ with $\alpha(k) \neq 0$. Hence $\gamma(G)A\alpha(k)R \neq 0$, since $A\alpha(k)R$ is a nonzero ideal. Therefore $\gamma(g)a\alpha(k) \neq 0$ for some $g \in G$ and $a \in A$. We define $\beta: F \to GA$ by $f \mapsto ga\alpha(f)$. Then

$$0 \neq \gamma(g)a\alpha(k) \in \gamma(G) \cap (\varphi + \gamma\beta)(F)$$

and, since $\gamma(G)$ is uniform, $\gamma(G)/\gamma(G) \cap (\varphi + \gamma\beta)(F)$ is a torsion R-module.

However,

$$\gamma(G)/\gamma(G)\cap(\varphi+\gamma\beta)(F) \simeq (\varphi+\gamma\beta)(F) + \gamma(G)/(\varphi+\gamma\beta)(F) \simeq \varphi(F) + \gamma(G)/(\varphi+\gamma\beta)(F).$$

Hence $\rho(M/(\varphi+\gamma\beta)(F)) = \rho(M/\varphi(F) + \gamma(G))$, so β is as required. $\square$

6.17 The connection with stable generation is made apparent next.

Corollary. *Let* $M = \sum_{i=1}^{s} m_i R$ *and* $A \triangleleft R$ *with* $A \not\subseteq P$. *Then there exists* $a_2, \ldots, a_s \in A$ *such that, for* $c = m_1 + \sum_{i=2}^{s} m_i a_i$,

$$\rho_P(M/cR) = \max\{0, \rho_P(M) - \rho_P(R)\}$$

and

$$g(M/cR, P) = \max\{0, g(M, P) - 1\}.$$

Proof. We apply 6.16 with $F = R$, $G = R^{s-1}$, $\varphi: 1 \mapsto m_1$ and $\gamma: e_i \mapsto m_{i+1}$ for $i = 1, \ldots, s-1$. $\square$

6.18 One can extend 6.16 to deal with finitely many modules.

Corollary. *Let* F, G_j *be right* R-*modules with* F *being projective and let* $\varphi: F \to M$, $\gamma_j: G_j \to M$, *for* $j = 1, \ldots, s$, *be given homomorphisms. Let* $P \in \operatorname{Spec} R$ *and* $A \triangleleft R$ *with* $A \not\subseteq P$. *Then there exist* $\beta_j: F \to G_j A$ *such that*

$$\rho_P(M/(\varphi + \sum \gamma_j \beta_j)(F)) = \max\{\rho_P(M/\varphi(F) + \sum \gamma_j(G_j)), \rho_P(M) - \rho_P(F)\}.$$

Proof. Take $G = G_1 \oplus \cdots \oplus G_s$ and $\gamma = (\gamma_1, \ldots, \gamma_s)$ and apply 6.16. $\square$

6.19 One can also extend 6.16 to deal with a finite set of prime ideals $\{P_1, \ldots, P_n\}$ say. Note that they can be so ordered that $\bigcap_{i=1}^{r-1} P_i \not\subseteq P_r$ for $r = 1, \ldots, n$. We state and prove the result for a single module G for clarity's sake; but, as before, it can readily be extended to any finite collection $G_1, \ldots, G_s$.

Corollary. *Let* $\Lambda = \{P_1, \ldots, P_n\}$ *be a finite set of prime ideals. Let* F, G *be right* R-*modules with* F *being projective and let* $\varphi: F \to M$, $\gamma: G \to M$ *be given. Then there exists* $\beta: F \to G$ *such that, for all* $P \in \Lambda$,

$$\rho_P(M/(\varphi + \gamma\beta)(F)) = \max\{\rho_P(M/\varphi(F) + \gamma(G)), \rho_P(M) - \rho_P(F)\}.$$

Proof. In 6.16 it is shown that this holds when $n = 1$, taking $A = R$. Suppose the primes P_i are ordered as above. The proof proceeds by induction on n. So we suppose there exists a map β as required for $P_1, \ldots, P_{n-1}$, and let $A = \bigcap_{i=1}^{n-1} P_i$ and $P = P_n$. Applying 6.16, we can choose $\beta': F \to GA$ such that

$$\rho_P(M/(\varphi + \gamma\beta + \gamma\beta')(F)) = \max\{\rho_P(M/(\varphi+\gamma\beta)(F) + \gamma(G)), \rho_P(M) - \rho_P(F)\}$$
$$= \max\{\rho_P(M/\varphi(F) + \gamma(G)), \rho_P(M) - \rho_P(F)\}.$$

Since $\gamma\beta'(F) \subseteq MA$, this also remains true for $P_1, \ldots, P_{n-1}$. $\square$

Of course, this result implies that

$$g(M/(\varphi + \gamma\beta)(F), P) = \max\{g(M/\varphi(F) + \gamma(G), P), g(M, P) - g(F, P)\}.$$

§7 Stability and Cancellation

7.1 Throughout this section R denotes a right Noetherian ring of finite Krull dimension and M is a finitely generated right R-module. The main result provides an upper bound for sr M in terms of the local behaviour of M at J-prime ideals. From this it will be deduced that if M is a large projective module then $M \simeq M' \oplus R$ and that if $M \oplus R \simeq N \oplus R$ then $M \simeq N$.

7.2 We need one further piece of notation which combines the two types of local information introduced in §6. If $P \in \operatorname{Spec} R$, then

$$b(M, P) = \begin{cases} g(M, P) + \mathcal{K}(R/P) & \text{if } g(M, P) \neq 0 \\ B(M, P) + 1 & \text{if } g(M, P) = 0, \end{cases}$$

and if $X \subseteq \operatorname{Spec} R$ then

$$b(M, X) = \sup\{b(M, P) | P \in X\}$$

with the convention that $b(M, \emptyset) = 0$.

7.3 The following facts are easily deduced from 6.4 and 6.14.

Lemma. (i) $B(M, P) + 1 \leq g(M, P) + \mathcal{K}(R/P)$.
(ii) $b(M, P) \leq \mathcal{K}(R/P)$ if and only if $g(M, P) = 0$.
(iii) $b(M, P) \geq 1$ if and only if $B(M, P) \geq 0$.
(iv) If $g(M, P) \leq 1$ then $b(M, P) = B(M, P) + 1$.
(v) If M has s generators then $b(M, P) \leq s + \mathcal{K}(R)$. □

The partition of the range of values of $b(M, P)$ provided by (ii) plays an important role in what follows.

7.4 When R is commutative or FBN, 6.3(v) shows that if $g(M, P) = 0$ then $B(M, P) = -1$. In that case it follows that

$$b(M, P) = \begin{cases} g(M, P) + \mathcal{K}(R/P) & \text{if } g(M, P) \neq 0, \\ 0 & \text{if } g(M, P) = 0, \end{cases}$$

this being the standard definition of $b(M, P)$ when R is commutative.

In the commutative case there are some well-known results and it seems advisable to start by mentioning them briefly. The Forster–Swan theorem, as improved by Eisenbud-Evans, asserts that

$$\operatorname{sr} M \leq b(M, X) \leq \sup\{g(M, P) + \mathcal{K}(R/P) | P \in X\},$$

where $X = J$-Spec R. Serre's theorem, that a large projective module has a free summand, and Bass' theorem, which allows free summands to be cancelled, are consequences of this.

The aim of this section is to extend these results to the noncommutative case.

7.5 First comes a stable generation result.

Proposition. *Let $M = \sum_{i=1}^{s} m_i R$ and $\Lambda \subseteq \operatorname{Spec} R$ be finite. Then there exists $r_2, \ldots, r_s \in R$ such that, for each $P \in \Lambda$, either $b(M, P) = 0$ or else $b(M/cR, P) < b(M, P)$, where $c = m_1 + \sum_{i=2}^{s} m_i r_i$.*

Proof. If $\Lambda = \emptyset$ there is nothing to prove. Suppose that $\Lambda = \{P_1, \ldots, P_n\}$ with $n \geq 1$, with the P_i ordered as in the proof of 6.19, and assume, as an inductive hypothesis, that the result holds for $\Lambda' = \{P_1, \ldots, P_{n-1}\}$. Thus there exist $r'_2, \ldots, r'_s \in R$ making the claims true for each $P \in \Lambda'$, using the element $c' = m_1 + \sum_{i=2}^{s} m_i r'_i$.

If $b(M, P_n) = 0$ there is nothing more to prove. Otherwise let $A = \bigcap_{j=1}^{n-1} P_j$. Using 6.17 if $g(M, P_n) \neq 0$, or 6.12 if $g(M, P_n) = 0$, there exist $a_2, \ldots, a_s \in A$ such that $b(M/cR, P_n) < b(M, P_n)$, where

$$c = c' + \sum_{i=2}^{s} m_i a_i = m_1 + \sum_{i=2}^{s} m_i r_i \quad \text{with } r_i = r'_i + a_i.$$

Since $a_i \in P_j$ for each i and for each $j \in \{1, \ldots, n-1\}$, it is clear that the elements r_i are as required. □

7.6 Next we note an elementary fact which is easily checked by considering separately the cases when $g(M, P) = 0$ and when $g(M, P) > 0$.

Lemma. *Let $X \subseteq \operatorname{Spec} R$, $X \neq \emptyset$, and $\Lambda = \{P \in X \mid b(M, P) = b(M, X)\}$, and suppose that $c \in M$ is such that $b(M/cR, P) < b(M, P)$ for each $P \in \Lambda$. Then $b(M/cR, X) < b(M, X)$.* □

7.7 The crucial step in our argument comes next.

Proposition. *If $X = J$-Spec R and $M \neq 0$ then the set $\{P \in X \mid b(M, P) = b(M, X)\}$ is finite.*

Proof. Note first, by 6.5, that $B(M, X) \geq 0$ and so $b(M, X) \geq 1$.

Now suppose that the result is false. This implies, using the notation of 6.9, that the set

$$\{P \in X_t \mid b(M, P) \geq \max(1, b(M, X_{t+1}))\}$$

is infinite when $t = 0$. We denote this set by $\Lambda(t, M)$ or $\Lambda(t)$ for short. We now

11.7.8   *Stability and cancellation*

choose M so as to obtain the largest possible value of t for which the set $\Lambda(t, M)$ is infinite and, subject to that, to obtain the smallest possible value of $s = b(M, X_{t+1})$.

Note that, if we discover a module M' such that $\Lambda(t, M')$ is infinite and $b(M', X_{t+1}) < s$, then a contradiction ensues and the proof is complete.

Suppose for a moment that $b(M, X_{t+1}) = 0$. Then 7.3(iii) shows that $B(M, X_{t+1}) = -1$. Also, $\Lambda(t) \cap X_{t+1} = \varnothing$. However, for all $P \in \Lambda(t)$, $b(M, P) \geq 1$ and so, by 7.3(iii), $B(M, P) \geq 0$. Thus $\Lambda(t) \subseteq Y$, as described in 6.9, this being a contradiction to 6.9. We conclude that $b(M, X_{t+1}) \geq 1$.

Next, let $P \in \Lambda(t)$ be such that $g(M, P) \leq 1$. Then, by 7.3(iv) and the definition of $\Lambda(t)$,

$$B(M, P) + 1 = b(M, P) \geq \max\{1, b(M, X_{t+1})\}$$

and so $B(M, P) \geq \max\{0, B(M, X_{t+1})\}$. Hence, by 6.9, the set

$$\{P \in \Lambda(t), P \notin X_{t+1} | g(M, P) \leq 1\}$$

is finite.

The maximality of t, together with the fact that $b(M, X_{t+1}) \geq 1$, shows that the set

$$\{P \in X_{t+1} | b(M, P) = b(M, X_{t+1})\}$$

is finite. This implies, by 7.5 and 7.6, that there exists $c \in M$ such that

$$b(M/cR, X_{t+1}) < b(M, X_{t+1}) = s;$$

and it also implies that the set

$$\Gamma = \{P \in X_t \setminus X_{t+1} | g(M, P) \geq 2, b(M, P) \geq \max\{1, b(M, X_{t+1})\}\}$$

is infinite.

Finally, note that, if $P \in \Gamma$, then $g(M/cR, P) \geq g(M, P) - 1 \geq 1$, and hence

$$b(M/cR, P) \geq b(M, P) - 1 \geq b(M, X_{t+1}) - 1 \geq b(M/cR, X_{t+1}).$$

Thus $\Gamma \subseteq \Lambda(t, M/cR)$ and so $M' = M/cR$ is as promised. □

7.8 At this point it becomes clear why the basic dimension term is required in the definition of $b(M, P)$. For suppose one defined (cf. 7.4)

$$b'(M, P) = \begin{cases} g(M, P) + K(R/P) & \text{if } g(M, P) \neq 0, \\ 0 & \text{if } g(M, P) = 0, \end{cases}$$

and $b'(M, X) = \sup\{b'(M, P) | P \in X\}$. Then the example below shows that the analogue of 7.7 would fail.

Example. Let $R = A_1(\mathbb{Z})$, the first Weyl algebra over $\mathbb{Z}$, let $M = R/xR$ and $X = \operatorname{Spec} R = J\text{-Spec} R$ (by 9.4.22). Then:

(i) $b'(M, X) = 2$;
(ii) $\{P \in X \mid b'(M, P) = 2\}$ is infinite;
(iii) $b(M, X) = B(M, X) + 1 = 3$.

Proof. First we discuss the prime ideals of R. Since $A_1(\mathbb{Q})$ is simple, each nonzero ideal of R meets $\mathbb{Z}$. It follows that the height 1 primes of R have the form pR, $0 \neq p$ a prime of $\mathbb{Z}$. Now $R/pR \simeq A_1(\mathbb{Z}/(p))$ which has centre $\mathbb{Z}/(p)[x^p, y^p]$ and it is easy to check that each ideal of $A_1(\mathbb{Z}/(p))$ is the extension of an ideal of the centre. In particular x^p and x^p, y^p generate prime ideals of R.

Next we calculate Krull dimensions. It is clear from 6.5.4 that $K(R) \leq 3$. It follows from 6.3.11(ii) that the factors of R by the prime ideals in the chain

$$0 \subset pR \subset pR + x^pR \subset pR + x^pR + y^pR$$

must have Krull dimensions, 3, 2, 1 and 0 respectively.

Now consider $b'(M, P)$. If $P = 0$ or if P has height 1 then M/MP is torsion over R/P and hence $b'(M, P) = 0$. For any other prime it is clear from 7.3(v) that $b'(M, P) \leq 2$. Hence $b'(M, X) \leq 2$. However, if $P = pR + x^pR$ it is easy to check, for $i = 1, \ldots, p - 1$, that

$$M/MP = R/xR + pR \simeq x^iR + pR/x^{i+1}R + pR \simeq x^{p-1}R + pR/P$$

and that $x^{p-1}R + pR/P$ is a uniform right ideal of R/P. Hence $\rho_P(M) = 1$, $\rho_P(R) = p$ and $b'(M, P) = 2$. This establishes (i), (ii).

Finally, we turn to (iii). Since M_R is torsion it is clear, from 6.4(i), that $B(M, P) \leq 2$ for all $P \in \operatorname{Spec} R$. Thus we need only show that $B(M, 0) \geq 2$. However, M is 2-critical; so if $M = M_s \supset M_{s-1} \supset \cdots \supset M_0 = 0$ is a basic composition series then $K(M_1) = 2$. Also M is $\mathbb{Z}$-torsion-free; so $(\operatorname{ann} M_1) \cap \mathbb{Z} = 0$ and hence $\operatorname{ann} M_1 = 0$. Thus $B(M, 0) = 2$ as required. $\square$

7.9 The connection with stable rank is now made.

Theorem. *If $X = J\text{-Spec } R$ then $\operatorname{sr} M \leq b(M, X)$.*

Proof. The proof is by induction on $b(M, X)$. Of course if $b(M, X) = 0$ then $B(M, X) = -1$ and so, by 6.5, $M = 0$.

Suppose next that $0 \neq M = \sum_{i=1}^{s} m_i R$ with $s > b(M, X) > 0$. By 7.7, the set

$$\Lambda = \{P \in X \mid b(M, P) = b(M, X)\}$$

is finite. Hence, by 7.5 and 7.6, there exist $r_2, \ldots, r_s \in R$ such that $b(M/cR, X) < b(M, X)$, where $c = m_1 + \sum_{i=2}^{s} m_i r_i$. The inductive hypothesis can now be applied to M/cR, noting that it is generated by the images of $m_2, \ldots, m_s$. The result follows easily. $\square$

7.10 The other promised result concerning $\operatorname{sr} M$ is an immediate consequence.

11.7.12 *Stability and cancellation* 411

Corollary. (i) sr $M \leqslant \sup \{g(M, P) + \mathcal{K}(R/P) | P \in J\text{-Spec } R\}$.
(ii) sr $M \leqslant \sup \{g(M, P) + \mathcal{K}(R/J(R)) | P \text{ primitive}\}$.

Proof. (i) This follows directly from 7.9 and 7.3(i).
(ii) It is evident from (i) that

$$\text{sr } M \leqslant \sup \{g(M, P) + \mathcal{K}(R/J(R)) | P \in J\text{-Spec } R\}.$$

We choose $P \in J\text{-Spec } R$ to achieve this supremum. If P is primitive the result is clear. So suppose P is not primitive. Let $\varepsilon = 1/\rho_P(R)$ and let $I \supset P$ be the ideal given by 4.6.10; so, for each $Q \in \text{Spec } R$ with $Q \supset P$, $Q \not\supset I$,

$$|\rho_Q(M)/\rho_Q(R) - \rho_P(M)/\rho_P(R)| < \varepsilon$$

and hence $g(M, Q) = g(M, P)$. Since P is a J-prime, Q can be chosen to be primitive, as required. □

7.11 The bounds given in 7.9 and 7.10 cannot in general be lowered. As an easy example, suppose that $M = I$, a nonprincipal ideal of a Dedekind domain R. Then $g(M, P) = 1$ for each $P \in X$ and so $b(M, X) = 2$. Since M is not cyclic, it follows that sr $M = 2$. Similar examples for rings of higher Krull dimensions are given in [Swan 77].

We note also that 7.10 applied to R itself shows, as in 6.7.4, that sr $R \leqslant 1 + \mathcal{K}(R/J(R))$.

7.12 The next result, which prepares the way for the other theorems promised earlier, involves the rational number $\rho_P(M)/\rho_P(R)$; this is denoted by $\hat{g}(M, P)$.

Lemma. Let $X = J\text{-Spec } R$ and $\mathcal{K}(R/J(R))_R = n$. Suppose that M_R is projective, $\vartheta_1 \in \text{Hom}(M, R)$, $\hat{g}(M, P) \geqslant n + 1$ for all P in X, and $b(R/\vartheta_1(M), X) \leqslant n$. Then, for any $r \leqslant n + 1$, there exist $\vartheta_2, \ldots, \vartheta_r \in \text{Hom}(M, R)$ such that $b(R^r/\vartheta(M), X) \leqslant n$, where $\vartheta = (\vartheta_1, \vartheta_2, \ldots, \vartheta_r) \in \text{Hom}(M, R^r)$.

Proof. The proof is by induction on r. The case $r = 1$ is part of the hypothesis. So we suppose the result holds for some $r \leqslant n$ and check its validity for $r + 1$.

Thus we assume that $\vartheta_2, \ldots, \vartheta_r$ are as described and, if $M' = R^r/\vartheta(M)$, that $b(M', X) \leqslant n$. It is clear, using 7.7, that the set

$$\Gamma = \{P \in X | b(M', P) = n\} \cup \{P \in X | \mathcal{K}(R/P) = n\}$$

is finite. We now apply 6.19 to the two maps $\varphi: M \to R^{r+1}$ and $\gamma: R \to R^{r+1}$ where $\varphi = (\vartheta, 0)$ and γ is the canonical map to the last component. This provides $\beta: M \to R$ such that, for all $P \in \Gamma$,

$$\rho_P(R^{r+1}/(\varphi + \gamma\beta)(M)) = \max \{\rho_P(R^{r+1}/\varphi(M) + \gamma(R)), \rho_P(R^{r+1}) - \rho_P(M)\}.$$

However, $\rho_P(R^{r+1}) - \rho_P(M) < 0$ since $\hat{g}(M, P) \geq n + 1$; so we deduce that
$$g(R^{r+1}/(\varphi + \gamma\beta)(M), P) = g(R^{r+1}/\varphi(M) + \gamma(R), P) = g(R^r/\vartheta(M), P).$$
So if $\vartheta_{r+1} = \beta$ and $\psi = (\vartheta_1, \ldots, \vartheta_{r+1})$ then, using 7.3,
$$b(R^{r+1}/\psi(M), P) \leq n \quad \text{for all } P \in \Gamma.$$
Now suppose that $P \in X \setminus \Gamma$; then $b(R^r/\vartheta(M), P) < n$. The homomorphism
$$R^{r+1}/\psi(M) \to R^r/\vartheta(M)$$
induced by factoring out the last component of R^{r+1} has a kernel, K say, with $g(K, P) \leq 1$. Therefore $g(R^{r+1}/\psi(M), P) \leq g(R^r/\vartheta(M), P) + 1$.

Hence $b(R^{r+1}/\psi(M), P) \leq n$ for all $P \in X$, as required. □

7.13 We now draw the desired conclusions.

Theorem. *Let R be right Noetherian, M_R be finitely generated projective and $\hat{g}(M, P) \geq 1 + K(R/J(R))$ for each $P \in J\text{-Spec } R$. Then:*
(i) *M is B-permeated by R;*
(ii) *('Serre's theorem') $M \simeq M' \oplus R$ for some M'_R;*
(iii) *('Bass' theorem') if $M \oplus R \simeq N \oplus R$ for some N_R then $M \simeq N$.*

Proof. (i) Suppose that $\varphi: M \to R$ and $a \in R$ are such that $\varphi(M) + aR = R$; i.e. the map $(\varphi, \gamma): M \oplus R \to R$, where γ is left multiplication by a, is a split epimorphism. It is required to find $\beta': M \to R$ so that $(\varphi + \gamma\beta')(M) = R$.

Let $K(R/J(R)) = n$, $X = J\text{-Spec } R$, and $\Lambda = \{P \in X \mid K(R/P) = n\}$. Then Λ is a finite set. Moreover, $\hat{g}(M, P) \geq 1 = \hat{g}(R, P)$ for each $P \in \Lambda$. Therefore, by 6.19, there exists $\beta: M \to R$ such that $g(R/(\varphi + \gamma\beta)(M), P) = 0$ for all $P \in \Lambda$. Let $\vartheta_1 = \varphi + \gamma\beta$; then 7.3 shows that $b(R/\vartheta_1(M), P) \leq n$ for all P in X. By 7.12 there exist $\vartheta_2, \ldots, \vartheta_{n+1}$ such that $b(R^{n+1}/\vartheta(M), X) \leq n$, where $\vartheta = (\vartheta_1, \vartheta_2, \ldots, \vartheta_{n+1})$. Hence, by 7.9, $\text{sr}(R^{n+1}/\vartheta(M)) \leq n$.

By construction $R^{n+1} = \vartheta(M) + \sum_{i=2}^{n+1} e_i R + e_1 a R$, where the e_i are the usual basis vectors. Therefore
$$R^{n+1} = \vartheta(M) + \sum_{i=2}^{n+1} (e_i + e_1 a r_i) R$$
for some $r_i \in R$. In particular
$$e_1 = \vartheta(m) + \sum_{i=2}^{n+1} (e_i + e_1 a r_i) s_i$$
for some $m \in M$, $s_i \in R$. Therefore
$$1 = \vartheta_1(m) + \sum_{2}^{n+1} a r_i s_i$$

and $0 = \vartheta_i(m) + s_i$ for $2 \leq i \leq n+1$. It follows that, if $\beta' = \beta - \sum_{i=2}^{n+1} r_i \vartheta_i$, then $1 = (\varphi + \gamma \beta')(m)$ as required.
(ii) The map $\varphi + \gamma \beta' : M \to R$ is a split epimorphism.
(iii) This is immediate from (i) and 4.4. □

7.14 As an illustration we note one consequence which is related to 5.7.8, 5.7.18 and 7.11.6.

Corollary. *If R is a simple right Noetherian domain with $\mathcal{K}(R_R) = 1$ and M is projective and nonzero, then $M \simeq I \oplus R^n$ for some $0 \neq I \lhd_r R$. Moreover, $I \oplus R$ is uniquely determined, up to isomorphism, by M.*

Proof. Straightforward, using Proposition 3.4.3(iv). □

§8 Additional Remarks

8.0 (a) As indicated above, this theory extends some well-known commutative results inspired by Serre's conjecture. The verification of Serre's conjecture, cited as 2.2, is described in [Lam **78**].

Nevertheless, the results here are not an extension of 2.2 since, as §2 demonstrates, its noncommutative analogue is false. Rather they follow the approach described in [Bass **64**] and the [Eisenbud and Evans **73**] version of the Forster–Swan theorem.

The extension of some of these results to noncommutative Noetherian rings was begun, for simple and then ideal invariant rings, by [Stafford **77, 78a**] and, for FBN rings, by [Warfield **80**]. The definitive results for Noetherian rings were established in [Stafford **81**]. The account given here follows [Coutinho **86**] who has refined Stafford's ideas, in particular by introducing the basic dimension. Thereby the results now apply to right Noetherian rings.

8.1 The terminology 'general linear rank' is not well established. Related results appear in [Rieffel **83**].

8.2 (a) The main source for this section is [Stafford **85a**]. However, 2.9 is due to [Ojanguren and Sridharan **71**] and 2.12 (and a corresponding result for $R[x, x^{-1}; \sigma]$) to [Bonang **P**].
(b) Other interesting examples of projective modules appear in [Swan **62**] and [Hodges and Osterburg **87**].

8.3 (a) The broad outline here echoes [Bass **64**], but has some novel features, especially 3.12 and 3.13.
(b) 3.7(ii) and 3.8 are due to [Stafford **77, 81**] and 3.4 to [Vaserstein **71**].

8.4 (a) This too is in the spirit of [Bass **64**]. 4.8 is due to [Warfield **80a**] although our approach is new. 4.9 is due to [Evans **73**].
(b) The terminology used is new.

8.5 (a) A good source for the topics in the first half of the section is [Suslin **81**] where it is noted that 5.8(ii) is due to Vaserstein.

(b) The second half, concerning matrices, originates from [Vaserstein **71**] where, in particular, 5.17 appears. A module-theoretic approach appeared in [Warfield **80a**]. This account is largely new.

8.6 (a) Basic dimension and its properties come from [Coutinho **86**]. We note that there is an alternative formulation of 6.2 which avoids critical modules. Namely $B(M, 0)$ is the least integer $t \geq -1$ such that there exists a chain of submodules

$$M = M_n \supseteq M_{n-1} \supseteq \cdots \supseteq M_0 = 0$$

with $\mathcal{K}(M_i/M_{i-1}) \leq t$ for all i such that M_i/M_{i-1} is faithful.

(b) The definition of $g(M, P)$ and this general approach comes from Warfield. These results are proved for FBN rings in [Warfield **80**] and for Noetherian rings in [Stafford **81**].

8.7 (a) The Forster–Swan theorem comes from [Forster **64**] and [Swan **67**] and was extended by [Eisenbud and Evans **73**]; see [Kunz **85**] for an account.

(b) This section again largely follows [Coutinho **86**] which simplified and extended to right Noetherian rings the major results of [Stafford **81, 82a**]. However, the definition of $b(M, P)$ in 7.2 differs from that in [Stafford **81**] where the formula denoted here by $b'(M, P)$ is used. The advantage of this change is demonstrated by 7.8.

(c) The conclusion of 7.13 is valid with the weakened hypothesis that $\hat{g}(M, P) \geq 1 + \mathcal{K}(R/P)$ for each $P \in J\text{-Spec } R$; see [Coutinho **86**].

(d) The notion of a basic element, introduced by [Eisenbud and Evans **73**], has been extended to right Noetherian rings by [Coutinho **86**] following results for FBN rings by [Chakravarti **85**].

(e) Cancellation is possible for some nonprojective modules; see [Stafford **79**].

(f) A cancellation theorem for certain C^*-algebras appears in [Rieffel **83a**].

Chapter 12
K_0 AND EXTENSION RINGS

Chapter 11 was concerned with the behaviour of finitely generated projective modules over a ring R. One would like to be able to extend such knowledge from R to an extension ring S, such as $R[x; \delta]$ for example. However the attractive idea of describing all finitely generated projective S-modules in terms of those over R is clearly unrealistic. For example, even for the simple case when $R = k[y]$ and $S = A_1(k) = k[y][x; d/dy]$ it is known that S has infinitely many non-isomorphic projective right ideals – whereas, of course, those of R are all free.

The more modest approach followed here starts by identifying two finitely generated projectives P, Q if $P \oplus R^n \simeq Q \oplus R^n$ for some n. The equivalence classes $[P]$ generate an abelian group denoted by $K_0(R)$. Then the aim becomes to calculate $K_0(S)$ in terms of $K_0(R)$, and this is carried out for most of the types of extension ring discussed in earlier chapters.

The first section defines $K_0(R)$ and describes its behaviour under various types of homomorphism. Then §2 outlines the relationship between the finitely generated projective-graded modules over a graded ring B and over its degree 0 subring B_0. This is the key to the use of filtered and graded methods, in §3, to describe $K_0(S)$ for many rings S, thereby obtaining several noncommutative analogues of the Hilbert syzygy theorem.

Up to that point the methods used are fairly elementary. However, they are not enough to handle the cases when $S = R[x, x^{-1}; \sigma]$ or when $S = R * U(g)$. These two are left until §§5 and 6 respectively, since one first needs to extend the notion of K_0 from rings to certain categories C. This is the burden of §4, together with results describing the effect on $K_0(C)$ of filtration, resolution and localization.

The final section calculates $K_0(S)$ for some simple Noetherian rings S and uses this to show that they need not be matrix rings over, or even Morita equivalent to, integral domains.

As the reader will be aware, or will surely guess, there is a whole sequence of groups $K_n(R)$. It is true that some of the results in this chapter can be extended to values of n higher than 0—but, not, it seems, by such elementary techniques. We make no further reference to these, except in the notes at the end of the chapter.

§1 K_0 of a Ring

1.1 This section describes the group $K_0(R)$, its elementary properties, and its behaviour under ring homomorphisms, especially localizations, and under matrix extensions.

1.2 Given a ring R, let $P_f(R)$ be the category of finitely generated projective right R-modules and, for each $P \in P_f(R)$, let $\langle P \rangle$ be its isomorphism class. Then $K_0(R)$ is the abelian group with generators $[P]$, one for each isomorphism class $\langle P \rangle$ with $P \in P_f(R)$, and with relations $[P] = [P'] + [P'']$ for each short exact sequence $0 \to P' \to P \to P'' \to 0$ or, equivalently, for each decomposition $P \simeq P' \oplus P''$. This is the **Grothendieck group** of R.

1.3 It is easy to see that each element of $K_0(R)$ has the form $[P] - [Q]$ for some $P, Q \in P_f(R)$. Those elements of the form $[P]$ comprise a subsemigroup denoted by $C(R)$ and called the **positive cone** of $K_0(R)$.

1.4 Lemma. *If $P, Q \in P_f(R)$ then the following conditions are equivalent:*
 (i) $[P] = [Q]$ in $K_0(R)$;
 (ii) $P \oplus D \simeq Q \oplus D$ for some $D \in P_f(R)$;
 (iii) $P \oplus R^t \simeq Q \oplus R^t$ for some t.

Proof. Straightforward. □

If P, Q are related as above, they are said to be **stably isomorphic**.

1.5 Corollary. (i) $[P] = [R^s]$ in $K_0(R)$ if and only if P is stably free of rank s;
 (ii) $K_0(R) = \mathbb{Z}[R]$, *the cyclic group generated by* $[R]$, *if and only if each* $P \in P_f(R)$ *is stably free.* □

Note that $\mathbb{Z}[R]$, the subgroup generated by $[R]$, is free provided that R has the invariant basis property. The quotient group $K_0(R)/\mathbb{Z}[R]$ is called the **projective class group** $\tilde{K}_0(R)$.

1.6 It is clear from 1.5(ii) that if $R = \mathbb{Z}$, or R is a division ring, or $R = A_1(k)$ with k a field of characteristic 0, then, by 7.11.6, $K_0(R) = \mathbb{Z}[R] \simeq \mathbb{Z}$ with $[R] \mapsto 1$. Therefore $\tilde{K}_0(R)$ is trivial in these cases.

As another example, let R be a commutative Dedekind domain. Its **ideal class group** $\mathscr{I}(R)$ is the abelian group whose generators are the isomorphism classes $\langle I \rangle$ of fractional ideals I of R with the operation $\langle I \rangle + \langle J \rangle = \langle IJ \rangle$. Recall from 5.7.18 that each $P \in P_f R$ has the form $R^n \oplus I$ for some I and that $I \oplus J \simeq R \oplus IJ$. So there is an epimorphism $\alpha : K_0(R) \to \mathscr{I}(R)$ given by $\alpha([R^n \oplus I]) = \langle I \rangle$. Since $\mathbb{Z}[R] = \ker \alpha$, then $\mathscr{I}(R) \simeq \tilde{K}_0(R)$. In fact α is split by the map $\beta : \mathscr{I}(R) \to K_0(R)$, $\langle I \rangle \mapsto [I] - [R]$ and so $K_0(R) \simeq \mathbb{Z}[R] \oplus \mathscr{I}(R)$.

12.1.8 K_0 of a ring   417

In particular suppose that $R = \mathbb{R}[X, Y]/(X^2 + Y^2 - 1)$ and $S = \mathbb{C}[X, Y]/(X^2 + Y^2 - 1)$. It was noted in 7.8.14(i) that R is a Dedekind domain and S a principal ideal domain; indeed $S = \mathbb{C}[z, z^{-1}]$. We will show now that $K_0(R) \simeq \mathbb{Z} \oplus (\mathbb{Z}/2\mathbb{Z})$.

To see this, first consider any ideal I of S of the form $I = \sum f_i S$, $f_i \in R$. If $g \in I \cap R$ and $\bar{\ }$ denotes a complex conjugate, then

$$g = \sum f_i s_i = \sum f_i \bar{s}_i = \tfrac{1}{2} \sum f_i (s_i + \bar{s}_i)$$

and so $I \cap R = \sum f_i R$ and $I = (I \cap R)S$.

Next let M be a maximal ideal of S. Recall, for example by 10.2.10, that $M \cap R$ is a maximal ideal of R and that each maximal ideal of R takes this form. Now M takes the form $(x - a, y - b)$ with $a, b \in \mathbb{C}$ or, equally, the form $z - \alpha$ with $\alpha \in \mathbb{C}$. We consider two cases.

If at least one of $a, b \notin \mathbb{R}$ then $M \neq \bar{M} = (x - \bar{a}, y - \bar{b})$ and so

$$M \cap \bar{M} = M\bar{M} = (z - \alpha)(\bar{z} - \bar{\alpha})S.$$

Hence $M \cap R = M \cap \bar{M} \cap R = (z - \alpha)(\bar{z} - \bar{\alpha})R$ and so is principal.

On the other hand, suppose $a, b \in \mathbb{R}$. Note first that $M \cap R$ cannot be principal. For if $M \cap R = fR$ then $M = (M \cap R)S = fS$ and so $f = (z - \alpha)z^n \beta$ with $n \in \mathbb{Z}$, $0 \neq \beta \in \mathbb{C}$. But then $f \neq \bar{f}$ and so $f \notin R$. Next let $M' = (\bar{z} - i)S$. One can check that $MM' = (ax + (b - 1)y)S$. However,

$$(M \cap R)(M' \cap R) = (M \cap R) \cap (M' \cap R) = (M \cap M') \cap R = MM' \cap R.$$

Hence $(M \cap R)(M' \cap R)$ is principal.

These facts combine to show that $\mathscr{I}(R) \simeq \mathbb{Z}/2\mathbb{Z}$ as claimed.

1.7 One can characterize $K_0(R)$ by a universal property. A function f from isomorphism classes in $\boldsymbol{P}_f R$ to an abelian group A is **additive** if $f(\langle P \rangle) = f(\langle P' \rangle) + f(\langle P'' \rangle)$ whenever there is a short exact sequence $0 \to P' \to P \to P'' \to 0$. The map $\langle P \rangle \mapsto [P] \in K_0(R)$ is the universal additive function to an abelian group; i.e. f factors through $K_0(R)$.

1.8 Let $\theta : R \to S$ be a ring homomorphism. If $P \simeq P' \oplus P''$ in $\boldsymbol{P}_f R$ then $P \otimes_R S \simeq (P' \otimes_R S) \oplus (P'' \otimes_R S)$ in $\boldsymbol{P}_f(S)$. The universality described above shows that θ induces the group homomorphism $\theta_* : K_0(R) \to K_0(S)$ given by $[P] \mapsto [P \otimes_R S]$ and then $K_0(\ )$ is a functor. Clearly, θ_* restricts to a semigroup homomorphism $\theta_* : C(R) \to C(S)$. Evidently these maps are injective if and only if $P \otimes_R S$ being stably isomorphic to $P' \otimes_R S$ implies that P is stably isomorphic to P'. Surjectivity is a little more complicated.

Proposition. *Let $\theta : R \to S$ be a ring homomorphism. Then:*

(i) $\theta_* : K_0(R) \to K_0(S)$ *is surjective if and only if given $Q \in \boldsymbol{P}_f(S)$, there exist $P \in \boldsymbol{P}_f(R)$ and $n \in \mathbb{N}$ such that $P \otimes_R S \simeq Q \oplus S^n$;*

(ii) $\theta_*: C(R) \to C(S)$ is surjective if and only if given $Q \in P_f S$ there exists $P \in P_f R$ and $n \in \mathbb{N}$ such that $(P \otimes_R S) \oplus S^n \simeq Q \oplus S^n$. □

One example of (ii) occurs in 7.11.6.

1.9 The next few results concern some special types of ring homomorphism.

Proposition. *If there are homomorphisms $\theta: R \to S$ and $\phi: S \to R$ such that $\phi\theta = 1_R$ then $\phi_* \theta_*$ is the identity on $K_0(R)$ and so $K_0(R)$ is a direct summand of $K_0(S)$.* □

1.10 Corollary. *If B is a graded ring and B_0 is the degree zero subring, then $K_0(B_0)$ is a direct summand of $K_0(B)$.* □

1.11 There is another situation similar to 1.9. Suppose that there is a nonzero homomorphism $R \to D$ with D being a division ring. Then the maps $\mathbb{Z} \to R \to D$ induce maps $K_0(\mathbb{Z}) \to K_0(R) \to K_0(D)$, i.e. $\mathbb{Z} \to K_0(R) \to \mathbb{Z}$ with the composed map being $1_\mathbb{Z}$. So $K_0(R) = \mathbb{Z}[R] \oplus T$ with $T \simeq K_0(R)/\mathbb{Z}[R]$, the projective class group.

1.12 Proposition. *Let R be a right regular, right Noetherian ring and let $\mathscr{S}$ be a right denominator set in R.*
(i) The map $R \to R_\mathscr{S}$ induces a surjection $K_0(R) \to K_0(R_\mathscr{S})$.
(ii) If every $P \in P_f(R)$ is stably free then every $Q \in P_f(R_\mathscr{S})$ is stably free.

Proof. (i) If $Q \in P_f(R_\mathscr{S})$ then $Q = MR_\mathscr{S}$ for some finitely generated R-submodule M of Q. Let

$$0 \to P_n \to P_{n-1} \to \cdots \to P_0 \to M \to 0$$

be a resolution of M by $P_i \in P_f(R)$. Then

$$0 \to P_n \otimes R_\mathscr{S} \to P_{n-1} \otimes R_\mathscr{S} \to \cdots \to P_0 \otimes R_\mathscr{S} \to M \otimes R_\mathscr{S} \to 0$$

is a projective resolution, and, by 7.4.2, $M \otimes R_\mathscr{S} \simeq MR_\mathscr{S} = Q$. Therefore

$$[Q] = \sum_{i=0}^{n} (-1)^i [P_i \otimes_R R_\mathscr{S}]$$

which is in the image of $K_0(R)$.
(ii) Clear from (i). □

There are examples to show that one does need the hypothesis that R is right regular.

1.13 Finally, we turn to matrix rings and Morita equivalent rings.

12.2.5   *Projective-graded modules*   419

Proposition. (i) *If* $R \overset{M}{\sim} S$ *via the progenerator* $_RP_S$ *then* $K_0(R) \simeq K_0(S)$ *with* $[R] \mapsto [P]$.
(ii) *If* $S = M_n(R)$ *then* $K_0(R) \simeq K_0(S)$ *with* $n[R] \mapsto [S]$.

Proof. (i) Clear.
(ii) Apply (i) taking $P = e_{11}S$.   □

The map θ_* induced by the diagonal embedding $\theta \colon R \to M_n(R)$ sends $[R]$ to $[S]$, of course.

§2   Projective-Graded Modules

2.1 Many of the rings R described in Chapter 1 have filtrations which give well-behaved associated graded rings, and this can be used in determining $K_0(R)$. The first step in this process is to investigate the relationship between the projective-graded modules over a graded ring B and over its degree 0 subring B_0; see 7.6.6. That is the aim of this section.

2.2 It is necessary first to view B_0 as a graded ring. In fact any ring R may be regarded as a graded ring by setting $R_0 = R$ and $R_n = 0$ for $n \neq 0$. We call this the **trivial grading** of R. In such a case a graded R-module is simply a direct sum of submodules, $M = \bigoplus_{n=0}^{\infty} M_n$, with M_n being the nth component of M.

2.3 If B, C are graded rings, a **graded homomorphism** $\theta \colon B \to C$ is a ring homomorphism such that $\theta(B_n) \subseteq C_n$ for all n.

There are two relevant examples to note, both concerning the degree zero subring B_0 of B with its trivial grading. Let $I = \bigoplus_{n=1}^{\infty} B_n$, this being the **augmentation ideal** of B. Then the projection $\pi \colon B \to B/I \simeq B_0$ and the injection $\iota \colon B_0 \to B$ are both graded homomorphisms.

2.4 Let $\boldsymbol{P}_{\text{gr}}(R)$ denote the category of projective-graded right R-modules over a graded ring R, this being a subcategory of $\boldsymbol{M}_{\text{gr}}(R)$; see 7.6.4. Our next step is to describe some functors between $\boldsymbol{P}_{\text{gr}}(B)$ and $\boldsymbol{P}_{\text{gr}}(B_0)$ and between $\boldsymbol{M}_{\text{gr}}(B)$ and $\boldsymbol{M}_{\text{gr}}(B_0)$.

2.5 Let $M = \bigoplus_{n=0}^{\infty} M_n$ be a graded B-module. Then each M_n is a B_0-module, so M may be viewed as a graded B_0-module by restriction. This gives the restriction functor $F \colon \boldsymbol{M}_{\text{gr}}(B) \to \boldsymbol{M}_{\text{gr}}(B_0)$.

Next, given $M \in \boldsymbol{M}_{\text{gr}}(B)$, note that $M \otimes_B B_0 \simeq M/MI \in \boldsymbol{M}_{\text{gr}}(B)$ and $F(M/MI) \in \boldsymbol{M}_{\text{gr}}(B_0)$. This gives a functor $Q \colon \boldsymbol{M}_{\text{gr}}(B) \to \boldsymbol{M}_{\text{gr}}(B_0)$ with $Q(M) = M \otimes_B B_0 \simeq M/MI$.

Finally, we define a functor $T \colon \boldsymbol{M}_{\text{gr}}(B_0) \to \boldsymbol{M}_{\text{gr}}(B)$. If $M \in \boldsymbol{M}_{\text{gr}}(B_0)$ then $T(M) = M \otimes_{B_0} B$ where this is given the **product grading**; namely

$$(M \otimes B)_n = \sum \{M_i \otimes B_j \mid i + j = n\}.$$

2.6 We note using 7.6.6 that Q restricts to a functor from $\boldsymbol{P}_{gr}(B)$ to $\boldsymbol{P}_{gr}(B_0)$, and that T restricts to a functor from $\boldsymbol{P}_{gr}(B_0)$ to $\boldsymbol{P}_{gr}(B)$. The same notation will be used for these restrictions.

Lemma. *Q is a right exact functor which commutes with direct sums both on $\boldsymbol{M}_{gr}(B)$ and $\boldsymbol{P}_{gr}(B)$. Moreover, if $M \neq 0$ then $Q(M) \neq 0$.*

Proof. The former claim is clear since tensoring is right exact. As for the latter, let n be minimal such that $M_n \neq 0$. Then M_n embeds in M/MI. □

Similar claims are valid for T.

2.7 The relationship between Q and T is explored next.

Proposition. *QT is naturally isomorphic to the identity functor on $\boldsymbol{M}_{gr}(B_0)$ and on $\boldsymbol{P}_{gr}(B_0)$.*

Proof. The isomorphism $M \simeq M \otimes_{B_0} B \otimes_B B_0$ is a natural isomorphism of graded modules. □

2.8 The next result establishes a weaker property for TQ, but its proof is relatively complicated.

Theorem. *If $M \in \boldsymbol{P}_{gr}(B)$ then $TQ(M) \simeq M$ in $\boldsymbol{P}_{gr}(B)$.*

Proof. The map $\pi: F(M) = M \to M/MI = Q(M)$ is a graded homomorphism in $\boldsymbol{M}_{gr}(B_0)$. Since $Q(M) \in \boldsymbol{P}_{gr}(B_0)$ there is a graded homomorphism $\gamma: M/MI \to M$ splitting π. (The lack of a natural choice of γ explains why the proof does not establish a natural isomorphism between TQ and the identity functor.)

Next let μ be the multiplication map $\mu: TF(M) = M \otimes_{B_0} B \to M \otimes_B B \simeq M$ and note that μ is a graded homomorphism in $\boldsymbol{P}_{gr}(B)$. Therefore the composite map

$$v: TQ(M) = (M/MI) \otimes_{B_0} B \xrightarrow{\gamma \otimes 1} M \otimes_{B_0} B \xrightarrow{\mu} M$$

is a graded homomorphism in $\boldsymbol{P}_{gr}(B)$. The aim now is to show that v is an isomorphism.

To see this we first apply Q to the constituents of v, getting

$$\begin{array}{ccccc}
QTQ(M) & \xrightarrow{\gamma \otimes 1 \otimes 1} & M \otimes_{B_0} B \otimes_B B_0 & \xrightarrow{\mu \otimes 1} & Q(M) \\
\downarrow \simeq & & \downarrow \simeq & & \| \\
Q(M) & \xrightarrow{\gamma} & M & \xrightarrow{\pi} & Q(M)
\end{array}$$

12.2.10   *Projective-graded modules*

where it is easily checked that the maps in the bottom row are indeed γ and π. Since $\pi\gamma = 1_{Q(M)}$, then $Q(v)$ is an isomorphism in $\boldsymbol{P}_{\mathrm{gr}}(B_0)$.

Next apply Q to the exact sequence

$$TQ(M) \xrightarrow{v} M \longrightarrow \mathrm{coker}\, v \longrightarrow 0.$$

By 2.6 this gives an exact sequence

$$QTQ(M) \xrightarrow{Q(v)} Q(M) \longrightarrow Q(\mathrm{coker}\, v) \longrightarrow 0$$

and, since $Q(v)$ is an isomorphism, $Q(\mathrm{coker}\, v) = 0$. Therefore, by 2.6, $\mathrm{coker}\, v = 0$ and so $v: TQ(M) \to M$ is a surjection. Since M is projective-graded, there is a graded homomorphism $\lambda: M \to TQ(M)$ splitting the short exact sequence

$$0 \longrightarrow \ker v \longrightarrow TQ(M) \xrightarrow{v} M \longrightarrow 0.$$

Hence, using 2.6, the sequence

$$0 \longrightarrow Q(\ker v) \longrightarrow QTQ(M) \xrightarrow{Q(v)} Q(M) \longrightarrow 0$$

is exact. However, $Q(v)$ is an isomorphism, so $Q(\ker v) = 0$ and hence, by 2.6, $\ker v = 0$. Thus v is an isomorphism.   □

2.9 Corollary. *Let B be a graded ring.*
(i) *Each projective-graded B-module M is isomorphic as a graded module to $P \otimes_{B_0} B$ for some $P \in \boldsymbol{P}_{\mathrm{gr}}(B_0)$; and if M is finitely generated then so is P.*
(ii) *If all projective B_0-modules are free then all projective-graded B-modules are free-graded.*

Proof. (i) Clear from 2.8.
(ii) If $P = \oplus P_n$ is a projective-graded B_0-module then each P_n is a projective B_0-module and so is free. Now apply (i).   □

2.10 If there is a split short exact sequence of graded B-homomorphisms

$$0 \to F_1 \to F_0 \to M \to 0$$

in which F_0 and F_1 are free-graded, M will be called **stably free-graded**.

Corollary. *Let B be a graded ring and suppose that all finitely generated projective B_0-modules are stably free. Then all finitely generated projective-graded B-modules are stably free-graded.*

Proof. Let M, P be as in 2.9 and note that since P_{B_0} is finitely generated projective so too are all P_n, and only finitely many P_n can be nonzero. For each such n there is a short exact sequence

$$0 \to F_{n1} \to F_{n0} \to P_n \to 0$$

in which F_{n1} and F_{n0} are free of finite rank. Hence

$$0 \to \bigoplus_n F_{n1} \to \bigoplus_n F_{n0} \to P \to 0$$

is an exact sequence in $M_{gr}(B_0)$ whose first two terms are free-graded of finite rank. Then

$$0 \to F_1 \to F_0 \to M \to 0$$

is an exact sequence in $M_{gr}(B)$, where $F_i = (\bigoplus_n F_{ni}) \otimes_{B_0} B$ for $i = 0, 1$ and so F_i is free-graded of finite rank. □

2.11 It is not, however, true that under such circumstances each projective B-module Q is extended from some projective B_0-module P; i.e. $Q \simeq P \otimes_{B_0} B$. For example, let $B = k[y][x; \sigma]$, where k is a field and $\sigma(y) = y + 1$. Then

$$(x + 1)(y + 1) - y(x + 1) = 1$$

and so $yB + (x + 1)B = B$. Therefore, by 11.2.7, $yB \cap (x + 1)B$ is stably free but not free. However, if B is graded by degree in x then $B_0 = k[y]$ over which all projectives are free. Hence, by 2.9, $yB \cap (x + 1)B$ is not extended from a projective B_0-module.

§3 Filtrations and the Syzygy Theorem

3.1 In this section the results of §2 are applied to filtered rings via their associated graded rings. Thereby two versions of Hilbert's syzygy theorem are proved for noncommutative rings.

Hilbert's syzygy theorem states that if $R = k[x_1, \ldots, x_n]$, the commutative polynomial ring over a field k, then every finitely generated R-module has a finite free resolution. Another version states that all finitely generated projective R-modules are stably free. These results can be viewed as lifting these properties from k to R.

Two such lifting results are proved in this section, even though they will be superseded by a more general theorem in §6; for they are both relatively straightforward to obtain and have immediate application to several of the extension rings described in Chapter 1.

We will use the notation $\{S_n\}$ for the filtration of a filtered ring S.

3.2 The first result is an easy consequence of earlier work.

12.3.4   Filtrations and the syzygy theorem

Theorem. *Let S be a filtered ring such that $\operatorname{gr} S$ is right Noetherian, right regular, and flat as a left S_0-module. Suppose also that every finitely generated projective right S_0-module is stably free; then the same is true for S.*

Proof. Let M_S be finitely generated. By 7.6.11 it can be arranged that $\operatorname{gr} M$ is finitely generated over $\operatorname{gr} S$ and so, using 2.10 and 11.1.6, $\operatorname{gr} M$ has a finite resolution by finitely generated free-graded $\operatorname{gr} S$-modules. Hence, by 7.6.17, M_S has a finite free resolution and 11.1.6 can be applied again. □

Note that the hypotheses imply, by 1.6.9 and 7.6.17, that S is a right Noetherian, right regular ring.

3.3 Corollary. *Let R be a right Noetherian, right regular ring whose finitely generated projective right modules are all stably free. Let S be one of the following rings:*
(i) $A_n(R)$;
(ii) $R*U(\mathbf{g})$ where R is a k-algebra and $\mathbf{g}$ is a finite dimensional k-Lie algebra over a commutative ring k;
(iii) $R[x; \sigma, \delta]$ with σ an automorphism;
(iv) $R[x, x^{-1}; \sigma]$;
(v) $R*G$ with G poly-(infinite cyclic).
Then every finitely generated projective right S-module is stably free.

Proof. (i), (ii) A filtration can be chosen to make $\operatorname{gr} S \simeq R[z_1, \ldots, z_n]$, where the z_i are central indeterminates.
   (iii) $\operatorname{gr} S \simeq R[x; \sigma]$.
   (iv) S is a localization of $R[x; \sigma]$; so one can apply (iii) together with 1.12.
   (v) S can be constructed from R by a sequence of extensions as in (iv). □

The comments in 2.11 show that one cannot deduce that all finitely generated projective right S-modules are free granted the same condition on R.

3.4 Next comes the second version of the syzygy theorem, which evidently extends the first.

Theorem. *Let S be a filtered ring such that $\operatorname{gr} S$ is right Noetherian, right regular, and flat as a left S_0-module. Then any finitely generated right S-module has a finite resolution by finitely generated projectives which are extended from S_0.*

Proof. Let M_S be finitely generated. Arguing as in the proof of 3.2, one can obtain resolutions

$$0 \to P \to F_n \to \cdots \to F_0 \to M \to 0$$
$$0 \to \operatorname{gr} P \to \operatorname{gr} F_n \to \cdots \to \operatorname{gr} F_0 \to \operatorname{gr} M \to 0$$

in which each F_i is a free S-module of finite rank and P_S is a finitely generated projective, and each $\operatorname{gr} F_i$ is a free-graded gr S-module of finite rank and gr P is a finitely generated projective gr S-module. By 2.9, $\operatorname{gr} P \simeq Q \otimes_{S_0} \operatorname{gr} S$ with Q being a finitely generated projective-graded S_0-module. It is sufficient now to prove that $P \simeq Q \otimes_{S_0} S$.

To ease notation, we denote gr S by B; so $S_0 = B_0$. Suppose that, with the usual notation for graded modules, $Q = \bigoplus_n Q_n$. The S_0-module embedding ϕ_n: $Q_n \to \oplus \{Q_i \otimes B_j | i + j = n\} = (\operatorname{gr} P)_n = P_n/P_{n-1}$ can be lifted to ψ_n as shown

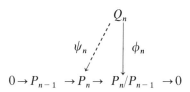

since Q_n is projective over S_0. Let $\psi = \oplus \psi_n \colon Q \to P$; then ψ is an S_0-module embedding and is a filtered homomorphism of filtered S_0-modules. Finally, define $\theta \colon Q \otimes_{S_0} S \to P$ by $\theta(q \otimes s) = \psi(q)s$, and give $Q \otimes S$ the product filtration with nth filtration submodule

$$\sum \{Q_i \otimes S_j | i + j = n\}.$$

One can check that θ is a homomorphism of filtered S-modules. The proof will be completed by showing θ to be an isomorphism.

To that end, we first consider $\operatorname{gr} \theta \colon \operatorname{gr}(Q \otimes S) \to \operatorname{gr} P$. Note that $Q \otimes S \simeq \bigoplus_n (Q_n \otimes S)$. Now the $(n + m)$th component of $\operatorname{gr}(Q_n \otimes S)$ is $(Q_n \otimes S_m)/(Q_n \otimes S_{m-1})$. However, since Q is a projective-graded right S_0-module so too is Q_n. Therefore

$$(Q_n \otimes S_m)/(Q_n \otimes S_{m-1}) \simeq Q_n \otimes (S_m/S_{m-1}) = Q_n \otimes B_m.$$

One can now check that $\operatorname{gr}(Q \otimes_{S_0} S) \simeq Q \otimes_{S_0} B$.

Consider now the composite map

$$\alpha \colon Q \otimes B \xrightarrow{\sim} \operatorname{gr}(Q \otimes S) \xrightarrow{\operatorname{gr} \theta} \operatorname{gr} P \xrightarrow{\sim} Q \otimes B.$$

If $q \in Q_n$ then

$$\alpha \colon q \otimes 1 \mapsto \overline{q \otimes 1} \mapsto \overline{\psi_n(q)} = \phi_n(q) \mapsto q \otimes 1.$$

So α is the identity map and $\operatorname{gr} \theta$ is an isomorphism. Therefore, by 7.6.14, θ is an isomorphism. □

3.5 This section ends by interpreting some of the preceding results in terms of K_0.

Corollary. (i) *Let S be a filtered ring such that $\operatorname{gr} S$ is right Noetherian, right regular and flat as a left S_0-module. Then the map $K_0(S_0) \to K_0(S)$ is surjective.*

(ii) *Furthermore, if either (a) $K_0(S_0) = \mathbb{Z}[S_0]$, or (b) S is graded, with the filtration on S being the corresponding one, then the map $K_0(S_0) \to K_0(S)$ is an isomorphism.*

Proof. (i) This comes from 3.4.
(ii)(a) This is covered by 3.2.
(ii)(b) 1.10 applies. □

The reasonable conjecture that the surjection above is always an isomorphism will be verified in §6.

3.6 Finally, we note two specific examples where this can be verified directly.

Corollary. *Let R be a right Noetherian right regular ring. If (i) $S = R[x; \sigma]$ for some automorphism σ, or (ii) $S = R[x, x^{-1}]$, then the map $K_0(R) \to K_0(S)$ is an isomorphism.*

Proof. (i) Here $R[x; \sigma]$ is graded and 3.5 can be applied.

(ii) Note first, by 1.9, that $K_0(R)$ embeds in $K_0(S)$ as a direct summand. However, $K_0(R) \simeq K_0(R[x])$ by (i), and $K_0(R[x]) \to K_0(R[x, x^{-1}])$ is a surjection by 1.12. Hence $K_0(R) \simeq K_0(S)$. □

We briefly consider what the corresponding result for $T = R[x, x^{-1}; \sigma]$ should be. The inclusion $\theta: R \to T$ factors as $R \hookrightarrow R[x; \sigma] \hookrightarrow T$. In the induced sequence

$$K_0(R) \to K_0(R[x; \sigma]) \to K_0(T)$$

the maps are, respectively, an isomorphism by 3.6(i), and a surjection by 1.12. However, we must expect there to be a kernel since if $P \in \boldsymbol{P}_f R$ then it is easy to show (see 7.3.4 and 5.4(i)) that $P^\sigma \otimes_R T \simeq P \otimes_R T$ and so $[P] - [P^\sigma] \in \ker \theta_*$.

The main result of §5 is that $\ker \theta_*$ is generated by the elements $[P] - [P^\sigma]$. The reader who is prepared to accept this and the result conjectured in 3.5 can ignore the next three sections apart from noting the statements of Theorems 5.5 and 6.13.

§4 K_0 of Module Categories

4.1 In order to prove the results mentioned at the end of §3, some generalization is required of the ideas introduced earlier. In particular it is necessary to define and describe K_0 of certain categories. The specific categories to which this will be applied are those already encountered, such as MR for a ring R or $\boldsymbol{P}_{\mathrm{gr}} B$ for a graded ring B.

4.2 The basic situation to be dealt with involves an abelian category A and a full additive subcategory B. In particular that means that in A maps have kernels and cokernels, and that in both A and B finite direct sums exist. If, further,
(i) whenever $0 \to M' \to M \to M'' \to 0$ is a short exact sequence in A such that M and M'' belong to B then $M' \in B$, and
(ii) the collection of isomorphism classes $\langle M \rangle$ of objects M of B forms a set, then B is an **admissible** subcategory of A.

For example, $P_f R$ is an admissible subcategory of MR for any ring R; and $M_{f\,gr}B$ is an admissible subcategory of $M_{gr}B$ for any right Noetherian graded ring B.

4.3 Let B be an admissible subcategory of A. The **Grothendieck group** $K_0(B)$ is the abelian group with generators $[M]$, one for each isomorphism class $\langle M \rangle$ in B, and relations $[M] = [M'] + [M'']$ whenever there is a sequence $0 \to M' \to M \to M'' \to 0$ in B which is exact in A. For brevity we call this an exact sequence in B.

For example, if $A = MR$ and $B = P_f R$ then $K_0(B)$ is just $K_0(R)$. So this definition is an extension of the earlier one.

4.4 In the same way as in 1.7, the map $\langle M \rangle \mapsto [M]$ is additive. It is the universal additive function from isomorphism classes in B to abelian groups.

4.5 We note the following facts, all easy extensions of similar results for $K_0(R)$.

Lemma. *Let B be an admissible subcategory of A.*
(i) *If $0 \to M_n \to \cdots \to M_1 \to M_0 \to 0$ is an exact sequence in B then $\sum_{i=0}^{n} (-1)^i [M_i] = 0$.*
(ii) *If there is a finite chain in A*
$$M = M_0 \supseteq M_1 \supseteq \cdots \supseteq M_{n+1} = 0$$
in which each M_i and each M_i/M_{i+1} is in B then $[M] = \sum_{i=0}^{n} [M_i/M_{i+1}]$.
(iii) *If $M_1, M_2 \in B$ then $[M_1 \oplus M_2] = [M_1] + [M_2]$.*
(iv) *Each element of $K_0(B)$ can be written in the form $[M] - [N]$ for some $M, N \in B$.* □

However, unlike $K_0(R)$, it is not in general true that if $[M] = [N]$ in $K_0(B)$ then $M \oplus C \simeq N \oplus C$ for some $C \in B$. For example, take $A = B = M_f(\mathbb{Z}/(4))$, $M = \mathbb{Z}/(4)$ and $N = \mathbb{Z}/(2) \oplus \mathbb{Z}/(2)$.

4.6 Suppose, for all i in some set I, that B_i is an admissible subcategory of A_i. Then $B = \bigoplus_i B_i$ is an admissible subcategory of $A = \bigoplus_i A_i$ and a routine calculation shows:

Proposition. $K_0(\bigoplus_i B_i) = \bigoplus_i K_0(B_i)$. □

4.7 Of course there is always a homomorphism $\alpha: K_0(B) \to K_0(A)$ induced by the inclusion of B in A. The next result describes a special case in which α is an isomorphism. It includes, for example, the case when A is the category of modules of finite length and B is the subcategory of semisimple modules.

Theorem. *Let B be an admissible subcategory of the abelian category A such that:*
(i) *if $0 \to M' \to M \to M'' \to 0$ is exact in A and $M \in B$ then $M', M'' \in B$; and*
(ii) *each $M \in A$ has a finite filtration*

$$M = M_0 \supseteq M_1 \supseteq \cdots \supseteq M_{n+1} = 0$$

for some n with $M_i/M_{i+1} \in B$ for $0 \leqslant i \leqslant n$.
Then the induced homomorphism $\alpha: K_0(B) \to K_0(A)$ is an isomorphism.

Proof. Given $M \in A$ with a finite filtration as above, $[M] = \sum_{i=0}^{n} [M_i/M_{i+1}]$ in $K_0(A)$ by 4.5(ii). On the one hand, this shows that α is surjective. On the other hand, this can be used to define a map f from isomorphism classes $\langle M \rangle$ of objects of A to $K_0(B)$ by $f(\langle M \rangle) = \sum_{i=0}^{n} [M_i/M_{i+1}]$, the latter being interpreted in $K_0(B)$, of course.

First we show that f is well defined. Since any two finite filtrations in A have equivalent refinements, it is enough to show that $f(\langle M \rangle)$ is unchanged by a refinement of the filtration. By induction, it will suffice to consider the insertion of just one extra term, say $M_j \supset M' \supset M_{j+1}$. However, there is a short exact sequence in A

$$0 \to M'/M_{j+1} \to M_j/M_{j+1} \to M_j/M' \to 0$$

and $M_j/M_{j+1} \in B$. Therefore, by hypothesis, the other terms also belong to B and so

$$[M_j/M_{j+1}] = [M_j/M'] + [M'/M_{j+1}]$$

in $K_0(B)$. Thus f is indeed well defined.

It is clear that f is additive and so, by the universal property of K_0, the corresponding map $\beta: K_0(A) \to K_0(B)$, $[M] \mapsto \sum [M_i/M_{i+1}]$, is a homomorphism. Finally, if $M \in B$ the filtration $M \supseteq 0$ in A shows that $\beta\alpha = 1$, so β is inverse to α. □

4.8 The next result concerns three categories $A \supseteq B \supseteq P$, where A is an abelian category, B an admissible subcategory and P is the full subcategory of all projective objects of A. (This will be applied, later, in the case where $A = B = M_f R$, and $P = P_f R$ with R being a right Noetherian, right regular ring, and in similar cases over graded rings.)

Theorem. *If A, B, P are as above and if each object in B has a finite projective resolution, then the homomorphism $\alpha: K_0(P) \to K_0(B)$ induced by the inclusion of P in B is an isomorphism.*

Proof. Let $M \in \boldsymbol{B}$; so there is a projective resolution

$$0 \to P_n \to \cdots \to P_0 \to M \to 0$$

with $P_j \in \boldsymbol{P}$ for all j. Hence, by 4.5(i), $[M] = \sum_{i=0}^{n} (-1)^i [P_i]$ in $K_0(\boldsymbol{B})$. Therefore α is surjective. As in the proof of 4.7 we will define $\beta: K_0(\boldsymbol{B}) \to K_0(\boldsymbol{P})$ by means of the map

$$f: \langle M \rangle \mapsto \sum_{i=0}^{n} (-1)^i [P_i]$$

from isomorphism classes in $\boldsymbol{B}$ to $K_0(\boldsymbol{P})$ and we will show that β is inverse to α.

The first step is to check that f is well defined. Suppose that

$$0 \to P'_n \to \cdots \to P'_0 \to M \to 0$$

is another projective resolution, arranged to have the same length by the adjunction of extra 0's if required. The long Schanuel lemma, 7.1.2, shows that

$$P_n \oplus P'_{n-1} \oplus P_{n-2} \oplus \cdots \simeq P'_n \oplus P_{n-1} \oplus P'_{n-2} \oplus \cdots$$

and so, in $K_0(\boldsymbol{P})$, $\sum_i (-1)^i [P_i] = \sum_i (-1)^i [P'_i]$ as required.

Next, to check that f is additive, suppose that $0 \to M' \to M \to M'' \to 0$ is exact in $\boldsymbol{B}$ and that there are projective resolutions

$$0 \to P'_n \to \cdots \to P'_0 \to M' \to 0$$
$$0 \to P''_n \to \cdots \to P''_0 \to M'' \to 0.$$

By [Cartan and Eilenberg **56**, Chap. 1, Proposition 2.5], there is then a projective resolution

$$0 \to P'_n \oplus P''_n \to \cdots \to P'_0 \oplus P''_0 \to M \to 0$$

and so $f(\langle M \rangle) = f(\langle M' \rangle) + f(\langle M'' \rangle)$ and f is additive.

It follows by universality that $\beta: [M] \mapsto \sum (-1)^i [P_i]$ is a homomorphism from $K_0(\boldsymbol{B})$ to $K_0(\boldsymbol{P})$, and the exact sequence $0 \to P \to P \to 0$, for $P \in \boldsymbol{P}$, makes clear that $\beta \alpha = 1$ as required. $\square$

4.9 Finally, we turn to localization. Here it seems best to be specific about the categories involved. First consider the case where R is a right regular right Noetherian ring and $\mathscr{S}$ is a right denominator set in R. Let $A = M_f R$, $B = M_f(R_{\mathscr{S}})$, F be the exact functor

$$- \otimes_R R_{\mathscr{S}} : \boldsymbol{A} \to \boldsymbol{B}$$

and C the full subcategory of A whose objects are those $M \in \boldsymbol{A}$ with $F(M) = 0$. Since F is exact, it induces a group homomorphism $\beta: K_0(\boldsymbol{A}) \to K_0(\boldsymbol{B})$.

12.5.1 *Skew Laurent extensions* 429

Theorem. *The sequence*

$$K_0(C) \xrightarrow{\alpha} K_0(A) \xrightarrow{\beta} K_0(B) \to 0,$$

in which α is induced by the inclusion of C in A and β by F, is exact.

4.10 Before proving 4.9 we note that the same proof, with minor changes of wording, can be applied to the case when R is graded and $\mathscr{S}$ consists of homogeneous elements. One needs the notion of a $\mathbb{Z}$-**graded ring**. As noted in 10.3.18, this is defined as in 1.6.3, but with the subgroups indexed by $n\in\mathbb{Z}$ rather than $n\in\mathbb{N}$. Similarly, one defines $\mathbb{Z}$-**graded modules** etc.

It is easy to see that the grading on R extends to a $\mathbb{Z}$-grading on $R_\mathscr{S}$. For example, if $R = k[x;\sigma]$ with $R_n = x^n k$, $n\in\mathbb{N}$, and $\mathscr{S} = \{x^n | n\in\mathbb{N}\}$, then $R_\mathscr{S} = k[x, x^{-1}; \sigma]$ and $(R_\mathscr{S})_n = x^n k$, $n\in\mathbb{Z}$.

The categories concerned here are $A = M_{\text{f gr}} R$ and $B = M_{\text{f gr}}(R_\mathscr{S})$, the latter denoting the category of finitely generated $\mathbb{Z}$-graded $R_\mathscr{S}$-modules.

4.11 *Proof* (of 4.9) The sequence is clearly a complex. Furthermore, if $N\in B$ then $N \simeq M_\mathscr{S} = F(M)$ for some $M\in A$, so the sequence is exact at $K_0(B)$.

It remains to consider $\ker \beta$. By 4.8 we can identify $K_0(A)$ with $K_0(P_{\text{f}} R)$ and $K_0(B)$ with $K_0(P_{\text{f}}(R_\mathscr{S}))$. Then an element of $\ker \beta$ has the form $[P] - [Q]$, where $P, Q \in P_{\text{f}} R$ and $[P_\mathscr{S}] = [Q_\mathscr{S}]$. That implies that $P_\mathscr{S}$ and $Q_\mathscr{S}$ are stably isomorphic. Now modify P and Q, by adding a suitable free module of finite rank, to make $P_\mathscr{S} \simeq Q_\mathscr{S}$; this will not change $[P] - [Q]$. We identity $P_\mathscr{S}$ with $Q_\mathscr{S}$ and consider the maps $\gamma_1: P \to P_\mathscr{S}$, $p \mapsto p \otimes 1$ and $\gamma_2: Q \to Q_\mathscr{S} = P_\mathscr{S}$, $q \mapsto q \otimes 1$.

It is clear that $\ker \gamma_1$ and $\ker \gamma_2 \in C$; but so too are $L/\operatorname{im} \gamma_1$ and $L/\operatorname{im} \gamma_2$, where $L = \operatorname{im} \gamma_1 + \operatorname{im} \gamma_2$. Evidently

$$[P] = [\ker \gamma_1] + [\operatorname{im} \gamma_1] \quad \text{and} \quad [L] = [\operatorname{im} \gamma_1] + [L/\operatorname{im} \gamma_1]$$

and so

$$[P] = [\ker \gamma_1] + [L] - [L/\operatorname{im} \gamma_1].$$

A similar equation holds for $[Q]$; and so

$$[P] - [Q] = [\ker \gamma_1] - [\ker \gamma_2] - [L/\operatorname{im} \gamma_1] + [L/\operatorname{im} \gamma_2] \in \operatorname{im} \alpha. \qquad \square$$

§5 Skew Laurent Extensions

5.1 In this section the results of §4 are applied to prove the fact mentioned in 3.6, namely that the map $K_0(R) \to K_0(R[x, x^{-1}; \sigma])$ is surjective with an easily specified kernel.

Throughout the section we let $T = R[x, x^{-1}; \sigma]$ and $S = R[x; \sigma]$.

5.2 In 7.3.4, given a module M_R a new R-module M^σ was defined via $m^\sigma r = (m\sigma^{-1}(r))^\sigma$. In the same way, given a left R-module N, one can define another left R-module $^\sigma N$ by $r(^\sigma n) = {^\sigma}(\sigma(r)n)$. The left R-module $^\sigma R$, with the ordinary right R-module structure, will play an important role now.

5.3 One can easily verify

Lemma. $M \otimes_R (^\sigma R) \simeq M^\sigma$. □

We will write $M \otimes_\sigma R$ rather than $M \otimes_R (^\sigma R)$ henceforth. The functor $MR \to MR$ obtained by mapping M to $M \otimes_\sigma R$ induces an endomorphism of $K_0(R)$ which coincides with the induced map σ_*. Furthermore, if R is right Noetherian then σ also induces an endomorphism, again named σ_*, of $K_0(M_f R)$.

5.4 The left R-module structure on T given by the embedding $R \hookrightarrow T$ is involved in the next result.

Proposition. (i) If $M \in M_f R$, $n \in \mathbb{Z}$, and $\phi = \sigma^n$, then $M \otimes_R T \simeq M^\phi \otimes_R T \in M_f T$.
(ii) If R is right Noetherian then there is a complex

$$K_0(M_f R) \xrightarrow{1-\sigma_*} K_0(M_f R) \xrightarrow{\alpha} K_0(M_f T)$$

with α being induced by the embedding $R \hookrightarrow T$.

Proof. (i) The map $\sum m_i^\phi \otimes x^i \mapsto \sum m_i \otimes x^{i+n}$ gives the isomorphism.
(ii) Since $_R T$ is free, the functor $- \otimes_R T$ is exact, so α exists. Note that $(1-\sigma_*)[M] = [M] - [M^\sigma]$ which, by (i), belongs to $\ker \alpha$. So $\text{im}(1-\sigma_*) \subseteq \ker \alpha$. □

5.5 If R is a right Noetherian, right regular ring then, by 4.8, $K_0(M_f R) \simeq K_0(R)$. By 7.7.5, T also is right Noetherian, right regular, so $K_0(M_f T) \simeq K_0(T)$. Therefore, by 5.4, there is a complex $K_0(R) \xrightarrow{1-\sigma_*} K_0(R) \xrightarrow{\alpha} K_0(T)$. The next result goes further.

Theorem. If R is a right Noetherian right regular ring then the sequence

$$K_0(R) \xrightarrow{1-\sigma_*} K_0(R) \xrightarrow{\alpha} K_0(T) \to 0$$

is exact.

Proof. T is the localization of $S = R[x; \sigma]$ with respect to $\mathcal{S} = \{x^n | n \geq 0\}$. By 4.9, there is an exact sequence

$$K_0(C) \to K_0(M_f S) \to K_0(M_f T) \to 0$$

12.5.6 Skew Laurent extensions

where C is the category of finitely generated $\mathcal{S}$-torsion modules over S.

The first step is to determine $K_0(C)$. Note that, since $xS = Sx$, if $M \in C$ then $Mx^n = 0$ for some $n \geq 1$. Thus M_S has a finite filtration with factors all in $M_f R$. However, there is an inclusion $M_f R \to C$ given by regarding an R-module as a S-module annihilated by x. This, together with 4.7, shows that $K_0(C) \simeq K_0(M_f R) \simeq K_0(R)$.

By 3.6, $K_0(S) \simeq K_0(R)$. Hence the exact sequence above can be rewritten in the form

$$K_0(R) \xrightarrow{\beta} K_0(R) \xrightarrow{\gamma} K_0(T) \longrightarrow 0.$$

It remains only to identify the maps β, γ here.

The map γ is induced by the embeddings $R \hookrightarrow S \hookrightarrow T$; so $\gamma = \alpha$ as required. The map β comes from regarding $P \in P_f R$ as an S-module annihilated by x and then using the isomorphism $K_0(S) \simeq K_0(R)$ given by 3.6. Note, however, that

$$0 \to xS \to S \to S/xS \to 0$$

is an exact sequence of (R, S)-bimodules. Therefore

$$0 \to P \otimes_R xS \to P \otimes_R S \to P \otimes_R (S/xS) \to 0$$

is exact. Now $P \otimes_R (S/xS)$ is P, regarded as an S-module annihilated by x; and $P \otimes_R xS \simeq P^\sigma \otimes_R S$. It follows that $\beta([P]) = [P] - [P^\sigma]$ and so $\beta = 1 - \sigma_*$ as claimed. □

5.6 It is perhaps worth noting that if $[P] - [Q] \in \ker \alpha$, i.e. if $[P \otimes_R T] = [Q \otimes_R T]$, it need not be true that Q is stably isomorphic to P^{σ^n} for some n. This is made clear by the next example.

Example. Take R to be the ring of integers in the algebraic number field $\mathbb{Q}(\sqrt{-71})$. In that case [H. Cohn 80, Table 3, p. 267] shows that the ideal class group is cyclic of order 7, generated by $\langle I \rangle$ say; and the automorphism σ of R induced by the field automorphism $a + b\sqrt{-71} \mapsto a - b\sqrt{-71}$ has order 2 and maps $\langle I \rangle$ to $\langle I^6 \rangle$. The isomorphism $K_0(R) \simeq \mathbb{Z}[R] \oplus \mathscr{I}(R)$ given by 1.6 makes the exact sequence of Theorem 5.5 become

$$\mathbb{Z}[R] \oplus \mathbb{Z}/(7) \xrightarrow{1-\sigma_*} \mathbb{Z}[R] \oplus \mathbb{Z}/(7) \xrightarrow{\alpha} K_0(T) \to 0$$

with each $\mathbb{Z}/(7)$ being the cyclic subgroup of $K_0(R)$ generated by $[I] - [R]$. It follows that

$$(1 - \sigma_*)([I] - [R]) = ([I] - [R]) - ([I^6] - [R])$$
$$= ([I] - [R]) + ([I] - [R]) = 2([I] - [R]).$$

However, $2([I] - [R])$ is also a generator for $\mathbb{Z}/(7)$, so the restriction of α to $\mathbb{Z}[R]$ must be a surjection onto $K_0(T)$. Hence all finitely generated projectives over T are stably free and, in particular, $[I^n \otimes_R T] = [I^m \otimes_R T] = [T]$ for all m, n.

§6 Filtered Rings

6.1 If S is a filtered ring such that $\operatorname{gr} S$ is right Noetherian, right regular and flat as a left S_0-module, then $K_0(S_0) \simeq K_0(S)$. This result was mentioned in 3.5 and its proof takes up the major part of this section. The basic idea of the proof involves the introduction and investigation of a graded ring A built from S and having $A_0 = S_0$.

6.2 Throughout the section S is a filtered ring, $B = \operatorname{gr} S$, and $R = S_0 = B_0$. The graded ring A is constructed next, as a subring of the polynomial ring $S[z]$, where z is a central indeterminate; namely $A = \bigoplus_n S_n z^n$. The grading is by degree in z; so $A_n = S_n z^n$.

6.3 The properties of A are not too complicated to obtain. The first to be considered is flatness.

Lemma. *If $_RB$ is flat then $_RS$ and $_RA$ are flat.*

Proof. Since $B = \bigoplus_n B_n$, each $_R(B_n)$ is flat. Now $B_n = S_n/S_{n-1}$ and induction shows that each $_R(S_n)$ is flat. However, $_RA \simeq \bigoplus_n S_n$ and $S = \bigcup S_n$; so by 7.1.5 $_RA$ and $_RS$ are flat. □

6.4 Note next that A has a filtration $\{S_n[z] \cap A\}$ and that $B[z]$ has a grading $\{B_n[z]\}$. (This filtration of A is not the one induced by the grading of A, and the grading of $B[z]$ is not by degree in z.)

Proposition. $\operatorname{gr} A \simeq B[z]$ *as a graded ring.*

Proof.
$$(\operatorname{gr} A)_n = (S_n[z] \cap A)/(S_{n-1}[z] \cap A) = \sum_{i \geq 0} (S_n/S_{n-1})z^{n+i} = B_n[z]z^n.$$
But the ring $B_0 \oplus B_1 z \oplus \cdots \oplus B_n z^n \oplus \cdots$, graded by degree in z, is isomorphic, as a graded ring, to B. Hence $\operatorname{gr} A \simeq B[z]$. □

6.5 This gives further properties of A.

Corollary. *If B is a right Noetherian, right regular ring then so too are S and A.*

Proof. By 7.7.5, $B[z]$ is right Noetherian and right regular. Then 7.7.4 deals with S and A. □

6.6 Of course $B[z]/zB[z] \simeq B[z]/(z-1)B[z] \simeq B$. There are similar isomorphisms for A.

Proposition. (i) zA is a graded ideal of A and $A/zA \simeq B$ as a graded ring.
(ii) $(z-1)A \triangleleft A$ and $A/(z-1)A \simeq S$.

Proof. Direct from the definitions. □

6.7 As in the proof of 5.5, use will be made of localization. If A is localized at the powers of z then $A_z \simeq S[z, z^{-1}]$. As in 4.10, A_z is $\mathbb{Z}$-graded and $M_{\text{f gr}}(A_z)$ is the category of finitely generated $\mathbb{Z}$-graded modules over A_z.

6.8 We now turn to the various categories involved and the homomorphisms between their Grothendieck groups.

Proposition. *The categories $M_{\text{f gr}} A_z$ and $M_{\text{f}} S$ are equivalent.*

Proof. Let G, H be the functors

$$G: M_{\text{f}} S \to M_{\text{f gr}} A_z, \qquad N \mapsto N \otimes_S A_z,$$

$$H: M_{\text{f gr}} A_z \to M_{\text{f}} S, \qquad M \mapsto M \otimes_{A_z} S$$

where A_z is viewed as an S-module via the embedding $S \hookrightarrow S[z, z^{-1}] = A_z$ and S is viewed as an A_z-module via the isomorphism $S \simeq S[z, z^{-1}]/(z-1)$.

Since the composed homomorphism

$$S \to S[z, z^{-1}] \to S[z, z^{-1}]/(z-1)$$

is the identity, $HG \simeq 1$. On the other hand, if $M \in M_{\text{f gr}} A_z$, so $M = \bigoplus_{n \in \mathbb{Z}} M_n$, then $M_i = M_j z^{i-j}$ and $M = M_0 \oplus M(z-1)$. Therefore $M_0 \simeq M/M(z-1) \simeq H(M)$, and $M \simeq GH(M)$. Since any graded homomorphism in $M_{\text{f gr}} A_z$ is determined by its degree 0 component it follows that $GH \simeq 1$. □

6.9 The next result is an immediate consequence of 6.3.

Corollary. *If B is right Noetherian and $_R B$ is flat then the inclusions $R \hookrightarrow S$ and $R \hookrightarrow A$ induce maps $K_0(M_{\text{f}} R) \to K_0(M_{\text{f}} S)$ and $K_0(M_{\text{f}} R) \to K_0(M_{\text{f}} A)$.* □

6.10 As noted in 2.2, R can be given the trivial grading.

Proposition. (i) $P_{\text{f gr}} R \simeq \bigoplus_{n \in \mathbb{N}} P_{\text{f}} R$.
(ii) $K_0(P_{\text{f gr}} R) \simeq \bigoplus_{n \in \mathbb{N}} K_0(R) \simeq K_0(R) \otimes_{\mathbb{Z}} \mathbb{Z}[t]$, *where the powers of t correspond to the different degrees.*

Proof. (i) Clear.
(ii) This is a consequence of 4.6. □

6.11 The trivial grading on R makes the inclusion $R \hookrightarrow B$ a graded homomorphism.

Proposition. (i) *The inclusion $\iota: R \hookrightarrow B$ induces an isomorphism $\iota_*: K_0(P_{\mathrm{f\,gr}}R) \to K_0(P_{\mathrm{f\,gr}}B)$.*
(ii) *If B is a right Noetherian, right regular ring, it also induces an isomorphism $\iota_*: K_0(P_{\mathrm{f\,gr}}R) \to K_0(M_{\mathrm{f\,gr}}B)$.*

Proof. (i) The composition $R \hookrightarrow B \to R$ is the identity map; so ι_* is clearly injective. However, if $P \in P_{\mathrm{f\,gr}}B$ then $P \simeq P' \otimes_R B$ for some $P' \in P_{\mathrm{f\,gr}}R$, by 2.9. Hence $[P] \in \operatorname{im} \iota_*$.
(ii) Theorem 4.8 shows that $K_0(M_{\mathrm{f\,gr}}B) \simeq K_0(P_{\mathrm{f\,gr}}B)$. □

6.12 Drawing these results together we get

Corollary. *Let B be right Noetherian and right regular and let $_R B$ be flat. Then there are isomorphisms*

$$K_0(R) \otimes_{\mathbb{Z}} \mathbb{Z}[t] \xrightarrow{\sim} K_0(P_{\mathrm{f\,gr}}R) \xrightarrow{\sim} K_0(P_{\mathrm{f\,gr}}B) \xrightarrow{\sim} K_0(M_{\mathrm{f\,gr}}B)$$

and

$$K_0(R) \otimes_{\mathbb{Z}} \mathbb{Z}[t] \xrightarrow{\sim} K_0(P_{\mathrm{f\,gr}}R) \xrightarrow{\sim} K_0(P_{\mathrm{f\,gr}}A) \xrightarrow{\sim} K_0(M_{\mathrm{f\,gr}}A).$$

Their composites map $[P] \otimes 1$ to $[P \otimes_R B]$ and $[P \otimes_R A]$ respectively.

Proof. The first isomorphism in each collection is given by 6.10. The second comes from 6.11 applied to B or to A, this being legitimate since A satisfies the conditions imposed on B in 6.11. The final isomorphism comes from Theorem 4.8. □

6.13 That completes the preliminaries. The proof of the main theorem which follows is rather lengthy. It is along similar lines to the proof of 5.5 and a comparison with that proof might be found helpful.

Theorem. (Quillen's theorem) *Let S be a filtered ring, $B = \operatorname{gr} S$, and $R = S_0 = B_0$. Suppose that B is right Noetherian and right regular, and that $_R B$ is flat. Then the inclusion $R \hookrightarrow S$ induces an isomorphism $K_0(R) \xrightarrow{\sim} K_0(S)$.*

Proof. Consider the localization A_z of A, where A is as before. Since A is right regular and right Noetherian, by 6.5, and $\mathscr{S} = \{z^n | n \in \mathbb{N}\}$ is homogeneous, it follows from 4.10 that there is an exact sequence

$$K_0(C) \xrightarrow{\alpha} K_0(M_{\mathrm{f\,gr}}A) \xrightarrow{\beta} K_0(M_{\mathrm{f\,gr}}A_z) \to 0. \qquad (*)$$

The rest of the proof is merely a detailed analysis of this sequence.

12.6.14 *Filtered rings*

First consider C. If $M \in C$ then $Mz^n = 0$ for some n. Since $B \simeq A/zA$, it follows that M has a finite filtration by B-modules. Moreover, $M_{\text{f gr}}B \hookrightarrow C$, by viewing each B-module as an A-module annihilated by z. Then 4.7, together with 6.12, implies that

$$K_0(C) \simeq K_0(M_{\text{f gr}}B) \simeq K_0(R) \otimes \mathbb{Z}[t].$$

Furthermore, 6.12 shows that

$$K_0(M_{\text{f gr}}A) \simeq K_0(R) \otimes \mathbb{Z}[t]$$

and 6.8 allied with 4.8 implies that

$$K_0(M_{\text{f gr}}A_z) \simeq K_0(M_f S) \simeq K_0(S).$$

Hence (*) can be rewritten as

$$K_0(R) \otimes \mathbb{Z}[t] \xrightarrow{\alpha'} K_0(R) \otimes \mathbb{Z}[t] \xrightarrow{\beta'} K_0(S) \to 0. \quad (**)$$

To analyse α', suppose $M \in P_f R$. If M is given the trivial grading then $M \otimes_R B \in M_{\text{f gr}}B$ and $[M] \otimes 1 \mapsto [M \otimes_R B]$ under the isomorphism $K_0(R) \otimes \mathbb{Z}[t] \xrightarrow{\sim} K_0(M_{\text{f gr}}B)$. However, the short exact sequence

$$0 \to zA \to A \to B \to 0$$

of graded right A-modules is also one of left R-modules. Therefore, since M_R is projective,

$$0 \to M \otimes_R zA \to M \otimes_R A \to M \otimes_R B \to 0$$

is an exact sequence. Thus in $K_0(M_{\text{f gr}}A)$

$$[M \otimes_R B] = [M \otimes_R A] - [M \otimes_R zA].$$

The isomorphism $K_0(M_{\text{f gr}}A) \xrightarrow{\sim} K_0(R) \otimes \mathbb{Z}[t]$ maps $[M \otimes_R A]$ to $[M] \otimes 1$ and $[M \otimes_R zA]$ to $[M] \otimes t$. Consequently,

$$\alpha'([M] \otimes 1) = [M] \otimes 1 - [M] \otimes t,$$

and so α' is simply right multiplication by $1 - t$. Therefore coker $\alpha' \simeq K_0(R)$ and so, from (**), $K_0(R) \simeq K_0(S)$.

It remains to check that this isomorphism is induced by the inclusion of R in S, and that is easily done making use of the fact that the composite map

$$R \hookrightarrow A \hookrightarrow A_z \twoheadrightarrow A_z/(z-1)A_z \xrightarrow{\sim} S$$

is the inclusion. □

6.14 Having proved the theorem, we now draw the conclusions we were aiming at.

Corollary. *Let R be a right Noetherian, right regular ring and let $S = A_n(R)$, or $S = R * U(g)$, where R is a k-algebra and g is a k-Lie algebra of finite dimension. Then the inclusion of R in S induces an isomorphism $K_0(R) \simeq K_0(S)$.*

Proof. By 1.7.13 and 1.7.14, $\operatorname{gr} S \simeq R[x_1, \ldots, x_n]$, the polynomial ring over R in a finite set of central indeterminates. Hence, by 7.7.5, $\operatorname{gr} S$ is right Noetherian and right regular, so 6.13 applies. □

6.15 It is perhaps worth noting that, under the hypotheses of Theorem 6.13, R is necessarily right Noetherian, because there is an epimorphism $B \to R$. Moreover, if $\operatorname{fd} B_R < \infty$ it is easily deduced as in 7.2.8 that R is right regular. (Presumably this is true without the additional hypothesis.)

§7 Applications to Simple Rings

7.1 In this section we will investigate $P_f S$ and calculate $K_0(S)$ for some simple Noetherian rings S, most of them being crossed products of a ring by a finite group. As a consequence some deductions will be made about their ring-theoretic structure.

7.2 Before starting the calculations, it seems sensible to describe the consequences. They relate to two known facts. The first is that a simple Artinian ring is always a matrix ring over a division ring; and the second, using 5.2.12, is that a simple hereditary Noetherian ring is always Morita equivalent to an integral domain with the same properties. It is reasonable to ask whether the latter rings are actually matrix rings over an integral domain and, more generally, whether every simple Noetherian ring is Morita equivalent to an integral domain. The consequences mentioned above are that these two questions both have negative answers.

7.3 The plan of the section will be to show first how knowledge of $K_0(S)$ and $P_f S$ can lead to such conclusions. Next $K_0(S)$ will be calculated for a few examples which will thereby be seen not to be matrix rings over integral domains. There will then be some results about crossed products by finite groups leading to the examples which are not Morita equivalent to integral domains.

7.4 The first result is immediate from 1.13.

Lemma. *For any ring S, if $[S]$ is not divisible by n in $K_0(S)$, then S is not an $n \times n$ matrix ring.* □

7.5 The key to dealing with Morita equivalence lies in the next result.

Proposition. *A ring S is Morita equivalent to a simple right Noetherian domain R if and only if S is a simple right Noetherian ring with a projective uniform right ideal P such that $\operatorname{End}(P_S) \simeq R$.*

Proof. ($\Rightarrow$) Here S is simple and right Noetherian since $S \overset{M}{\sim} R$. Let ${}_R P_S$ be the progenerator giving the Morita equivalence. Since R is an integral domain, P_S is uniform, by 3.4.5. It is then clear from 3.4.3 that P_S embeds in S.

($\Leftarrow$) Suppose S has a projective uniform right ideal P. Since S is a simple ring, $SP = S$; and so P_S is a progenerator. Therefore $R \overset{M}{\sim} S$ and R is simple and right Noetherian. Then 3.3.5 shows that R is an integral domain. □

7.6 Corollary. *If S is a simple right Noetherian ring such that there is no uniform module in $P_f S$, then S is not Morita equivalent to an integral domain.* □

7.7 The first few examples will pertain to 7.4.

Example. *Let $S = A_1(k) \# G$ as in 7.8.14(ii). Then:*
(i) *S is a simple Dedekind prime ring with $\operatorname{u\,dim} S = n$;*
(ii) *$K_0(S) \simeq \mathbb{Z}^n$ with $[S] \mapsto (1, 1, \ldots, 1)$;*
(iii) *S is not a matrix ring over a domain.*

Proof. (i) This is shown in 7.8.14(ii).
  (ii) S has a filtration $\{S_n\}$ given by total degree in x and y, i.e.

$$S_n = \{\sum_{i,j,h} a_{ijh} x^i y^j \sigma^h \mid a_{ijh} \in k, i + j \leq n\}.$$

Then $\operatorname{gr} S = kG[\bar{x}, \bar{y}]$ with the relations $\bar{x}\bar{y} - \bar{y}\bar{x} = 0$, $\bar{x}\sigma = \sigma\varepsilon\bar{x}$ and $\bar{y}\sigma = \sigma\varepsilon^{-1}\bar{y}$. Hence $\operatorname{gr} S$ can be viewed as an iterated skew polynomial ring $kG[\bar{y}; \sigma_1][\bar{x}; \sigma_2]$; so, by 7.5.3, and 1.2.9, $\operatorname{gr} S$ is a Noetherian ring of global dimension 2. Of course, $\operatorname{gr} S$ is free over $S_0 = kG$. So 6.13 shows that $K_0(S) \simeq K_0(kG)$ with $[S] \mapsto [kG]$.

However, kG is a semisimple Artinian ring, by Maschke's theorem, and is commutative. Hence kG is a direct sum of fields. Furthermore, kG has n distinct simple modules each of dimension 1 over k, namely k with σ acting as ε^i for $0 \leq i \leq n - 1$. Hence $kG \simeq k^n$ and so $K_0(kG) \simeq \mathbb{Z}^n$ with $[kG] \mapsto (1, 1, \ldots, 1)$.

  (iii) It is clear from (ii) and 7.4 that S is only a 1×1 matrix ring and yet S is not an integral domain. □

7.8 Example. *Let $R = \mathbb{R}[X, Y]/(X^2 + Y^2 - 1)$, the real coordinate ring of $\mathbb{S}^1$, and let σ be the automorphism of R induced by rotation of $\mathbb{S}^1$ through $2\pi\alpha$ for some irrational number α, $0 < \alpha < 1$. Let $S = R[t, t^{-1}; \sigma]$ and $T = \operatorname{End}(S \oplus AS)_S$ for some non-principal ideal A of R (as provided by 1.6). Then:*
(i) *S is a simple Dedekind domain and T is a simple Dedekind prime ring with $\operatorname{u\,dim} T = 2$;*

(ii) $K_0(T) \simeq K_0(S) \simeq \mathbb{Z} \oplus \mathbb{Z}/(2)$ with $[T] \mapsto (2, \bar{1})$;
(iii) T is not a matrix ring over a domain.

Proof. (i) Note that $S \otimes \mathbb{C} \simeq \mathbb{C}[z, z^{-1}][t, t^{-1}; \sigma] = P_\lambda(\mathbb{C})$, where $z = x + iy$ and $\lambda = e^{2\pi \alpha i}$. By 7.11.3, $S \otimes \mathbb{C}$ is a simple Dedekind domain so $K(S \otimes \mathbb{C}) = 1$. Hence by 6.5.3, $K(S) = 1$. However, R is a Dedekind domain, by 7.8.14(i), and then 7.11.2 shows that S is a simple Dedekind domain. Evidently $S \oplus AS$ is a progenerator and $\text{u dim}(S \oplus AS) = 2$. So T is a simple Dedekind prime ring, by 5.2.10, and $\text{u dim } T = 2$, by 3.4.5.
  (ii) Since $T \overset{M}{\sim} S$ via the progenerator $S \oplus AS$, then $K_0(T) \simeq K_0(S)$ via $[T] \mapsto [S \oplus AS]$. Note, however, that $^\sigma R$, as defined in 5.2, is isomorphic to R. Hence, in 5.5, $\sigma_* = 1$ and so $K_0(S) \simeq K_0(R)$ with $[S \oplus AS] \mapsto [R \oplus A]$. However, as noted in 1.6, $K_0(R) \simeq \mathbb{Z} \oplus \mathbb{Z}/(2)$ and then $[R] \mapsto (1, 0)$, $[A] \mapsto (1, \bar{1})$.
  (iii) Clear from (i), (ii) and 7.4. □

7.9 We note that if S, in the above example, is replaced by the ring $R[t; \delta]$, where $\delta = -y \partial/\partial x + x \partial/\partial y$ (that being the derivation corresponding to the unit tangent vector field on $\mathbb{S}^1$) then the same conclusions hold for similar reasons. This example also appears in 15.3.13.

7.10 The remainder of the section aims towards the second question raised in 7.2 and will make use of 7.6. The example concerned takes the form $S = R \# G$ with R a simple Noetherian domain and G a finite p-group. Before it, some preparatory results are required.

7.11 Lemma. *Let k be a field of characteristic p and let G be a finite p-group. Then kG has a chain of ideals*

$$0 = A_0 \subset A_1 \subset \cdots \subset A_n = kG,$$

where $n = |G|$ and each A_i/A_{i-1} is the one-dimensional trivial kG-module.

Proof. Being a p-group, G has a nontrivial centre C; and if $g \in C$ then $1 - g$ is central and nilpotent in kG, since $(1 - g)^m = 0$ if $m = p^a = \text{order } g$. Let I be the nilpotent ideal of kG generated by $\{1 - g | g \in C\}$ and note that $kG/I \simeq kG_1$ where $G_1 = G/C$.
  Induction on $|G|$ shows that the augmentation ideal of G is nilpotent, and the result is then immediate. □

7.12 Let $S = R \# G$ with R a ring and G any finite group. Then S_R is free. Hence there are functors $\mathbf{P}_f S \to \mathbf{P}_f R$, $M_S \mapsto M_R$ and $\mathbf{P}_f R \to \mathbf{P}_f S$, $N_R \mapsto N \otimes_R S$. We let F denote the composite

$$F: \mathbf{P}_f S \to \mathbf{P}_f S, \quad M_S \mapsto (M_R) \otimes_R S.$$

12.7.13

Recall, from 7.8.5 and 7.8.6, that, if $f = \sum \{g | g \in G\}$ then $R_S \simeq fS_S$ and $_SR \simeq _SSf$. Further, if $S = SfS$ (which, of course, is necessarily true when S is simple) then $S \simeq {}_S(R \otimes_{R^G} R)_S$.

Proposition. *Suppose $S = SfS$ and $M \in P_f S$.*
(i) $F(M) \simeq M \otimes_S (R \otimes_{R^G} S)$.
(ii) *If, in addition, R is a k-algebra, with k a field of characteristic p, and G is a finite p-group with $|G| = n$ then $F(M) \simeq M^n$.*

Proof. (i) $M_R \simeq M \otimes_S S_R$ and so
$$F(M) \simeq M \otimes_S (R \otimes_{R^G} R) \otimes_R S \simeq M \otimes_S (R \otimes_{R^G} S).$$

(ii) Let $\{A_i\}$ be as in 7.11. Since G acts trivially on the factors A_i/A_{i-1} an easy induction shows that $A_i S = A_i R = B_i$ say. Moreover, since $S \simeq kG \otimes_k R$ as a k-vector space, the map $A_i \otimes_k R \to A_i R = B_i$ is an isomorphism. Of course $A_i/A_{i-1} \otimes_k R \simeq R$. Hence the short exact sequence
$$0 \to A_{i-1} \to A_i \to A_i/A_{i-1} \to 0$$
gives another short exact sequence
$$0 \to B_{i-1} \to B_i \to R \to 0.$$
This is a sequence of (R^G, R)-bimodules and, in fact, of (R^G, S)-bimodules using the right S-module structure of R.

Recall next that R_{R^G} is projective by 3.5.4 and 7.8.6. Therefore, letting $C_i = R \otimes_{R^G} B_i$, we see that
$$0 \to C_{i-1} \to C_i \to R \otimes_{R^G} R \to 0$$
is an exact sequence of S-bimodules. Hence
$$0 \to M \otimes_S C_{i-1} \to M \otimes_S C_i \to M \otimes_S (R \otimes_{R^G} R) \to 0$$
is an exact sequence of right S-modules. However, $M \otimes_S (R \otimes_{R^G} R) \simeq M \otimes_S S \simeq M$ and so the sequence splits. Hence $M \otimes_S C_i \simeq (M \otimes_S C_{i-1}) \oplus M$. By induction, $M \otimes C_n \simeq M^n$ and thus
$$F(M) \simeq M \otimes_S (R \otimes_{R^G} S) \simeq M \otimes_S (R \otimes_{R^G} B_n) \simeq M \otimes C_n \simeq M^n. \quad \square$$

This means, of course, that in case (ii) the map $K_0(S) \to K_0(S)$ induced by F is simply multiplication by $|G|$.

7.13 To deal with the example below, the notion of a **trace function**, or **trace**, on a ring R will be useful. It is any abelian group homomorphism t, from R to an abelian group A satisfying $t(r_1 r_2) = t(r_2 r_1)$ for all $r_1, r_2 \in R$. The traces of interest here are those with $t(1) \neq 0$, these being called **proper**.

There is an easy criterion for the existence of such a trace which involves the **universal trace function**. This is constructed by letting $A = R/[R, R]$, where $[R, R]$ is the additive subgroup of R generated by all the elements $r_1 r_2 - r_2 r_1$ of R, and letting t be the natural map $R \to A$. Evidently R has a proper trace if and only if $1 \notin [R, R]$.

7.14 We now describe some cases where proper traces exist.

Lemma. *Suppose R has a proper trace $t: R \to A$.*
(i) *There is a trace $t_n: M_n(R) \to A$ given by $t_n([r_{ij}]) = \sum_i t(r_{ii})$. This is proper if and only if $nt(1) \neq 0$.*
(ii) *For any group G acting on R, if $t(r^g) = t(r)$ for all $r \in R$, $g \in G$ then there is a proper trace t' on $R \# G$ given by $t'(\sum g r_g) = t(r_1)$.*

Proof. (i) This is easily checked.
(ii) Note that the coefficient of 1 in $(\sum g r_g)(\sum h s_h)$ is $\sum \{(r_g)^h s_h \mid h = g^{-1}\}$. However, if $h = g^{-1}$ then
$$t((r_g)^h s_h) = t(r_g(s_h)^g) = t((s_h)^g r_g).$$
Hence t' is a trace function which evidently is proper. □

7.15 The connection with $P_f S$ comes next.

Theorem. *Let p be a prime number. Let R be a simple Noetherian domain such that $pR = 0$ and $K_0(R) = \mathbb{Z}[R]$. Let G be a finite p-group of outer automorphisms of R with $G \neq 1$. Let $S = R \# G$ and suppose that S has a proper trace. Then p divides $u \dim M$ for all $M \in P_f S$.*

Proof. Recall from 7.8.12 and 7.8.13 that S is a simple Noetherian ring and $u \dim S = n$ where $n = |G|$. Let $M \in P_f S$; then $[M_R] = [R^m]$ for some m and so, under the map $K_0(S) \to K_0(S)$ induced by the functor F of 7.12, $[M] \mapsto [S^m]$. However, by 7.12, $[M] \mapsto [M^n]$; so $[M^n] = [S^m]$ and hence
$$M^n \oplus S^r \simeq S^m \oplus S^r$$
for some r which may be increased to make r a multiple of p. Note also that since $u \dim S = n$ then $u \dim M = m$.

Suppose now that $p \nmid m$, and so $p \nmid m + r$. Let $V_S = M^{n/p} \oplus S^{r/p}$. Then $V^p \simeq S^{m+r}$ and therefore $M_p(\text{End } V) \simeq M_{m+r}(S)$. By 7.14(i), $M_{m+r}(S)$ has a proper trace, t say. But then, in $M_p(\text{End } V)$,
$$t(e_{11}) = t(e_{1i} e_{i1}) = t(e_{i1} e_{1i}) = t(e_{ii})$$
and so $t(1) = pt(e_{11}) = 0$, a contradiction. Hence $p \mid m$. □

7.16 Example. Let k be a field of characteristic 2 containing an element λ which is not a root of 1. Let $R = P_\lambda(k)$ and let θ be the k-automorphism of R such that $\theta(x) = x^{-1}$, $\theta(y) = y^{-1}$. Let $G = \{1, \theta\}$ and $S = R\#G$. Then:
(i) S is a simple Noetherian ring and $\operatorname{u\,dim} S = 2$;
(ii) S has no proper idempotents;
(iii) S is not Morita equivalent to an integral domain.

Proof. One can check that the hypotheses of 7.15 are satisfied, using 1.8.7 to see that R is a simple Noetherian domain, 3.3 to see that $K_0(R) = \mathbb{Z}[R]$, a degree argument to see that θ is outer and an iterated application of 7.14(ii) to see that S has a proper trace. As noted above, S is simple Noetherian and $\operatorname{u\,dim} S = 2$. The fact that $eS \in P_f S$ whenever $e = e^2 \in S$ yields (ii); and (iii) follows from 7.6. □

7.17 We note that an example in characteristic p can be obtained in a similar fashion. Take k to be a field of characteristic p containing an element λ which is not a root of 1. Let $R = P_\lambda(k) \otimes \cdots \otimes P_\lambda(k)$, with p factors, and with generators $x_1, y_1, \ldots, x_p, y_p$, and θ be defined by $x_i \mapsto x_{i+1}$, $y_i \mapsto y_{i+1}$, taking suffices mod p. Then $S = R\#\langle\theta\rangle$ is a simple Noetherian ring, $\operatorname{u\,dim} S = p$, S has no proper idempotents and S is not Morita equivalent to an integral domain.

§8 Additional Remarks

8.0 (a) For surveys of algebraic K-theory see [Bass **74, 75**], [Quillen **75**] and [Lam and Siu **75**]. For fuller accounts, including historical references and the higher K-groups, see [Bass **68**], [Berrick **82**], [Milnor **71**], [Silvester **81**], and [Swan **68**]. For K-theory in C^*-algebras, see [Rieffel **85**].
  (b) The fact that $A_1(k)$ has infinitely many non-isomorphic right ideals appears in [Webber **71**]. Related results, about their endomorphism rings, appears in [Stafford **87**].

8.1 Related to the Grothendieck group $K_0(R)$ is the **state space**; see [Goodearl and Warfield **81**].

8.2 2.9 is due to Swan; see [Bass **68**, 12.3.3].

8.3 (a) The special case of 3.3, that if k is a field then finitely generated projective $k[x_1, \ldots, x_n]$-modules are stably free, was noted in [Serre **58**]. That the arguments there extend to give 3.3 became 'folklore' once the more general result 6.13, Quillen's theorem, was proved.
  (b) 3.4 is a special case of 6.13 but this elementary proof is new.
  (c) 3.6(i), for $S = R[x]$, is due to [Serre **58**]; see [Bass **75**]. There are examples, see [Lam **78**, p. 64], to show the result fails if R is not regular.

8.4 Related results can be found in [Bass **74**].

8.5 (a) 5.5 is a special case of results of [Farrell and Hsiang **70**] and [Siebenmann **70**]; this proof follows that suggested in [Quillen **73**].

(b) The Grothendieck group of a polycyclic by finite group ring is discussed in [Farrell and Hsiang **81**] and [Brown, Howie, and Lorenz **P**].

8.6 The results and proofs in this section are the case when $n=0$ of results from [Quillen **73**]; see also [Grayson **80**].

8.7 (a) 7.5 is due to [Hart **67**]; see also [Hart and Robson **70**].

(b) 7.7 is an example of [Zalesskii and Neroslavskii **77**], (see also [Goodearl **79**]). The idea of using K_0 to establish that a given ring is not a matrix ring comes from [Hart **80**] and concerned Example 7.8. This proof of 7.7(iii) is due to [Goodearl **84**].

(c) 7.10–7.17 are based on [Lorenz **85**]. The first proof that a simple Noetherian ring need not be Morita equivalent to a domain was in [Stafford **78b**], the example being as in 7.7. See also [Stafford **79a**].

Part IV
EXAMPLES

Chapter 13

POLYNOMIAL IDENTITY RINGS

A polynomial identity ring, or PI ring for short, is defined, loosely, as a ring all of whose elements satisfy some polynomial identity. What this means precisely will be explained in §1. However, the trite observation that if R is a commutative ring and if $f(x_1, x_2) = x_1 x_2 - x_2 x_1 \in \mathbb{Z}\langle x_1, x_2 \rangle$ then R satisfies f (i.e. $f(r_1, r_2) = 0$ for all $r_1, r_2 \in R$) shows that commutative rings are PI rings. It will be shown that any ring finite as a module over its centre, such as a matrix ring over a commutative ring, is a PI ring. Moreover, any subring or homomorphic image of a PI ring is a PI ring. Thus PI rings occur in abundance.

The concern of this chapter is the structure and properties of PI rings. The major theme throughout is that prime PI rings are closely linked with their centres and general PI rings are much more like commutative rings than are other noncommutative rings. This latter claim is highlighted by the verification, for PI rings, of the Köthe conjecture in §2 and, for affine PI algebras over a field, of the Kurosch conjecture, in §8, and the Nullstellensatz and the catenary property for prime ideals in §10.

The basic structure theorems concern primitive and prime PI rings. In §3 it is shown that the former are central simple algebras. The proof that the latter are all orders in central simple algebras appears only in §6, although indications appear earlier.

In §§7 and 8 special features related to Azumaya algebras and rings integral over central subrings are demonstrated. The next section describes the trace ring and uses it to complete the discussion, begun in Chapter 5, of Dedekind prime rings. Finally, in §10, these results are used in the study of affine algebras over a field, with results as mentioned above.

§1 Polynomial Identities

1.1 This section makes precise the notion of a polynomial identity and of a PI ring. It establishes that the class of PI rings includes all subrings of matrix rings over commutative rings and it describes some universal examples of PI rings.

1.2 The identities which concern us here come from polynomials f in the free algebra $\mathbb{Z}\langle x_1,\ldots,x_n\rangle$ for some n, or, equivalently, polynomials in $\mathbb{Z}\langle x_1, x_2,\ldots\rangle = \mathbb{Z}\langle x\rangle$ for short.

Let $f(x_1,\ldots,x_n)\in\mathbb{Z}\langle x\rangle$. Then a ring R **satisfies** f, and f is a **polynomial identity** of R, if $f(r_1,\ldots,r_n)=0$ for all $r_i\in R$. This is, of course, equivalent to insisting that $f\in\ker\vartheta$ for every ring homomorphism $\vartheta:\mathbb{Z}\langle x\rangle\to R$. The **degree** of the identity f is its usual degree as an element of $\mathbb{Z}\langle x\rangle$; and f is **monic** if at least one of the words of highest degree in the support of f has coefficient 1.

1.3 One identity of particular importance is

$$s_n = \sum_{\sigma\in S_n} (\operatorname{sgn}\sigma)x_{\sigma(1)}x_{\sigma(2)}\cdots x_{\sigma(n)},$$

where S_n is the symmetric group. This is called the n**th standard identity**. Evidently s_n has degree n and is monic.

Of course, s_2 is the identity $x_1x_2 - x_2x_1$ cited in the introduction to this chapter.

1.4 As usual, we normally will assume that a PI ring R has a 1. However, it is useful, on occasion, to allow this restriction to lapse. In that case, our attention is focused on elements $f\in\mathbb{Z}\langle x\rangle$ which have zero constant term.

1.5 Next comes a cautionary, albeit trivial, example. Suppose that k is a field of characteristic 2 and $R = k\langle y_1,\ldots,y_n\rangle$. Then R satisfies the identity $2x_1\in\mathbb{Z}\langle x\rangle$, but for trivial reasons. Bearing in mind the general aim to show that PI rings are close to commutative, it is clear that this example must be excluded.

1.6 Therefore a **polynomial identity ring**, always abbreviated henceforth to **PI ring**, is defined to be a ring which satisfies some monic polynomial in $\mathbb{Z}\langle x\rangle$.

Naturally, this includes all commutative rings!

1.7 It is easy to check that the class of PI rings is closed under certain operations.

Lemma. (i) *Any subring or homomorphic image of a PI ring is a PI ring.*

(ii) *A finite direct product of PI rings is a PI ring.*

(iii) *If $\{R_i|i\in I\}$ is a set of PI rings all satisfying the same identity f then $\prod R_i$ does so too.*

(iv) *If N is a nilpotent ideal of R and R/N is a PI ring then R is a PI ring.*

Proof. (i) Clear, using the same monic identity.

(ii) If $f_1,\ldots,f_m\in\mathbb{Z}\langle x\rangle$ are monic identities of $R_1,\ldots,R_m$ respectively then $f_1f_2\cdots f_m$ is a monic identity of $\prod R_i$.

(iii) Clear.

(iv) If f is a monic identity of R/N and $N^t = 0$ then f^t is a monic identity of R. □

1.8 For many purposes it is useful to refine the type of identity considered. A **multilinear polynomial of degree** n is a nonzero element $f \in \mathbb{Z}\langle x_1, \ldots, x_n \rangle$ taking the form

$$f = \sum_{\sigma \in S_n} a_\sigma x_{\sigma(1)} \cdots x_{\sigma(n)}$$

with each $a_\sigma \in \mathbb{Z}$. This means, of course, that

$$f(c_1 x_1, \ldots, c_n x_n) = c_1 c_2 \cdots c_n f(x_1, \ldots, x_n)$$

for all $c_i \in \mathbb{Z}$.

For example, the standard identity s_n defined in 1.3 is multilinear of degree n.

1.9 The next result shows that a PI ring could, equally well, be defined as a ring which satisfies some monic multilinear identity.

Proposition. *If R satisfies an identity f of degree d then R also satisfies a multilinear identity g of degree at most d. Furthermore, each coefficient in g is also a coefficient in f; and if f is monic, so too is g.*

Proof. The words in the support of f can be partitioned into those in which x_1 appears and the others. This decomposes $f(x_1, \ldots, x_n)$ into a sum

$$f_1(x_1, \ldots, x_n) + f_2(x_2, \ldots, x_n).$$

Setting $x_1 = 0$ makes clear that f_2 and hence also f_1, is an identity of R. Of course, if f is monic then so is at least one of f_1 and f_2. An easy induction now allows us to assume that each x_i occurs in each word in the support of f, although the number of indeterminates and the set of coefficients may have been reduced.

If f is not already multilinear then the words of highest degree in the support of f cannot be multilinear. So some indeterminate, x_1 say, occurs in these words with maximal degree m say. Consider then the polynomial

$$g(x_1, \ldots, x_{n+1}) = f(x_1 + x_{n+1}, x_2, \ldots, x_n) \\ - f(x_1, x_2, \ldots, x_n) - f(x_{n+1}, x_2, \ldots, x_n).$$

A simple calculation demonstrates that the words in the support of g all come from those words w in the support of f which have degree at least 2 in x_1; they are the words obtained by replacing some, but not all, of the x_1's in w by x_{n+1}. They retain the same coefficient as w; and, if f is monic, then g is monic.

A straightforward induction completes the proof. □

The process used above in the passage from f to g is called **linearization**.

1.10 One advantage enjoyed by multilinear identities is noted next.

Lemma. *Let $\mathscr{A}$ be a set of generators of a ring R viewed as a module over its centre. Suppose that $f(x_1,\ldots,x_n)$ is multilinear and that $f(a_1,\ldots,a_n) = 0$ for all $a_i \in \mathscr{A}$. Then R satisfies f.*

Proof. Each element of R is a finite sum of elements ca, with $a \in \mathscr{A}$, $c \in Z(R)$. However,
$$f(c_1 a_1, \ldots, c_n a_n) = c_1 \cdots c_n f(a_1, \ldots, a_n) = 0.$$
Therefore, by linearity, R satisfies f. □

1.11 One immediate consequence concerns any **central extension** S of a ring R, this meaning that $S \supseteq R$ with S being generated, as a ring, by R and the centre of S.

Corollary. *Let S be a central extension of R.*
(i) If R satisfies a multilinear identity f then so does S.
(ii) If R is a PI ring then S is a PI ring.

Proof. Clear from 1.10 and 1.9. □

1.12 Recall that a multilinear element $f \in \mathbb{Z}\langle x_1, \ldots, x_n \rangle$ is **alternating** if, for all $i, j \in \{1, \ldots, n\}$ with $i \neq j$, $f(x_1, \ldots, x_{j-1}, x_i, x_{j+1}, \ldots, x_n) = 0$. This is equivalent to requiring that the interchange of any two indeterminates in f gives $-f$.

Once again s_n serves as an example.

1.13 A second consequence of 1.10 establishes, in (ii), a claim made in the introduction to the chapter, and thereby provides a large class of examples of PI rings.

Corollary. *(i) Let R be a ring finitely generated, by m elements, as a module over its centre. Then R satisfies every alternating multilinear polynomial of degree $m+1$ or more. In particular R satisfies s_n for $n \geq m+1$.*
(ii) If A is a commutative ring then $M_n(A)$ satisfies s_t for all $t \geq n^2 + 1$.
(iii) If R is finitely generated as a right module over a commutative subring A then R is a PI ring.

Proof. (i) This is immediate from 1.10 and 1.12.
(ii) Clear from (i).
(iii) By 8.2.8, R is a subquotient of $M_m(A)$ for some m. □

This result will be extended and improved in later sections.

1.14 We end this section by discussing some universal examples of PI rings.

Let $F = \mathbb{Z}\langle x \rangle$ where $x = \{x_i | i \in I\}$. An ideal H of F is a **T-ideal** if $\varphi(H) \subseteq H$ for all endomorphisms φ of F.

Proposition. (i) *If R is a PI ring then the set J of all identities of R in F is a T-ideal. Indeed, J is the intersection of the kernels of all the homomorphisms from F to R.*

(ii) *If H is any T-ideal of F then each element of H is an identity of the ring F/H. In particular, if H contains a monic polynomial then F/H is a PI ring.*

Proof. (i) Clear; but note that if the index set I is small then J might be 0.

(ii) Let $f = f(x) \in H$ and suppose $x_1, \ldots, x_m$ are the indeterminates appearing in f. Let $g_1, \ldots, g_m \in F$. Define $\varphi: F \to F$ by $x_i \mapsto g_i$ for $i = 1, \ldots, m$ and $x_i \mapsto x_i$ otherwise. By hypothesis $\varphi(f) \in H$ and so $f(g_1, \ldots, g_m) \in H$. Hence f is satisfied by F/H. □

Thus F/H is universal for PI rings satisfying the identities in H in the sense that, if R is such a ring, then for any set map $\vartheta: \{x_i | i \in I\} \to R$ there is a commutative diagram

with $\bar{\vartheta}$ a uniquely specified ring homomorphism.

In particular, if I is sufficiently large that R is a homomorphic image of F then R is a homomorphic image of F/H.

1.15 In practice it is often convenient to work with a K-algebra R over some commutative ring K. In that case one considers the free K-algebra $K\langle x \rangle$ and uses only K-algebra endomorphisms and homomorphisms. The corresponding ring $K\langle x \rangle/H$ is then a universal PI K-algebra. The elements of H are called **K-algebra polynomial identities** of R. Note that we still insist, however, that H should contain a monic polynomial with integer coefficients.

1.16 For later use, a slight refinement of 1.14 is needed. An identity $f \in F$ of a PI ring R is called **stable** if f is also an identity of $R \otimes_\mathbb{Z} A$ for every commutative ring A. The identity $f(x) = x^3 - x$ on $\mathbb{Z}/(2)$ is an example of an identity which fails to be stable—taking $A = \mathbb{Z}[t]$ for example, with t being a central indeterminate.

1.17 Proposition. *Let I be the set of all stable identities of R in F. Then:*

(i) *I is a T-ideal of F;*

(ii) I contains all multilinear identities of R in F;
(iii) I is a graded ideal of F under its standard grading;
(iv) F/I satisfies all identities in I.

Proof. (i) Clear.

(ii) This follows from 1.11.

(iii) Let $f = f(x_1,\ldots,x_m) \in I$. By hypothesis, f is also an identity for $R[t]$, with t a central indeterminate, since $R[t] \simeq R \otimes_{\mathbb{Z}} \mathbb{Z}[t]$. Thus, for all $r_1,\ldots,r_m \in R$,

$$0 = f(tr_1,\ldots,tr_m) = \sum_i t^i f_i(r_1,\ldots,r_m),$$

where f_i is the ith homogeneous component of f. Hence f_i is an identity of R, as required.

(iv) This follows from (i) and 1.14 (ii). □

Of course, F/I can also be viewed as a universal object.

1.18 Finally we turn to matrix rings over commutative rings, since these play a major role in this chapter. We fix a commutative ring K, an integer n and an index set I, and consider the free algebra $F = K\langle x_i | i \in I \rangle = K\langle x \rangle$. Let G be the intersection of the kernels of all K-algebra homomorphisms from F to $M_n(A)$, as A varies over all commutative K-algebras. Note that if I is large enough then $G \neq 0$ since $s_{n^2+1} \in G$. Evidently $G \triangleleft F$ and F/G is universal in the sense that, given any set map $\vartheta: \{x_i | i \in I\} \to M_n(A)$ for some A, there is a commutative diagram

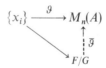

with a unique K-algebra homomorphism $\bar{\vartheta}$.

In particular, any K-subalgebra of $M_n(A)$ with a generating set whose cardinality is no more than that of I is a homomorphic image of F/G.

Of course, if I is finite then F/G is an affine K-algebra.

1.19 There is another description of the K-algebra F/G above which reflects the observation that the K-algebra universal for subalgebras of $n \times n$ matrices over K-algebras ought, itself, to take that form.

For simplicity, consider the affine case, say with m generators. Let A be the commutative polynomial ring over K on mn^2 indeterminates

$$\{a_{ij}^k | i,j \in \{1,\ldots,n\}, k \in \{1,\ldots,m\}\};$$

we write $A = K[\mathbf{a}]$. In the matrix ring $M_n(A)$, let X_k be the matrix (a_{ij}^k). This is called a **generic** $n \times n$ **matrix**; and the K-subalgebra of $M_n(A)$ generated by $X_1, \ldots, X_m$ is called the K-**algebra of** m **generic** $n \times n$ **matrices** and is written $K[X_1, \ldots, X_m]$.

1.20 Proposition. *If F, G are as in 1.18, with the index set $I = \{1, 2, \ldots, m\}$, then $F/G \simeq K[X_1, \ldots, X_m]$ as a K-algebra.*

Proof. It is sufficient to show that $F[X_1, \ldots, X_m]$ has the universal property of F/G. Let $r_1, \ldots, r_m$ be $n \times n$ matrices over some commutative K-algebra B and let $\vartheta : x_i \mapsto r_i$ be the given set map. There is a unique K-algebra homomorphism $K[\mathbf{a}] \to B$, given by mapping each a_{ij}^k to the (i,j) entry of r_k. This extends to a K-algebra homomorphism $M_n(K[\mathbf{a}]) \to M_n(B)$ and then restricts to a K-algebra homomorphism $K[X_1, \ldots, X_m] \to K[r_1, \ldots, r_m]$. The latter is the required map $\bar{\vartheta}$ and clearly is unique. □

1.21 We note the rather surprising fact that although the X_i are all matrices, nevertheless if K is an integral domain then so also is the generic matrix ring. This will not be used here and its proof is omitted. It can be found in [Cohn 77, 12.6.2] or [Jacobson 75, p. 90].

§2 Nilpotence

2.1 In this section we present the first evidence that PI rings behave a little like commutative rings. The main result demonstrates the truth, for PI rings, of the Köthe conjecture.

2.2 A PI ring R is said to have **minimal degree** m if m is the least possible degree of a monic polynomial identity of R. This identity can, by 1.9, be chosen to be multilinear.

2.3 There will be some discussion about the minimal degree of certain rings later, although in this section it will simply provide the means of proof by induction. Note that the least possible value the minimal degree can attain is 2; and the monic multilinear polynomial concerned must then be $x_1 x_2 - n x_2 x_1$ for some $n \in \mathbb{Z}$.

2.4 We note one more rather obvious fact.

Lemma. *Let $g(x_1, \ldots, x_m)$ be a monic multilinear identity whose monic term is $x_1 x_2 \cdots x_m$. Then $g = g_1(x_1, \ldots, x_{m-1}) x_m + g_2(x_1, \ldots, x_m)$, where g_1 is a monic multilinear polynomial and g_2 is multilinear with no word ending in x_m.* □

2.5 The verification of the Köthe conjecture is one case in which, as mentioned in 1.4, it is helpful to work with rings not necessarily having a 1.

Theorem. (i) *A nonzero PI ring with a nonzero nil right ideal has a nonzero nilpotent ideal.*
(ii) *A nonzero nil PI ring has a nonzero nilpotent ideal.*

Proof. It is evident that (i) and (ii) are equivalent. We aim at proving (ii); and that will follow if it is shown that the nil ring R contains a nonzero nilpotent right ideal B, since then $A = RB + B$ is a nonzero nilpotent ideal of R.

The proof proceeds by induction on the minimal degree m of R. If $m = 2$ then, as noted in 2.3, R satisfies $x_1 x_2 - n x_2 x_1$ for some $n \in \mathbb{Z}$. Choose $0 \neq b \in R$ with $b^2 = 0$; then $bRb = nb^2 R = 0$. So if $B = b\mathbb{Z} + bR$ then $B^2 = 0$ as required.

Next, suppose the result holds for all smaller minimal degrees, and again choose $0 \neq b \in R$ with $b^2 = 0$. If $bR = 0$ then $B = b\mathbb{Z}$ is a nilpotent right ideal of R. If $bR \neq 0$ and R satisfies $g(x_1, \ldots, x_m)$, as in 2.4, let g_1, g_2 be as there described; so $g = g_1 x_m + g_2$. Note that for all $r_1, \ldots, r_{m-1} \in R$

$$g_2(br_1, \ldots, br_{m-1}, b) = 0$$

since each word in g_2 involves an x_m before some other x_i. Therefore

$$g_1(br_1, \ldots, br_{m-1})b = 0.$$

Let $W = \{r \in bR \mid rbR = 0\}$. This is an ideal of the ring bR; and we have established that g_1 is a monic multilinear identity of degree $m - 1$ satisfied by bR/W. If $W = bR$ then bR is a nilpotent right ideal of R. Otherwise, by induction, bR/W contains a nonzero nilpotent right ideal. Hence there is a right ideal I of bR, strictly containing W, with $I^2 \subseteq W$. Now $0 \neq IbR \triangleleft_r bR$, since $I \not\subseteq W$ and

$$(IbR)^2 \subseteq I^2 bR \subseteq WbR = 0. \qquad \square$$

2.6 Corollary. (i) *A semiprime PI ring contains no nonzero nil right ideals.*
(ii) *The prime radical of a PI ring is its maximal nil right ideal.* $\square$

2.7 One consequence of 2.6 (i) which will be used later concerns polynomial rings.

Proposition. (i) *If R is a ring with no nil right ideals then $R[x]$, the polynomial ring in a central indeterminate, is semiprimitive.*
(ii) *If R is a semiprime PI ring then $R[x]$ is semiprimitive.*

Proof. (i) Suppose J is the Jacobson radical of $R[x]$ and $J \neq 0$. Pick $0 \neq p = \sum_0^n p_i x^i \in J$ having the fewest nonzero coefficients. It follows, by considering $p_i p - p p_i$, that the coefficients of p commute with each other.

Now $1 - px$ is a unit, with inverse $q = \sum_0^m q_i x^i$ say. Then $(1 - px)q = 1$ and so

$q = 1 + pxq = 1 + px(1 + pxq)$. It follows that

$$q = 1 + px + p^2x^2 + \cdots + p^k x^k + p^{k+1} x^{k+1} q.$$

Choose k so that $(n+1)k \geq m$. Then no term occurring in $q, 1, px, \ldots, p^k x^k$ has degree greater than $nk + k = (n+1)k$, whereas in $p^{k+1} x^{k+1} q$ there is the term

$$p_n^{k+1} q_m x^{(n+1)(k+1)+m}.$$

It follows that

$$p_n^{k+1} q_m = 0 = p_n^{k+2} q_{m-1} = \cdots = p_n^{k+m+1} q_0$$

and so $p_n^{k+m+1} q = 0$. However, q is a unit, so p_n is a nilpotent element.

The same argument can be applied to any nonzero element of $p_n R$. Thus $p_n R$ is a nil right ideal—a contradiction.

(ii) Clear from (i) and 2.6. □

2.8 We end this section with a result demonstrating a weak version of being bounded.

Proposition. *Let R be a semiprime PI ring and let $a \in R$. Then, for $n \gg 0$, $a^n R + \mathrm{r\,ann}(a^n)$ contains a nonzero ideal of R.*

Proof. This is clear if a is nilpotent. Otherwise choose n large enough that the minimal degree of $a^n R$ is as small as possible, and let $g(x_1, \ldots, x_m)$ be the corresponding monic multilinear identity. As in 2.4 we may suppose that

$$g = x_1 g_1(x_2, \ldots, x_m) + g_2(x_1, \ldots, x_m)$$

with g_1 monic and multilinear, and g_2 multilinear with no word starting with x_1. For any $r_1, \ldots, r_m \in R$

$$\begin{aligned}
0 &= g(a^n r_1, a^{2n} r_2, \ldots, a^{2n} r_m) \\
&= a^n r_1 g_1(a^{2n} r_2, \ldots, a^{2n} r_m) + g_2(a^n r_1, a^{2n} r_2, \ldots, a^{2n} r_m) \\
&= a^n r_1 s + a^{2n} t,
\end{aligned}$$

where $s, t \in R$. By hypothesis, g_1 is not satisfied by $a^{2n} R$. So there exist $r_2, \ldots, r_m$ such that $s \neq 0$. In that case $0 = a^n(r_1 s + a^n t)$ and so $r_1 s \in a^n R + \mathrm{r\,ann}(a^n)$. However, r_1 is arbitrary; so $0 \neq RsR \subseteq a^n R + \mathrm{r\,ann}(a^n)$. □

2.9 Corollary. *If R is a semiprime PI ring and $\mathrm{r\,ann}\, a = 0$, for some $a \in R$, then aR contains a nonzero ideal of R.* □

2.10 This has an interesting consequence which will, however, be superseded by 6.5.

Corollary. *If R is an integral domain and a PI ring then R is an Ore domain.*

Proof. Let a,b be nonzero elements of R. By 2.9, $aR \supseteq I$ with $0 \neq I \triangleleft R$. Hence $0 \neq bI \subseteq I \subseteq aR$. Thus, if $0 \neq a' \in I$ then $ab' = ba'$ for some $b' \in R$. Hence R is right Ore and, by symmetry, left Ore. □

§3 Central Simple Algebras

3.1 The main result in this section, which is crucial to the theory of PI rings, is that primitive PI rings are all central simple algebras.

3.2 We saw, in 1.13, that if A is any commutative ring then $M_n(A)$ satisfies s_{n^2+1}. A result in the opposite direction is needed for what follows.

Proposition. *If A is a commutative ring then $M_n(A)$ satisfies no monic identity of degree less than 2n.*

Proof. By 1.9 we need only consider a monic multilinear identity, say

$$f = x_1 x_2 \cdots x_t + \sum_{\sigma \neq 1} a_\sigma x_{\sigma(1)} \cdots x_{\sigma(t)}$$

with $t \leq 2n - 1$. Choose $r_1, \ldots, r_t$ to be the first t matrix units in the sequence

$$e_{11}, e_{12}, e_{22}, e_{23}, e_{33}, \ldots, e_{nn}.$$

Then it is clear that $f(r_1, \ldots, r_t) = r_1 r_2 \cdots r_t \neq 0$. □

3.3 The comparison of 3.2 with 1.13 suggests an obvious question which is answered by the next result.

Theorem. (i) *If a PI ring satisfies s_m then it satisfies s_t for all $t \geq m$.*
(ii) (Amitsur–Levitzki theorem) *If A is a commutative ring then $M_n(A)$ satisfies s_{2n}.*

Proof. (i) Collecting together those monomials beginning with the same indeterminate gives

$$s_{m+1}(x_1, \ldots, x_{m+1}) = \sum_{i=1}^{m+1} (-1)^{i-1} x_i s_m(x_1, \ldots, \hat{x}_i, \ldots, x_{m+1})$$

where $\hat{x}_i$ indicates the omission of x_i. Hence the ring satisfies s_{m+1}; and induction completes the proof.

(ii) The proof of (ii) is combinatorial and is omitted. Proofs are available in [Cohn 77, 12.5.11], [Rowen 80, 1.4.1] or [Bollobas 79, p. 20]. □

13.3.8 *Central simple algebras* 455

The only use made of (ii) in this chapter will be to give precise formulae, rather than inequalities, when discussing degrees.

3.4 We now turn towards central simple algebras starting, however, a little more generally.

Lemma. *Let $R = M_t(D)$ for some division ring D. Let H be a maximal subfield of D, C be the centre of D and V_R be a simple module; so $D \simeq \text{End } V_R$.*
(i) *$S = R \otimes_C H$ is simple and V is a simple S-module with $\text{End } V_S \simeq H$.*
(ii) *If V has dimension $m < \infty$ over H then $S \simeq M_m(H)$, $\dim D_H = \dim H_C = m/t$ and $\dim R_C = m^2$.*

Proof. (i) 9.6.9 shows that S is simple. Now V is given an S-module structure via $v(r \otimes h) = hvr$ and clearly is simple. It is easy to check that $\text{End } V_S$ is isomorphic to the centralizer of H in D, i.e. H itself.

(ii) That $S \simeq M_m(H)$ is clear. However, $\dim S_H = \dim R_C$, so $\dim R_C = m^2 = t^2 \dim D_C$. Since $V \simeq D^t \simeq H^m$, it follows that $\dim D_H = m/t = \dim H_C$. □

3.5 This applies, of course, to the case when R is any central simple algebra with centre C, giving (i) below; and (ii) is then immediate from 1.11.

Corollary. *Let $R = M_t(D)$ be a central simple algebra, where D is a division ring with centre C. Let H be a maximal subfield of D. Then:*
(i) *$D \otimes_C H \simeq M_m(H)$, where $m^2 = [D:C]$, and $R \otimes_C H \simeq M_n(H)$, where $n^2 = (mt)^2 = [R:C]$;*
(ii) *R satisfies s_{2n} and does not satisfy s_{2n-1}.* □

3.6 For a central simple algebra R as above, it is customary to call n the **PI degree** of R. Of course, n^2 is the rank of R over its centre. Also $n = \frac{1}{2}d$, where d is the minimal degree of R.

Any field H containing C and such that $R \otimes_C H \simeq M_n(H)$ is called a **splitting field** for R.

3.7 We now turn towards primitive rings, starting with a straightforward deduction from the density theorem (0.3.7).

Lemma. *Let R be a primitive ring with a faithful simple module V_R and let $D = \text{End } V_R$. Then either $\dim_D V = n$ and $R \simeq M_n(D)$ for some $n < \infty$, or else, for each $n < \infty$, there is a subring S_n of R together with a surjective ring homomorphism $S_n \to M_n(D)$.* □

3.8 The preceding results combine to yield the main result of this section.

Theorem. (Kaplansky's theorem) *If R is a primitive PI ring of minimal degree d then R is a central simple algebra of dimension $(d/2)^2$ over its centre.*

Proof. The latter option in 3.7 is not tenable here. For if C is the centre of D then 3.2 shows that the minimal degree of $M_n(C)$ is at least $2n$, and hence the same is true of $M_n(D)$, of S_n and so of R.

Therefore the former option applies. Thus R has a faithful simple module V with $\dim_D V = n < \infty$ and $R \simeq M_n(D)$. In particular, R is simple. Let H be a maximal subfield of D. By 3.4(i), V is a simple faithful $R \otimes_C H$-module with $\operatorname{End} V_{R \otimes H} \simeq H$. Moreover, $R \otimes_C H$ is simple, by 3.4(i), and a PI ring, by 1.11. The preceding paragraph applies to show that $_H V$ is finite dimensional. By 3.4(ii) so too is R_C; thus R is a central simple algebra. The rest is clear from 3.5. □

3.9 This has an immediate corollary, which can, however, be proved more directly.

Corollary. *If $R' \subseteq R$ are simple rings and R is a central simple algebra then so too is R'.*

Proof. R is a PI ring, so R' is too. □

Other consequences are described in the next section.

§4 Embeddings and Matrix Rings

4.1 In §3 it was shown that primitive PI rings, being central simple algebras, embed in matrix rings over fields and so satisfy some standard identity s_t. This section extends these results to much wider classes of PI rings.

4.2 The first step concerns embeddings.

Theorem. *Let R be a semiprime PI ring of minimal degree d.*
(a) *There is an embedding, not preserving 1, $R \hookrightarrow M_n(A)$, where $n = d/2$ and A is a product of fields; and if R is prime then A can be chosen to be a field.*
(b) *There is a similar embedding, preserving 1, with $n = (d/2)!$.*

Proof. By using 2.7, there is no loss in assuming that R is semiprimitive. Then $R \hookrightarrow \prod R/P_i$, where P_i ranges over the primitive ideals of R. By 3.5 and 3.8, each $R/P_i \hookrightarrow M_{n_i}(H_i)$, where n_i is the PI degree of R/P_i and H_i is a splitting field. Since $n_i \leq d/2$ then $R/P_i \hookrightarrow M_n(H_i)$, (a) as the top corner with $n = d/2$, or (b) via the identification $M_n(H_i) \simeq M_{n/n_i}(M_{n_i}(H_i))$ with $n = (d/2)!$. Hence $R \hookrightarrow M_n(A)$ where $A = \prod H_i$.

Finally, if R is prime, choose $I \triangleleft A$ maximal with respect to having $M_n(I) \cap R = 0$. Evidently I is prime and
$$R \hookrightarrow M_n(A/I) \hookrightarrow M_n(F)$$
where F is the field of fractions of A/I. □

4.3 It is useful to investigate the effect of this embedding on the centre of R.

Proposition. *Let $K \subseteq C$, the centre of R, with R as in 4.2. There is an embedding $K \hookrightarrow A \subseteq M_n(A)$ which preserves the action of K on R.*

Proof. As in 4.2, we may assume that R is semiprimitive. Let $\vartheta_i : K \to R/P_i$ be the restriction of the map $R \to R/P_i$. Then
$$\text{im } \vartheta_i \subseteq Z(R/P_i) \hookrightarrow Z((R/P_i) \otimes H_i) = Z(M_{n_i}(H_i)) = H_i.$$
Since $\bigcap \ker \vartheta_i = 0$ it follows that $K \hookrightarrow A$. This embedding preserves the action of K on R. □

The reader may have noted that the image of K in $M_n(A)$ provided above does not necessarily coincide with the image given in 4.2(a) by $K \subseteq R \to M_n(A)$.

4.4 This applies especially to affine K-algebras.

Corollary. *Let R be a semiprime PI ring of minimal degree $2n$ which is affine, with s generators, over some subring K of its centre. Then:*
(i) $R \hookrightarrow M_n(A)$, where A is a commutative affine K-algebra with $n^2 s$ generators;
(ii) if K is Noetherian then R has finitely many minimal primes.

Proof. (i) Suppose that $R = K[r_1, \ldots, r_s]$ and that $r_k = (a_{ij}^k)$ is the matrix representation of r_k in $M_n(A)$ given by 4.2. Then $R \hookrightarrow M_n(K[a_{ij}^k])$.
(ii) Evidently $M_n(K[a_{ij}^k])$ is Noetherian and so satisfies the a.c.c. on annihilator ideals. Therefore the same is true of R. Hence 2.2.15 shows that R has finitely many minimal primes. □

By passing to $R/N(R)$, it is clear that (ii) is valid when R is not necessarily semiprime. This is the first appearance of a phenomenon that recurs in this chapter, namely that being affine over a central Noetherian ring K gives R many of the properties of a Noetherian ring; see 9.13, for example.

4.5 It is not, however, true that every PI ring can be embedded in a matrix ring over a commutative ring. To see this, let k be a field of characteristic zero and R be the **exterior algebra** on a countable dimensional vector space over k with basis $e_1, e_2, \ldots$. Thus R has generators $e_1, e_2, \ldots$ and relations $e_i e_j = -e_j e_i$ for all i, j (which imply, of course, $e_i^2 = 0$ for all i).

Proposition. (i) *The exterior algebra R does not satisfy s_n for any n.*
(ii) *R cannot be embedded in a matrix ring over a commutative ring.*
(iii) *R is a PI ring, satisfying the identity $[[x,y],z]$.*

Proof. (i) $s_n(e_1,\ldots,e_n) = \sum \text{sgn}\,\sigma\,\text{sgn}\,\sigma e_1 e_2 \cdots e_n = n!\, e_1 \cdots e_n \neq 0$.
(ii) Clear from (i) and 1.13(ii).
(iii) If $w = e_{i(1)} \cdots e_{i(r)}$ is a monomial then $we_j = 0$ if $j \in \{i(1),\ldots,i(r)\}$, and otherwise $we_j = (-1)^r e_j w$. Thus w is central in R if and only if r is even.

We claim now that if $u,v \in R$ then $[u,v]$ is central—noting that this will complete the proof. To check this it is enough to consider the case when u,v are both monomials. If either one has even length then $[u,v] = 0$; and if both have odd length then $[u,v] = 2uv$ which has even length, so is central. □

4.6 There are some interesting consequences of 4.2, the first of which suggests, correctly, that prime PI rings might always be Goldie rings.

Corollary. *A prime PI ring R satisfies the a.c.c. on annihilators.*

Proof. Clear from 4.2. □

The proof that R is necessarily a Goldie ring is deferred to §6 where a more precise result will be obtained.

4.7 Corollary. (i) *A semiprime PI ring of minimal degree d satisfies s_d.*
(ii) *If R is a semiprime PI ring then so too is $M_t(R)$.*

Proof. (i) This follows from 3.3 and 4.2.
(ii) Since $R \hookrightarrow M_n(A)$ then $M_t(R) \hookrightarrow M_{nt}(A)$. □

4.8 Both (i) and (ii) above have extensions to non-semiprime rings.

Theorem. *If R is a PI ring and $t \in \mathbb{N}$, then $M_t(R)$ satisfies $(s_d)^n$ for some n,d. In particular:*
(i) *every PI ring satisfies some power of a standard identity;*
(ii) *if R is a PI ring then so too is $M_t(R)$.*

Proof. It is evident that (i) and (ii) are consequences of the main result; so we need only consider that. Let N be the prime radical of R. Since prime ideals of $M_t(R)$ are precisely of the form $M_t(P)$ for $P \in \text{Spec}\,R$ (by 3.6.2, 3.6.3 for example) it is clear that $M_t(N)$ is the prime radical of $M_t(R)$.

By 4.7, $M_t(R)/M_t(N)$ must satisfy s_d for some d. Thus if $\alpha = (a_1,\ldots,a_d) \in M_t(R)^d$ then $s_d(\alpha) \in M_t(N)$ which is nil; hence $s_d(\alpha)^n = 0$ for some $n = n(\alpha)$. It is enough to show that an n can be chosen, independent of α, with this property.

13.5.2 Central polynomials

For each $\alpha \in M_t(R)^d$, we choose a copy R_α of R, and set $T = \prod_\alpha R_\alpha$. Evidently T satisfies any identity satisfied by R; so T is a PI ring. Note also that

$$(M_t(T))^d \simeq \prod_\alpha (M_t(R_\alpha))^d.$$

Let $(b_1, \ldots, b_d) \in (M_t(T))^d$ correspond to the element $\prod \alpha \in \prod (M_t(R_\alpha))^d$. The argument above, applied to T, ensures that $s_d(b_1, \ldots, b_d)^n = 0$ for some n. The component of this expression in $M_t(R_\alpha)$ is $s_d(\alpha)^n$; so this fixed value of n is as required. □

4.9 We note some consequences of 4.8.

Corollary. *Let R be a PI ring.*
(i) If S is an extension ring of R with S_R finitely generated then S is a PI ring.
(ii) If M_R is finitely generated then $\mathrm{End}\, M$ is a PI ring.

Proof. In each case 8.2.8 demonstrates that the ring concerned is a subquotient of $M_t(R)$ for some t. □

4.10 A result of [Regev 72], or see [Rowen 80, 6.1.1], shows that if R, S are PI rings then so too is $R \otimes_\mathbb{Z} S$. This appears to require a considerable combinatorial proof. We content ourselves here with a more special result.

Corollary. *If R, S are PI k-algebras over a field k and R is semiprime, then $R \otimes_k S$ is a PI k-algebra.*

Proof. We know, by 4.2, that $R \hookrightarrow M_n(A)$ for some commutative ring A. Then

$$R \otimes_k S \hookrightarrow M_n(A) \otimes_k S \simeq A \otimes_k M_n(k) \otimes_k S \simeq A \otimes_k M_n(S).$$

By 4.8, $M_n(S)$ is a PI ring; and hence, by 1.11, its central extension $A \otimes_k M_n(S)$ is a PI ring too. □

§5 Central Polynomials

5.1 It is clear from 4.5 that the exterior algebra R has the property that $ab - ba$ is central for all $a, b \in R$, but is not always zero. This will be interpreted here as showing that R has a 'central polynomial'. Indeed the point of this section is that a similar 'central polynomial' exists for any matrix ring $M_n(A)$ over a commutative ring A.

5.2 First some terminology is required. Let R be a ring and let $g \in \mathbb{Z}\langle x \rangle$, say $g = g(x_1, \ldots, x_t)$. Each element $g(r_1, \ldots, r_t)$, with $r_i \in R$, is called an **evaluation** of g; the set of all evaluations is denoted by $g(R)$; and the additive subgroup of R they generate is $g(R)^+$.

If $g = a + f$, where $a \in \mathbb{Z}$ and f is a polynomial identity of R, then it is automatic that $g(R) \subseteq C$, the centre of R. Any *other* polynomial g such that $g(R) \subseteq C$ is called a **central polynomial** for R; it differs from the trivial examples above by having at least two distinct evaluations.

5.3 It was noted, in 5.1, that $g = x_1 x_2 - x_2 x_1$ is a central polynomial for the exterior algebra. Our main interest here, however, is in the case when $R = M_n(A)$ for some commutative ring A.

Example. $(x_1 x_2 - x_2 x_1)^2$ is a central polynomial for $R = M_2(A)$.

Proof. Let $r_1, r_2 \in R$ and note that $\mathrm{trace}(r_1 r_2 - r_2 r_1) = 0$. However, for any $r \in R$, the Cayley–Hamilton theorem asserts that $r^2 - (\mathrm{trace}\, r)r + (\det r)1 = 0$. In particular, $(r_1 r_2 - r_2 r_1)^2$ is a scalar; so every evaluation of $(x_1 x_2 - x_2 x_1)^2$ is central. Taking $x_1 = e_{12}$ and $x_2 = e_{21}$ shows it is non-trivial. $\square$

5.4 The process of obtaining central polynomials for larger sizes of matrix rings involves multilinear polynomials of a special type. Recall, from 1.12, that the standard polynomials s_d are alternating. A more general notion is that of being alternating in a certain subset of the variables. More precisely, if $f \in \mathbb{Z}\langle x \rangle$ has the property that the interchange of any two of the variables $x_1, \ldots, x_t$ changes f to $-f$, then f will be called t-**alternating.** If

$$f = f(x_1, \ldots, x_t, x_{t+1}, \ldots, x_{t+u})$$

it is sometimes convenient to rewrite x_{t+i} as y_i, getting

$$f = f(x_1, \ldots, x_t, y_1, \ldots, y_u),$$

thus distinguishing the alternating part.

5.5 We now wish to describe some t-alternating multilinear polynomials. Let t be an integer with $t \geq 2$. We define

$$c_{2t-1}(x_1, \ldots, x_t, y_1, \ldots, y_{t-1}) = \sum_{\sigma \in S_t} \mathrm{sgn}\,\sigma\, x_{\sigma(1)} y_1 x_{\sigma(2)} y_2 \cdots y_{t-1} x_{\sigma(t)}$$

and

$$c_{2t}(x_1, \ldots, x_t, y_1, \ldots, y_t) = (c_{2t-1}(x_1, \ldots, y_{t-1})) y_t.$$

These polynomials, which are obtained from the standard polynomial s_t by interspersing extra variables, are called the **Capelli polynomials**.

5.6 Lemma. (i) c_{2t-1} and c_{2t} are multilinear and t-alternating;

(ii) $$c_{2t+1} = \sum_{i=1}^{t+1} (-1)^{t+1-i} c_{2t}(x_1, \ldots, \hat{x}_i, \ldots, y_t) x_i.$$

Proof. (i) Clear.
(ii) Straightforward (cf. 3.3). $\square$

13.5.8   Central polynomials   461

5.7 Proposition. *If $t = n^2$ and $R = M_n(A)$ for some commutative ring A, then $c_{2t}(R)$ contains all the matrix units of R and $c_{2t}(R)^+ = R$.*

Proof. First substitute for $x_1, \ldots, x_t$ in c_{2t-1} the matrix units

$$e_{11}, e_{12}, \ldots, e_{1n}, e_{21}, \ldots, e_{nn},$$

and for $y_1, \ldots, y_{t-1}$ the matrix units

$$e_{11}, e_{21}, \ldots, e_{n1}, e_{12}, \ldots, e_{n-1\,n}.$$

The only monomial in c_{2t-1} to give a nonzero evaluation is $x_1 y_1 x_2 \cdots x_{t-1} y_{t-1} x_t$ since, in a nonzero evaluation, the values of y_i and y_{i+1} determine the value of the letter lying between them. Thus $e_{1n} \in c_{2t-1}(R)$. By symmetry, $e_{ij} \in c_{2t-1}(R)$ for all $i \neq j$ and then it is clear that all matrix units are in $c_{2t}(R)$. Finally, since $c_{2t}(R)$ is A-multilinear, it follows that $c_{2t}(R)^+ = R$.   □

5.8 We now investigate several identities concerning such polynomials.

Proposition. *Let $f = f(x_1, \ldots, x_t, y_1, \ldots, y_u)$ be multilinear and t-alternating, let $R = M_n(A)$ with $t = n^2$, let β be an A-linear transformation of the n^2-dimensional free A-module R, and let $v_1, \ldots, v_t, w_1, \ldots, w_u \in R$. Then:*
  (i) $f(\beta v_1, \ldots, \beta v_t, w_1, \ldots, w_u) = \det \beta \, f(v_1, \ldots, w_u)$;
  (ii) $f((\lambda I - \beta) v_1, \ldots, (\lambda I - \beta) v_t, w_1, \ldots, w_u) = \det(\lambda I - \beta) f(v_1, \ldots, w_u)$, where λ is a central indeterminate over A;
  (iii) *if* $\det(\lambda I - \beta) = \sum_{i=0}^{t} (-1)^i a_i \lambda^{t-i}$, where $a_i \in A$ and $a_0 = 1$, then for $1 \leq k \leq t$,

  $$a_k f(v_1, \ldots, w_u) = \sum f(\beta^{i(1)} v_1, \ldots, \beta^{i(t)} v_t, w_1, \ldots, w_u),$$

  where the summation is over $\{i(1), \ldots, i(t)\}$ such that $\sum_j i(j) = k$ and each $i(j)$ is either 0 or 1;
  (iv) $(\operatorname{trace} \beta) f(v_1, \ldots, w_u) = \sum_{i=1}^{t} f(v_1, \ldots, \beta v_i, \ldots, w_u)$;
  (v) *if $r, s \in R$ then*

  $$(\operatorname{trace} r)(\operatorname{trace} s) f(v_1, \ldots, w_u) = \sum_{i=1}^{t} f(v_1, \ldots, r v_i s, \ldots, w_u).$$

Proof. (i) It is easy to see that it is enough to prove this in the case that $\{v_1, \ldots, v_t\}$ is a basis for R over A. So then $\beta v_j = \sum_{i=1}^{t} b_{ij} v_i$ with $b_{ij} \in A$ and then

$$\begin{aligned}
&f(\beta v_1, \ldots, \beta v_t, w_1, \ldots, w_u) \\
&= f(\sum b_{i1} v_i, \ldots, \sum b_{it} v_i, w_1, \ldots, w_u) \\
&= \sum_{\sigma \in S_t} b_{\sigma(1)1} \cdots b_{\sigma(t)t} f(v_{\sigma(1)}, \ldots, v_{\sigma(t)}, w_1, \ldots, w_u) \\
&= \sum_\sigma \operatorname{sgn} \sigma \, b_{\sigma(1)1} \cdots b_{\sigma(t)t} f(v_1, \ldots, v_t, w_1, \ldots, w_u) \\
&= \det \beta \, f(v_1, \ldots, v_t, w_1, \ldots, w_u).
\end{aligned}$$

(ii) This follows from (i) by considering $\lambda I - \beta$ as an $A[\lambda]$-linear transformation of $M_n(A[\lambda])$.

(iii) Compare the coefficients of λ^{t-k} in (ii).

(iv) Take $k = 1$ in (iii).

(v) We apply (iv) with β being the map $x \mapsto rxs$ for $x \in R$. If $r = (r_{ij})$, $s = (s_{ij})$, and $\{e_{ij}\}$ are the usual matrix units in R then the coefficient of e_{ij} in βe_{ij} is $r_{ii}s_{jj}$. Hence trace $\beta = \sum_{i,j} r_{ii}s_{jj} = (\operatorname{trace} r)(\operatorname{trace} s)$. □

5.9 Continuing with the notation of 5.8, for each i write

$$f(x_1, \ldots, x_t, y_1, \ldots, y_u) = \sum_j f'_{ij} x_i f''_{ij},$$

where

$$f'_{ij} = f'_{ij}(x_1, \ldots, \hat{x}_i, \ldots, y_u) \quad \text{and} \quad f''_{ij} = f''_{ij}(x_1, \ldots, \hat{x}_i, \ldots, y_u).$$

Let x' be a further indeterminate, $\nabla_i f = \sum_j f''_{ij} f'_{ij} x' x_i$ and $\nabla f = \sum_i \nabla_i f$; so $\nabla f = \nabla f(x_1, \ldots, x_t, y_1, \ldots, y_u, x')$.

Proposition. $\nabla f(v_1, \ldots, v_t, w_1, \ldots, w_u, r) = (\operatorname{trace} r)\operatorname{trace}(f(v_1, \ldots, w_u))$.

Proof. Let

$$r'_{ij} = f'_{ij}(v_1, \ldots, \hat{v}_i, \ldots, w_u),$$
$$r''_{ij} = f''_{ij}(v_1, \ldots, \hat{v}_i, \ldots, w_u),$$

and let $s \in R$. By 5.8(v),

$$(\operatorname{trace} r)(\operatorname{trace} s) f(v_1, \ldots, w_u) = \sum_{i=1}^t f(v_1, \ldots, rv_i s, \ldots, w_u) = \sum_{i=1}^t \sum_j r'_{ij} rv_i s r''_{ij}.$$

Therefore

$$\operatorname{trace}[(\operatorname{trace} r)(\operatorname{trace} s) f(v_1, \ldots, w_u)] = \sum_{i=1}^t \sum_j \operatorname{trace}(r'_{ij} rv_i s r''_{ij})$$
$$= \sum_i \sum_j \operatorname{trace}(r''_{ij} r'_{ij} rv_i s).$$

Hence

$$\operatorname{trace}([(\operatorname{trace} r)\operatorname{trace}(f(v_1, \ldots, w_u)) - \nabla f(v_1, \ldots, v_t, w_1, \ldots, w_u, r)]s) = 0.$$

Taking s to be each matrix unit in turn shows that the claimed result holds. □

5.10 Suppose that f satisfies one further condition, namely that $f(R)^+ = R$. Then 5.9 makes clear that ∇f is a central polynomial with $\nabla f(R)^+ = A$.

Corollary. ∇c_{2t} is a central polynomial for $M_n(A)$, where $t = n^2$ and A is any commutative ring. Furthermore, $\nabla c_{2t}(M_n(A))^+ = A$.

Proof. This follows readily from the remarks above, together with 5.6 and 5.7. □

5.11 Whilst ∇c_{2t} is evidently multilinear it is not clear whether it is t-alternating. To arrange this extra property, we define g_n to be the polynomial obtained from ∇c_{2t} by replacing x_1 throughout by $c_{2t}(u_1,\ldots,u_t,v_1,\ldots,v_t)$, these being further indeterminates. This is the polynomial we have been seeking.

Corollary. $g_n(u_1,\ldots,u_t,v_1,\ldots,v_t,x_2,\ldots,x_t,y_1,\ldots,y_t,x')$ is a t-alternating multilinear central polynomial for $M_n(A)$, where A is any commutative ring and $t = n^2$; and $g_n(M_n(A))^+ = A$.

Proof. Clear, using 5.7. □

In later applications the different indeterminates will not be distinguished as above, except that the first t are alternating.

5.12 We end this section by proving some simple results concerning any n^2-alternating multilinear central polynomial f of $M_n(A)$, for all commutative rings A. These will, of course, be applied to g_n.

Proposition. *Let f be as above. Then:*
(i) $f(M_n(A))^+ = A$;
(ii) *if $m < n$ then $f(M_m(A)) = 0$.*

Proof. (i) Note first that $f(M_n(A))^+$ is an ideal, I say, of A. If $I \neq A$ then A/I is a (nonzero) commutative ring yet $f(M_n(A/I)) = 0$. This contradicts f being a central polynomial for $M_n(A/I)$.
(ii) Consider the embedding, not preserving 1, of $M_m(A)$ into the top left-hand corner of $M_n(A)$. Then $f(M_m(A))$ must contain only scalar matrices of $M_n(A)$ belonging to $M_m(A)$—namely 0. □

5.13 There is another explanation of (ii) above in terms of bases for $M_n(A)$ over A. Suppose $f = f(x_1,\ldots,x_d)$, where $d \geq t = n^2$.

Lemma. *Let $v_1,\ldots,v_t \in M_n(A)$ and let Λ be a generating set for $M_n(A)$ over A.*
(i) *If $\{v_1,\ldots,v_t\}$ is a basis for $M_n(A)$ over A then there exist $v_{t+1},\ldots,v_d \in \Lambda$ such that $f(v_1,\ldots,v_d) \neq 0$.*
(ii) *If some v_i is a linear combination of $\{v_1,\ldots,\hat{v}_i,\ldots,v_t\}$ then $f(v_1,\ldots,v_d) = 0$ for all choices of $v_{t+1},\ldots,v_d \in M_n(A)$.*

Proof. Clear. □

5.14 Now we define a new polynomial, namely

$$f'(x_0, x_1, \ldots, x_d) = \sum_{i=0}^{t}(-1)^i f(x_0, x_1, \ldots, \hat{x}_i, \ldots, x_d) x_i.$$

Lemma. (i) f' is multilinear and $t+1$-alternating.
(ii) f' is a polynomial identity for $M_n(A)$.

Proof. (i) Clear from the definition.
(ii) Clear from (i) and 1.10. □

5.15 Corollary. *If $v_0, v_1, \ldots, v_d \in M_n(A)$, then*

$$f(v_1, \ldots, v_d) v_0 = \sum_{i=1}^{t}(-1)^{i-1} f(v_0, v_1, \ldots, \hat{v}_i, \ldots, v_d) v_i.$$

Proof. By 5.14, $f'(v_0, v_1, \ldots, v_d) = 0$. □

§6 Semiprime Rings and Central polynomials

6.1 The fact that a semiprime PI ring R embeds in a matrix ring over a commutative ring suggests that the results of §5 should have application to R. That is the point of this section. It will be shown that R has a central polynomial and that each nonzero ideal of R contains nonzero central elements. This leads to results connecting the properties of R and of its centre.

However, the major result proved here is that each prime PI ring is an order in a central simple algebra. The fact that this result fails to extend to semiprime PI rings causes us to concentrate our attention on the prime case, with most of the results mentioned above being proved first in that case.

6.2 We start with an elementary result.

Proposition. *Let A be a commutative ring and let R be a subring of $M_n(A)$ such that $RA = M_n(A)$. If f is a multilinear central polynomial for $M_n(A)$ then f is a central polynomial for R.*

Proof. Note first that $f(R) \subseteq f(M_n(A)) \cap R \subseteq Z(R)$, the centre of R. However, f cannot be constant on R since then, using 1.11, it would also be constant on the central extension $M_n(A)$. Thus f is a central polynomial for R. □

6.3 Corollary. (i) *Every central simple algebra R has a central polynomial, namely g_n where $n = \text{PI deg } R$.*
(ii) *Each classical order in the central simple algebra R has this same central polynomial.*

13.6.6   *Semiprime rings and central polynomials*   465

Proof. (i) By 3.5, $R \subseteq M_n(H)$ for some splitting field H and also $M_n(H) = RH$. The result is now clear from 5.11 and 6.2.

(ii) Clear from the definition, 5.3.5, using (i) and 6.2. □

6.4 This, together with 3.8, shows that each primitive PI ring has a central polynomial. We now take the first step towards extending this result to prime and semiprime rings.

Theorem. *Let R be a semiprime PI ring with centre C and let $0 \neq I \triangleleft R$. Then, for some m, $0 \neq g_m(I) \subseteq I \cap C$.*

Proof. Note first that $I[x]$ is a nonzero ideal of the semiprimitive PI ring $R[x]$ and that $C[x]$ is the centre of $R[x]$. So if $0 \neq g_m(I[x]) \subseteq I[x] \cap C[x]$ then the result follows.

This means that we may as well suppose that R is semiprimitive. If P is a primitive (and hence maximal) ideal of R then the image of I in R/P is either R/P or zero. Now R/P has g_m as a central polynomial for some m. The values of m, as P varies, are bounded; for $R \hookrightarrow M_d(A)$ for some d and some commutative ring A, and so, by 5.12, $g_m(R) = 0$ whenever $m \geqslant d + 1$.

Thus there is a largest value of m such that g_m is a central polynomial for some primitive factor R/P in which I has a nontrivial image. It follows directly that $0 \neq g_m(I) \subseteq I \cap C$. □

We note that this result is valid even if R has no 1. One can see that either by checking that the existence of a 1 was not needed in the earlier results or else, more easily, by adjoining a 1 to R as in 2.3.9 and noting that each ideal of R is also one of R^1.

6.5 We now turn to the definitive results for prime PI rings.

Theorem. (Posner's theorem) *Let R be a prime PI ring with centre C. Let $\mathscr{S} = C \backslash \{0\}$, $Q = R_\mathscr{S}$, and $Z = C_\mathscr{S}$, the quotient field of C. Then Q is a central simple algebra with centre Z, R is an order in Q and $Q = RZ$.*

Proof. Since C is an integral domain, $R_\mathscr{S}$ exists. Moreover, by 6.4, each nonzero ideal of R meets $\mathscr{S}$. Thus it is clear from 2.1.16 that Q is simple. However, Q is a central extension of R and so is a PI ring. Hence 3.8 shows that Q is a central simple algebra. It is clear that Z is its centre and that $Q = RZ$; and it is easily verified that R is an order in Q. □

6.6 Corollary. (i) *A prime PI ring is a bounded Goldie ring.*

(ii) *A ring R is a prime PI ring if and only if it is an order in a central simple algebra.*

(iii) *A Noetherian PI ring R is FBN and* $K(R) = \dim R$.

Proof. Clear, using 6.4.8. □

6.7 If R, Q are as in 6.5, we extend the definition of **PI degree** by setting PI deg R = PI deg Q.

Corollary. *Let R be a prime PI ring with* PI deg $R = n$. *Then*:
(i) $g_n(I) \neq 0$ *for each* $0 \neq I \lhd R$;
(ii) *if* $\mathcal{T} = g_n(R)\setminus\{0\}$ *then* $Q = R_{\mathcal{T}}$;
(iii) g_n *is a central polynomial for R, but* g_m *is a polynomial identity for R if* $m > n$;
(iv) s_{2n} *is an identity for R, but* s_{2n-1} *is not*;
(v) $R \hookrightarrow M_n(H)$ *for some field H, but* $R \not\hookrightarrow M_m(A)$ *for any commutative ring A if* $m < n$.

Proof. These are straightforward consequences of the earlier results. □

6.8 Many of these results extend easily to the semiprime case.

Corollary. *Let R be a semiprime PI ring. Then there is an integer n such that if* $m > n$ *then* g_n *is a central polynomial,* g_m *and* s_{2n} *are identities and* s_{2n-1} *is not an identity.*

Proof. Choose the largest n such that n = PI degree R/P for some minimal prime P of R. □

6.9 Any infinite direct product of fields demonstrates that a semiprime PI ring need not be a Goldie ring. However, 6.5 together with 3.2.5 gives the next result.

Corollary. *If a semiprime PI ring R has only finitely many minimal primes then R is a Goldie ring.* □

This applies, in particular, to any semiprime PI ring which is affine over a Noetherian subring of its centre, by 4.4.

6.10 Next, another result special to the prime case.

Theorem. *Let R be a prime PI ring with centre C and with* PI deg $R = n$. *Then there is a C-submodule F of R which is free of rank* n^2 *together with an R-monomorphism* $\varphi : R_R \to R_R$ *such that* $\varphi(R) \subseteq F$.

Proof. Let H be a splitting field for Q, the quotient ring of R; so $RH = M_n(H)$. We apply 5.15, with $f = g_n$, by choosing $v_1, \ldots, v_d \in R$ so that $\{v_1, \ldots, v_t\}$ is a basis for $M_n(H)$ over H and $f(v_1,\ldots,v_d) \neq 0$, this being possible by 5.13. Then, if $c = f(v_1,\ldots,v_d)$, we see from 5.15 that $cR \subseteq \bigoplus_{i=1}^{t} Cv_i$. Evidently $R \simeq cR$ as a right R-module. □

We note that the composite map

$$R \hookrightarrow F \simeq C^t \hookrightarrow R^t$$

is also a monomorphism of (R, R)-bimodules, since c is central.

6.11 We recall that 6.5, 6.6 and 6.10 were quoted and used in Chapter 5—see 5.3.9 and 5.3.10. Another result, akin to 5.3.14, is an immediate consequence of 6.10 together with 10.1.11.

Proposition. *If R is a prime PI ring with centre C then the following are equivalent:*
(i) *C is Noetherian;*
(ii) *R_C is finitely generated and R is right Noetherian.* □

6.12 We note that 5.3.7 (iii) demonstrates that the centre of a Noetherian prime PI ring need not be Noetherian.

6.13 One can extend 6.11 to the semiprime case. First recall that if R is a semiprime PI ring with finitely many minimal primes, $P_1,\ldots,P_m$ say, then not only is R and each R/P_i a Goldie ring but also, by 3.2.4, there are central idempotents e_i of $Q(R)$ such that $R \subseteq \bigoplus e_i R$ and $e_i R \simeq R/P_i$. Furthermore, each ideal $e_i R \cap R$ of R must, by 6.4, contain a nonzero central element, a_i say, which is thus regular in $e_i R$. It follows that $a = \sum a_i$ is a central regular element of R. Note that $a(\bigoplus e_i R) \subseteq R$.

6.14 Corollary. *If R is a semiprime PI ring with centre C then the following are equivalent:*
(i) *C is Noetherian;*
(ii) *R_C is finitely generated and R is right Noetherian.*

Proof. Once again 10.1.11 shows that (ii)⇒(i). So suppose C is Noetherian. Since each ideal of R meets C, R must have finite uniform dimension as an (R, R)-bimodule. Therefore, by 2.2.15, R has finitely many minimal primes. Let C_i be the centre of $e_i R$, using the notation of 6.13. Then $a(\bigoplus C_i) \subseteq C$; so clearly each C_i is Noetherian. Therefore, by 6.11, $e_i R$ is finitely generated over C_i and is Noetherian. The result now follows. □

6.15 Of course, in the circumstances of 6.14, it follows also that R is left Noetherian. This, however, is part of a general phenomenon.

Theorem. *If R is a semiprime PI ring then the following are equivalent:*
(i) *R satisfies the a.c.c. for ideals;*
(ii) *R is right Noetherian;*
(iii) *R is left Noetherian.*

Proof. Evidently it is enough to prove that (i) ⇒ (ii). Furthermore, the argument used in 6.14 demonstrates that it will be enough to prove the result when R is prime.

To do so, we make use of the map $\varphi: R \hookrightarrow C^t \subseteq R^t$ given by 6.10. To show that R is right Noetherian we aim to show that if $s_1, s_2, \ldots$ is a sequence of elements of R then, for $m \gg 0$, $s_m \in \sum_{i=1}^{m-1} s_i R$.

However, R^t inherits from R the a.c.c. on (R, R)-sub-bimodules; therefore, for $m \gg 0$,

$$\varphi(s_m) \in \sum_{i=1}^{m-1} R\varphi(s_i)R.$$

But $\varphi(s_i) \in C^t$ and so $R\varphi(s_i)R = \varphi(s_i)R$ since, as noted after 6.10, φ is a bimodule monomorphism. Therefore $\varphi(s_m) = \sum_{i=1}^{m-1} \varphi(s_i) r_i$ and thus, by 6.10 again, $s_m = \sum s_i r_i$. □

The semiprime hypothesis here is necessary as 1.1.9 shows.

§7 Prime Ideals and Azumaya Algebras

7.1 If R is a prime PI ring and $P \in \operatorname{Spec} R$ then we will see, in 7.2, that PI deg $R/P \leq$ PI deg R. The case where equality occurs is especially important; indeed if this holds for all prime ideals of R then R must be an Azumaya algebra. This class of algebras and these results are outlined next.

7.2 First the result mentioned above.

Lemma. *Let R be a prime PI ring with* PI deg $R = n$ *and let* $P \in \operatorname{Spec} R$.
(i) PI deg $R/P \leq$ PI deg R *with equality if and only if* $g_n(R) \not\subseteq P$.
(ii) *If* PI deg $R/P =$ PI deg R *for all* $P \in \operatorname{Spec} R$ *then* $g_n(R)R = R$.

Proof. This is clear from 6.7 (i) since if $g_m(R/P) \neq 0$ then $g_m(R) \not\subseteq P$. □

7.3 A prime ideal P of a prime PI ring R is **regular** if PI deg $R/P =$ PI deg R. Not all primes are regular: consider, for example

$$R = \begin{bmatrix} \mathbb{Z} & 2\mathbb{Z} \\ \mathbb{Z} & \mathbb{Z} \end{bmatrix} \quad \text{and} \quad P = \begin{bmatrix} 2\mathbb{Z} & 2\mathbb{Z} \\ \mathbb{Z} & \mathbb{Z} \end{bmatrix},$$

where PI deg $R = 2$, PI deg $R/P = 1$.

7.4 There is a straightforward way of removing non-regular primes of a prime PI ring R. Let $C = Z(R)$ and note that if $\mathscr{S}$ is any multiplicatively closed subset of $C \backslash \{0\}$ then $R_{\mathscr{S}}$ exists, is an order in $Q(R)$ and is a prime PI ring. If $\mathscr{S} = \{c^n | n \in \mathbb{N}\}$ we write R_c rather than $R_{\mathscr{S}}$.

Proposition. *Let R be a prime PI ring with PI $\deg R = n$ and let $0 \neq c \in g_n(R)$. Then in R_c each prime ideal is regular.*

Proof. Suppose $P \in \operatorname{Spec} R_c$ is such that PI $\deg R_c/P < n$. Then g_n must be a polynomial identity for R_c/P and so $c \in P$. This is a contradiction, since $c^{-1} \in R_c$. □

This proof applies, of course, to $R_{\mathscr{S}}$ for any $\mathscr{S}$ containing c. This will be used in the next result.

7.5 With R and C as above, let P be a regular prime of R. The set $\mathscr{S} = \{a \in C | a \notin P\}$ gives a localization $R_{\mathscr{S}}$ which we prefer to write as R_P.

Proposition. *With the notation above, R_P is a prime PI ring, PR_P is its Jacobson radical and unique maximal ideal, $R_P/PR_P \simeq Q(R/P)$ and all prime ideals of R_P are regular.*

Proof. The remarks in 7.4 make clear that all primes of R_P are regular since, of course, $g_n(R) \not\subseteq P$. Indeed, it is clear from 6.7 that, if $I \triangleleft R$ and $I \supset P$, then $I \cap \mathscr{S} \neq \emptyset$. This shows that PR_P is the unique maximal ideal of R_P. Thus R_P/PR_P is primitive and hence a central simple algebra. Hence $R_P/PR_P \simeq Q(R/P)$. □

7.6 Next we aim towards Azumaya algebras. For now, let R be any ring, with centre C, and let $E = \operatorname{End} R_C$; then R is an (E, C)-bimodule. If $a, b \in R$ then the map $R \to R$ via $r \mapsto arb$ belongs to E. Thus there is a ring homomorphism $\vartheta: R \otimes_C R^{\operatorname{op}} \to E$ given by

$$a \otimes b \mapsto (r \mapsto arb).$$

The ring R is an **Azumaya algebra**, over its centre C, provided:
(i) R_C is finitely generated and projective; and
(ii) $\vartheta: R \otimes_C R^{\operatorname{op}} \to E$ is an isomorphism.

7.7 We note some immediate examples and consequences.

Proposition. *(i) Any Azumaya algebra is a PI ring.*
  (ii) Each central simple algebra R is an Azumaya algebra.
  (iii) If A is a commutative ring then $M_n(A)$ is an Azumaya algebra.

Proof. (i) The fact that R_C is finitely generated means that 1.13 applies.

(ii) By 9.6.9, $R \otimes R^{op}$ is simple, and ϑ is not trivial; hence ϑ is injective. Counting the dimensions over C make clear that ϑ is also surjective. The rest is trivial.

(iii) Consideration of matrix units makes clear that ϑ is an isomorphism, and evidently $M_n(A)$ is finitely generated and projective. □

7.8 As this result suggests, the theory of Azumaya algebras is a natural extension of the theory of central simple algebras. Here we will restrict ourselves to the facts we require; a fuller account is given in [DeMeyer and Ingraham **70**], where they are termed **central separable algebras**.

7.9 Proposition. *If R is an Azumaya algebra with centre C then there is a $(1,1)$-correspondence between ideals I of R and H of C given by $I \mapsto I \cap C$, $H \mapsto HR$.*

Proof. Since R_C is finitely generated and projective, the dual basis lemma, 3.5.2, shows that $RR^* = E = \operatorname{End} R_C$, where $R^* = \operatorname{Hom}_C(R, C)$. In particular $1 = \sum_{i=1}^n r_i \gamma_i$ for some $r_i \in R$, $\gamma_i \in R^*$. However, $R^* \subseteq E \simeq R \otimes R^{op}$. Identifying R^* with its image in $R \otimes R^{op}$ enables us to write each γ_i as $\sum_j p_{ij} \otimes q_{ij}$ with $p_{ij}, q_{ij} \in R$; thus $\gamma_i(r) = \sum p_{ij} r q_{ij}$ for each $r \in R$. Furthermore, since $1 = \sum r_i \gamma_i(1)$, then $R = \sum R \gamma_i(1) = R \sum C \gamma_i(1)$. However, R_C is projective; so $\sum C \gamma_i(1) = C$ and therefore $1 = \sum b_i \gamma_i(1)$ for some $b_i \in C$.

We now turn to the $(1,1)$-correspondence. First note that if $a \in I$ then $a = \sum r_i \gamma_i(a)$ and yet

$$\gamma_i(a) = \sum p_{ij} a q_{ij} \in RaR \cap C \subseteq I \cap C.$$

Hence $a \in R(I \cap C) = (I \cap C)R$ and so $I = (I \cap C)R$.

Next suppose that $a \in HR \cap C$; so $a = \sum h_j s_j$ with $h_j \in H$, $s_j \in R$. Therefore

$$a = \sum_i b_i \gamma_i(1)a = \sum_i b_i \gamma_i(a)$$
$$= \sum_{i,j} b_i \gamma_i(h_j s_j) = \sum_{i,j} b_i \gamma_i(s_j) h_j \in H.$$

This shows that $HR \cap C = H$.

It follows directly that there is a $(1,1)$-correspondence as claimed. □

Note that this correspondence preserves products and hence primeness.

7.10 Corollary. *If R is an Azumaya algebra with centre C then the following are equivalent:*

(i) *R has a.c.c. on ideals;*

(ii) R is right and left Noetherian;
(iii) C is Noetherian.

Proof. This follows directly from 7.9. □

7.11 Proposition. *Let R be an Azumaya algebra with centre C and let M be a maximal ideal of C. Then R/MR is a central simple algebra with centre C/M.*

Proof. By 7.9, MR is a maximal ideal of R and, by 7.7, R/MR is a PI ring. Thus R/MR is a central simple algebra. If $\bar{C}=C/M$ and $\bar{R}=R/MR$, then $\bar{C} \subseteq Z(\bar{R}) \subseteq \bar{R}$, and $\bar{C}, Z(\bar{R})$ are both fields. It remains to check that $\bar{C} = Z(\bar{R})$.

However, there is a commutative diagram

$$\begin{array}{ccccccc} \bar{R} \otimes_{\bar{C}} \bar{R}^{op} & \xrightarrow{\sim} & (R \otimes_C R^{op}) \otimes_C \bar{C} & \xrightarrow{\sim} & (\text{End } R_C) \otimes_C \bar{C} & \xrightarrow{\sim} & \text{End } \bar{R}_{\bar{C}} \\ \| & & & & & & \| \\ \bar{R} \otimes_{\bar{C}} \bar{R}^{op} & \longrightarrow & \bar{R} \otimes_{Z(\bar{R})} \bar{R}^{op} & \longrightarrow & \text{End } \bar{R}_{Z(\bar{R})} & \hookrightarrow & \text{End } \bar{R}_{\bar{C}} \end{array}$$

from which it follows that $\text{End } \bar{R}_{Z(\bar{R})} \simeq \text{End } \bar{R}_{\bar{C}}$. This ensures that $\bar{R}$ has the same rank over each field and hence $Z(\bar{R}) = \bar{C}$. □

7.12 Continuing with the notation of 7.11, note that $C \backslash M$ is an Ore set; so one can form localizations R_M and C_M. Then R_M is a finitely generated projective module over the commutative local ring C_M and so is free.

Lemma. $R/MR \simeq R_M/MR_M$.

Proof. By 7.11, R/MR is a central simple algebra of which R_M/MR_M is a localization. □

7.13 Again let R be an Azumaya algebra over C. Suppose, for each maximal ideal M of C, that R/MR has the same rank over C/M, say n^2. Then R is called an **Azumaya algebra of rank** n^2. This is equivalent, by 7.12, to R_M being of rank n^2 over C_M for each M.

7.14 We can now give the main result of this section.

Theorem. (Artin–Procesi theorem) *If R is a prime ring then the following conditions are equivalent:*
(i) *R is an Azumaya algebra of rank n^2;*
(ii) *R is a prime PI ring whose prime ideals are all regular, with $\text{PI deg } R = n$;*
(iii) *$g_n(R)R = R$.*

Proof. (i)$\Rightarrow$(ii) By 7.7, R is a PI ring; and, by 7.13, R_M is free of rank n^2 over C_M for each maximal ideal M of R. It follows that $Q(R)$ has rank n^2 over $Q(C)$

and so $Q(R)$ and R have PI degree n. Furthermore, R/MR is of rank n^2 over its centre and so, by 3.6, PI deg $R/MR = n$. Finally, if $P \in \operatorname{Spec} R$, then $P \subseteq MR$ for some M, by 7.9. Then PI deg $R \geqslant$ PI deg $R/P \geqslant$ PI deg R/MR and therefore PI deg $R/P = n$.

(ii) $\Rightarrow$ (iii) See 7.2.

(iii) $\Rightarrow$ (i) Let $t = n^2$ and $g_n = g_n(x_1, \ldots, x_t, y_1, \ldots, y_s)$. Since $g_n(R)R = R$ there exist $a_{ij}, b_{ij}, c_j \in R$ such that

$$1 = \sum_j g_n(a_{1j}, \ldots, a_{tj}, b_{1j}, \ldots, b_{sj})c_j = \sum_j g_n(a_{ij}, b_{ij})c_j$$

for short. Define $h_{ij} \in \operatorname{Hom}_C(R, C)$ by

$$h_{ij}(r) = g_n(r, a_{1j}, \ldots, \hat{a}_{ij}, \ldots, a_{tj}, b_{1j}, \ldots, b_{sj}).$$

By 5.15,

$$g_n(a_{ij}, b_{ij})r = \sum_{i=1}^{t} (-1)^{i-1} h_{ij}(r) a_{ij}$$

and so

$$r = r1 = \sum_j r g_n(a_{ij}, b_{ij})c_j = \sum_j g_n(a_{ij}, b_{ij})r c_j$$
$$= \sum_{i,j} h_{ij}(r)(-1)^{i-1} a_{ij} c_j.$$

Thus $1 = \sum_{i,j} d_{ij} h_{ij} \in RR^* = E$, where $d_{ij} = (-1)^{i-1} a_{ij} c_j$. Hence R_C is finitely generated and projective, by 3.5.2, and $E = \sum Rh_{ij}$.

Let $\vartheta: R \otimes R^{\operatorname{op}} \to E$ be the canonical map. By the definition of h_{ij}, there exist $e_{ijk}, f_{ijk} \in R$ such that, for each $r \in R$,

$$h_{ij}(r) = \sum_k e_{ijk} r f_{ijk} = \vartheta(\sum_k e_{ijk} \otimes f_{ijk})(r).$$

Hence, for any $a \in R$,

$$ah_{ij} = \vartheta(\sum a e_{ijk} \otimes f_{ijk})$$

and so $E = \sum Rh_{ij} = \vartheta(R \otimes R^{\operatorname{op}})$. Thus ϑ is surjective.

Finally, suppose $\vartheta(\sum_p a_p \otimes b_p) = 0$ for some $a_p, b_p \in R$; i.e. $\sum_p a_p r b_p = 0$ for each $r \in R$. Then

$$\sum_p a_p \otimes b_p = \sum_p (\sum_{i,j} d_{ij} h_{ij}(a_p) \otimes b_p) = \sum_{i,j} d_{ij} \otimes \sum_p h_{ij}(a_p) b_p$$
$$= \sum_{i,j} d_{ij} \otimes \sum_{p,k} e_{ijk} a_p f_{ijk} b_p = 0.$$

Thus ϑ is injective and R is Azumaya. It is clear that R has rank n^2. $\square$

7.15 To illustrate the usefulness of 7.14, we now give an application to Noetherian PI rings—using only a weak Noetherian condition.

Proposition. *If R is a PI ring with a.c.c. on ideals, then:*
(i) *each prime ideal of R has finite height;*
(ii) *R has finitely many idempotent prime ideals.*

Proof. (i) We can assume, by Noetherian induction, that the result holds for proper factor rings of R. Since all prime ideals contain the prime radical, we may suppose that R is semiprime and hence Noetherian, by 6.15. In that case 2.2.15 ensures that R has finitely many minimal primes; so we may suppose R to be prime, of PI degree n say. Choose $0 \neq c \in g_n(R)$.

By assumption, the result is true in R/cR. However, it is also true in R_c; for R_c is an Azumaya algebra over its centre which, by 7.10, must be Noetherian. Now each prime ideal in a commutative Noetherian ring has finite height (see [Kaplansky 70] or [Atiyah and Macdonald 69]) and the (1, 1)-correspondence given by 7.9 transfers this property to R_c.

Now let $P \in \operatorname{Spec} R$. If $c \notin P$ then any chain of primes inside P is preserved inside P_c; so height $P \leq$ height P_c. Next suppose $c \in P$. Then 4.1.12 shows that height $P \leq$ height $P/(c) + 1$. In either case, P has finite height.

(ii) As in (i), it is enough to deal with the case when R is prime and Noetherian; and we may suppose the result holds for R/cR where $0 \neq c \in g_n(R)$, as in (i). The (1, 1)-correspondence given by 7.9, together with the fact that a commutative Noetherian domain cannot have proper idempotent ideals other than 0, shows that R_c has no idempotent prime ideals. Thus each idempotent prime ideal of R contains cR. □

§8 Integral Extensions and Prime Rings

8.1 Throughout this section R will denote a PI ring and K a subring of its centre.

Much of the section concerns the case when R is integral over K. For commutative rings, there is a well-known theory which, amongst other things, links the prime ideals of R and K in a fashion similar to that described, in Chapter 10, for finite normalizing extensions. All this theory concerning prime ideals extends to the noncommutative PI case. This is a little surprising in the light of the failure of some basic facts about integrality to extend. However, close variants do extend and act as replacements.

8.2 It will be helpful to recall some of the elementary commutative results which will be needed.

Proposition. *Let $K \subseteq R \subseteq S$ be commutative rings.*
(i) *If M_K is a finitely generated module, $p \triangleleft K$, $\varphi \in \operatorname{End} M$ and $\varphi(M) \subseteq Mp$ then, for some $p_i \in p$,*

$$\varphi^n + p_{n-1}\varphi^{n-1} + \cdots + p_0 = 0.$$

(ii) R is affine over K and integral over K if and only if R_K is finitely generated.
(iii) The elements of R integral over K form a subring of R.
(iv) If S is integral over R, and R over K, then S is integral over K.

Proof. See for example [Atiyah and Macdonald **69**, 2.4, 5.1, 5.3, 5.4]. □

8.3 Before any positive results, consider the following cautionary example.

Example. Let K be any commutative ring and R be the K-subalgebra of $M_2(K[x])$ generated by xe_{12} and xe_{21}. These generators are integral over K but R is not. Indeed, neither the sum nor the product of the generators is integral.

Proof. Here $(xe_{12})^2 = 0 = (xe_{21})^2$ yet the elements $xe_{12}xe_{21} = x^2 e_{11}$, $xe_{21}xe_{12} = x^2 e_{22}$, and $xe_{12} + xe_{21}$ are readily checked not to be integral. □

8.4 This demonstrates how easily such results can fail to extend. However, some facts are immediate.

Lemma. (i) If $r \in R$ is integral and $c \in K$ then cr is integral.
(ii) If R_K is finitely generated then R is affine and integral over K.

Proof. (i) Clear.
(ii) Of course R is affine over K. If $r \in R$, let φ be left multiplication by r; so $\varphi \in \mathrm{End}(R_K)$. By 8.2 (i), with $p = K$,

$$\varphi^n + p_{n-1}\varphi^{n-1} + \cdots + p_0 = 0$$

for some $p_i \in K$. Hence

$$(r^n + p_{n-1}r^{n-1} + \cdots + p_0)R = 0$$

and so

$$r^n + p_{n-1}r^{n-1} + \cdots + p_0 = 0.$$ □

8.5 We now aim towards the converse of 8.4(ii), thus extending 8.2(ii). This depends upon an elementary fact concerning nilpotency.

A ring, without 1, is **locally nilpotent** if every finitely generated subring is nilpotent. It is easily seen, using Zorn's lemma, that any ring S contains a unique ideal A maximal amongst locally nilpotent ideals of S; and that then S/A has no nonzero locally nilpotent ideal. That implies, of course, that S/A is semiprime.

Proposition. If S is a ring, not necessarily with 1, and S is not locally nilpotent, then S has a proper prime ideal.

Proof. With the notation above, $A \neq S$; and so S/A is semiprime. Thus A is an intersection of prime ideals. □

13.8.8   *Integral extensions and prime rings*

8.6 Note that, in the commutative case, to determine that an affine K-algebra R is integral, it is enough to determine that its generators are integral. The result to be obtained relies not only upon the generators being integral but also upon certain monomials in those generators being integral too. The next result helps to delineate these monomials.

Lemma. *Let k be a field, $a_1, \ldots, a_p \in M_n(k)$, and S be the k-subspace of $M_n(k)$ spanned by all monomials of degree at least 1 in the elements a_i. Then S is spanned by those monomials of degree n^2 or less.*

Proof. Note that $\dim S \leq n^2$. Hence, if S_j is the subspace spanned by monomials of degree at most j, then it is clear that $\dim S_j = \dim S$ for some $j \leq n^2$; and so $S_{n^2} = S$. □

8.7 Before discussing integrality we prove a more special result about nilpotency.

Proposition. *Let $R = K[a_1, \ldots, a_p]$ be affine over K. Suppose that the minimal degree of R is d and that all monomials in the a_i, of degree $(d/2)^2$ or less, are nilpotent. Then the ideal $I = \sum a_i R$ is nilpotent.*

Proof. Suppose not. Then the subring, without 1, generated by $\{a_1, \ldots, a_p\}$ is not nilpotent. Thus I, viewed as a ring, is not locally nilpotent and so, by 8.5, contains a prime ideal, P say.

By 6.7, I/P embeds in $M_n(H)$ for some field H with $n \leq d/2$ and $(I/P)H = M_n(H)$. Now 8.6 shows that the monomials of degree at most n^2 span $M_n(H)$. However, each is nilpotent so must have trace 0. This is a contradiction. □

8.8 Next comes the result we have been aiming at.

Theorem. *Let R be K-affine, generated by $\{a_1, \ldots, a_p\}$. Let d be the minimal degree of R and suppose that each monomial in the generators, of degree at most $(d/2)^2$, is integral over K. Then R is a finitely generated K-module.*

Proof. By 1.9 and 1.17 we know that R satisfies some nonzero stable identity of degree d; so by introducing extra generators a_i all equal to zero, it can be arranged that p is large enough for there to be a stable identity in $F = \mathbb{Z}\langle x_1, \ldots, x_p \rangle$. As in 1.17, let I be the ideal of F consisting of all stable identities of R, and let $S = F/I$. There are maps $F \to S \to R$ mapping $x_i \mapsto b_i \mapsto a_i$ say, for each $i \leq p$. Recall from 1.17 that S is a PI ring with minimal degree at most d and that S is graded.

If $w = w(x)$, a word in the generators of F, we will write $w(b)$, $w(a)$ for its image in S, R respectively. We let $\mathscr{W}_j(F)$ be the set of all words in F of degree at most j, with $\mathscr{W} = \bigcup \mathscr{W}_j$; and let $\mathscr{W}_j(S)$, $\mathscr{W}_j(R)$ etc., be their images.

First consider $\mathscr{W}_q(R)$ with $q = (d/2)^2$. By hypothesis, there is an integer m such that each word in $\mathscr{W}_q(R)$ satisfies a monic polynomial of degree at most m.

Now consider S. Let I be the ideal of S generated by $\mathscr{W}(S)$ and J the ideal generated by the mth powers of words in $\mathscr{W}_q(S)$. It follows from 8.7 that I/J is nilpotent, say with $I^n \subseteq J$.

As a final piece of notation, let U_j denote the K-submodule of R generated by $\mathscr{W}_j(R)$. We aim to prove that $U_n = R$, thus establishing the result.

Of course, $\bigcup U_j = R$; so it will suffice to show that, if $j \geq n$ and $U_j = U_n$, then $U_{j+1} = U_n$. Let $w(\boldsymbol{a}) \in \mathscr{W}_{j+1}(R)$ and consider $w(\boldsymbol{b})$: we may suppose that w has degree $j+1 > n$ and so $w(\boldsymbol{b}) \in J$. Thus

$$w(\boldsymbol{b}) = \sum s_i w_i(\boldsymbol{b})^m s_i'$$

with $s_i, s_i' \in S$, $w_i \in \mathscr{W}_q(F)$. Since S is graded and $w(\boldsymbol{b})$ has degree $\leq j+1$, we may assume that each $s_i w_i(\boldsymbol{b})^m s_i'$ is homogeneous of degree at most $j+1$. If r_i, r_i' are the images of s_i, s_i' in R, then

$$w(\boldsymbol{a}) = \sum r_i w_i(\boldsymbol{a})^m r_i'.$$

Now $w_i(\boldsymbol{a})^m$ can be written as a K-linear combination of words of smaller degree, because $w_i(\boldsymbol{a})$ is integral. Consequently, $w(\boldsymbol{a})$ is the image of an element in S of degree at most j; thus $w(\boldsymbol{a}) \in U_j$ as required. □

8.9 We note two immediate consequences. The first extends 8.2(ii), as promised; and the second solves, for PI rings, the Kurosch problem.

Corollary. (i) R is affine and integral over K if and only if R_K is finitely generated.
(ii) R is affine and algebraic over a field k if and only if R_k is finite dimensional. □

8.10 One can also use 8.8 to obtain a result related to 8.2(iii).

Corollary. If R has a set of generators, as a K-algebra, such that all monomials in those generators are integral over K then R is integral over K.

Proof. If $r \in R$ then $r \in R' = K[a_1, \ldots, a_n]$ for some finite subset $\{a_1, \ldots, a_n\}$ of the given generating set. By 8.8, R' is a finitely generated K-module and so, by 8.4, is integral. Hence r is integral. □

8.11 Although 8.10 is sufficient for later application, there is a better result which will be proved next. This needs some results concerning the trace, determinant and characteristic polynomial of a matrix over a field H.

Proposition. (a) If $\{r_1, \ldots, r_{n^2}\}$ is a basis for $M_n(H)$ over H then $\Delta = \det(\operatorname{tr}(r_i r_j)) \neq 0$.
(b) Let K be a subring of H and $s \in M_n(H)$.
   (i) If s is integral over K, then the coefficients of the minimal polynomial and characteristic polynomial of s over H are integral over K.

(ii) *If s is a K-linear combination of elements integral over K then* tr(s) *is integral over K.*

Proof. (a) This is a well-known fact about the discriminant. See [Cohn 77, pp. 201 and 461] for example.

(b) (i) Let $f(x) \in K[x]$ be the monic polynomial satisfied by s, and $g(x) \in H[x]$ the minimal polynomial. Of course, $g(x)|f(x)$ in $H[x]$, so each root of $g(x)$ satisfies $f(x)$ and is integral over K. The coefficients in $g(x)$ are all generated by these roots and therefore, by 8.2(iii), they too are integral over K. Since the roots of the characteristic polynomial are also roots of $g(x)$, the same argument applies there.

(ii) This follows from (i), since the trace is K-linear. □

8.12 Theorem. *Let $R = \sum a_i K$ with each a_i being integral over K. Then R is integral over K.*

Proof. Suppose, to the contrary, that $c \in R$ is not integral over K. Let

$$\mathscr{S} = \{f(c) | f(x) \text{ is monic}, f(x) \in K[x]\}.$$

Then $0 \notin \mathscr{S}$ and so, by Zorn's lemma, there exists an ideal P of R maximal with respect to having $P \cap \mathscr{S} = \emptyset$. By passing to R/P and $K/P \cap K$ we may suppose that R is prime, K is an integral domain and each nonzero ideal of R meets $\mathscr{S}$.

Since R is a prime PI ring, $R \subseteq M_n(H)$ with $RH = M_n(H)$ and $n = \text{PI deg } R$. In particular R contains a basis $r_1, \ldots, r_{n^2}$ of $M_n(H)$ over H. We let Δ be as in 8.11(a). Note, by hypothesis and 8.11(b), that $\text{tr}(r_i r_j)$ is integral over K for each i, j. Hence 8.2(iii) shows that Δ is integral over K.

For any $r \in R$ we can write $rr_i = \sum_j a_{ij} r_j$ with $a_{ij} \in H$ and so $rr_i r_k = \sum_j a_{ij} r_j r_k$. Therefore $\text{tr}(rr_i r_k) = \sum_j a_{ij} \text{tr}(r_j r_k)$. It follows, by Cramer's rule, that $\Delta a_{ij} = \Delta_{ij}$ where, once again, Δ_{ij} is the determinant of a matrix of traces of elements of R. Hence, as before, Δ_{ij} is integral over K. However, since

$$\Delta r r_i = \sum \Delta a_{ij} r_j = \sum \Delta_{ij} r_j,$$

we see that Δr satisfies the polynomial $\det(xI - (\Delta_{ij}))$. The coefficients of this polynomial are all integral over K and hence, by 8.2(iv), Δr is integral over K.

This establishes that each element in ΔR is integral. However, each nonzero ideal of R contains some monic polynomial over K in c, $f(c)$ say. The fact that $f(c)$ is integral ensures that c is integral—which completes the proof. □

8.13 Next comes a result extending 8.2(i).

Lemma. *Let R be integral over K, p be an ideal of K and $a \in pR$. Then a satisfies an equation*

$$a^n + p_{n-1} a^{n-1} + \cdots + p_0 = 0$$

for some $p_i \in p$.

Proof. Let $a = \sum_{i=1}^{m} q_i r_i$ where $q_i \in p$, $r_i \in R$, and let $S = K[r_1, \ldots, r_m]$. By 8.8, S_K is finitely generated and clearly $aS \subseteq pS$. Hence 8.2(i) applies, with φ being left multiplication by a. □

8.14 The relationship between prime ideals of R and K can now be established. The results are more precise than those in Chapter 10, and they form a complete extension of the commutative theory.

Theorem. *Let R be integral over K.*
  (i) (Lying over) If $p \in \operatorname{Spec} K$, then there exists $P \in \operatorname{Spec} R$ such that $P \cap K = p$.
  (ii) (Going up) If $p, q \in \operatorname{Spec} K$ with $p \subset q$ and $P \in \operatorname{Spec} R$ with $P \cap K = p$, then there exists $Q \in \operatorname{Spec} R$ with $P \subset Q$ and $Q \cap K = q$.
  (iii) (Incomparability) If $P, Q \in \operatorname{Spec} R$ with $P \subset Q$ then $P \cap K \subset Q \cap K$.
  (iv) (Going down) Suppose that R is prime and K is integrally closed. Let $p, q \in \operatorname{Spec} K$ with $p \subset q$ and $Q \in \operatorname{Spec} R$ with $Q \cap K = q$. Then there exists $P \in \operatorname{Spec} R$ with $P \subset Q$ and $P \cap K = p$.

Proof. (i) If $a \in pR \cap K$, then 8.13 shows that $a^n \in p$. Hence $pR \cap K = p$. Amongst all ideals I of R such that $I \cap K = p$ there is a maximal one, P say; and it is clear that P is prime.

(ii) This follows from (i) by passing to R/P and K/p.

(iii) As in (ii), we may as well assume that $P = 0$ and so $p = 0$. In that case, R is a prime PI ring, and so Q contains a nonzero central element c say. Let $x^n + a_{n-1}x^{n-1} + \cdots + a_0$ be a monic polynomial of least possible degree satisfied by c. Since c is regular, $a_0 \neq 0$. Hence $a_0 = -(c^n + a_{n-1}c^{n-1} + \cdots + a_1 c) \in Q \cap R$ and thus $Q \cap R \neq 0$.

(iv) The main step in this proof shows that if $\mathscr{S}$ is the multiplicatively closed set
$$\mathscr{S} = \{cr \mid 0 \neq c \in K \setminus p, \, r \in \mathscr{C}_R(Q)\}$$
then $pR \cap \mathscr{S} = \varnothing$.

Suppose, to the contrary, that $cr \in pR \cap \mathscr{S}$. By 8.13, there is a monic polynomial $f(x) \in K[x]$, of degree m say, with nonleading coefficients in p, such that $f(cr) = 0$. Let $g(x)$ be the minimal polynomial for r over Z, the centre of the quotient ring of R. If $\deg g(x) = n$ then $h(x) = c^n g(c^{-1}x)$ is the minimal polynomial for cr. As in 8.11(b) one sees that the coefficients of $g(x), h(x)$ are integral over K. Therefore $g(x), h(x) \in K[x]$. Evidently $h(x) \mid f(x)$ in $Z[x]$ and hence also in $K[x]$; say $f(x) = h(x)k(x)$.

Let $\bar{K} = K/p$ and $\bar{Z}$ be the quotient field of $\bar{K}$. In $\bar{K}[x]$, $\bar{f}(x) = x^m = \bar{h}(x)\bar{k}(x)$. The uniqueness of factorization in $\bar{Z}[x]$ implies that $\bar{h}(x) = x^t$ for some t. Therefore, in $h(x)$, all but the leading coefficient belong to p. Since $c \notin p$ it is then clear that the same is true of $g(x)$. Therefore $r^n \in pR \subseteq Q$; but this contradicts the hypothesis that $r \in \mathscr{C}_R(Q)$.

We have, therefore, proved that $pR \cap \mathcal{S} = \emptyset$. Hence an ideal P of R can be chosen with $P \supseteq pR$ and maximal with respect to having $P \cap \mathcal{S} = \emptyset$. As usual, P is prime. Also $P \subseteq Q$ since otherwise $P \cap \mathscr{C}(Q) \neq \emptyset$ and yet $P \cap \mathscr{C}(Q) \subseteq P \cap \mathcal{S}$. Finally, $P \cap K \subseteq p$ since $K \backslash p \subseteq \mathcal{S}$. □

8.15 This section ends with a consequence which will be used later.

Proposition. *Let R be a Noetherian prime PI ring which is integral over K, and let $P \in \operatorname{Spec} R$ with height $P \geq 2$. Then there is an infinite set of primes $P_i \in \operatorname{Spec} R$ such that $P_i \subset P$, $\bigcap P_i = 0$, height $P_i = 1$, and $\operatorname{PI} \deg R/P_i = \operatorname{PI} \deg R$.*

Proof. First localize with respect to $K \backslash p$, where $p = P \cap K$. The hypotheses on R are preserved; and the result, for R, is an easy consequence of the result for the localized ring.

Thus there is no loss in supposing that p is the unique maximal ideal of K; so p contains only central non-units of R. Let $0 \neq a \in p$. By 4.1.11, P cannot be minimal over a; so $P \supset P' \supseteq (a)$ for some $P' \in \operatorname{Spec} R$, height $P' = 1$.

Suppose for a moment that, as a varies, only finitely many height 1 primes $P_1, \ldots, P_m$ occur. Then $p \subseteq \bigcup_{i=1}^{m} (P_i \cap K)$ and so, by a well-known commutative result, [Atiyah and Macdonald **69**, 1.11], it follows that $p \subseteq P_i \cap K$ for some i. However, that contradicts 8.14(iii). We conclude that infinitely many height 1 primes occur, say P_i, $i \in I$.

Note next that if $\bigcap P_i \neq 0$ then each P_i would be a minimal prime of $\bigcap P_i$. However, $\bigcap P_i$, being a semiprime ideal in a Noetherian ring, has only finitely many minimal primes. Therefore $\bigcap P_i = 0$.

Let $J = \{i \in I \mid \operatorname{PI} \deg R/P_i = \operatorname{PI} \deg R\}$ and $J' = I \backslash J$. Evidently

$$\operatorname{PI} \deg (R/\bigcap \{P_i \mid i \in J'\}) < \operatorname{PI} \deg R,$$

so $\bigcap \{P_i \mid i \in J'\} \neq 0$. Therefore $\bigcap \{P_i \mid i \in J\} = 0$, showing that the set $\{P_i \mid i \in J\}$ is as required. □

§9 The Trace Ring and Maximal Orders

9.1 Throughout this section R will denote a prime PI ring with centre C. A type of closure operation on R will be introduced. This produces the trace ring, which is integral over its centre and is an equivalent order to R in the quotient ring of R; indeed it shares a nonzero ideal with R. It is used here to prove some results, mentioned in Chapter 5, §3, about Dedekind prime rings and hereditary Noetherian rings. In the next section it will be used in the study of prime PI rings affine over a field k.

9.2 The quotient ring of R, Q say, is a central simple algebra of dimension n^2 over its centre Z, where $n = \operatorname{PI} \deg R$. Therefore each $r \in R$ is represented, by left multiplication on Q, as a linear transformation of Q over Z. This has a characteristic polynomial, $f(x)$ say, and it is the coefficients of $f(x)$ which are of interest here. Indeed, we let T be the subring of Z generated, over C, by all such coefficients, as r varies throughout R. It is clear that TR is a subring of Q containing R. This is called the **trace ring**, or **characteristic closure**, of R.

9.3 The other coefficients in the characteristic polynomials can be written in terms of traces provided that $\mathbb{Q} \subseteq C$. Thus in that case one can view T as being generated by the traces. This explains the name.

9.4 The reader should be warned that the definition of trace ring here is not the same as in much of the literature. The other definition uses the reduced characteristic polynomial $g(x)$, this being the characteristic polynomial of r, viewed as an $n \times n$ matrix via a splitting field of Q. In fact $f(x) = (g(x))^n$, so the comments in 9.3 show that when $\mathbb{Q} \subseteq C$ the two trace rings coincide. In general, the one used here is the smaller of the two.

The advantage of adopting the definition given in 9.2 is that then 5.8 can easily be used below.

9.5 We note one advantage of TR.

Proposition. *TR is integral over T.*

Proof. This is clear from 8.10 or 8.12. □

Thus the results of §8 are applicable to TR.

9.6 In using TR in the study of R the evaluations of the central polynomial g_n, as defined in 5.11, provide a link.

Proposition. *Let $0 \neq I \triangleleft R$. Then:*
(i) $0 \neq g_n(I)^+ \triangleleft T$;
(ii) $g_n(I)R$ is a nonzero ideal of both R and TR.

Proof. (i) Let $t = n^2$, $g_n = g_n(x_1, \ldots, x_d)$ with $d > t$, and $v_1, \ldots, v_d \in I$. By 5.8(iii),
$$a_k g_n(v_1, \ldots, v_d) \in g_n(I)^+$$
for each coefficient a_k in the characteristic polynomial of any $a \in R$. Hence $g_n(I)^+ \triangleleft T$.

(ii) This follows directly from (i). □

9.7 Corollary. *(i) R and TR are equivalent orders in Q.*

(ii) There is a $(1,1)$-correspondence between regular primes P of TR and regular primes of R given by $P \mapsto P \cap R$.

Proof. (i) Combine 9.6 with 3.1.6.

(ii) By 3.6.3(iii), there is a correspondence, as described, between the prime ideals of R and TR which do not contain $g_n(R)R$. By 7.2, these are the regular primes. □

9.8 There are circumstances under which $R = TR$.

Proposition. (i) If R is a maximal order in Q then $R = TR$.
(ii) If R is an Azumaya algebra over C then $R = TR$.

Proof. (i) This is clear from 9.7(i).
(ii) By 7.14, $g_n(R)R = R$. This combined with 9.6(ii) shows that $R = TR$. □

9.9 In general, as the next example shows, these rings are not equal.

Example. Let S be the commutative ring $\mathbb{Q}[x, y, y^{-1}]$ and

$$R = \begin{bmatrix} S & xS \\ S & \mathbb{Q}[y] + xS \end{bmatrix} \subseteq M_2(S).$$

Then $T = S$ and $TR = M_2(S) \neq R$.

Proof. The characteristic polynomial of $y^{-1}e_{11} + e_{22}$ demonstrates easily that $y^{-1} \in T$ and so $T \supseteq S$. However, S is integrally closed and so $M_2(S)$ is a maximal order. Therefore, by 9.7(i), $TR = M_2(S)$ and so $T = S$. □

9.10 We next investigate when TR is Noetherian. This requires a well-known result.

Lemma. (Artin–Tate lemma) Let $A \subseteq B \subseteq S$ be rings such that A, B are central subrings of S with S being an affine A-algebra and a finitely generated B-module.
(i) There exists an affine A-subalgebra B' of B such that S is a finitely generated B'-module.
(ii) If either A is Noetherian or B is a direct summand of S_B, then B is a finitely generated B'-module and an affine A-algebra.

Proof. (i) Let $S = A[s_1, \ldots, s_m] = \sum_{j=1}^n u_j B$. Then $s_i = \sum_j u_j b_{ij}$ and $u_i u_j = \sum_k u_k b_{ijk}$ for b_{ij}, $b_{ijk} \in B$. Let $B' = A[b_{ij}, b_{ijk}]$; then $S = \sum_j u_j B'$ as required.

(ii) If A is Noetherian then so is B'. Hence S and B are Noetherian B'-modules and the rest is clear.

On the other hand, if B is a direct summand of S_B then $B_{B'}$ is finitely generated and so is A-affine. □

9.11 Proposition. (i) If R is Noetherian then TR is Noetherian TR_R is finitely generated.

(ii) *Suppose that R is an affine K-algebra with K being a Noetherian subring of C. Then each of TR, T and the centre of TR is a Noetherian affine K-algebra, and TR_T is finitely generated.*

Proof. (i) If $0 \neq c \in g_n(R)$ then $TR \simeq cTR \subseteq R$. The rest is clear.

(ii) Suppose $R = K[r_1, \ldots, r_m]$ and hence $TR = T[r_1, \ldots, r_m]$. By 9.5 and 8.9, TR is a finitely generated T-module. We choose a finite generating set $\{u_j\}$ which includes 1 and each r_i; and then note, by 9.10, that there is an affine K-subalgebra T' of T, necessarily Noetherian, such that $TR_{T'}$ is finitely generated.

Let $R' = \sum_j u_j T'$. By the construction of T', as in 9.10, it is clear that $R \subseteq R' \subseteq TR$. The proof of (i) shows that $TR \simeq cTR \subseteq R'$ for some c; and thus $TR_{T'}$ is finitely generated. The same follows for T and the centre of TR. The rest is clear. □

9.12 The rest of this section illustrates the use of the trace ring. First recall from 9.3.9 that a commutative integral domain C is a G-domain if, using the notation of 7.4, $C_c = Q(C)$, the quotient field, for some $0 \neq c \in C$. In the same vein, a prime PI ring R is a **G-ring** if $R_c = Q(R)$ with $0 \neq c \in C$. Of course, this is equivalent to saying that all nonzero primes of R contain c.

Proposition. *If R is Noetherian and a G-ring then R has only finitely many nonzero primes all of which are maximal.*

Proof. By 9.11 and 9.5, we know that TR is Noetherian and integral over T. However, TR is also a G-ring, of course; so 8.15 implies that nonzero primes of TR must have height 1. Using 9.11 again, we see that $TR = \sum_{i=1}^{m} t_i R$. Thus 10.2.10 can be applied to deduce that nonzero primes of R all have height 1. Since each must therefore be minimal over c, they must be finite in number. □

9.13 The following result is related to 7.15(ii). A result similarly related to 7.15(i) will be proved in 10.10.

Proposition. *If R is a PI ring affine over a Noetherian subring K of its centre, then R has only finitely many idempotent prime ideals.*

Proof. By 4.4, R has only finitely many minimal prime ideals. Therefore it is enough to deal with the case when R is prime with $\text{PI deg } R = n$, say. Let $0 \neq c \in g_n(R)$. Then 7.4 shows that R_c is an Azumaya algebra over its centre and hence, by 9.8, R_c is its own trace ring. However, R_c is affine over K and so, by 9.11(ii), its centre is Noetherian and thus has no nonzero idempotent ideals. By 7.9, the same is true of R_c. Hence every idempotent prime ideal of R contains c. An induction on PI degree completes the argument. □

9.14 The next few results complete the discussions begun in §3 of Chapter 5. The first is an analogue of Noether's characterization of commutative Dedekind domains as Noetherian integrally closed domains of dimension 1.

Theorem. *Let R be a prime PI ring which is Noetherian, a maximal order and has classical Krull dimension 1. Then the centre C of R is a Dedekind domain and R is a maximal classical C-order and a Dedekind prime ring.*

Proof. By 9.5 and 9.8, R is integral over C. Hence, by 8.14, the classical Krull dimension of C must be 1. However, by 5.1.10, C is a Krull domain. These two facts imply that C is a Dedekind domain (e.g. by [Bourbaki **61–65**, Chapter 7, §2] or [Kaplansky **70**, §2.4 Exercise 2]). We can thus apply 6.11 to deduce that R is a finitely generated C-module, and 6.5 to check that $Q = RZ$ with Z the quotient field of C. Thus R is a C-order and so, by 5.3.13 and 5.3.14 is a maximal classical C-order. Then 5.3.16 asserts that R is a Dedekind prime ring. □

9.15 Corollary. *Let R be a prime PI ring which is an Asano order. Then R is a maximal classical C-order over the Dedekind domain C.*

Proof. We know, by 5.2.6, that R is a maximal order and that each ideal is finitely generated. By 6.15, R must be Noetherian; and by 5.2.9, R has classical Krull dimension 1. Therefore 9.14 applies. □

9.16 Of course, 9.15 applies, in particular, to a PI ring which is a Dedekind prime ring. We end this section by extending the result to hereditary Noetherian prime rings.

Theorem. *If R is a hereditary Noetherian prime PI ring, then the centre C is a Dedekind domain and R is a classical C-order.*

Proof. By 7.15, R has only finitely many idempotent primes ideals. Thus 5.6.12 shows that R has a minimal nonzero idempotent ideal and so, by 5.6.8, is obtained from a Dedekind prime ring, S say, by a finite iteration of forming idealizers at generative isomaximal right ideals. It is elementary to check that the centre of R and of S coincide. Therefore, by 9.15, C is a Dedekind domain and S is a C-order. Hence, by 5.3.11 and 5.3.14, R is a classical C-order. □

§10 Affine k-algebras

10.1 Earlier results, like 8.2.14 for instance, have illustrated how well behaved are commutative affine k-algebras over a field k. In this section it will be shown that the same is true for PI rings which are affine k-algebras. In particular, the Nullstellensatz and the catenary property for prime ideals are satisfied.

Throughout the section, k is a field and R is a PI ring with centre C.

10.2 We start, however, by noting some examples which demonstrate serious differences from the commutative case.

First consider the ring R described in 9.9. It is easy to check that R is $\mathbb{Q}$-affine, with generators $\{e_{11}x, e_{11}y, e_{11}y^{-1}, e_{21}, e_{12}x, e_{22}, e_{22}y\}$; and it has already been noted that R is a prime PI ring. However, neither R nor its centre C is Noetherian; for $e_{22}Re_{22} \simeq C = \mathbb{Q}[y] + xS$ and $xS/x^2S \simeq \mathbb{Q}[y, y^{-1}]$ which is not a finitely generated $\mathbb{Q}[y]$-module. Thus C is not Noetherian and, by 1.1.7, R is not Noetherian.

Next consider the ring R described in 5.3.7(iii). This is not k-affine; but it has a k-affine subalgebra S which, like R, is a Noetherian prime PI ring whose centre is not Noetherian, namely

$$S = \begin{bmatrix} A^\sigma + tA[t] & tA[t] \\ tA[t] & A^\tau + tA[t] \end{bmatrix},$$

where $A = k[x, y]$. To see that S is k-affine, note first that $A[t]$ is k-affine and is a finitely generated module over each of $A^\sigma + tA[t]$ and $A^\tau + tA[t]$. Hence, by 9.10, each of these subalgebras is affine. It follows readily that S is k-affine. The same argument as in 5.3.7 yields the other properties of S and its centre.

10.3 In obtaining a Nullstellensatz, the terminology of Chapter 9, §1 is used. Recall that the commutative result asserts that k-affine fields are finite dimensional over k—an immediate consequence being that commutative Artinian k-affine rings are finite dimensional.

Theorem. *Let R be a PI affine k-algebra. Then:*
(i) *if M_R is a simple module then M_k is finite dimensional;*
(ii) *R satisfies the Nullstellensatz;*
(iii) *R is a Jacobson ring;*
(iv) *if R is Artinian then R_k is finite dimensional.*

Proof. (i) By passing to $R/\mathrm{ann}\, M$ we may suppose R to be primitive and hence a central simple algebra over its centre C. By 9.10, C is affine over k and so, being a field, has finite dimension over k. Hence R_k and M_k are finite dimensional too.

(ii) Since M_k is finite dimensional so too is $\mathrm{End}\, M_R$. Thus R has the endomorphism property. Since $R[x]$ satisfies the same hypotheses as R, the same is true of $R[x]$. Hence, by 9.1.6, R satisfies the Nullstellensatz.

(iii) This is clear from (ii) and 9.1.2.

(iv) By (i) each simple composition factor of R_R is finite dimensional; hence so too is R. □

10.4 One can extend 10.3 to cover algebras over commutative Jacobson rings, obtaining a result similar to 9.4.21.

Theorem. *Let K be a commutative Jacobson ring and R be an affine PI K-algebra. Then:*
 (i) *if M_R is a simple module then $\operatorname{ann}_K M$ is a maximal ideal and M is finite dimensional over the field $K/\operatorname{ann}_K M$;*
 (ii) *R satisfies the Nullstellensatz over K;*
 (iii) *R is a Jacobson ring.*

Proof. (i) By passing to factor rings, we may assume that M_R is faithful and so R is primitive. Then R is a central simple algebra, finite dimensional over its centre C, which is a field. By 9.10, C is K-affine and so, by 9.3.10 and 9.4.20, K must be a field. The commutative Nullstellensatz now shows that C_K is finite dimensional. Hence so too are R_K and M_K.
 (ii) We see from (i), applied to $R[x]$, that $R[x]$ satisfies the endomorphism property over K. Hence, by 9.2.6, R satisfies the Nullstellensatz.
 (iii) This is clear from (ii) and 9.1.2. □

10.5 Now we turn to the prime ideal structure. If R is a prime k-algebra and Z is the centre of $Q(R)$ then the **transcendence degree** of R, $\operatorname{tr deg} R$, is defined to be $\operatorname{tr deg} Z$.

Note, as an immediate consequence, that all orders in $Q(R)$ have the same transcendence degree. Therefore, in particular, $\operatorname{tr deg} TR = \operatorname{tr deg} R$.

Note also that since $Z = Q(C)$ then $\operatorname{tr deg} R = \operatorname{tr deg} C$.

10.6 The next result is an extension of 8.2.14, that being the corresponding fact for commutative algebras.

Proposition. *If R is a prime PI affine k-algebra then $\dim R = \operatorname{tr deg} R = \operatorname{GK}(R)$.*

Proof. First consider the case when R is an Azumaya algebra over C, bearing in mind, by 9.10, that C is k-affine. By 8.2.14

$$\dim C = \operatorname{tr deg} C = \operatorname{GK}(C).$$

However, $\operatorname{GK}(R) = \operatorname{GK}(C)$ since R_C is finitely generated; $\operatorname{tr deg} R = \operatorname{tr deg} C$ by definition; and $\dim R = \dim C$ by 7.9. Thus the result holds in this case.

Now consider the general case. Choose $0 \neq c \in g_n(R)$, where $n = \operatorname{PI deg}(R)$, and recall, from 7.4, that R_c is an Azumaya algebra. By definition, $\operatorname{tr deg} R = \operatorname{tr deg} R_c$; and 8.2.13 shows that $\operatorname{GK}(R) = \operatorname{GK}(R_c)$. Also, it is clear that $\dim R \geqslant \dim R_c$ yet, by 8.3.6, $\dim R \leqslant \operatorname{GK}(R)$. □

10.7 This enables us to prove a result related to 7.15(i).

Proposition. *Let R be a PI affine K-algebra, with s generators, where K is a Noetherian subring of C and let m be the minimal degree of R. Then:*

(i) any chain of primes in R sharing the same intersection with K has length at most $(\frac{1}{2}m)^2 s$;
(ii) if $P \in \operatorname{Spec} R$, then $\operatorname{ht} P \leqslant (1 + (\frac{1}{2}m)^2 s)(1 + \operatorname{ht}(P \cap K)) - 1$;
(iii) $\dim R \leqslant (1 + (\frac{1}{2}m)^2 s)(1 + \dim K) - 1$;
(iv) R satisfies the a.c.c. for prime ideals.

Proof. (i) Let $P_1 \subset \cdots \subset P_t$ with $P_i \cap K = p$ for each i, and let $\mathscr{S} = K \backslash p$. Then $(R/P_1)_{\mathscr{S}}$ is an affine algebra, over the field $Q(K/p)$, with s generators, and it has the chain of prime ideals $(P_i)_{\mathscr{S}}$. However, $\dim(R/P_1)_{\mathscr{S}} = \operatorname{GK}(R/P_1)_{\mathscr{S}} \leqslant (\frac{1}{2}m)^2 s$ by 4.4 and 10.6.

(ii), (iii), (iv) These are now clear. □

Using the König graph theorem, one can deduce that R satisfies the a.c.c for semiprime ideals—see [Gordon and Robson 73, 7.7].

10.8 One can also draw some conclusions regarding prime radicals.

Lemma. *Let R be a prime PI ring whose trace ring TR is Noetherian, and let H be an ideal of both R and TR. Then $N_R(H) = N_{TR}(H) \cap R$.*

Proof. Since TR is Noetherian, $N_{TR}(H)/H$ is nilpotent and hence $N_R(H) \supseteq N_{TR}(H) \cap R$. Now consider the embedding
$$\bar{R} = R/(N_{TR}(H) \cap R) \hookrightarrow TR/N_{TR}(H) = \overline{TR}.$$
Since $\overline{TR}$ is a centralizing extension of $\bar{R}$, each nilpotent ideal I of $\bar{R}$ extends to a nilpotent ideal $I\overline{TR}$ of $\overline{TR}$. Therefore $\bar{R}$ has no nilpotent ideals and is semiprime. Thus $N_R(H) = N_{TR}(H) \cap R$. □

10.9 Theorem. (i) *Let R be a prime PI affine K-algebra where K is a Noetherian subring of C. Then, in each factor ring $\bar{R}$ of R, the prime radical is nilpotent.*
(ii) *If, further, K is a Jacobson ring, then $J(\bar{R})$ is nilpotent.*

Proof. (i) We proceed by induction on the classical Krull dimension, $\dim R = 0$ being trivial. Let $0 \neq I \triangleleft R$ and $H = g_n(I)R$. By 9.6, $0 \neq H \triangleleft TR$ and, by 9.11, TR is Noetherian; so 10.8 applies. In particular $N_R(H)^h \subseteq H$ for some h.

Let $P_1, \ldots, P_t$ be the minimal primes of H, these being finite in number, by 4.4. By induction,
$$(N_{R/P_i}(I + P_i/P_i))^{r_i} \subseteq I + P_i/P_i$$
for some r_i, since $\dim R/P_i < \dim R$. Hence $N_R(I)^{r_i} \subseteq I + P_i$ and so
$$\left(\prod_{i=1}^t N_R(I)^{r_i}\right)^h \subseteq \left(\prod_i (I + P_i)\right)^h \subseteq I + \left(\bigcap_i P_i\right)^h \subseteq I.$$

(ii) Clear from (i) and 10.4. □

10.10 This theorem applies, in particular, to $K[X_1,\ldots,X_m]$, the K-algebra of m generic $n \times n$ matrices, since, as noted in 1.21, it is an integral domain. (But see also 11.10.)

10.11 Next comes a preparatory result.

Lemma. *Let $R \subseteq S$ be prime PI affine k-algebras, with S being a central extension of R such that if $0 \neq I \triangleleft S$ then $0 \neq I \cap R$. Then $\operatorname{tr deg} R = \operatorname{tr deg} S$.*

Proof. Note that C is central in S. The hypothesis, together with 6.4, shows that each I, as above, meets C nontrivially. Hence localization of S at $C \backslash \{0\}$ yields $Q(S)$. Thus $Q(S)$ is affine over Z, the centre of $Q(R)$. Moreover, $Q(S)$ is finite dimensional over its centre, Z' say. Hence, by 9.10, Z' is affine over Z. Therefore, by the Nullstellensatz, Z' is finite dimensional over Z and so $\operatorname{tr deg} Z' = \operatorname{tr deg} Z$. □

10.12 We now prove a result, concerning height of primes, which yields the catenary property. Its proof depends upon the commutative case, see [Kaplansky **70**, Theorem 151] or [Zariski and Samuel **60**, Chapter 7, §7].

Theorem. (Schelter's theorem) *Let R be a prime PI affine k-algebra and $P \in \operatorname{Spec} R$. Then $\dim R = \operatorname{ht} P + \dim R/P$.*

Proof. Let $n = \operatorname{PI deg} R$ and $0 \neq c \in g_n(R)$. The proof involves both R_c and TR, the trace ring. Note first, by 7.4, that R_c is an Azumaya algebra over its centre. Therefore, by 9.10, the centre is k-affine, and so the commutative result applies. The correspondence given by 7.9 then shows that the result holds for R_c.

Next consider TR. By 9.11, T is affine yet TR is integral over T. Thus the result is true for T and hence, by 8.14, for TR.

Finally, we turn to R itself. Induction on height shows that it is enough to prove the result when $\operatorname{ht} P = 1$. If P is a regular prime, we choose $0 \neq c \in g_n(R)$, $c \notin P$. Evidently P_c is a height 1 prime of R_c and so, as noted above, $\dim R_c = 1 + \dim R_c/P_c$. However, $\operatorname{tr deg} R_c = \operatorname{tr deg} R$, and also $\operatorname{tr deg}(R_c/P_c) = \operatorname{tr deg}(R/P)$ since $Q(R_c/P_c) = Q(R/P)$. These facts, combined with 10.6, show that $\dim R = 1 + \dim R/P$ as required.

On the other hand, if P is a nonregular prime then $P \supseteq g_n(R)R$ which, by 9.6, is an ideal of TR. By Zorn's lemma there is an ideal P' say, of TR which contains $g_n(R)R$ and is maximal with respect to having $P' \cap R \subseteq P$. One can check that P' is prime and so, since TR is a central extension of R, $P' \cap R$ is prime. Evidently $P' \cap R \neq 0$, so $P' \cap R = P$.

Now $R/P \hookrightarrow TR/P'$ and 10.11 applies to show that $\operatorname{tr deg} R/P = \operatorname{tr deg} TR/P'$. As noted in 10.5, $\operatorname{tr deg} TR = \operatorname{tr deg} R$. Thus if $\operatorname{ht} P' = 1$ then $\dim TR = 1 + \dim TR/P'$ and so, using 10.6, $\dim R = 1 + \dim R/P$.

Hence it remains only to show that $\operatorname{ht} P' = 1$. Suppose, to the contrary, that ht $P' > 1$; then 8.15 provides a regular prime $P'' \subset P'$ with ht $P'' = 1$. Since $P'' \neq 0$, we see that $0 \neq g_n(R)RP'' \subseteq P'' \cap R$ and hence $P'' \cap R = P$. Therefore TR/P'' is a central extension of R/P and hence has the same PI degree. This means that P must be regular, which is a contradiction. □

10.13 Recall that a chain is **saturated** if no additional term can be inserted. A ring R has the **catenary property** if given any $P, P' \in \operatorname{Spec} R$ with $P \supseteq P'$, any two saturated chains of primes between P and P' have the same length.

Corollary. *Each PI affine k-algebra has the catenary property.*

Proof. This is easily deduced from 10.12. □

§11 Additional Remarks

11.0 (a) Although the invention of PI rings is sometimes ascribed to [Kaplansky **48**], there were earlier results; see [Amitsur **74**] for a historical survey.

(b) The account here which is influenced by [Small **80**] is essentially self-contained, the proofs omitted not being strictly necessary for our purposes. For fuller accounts the reader is referred to [Rowen **80**], [Jacobson **75**] and [Procesi **73**].

11.1 (a) This is a straightforward account of basic results. The ring of generic matrices plays only a small part in this account. However, it has great importance elsewhere. It arose in [Amitsur **55**], where it was shown to be an integral domain; our description follows [Procesi **73**]. It is used in the proof, by [Amitsur **72**], that division algebras need not be crossed products; accounts of this are available in [Pierce **82**], [Jacobson **75**] and [Rowen **80**].

(b) Homological properties of generic matrices are discussed in [Coutinho **85**], [Le Bruyn **87**], [Small and Stafford **85**]. The relationship with invariants is described in [Formanek **84**].

11.2 (a) This material is derived from [Amitsur **56a**] and [Levitzki **50**].

(b) There is a general technique, analogous to 2.7(ii), which converts prime ideals into primitive ideals. It involves the construction of formal power series and polynomials in large numbers of noncommuting indeterminates; see [Passman **81**].

11.3 Theorem 3.8 comes from [Kaplansky **48**].

11.4 4.2 is due to [Amitsur **52**], 4.5(i) to [Cohn **77**, p. 466], 4.8(i) to [Amitsur **53**] and 4.8(ii) to [Procesi and Small **68**].

11.5 The existence of central polynomials was established by [Formanek **72**] and by [Razmyslov **73**]. This account, which is based on the latter, follows [Rowen **80**].

11.6 6.4 is due to [Rowen **73**], 6.5 to [Posner **60**], 6.10 and 6.11 to [Formanek **74**] and 6.15 to [Cauchon **76a**]

11.7 (a) 7.5 is due to [Small **71**], 7.14 to [Artin **69**] and [Procesi **72**]. Note, however, that the general Artin–Procesi theorem (see [Rowen **80**]) does not require R to be prime.

13.11.10 *Additional remarks*

(b) The relationship between the PI degrees of a prime PI ring, of its prime factors and of its subrings is discussed in [Bergman and Small **75**]. Further results about localization appear in [Braun and Small **86**], [Braun and Warfield **P**].

11.8 (a) 8.8 was proved by [Sirsov **57**], 8.12 by [Amitsur and Small **80**] and 8.14 by [Blair **73**].

(b) There is a more general notion of integrality defined in [Schelter **76**] which applies well to PI rings; see [Paré and Schelter **78**] and [Artin and Schelter **81**].

11.9 (a) The main results of this section are due to [Schelter **78**] although this account is influenced by [Amitsur and Small **80**].

(b) The trace ring, as described in 9.4, was introduced by [Schelter **78**] and [Razmyslov **74**]. The one defined in 9.2 comes from [Rowen **80**]. In [Amitsur **80**] it is shown, for either, that the trace ring of TR is TR.

(c) Further comments on G-rings come in 14.5.6.

(d) 9.10 comes from [Artin and Tate **51**], and 9.14 from [Asano **49**]. 9.13 and 9.16 come from [Robson and Small **74, 76**]. Further results about centres of hereditary PI rings can be found in [Jondrup **81**] and [Page **84**]. See also [Chatters and Hajarnavis **P**].

11.10 (a) 10.3 and 10.4 are due to [Amitsur and Procesi **66**]. 10.7 combines results of [Procesi **67**], [Markov **73**] and [Malliavin **76**]. 10.9 comes from [Schelter **78**] and Razmyslov **74**], and 10.12 from [Schelter **78**].

(b) 10.9 has been improved by [Braun **84**] which shows that if Λ is a commutative Noetherian ring and R is a PI ring affine over Λ then $N(R)$ is nilpotent. An example of [Golod and Shafarevich **64**], see [Herstein **68**], shows that if R is an affine k-algebra, for some field k, then $N(R)$ need not be nilpotent; so the hypothesis that R is a PI ring is important here.

(c) [Braun **84**] also proves that if R is a PI ring and an affine k-algebra then (i) $J(R)$ is nilpotent, and (ii) R is a homomorphic image of a generic matrix ring.

(d) The bound on dim R provided by 10.7 is not optimal; compare [Procesi **73**, 8.3.2].

Chapter 14
ENVELOPING ALGEBRAS OF LIE ALGEBRAS

The theory of enveloping algebras of Lie algebras g over a field k and of their irreducible representations is far too substantial to describe in one chapter. This chapter describes only one portion of this theory, a portion which is closely related to the earlier chapters and which, it is hoped, can serve as an introduction to this subject.

The overall theory splits in several ways. One split is determined merely by the characteristic of k. If this is finite one can check (see [Jacobson **62**, Chapter 6, Lemma 5] for example) that $U(g)$ is a finite module over its centre; and then the methods of Chapter 13 apply. This chapter, therefore, concentrates almost exclusively on the case when char $k = 0$.

Note next that a Lie algebra is a split extension of its solvable radical by a semisimple subalgebra. This leads the theory to consider separately the cases when g is solvable and when g is semisimple. It is the former case that is dealt with here.

That is not intended to suggest that the study of representations of a semisimple Lie algebra is not connected with topics met earlier in this book. Indeed, in classifying the primitive ideals P of $U(g)$, the quotient ring $Q(U(g)/P)$ and u dim $U(g)/P$ (known in this context as the **Goldie rank**) are important invariants. There are polynomials, the **Goldie rank polynomials**, whose evaluations give these ranks, and the additivity principle is used in demonstrating their existence. However, an account just of this theory would be much too large for a chapter.

Therefore, this chapter studies the enveloping algebra $U(g)$ when g is solvable and k has characteristic 0. It concentrates on the ring-theoretic aspects, aiming to describe, as precisely as possible, the structure of each prime factor ring R, with special emphasis on the primitive case.

The chapter opens with a section outlining the basic results about Lie algebras which are needed here. In particular, it notes the distinction, when k is not algebraically closed, between g being solvable or completely solvable. In practice the chapter concentrates on the latter case. But §5 shows how some of this can be extended to the solvable case.

14.1.5 Basics

There is, in §2, an analysis of prime ideals of $U(g)$ along the lines met in Chapter 10. It is shown, in particular, that they are all completely prime. Now let R denote a prime factor ring of $U(g)$. In §3 $E(R)$, the set of eigenvectors which is at the heart of the chapter, is described. Indeed what is shown, in the later sections, is that the structure of R, after localizing at one, or at all, of the elements of E, has a very special form. This is described in §8 with the main structure theorem appearing in §9. However, the special cases when g is nilpotent or g is algebraic are more easily dealt with, this being shown in §§6 and 7 respectively. The results obtained can be summed up, roughly, as asserting that R is closely related to a Weyl algebra.

Throughout this chapter, k will denote a field of characteristic 0, g a finite dimensional k-Lie algebra and $U = U(g)$. The fact, given by 1.7.4, that U is Noetherian will be used without comment.

§1 Basics

1.1 This section reviews some of the basic definitions and results concerning Lie algebras and their enveloping algebras.

1.2 Recall from 1.7.1 that g has a product $(x, y) \mapsto [x, y]$. This immediately yields the **adjoint representation** of g; namely, each $y \in g$ is represented by the linear transformation $g \to g$ given by $x \mapsto [x, y]$. This linear transformation is denoted by ad y; so $[x, y] = (x)$ ad y.

Evidently the kernel of this representation is the centre of g; and the **ideals** of g are the k-subspaces of g closed under all ad y, $y \in g$.

1.3 A **right g-module** is a vector space M together with a k-bilinear map $M \times g \to M$, $(m, x) \mapsto [m, x]$ such that

$$[m, [x, y]] = [[m, x], y] - [[m, y], x]$$

for each $m \in M$, $x, y \in g$.

For example, g is itself a right g-module under its own product; and its ideals are the submodules.

1.4 If M is a g-module then the action of g on M gives a representation of g which, by the universal property of $U(g)$ described in 1.7.2, extends to an action of $U(g)$ on M. This makes M into a $U(g)$-module. For example if $m \in M$, $x, y \in g$, and xy is the monomial in $U(g)$ then $m(xy) = [[m, x], y]$.

Conversely, if M is any $U(g)$-module then M is also a g-module by restriction; so $[m, x] = mx$. Thus these two terms are interchangeable.

1.5 g is **abelian** if $[x, y] = 0$ for each $x, y \in g$. This is the same as asserting that the adjoint representation of g is zero; or that $U(g)$ is commutative.

1.6 More generally if g has a chain of ideals

$$0 = g_0 \subset g_1 \subset \cdots \subset g_m = g$$

with $[g_{i+1}, g] \subseteq g_i$ for each i then g is **nilpotent**. This condition asserts that g acts trivially on g_{i+1}/g_i; and by refining the chain it can be arranged that each factor g_{i+1}/g_i is one-dimensional.

Note that the adjoint representation, with respect to the appropriate basis, is then by strictly lower triangular matrices. Of course, the set of all strictly lower triangular $n \times n$ matrices over k forms a nilpotent Lie algebra.

1.7 We recall, from 6.6.2, that g is **solvable** if it has a k-basis $\{x_1, \ldots, x_n\}$ such that $kx_1 + \cdots + kx_i \triangleleft kx_1 + \cdots + kx_{i+1}$ for each i. Evidently this is equivalent to requiring that g have a chain of Lie subalgebras

$$0 = g_0 \triangleleft g_1 \triangleleft \cdots \triangleleft g_m = g$$

with each an ideal in the next and each factor being abelian.

1.8 We also recall, from 7.5.7, that g is **completely solvable** if, in the notation of 1.7, it can also be arranged that each $kx_1 + \cdots + kx_i$ is an ideal of g or, equivalently, that each g_i is an ideal of g and each g_{i+1}/g_i is one-dimensional over k.

Note that the adjoint representation of g, with respect to this basis, is by lower triangular matrices. Also, the set of all lower triangular matrices over k forms a completely solvable Lie algebra.

1.9 It is elementary to check the following implications of properties of g:

abelian $\Rightarrow$ nilpotent $\Rightarrow$ completely solvable $\Rightarrow$ solvable.

1.10 Since char $k = 0$, Lie's theorem, as mentioned in 7.5.7, can be applied. It asserts that if k is algebraically closed and g is solvable then g is completely solvable. This will be proved in 5.3 as part of a more general result dealing with the case when k is not algebraically closed. It will also be proved there that if g is solvable then $[g, g]$ is nilpotent. However, when g is completely solvable this is made clear by the adjoint representation of g as triangular matrices, together with the fact that the kernel of this representation is the centre.

1.11 We should, perhaps, provide an example of a solvable Lie algebra which is not completely solvable. Let g be the three-dimensional $\mathbb{R}$-Lie algebra with basis x, y, z and multiplication

$$[y, x] = -z, \qquad [z, x] = y, \qquad [y, z] = 0.$$

Then $ky + kz$ is the only proper nonzero ideal of g, and so g is as required.

14.1.16 Basics

1.12 A Lie algebra g is said to **act semisimply** on a module M when M is a semisimple g-module. If this is true for every finite dimensional g-module then g is **semisimple**. (This is equivalent to the property that g contains no nonzero ideal which is solvable.)

As mentioned in the introduction, this chapter will not discuss enveloping algebras of semisimple Lie algebras but will concentrate, rather, on completely solvable Lie algebras. The last few results of this section aim in that direction.

1.13 A finite dimensional g-module M is called **triangular** (or **triangularizable** or **trigonalizable**) if it has a chain of submodules

$$0 = M_0 \subset M_1 \subset \cdots \subset M_n = M$$

with each factor having dimension 1 over k. If, in addition, g annihilates each factor then M is **strictly triangular**, although the alternative terminology that g is **acting nilpotently** on M will also be used.

More generally, an infinite dimensional g-module M is **triangular** (or **strictly triangular**) if each element belongs to a finite dimensional submodule with that property; and, in the latter case, g is said to **act locally nilpotently** on M.

Note that these properties are retained by submodules, factor modules and sums of such modules.

Lemma. *If M is a triangular g-module then $[g,g]$ acts locally nilpotently on M.*

Proof. The commutator of two triangular matrices is strictly triangular. □

1.14 It is clear from the definitions above that g is completely solvable if and only if g is a triangular g-module via the adjoint representation; and that g is nilpotent if and only if g is strictly triangular.

1.15 Let M be a nonzero g-module. The nonzero elements $m \in M$ such that $km \triangleleft M$ are called the **eigenvectors** (or **semi-invariants**) of g in M. The set of eigenvectors is denoted by $E(M; g)$ or simply $E(M)$. Those eigenvectors which also have $mg = 0$ are the **invariants** of g, and M^g denotes the set of these.

Of course if M is triangular then $E(M) \neq \emptyset$, and if M is strictly triangular then $M^g \neq \emptyset$.

1.16 If $m \in M$ is an eigenvector then, for each $x \in g$, there is an element $\lambda(x) \in k$ such that $[m, x] = \lambda(x)m$. This gives a linear form $\lambda \in g^*$, i.e. a k-linear map from g to k. This is called an **eigenvalue** of g on M, and is zero if m is an invariant.

Given any linear form λ on g we let

$$M_\lambda = M_\lambda(g) = \{m \in M \mid [m, x] = \lambda(x)m \quad \text{for all } x \in g\}.$$

This is a subset of $E(M) \cup \{0\}$. In particular $M_0 = M^g \cup \{0\}$.

1.17 Lemma. *Let M be a nonzero g-module, $0 \neq m \in M_\lambda$ and $g' = \ker \lambda$. Then $[g,g] \subseteq g'$ and $g' \triangleleft g$; and if $\lambda \neq 0$ then $\operatorname{codim} g' = 1$.*

Proof. If $x, y \in g$ then
$$[m,[y,x]] = [[m,y],x] - [[m,x],y]$$
and so
$$\lambda([y,x])m = \lambda(y)\lambda(x)m - \lambda(x)\lambda(y)m = 0.$$
Thus $[g,g] \subseteq g'$ and so $g' \triangleleft g$. Finally, $\dim(g/g') = 1$ since $\dim(\operatorname{im} \lambda) = 1$. □

1.18 Proposition. *Let g be completely solvable, M be a nonzero triangular g-module, $a \triangleleft g$ and $\lambda \in a^*$ be an eigenvalue of a on M. Then:*
(i) *$M_\lambda(a)$ is a g-submodule of M;*
(ii) *there is an eigenvalue μ of g whose restriction to a is λ.*

Proof. (i) By 1.13, $[g,a]$ acts locally nilpotently on M and so $\lambda([g,a]) = 0$. Now suppose that $m \in M_\lambda(a)$, $a \in a$ and $x \in g$. Then
$$\begin{aligned}[] [[m,x],a] &= [[m,a],x] + [m,[x,a]] \\ &= \lambda(a)[m,x] + \lambda([x,a])m \\ &= \lambda(a)[m,x]. \end{aligned}$$
Hence $[m,x] \in M_\lambda(a)$, as required.

(ii) It follows from 1.15 that $M_\lambda(a)$ contains an eigenvector m' for g, with eigenvalue μ say. Evidently μ, restricted to a, equals λ. □

1.19 It is clear from the definition of $M_\lambda(g)$ that $m \in M_\lambda(g)$ if and only if $m(\operatorname{ad} x - \lambda(x)) = 0$ for all $x \in g$. More generally, the **weight space** of M corresponding to λ is defined to be
$$M^\lambda = M^\lambda(g)$$
$$= \{m \in M \mid \text{for each } x \in g, m(\operatorname{ad} x - \lambda(x))^{i(x)} = 0 \text{ for some } i(x) \in \mathbb{N}\}.$$

We note one well known fact from linear algebra.

Proposition. *If g is abelian and M is triangular then $M = \bigoplus_{\lambda \in g^*} M^\lambda$.* □

1.20 We note one further elementary fact from linear algebra which will be of use.

Lemma. *Let V be a finite dimensional k-space and A a subspace of V^* such that no nonzero $v \in V$ is annihilated by all $a \in A$. Then $A = V^*$.* □

§2 Prime Ideals

2.1 One of the omissions from Chapter 10 was an account of the relationship between Spec R and Spec S, where R is a k-algebra and $S = R * U(g)$. That omission is, in part, repaired here with, in particular, an analogue of the weak incomparability result described in 10.6.6.

These results are used to show that if g is completely solvable then each prime ideal of $U(g)$ is completely prime.

2.2 The section starts with some elementary results concerning derivations.

Lemma. *Let R be a k-algebra with a k-derivation δ and let $\mathscr{S}$ be a right denominator set in R. Then δ extends uniquely to a k-derivation of $R_\mathscr{S}$.*

Proof. We check first that the ideal ass $\mathscr{S}$, defined in 2.1.3, is δ-stable. Let $r \in \text{ass } \mathscr{S}$; so $rs = 0$ for some $s \in \mathscr{S}$. Therefore

$$0 = \delta(rs) = \delta(r)s + r\delta(s).$$

However, $r\delta(s) \in \text{ass } \mathscr{S}$ and thus there exists $t \in \mathscr{S}$ with $r\delta(s)t = 0$. Hence $\delta(r)st = 0$ and so $\delta(r) \in \text{ass } \mathscr{S}$.

This shows that there is no loss in supposing that ass $\mathscr{S} = 0$. Since $ss^{-1} = 1$, if δ is to be extended then

$$0 = \delta(ss^{-1}) = \delta(s)s^{-1} + s\delta(s^{-1})$$

and so $\delta(s^{-1})$ must equal $-s^{-1}\delta(s)s^{-1}$. We therefore define $\delta(rs^{-1}) = \delta(r)s^{-1} - rs^{-1}\delta(s)s^{-1}$. It can be checked that this is a k-derivation of $R_\mathscr{S}$. □

2.3 Proposition. *Let R be a k-algebra with a k-derivation δ. Then the prime radical and the minimal primes of R are δ-stable.*

Proof. First recall Leibniz's formula, namely

$$\delta^n(r_1 r_2 \cdots r_t) = \sum \frac{n!}{n_1! n_2! \cdots n_t!} \delta^{n_1}(r_1) \delta^{n_2}(r_2) \cdots \delta^{n_t}(r_t)$$

with the summation being over all t-tuples $(n_1, \ldots, n_t)$ such that $n = n_1 + \cdots + n_t$.

Let P be a minimal prime and set

$$P' = \{r \in R \mid \delta^n(r) \in P \text{ for all } n \geq 0\}$$

noting that $P' \subseteq P$. Suppose first that $P' \subset P$, and so P' cannot be prime. Choose $a, b \in R \setminus P'$ with $aRb \subseteq P'$ and choose the least possible $s, t \in \mathbb{N}$ such that $\delta^s(a) \notin P$, $\delta^t(b) \notin P$. Finally, pick $c \in R$ such that $\delta^s(a) c \delta^t(b) \notin P$.

Now consider the expansion of $\delta^{s+t}(acb)$ using Leibniz's formula. Exactly one term, namely $((s+t)!/s!t!)\delta^s(a)c\delta^t(b)$ fails to belong to P. Hence $\delta^{s+t}(acb) \notin P$ and so $acb \notin P'$, contradicting the inclusion $aRb \subseteq P'$.

It follows that $P' = P$ and so P is δ-stable. Hence the prime radical, which is the intersection of the minimal primes, is δ-stable too. □

2.4 We now turn to the study of ideals in crossed products $S = R * U(g)$ where R is a k-algebra. (This, of course, includes the case where $S = R[x; \delta]$.) The definition, 1.7.12, requires that each element of g induces a k-derivation of R, and it also induces one of S. An ideal which is stable under all these derivations is called **g-stable**. Of course, each ideal of S is g-stable.

Lemma. *Let R be a k-algebra and $S = R * U(g)$.*
(i) If $I \triangleleft S$ then $I \cap R \triangleleft R$ and is g-stable.
(ii) Conversely, if $J \triangleleft R$ and is g-stable then $JS \triangleleft S$. Moreover, $S/JS \simeq (R/J) * U(g)$.

Proof. This is easily checked from the definitions. □

2.5 The behaviour of prime ideals is similar to that described in 10.5.6.

Proposition. *Let R be a right Noetherian k-algebra and $S = R * U(g)$.*
(i) (Cutting down) If $P \in \operatorname{Spec} S$ then $P \cap R \in \operatorname{Spec} R$ and is g-stable.
(ii) (Lying over) If $A \in \operatorname{Spec} R$ and is g-stable then $AS \in \operatorname{Spec} S$.

Proof. (i) Using 2.4, we may, by passing to $S/(P \cap R)S$, assume that $P \cap R = 0$. We need to demonstrate that R is prime. Note first, by 2.3, that $N(R)$ is g-stable and so, by 2.4, $N(R)S \triangleleft S$. Moreover, since $N(R)$ is nilpotent, so also is $N(R)S$. Hence $N(R)S \subseteq P$ and $N(R) \subseteq P \cap R = 0$. So certainly R is semiprime.

Similarly, if $P_1, \ldots, P_t$ are the minimal primes of R then each $P_i S$ is an ideal of S and, since $P_1 P_2 \cdots P_t = 0$, then $(P_1 S)(P_2 S) \cdots (P_t S) = 0$. It follows that some $P_i S \subseteq P$ and hence $P_i = 0$. Therefore R is prime.

(ii) This time 2.4 enables us to assume that $A = 0$ and hence that R is prime. We order the standard monomials, in a basis of g, as in 9.4.15; i.e. by total degree and then lexicographically. If a_1, a_2 are the coefficients of the leading terms of two nonzero elements s_1, s_2 of S, then $a_1 R a_2 \neq 0$ and so $s_1 R s_2 \neq 0$. Therefore S is prime. □

2.6 Corollary. *Let a be an ideal of g.*
(i) If $P \in \operatorname{Spec} U(g)$ then $P \cap U(a) \in \operatorname{Spec} U(a)$ and is g-stable.
(ii) If $A \in \operatorname{Spec} U(a)$ and is g-stable then $AU(g) \in \operatorname{Spec} U(g)$.

Proof. $U(g) \simeq U(a) * U(g/a)$. □

2.7 The next result about prime ideals of $R*U(g)$ relies upon a localization.

Lemma. *Let R be a k-algebra, $\mathscr{S}$ a right denominator set of regular elements of R and $R*U(g)$ a crossed product. Then $\mathscr{S}$ is a right denominator set of regular elements of $R*U(g)$ and there is a crossed product $R_{\mathscr{S}}*U(g) \simeq (R*U(g))_{\mathscr{S}}$.*

Proof. Since $R*U(g)$ is free over R, it is clear that the elements of $\mathscr{S}$ are regular in $R*U(g)$. Now let $s \in \mathscr{S}, f \in R*U(g)$. With respect to the ordering of monomials, used above in 2.5, let $a \in R$ be the leading coefficient. Choose $s_1 \in \mathscr{S}$ so that $as_1 \in sR$. Then the leading term of fs_1 is a multiple of s, say sf_1; and, of course, the leading monomial of $fs_1 - sf_1$ is lower in the ordering than that of f. By induction, there exists $s_2 \in \mathscr{S}$ such that $(fs_1 - sf_1)s_2 \in s(R*U(g))$ and so $f(s_1 s_2) \in s(R*U(g))$. This establishes the right Ore condition. The rest is clear. □

2.8 Theorem. *Let R be a right Noetherian k-algebra, $S = R*U(g)$, and $P_0 \subset P_1 \subset \cdots \subset P_m$ be a chain of prime ideals of S such that $P_0 \cap R = P_m \cap R$. Then $m \leq \dim g$.*

Proof. Using 2.6 we reduce immediately to the case when R is prime and $P_0 \cap R = 0$. Let $Q = Q(R)$; then 2.7 shows that S has a localization $T = Q*U(g)$. Further, since Q is simple Artinian, 6.5.7 asserts that $\mathcal{K}(T) \leq \dim g$. Hence no chain of primes in T has length greater than $\dim g$. However, 2.1.16 gives a $(1,1)$-correspondence between primes of T and primes P of S with $P \cap R = 0$. □

2.9 This provides an analogue of 10.6.6.

Corollary. *If R is right Noetherian, $S = R[x;\delta]$ and $P_0 \subset P_1 \subseteq P_2$ are prime ideals of S with $P_0 \cap R = P_2 \cap R$ then $P_1 = P_2$.* □

2.10 We need next to analyse under what conditions one gets two comparable prime ideals P_0, P_1 in $R[x;\delta]$ having the same intersection with R. Once again it is enough to consider the case when $P_0 = 0$.

Lemma. *Let R be a prime right Noetherian k-algebra with $Q = Q(R)$ and $Z = Z(Q)$ and let $S = R[x;\delta]$ and $T = Q[x;\delta]$. Suppose that there exists $0 \neq P \in \operatorname{Spec} S$ with $P \cap R = 0$. Then:*
  (i) *δ is inner on Q and $T = Q[y]$ for some central indeterminate y;*
  (ii) *$PT = fT$ for some irreducible monic polynomial $f \in Z[y]$ and $T/PT \simeq Q \otimes_Z F$, where F is the field $Z[y]/(f)$;*
  (iii) *if RF denotes the image in $Q \otimes_Z F$ of $R \otimes_k F$, then $Q \otimes_Z F$ is a localization of RF with respect to the regular elements $\mathscr{C}_R(0) \otimes 1$, and so RF is prime.*

Proof. (i) Clear from 1.8.4.

(ii) $Q = M_n(D)$ for some division ring D and so $T = M_n(D[y])$. However, $PT \in \operatorname{Spec} T$ and so $PT = M_n(A)$ for some $A \in \operatorname{Spec} D[y]$. Then 9.6.3 applies to give f. Also

$$T/PT = Q[y]/fQ[y] \simeq Q \otimes_Z (Z[y]/(f)).$$

(iii) Clear. □

It is easy to see that (i) is also sufficient for the existence of a prime P as described.

2.11 The main result of this section comes next. Its proof uses 2.10 in making the inductive step.

Theorem. *If g is completely solvable then each prime ideal of $U(g)$ is completely prime.*

Proof. This is clear when $\dim g = 0$. We now proceed by induction on $\dim g$, assuming the result for algebras of smaller dimension over any field of characteristic 0.

Since g is completely solvable, it has an ideal, a say, of codimension 1 and, as noted in 7.5.7, then $U(g) \simeq U(a)[x; \delta]$ for some x, δ. Let $A \in \operatorname{Spec} U(g)$ and $A' = A \cap U(a)$. Then, by 2.6, the rings $R = U(a)/A'$ and $S = U(g)/A'U(g)$ are prime; and, of course, the image of A in S is a prime ideal, P say.

By induction, R is an integral domain and hence so too is S, since $S \simeq R[x; \delta]$. So if $P = 0$, and so $A = A'U(g)$, the proof is complete. Otherwise 2.10 applies. In that case, consider the composite map

$$\varphi \colon U(a) \otimes_k F \to R \otimes_k F \to RF$$

with RF as in 2.10(iii). Note first that it is surjective; so $\ker \varphi$ is prime. On the other hand, $U(a) \otimes_k F \simeq U(a \otimes_k F)$ and so, by the inductive hypothesis, $\ker \varphi$ is completely prime. It follows, in turn, that RF, $Q \otimes_Z F$ and T/PT are integral domains, and thus the subring S/P of T/PT is as well. □

We will see, in 5.5, that this result is also valid when g is solvable.

§3 Eigenvalues and Prime Ideals

3.1 The results in this section describe more fully the prime and simple factor rings R of $U(g)$ when g is completely solvable. This depends upon a study of eigenvectors.

3.2 Viewed as a right U-module, $U = U(g)$ is also, by restriction, a g-module. However, it has another g-module structure given by the adjoint representation

14.3.4

of g; namely

$$[u, x] = ux - xu \quad \text{for } u \in U, x \in g.$$

In particular, each ideal of U is g-stable and so is a g-submodule under this action. This, together with 2.2, yields

Lemma. *If R is a factor ring of U and $\mathcal{S}$ is a right denominator set in R, then both R and $R_\mathcal{S}$ inherit the adjoint g-module structure from U, and each of their ideals is a g-submodule.* □

Henceforth when a ring R or $R_\mathcal{S}$ as above is viewed as a g-module it will invariably be via the adjoint representation. Occasionally, for emphasis, it will be described briefly as an ad g**-module**.

3.3 We saw in 1.14 the connection between completely solvable Lie algebras and triangular modules. This extends to U.

Proposition. *If g is completely solvable (or nilpotent) then $U(g)$ is a triangular (or strictly triangular) g-module under the adjoint representation.*

Proof. Let $x_1, \ldots, x_n$ be a basis for g as described in 1.8 (or 1.6). We order the standard monomials x^m as in 2.5 and note, for $y \in g$, that $[x^m, y] = cx^m + u$, where $c \in k$ and u is a linear combination of monomials which precede x^m in the ordering; and if g is nilpotent then $c = 0$. The result follows easily. □

This result is also valid for U/I, by 1.13.

3.4 The triangular structure provides eigenvectors in $E(R)$, by 1.15; and, as we will now see, they are normal elements of the ring. Part (iv) below was mentioned earlier in 4.2.7.

Lemma. *Let g be completely solvable, R be a factor ring of $U(g)$, and $\mathcal{S}$ be a right denominator set in R. Then:*
 (i) if $0 \neq I \triangleleft R_\mathcal{S}$ then $I \cap E(R) \neq \emptyset$;
 (ii) if $a \in E(R_\mathcal{S})$ then $aR = Ra$;
 (iii) $R^g = Z(R) \backslash \{0\}$;
 (iv) each ideal of R has a normalizing sequence of generators; and if g is nilpotent, a centralizing sequence.

Proof. (i) Clear from above.
 (ii) If $\lambda \in g^*$ is the corresponding eigenvalue then $[a, x] = \lambda(x)a$ for each x in g. Thus $a\bar{x} = \bar{x}a + \lambda(x)a$ and so $aR = Ra$.
 (iii) Clear from the proof of (ii) since $\lambda = 0$.
 (iv) This follows by induction using (i), (ii) and (iii). □

3.5 Naturally, the analysis of eigenvectors which follows is similar to the analysis of normal elements undertaken in Chapter 10. As in that case, it is not true that $E(R)$ is closed under addition; but some useful properties are immediate.

Lemma. *Let g, R, $\mathcal{S}$ be as in 3.4, and let a, $b \in E(R_\mathcal{S})$ with eigenvalues λ, μ respectively.*
(i) *If $ab \neq 0$ then $ab \in E(R_\mathcal{S})$ with eigenvalue $\lambda + \mu$.*
(ii) *If $\lambda = \mu$ and $a + b \neq 0$ then $a + b \in E(R_\mathcal{S})$ with eigenvalue λ.* □

3.6 It is now easy to characterize prime and simple factors of $U(g)$.

Proposition. *Let g be completely solvable and R be a factor ring of $U(g)$. Then:*
(i) *R is prime if and only if $E(R)$ is multiplicatively closed;*
(ii) *R is simple if and only if $E(R)$ is a multiplicative group;*
(iii) *if R is prime and $\mathcal{S} = E(R)$ then $R_\mathcal{S}$ is a simple ring and is a triangular g-module;*
(iv) *if R is prime, $Q = Q(R)$ and $e \in E(Q)$ then $e = ab^{-1}$ with a, $b \in E(R)$.*

Proof. (i) It follows from 3.4(i), (ii) that R is prime if and only if $a, b \in E(R)$ implies $ab \neq 0$; and then 3.5(i) completes the argument.

(ii) Evidently R is simple if and only if each $a \in E(R)$ is a unit. However, if $a \in E(R)$ with eigenvalue λ and $a^{-1} \in R$ one can check that $a^{-1} \in E(R)$ with eigenvalue $-\lambda$.

(iii) It is easily verified that $E(R)$ is a right denominator set of regular elements. Then 3.4(i) shows that $R_\mathcal{S}$ is simple, and the rest is clear.

(iv) Let $B = \{r \in R | er \in R\}$. Then $0 \neq B \triangleleft R$ and so $E(B) \neq \emptyset$. Choose $b \in E(B)$ and note that $eb = a \in E(R)$, by 3.5. □

The fact that $R_\mathcal{S}$, in (iii), is triangular does rely upon $\mathcal{S}$ consisting of eigenvectors. For example, take $R = A_1(k)$, this being, as noted in 1.7.7, a homomorphic image of a three-dimensional nilpotent Lie algebra g. Then R is a triangular g-module but R_y, its localization at the powers of y, is not, since y^{-1} does not belong to any finite dimensional g-submodule.

3.7 This is easily adapted to the nilpotent case.

Corollary. *Let g be nilpotent and R be a factor ring of $U(g)$. Then:*
(i) *R is prime if and only if $Z(R)$ is an integral domain;*
(ii) *R is simple if and only if $Z(R)$ is a field;*
(iii) *if R is prime and $\mathcal{S} = R^g$ then $R_\mathcal{S}$ is simple.* □

3.8 Next, some of these ideas are used to study an example mentioned earlier, in 8.6.10. First, however, note that there is a canonical surjection, called the

14.4.1   *Primitive ideals*   501

augmentation map, $U(g) \to k$ given by $x \mapsto 0$ for all $x \in g$. Its kernel is the **augmentation ideal**.

Lemma. *If $\vartheta: U(g) \to k$ is a surjective homomorphism then there a Lie algebra $h \subseteq U(g)$ such that $U(g) = U(h)$ and $\ker \vartheta$ is the augmentation ideal of $U(h)$.*

Proof. Set $h = \{x - \vartheta(x) | x \in g\}$.   □

3.9 Proposition. *The algebra $R = k[y][x; -y^2 d/dy]$ is not almost commutative.*

Proof. By extending the field we can assume k to be algebraically closed. By 8.4.3, we need to show that R is not a homomorphic image of some $U(g)$. Suppose to the contrary that $\psi: U(g) \to R$ is surjective. Let $P = \ker \psi$ and let Q be the inverse image in $U(g)$ of the ideal $I = xR + yR$. One verifies easily that $I^n = \sum_{i=0}^n x^i y^{n-i} R$; so $\bigcap I^n = 0$ and $\bigcap Q^n \subseteq P$.

By 3.8, we may suppose Q to be the augmentation ideal of g. So, if s is any semisimple subalgebra of g, then

$$s = [s, s] \subseteq Q^2$$

and so, by induction, $s \subseteq \bigcap Q^n \subseteq P$. Therefore we may assume g to be solvable and hence, by 1.10, completely solvable.

Note next that $R_y \simeq A_1'(k)$ which is simple. However, 3.4 shows that Q contains a nonzero normal element, e say, which by 2.1.16 generates an ideal of R_y. It follows that e is a unit of R_y and so has the form ay^m for some $a \in k$, $m \geqslant 0$. However, one can check that

$$k[y] \supseteq \{w \in R_y | [y^m, w] \in ky^m\}$$

which, in turn, contains the image of g. It follows that $R = k[y]$, a contradiction.   □

Of course the same argument can be used for the algebra $k[y][x; -y^n d/dy]$ for any $n \geqslant 2$.

§4 Primitive ideals

4.1 Throughout this section g is completely solvable and R is a prime factor ring of $U(g)$ and hence is an integral domain. Other notations will be established and accumulated.

The main result provides a description of a localization R_e of R with e being a particular eigenvector. This description will be refined substantially throughout later sections; but it suffices, here, to characterize the primitive factor rings amongst the primes. Indeed the following three conditions upon R will be seen to be equivalent:

(i) *R* is primitive;
(ii) there is a normal element $e \in R$ such that R_e is simple;
(iii) $Z(Q(R))$ is a finite dimensional field extension of k.

The reader is recommended to compare this with the commutative Nullstellensatz which applies, of course, if **g** is abelian. Whilst this comparison might suggest that primitive factor rings should be simple, and so $e = 1$, examples will be given to counter this.

4.2 These results rely upon an analysis of $E(R)$. We start with an illustrative example which will be cited throughout the section.

Example. *Let* **g** *be the three-dimensional completely solvable Lie algebra with basis x, y, z and products $[y, z] = 0, [y, x] = y$, and $[z, x] = \lambda z$, where $\lambda \in k$ and $\lambda \neq 0$.*
(i) *If $\lambda \in k \backslash \mathbb{Q}$ then each distinct monomial in y, z is an eigenvector for a different eigenvalue,*

$$E(U) = \{cy^i z^j | c \in k \backslash \{0\}, i, j \in \mathbb{N}\}$$

and $Z(U) = k$.
(ii) *If $\lambda \in \mathbb{Q}$ then there are distinct monomials having the same eigenvalue; and then their k-linear combinations belong to $E(U)$.*
(iii) *If $\lambda \in \mathbb{Q}$ and $\lambda < 0$ then $Z(U) \supset k$; but if $\lambda > 0$ then $Z(U) = k$.*

Proof. (i) The element $y^i z^j$ corresponds to the eigenvalue $x \mapsto i + j\lambda, y \mapsto 0, z \mapsto 0$. The claims follow easily.

(ii), (iii) If $\lambda < 0$, say $\lambda = -p/q$, then $y^p z^q$ corresponds to the eigenvalue 0 and so is central. If $\lambda > 0$ with $\lambda = p/q$ then, for example, $y^{2p} z^q$ and $y^p z^{2q}$ have the same eigenvalue; but one can check that $Z(U) = k$. □

4.3 The analysis below depends upon a reduction to subalgebras, the first step of which is covered next.

Proposition. *Let $e \in E(R; \mathbf{g})$ with nonzero eigenvalue λ, $\mathbf{g}' = \ker \lambda$, R' be the image of $U(\mathbf{g}')$ in R, and $\mathcal{S} = R' \backslash \{0\}$. Then:*
(i) $R = R'[x; \delta]$ *where $x \in \mathbf{g} \backslash \mathbf{g}'$ and $\delta = \operatorname{ad} x$;*
(ii) $R_{\mathcal{S}}$ *is simple;*
(iii) $E(R; \mathbf{g}) = E(R'; \mathbf{g}) \subseteq E(R'; \mathbf{g}')$.

Proof. Note first, by 1.17, that $\mathbf{g}'$ is an ideal of codimension 1 in $\mathbf{g}$ and so $U(\mathbf{g}) = U(\mathbf{g}')[x; \delta]$, as in 7.5.7. Therefore $R = R'[x; \delta]/I$, where I is a completely prime ideal with $I \cap R' = 0$. Since $R' \subseteq R$, then R' is an integral domain with quotient division ring $D = R'_{\mathcal{S}}$. Evidently $D[x; \delta]$ is the localization of $R'[x; \delta]$ at $\mathcal{S}$; and 2.1.16 shows that $I_{\mathcal{S}} \in \operatorname{Spec}(D[x; \delta])$ and $R_{\mathcal{S}} = D[x; \delta]/I_{\mathcal{S}}$.

(i), (ii) Clearly it is enough to show that $D[x; \delta]$ is simple, since then I must

14.4.7 *Primitive ideals* 503

be zero. Suppose, to the contrary, that $D[x;\delta]$ is not simple. Then 1.8.4 implies that δ is inner and so $D[x;\delta] = D[y]$ for some central indeterminate y. However e centralizes R', by definition of g', and so $e \in Z(D[y]) \cap R = Z(R)$. Therefore $\lambda = 0$, a contradiction.

(iii) Let $b \in E(R) = E(R;g)$. Then $0 \neq bR \triangleleft R$ and, by 2.1.16, $bR_{\mathscr{G}} \triangleleft R_{\mathscr{G}}$. Hence, by (ii), b must be a unit in $R_{\mathscr{G}}$ and so $b \in D$. Therefore $b \in D \cap R = R'$ and so $b \in E(R';g) \subseteq E(R';g')$. □

4.4 Let $g\hat{}\, = \bigcap \{\ker \lambda \,|\, \lambda$ an eigenvalue of g on $R\}$, and let $R\hat{}\,$ be the image of $U(g\hat{}\,)$ in R. Let $(g')\hat{}\,$ and $(R')\hat{}\,$ be defined likewise.

Lemma. (i) $[g,g] \subseteq g\hat{}\,$ and so $g/g\hat{}\,$ is abelian.
(ii) $g\hat{}\, = (g')\hat{}\, = \bigcap \{\ker \mu \,|\, \mu$ an eigenvalue of g on $R'\}$.
(iii) $R\hat{}\, = (R')\hat{}\,$.

Proof. (i) This is immediate from 1.17.
(ii) By 1.18, each eigenvalue of g' on R' is the restriction of an eigenvalue of g on R' and hence of g on R.
(iii) Clear. □

4.5 Theorem. (i) $R = R\hat{}\,[x_1;\delta_1]\cdots[x_m;\delta_m]$ with $g = g\hat{}\, + \sum_{i=1}^{m} kx_i$ and $\delta_i = \mathrm{ad}\, x_i$.
(ii) $E(R) \subseteq E(R\hat{}\,) = Z(R\hat{}\,)\setminus\{0\}$.

Proof. Both (i) and the containment $E(R) \subseteq E(R\hat{}\,)$ are clear, by induction, using 4.3 and 4.4. By construction, it is clear that the only eigenvalue of $R\hat{}\,$ is zero, and so $E(R\hat{}\,) = Z(R\hat{}\,)\setminus\{0\}$. □

4.6 The k-subalgebra of R generated by $E(R)$ is the **semicentre** of R, denoted by $Sz(R)$. (This corresponds, roughly, to the algebra of the group, defined in 10.3.17.)

Corollary. $Sz(R)$ *is a commutative integral domain. Furthermore,*

$$Z(Q(R)) \subseteq Q(Sz(R)) \subseteq Q(Z(R\hat{}\,)) = Z(Q(R\hat{}\,)).$$

Proof. $Z(R\hat{}\,)$ is a commutative integral domain, and so, by 4.5(ii), $Sz(R)$ is as well and $Q(Sz(R)) \subseteq Q(Z(R\hat{}\,))$.

Note next, by 3.6(iv), that each $e \in E(Q(R))$ takes the form ab^{-1} with $a, b \in E(R)$. Since $Q(R)^g = Z(Q(R))\setminus\{0\}$ we see that $Z(Q(R)) \subseteq Q(Sz(R))$. Similarly, $Z(Q(R\hat{}\,)) \subseteq Q(Sz(R\hat{}\,))$. However,

$$Q(Sz(R\hat{}\,)) = Q(Z(R\hat{}\,)) \subseteq Z(Q(R\hat{}\,)). \qquad \square$$

4.7 It is not always the case that $Z(Q(R)) = Q(Z(R))$. This is clear in 4.2 in the case when $\lambda \in \mathbb{Q}$, $\lambda > 0$. For if $\lambda = p/q$ then $y^p z^{-q} \in Z(Q(U)) \setminus Q(Z(U))$. Thus, in this case, $Z(Q(U))$ is an infinite dimensional extension of $Q(Z(U))$.

On the other hand, if $\lambda \in k \setminus \mathbb{Q}$ then $U \subseteq k[y, y^{-1}, z, z^{-1}][x; \delta] = S$ say. However, S is one of the simple algebras described in 1.8.6 and so, by 2.1.16, $k = Z(S) = Z(Q(S)) = Z(Q(U))$.

4.8 If $\mathscr{S}$ is a right denominator set in R, then the collection $\Lambda(R_{\mathscr{S}}; g)$ of eigenvalues of g on $R_{\mathscr{S}}$ is a subset of g^*. The subgroup this generates will be denoted by $\Gamma(R_{\mathscr{S}}; g)$. If $\mathscr{S} = E(R)$ it is clear from 3.5 and 3.6(iv) that $\Lambda(R_{\mathscr{S}}; g) = \Gamma(R_{\mathscr{S}}; g) = \Gamma(R; g)$.

We aim now to discuss the nature of $\Gamma(R; g)$; to find $e \in E(R)$ such that $\Lambda(R_e; g) = \Gamma(R_e; g)$; and then use this to describe the semicentre of R_e.

4.9 In the example described in 4.2, if $\lambda \in k \setminus \mathbb{Q}$ then $\Lambda(U; g) = \{i + j\lambda \mid i, j \in \mathbb{N}\}$ which is a free abelian semigroup of rank 2, and $\Gamma(U; g)$ is a free abelian group of rank 2. If we choose $e = yz$ it is clear that R_e is the ring S described in 4.7 and that $\Lambda(R_e; g) = \Gamma(R; g) = \Gamma(R_e; g)$.

If $\lambda \in \mathbb{Q}$, with $\lambda = p/q$ and p, q relatively prime, then $\Gamma(U; g)$ is free abelian of rank 1 generated by the eigenvalue $y \mapsto 0$, $z \mapsto 0$, $x \mapsto 1/q$, and the same element e may be chosen.

4.10 Let
$$0 = g_0 \subset g_1 \subset \cdots \subset g_n = g$$
be a chain of ideals with $\dim(g_i/g_{i-1}) = 1$ for each i. There is, for each i, a corresponding eigenvalue, λ_i say; and the Jordan–Hölder theorem shows that the collection $\{\lambda_1, \ldots, \lambda_n\}$ is independent of the choice of chains of ideals. These eigenvalues are called the **Jordan–Hölder weights** of g on g.

Lemma. $\Lambda(R; g) \subseteq \sum \mathbb{N} \lambda_i$.

Proof. It is enough to prove the result for U. We order the standard monomials as usual and choose one, say x^m with $m = (m_1, \ldots, m_n)$. Let M and M' be the subspaces of U spanned, respectively, by all monomials $w \leq x^m$ and by all $w < x^m$. Then M/M' is a g-module of dimension 1 and it has the eigenvalue $m_1 \lambda_1 + \cdots + m_n \lambda_n$. The Jordan–Hölder theorem can now be used to show that all eigenvalues are of this form. □

4.11 Since $\operatorname{char} k = 0$ it is clear that g^* is $\mathbb{Z}$-torsion-free. By 4.10, $\Gamma = \Gamma(R; g)$ is a subgroup of $\sum \mathbb{Z} \lambda_i$ and so is free of finite rank, say with a basis $\gamma_1, \ldots, \gamma_r$. There must, perforce, be eigenvalues $\lambda_1, \ldots, \lambda_s \in \Lambda(R; g)$ and $n_{ij} \in \mathbb{Z}$ such that

$$\gamma_i = \sum_j n_{ij} \lambda_j \quad \text{for } i = 1, \ldots, r.$$

For each λ_j, let e_j be a corresponding eigenvector in R; and let $e = e_1 e_2 \cdots e_s$. This is the element we have been seeking.

In the group of units of R_e generated by $e_1, \ldots, e_s$ there are words $w_1, \ldots, w_r$ which are eigenvectors corresponding, respectively, to the eigenvalues $\gamma_1, \ldots, \gamma_r$. We fix these and let W be the multiplicative subgroup they generate.

4.12 Theorem. (i) W is a free abelian group of rank r.
  (ii) There is a $(1,1)$-correspondence between eigenvalues of R_e and elements of W.
  (iii) $\Lambda(R_e; g) = \Gamma(R_e; g) = \Gamma(R; g) = \Gamma(R_E; g)$.
  (iv) If $w \in W$ then $wZ(R_e) \setminus \{0\}$ is the set of all eigenvectors in R_e having the same eigenvalue as w.
  (v) $Sz(R_e) = Z(R_e)[w_1, w_1^{-1}, \ldots, w_r, w_r^{-1}]$, this being a commutative Laurent polynomial ring.
  (vi) $Z(Q(R)) = Q(Z(R_e))$.

Proof. (i), (ii), (iii) Clear.
  (iv) If a, w have the same eigenvalue then aw^{-1} has eigenvalue 0 and so belongs to $Z(R_e)$. Hence $a \in wZ(R_e)$.
  (v) Since $\hat{g}$ acts trivially on $Sz(R_e)$, under the adjoint action, then $Sz(R_e)$ is a module over $g/\hat{g}$. Thus 1.19, combined with (iv), shows that $Sz(R_e)$ is free on W over $Z(R_e)$.
  (vi) If $q \in Z(Q(R))$ then $q = ab^{-1}$ with $a, b \in E(R)$ and a, b having the same eigenvalue. Thus there exists $w \in W$ such that aw and bw are both central in R_e. Hence $q \in Q(Z(R_e))$. □

Looking again at 4.2, and taking $e = yz$, one sees that, if $\lambda \in k \setminus \mathbb{Q}$, then

$$Sz(U_e) = k[y, y^{-1}, z, z^{-1}]$$

with $k = Z(U_e)$, as predicted by (v) above. Similarly, if $\lambda = p/q$ with p, q coprime then

$$Sz(U_e) = Z(U_e)[w, w^{-1}],$$

where $w = y^i z^j$ and $i + \lambda j = 1/q$.

4.13 With this detailed information we can now give the characterization of primitive factors.

Theorem. *Let g be completely solvable and R be a prime factor ring of $U(g)$. The following conditions are equivalent:*
  (i) R *is primitive*;
  (ii) R_e *is simple for some $e \in E(R)$*;
  (iii) $Z(Q(R))$ *is a finite dimensional field extension of k*.

Proof. (i)⇒(iii) Let M be a simple faithful R-module. Since $aR = Ra$ for all $a \in E(R)$, it follows that $M = Ma$ and so M is also a simple R_E-module. By 3.6, R_E is simple and so, by 2.1.16, $Z(Q(R)) = Z(R_E)$. However, $Z(R_E) \hookrightarrow \operatorname{End} M_R$ which, by 9.5.5, is finite dimensional over k.

(iii)⇒(ii) By 4.12(vi), $Z(Q(R)) = Q(Z(R_e))$ with e as specified. Therefore $Z(R_e)$ must be finite dimensional over k, so is a field. But then it follows from 4.12(v) that each eigenvector of g in R_e is a unit and so, by 3.6(iii), R_e is simple.

(ii)⇒(i) By 9.4.22, $J(R) = 0$ and yet, with e as specified, each nonzero prime ideal of R must contain e. Hence R must be primitive. □

Note, as in 3.6, that $R_e = R_E$ since R_e is simple.

4.14 Corollary. *With g as in 4.2, $U(g)$ is primitive if and only if $\lambda \in k \backslash \mathbb{Q}$. In that case*
$$U_e = U_E = k[y, y^{-1}, z, z^{-1}][x; \delta].$$

Proof. This is clear from 4.13, 4.7 and 4.9. □

Note, in particular, that $U(g)$ can be primitive without being simple.

4.15 However, when g is nilpotent this cannot happen.

Theorem. *Let g be nilpotent and R be a factor ring of $U(g)$. Then R is primitive if and only if R is simple.*

Proof. Let R be primitive; then $Z(R)$ must, by 4.13, be finite dimensional over k and hence be a field. Therefore 3.7(ii) applies to show R is simple. □

§5 The Solvable Case

5.1 As noted in 1.10, if k is algebraically closed and g is solvable then g is completely solvable. This is an immediate consequence of a result we prove next, namely that if k is any field of characteristic 0 and g is solvable then there is a finite extension field k' of k such that $g' = g \otimes k'$ is completely solvable. A connection between $\operatorname{Spec} U(g)$ and $\operatorname{Spec} U(g')$ is then outlined and thereby some earlier results are extended to the solvable case.

5.2 The first, preparatory, result extends 1.18.

Lemma. *Let M be a finite dimensional g-module, $a \triangleleft g$ and $\lambda \in a^*$. Then M_λ is a g-submodule of M.*

Proof. Let $m = m_0 \in M_\lambda$ and $x \in g$. We need to check that $[m, x] \in M_\lambda$. Let

14.5.4   The solvable case

$m_1 = [m, x]$ and $m_{i+1} = [m_i, x]$ for $i = 1, 2, \ldots$. Since M is finite dimensional, there exists j such that

$$m_{j+1} \in km_0 + \cdots + km_j = M'$$

say. Hence $[M', x] \subseteq M'$.

Suppose we know, for some $i > 0$, that $[m_i, a] \in \lambda(a)m_i + km_{i-1} + \cdots + km_0$ for all $a \in \mathbf{a}$. This is, of course, true when $i = 0$; and we aim to prove it true for $i + 1$ and hence, by induction, for $j + 1$.

To see this note that, since $[x, a] \in \mathbf{a}$,

$$[m_{i+1}, a] = [[m_i, x], a] = [[m_i, a], x] + [m_i, [x, a]]$$
$$\in [(\lambda(a)m_i + km_{i-1} + \cdots + km_0), x] + km_i + \cdots + km_0$$
$$\subseteq \lambda(a)m_{i+1} + km_i + \cdots + km_0$$

as required.

It follows that M' is an $\mathbf{a}$-module and that the trace of ad a, restricted to M', is $(j+1)\lambda(a)$. This can be applied to $[x, a] \in \mathbf{a}$ whose trace, when restricted to M', is clearly 0. Hence $\lambda([x, a]) = 0$ and the proof of 1.18(i) can now be used. □

5.3 Theorem. *If* $\mathbf{g}$ *is solvable then*:
(i) *there is a finite extension* k' *of* k *such that* $\mathbf{g}' = \mathbf{g} \otimes k'$ *is completely solvable*;
(ii) $[\mathbf{g}, \mathbf{g}]$ *is nilpotent*;
(iii) *(Lie's theorem) if* k *is algebraically closed then* $\mathbf{g}$ *is completely solvable.*

Proof. (i) Let $x_1, \ldots, x_n$ be a basis for $\mathbf{g}$ as described in 1.7 and let $\mathbf{a} = kx_1 + \cdots + kx_{n-1}$. This is an ideal of $\mathbf{g}$ and, by induction on dimension, the result holds for $\mathbf{a}$. We choose k' to be a finite extension making $\mathbf{a} \otimes k'$ completely solvable and such that all the eigenvalues of ad x_n belong to k'. We now claim that $\mathbf{g}' = \mathbf{g} \otimes k'$ is completely solvable.

We know that $M = \mathbf{a} \otimes k'$ contains an $\mathbf{a} \otimes k'$-eigenvector y with eigenvalue λ, say. By 5.2 M_λ is a $\mathbf{g}'$-submodule of M and so $[M_\lambda, x_n] \subseteq M_\lambda$. By the choice of k', there is an eigenvector y' for x_n in M_λ, and then $ky' \lhd \mathbf{g}'$. By passing to $\mathbf{g}'/ky'$ and repeating this process it follows that $\mathbf{g}'$ is completely solvable.

(ii) This follows from (i) and 1.10.
(iii) Clear from (i). □

5.4 Throughout the rest of this section, let $\mathbf{g}, \mathbf{g}', k, k'$ be as in 5.3; and let $U = U(\mathbf{g})$, $U' = U(\mathbf{g}')$.

Of course $U' \simeq U \otimes_k k'$ so U' is a finite centralizing extension of U. Thus the results of Chapter 10 (which, in these circumstances, have rather easier proofs) can be applied.

Proposition. (i) (Cutting down) *If $P' \in \operatorname{Spec} U'$ then $P' \cap U \in \operatorname{Spec} U$.*
(ii) (Lying over) *If $P \in \operatorname{Spec} U$ then there exists $P' \in \operatorname{Spec} U'$ with $P' \cap U = P$.*

Proof. (i) See 10.2.4.
(ii) By 10.2.9, there exists P' such that P is a minimal prime of $P' \cap U$. Therefore $P = P' \cap U$ by (i). □

5.5 Corollary. *Each prime ideal of $U(g)$ is completely prime.*

Proof. If $P \in \operatorname{Spec} U$, let $P' \in \operatorname{Spec} U'$ be as in 5.4(ii). Then $U/P \hookrightarrow U'/P'$ which, by 2.11, is an integral domain. □

5.6 We now extend a notion introduced, for prime PI rings, in 13.9.12. A prime ring S is called a **G-ring** if the intersection of its nonzero prime ideals is nonzero.

The relevance here is that when g is completely solvable and R is a prime factor ring of $U(g)$ then each nonzero ideal of R contains an eigenvector, by 3.4. Thus 4.13(ii) simply asserts that the ring R there is a G-ring.

5.7 Further facts relating P and P' are required. Let $R = U/P$, $Q = Q(R)$ and $Z = Z(Q)$; and R', Q', Z' be the corresponding algebras for P'.

Proposition. *Let $P \in \operatorname{Spec} U$, $P' \in \operatorname{Spec} U'$ with $P' \cap U = P$. Then:*
(i) *R is primitive if and only if R' is primitive;*
(ii) *R is a G-ring if and only if R' is a G-ring;*
(iii) *Z is finite dimensional over k if and only if Z' is finite dimensional over k'.*

Proof. (i) Clear from 10.2.16, using 10.2.12 and 10.2.14.
(ii) Let A, A' denote respectively the intersections of the prime ideals Q, Q' of U, U' strictly containing P, P'. It is clear from 5.4 that $A = A' \cap U$. Thus if $A \supset P$ then $A' \supset P'$. Conversely, if $A' \supset P'$ then incomparability (which holds by 10.2.12 and 10.2.13) shows that $A \supset P$.
(iii) Of course $Z \subseteq Z'$; so if Z' is finite over k' which is finite over k then Z is finite over k. For the converse, consider the multiplication map

$$Q \otimes_k k' \to Q'.$$

This is a k-algebra homomorphism and its image, Qk', is a finitely generated Q-module and so is a subdivision ring of Q'. However, it contains R' and thus equals Q'. Now

$$Q \otimes_k k' = Q \otimes_Z Z \otimes_k k'$$

and so, by 9.6.9, each ideal has the form $Q \otimes I$ with $I \triangleleft Z \otimes_k k'$. Therefore

$$Q' \simeq Q \otimes_Z (Z \otimes k'/I)$$

and so $Z' \simeq Z \otimes k'/I$. Thus if Z is finite dimensional over k then Z' is over k'. □

5.8 When these results are combined with 4.13 one quickly deduces

Theorem. *Let g be solvable and R be a prime factor ring of $U(g)$. The following conditions are equivalent*:
  (i) *R is primitive*;
  (ii) *R is a G-ring*;
  (iii) *$Z(Q(R))$ is a finite dimensional field extension of k.* □

5.9 Evidently the close relationship between R and R' indicates that the structure of R is 'close' to that of R'. How, precisely, the more detailed results can be applied to R has not yet been completely understood. For further comment see 15.3.14.

§6 When Eigenvectors are Central

6.1 The next few sections extend and refine the information obtained in §4 about R_e and R_E. We know, from 1.7.7 for example, that $A_n(k)$ can occur as a factor ring R of $U(g)$ and then $R = R_e = R_E = A_n(k)$. We also know, from 4.14, that R_E and R_e can take a different form. The next four sections will demonstrate that these examples, together, give a reasonable indication of the general situation.

6.2 Throughout this section g is completely solvable and R is a prime factor ring of $U(g)$. Furthermore it will be assumed that $E(R)$ is central; so $E(R) = R^g = Z(R) \backslash \{0\}$. This is, of course, what happens whenever g is nilpotent; and it is also what happens for the algebra $R^{\wedge}$ described in 4.5. In fact the results of this section about the structure of R in this special case will, in later sections, be combined with those of §4 in order to describe the general case. The main consequence obtained in this section, however, is that R_E must, in this special case, necessarily be a Weyl algebra.

6.3 It is possible for these conditions to be satisfied by $R = U(g)$ without g being nilpotent. The simplest example is the four-dimensional 'diamond algebra' g which has a basis $\{t, x, y, z\}$ and $[t, x] = x$, $[t, x] = -y$, $[x, y] = z$ and z central. The fact that $U(g)$ satisfies these conditions will be demonstrated in 6.6. For now we note that

$$U_z \simeq (k[z, z^{-1}] \otimes_k A_1(k))[t; \delta]$$

as is easily seen by taking x and yz^{-1} as generators for $A_1(k)$.

6.4 If B is an algebra and δ a derivation of B, then δ is **locally nilpotent** if, given $b \in B$, there exists n such that $\delta^n(b) = 0$.

Lemma. *Let B be a k-algebra, δ be a locally nilpotent k-derivation of B, and $C = \ker \delta$. Suppose that $\delta(y) = 1$, for some $y \in Z(B)$. Then:*
(i) $B = C[y]$ which is a polynomial ring over C in the central indeterminate y;
(ii) $C \simeq B/By$;
(iii) $B[x; \delta] \simeq C \otimes_k A_1(k)$.

Proof. Suppose that $c_0 + c_1 y + \cdots + c_m y^m = 0$. Applying δ^m gives $m! c_m = 0$ and hence $c_m = 0$. Thus y is an indeterminate over C. To see that $B = C[y]$, let $b \in B$. We know that $\delta^n(b) = 0$ for some n. To prove that $b \in C[y]$, induction on n will be used. Of course, when $n = 1$ then $\delta(b) = 0$ and so $b \in C$. In general one can check that

$$b - \delta(b)y + \frac{1}{2!}\delta^2(b)y^2 - \cdots + (-1)^{n-1}\frac{1}{(n-1)!}\delta^{n-1}(b)y^{n-1} \in \ker \delta = C,$$

and the result follows immediately. □

6.5 In the next result, and later, it is convenient to write b rather than $b \otimes 1$ and a rather than $1 \otimes a$ in $B \otimes_k A$.

Lemma. *Let B be a k-algebra, δ be a k-derivation on $S = B \otimes_k A_n(k)$ and $T = S[t; \delta]$. There exists $s \in S$ such that, if $\delta' = \delta - \operatorname{ad} s$, then:*
(i) δ' restricted to B is a derivation on B;
(ii) δ' restricted to $A_n(k)$ is zero;
(iii) $T = B[t'; \delta'] \otimes_k A_n(k)$, where $t' = t - s$.

Proof. (i), (ii) By writing $S = B \otimes A_{n-1}(k) \otimes A_1(k)$ and using induction, it is sufficient to prove the result for the case $n = 1$. Since $[x, y] = 1$, then $[\delta(x), y] + [x, \delta(y)] = 0$.
Now $\delta(x) = \sum_{i,j} b_{ij} x^i y^j$ with $b_{ij} \in B$. Let

$$s_1 = \sum_{i,j} \frac{1}{j+1} b_{ij} x^i y^{j+1} \quad \text{and} \quad \delta_1 = \delta - \operatorname{ad} s_1.$$

Then $[x, s_1] = \delta(x)$, and $\delta_1(x) = 0$.
However, $[x, y] = 1$ and so $[\delta_1(x), y] + [x, \delta_1(y)] = 0$ and therefore $[x, \delta_1(y)] = 0$. It follows that $\delta_1(y) \in B[x]$, with $\delta_1(y) = \sum b_j x^j$ say. Let

$$s_2 = -\sum \frac{1}{j+1} b_j x^{j+1}, \quad s = s_1 + s_2 \quad \text{and} \quad \delta' = \delta - \operatorname{ad} s.$$

Then

$$\delta'(x) = \delta_1(x) = 0 \quad \text{and} \quad \delta'(y) = \delta_1(y) - [y, s_2] = 0.$$

Thus $\delta'(A_1(k)) = 0$. Now let $b \in B$, $a \in A_1$; then $[b, a] = 0$ and so
$$0 = \delta'[b, a] = [\delta'(b), a] + [b, \delta'(a)] = [\delta'(b), a].$$
Hence $\delta'(b)$ centralizes $A_1(k)$, so $\delta'(b) \in B$.
(iii) Clear. □

Note that if B is commutative then the new derivation coincides with δ on B.

6.6 If 6.5 is applied to the algebra
$$(k[z, z^{-1}] \otimes_k A_1(k))[t; \delta]$$
described in 6.3 one sees that
$$U_z \simeq k[z, z^{-1}, t] \otimes_k A_1(k).$$
This will be shown, in 6.8(i), to be a typical situation.

Let F be the quotient field of $k[z, z^{-1}, t]$. Then $F \otimes_k A_1(k) \simeq A_1(F)$ which is simple. It follows that $E(U) \subseteq F$ and so $E(U)$ is central. This establishes the claim, in 6.3, that U satisfies the conditions on R assumed in this section.

6.7 One further preliminary result is required.

Lemma. *Let A, B be k-algebras. Then $A \otimes_k B$ is k-affine if and only if both A and B are k-affine.*

Proof. ($\Leftarrow$) Clear.
($\Rightarrow$) Let $A \otimes B$ be generated by $\{f_i\}$, $i \in \{1, \ldots, n\}$, with $f_i = \sum_j a_{ij} \otimes b_{ij}$. Then A is generated by $\{a_{ij}\}$ and B by $\{b_{ij}\}$. □

6.8 After these preliminaries the main result of this section can be proved. It describes R_e and hence R_E. For later use, both in respect of $R^{\hat{}}$, as in 4.5, and in the algebraic case considered in §7, it gives, in (iii) and (iv), some extra details concerning related Lie algebras **h** and **s**.

Theorem. *Let **g** be completely solvable and $R = U/P$, a prime factor ring of $U(\mathbf{g})$, with $Sz(R) = Z(R)$. Suppose also that $\mathbf{g} \lhd \mathbf{h}$ with **h** completely solvable, that P is **h**-stable and that **s** is a subalgebra of **h** over which **g** is a semisimple module. Then:*
(i) *there exists $e \in Z(R)$ such that $R_e \simeq Z(R)_e \otimes_k A_p(k)$ for some p;*
(ii) *$Z(R_e) = Z(R)_e$ and is k-affine;*
(iii) *e can be chosen to be an **h**-eigenvector;*
(iv) *the generators x_i, y_i of $A_p(k)$ in R_e can be chosen to be **s**-eigenvectors.*

Proof. Note first that (ii) is an immediate consequence of (i) together with 6.7. The rest of the proof uses induction on dim **g**, the case $\mathbf{g} = 0$ being clear.

(i) Let g' be an h-submodule of g of codimension 1, and let R' be the image of $U(g')$ in R. Let Z and Z' denote the centres of R and R' respectively. Since $Sz(R; g) = Z$, 1.18 shows that 0 is the only g'-eigenvalue of R' and thus $Sz(R'; g') = Z'$. Therefore, by the induction hypothesis, there is an h-eigenvector $f \in Z'$ such that $R'_f = Z'_f \otimes A_q(k)$ for some q, with the generators x_i, y_i of $A_q(k)$ being s-eigenvectors.

Choose any $x \in g \setminus g'$ and let $\delta = \operatorname{ad} x$. Since f is an h-eigenvector it is, of course, a g-eigenvector. Therefore its eigenvalue in g^* is zero and $\delta(f) = 0$. However, $U(g) = U(g')[x; \delta]$; so R_f is a homomorphic image of

$$R'_f[x; \delta] = (Z'_f \otimes A_q(k))[x; \delta].$$

Since $\delta(f) = 0$ it follows that $f \in Z$.

By 6.5, there exists $r \in R'_f$ such that

$$(Z'_f \otimes A_q(k))[x; \delta] \simeq Z'_f[x'; \delta] \otimes A_q(k),$$

where $x' = x - r$ and $\delta = \operatorname{ad} x' = \operatorname{ad} x$ on Z'_f. Since $A_q(k)$ is simple, 9.6.9 shows that each ideal of $Z'_f[x'; \delta] \otimes A_q(k)$ has the form $I \otimes A_q(k)$ with $I \triangleleft Z'_f[x'; \delta]$. Thus

$$R_f \simeq (Z'_f[x'; \delta]/I) \otimes A_q(k).$$

We now consider the action of the derivation δ on Z'_f and on Z'. Note that Z' is a triangular g-module. If it had a g-eigenvector with nonzero eigenvalue this would belong to $Sz(R) \setminus Z(R)$ which, by hypothesis, is empty. It follows that g, and hence δ, must act locally nilpotently on Z'. Furthermore, δ is locally nilpotent on Z'_f since $\delta(f) = 0$.

One can check directly that $\delta(Z')$ is an h-submodule of Z', using the facts that $g = g' + kx$ and $[Z', g'] = 0$.

We now consider separately two cases. First, suppose $\delta(Z') = 0$. Then the isomorphism above becomes

$$R_f \simeq (Z'_f[x']/I) \otimes A_q(k).$$

It follows that

$$Z(R_f) = Z'_f[x']/I \subseteq Z(R)_f \subseteq Z(R_f)$$

and (i) is established in this case.

Secondly, suppose $\delta(Z') \neq 0$. Then, since R' is a triangular h-module, $\delta(Z')$ contains an h-eigenvector, v say. Of course v is also a g-eigenvector; so $v \in Sz(R) = Z(R)$ and thus $\delta(v) = 0$. One can check directly that the nonzero k-subspace

$$V = \{u \in Z' \mid \delta(u) \in kv\}$$

is also an h-submodule and so is s-stable. Since s acts semisimply on g, it follows that V is a semisimple s-module and hence is spanned by s-eigenvectors. We choose $u \in V$ so that u is an s-eigenvector and $\delta(u) = v$.

Now let $e = fv$; evidently e is an h-eigenvector and $R'_e = (R'_f)_v$. Also let

$y = v^{-1}u \in Z'_e$ and note that $\delta(y) = 1$. Next, recall that δ is locally nilpotent on Z' and hence on Z'_e, since $\delta(e) = 0$. Therefore, by 6.4,

$$Z'_e[x'; \delta] \simeq C \otimes_k A_1(k)$$

with $C \simeq Z'_e/(y)$. Hence

$$R'_e[x; \delta] = (Z'_e[x'; \delta]) \otimes A_q(k) \simeq C \otimes A_{q+1}(k);$$

and (i) follows, as in the first case.

(iii) This has already been arranged.

(iv) We already have, by induction, that $x_1, \ldots, x_q, y_1, \ldots, y_q$ are s-eigenvectors, and we have arranged that $y = y_{q+1}$ is one. It remains to show that $x' = x - r$ could be chosen to be an s-eigenvector.

The fact that s acts semisimply on **g** means that x could be chosen to be an s-eigenvector, say with eigenvalue $\lambda \in s^*$. Now consider $r \in R'_f$; since R'_f is s-semisimple, r has a unique representation $r = \sum\{r^\alpha | \alpha \in s^*\}$, where r^α is an s-eigenvector, with eigenvalue α, or is zero. Let w be one of the elements $x_1, y_1, \ldots, x_q, y_q$. Since w is an s-eigenvector, with eigenvalue μ say, so too are wx, xw and $[w, x]$, with eigenvalue $\mu + \lambda$, and likewise $[w, r^\alpha]$ with eigenvalue $\mu + \alpha$, if nonzero.

Recall, however, that, by the choice of r,

$$[w, r] = \delta(w) = [w, x].$$

Hence

$$[w, r - r^\lambda] = [w, x - r^\lambda] = [w, x] - [w, r^\lambda]$$

which, if nonzero, is an s-eigenvector with eigenvalue $\mu + \lambda$; and

$$[w, r - r^\lambda] = [w, \sum\{r^\alpha | \alpha \neq \lambda\}] = \sum\{[w, r^\alpha] | \alpha \neq \lambda\}$$

each summand of which is either zero or else an s-eigenvector with eigenvalue $\mu + \alpha$ with $\alpha \neq \lambda$. It follows that $[w, r - r^\lambda]$ must be zero. Hence r^λ could be used rather than r; and then $x' = x - r^\lambda$ is an s-eigenvector. □

6.9 Corollary. *Let **g** be nilpotent and R be a prime factor ring of $U(\mathbf{g})$. Then*:
(i) $R_Z \simeq A_q(Q(Z))$, *where* $Z = Z(R)$;
(ii) R *is primitive if and only if* $R \simeq A_q(K)$ *for some field* K.

Proof. (i) Clear from 6.8(i).

(ii) If R is primitive then 4.15 shows that R is simple; so $Z = Q(Z)$ and (i) applies with $Z = K$. □

§7 When g is Algebraic

7.1 This section deals with another special case, namely when **g** is algebraic. Whilst the results here serve as a guide to what to expect in the general case, they are not used later.

7.2 An **algebraic** Lie algebra g is the Lie algebra of a (linear) affine algebraic group. The facts required here about a completely solvable algebraic Lie algebra g, which may be found in [Borho, Gabriel and Rentschler 73], are:
(i) $g = n \oplus s$, where n is the maximum nilpotent ideal of g and s is an abelian subalgebra which acts semisimply on n;
(ii) if $W(g)$ and $\Gamma(g)$ are, respectively, the k-subspace and the $\mathbb{Z}$-submodule of g^* spanned by the Jordan–Hölder weights of g on g (see 4.10) then $\dim_k W(g) = \mathrm{rank}_{\mathbb{Z}} \Gamma(g)$.

7.3 We note that then $\dim_k W' = \mathrm{rank}_{\mathbb{Z}} \Gamma'$ for any k-subspace W' and $\mathbb{Z}$-submodule Γ' of $\Gamma(g)$ such that $k\Gamma' = W'$. This is easily verified by choosing simultaneous bases for Γ' and $\Gamma(g)$ over $\mathbb{Z}$.

7.4 Examples. (i) The algebra g of lower triangular $n \times n$ matrices is algebraic and completely solvable. In this case n is the algebra of strictly lower triangular matrices, s the diagonal matrices and each e_{ij} is an s-eigenvector with eigenvalue

$$\begin{bmatrix} c_1 & & 0 \\ & \ddots & \\ 0 & & c_n \end{bmatrix} \mapsto c_j - c_i.$$

Therefore $\dim_k W(g) = \mathrm{rank}_{\mathbb{Z}} \Gamma(g) = n - 1$.

(ii) Let g be the Lie algebra described in 4.2. Then g is algebraic if and only if $\lambda \in \mathbb{Q}$. In this case $n = ky + kz$ and $s = kx$.

7.5 The decomposition $g = n \oplus s$ leads to a decomposition. Recall from 1.8.6 that $A'_1(k) = k[y, y^{-1}][x; \delta]$, where $\delta = -d/dy$. The decomposition we seek involves $A'_r(k)$, the tensor product of r copies of $A'_1(k)$. Since $A'_1(k)$ is simple, so too is $A'_r(k)$.

7.6 Theorem. *Let g be completely solvable and algebraic and R be a prime factor ring of $U(g)$. Then there exists $e \in E(R)$ such that*

$$R_e \simeq Z(R_e) \otimes A_p(k) \otimes A'_r(k).$$

The numbers r, p are uniquely determined; namely $r = \mathrm{rank}\, \Gamma(R)$ and $p + r = \mathcal{K}(R_E)$.

Proof. The proof makes use of $g\hat{}$ and $R\hat{}$ as defined in 4.4. Note first, by 7.3, that there is a $\mathbb{Z}$-basis $\gamma_1, \ldots, \gamma_r$ for $\Gamma(R; g)$ which is also a k-basis for $k\Gamma(R; g)$. Now $g\hat{} = \bigcap \{\ker \lambda \mid \lambda \in \Lambda(R; g)\}$ and $\Gamma(R; g) = \mathbb{Z}\Lambda(R; g)$; so it follows that $g\hat{} = \bigcap_{i=1}^{r} \ker \gamma_i$.

Now $g = n \oplus s$ and, of course, $g\hat{} \supseteq n$. Thus if t is a complement to $g\hat{} \cap s$ in s then g is the semidirect product of the ideal $g\hat{}$ with the abelian subalgebra t. Each eigenvalue of g on R annihilates $g\hat{}$ and so is uniquely the extension of an eigenvalue of t on R. It follows that the k-subspace of t^* spanned by the

14.8.1 The simple algebras $\mathscr{A}(V, \delta, \Gamma)$ 515

restriction to t of $\gamma_1, \ldots, \gamma_r$ has dimension r; and, by 1.20, it must equal t^*. Thus the restrictions of $\gamma_1, \ldots, \gamma_r$ form a basis for t^*.

Next consider $R\hat{}$. We know from 4.5(ii) that $Z(R\hat{}) = Sz(R\hat{})$. Therefore 6.8 shows that there exists $e \in E(R\hat{}; g) \subseteq Z(R\hat{})$ such that $R_e\hat{} = Z(R_e\hat{}) \otimes A_p$ for some p. Moreover, the generators $\{x_i, y_i\}$ of A_p can be chosen to be t-eigenvectors; and if α_i, β_i are the eigenvalues, in t^*, of x_i, y_i respectively then $\beta_i = -\alpha_i$ since $[x_i, y_i] = 1$.

We turn now to R_e. One can interpret 4.5(i) as saying that $R_e = R_e\hat{} \# U(t)$; hence $R_e = (Z(R_e\hat{}) \otimes A_p) \# U(t)$. For each $t \in t$, let $t' = t - \sum_{i=1}^{p} \alpha_i(t) x_i y_i$. Then the elements t' commute with each other and with each generator of A_p and hence centralize A_p. Moreover, t' induces the same derivation as t on $Z(R_e\hat{})$. Hence if $t' = \{t' | t \in t\}$ then t' is abelian and

$$R_e = (Z(R_e\hat{}) \# U(t')) \otimes A_p$$
$$\simeq (Z(R_e\hat{}) \# U(t)) \otimes A_p.$$

Finally, we must analyse $Z(R_e\hat{}) \# U(t)$. Since t acts semisimply on $R\hat{}$, then $Z(R\hat{})$ is spanned by t-eigenvectors. Of course $g\hat{}$ acts trivially on $Z(R\hat{})$; so these t-eigenvectors must be g-eigenvectors. This shows that $Z(R\hat{}) \subseteq Sz(R)$ and hence, by 4.5(ii), $Z(R\hat{}) = Sz(R)$.

However, 4.12(v) shows that, for some $f \in E(R; g)$

$$Sz(R_f) = Z(R_f) \otimes k[w_1, w_1^{-1}, \ldots, w_r, w_r^{-1}]$$

where each w_i has eigenvalue γ_i. By replacing e by ef we may suppose this holds for R_e rather than R_f. We saw above that the restriction to t of $\gamma_1, \ldots, \gamma_r$ formed a basis for t^*. We choose a dual basis $v_1, \ldots, v_r$ in t and let $t_i = w_i^{-1} v_i$. One can check that

$$[w_j, t_i] = \delta_{ij}, \quad [w_i, w_j] = 0 \text{ and } [t_i, t_j] = 0$$

for all i, j. However, then

$$Z(R_e\hat{}) \# U(t) \simeq (Z(R_e) \otimes k[w_1, w_1^{-1}, \ldots, w_r, w_r^{-1}])[t_1; \partial/\partial w_1] \cdots [t_r; \partial/\partial w_r]$$
$$= Z(R_e) \otimes A_r'.$$

It remains only to check that $K(R_E) = p + r$. Let $F = Q(Z(R_e))$ and $T = F \otimes A_p(k) \otimes A_r'(k)$. Note that T is simple and $R \subseteq T \subseteq R_E$. It follows that $T = R_E \simeq A_p(F) \otimes_F A_r'(F)$. It is easy to see, from 6.5.7, that $K(R_E) \geq p + r$. On the other hand, R_E is a localization of $A_{p+r}(F)$ and so, by 6.5.3 and 6.5.8, $K(R_E) \leq p + r$. □

§8 The Simple Algebras $\mathscr{A}(V, \delta, \Gamma)$

8.1. The preceding sections have shown that under suitable circumstances the simple rings R_E are built up from Weyl algebras and the rings $A_r'(k)$. In general, as will be shown in §9, the simple rings R_E are rather more complicated 'amalgams' of such rings. This section describes the appropriate types of algebra.

8.2 The ingredients required will be fixed throughout this section. They are a finite dimensional k-vector space V, an alternating bilinear form $\delta: V \times V \to k$ and a finitely generated subgroup Γ of the additive group $V^* = \mathrm{Hom}_k(V, k)$. Of course Γ is a free abelian group of finite rank.

8.3 The first step involves δ. Note that since δ is alternating V has a 'standard' basis, say $\{x_1, \ldots, x_p, y_1, \ldots, y_p, z_1, \ldots, z_h\}$ such that $\delta(x_i, y_j) = \delta_{ij}$ and

$$0 = \delta(x_i, x_j) = \delta(y_i, y_j) = \delta(x_i, z_j) = \delta(y_i, z_j)$$

for all i, j. The subspace $\sum k z_i$ is denoted by V^δ and equals $\{v \in V \mid \delta(v, V) = 0\}$.

Now let $S_\delta(V)$ be the k-algebra freely generated by V subject to the relations

$$v_1 v_2 - v_2 v_1 - \delta(v_1, v_2) 1 = 0$$

for all $v_1, v_2 \in V$. This is called the δ-**symmetric algebra** on V since, if $\delta = 0$, the ordinary symmetric algebra $S_0(V)$ is obtained. With respect to the standard basis above

$$S_\delta(V) = A_p(k) \otimes k[z_1, \ldots, z_h]$$

where $k[z_1, \ldots, z_h] = S_0(V^\delta)$ and so $S_\delta(V)$ is a Noetherian integral domain.

8.4 The second step brings in Γ. Each $\gamma \in \Gamma$ induces an automorphism ϑ_γ of $S_\delta(V)$ given by $v \mapsto v - \gamma(v)$ for each $v \in V$. The map $\Gamma \to \mathrm{Aut}(S_\delta(V))$ given by $\gamma \mapsto \vartheta_\gamma$ is an injective group homomorphism whose image we denote by G. We let

$$V^\Gamma = \bigcap \{\ker \gamma \mid \gamma \in \Gamma\}$$

and note that

$$V^\Gamma = \{v \in V \mid \vartheta_\gamma(v) = v \text{ for all } \vartheta_\gamma \in G\}.$$

Finally, we define the object of interest in this section, namely the skew group ring $S_\delta(V) \# G$ which is denoted here by $\mathscr{A}(V, \delta, \Gamma)$ or $\mathscr{A}$ for short. Thus if $g = \vartheta_\gamma$ and $v \in V$ then, in $\mathscr{A}(V, \delta, \Gamma)$, one has $[g, v] = \gamma(v) g$.

Since Γ is free, $\mathscr{A}$ can be viewed as an iterated skew Laurent polynomial extension of $S_\delta(V)$. When combined with the description of $S_\delta(V)$ in 8.3, this shows that $\mathscr{A}$ is a Noetherian integral domain. It also shows that its units are precisely the elements ag, where $g \in G$, $0 \neq a \in k$, from which it is clear that k is its maximal subfield.

8.5 We illustrate the definition by describing some special cases.

Examples. (i) Suppose $\dim V = 2p$, δ is non-singular on V and $\Gamma = \{0\}$. Then $\mathscr{A}(V, \delta, \Gamma) = S_\delta(V) = A_p(k)$.

14.8.9 *The simple algebras $\mathscr{A}(V, \delta, \Gamma)$* 517

(ii) Suppose dim $V = 1$, and so $\delta = 0$. If rank $\Gamma = 1$ then $S_\delta(V) \simeq k[z_1]$ and $\mathscr{A}(V, \delta, \Gamma) \simeq A_1'(k)$. Similarly, if rank $\Gamma = n$ then $\mathscr{A}(V, \delta, \Gamma)$ is isomorphic to a simple algebra described in 1.8.6.

As this last example indicates, the algebras constructed this way are a little more complicated than those obtained in §7.

8.6 Given two sets of ingredients $(V_1, \delta_1, \Gamma_1)$ and $(V_2, \delta_2, \Gamma_2)$ one can form, in the obvious way, the set $(V, \delta, \Gamma) = (V_1 \oplus V_2, \delta_1 \oplus \delta_2, \Gamma_1 \oplus \Gamma_2)$.

Lemma. $\mathscr{A}(V, \delta, \Gamma) = \mathscr{A}(V_1, \delta_1, \Gamma_1) \otimes \mathscr{A}(V_2, \delta_2, \Gamma_2)$. □

8.7 One connection with the problem at hand is easily made.

Proposition. *$\mathscr{A}(V, \delta, \Gamma)$ is a homomorphic image of $U(\mathfrak{g})$ for some completely solvable $\mathfrak{g}$, with $\mathfrak{g}$ of solvable length 2.*

Proof. Suppose Γ has rank r, and let g_i, g_i^{-1} be the generators of G, for $i = 1, \ldots, r$. Let A be the k-vector space spanned by $\{g_1, g_1^{-1}, \ldots, g_r, g_r^{-1}, 1\}$. Then $\mathfrak{g} = A + V$ is a Lie subalgebra of $\mathscr{A}$, A is an abelian ideal of $\mathfrak{g}$, $\mathfrak{g}/A$ is abelian and $\mathfrak{g}$ acts semisimply on A. Evidently $\mathscr{A}$ is a homomorphic image of $U(\mathfrak{g})$. □

8.8 The ideals of $\mathscr{A}(V, \delta, \Gamma)$ are easily described.

Proposition. *If $I \triangleleft \mathscr{A}(V, \delta, \Gamma)$ then I is generated by $I \cap k[z_1, \ldots, z_h]$.*

Proof. Note that $[V, V] \subseteq k$ and so if ad $V = \{\text{ad } v | v \in V\}$ then ad V is an abelian Lie algebra of derivations of $\mathscr{A}$. Further, the elements of G are ad V-eigenvectors; and if $g \in G$ corresponds to $\gamma \in \Gamma$, then γ is the eigenvalue of g.

For any $v \in V$, ad v acts locally nilpotently on $S_\delta(V)$. So $\mathscr{A}^\gamma = S_\delta(V)g$, where $\mathscr{A}^\gamma = \{a \in \mathscr{A} | \text{for each } v \in V \text{ there is an } n \text{ such that } a(\text{ad } v - \gamma(v)1)^n = 0\}$ as in 1.19. By 1.19, if W is an ad V-submodule of $\mathscr{A}$ then $W = \bigoplus_{g \in G}(W \cap S_\delta(V)g)$.

We apply this now to I, to see that I is generated by $I \cap S_\delta(V)$. However, $S_\delta(V) = A_p \otimes k[z_1, \ldots, z_h]$ and so, by 9.6.9, I is generated by $I \cap k[z_1, \ldots, z_h]$. □

8.9 It is now easy to determine when $\mathscr{A}$ is simple.

Theorem. *$\mathscr{A}(V, \delta, \Gamma)$ is simple if and only if $V^\delta \cap V^\Gamma = 0$.*

Proof. Any element in $V^\delta \cap V^\Gamma$ is a central non-unit of both $S_\delta(V)$ and $\mathscr{A}$. So if $V^\delta \cap V^\Gamma \neq 0$ then $\mathscr{A}$ is not simple.

Conversely, suppose $V^\delta \cap V^\Gamma = 0$ and consider the restriction to V^δ of Γ. This gives a subspace of $(V^\delta)^*$. If it is a proper subspace, then by 1.20 there exists $0 \neq v \in V^\delta$ such that $\gamma(v) = 0$ for all $\gamma \in \Gamma$; but then $v \in V^\delta \cap V^\Gamma$, a contradiction.

Hence we can choose $\gamma_1, \ldots, \gamma_h \in \Gamma$ so that their restrictions to V^δ are a basis for $(V^\delta)^*$. Further, we may choose $z_1, \ldots, z_h \in V^\delta$ to be a dual basis. Let $g_1, \ldots, g_h \in G$ correspond to $\gamma_1, \ldots, \gamma_h$. Then the subalgebra

$$\mathscr{B} = k[z_1, \ldots, z_h][g_1, g_1^{-1}, \ldots, g_h, g_h^{-1}]$$

of $\mathscr{A}$ is isomorphic to the simple algebra A'_h and contains $k[z_1, \ldots, z_h]$. Therefore if $0 \neq I \triangleleft \mathscr{A}$ then, by 8.8, $I \cap \mathscr{B} \neq 0$; and so $1 \in I$ and $I = \mathscr{A}$. Hence $\mathscr{A}$ is simple. □

8.10 This, combined with the preceding results, gives information about the general algebra $\mathscr{A}(V, \delta, \Gamma)$. Let V' be a complement in V to $V^\delta \cap V^\Gamma$, and let δ' and Γ' be the restrictions to V' of δ and Γ.

Corollary. (i) $\mathscr{A}(V, \delta, \Gamma) \simeq \mathscr{A}(V', \delta', \Gamma') \otimes_k S_0(V^\delta \cap V^\Gamma)$, *this being the tensor product of a simple algebra with a polynomial algebra.*
  (ii) $S_0(V^\delta \cap V^\Gamma)$ *is the centre of* $\mathscr{A}(V, \delta, \Gamma)$.
  (iii) *The ideals of* $\mathscr{A}(V, \delta, \Gamma)$ *are in* $(1,1)$-*correspondence with those of* $S_0(V^\delta \cap V^\Gamma)$.

Proof. (i) This is clear from 8.6 and 8.9.
  (ii) Since $\mathscr{A}(V', \delta', \Gamma')$ is simple, its centre is a field. The remarks in 8.4 show that its centre must be k, and then 9.6.9 applies.
  (iii) Clear from 9.6.9. □

8.11 Subalgebras $A_p(k)$ and $A'_h(k)$ of $\mathscr{A}$ have been identified above. The next result exhibits $\mathscr{A}$ as a subalgebra.

Proposition. *Let* rank $\Gamma = r$, p *be as in* 8.3 *and* $f = \dim(V^\delta \cap V^\Gamma)$. *Then*

$$\mathscr{A}(V, \delta, \Gamma) \hookrightarrow A_p(k) \otimes A'_r(k) \otimes k[t_1, \ldots, t_f].$$

Proof. It is clear from 8.10 that we may assume that $\mathscr{A}$ is simple and $f = 0$. If $g_1, \ldots, g_r$ are free generators of G and $v \in V$, then $[g_i, v] = \gamma_i(v)g_i$ where $\gamma_i \in \Gamma$ corresponds to $g_i \in G$. Now let $\{x'_i, y'_i\}$ be the generators of $A_p(k)$ and $\{r_i^{\pm 1}, s_i\}$ those of $A'_r(k)$; thus $[x'_i, y'_i] = 1$ and $[r_i, s_i] = r_i$, etc. We define a mapping

$$x_i \mapsto x'_i + \sum_j \gamma_j(x_i)s_j$$

$$y_i \mapsto y'_i + \sum_j \gamma_j(y_i)s_j$$

$$z_i \mapsto \sum_j \gamma_j(z_i)s_j$$

$$g_j \mapsto r_j$$

and note that, since this preserves all the relations on the generators of $\mathscr{A}$, it

14.8.13 The simple algebras $\mathscr{A}(V,\delta,\Gamma)$

gives a homomorphism from $\mathscr{A}$ into the given algebra. This is clearly nonzero and so, since $\mathscr{A}$ is simple, is injective. □

8.12 There is a useful alternative description of $\mathscr{A}(V,\delta,\Gamma)$.

Proposition. $\mathscr{A}(V,\delta,\Gamma) \simeq kG * U(V)$ where V is viewed as an abelian Lie algebra.

Proof. Choose a basis $\{v_i\}$ for V and $\{g_j\}$ for G. The monomials in $\{v_i, g_j, g_j^{-1}\}$ from a basis for $\mathscr{A}$ over k. Furthermore, $kG \subseteq \mathscr{A}$ and, for each i,j, $[v_i, g_j] \in kG$ and $[v_i, v_j] \in k$. Evidently the conditions in the definition, 1.7.12, are satisfied. □

8.13 This description helps one compute the various dimensions of $\mathscr{A}$. First, however, note that one can choose the standard basis, described in 8.3, starting with V^Γ. Thus we can arrange that

$$\{x_1,\ldots,x_q, y_1,\ldots,y_r, z_1,\ldots,z_j\}$$

is a basis for V^Γ where $q \leq r \leq p$ and $j \leq h$. Let $t = r + j$ (this being the maximal dimension of a subspace of V^Γ on which δ is zero).

Theorem. (i) $\mathrm{GK}(\mathscr{A}) = \mathrm{rank}\,\Gamma + \dim V$.
(ii) $K(A) = \mathrm{gld}\,\mathscr{A} = \mathrm{rank}\,\Gamma + t$.

Proof. (i) Since Γ is free abelian, 8.2.15 shows that $\mathrm{GK}(kG) = \mathrm{rank}\,\Gamma$. Then 8.2.10, combined with 8.12, gives the result.
(ii) Let V' be the subspace of V spanned by

$$\{x_{q+1},\ldots,x_p, y_{r+1},\ldots,y_p, z_{j+1},\ldots,z_h\}.$$

Since $V' \cap V^\Gamma = 0$, the argument in the proof of 8.9 shows that the restriction to V' of Γ spans $(V')^*$. Hence we can choose dual bases $\{\gamma_1,\ldots,\gamma_m\}$ for $(V')^*$ and $\{v_1,\ldots,v_m\}$ for V'.

Now let $\mathscr{B} = kG[y_1,\ldots,y_r, z_1,\ldots,z_j]$ and let $\dim \mathscr{B}$ denote either $K(\mathscr{B})$ or $\mathrm{gld}\,\mathscr{B}$. We know, by 6.5.4 and 7.5.3, that

$$\dim \mathscr{B} = \mathrm{rank}\,\Gamma + r + j = \mathrm{rank}\,\Gamma + t.$$

We can view $\mathscr{A}$ as being constructed as an iterated skew polynomial ring over $\mathscr{B}$, using the sequence of indeterminates

$$g_1^{-1}v_1,\ldots,g_m^{-1}v_m, x_1,\ldots,x_q,$$

where g_i corresponds to γ_i. However, the sequence

$$g_1,\ldots,g_m, y_1,\ldots,y_q$$

provides a central element at each stage to which 9.1.14 can be applied. Thus $\dim \mathscr{A} = \dim \mathscr{B}$. □

§9 The General Case

9.1 Throughout this section g is completely solvable and R is a prime factor ring of $U(g)$. The aim here, as in the algebraic case described in §7, is to elucidate the structure of R_e and R_E.

The arguments used in §7 depended heavily upon the decomposition $g = n \oplus s$ which, in particular, allowed the use of 6.8 in an inductive proof. In the general case such a decomposition is not available, and a more detailed analysis is required before R_e and R_E can be described.

We will, eventually, show that R_E is one of the simple algebras $\mathscr{A}(V, \delta, \Gamma)$ over an extension field K of k.

9.2 The notation throughout this analysis will be cumulative. It includes $g\hat{\,}$ and $R\hat{\,}$ as in 4.4. Applied to $R\hat{\,}$, 6.8 provides an eigenvector (for g as well as for $g\hat{\,}$), e_1 say. For R itself 4.11 together with 4.5(ii) provides another, e_2 say. Combining these in $e = e_1 e_2$, all the claims of 6.8, applied to $R_e\hat{\,}$, and 4.5 and 4.12 applied to R_e are valid. In particular this gives

Proposition. (i) $R_e\hat{\,} \simeq Z(R_e\hat{\,}) \otimes A_p(k)$ for some p.
(ii) $Z(R_e\hat{\,})$ is affine over k.
(iii) $R_e = R_e\hat{\,}[x_1; \delta_1] \cdots [x_m; \delta_m]$, where $x_1, \ldots, x_m$ are a basis of a complementary subspace to $g\hat{\,}$ in g. □

9.3 The strategy employed over the next few pages involves choosing generators of the k-algebra R_e in such a way that its structure becomes plain. The first step modifies the choice of $x_1, \ldots, x_m$. Note, by 6.5, that for each $i \in \{1, \ldots, m\}$, there exists $s_i \in R_e\hat{\,}$ such that $\delta_i - \text{ad } s_i$ is zero on $A_p(k)$ and a derivation on $Z(R_e\hat{\,})$. Moreover, as noted after 6.5, this latter derivation coincides with δ_i. Thus if each x_i is replaced by the element $x_i - s_i$ then the δ_i become trivial on $A_p(k)$.

Proposition. $R_e \simeq Z(R_e\hat{\,})[x_1; \delta_1] \cdots [x_m; \delta_m] \otimes A_p(k)$. □

9.4 Let $S = Z(R_e\hat{\,})[x_1; \delta_1] \cdots [x_m; \delta_m]$; so $R_e \simeq S \otimes A_p(k)$. We now concentrate attention on S. Note that under the adjoint action of g on $Z(R_e\hat{\,})$ the subalgebra $g\hat{\,}$ acts trivially. Thus $Z(R_e\hat{\,})$ is a $g/g\hat{\,}$ module.

9.5 We note that S is a crossed product.

Proposition. $S \simeq Z(R_e\hat{\,}) * U(g/g\hat{\,})$, where $g/g\hat{\,}$ is an abelian Lie algebra of dimension m.

Proof. It was noted in 4.4 that $g/g\hat{\,}$ is abelian; and since m is the dimension of a complementary subspace to $g\hat{\,}$ in g, it is clear that $\dim(g/g\hat{\,}) = m$. The definition of $g\hat{\,}$ ensures that $g/g\hat{\,}$ acts faithfully on $Z(R_e\hat{\,})$ and the result is now clear. □

14.9.8 *The general case* 521

In particular this means that we may take $\{x_1,\ldots,x_m\}$ to be a basis for $g/g\hat{}$. In the next few pages the choice of $x_1,\ldots,x_m$ will be refined several times. On each occasion this will involve replacing an x_i by $x_i + u_i$ for some $u_i \in Z(R_e\hat{})$. In effect this gives a different crossed product structure for $Z(R_e\hat{}) * U(g/g\hat{})$; but each x_i retains the same adjoint action on $Z(R_e\hat{})$.

9.6 As noted in 9.2, 4.11 and 4.12 can be applied. Thus

$$\Lambda(R_e; g) = \Gamma(R_e; g) = \Gamma(R; g) = \Gamma \text{ say,}$$

$\gamma_1, \ldots, \gamma_r$ is a $\mathbb{Z}$-basis for Γ, $w_1, \ldots, w_r$ are corresponding eigenvectors in R_e and W is the free abelian group of rank r generated by $w_1, \ldots, w_r$.

9.7 Since $Z(R_e\hat{})$ is, by 9.4, a $g/g\hat{}$ module and since $g/g\hat{}$ is abelian, there is, by 1.19, a decomposition of $Z(R_e\hat{})$ into weight spaces. Let C denote the weight space corresponding to the eigenvalue 0, this being a subalgebra, of course.

Proposition. *Let $\gamma \in \Gamma$ and let $w_\gamma \in W$ be an eigenvector for γ. Then:*
(i) *the γ-weight space is Cw_γ;*
(ii) $Z(R_e\hat{}) = C[w_1, w_1^{-1}, \ldots, w_r, w_r^{-1}]$, *the Laurent polynomial ring;*
(iii) C *is affine over k.*

Proof. (i) It is easy to check that an element a belongs to the weight space of γ if and only if $zw_\gamma^{-1} \in C$.

(ii) Since distinct monomials in the $w_i^{\pm 1}$ correspond to distinct eigenvalues, the weight space decomposition of $Z(R_e\hat{})$ is simply

$$Z(R_e\hat{}) = \oplus \{Cw | w \in W\}$$

and so $Z(R_e\hat{})$ is as claimed.

(iii) By 9.2(ii) $Z(R_e\hat{})$ is k-affine and, by (ii), C is a factor. □

9.8 The next step is to decompose C. In doing this, it is necessary to invert an element of C. Since that can be incorporated into e, as before, it will promptly be ignored.

Of course C is a $g/g\hat{}$-module. Let D be the submodule, and subalgebra, annihilated by $g/g\hat{}$.

Proposition. (i) *After possibly inverting an element of D and reordering the basis $x_1, \ldots, x_m$ of $g/g\hat{}$, there exist $y_1, \ldots, y_f \in C$ with $f \leq m$ such that $C = D[y_1, \ldots, y_f]$, a commutative polynomial ring, and (y_j) ad $x_i \in D$ for all i, j, with (y_i) ad $x_i = 1$ and (y_j) ad $x_i = 0$ whenever $i < j$.*

(ii) *Furthermore, D is affine over k and R_e is a free right and left D-module.*

Proof. (i) Note first that since $g/g\hat{}$ is abelian either $C = D$, in which case the result is clear, or else there exists some x_i, say x_1, and some $y_1 \in C \setminus D$ such that

$0 \neq (y_1)$ ad $x_1 \in D$. By inverting this element of D we may assume that (y_1) ad $x_1 = 1$. It follows from 6.4 that $C = C'[y_1]$ where $C' = \ker(\text{ad } x_1) \supseteq D$. This same process can now be repeated on the subalgebra C' to yield the result.

(ii) These follow immediately using 9.7, 9.3 and 6.7. □

9.9 The situation so far is that, with suitable choice of e,

$$R_e = D[y_1, \ldots, y_f][w_1, w_1^{-1}, \ldots, w_r, w_r^{-1}][x_1, \ldots, x_m] \otimes_k A_p(k),$$

where $D = Z(R_e)$. As things stand, the commutators $[x_i, x_j]$ belong to the ring

$$D[y_1, \ldots, y_f][w_1, w_1^{-1}, \ldots, w_r, w_r^{-1}] = T \text{ say}.$$

The next step is to modify the choice of $\{x_i\}$ so as to arrange that $[x_i, x_j] \in D$. This will be achieved by replacing x_i by $x_i - t_i$ for some $t_i \in T$ and so will not affect the derivation ad x_i on T. As a preliminary step we have

Lemma. *The elements $x_1, \ldots, x_m$ can be chosen so that*

$$[x_i, x_j] \in \bigcap \{\ker(\text{ad } x_h) | h = 1, \ldots, f\} = H \text{ say}$$

for all i, j, and $[x_i, x_j] = 0$ whenever $i \leq f$. Furthermore, $H \subseteq D[w_1, w_1^{-1}, \ldots, w_r, w_r^{-1}]$.

Proof. Note first that each weight space of $g/g^{\hat{}}$ in T is stable under ad x_1. Indeed if $w \in W$ has eigenvalue $\gamma \in \Gamma$ then ad x_1 coincides, on Cw, with $\partial/\partial y_1 + \gamma(x_1)1$. Thus ad x_1 is surjective on each weight space.

Suppose $[x_1, x_i] = r_i \in T$. Then $r_i = \sum \{r_{i\gamma} | \gamma \in \Gamma\}$ where $r_{i\gamma}$ is in the weight space of γ. The surjectivity ensures the existence of $s_{i\gamma}$ in that weight space with $(s_{i\gamma})$ ad $x_1 = r_{i\gamma}$. Therefore (s_i) ad $x_1 = r_i$ where $s_i = \sum s_{i\gamma}$. Hence if x_i is replaced by $x_i - s_i$ for $i = 2, \ldots, m$ then x_1 commutes with each new x_i.

Let $T' = D[y_2, \ldots, y_f][w_1, w_1^{-1}, \ldots, w_r, w_r^{-1}]$ and note that $\ker(\text{ad } x_1) \subseteq T'$ and that $x_2, \ldots, x_m \in \ker(\text{ad } x_1)$. The process can now be repeated with x_2 and T'. No change in x_1 will be required; and since $[x_2, x_j] \in \ker(\text{ad } x_1)$ the required modifications to $x_3, \ldots, x_m$ will not affect their commutators with x_1.

Repetition for $x_3, \ldots, x_m$ yields the result. □

9.10 The final modification of the x_i comes next.

Proposition. *The elements $x_1, \ldots, x_m$ can be chosen so that $[x_i, x_j] \in D$ for all i, j and $[x_i, x_j] = 0$ if $i \leq f$.*

Proof. Consider ad x_i acting on $D[w_1, w_1^{-1}, \ldots, w_r, w_r^{-1}]$, where $i \in \{f+1, \ldots, m\}$. If $w \in W$ and $\gamma \in \Gamma$ is the corresponding eigenvalue then (dw) ad $x_i = \gamma(x_i)dw$ for each $d \in D$. Therefore

$$\text{im}(\text{ad } x_i) \oplus \ker(\text{ad } x_i) = D[w_1, w_1^{-1}, \ldots, w_r, w_r^{-1}].$$

Hence, when the action is restricted to H, as above, one sees that

$$\operatorname{im}(\operatorname{ad} x_i) \oplus \ker(\operatorname{ad} x_i) = H.$$

Note that x_{f+1} acts semisimply on H. Hence, by replacing x_j by $x_j - a_j$ for some $a_j \in \operatorname{im}(\operatorname{ad} x_{f+1})$, it can be arranged that

$$[x_{f+1}, x_j] \in H \cap \ker(\operatorname{ad} x_{f+1})$$

for $j = f+2, \ldots, m$.

Repetition of this process for $x_{f+2}, \ldots, x_m$, each time altering all of $x_{f+1}, \ldots, x_m$ except for the element concerned, allows us to arrange that

$$[x_i, x_j] \in \bigcap \{\ker(\operatorname{ad} x_h) | h = 1, \ldots, m\}$$

for all i, j; i.e. $[x_i, x_j] \in Z(R_e) = D$. □

9.11 This process of adjusting the elements x_i has not affected the description

$$R_e = D[y_1, \ldots, y_f][w_1, w_1^{-1}, \ldots, w_r, w_r^{-1}][x_1; \delta_1] \cdots [x_m; \delta_m] \otimes A_p(k).$$

Thus R_e is still free over D.

9.12 We now aim towards the structure of R_E. Let K be the quotient field of D. Since $D = Z(R_e)$ then $K \otimes R_e$ is the localization of R_e with respect to $D \setminus \{0\}$. Equally, it is a localization of R with respect to a particular subset of E. In fact it will turn out that $K \otimes R_e = R_E$ and is a K-algebra $\mathcal{A}(V, \delta, \Gamma)$.

9.13 Let U be the free D-submodule of R_e with basis $\{y_1, \ldots, y_f, x_1, \ldots, x_m\}$ and U_0 the free k-submodule with the same basis. Let $V = K \otimes_D U = K \otimes_k U_0$.

Now let $\gamma \in \Gamma$ and let $w \in W$ be the corresponding eigenvector. Note that γ is defined on $\sum_{i=1}^m k x_i$, but can be extended to a linear form on U_0 by annihilating $y_1, \ldots, y_f$. Then for all $u \in U_0$, $[w, u] = \gamma(u) w$. It can be extended further, to say $\gamma' \in V^*$, by defining $\gamma'(\sum c_i \otimes u_i) = \sum c_i \gamma(u_i)$, where $c_i \in K$ and $u_i \in U_0$. The map $\gamma \mapsto \gamma'$ gives an injective group homomorphism $\Gamma \hookrightarrow V^*$.

The bracket product $[\ ,\ ]$ is a K-valued alternating bilinear form, δ say, on V. The algebra $S_\delta(V)$, as defined in 8.3, is then given by

$$S_\delta(V) = K \otimes_D D[y_1, \ldots, y_f][x_1; \delta_1] \cdots [x_m; \delta_m].$$

The words $w \in W$ acting on $S_\delta(V)$ yield the automorphisms ϑ_γ as in 8.4.

Proposition. *The K-algebra*

$$K \otimes_D D[y_1, \ldots, y_f][w_1, w_1^{-1}, \ldots, w_r, w_r^{-1}][x_1; \delta_1] \cdots [x_m; \delta_m]$$

is isomorphic to the K-algebra $\mathcal{A}(V, \delta, \Gamma)$. □

9.14 Lemma. $V^\delta \cap V^\Gamma = 0$.

Proof. If the given basis of V is partitioned as

$$\{y_1, \ldots, y_f\} \cup \{x_1, \ldots, x_f\} \cup \{x_{f+1}, \ldots, x_m\}$$

the resulting block matrix representation of δ takes the form

$$\begin{bmatrix} 0 & A & * \\ B & * & * \\ * & * & * \end{bmatrix} \text{ with } A = \begin{bmatrix} -1 & & * \\ & -1 & \\ & & \ddots \\ 0 & & -1 \end{bmatrix} \text{ and } B = \begin{bmatrix} 1 & & 0 \\ & 1 & \\ & & \ddots \\ * & & 1 \end{bmatrix},$$

where $*$ denotes unspecified entries. Since

$$\det \begin{bmatrix} 0 & A \\ B & * \end{bmatrix} \neq 0$$

it follows that $V^\delta \subseteq Kx_{f+1} + \cdots + Kx_m = Kx$, where $x = kx_{f+1} + \cdots + kx_m$.

The elements of x still have the same adjoint action on W as those of $\mathbf{g}/\mathbf{g}^{\hat{}}$. Thus none is annihilated by all $\gamma \in \Gamma$, and so, by 1.20, Γ, restricted to x, spans x^*. Dual bases can, therefore, be chosen for x and for this restriction. When extended to Kx it follows that $V^\Gamma \cap Kx = 0$. Hence $V^\delta \cap V^\Gamma = 0$. □

9.15 The structure of R_E can now be deduced.

Theorem. $R_E = K \otimes_D R_e$ and is a simple K-algebra $\mathscr{A}(V', \delta', \Gamma')$ for some V', δ', Γ'.

Proof. In 9.13 it is shown that

$$K \otimes_D R_e \simeq \mathscr{A}(V, \delta, \Gamma) \otimes_k A_p(k)$$
$$\simeq \mathscr{A}(V, \delta, \Gamma) \otimes_K A_p(K).$$

But $A_p(K) \simeq \mathscr{A}(V_1, \delta_1, \Gamma_1)$, by 8.5 and so, using 8.6,

$$K \otimes_D R_e \simeq \mathscr{A}(V', \delta', \Gamma').$$

By 9.14 and 8.9, the algebra $\mathscr{A}(V, \delta, \Gamma)$ is simple and thus $K \otimes R_e$ is simple. In particular all normal elements in $K \otimes R_e$ must be units; so it follows that $K \otimes R_e = R_E$. □

9.16 Combining this with 4.13 we obtain

Corollary. *If R is primitive then there exists $e \in E(R)$ such that R_e is simple and is isomorphic to an $\mathscr{A}(V, \delta, \Gamma)$ over an extension field K of k.* □

9.17 The final portion of this section considers the subring R_e of R_E once again. The aim is to view it as a version of $\mathscr{A}(V, \delta, \Gamma)$ but over D rather than over K.

14.9.19 *The general case*

After an appropriate choice of e, the proofs of §8 will be adapted to yield a $(1, 1)$-correspondence between the ideals of R_e and those of D.

The K-space V has a standard basis with respect to δ, say $\{y'_1, \ldots, y'_q, x'_1, \ldots, x'_q, z_1, \ldots, z_t\}$, giving $S_\delta(V)$ the form $A_q(K) \otimes_K K[z_1, \ldots, z_t]$. By modifying e appropriately it can be assumed that this basis belongs to U and is a basis for U over D. Therefore, letting $S_\delta(U)$ be the D-subalgebra of $S_\delta(V)$ generated by U, we have

Lemma. $S_\delta(U) \simeq A_q(k) \otimes_k D[z_1, \ldots, z_t]$. □

9.18 Now set $U^\delta = \{u \in U \mid \delta(u, U) = 0\}$. Then U^δ is the free D-module with basis $\{z_1, \ldots, z_t\}$; and $K \otimes_D U = V$ and $V^\delta = K \otimes_D U^\delta$. No element of V^δ is annihilated by all $\gamma \in \Gamma$; so, by 1.20, we can choose $\gamma_1, \ldots, \gamma_t \in \Gamma$ whose restrictions to V^δ are a basis for $(V^\delta)^*$ and a dual basis in V^δ. By modifying e again, this dual basis may be arranged to lie in U^δ. Without loss of generality, let it be $z_1, \ldots, z_t$. Let $w_1, \ldots, w_t$ be eigenvectors corresponding to the eigenvalues $\gamma_1, \ldots, \gamma_t$.

Lemma. $D[z_1, \ldots, z_t][w_1, w_1^{-1}, \ldots, w_t, w_t^{-1}] \simeq A'_t(k) \otimes_k D$.

Proof. Clear. □

9.19 The next result effectively extends 8.8 and 8.9 to this more general case, and the proof below mimics their proofs.

Theorem. *With the above choice of e, there is a $(1, 1)$-correspondence between ideals I of R_e and J of D given by $I \mapsto I \cap D$, $J \mapsto JR_e$. This correspondence preserves primeness.*

Proof. The factor $A_p(k)$ in R_e can, as in 9.11, safely be ignored; for ease of notation we will assume it is absent. Note that, since R_e is free over D then $JR_e \cap D = J$.

Note next that the commutator of any two elements of U_0 lies in D. Hence ad U_0 is an abelian Lie algebra of linear transformations of R_e and, moreover, ad U_0 acts locally nilpotently on $S_\delta(U)$. Also, if $w_\gamma \in W$ corresponds to $\gamma \in U_0^*$ the corresponding weight space in R_e is $S_\delta(U)w_\gamma$.

Now let $I \triangleleft R_e$. From above, $I = \bigoplus_\gamma (I \cap S_\delta(U))w_\gamma$ and so I is generated by $I \cap S_\delta(U)$. But then, by 9.17, I is generated by $I \cap D[z_1, \ldots, z_t]$. Therefore, in the notation of 9.18, I is generated by

$$I \cap D[z_1, \ldots, z_t][w_1, w_1^{-1}, \ldots, w_t, w_t^{-1}]$$

and hence by $I \cap D$ as required. □

Since D is, by 9.8, a commutative affine k-algebra, this shows that the ideal structure and the prime spectrum of R_e are those of such an algebra. This is sometimes described by saying that R is 'generically a commutative affine k-algebra'.

§10 Additional Remarks

10.0 (a) The general theory of enveloping algebras is covered by [Dixmier **77**] including much of the basic material in the early sections of this chapter. Other results are in [Joseph **81**].

(b) [Borho, Gabriel and Rentschler **73**] deal with the solvable case, and our account in §§5–7 is heavily influenced by it.

(c) The material in §§8–9 has not appeared previously in a book.

(d) The semisimple case is covered by [Jantzen **83**]; see also [Joseph **83, 84**], and [Borho **86**] together with the other papers in that volume.

10.1 No comment.

10.2 (a) Most of this appears in [Dixmier **77**]. However, 2.7 is new and 2.8 has a much shorter proof than the non-Noetherian version due to [Passman **Pb**].

(b) When k is algebraically closed there is a geometric aspect covered by [Dixmier **77**] and [Borho, Gabriel and Rentschler **73**]. Recall that if $g = \sum_{i=1}^{n} kx_i$ is abelian then the set of maximal ideals of $U(g)$ is in $(1,1)$-correspondence with g^* via

$$\lambda \mapsto (x_1 - \lambda(x_1), \ldots, x_n - \lambda(x_n)) \quad \text{for } \lambda \in g^*.$$

Similarly, if g is solvable there is a $(1,1)$-correspondence between the set of primitive ideals of $U(g)$ and g^*/G, the space of orbits of the adjoint algebraic group G of g.

(c) If k is algebraically closed and g is solvable then [Gabber **P**] proves that $U(g)$ has the catenary property, defined in 13.10.13; see also [Malliavin **79**] and [Lorenz **81a**].

10.3 3.4(iv) and 3.6 are due to [McConnell **68**].

10.4 4.13 is due to [Dixmier **77**, 4.5.7].

10.5 5.5 comes from [Gabriel **71**].

10.6, 10.7 The material in these sections is mainly due to Borho and to [McConnell **74a**]. However, 6.9 is a result of [Nouazé and Gabriel **67**].

10.8, 10.9 (a) The results in §§8 and 9 come from [McConnell **75, 77b**], but the proofs here are new and more elementary. 8.13 comes from [Tauvel **78**] and [McNaughton **77**]; see also [McConnell **77a, 82a**].

(b) The algebra $\mathscr{A}(V, \delta, \Gamma)$ determines the data V, δ and Γ up to an isomorphism of the data, i.e. an isomorphism of vector spaces and of groups preserving the actions and form, [McConnell **P**].

(c) [Heinicke **81a**] shows that, for any solvable Lie algebra g and any factor ring of $U(g)$, the left and right Krull dimensions are equal; see also 6.4.10 and [Brown and Smith **85**].

Chapter 15

RINGS OF DIFFERENTIAL OPERATORS ON ALGEBRAIC VARIETIES

The content of this chapter is most easily described via the Weyl algebra $A_n(k)$ over a field k of characteristic 0. This can be viewed as the ring of differential operators either on the algebraic variety k^n, or on its coordinate ring $k[y_1,\ldots,y_n]$. Earlier chapters have shown that $A_n(k)$ has many interesting properties. These can, to a large extent, be extended to the ring of differential operators on any nonsingular (i.e. smooth) irreducible affine algebraic variety V, or equivalently on its coordinate ring $A = A(V)$ which is then regular. This ring of differential operators is denoted by $\mathscr{D}(A)$. It will be shown under these circumstances that $\mathscr{D}(A)$ is a simple Noetherian domain, that $K_0(\mathscr{D}(A)) \simeq K_0(A)$ and that gld $\mathscr{D}(A) =$ dim $A = \frac{1}{2}\mathrm{GK}(\mathscr{D}(A)) = \mathcal{K}(\mathscr{D}(A))$.

The study of $\mathscr{D}(A)$ builds upon an investigation of the derivations and the differentials of the commutative ring A. In fact $\mathscr{D}(A)$ is approached here via the derivation ring $\Delta(A)$, this being the ring generated by A and its derivations. The relevance of $\Delta(A)$ becomes clear in §5, where $\mathscr{D}(A)$ is defined and is shown to coincide with $\Delta(A)$ when A is regular.

Some general aspects are studied in §1 with the main techniques being established in §2. This involves localization of $\Delta(A)$ at the powers of a single, carefully chosen element, together with a corresponding globalization. The localized rings are shown to be very close to Weyl algebras and it is this which enables one in later sections to extend results from $A_n(k)$ to $\Delta(A)$ and $\mathscr{D}(A)$. Whilst some hold generally, most rely heavily upon A being regular, as examples will indicate.

The presence of the commutative ring A at the heart of this theory leads, as we shall see, to a left–right symmetry in the arguments concerning $\Delta(A)$ when A is regular. It is convenient, on the whole, to concentrate on left modules in this chapter.

§1 Algebras Over a Ring

1.1 In this section the basic properties of the derivations, differentials and differential operators of a commutative K-algebra A, over a commutative ring K, are studied. **This notation will be kept fixed throughout the section,** and R will always denote the commutative polynomial K-algebra $K[y_1,\ldots,y_n]$.

1.2 The set of all K-derivations of A is denoted by $\operatorname{Der}_K A$. As noted earlier, the commutator of two derivations is again a derivation. This product gives $\operatorname{Der}_K A$ the structure of a K-Lie algebra, provided we allow such an algebra to be non-free over K. The first result notes an obvious A-module structure on $\operatorname{Der}_K A$ and a connection between the two products.

Lemma. Let $a, b \in A$, $\gamma, \delta \in \operatorname{Der}_K A$. Then:
(i) $\operatorname{Der}_K A$ is an A-module via $(a\delta)(b) = a\delta(b)$;
(ii) $[a\delta, b\gamma] = a\delta(b)\gamma - b\gamma(a)\delta + ab[\delta, \gamma]$. □

1.3 $\operatorname{Der}_K R$ is easily described.

Example. $\operatorname{Der}_K R$ *is a free R-module of rank n with basis* $\{\partial/\partial y_1, \ldots, \partial/\partial y_n\}$.

Proof. If $\delta \in \operatorname{Der}_K R$ and $f \in R$ then
$$\delta(f) = \sum (\partial f/\partial y_i)\delta(y_i).$$
□

1.4 The main object of study in this chapter is the K-subalgebra of $\operatorname{End}_K A$ generated by A and $\operatorname{Der}_K A$. This ring of differential operators on A is denoted by $\Delta(A)$ and is called the **derivation ring** of A. (In 5.2 a larger ring, the ring of differential operators $\mathscr{D}(A)$, will be defined and its relationship with $\Delta(A)$ described.)

We note that A is a cyclic left $\Delta(A)$-module generated by 1_A. If we let $I = \operatorname{ann}_{\Delta(A)}(1_A)$ then it is clear that $I \supseteq \Delta(A) \operatorname{Der}_K A$ and that $I \cap A = 0$. Hence
$$I = \Delta(A) \operatorname{Der}_K A \quad \text{and} \quad \Delta(A) = A \oplus \Delta(A) \operatorname{Der}_K A.$$
So $A \simeq \Delta(A)/\Delta(A) \operatorname{Der}_K A$.

1.5 Example. $\Delta(R) \simeq A_n(K)$, *the n-th Weyl algebra over K.*

Proof. $\Delta(R) = K[y_1,\ldots,y_n][\partial/\partial y_1,\ldots,\partial/\partial y_n]$ by 1.3. □

1.6 It is convenient, on occasion, to consider not only $\Delta(A)$ but also the K-subalgebras of $\Delta(A)$ generated by A and $\boldsymbol{d}$, where $\boldsymbol{d}$ is any A-submodule of $\operatorname{Der}_K A$ closed under the Lie product. This subalgebra will be denoted by $A[\boldsymbol{d}]$; so then $\Delta(A) = A[\operatorname{Der}_K A]$. The study of $\Delta(A)$ and $A[\boldsymbol{d}]$ depends upon a close study of $\operatorname{Der}_K A$ which we now begin.

1.7 It is convenient first to extend the notion of a K-derivation. If M is an A-module then a K-module homomorphism $\delta: A \to M$ is called a K-**derivation** of A to M if $\delta(ab) = a\delta(b) + b\delta(a)$ for all $a, b \in A$. The set of all such derivations is denoted by $\mathrm{Der}_K(A, M)$.

Of course $\mathrm{Der}_K(A, A) = \mathrm{Der}_K A$. As with $\mathrm{Der}_K A$, it is easily seen that $\mathrm{Der}_K(A, M)$ is an A-module via $(a\delta)(b) = a(\delta(b))$.

1.8 There is a universal derivation from A to a module, and it is easily described. Let F be the free A-module on symbols da, $a \in A$, and let N be the submodule of F generated by $d\alpha$, $d(a + b) - da - db$, and $d(ab) - a(db) - b(da)$ for $\alpha \in K$, $a, b \in A$. Then $\Omega = \Omega_K(A) = F/N$ is the **module of (Kähler) differentials** of A and the derivation $d = d_A: A \to \Omega_K(A)$ given by $d(a) = da$ is the **universal derivation** of A.

1.9 The universality of Ω and d is summed up next.

Proposition. *Given any A-module M and derivation $\delta \in \mathrm{Der}_K(A, M)$ there is a unique $\varphi \in \mathrm{Hom}_A(\Omega(A), M)$, such that $\delta = \varphi d$.*

Proof. Straightforward. □

1.10 There is a close relationship with $\mathrm{Der}_K A$.

Proposition. *(i) Let M be any A-module. The map $\mathrm{Hom}_A(\Omega(A), M) \to \mathrm{Der}_K(A, M)$ given by $\varphi \mapsto \varphi d$ is an isomorphism of A-modules.*
 (ii) $\mathrm{Der}_K A \simeq \mathrm{Hom}_A(\Omega(A), A) = \Omega(A)^*$.

Proof. (i) It is clear that the map is a well-defined A-homomorphism; and 1.9 shows it to be an isomorphism.
 (ii) Clear. □

1.11 Once again we use R as an example.

Example. *$\Omega_K(R)$ is a free R-module of rank n on the generators $dy_1, \ldots, dy_n$.*

Proof. As in 1.3, one sees that
$$df = \sum (\partial f/\partial y_i) dy_i$$
for each $f \in R$. Hence $\Omega(R)$ is generated over R by the dy_i. Now let M be a free R-module with basis $m_1, \ldots, m_n$ and define $\delta \in \mathrm{Der}_K(R, M)$ by $\delta(y_i) = m_i$ for each i. The map φ given by 1.9 maps dy_i to m_i; therefore the dy_i freely generate $\Omega(R)$. □

1.12 A simple consequence of 1.10 is often useful.

Corollary. *Suppose that $K \subseteq A$ and has no $\mathbb{Z}$-torsion, and that $\Omega_K(A)$ is free over A on a basis $\{da_1, \ldots, da_n\}$ with $a_i \in A$. Then:*
(i) *$\{a_1, \ldots, a_n\}$ is a set of indeterminates over K.*
(ii) *The derivations $\partial/\partial a_i$ extend uniquely from $K[a_1, \ldots, a_n]$ to A.*
(iii) *$\operatorname{Der}_K A$ is free on the basis $\{\partial/\partial a_1, \ldots, \partial/\partial a_n\}$.*
(iv) *Each derivation on A is the unique extension of a derivation from $K[a_1, \ldots, a_n]$ to A.*
(v) *If $\gamma, \delta \in \operatorname{Der}_K A$ and $[\gamma, \delta](a_i) = 0$ for each i then $[\gamma, \delta] = 0$.*

Proof. The dual basis lemma shows that $\operatorname{Der}_K A$ is free on derivations $\delta_1, \ldots, \delta_n$, these having the property that $\delta_i(a_i) = 1$, $\delta_i(a_j) = 0$ if $i \neq j$. Applying any δ_i to an equation showing K-dependency of $\{a_1, \ldots, a_n\}$ yields another equation of smaller degree. The choice of one of least degree demonstrates the truth of (i), and (ii), (iii) and (iv) follow directly with δ_i being the extension of $\partial/\partial a_i$. Finally (v) follows from (iv) since $[\gamma, \delta] = 0$ when restricted to $K[a_1, \ldots, a_n]$. □

1.13 The effect of homomorphisms on derivations and differentials is dealt with by the next few results. The duality given by 1.10 is used in conjunction with the following fact [Atiyah and Macdonald **69**, 2.9]:

A sequence of B-modules
$$N' \xrightarrow{\alpha} N \xrightarrow{\beta} N'' \longrightarrow 0$$
is exact if and only if the induced sequence
$$0 \to \operatorname{Hom}_B(N'', M) \xrightarrow{\sigma} \operatorname{Hom}_B(N, M) \xrightarrow{\tau} \operatorname{Hom}_B(N', M)$$
is exact for all ${}_B M$. Evidently α is a split injection if and only if τ is surjective for all ${}_B M$.

Lemma. *Let A, B be commutative K-algebras, $\psi: A \to B$ a K-algebra homomorphism and M a B-module. Then:*
(a) *$\operatorname{Hom}_B(B \otimes_A \Omega_K(A), M) \simeq \operatorname{Der}_K(A, M)$.*
(b) *There are exact sequences of B-modules*

(i) $0 \longrightarrow \operatorname{Der}_A(B, M) \xrightarrow{\sigma} \operatorname{Der}_K(B, M) \xrightarrow{\tau} \operatorname{Der}_K(A, M)$

(ii) $B \otimes_A \Omega_K(A) \xrightarrow{\alpha} \Omega_K(B) \xrightarrow{\beta} \Omega_A(B) \longrightarrow 0$

with α being a split injection if and only if τ is surjective for all ${}_B M$.

Proof. (a)
$$\operatorname{Hom}_B(B \otimes_A \Omega_K(A), M) \simeq \operatorname{Hom}_A(\Omega_K(A), \operatorname{Hom}_B(B, M)) \simeq \operatorname{Hom}_A(\Omega_K(A), M)$$
$$\simeq \operatorname{Der}_K(A, M)$$

using 1.10 for the final isomorphism.

(b) (i) The B-module structure on M makes $\operatorname{Der}_K(A, M)$ into a B-module. The rest is clear.

(ii) We define α, β by $\alpha(b \otimes da) = bd(\psi(a))$ and $\beta(bd(b')) = bd(b')$. Applied to the sequence in (ii), $\text{Hom}_B(-, M)$ yields the sequence (i), using (a). The rest now follows from the fact above. □

1.14 One obvious consequence is noted for later use.

Corollary. *If A, B, ψ are as above and if $\Omega_K(A) = 0$ then $\Omega_K(B) \simeq \Omega_A(B)$.* □

1.15. For factor rings, another more special result can be obtained.

Lemma. *Let B be a commutative K-algebra, $I \triangleleft B$, $A = B/I$ and M be an A-module. Then there are exact sequences of B-modules:*

(i) $0 \to \text{Der}_K(A, M) \xrightarrow{\alpha} \text{Der}_K(B, M) \xrightarrow{\beta} \text{Hom}_B(I, M)$;

(ii) $I/I^2 \xrightarrow{\beta'} A \otimes_B \Omega_K(B) \xrightarrow{\alpha'} \Omega_K(A) \to 0$.

Proof. (i) The map α is given by composition with the surjection $B \to A$ and β is restriction to I (which yields homomorphisms since $IM = 0$).

(ii) Let $\beta': x + I^2 \mapsto 1 \otimes dx$ and $\alpha': 1 \otimes db \mapsto d(b+I)$ for $x \in I$, $b \in B$. These are well defined; and when $\text{Hom}_B(-, M)$ is applied, as in 1.13, sequence (i) is obtained. □

1.16 Further information about the maps α, α' can be obtained in one important case. First a straightforward result is noted.

Lemma. *Let B be a commutative K-algebra, $I \triangleleft B$, M be a B-module, $N \triangleleft M$ with $IM \subseteq N$ and $\delta \in \text{Der}_K(B, M)$. Then δ induces a derivation from B/I to M/N if and only if $\delta(I) \subseteq N$.* □

1.17 The case when $B = R$, i.e. when A is a affine K-algebra, is now dealt with.

Proposition. *Let $A = R/I$ for some $I \triangleleft R$, let $\{f_j\}$ be a generating set for I, and let $z_i = y_i + I$, $i = 1, \ldots, n$.*
(i) (a) There is a surjection $\{\delta \in \text{Der}_K R \mid \delta(I) \subseteq I\} \to \text{Der}_K A$.
   (b) If K is Noetherian then $\text{Der}_K A$ is finitely generated.
(ii) $\Omega_K(A)$ is generated by $\{dz_1, \ldots, dz_n\}$ with the relations $\sum_i (\partial f_j / \partial y_i) dz_i = 0$ for all j.

Proof. (i)(a) Applying 1.15(i), with $B = R$ and $M = A$, gives an injection $\text{Der}_K A \hookrightarrow \text{Der}_K(R, A)$. The projection $R \to A$ gives a map $\text{Der}_K R \to \text{Der}_K(R, A)$. This map is surjective, for if $\delta \in \text{Der}_K(R, A)$ then δ is the image of $\gamma \in \text{Der}_K R$ defined by choosing $b_i \in R$ such that $\delta(y_i) = b_i + I$ and setting $\gamma(y_i) = b_i$. We now apply 1.16, with $B = M = R$, to give the result.

(b) All the maps involved are R-module homomorphisms and $\mathrm{Der}_K R$ is finitely generated. Therefore $\mathrm{Der}_K A$ is finitely generated over R and hence over A.

(ii) Putting $B = R$ in 1.15(ii) gives the exact sequence

$$I/I^2 \xrightarrow{\beta'} A \otimes \Omega_K(R) \xrightarrow{\alpha'} \Omega_K(A) \longrightarrow 0.$$

Using 1.11, $A \otimes \Omega_K(R)$ is free on $\{1 \otimes dy_i\}$; so $\Omega_K(A)$ is generated by $\{dz_i\}$. The relations are determined by $\mathrm{im}\,\beta'$ which is generated by the elements $1 \otimes df_j$. However 1.11 shows that $df_j = \sum(\partial f_j/\partial y_i) dy_i$. □

1.18 In discussing $\Delta(A)$ and, more generally, $A[d]$, filtered and graded techniques are useful. It is convenient first to recall the notion of a **symmetric algebra** $S_A(M)$ of an A-module M. This is a commutative A-algebra which comes equipped with an A-module homomorphism $M \to S_A(M)$ and is defined by the universal property that any A-module homomorphism from M to a commutative A-algebra B can be factored through a unique A-algebra homomorphism $S_A(M) \to B$. It is readily constructed as the factor of the polynomial algebra over A on generators $\{x_m | m \in M\}$ by the relations

$$\{\sum a_i x_{m_i} | \sum a_i m_i = 0, a_i \in A, m_i \in M\}.$$

In particular if $_A M$ is free of rank t then $S_A(M) = A[x_1, \ldots, x_t]$, the commutative polynomial algebra.

Note that the zero map $M \to A$ induces a homomorphism $\alpha: S_A(M) \to A$ whose kernel, I say, is called the **augmentation ideal**. Evidently α splits; i.e. $S_A(M) \simeq A \oplus I$ as an A-module. Also the algebra $A \oplus M$, with $M^2 = 0$, shows that $M \hookrightarrow S_A(M)$.

1.19 Now let d be any A-submodule of $\mathrm{Der}_K A$ closed under the Lie product. Then $A[d]$ has a standard filtration over A based on d as a generating set. Thus $A[d]_m$ is the A-submodule spanned by all product of at most m derivations from d.

Proposition. $\mathrm{gr}\,A[d]$ *is commutative, and there is a canonical surjection* $S_A(d) \to \mathrm{gr}\,A[d]$.

Proof. If $\delta_1, \delta_2 \in d$ and $a \in A$ then $\delta_1 a - a\delta_1 = \delta_1(a)$ and $\delta_1 \delta_2 - \delta_2 \delta_1 \in d$. This ensures that $\mathrm{gr}\,A[d]$ is commutative, and then the surjection is clear. □

There is an extension of this result in 4.5.

1.20 This has some interesting consequences.

Theorem. (a) *Suppose that d is finitely generated as an A-module. Then:*
  (i) $A[d]$ *is an almost centralizing extension of A;*
  (ii) $\mathrm{gr}\,A[d]$ *is a commutative affine A-algebra;*

(iii) If A is Noetherian then gr $A[\boldsymbol{d}]$ and $A[\boldsymbol{d}]$ are Noetherian.
(b) If K is Noetherian and A is K-affine then gr $\Delta(A)$ is a commutative affine K-algebra and gr $\Delta(A)$ and $\Delta(A)$ are Noetherian.

Proof. (a)(i) If $\delta_1, \ldots, \delta_s$ generate $\boldsymbol{d}$ over A they also generate $A[\boldsymbol{d}]$ over A. The relations noted above ensure that $A[\boldsymbol{d}]$ is almost centralizing.
(ii) This is immediate from 1.19.
(iii) Clear from (ii) and 1.6.9.
(b) Immediate from 1.17 (i) and (a). □

1.21 Even when $K = k$, a field, and A is k-affine one cannot deduce that $A[\boldsymbol{d}]$ is almost commutative; for the algebra in 14.3.9 has this form, with $A = k[y]$ and $\boldsymbol{d}$ generated by $y^2 \partial/\partial y$.

However 8.6.9 shows that in such circumstances $A[\boldsymbol{d}]$ and $\Delta(A)$ are somewhat commutative and thus 8.6.20 gives information about GK dimensions over these algebras. In particular we note

Corollary. *If K is a field and A is K-affine then $A[\boldsymbol{d}]$ and $\Delta(A)$ are right finitely partitive (for GK dimension).* □

1.22 The results of Chapter 9 also apply.

Corollary. *Let A be affine over K and $\boldsymbol{d}$ be finitely generated over A.*
(i) *$A[\boldsymbol{d}]$ and $\Delta(A)$ are constructible K-algebras.*
(ii) *If K is a Jacobson ring then $A[\boldsymbol{d}]$ and $\Delta(A)$ satisfy the Nullstellensatz over K.*
(iii) *If $K = k$, a field, then the endomorphism ring of any simple module over $A[\boldsymbol{d}]$ or $\Delta(A)$ is finite dimensional over k.*

Proof. (i) The inclusions $K \subseteq A \subseteq A[\boldsymbol{d}]$ show that $A[\boldsymbol{d}]$ is constructible; and setting $\boldsymbol{d} = \mathrm{Der}_K A$ deals with $\Delta(A)$.
(ii), (iii) Apply 9.4.21 and 9.5.5. □

1.23 Next we turn to localization. Once again we start by studying derivations and differentials.

Lemma. *Let $\mathscr{S}$ be a multiplicatively closed subset of A and $I = \mathrm{ass}_A \mathscr{S}$. Let M be an A-module, $N = \mathrm{ass}_M \mathscr{S}$ and $\delta \in \mathrm{Der}_K(A, M)$.*
(i) $\delta(I) \subseteq N$.
(ii) *δ induces a unique derivation in $\mathrm{Der}_K(A/I, M/N)$.*
(iii) *δ induces a unique derivation in $\mathrm{Der}_K(A_\mathscr{S}, M_\mathscr{S})$.*

Proof. Straightforward. □

1.24 Proposition. *Let K be Noetherian, A be K-affine, $\mathscr{S}$ be a m.c. subset of A, $\vartheta: R \to A$ a K-algebra surjection and M an A-module. Then:*

(i) $A_{\mathscr{S}} \otimes_A \mathrm{Der}_K(A, M) \simeq \mathrm{Der}_K(A_{\mathscr{S}}, M_{\mathscr{S}})$; and if $\mathrm{ass}_A \mathscr{S} = 0$ and $\mathrm{ass}_M \mathscr{S} = 0$ then

$$\mathrm{Der}_K(A, M) \simeq \{\delta \in \mathrm{Der}_K(A_{\mathscr{S}}, M_{\mathscr{S}}) \mid \delta(A) \subseteq M\};$$

(ii) there is an isomorphism

$$\varphi: A_{\mathscr{S}} \otimes_A \Omega_K(A) \to \Omega_K(A_{\mathscr{S}})$$

which makes the universal derivation for $A_{\mathscr{S}}$ correspond to the unique extension to $A_{\mathscr{S}}$ of φd_A.

Proof. (ii) By 1.23, the map $\tau: \mathrm{Der}_K(A_{\mathscr{S}}, M_{\mathscr{S}}) \to \mathrm{Der}_K(A, M_{\mathscr{S}})$ in 1.13(b) is surjective for each $M_{\mathscr{S}}$. Thus 1.13(b) shows the exactness of the sequence

$$0 \to A_{\mathscr{S}} \otimes \Omega_K(A) \to \Omega_K(A_{\mathscr{S}}) \to \Omega_A(A_{\mathscr{S}}) = 0.$$

(i) Since $\Omega_K(A)$ is finitely generated

$$A_{\mathscr{S}} \otimes \mathrm{Hom}_A(\Omega_K(A), M) \simeq \mathrm{Hom}_{A_{\mathscr{S}}}(A_{\mathscr{S}} \otimes \Omega_K(A), M_{\mathscr{S}}) \simeq \mathrm{Hom}_{A_{\mathscr{S}}}(\Omega_K(A_{\mathscr{S}}), M_{\mathscr{S}})$$

by (ii). So, from 1.10,

$$A_{\mathscr{S}} \otimes_A \mathrm{Der}_K(A, M) \simeq \mathrm{Der}_K(A_{\mathscr{S}}, M_{\mathscr{S}})$$

and the result follows. □

1.25 This is readily applied to $\Delta(A)$.

Theorem. *Let K be Noetherian, A be an affine K-algebra and $\mathscr{S}$ be a m.c. subset of A. Then the right, and left, quotient ring $\Delta(A)_{\mathscr{S}}$ exists and equals $\Delta(A_{\mathscr{S}})$; and if $\mathscr{S}$ consists of regular elements of A then $\Delta(A) \hookrightarrow \Delta(A_{\mathscr{S}})$.*

Proof. In 1.24(i) it is shown that $\Delta(A_{\mathscr{S}})$ is generated over $A_{\mathscr{S}}$ by the image in $\mathrm{Der}_K(A_{\mathscr{S}})$ of $\mathrm{Der}_K A$. The relations $\delta a - a\delta = \delta(a)$ can be used to write each element of $\Delta(A_{\mathscr{S}})$ in the form rs^{-1} or $s^{-1}r$ with $r \in \Delta(A)$, $s \in \mathscr{S}$. Thus, by the universal property of quotient rings, $\Delta(A_{\mathscr{S}})$ is the right and left quotient ring of $\Delta(A)$ with respect to $\mathscr{S}$. □

One can also deduce the same result whenever A is a localization of an affine K-algebra.

§2 Affine Algebras Over a Field

2.1 This section establishes the basic technique to be used in the investigation of $\Delta(A)$ in the case when A is an affine domain over a field of characteristic 0.

15.2.5 Affine algebras over a field

As mentioned in the introduction, the technique involves localization with respect to the powers of a single, carefully chosen element c. Indeed it can be arranged that $\Delta(A_c)$ is free of finite rank over a subalgebra $A_n(k)_c$ with $c \in k[y_1, \ldots, y_n]$ and $n = \dim A$. When A is regular, globalization is achieved by the choice of a finite number of elements c_i so that $\prod \Delta(A_{c_i})$ is faithfully flat over $\Delta(A)$.

2.2 It is convenient to fix some notation for this and the succeeding sections. We let A be an integral domain, affine over a field k of characteristic 0, with $\dim A = n$. Also $\{y_1, \ldots, y_n\}$ is a transcendence basis over k for the field of fractions L of A with each $y_i \in A$. Finally,

$$R = k[y_1, \ldots, y_n] \subseteq A \quad \text{and} \quad Q = Q(R) = k(y_1, \ldots, y_n) \subseteq L.$$

2.3 Once again the theory starts by investigating derivations and differentials, starting with a straightforward case.

Example. $\Omega_k(Q)$ *is a free Q-module of rank n on $dy_1, \ldots, dy_n$ and $\mathrm{Der}_k(Q)$ is free on $\partial/\partial y_1, \ldots, \partial/\partial y_n$.*

Proof. Combine 1.24 with 1.3 and 1.11. □

2.4 This is easily extended to finite field extensions of Q such as L.

Proposition. *Let $F \supseteq Q$ be a finite field extension. Then $\Omega_k(F)$ is a free F-module with basis $dy_1, \ldots, dy_n$ and $\mathrm{Der}_k(F)$ is free with a basis consisting of extensions to F of $\partial/\partial y_1, \ldots, \partial/\partial y_n$.*

Proof. Each $f \in F$ is algebraic over Q. The application of d_F to the relevant expression demonstrates that $d_F y_1, \ldots, d_F y_n$ span $\Omega_k(F)$. From 1.13(b) one sees that $\Omega_Q(F) = 0$; furthermore, one obtains a surjection $F \otimes_Q \Omega_k(Q) \to \Omega_k(F)$ which is an isomorphism provided each $\delta \in \mathrm{Der}(Q, {}_F M)$ extends to some $\eta \in \mathrm{Der}_k(F, {}_F M)$. However, $F = Q(f)$ for some $f \in F$, say with minimum polynomial $p(x) = \sum p_i x^i \in Q[x]$; so $p'(f) \neq 0$. The isomorphism $Q[x]/(p) \simeq F$ makes M a $Q[x]$-module. We define $\zeta \in \mathrm{Der}_k(Q[x], M)$ to extend δ and have $\zeta(x) = -p'(f)^{-1} \sum f^i \delta(p_i)$. Clearly $\zeta(p) = 0$ and so ζ induces η as required. This deals with $\Omega_k(F)$; and the duality provided by 1.10 then establishes the claims about $\mathrm{Der}_k F$. □

Note that 1.12(iv), (v) now applies to $\mathrm{Der}_k F$.

2.5 Corollary. $\Delta(L) = L[\partial/\partial y_1, \ldots, \partial/\partial y_n]$ *and contains a copy of $B_n(k)$ over which it is a free module of finite rank.*

Proof. It is clear from 2.4 and 1.12 that $\Delta(L)$ is generated by the commuting derivations $\partial/\partial y_1, \ldots, \partial/\partial y_n$. Hence

$$\Delta(L) \supseteq k(y_1, \ldots, y_n)[\partial/\partial y_1, \ldots, \partial/\partial y_n] \simeq B_n(k).$$

Since L is free over $k(y_1, \ldots, y_n)$ the rest follows. □

2.6 In fact a similar structure holds after inverting just a single, carefully chosen, element of R.

Theorem. *There exists $0 \neq c \in R \subseteq A$ such that $\Omega(A_c)$ is a free A_c-module with basis $\{dy_i\}$, Der A_c is a free A_c-module with basis $\{\partial/\partial y_i\}$,*

$$\Delta(A_c) = A_c[\partial/\partial y_1, \ldots, \partial/\partial y_n] \supseteq R_c[\partial/\partial y_1, \ldots, \partial/\partial y_n] = A_n(k)_c$$

and $\Delta(A_c)$ is free of finite rank, as a left or right module, over the subalgebra $A_n(k)_c$.

Proof. Note first that QA is finite dimensional over Q and hence is Artinian. Therefore QA is a field and so is equal to L. Thus if $a \in A$ then $a^{-1} = c^{-1}b$ for some $b \in A$, $c \in R$ and so $a^{-1} \in A_c$.

Let $M = Ady_1 + \cdots + Ady_n \subseteq \Omega(A)$. By 2.4 and 1.24 one sees that $L \otimes_A M = \Omega(L) = L \otimes_A \Omega(A)$ and hence $L \otimes_A (\Omega(A)/M) = 0$. Since $\Omega(A)$ is finitely generated, by 1.17, there exists $a \in A$ such that $A_a \otimes_A (\Omega(A)/M) = 0$ and, by the paragraph above, we may choose $a \in R$. It follows that $\Omega(A_a)$ is free over A_a on $dy_1, \ldots, dy_n$. Then duality, 1.10, shows that Der A_a is free on $\partial/\partial y_1, \ldots, \partial/\partial y_n$.

A basis for L over Q can be chosen from A; and it is easily arranged, by inverting some $b \in R$, that each k-algebra generator of A, and each product of one of these with a basis element, is an R_b-linear combination of that basis. That ensures that it forms a basis for A_b over R_b.

Finally, let $c = ab$. □

2.7 Before proceeding further we need a result about differentials which extends 1.15.

Proposition. *Let B be the localization of A at some maximal ideal P and let I be the maximal ideal of B. Then*

$$I/I^2 \simeq B/I \otimes_B \Omega_k(B).$$

Proof. Note first that $B/I^2 \simeq A/P^2$ which, by the Nullstellensatz, is a finite dimensional k-algebra. Therefore, by the Wedderburn principal theorem, see [Cohn 77, p. 386, Theorem 4], B/I^2 has a subfield F isomorphic to B/I such that $B/I^2 = F \oplus I/I^2$. Thus there is a map $B/I \to B/I^2$ to which 1.13 can be applied. This yields an exact sequence

$$B/I^2 \otimes_{B/I} \Omega_k(B/I) \to \Omega_k(B/I^2) \to \Omega_{B/I}(B/I^2) \to 0.$$

However, B/I is finite dimensional over k and so 2.4 shows that $\Omega_k(B/I) = 0$. We deduce that $\Omega_k(B/I^2) \simeq \Omega_{B/I}(B/I^2)$.

Note next the elementary fact that $\text{Der}_k(B, M) \simeq \text{Der}_k(B/I^2, M)$ for all B/I-modules M. Hence the canonical surjection

$$B/I \otimes_B \Omega_k(B) \to B/I \otimes_B \Omega_k(B/I^2)$$

is an isomorphism.

Next consider the exact sequence

$$I/I^2 \to B/I \otimes_B \Omega_k(B) \to \Omega_k(B/I) \to 0$$

provided by 1.15. The facts above give a surjection

$$I/I^2 \to B/I \otimes_B \Omega_{B/I}(B/I^2)$$

which we now show is injective. This is equivalent to proving that the map of dual vector spaces

$$\text{Hom}_{B/I}(B/I \otimes_B \Omega_{B/I}(B/I^2), B/I) \to \text{Hom}_{B/I}(I/I^2, B/I)$$

is surjective.

The left-hand term is isomorphic to $\text{Hom}_{B/I}(\Omega_{B/I}(B/I^2), B/I)$ and hence to $\text{Der}_{B/I}(B/I^2)$; and the map above sends $\delta \in \text{Der}_{B/I}(B/I^2)$ to $\alpha \in \text{Hom}_{B/I}(I/I^2, B/I)$ merely by restricting δ to I/I^2.

Finally, suppose $\alpha \in \text{Hom}_{B/I}(I/I^2, B/I)$; we seek δ. Using the decomposition $B/I^2 = F \oplus I/I^2$, if $g \in B/I^2$ then $g = g_1 + g_2$ with $g_1 \in F$, $g_2 \in I/I^2$. We define $\delta(g) = \alpha(g_2)$. This is as required. □

2.8 The reason for proving 2.7 is that it gives connections with regular rings. These were defined in 7.7.1; but when applied to a commutative Noetherian local ring B, the term means merely that the ring has finite global dimension, as in 7.1.15. Another characterization is useful here. Recall that if B is a local Noetherian integral domain with maximal ideal I then rank I/I^2, as a B/I-vector space, is the minimal cardinality of a set of generators for I, by Nakayama's lemma. Hence, by the generalized principal ideal theorem, cf. 4.1.13, rank $I/I^2 \geqslant \dim B$, the Krull dimension of B.

Proposition. (i) *A commutative Noetherian local integral domain B with maximal ideal I is regular if and only if rank $I/I^2 = \dim B$.*

(ii) *A commutative Noetherian integral domain C is regular if and only if C_P is regular for each maximal ideal P of C.*

Proof. See [Kunz 85, 7.2.4]. □

2.9 The connection between these different notions is made by the next result.

Theorem. *Let B be the localization of A at some maximal ideal and let I be the maximal ideal of B. Then the following are equivalent:*
 (i) *$\Omega_k(B)$ is free of rank n over B with a basis $db_1, \ldots, db_n$ where $b_1, \ldots, b_n$ is a minimal generating set for I.*
 (ii) *$\Omega_k(B)$ is free over B.*
 (iii) *B is regular.*

Proof. (i)⇒(ii) Trivial.
 (ii)⇒(iii) Note that L, the quotient field of A, is also the quotient field of B. By 1.24, $\Omega_k(L) \simeq L \otimes_B \Omega_k(B)$ and, by 2.4, $\Omega_k(L)$ is free of rank n. Thus $\Omega_k(B)$ must be free of rank n. Then 2.7 shows that rank $I/I^2 = n$ as required.
 (iii)⇒(i) By 2.7, $I/I^2 \simeq B/I \otimes_B \Omega_k(B)$. It follows that $\Omega_k(B)$ is spanned by the elements $db_1, \ldots, db_n$, and hence the same is true of $\Omega_k(L)$. But $\Omega_k(L)$ is free of rank n and so $db_1, \ldots, db_n$ must be a basis for $\Omega_k(L)$. Thus $\Omega_k(B)$ is freely generated by $db_1, \ldots, db_n$. □

2.10 Corollary. *If c is chosen as in 2.6 then A_c is regular.*

Proof. It is clear from 1.24 that $\Omega_k(B)$ is free for each localization B of A_c at a maximal ideal. Thus, by 2.9, each B is regular. □

2.11 The next few results aim towards the case when A is regular.

Corollary. *If A is regular then $\mathrm{Der}_k A$ and $\Omega_k(A)$ are finitely generated projective modules.*

Proof. 1.17 shows that $\Omega_k(A)$ is finitely generated; and 2.9 shows that $\Omega_k(A_P)$ is free for each maximal ideal P of A. However, $\Omega_k(A_P) \simeq A_P \otimes \Omega_k(A)$ and so, by 7.1.15, $\Omega_k(A)$ is projective. Since $\mathrm{Der}_k A = \Omega(A)^*$, the dual basis lemma, 3.5.2, shows that $\mathrm{Der}_k A$ is finitely generated projective also. □

2.12 In fact this characterizes regular rings.

Theorem. *A is regular if and only if $\Omega_k(A)$ is projective.*

Proof. (⇐) If P is any maximal ideal of A then $\Omega_k(A_P) \simeq A_P \otimes \Omega_k(A)$. Hence $\Omega_k(A_P)$ is projective and hence free. Then 2.9 shows that A_P is regular; so A is regular. □

2.13 The next result provides the globalization we need.

Theorem. *Let A be a regular domain.*
 (i) *Given any maximal ideal P of A there exists $c = c(P) \in A \setminus P$ such that $\Omega_k(A_c)$*

is free on $db_1,\ldots,db_n$ and $\mathrm{Der}_k(A_c)$ is free on $\partial/\partial b_1,\ldots,\partial/\partial b_n$ for some $b_i \in A$.

(ii) There is a finite subset $\{c_1,\ldots,c_t\}$ of $\{c(P)|P$ a maximal ideal of $A\}$ such that $\prod_{j=1}^{t} \Delta(A_{c_j})$ is left and right faithfully flat over $\Delta(A)$.

Proof. (i) Since A_P is regular, 2.9 applies, and the b_i can be chosen to belong to P. Thus $\Omega_k(A_P)$ is free on $db_1,\ldots,db_n$ say. It is easily arranged, as in the proof of 2.6, that $\Omega_k(A_c)$ is free on $db_1,\ldots,db_n$, this having the consequence, by duality, that $\mathrm{Der}_k A_c$ is free on $\partial/\partial b_1,\ldots,\partial/\partial b_n$.

(ii) Let $c_j = c(P_j)$ and let H be the ideal $\sum\{c_j A | P_j$ maximal$\}$. Note that H is not contained in any maximal ideal; so $H = A$ and $H = \sum_{j=1}^{t} c_j A$ for some t.

With this choice of $c_1,\ldots,c_t$, let I be a left ideal of $\Delta(A)$ such that $\Delta(A_{c_j}) \otimes (\Delta(A)/I) = 0$ for each j. Then there exists $m \gg 0$ such that $c_j^m \in I$ for each j. However, then $A = H^{tm} \subseteq I$ and so $\Delta(A) = I$ as required. The same argument applies to the right-hand side. □

We note that (i) above only requires that A_P be regular; and in that case A_c too will be regular.

2.14 For later use a result linking regularity with minors of a certain matrix is now proved. Since A is affine over k then $A \simeq T/H$, where T is a polynomial algebra $k[t_1,\ldots,t_m]$ and $H = (f_1,\ldots,f_s) \triangleleft T$ say. Let α denote the Jacobian matrix $(\partial f_i / \partial t_j)$. This can be viewed as a map $T^s \to T^m$. It induces a map $\beta: A^s \to A^m$ which extends to localizations of A; thus $\beta: L^s \to L^m$ also. And if $B = A_P$ for some maximal ideal P and $I = P_P$ then β induces a map $\gamma: (B/I)^s \to (B/I)^m$.

Proposition. (i) *If δ is a derivation of T leaving H invariant then, for each r, δ leaves invariant the ideal generated by H and all the $r \times r$ minors of α.*

(ii) $\mathrm{rank}_L \beta = m - n$

(iii) $\mathrm{rank}_{B/I} \gamma \leq m - n$ *with equality precisely when B is regular.*

Proof. (i) This is a straightforward calculation; see [Hart **74**].

(ii) It is clear from 1.17(ii) that $\Omega_k(A)$ has a free presentation

$$A^s \xrightarrow{\beta} A^m \longrightarrow \Omega_k(A) \longrightarrow 0$$

which, by 1.24, yields a free presentation

$$L^s \xrightarrow{\beta} L^m \longrightarrow \Omega_k(L) \longrightarrow 0$$

of $\Omega_k(L)$. Since $\mathrm{rank}\, \Omega_k(L) = n$ then $\mathrm{rank}\, \beta = m - n$.

(iii) Tensoring with B/I yields the exact sequence

$$(B/I)^s \xrightarrow{\gamma} (B/I)^m \longrightarrow \Omega_k(B)/I\Omega_k(B) \longrightarrow 0.$$

Thus rank (im γ) + rank ($\Omega_k(B)/I\Omega_k(B)$) = m where the ranks are over B/I. However, $\Omega_k(B)/I\Omega_k(B) \simeq I/I^2$, by 2.7; and rank $I/I^2 = n$ if and only if B is regular. The result follows. □

§3 Dimensions

3.1 We retain the fixed notation of §2 and begin to determine the properties of $\Delta(A)$. We already know, by 1.20, that $\Delta(A)$ is Noetherian and will see, shortly, that it is an integral domain which is simple when A is regular. The aim of this section, after establishing these basic properties, is to calculate the various dimensions of $\Delta(A)$. This aim is realized in full when A is regular, and in part in greater generality.

The strategy involved depends upon the single element localizations A_c described in 2.6 and 2.13. The proofs rely largely upon the fact that $\operatorname{Der}_k A_c$ is free on $\{\partial/\partial y_1, \ldots, \partial/\partial y_n\}$ for some choice of $\{y_1, \ldots, y_n\}$. In practice, therefore, some of the results take this as the working hypothesis. These are then applied, via globalization, to the general case, especially when A is regular.

3.2 The first result in part echoes 2.6.

Proposition. *Suppose that $\operatorname{Der}_k A$ is free on $\partial/\partial y_1, \ldots, \partial/\partial y_n$. Then:*
 (i) *$\Delta(A)$ can be obtained from A by a sequence of Ore extensions*

$$\Delta(A) = A[x_1; -\partial/\partial y_1] \cdots [x_n; -\partial/\partial y_n],$$

 where $x_i x_j = x_j x_i$ for all i, j;
 (ii) *$\Delta(A)$ is a Noetherian integral domain with an involution (i.e. anti-automorphism of order 2) which fixes A;*
 (iii) *$\Delta(A)$ is simple and A is a simple $\Delta(A)$-module;*
 (iv) *$A_n(k) \simeq k[y_1, \ldots, y_n][x_1; -\partial/\partial y_1] \cdots [x_n; -\partial/\partial y_n] \subseteq \Delta(A)$;*
 (v) *$\operatorname{GK}(\Delta(A)) = 2n$;*
 (vi) *$n \leq K(\Delta(A)) \leq 2n$;*
 (vii) *$n \leq \operatorname{gld} \Delta(A) \leq 2n = n + \operatorname{gld} A$, provided $\operatorname{gld} A < \infty$.*

Proof. (i) There is a surjection, ϑ say, from the iterated Ore extension onto $\Delta(A)$. Evidently $\ker \vartheta \cap A = 0$ and yet $\operatorname{ad} y_i$, applied to the Ore extension, acts as $\partial/\partial x_i$. These readily combine to show that $\ker \vartheta = 0$.

 (ii) It is clear from (i) and 1.20 that $\Delta(A)$ is a Noetherian domain. It is readily checked that the map $x_i \mapsto -x_i$, $a \mapsto a$ for each $a \in A$, $i \in \{1, \ldots, n\}$, is an involution.

 (iii) Let $0 \neq I \triangleleft \Delta(A)$. Applying $\operatorname{ad} y_i$, as in (i), one sees that $I \cap A \neq 0$. Hence $\dim(A/I \cap A) < \dim A$ and so

$$(I \cap A) \cap k[y_1, \ldots, y_n] \neq 0.$$

Applying $\operatorname{ad} x_i$ now shows that $I \cap k \neq 0$ and so $I = \Delta(A)$.

15.3.6   Dimensions

As noted in 1.4, A is a left $\Delta(A)$-module; and a submodule is just an ideal invariant under each ad x_i. Such an ideal generates an ideal of $\Delta(A)$; so A must be a simple module.

(iv) Obvious.

(v), (vi) Note that $\Delta(A) \simeq A * U(g)$ with g being the abelian k-Lie algebra spanned by $\partial/\partial y_1, \ldots, \partial/\partial y_n$; then 8.2.10 and 6.5.7 apply.

(vii) Use (i) combined with 7.1.15 and 7.5.3.   □

3.3 In 3.9 it will be shown that if $\text{Der}_k A$ is as described in 3.2, then gld $A < \infty$; so the extra hypothesis in 3.2(vii) is redundant.

3.4 Of course, the involution of $\Delta(A)$ provided by 3.2(ii) interchanges left ideals and right ideals, thus demonstrating the left–right symmetry of $\Delta(A)$. The earlier results concerning localizations $\Delta(A_c)$ and the faithfully flat overring $\prod \Delta(A_{c_i})$ are all valid on each side. The effect of this is that all the arguments which follow can be applied on either side. Note, however, that this demonstration of symmetry relies upon A being regular. Some related cautionary examples are mentioned later in 6.5.

3.5 Making use of the results of Chapter 9, we can improve upon 3.2(vi), (vii).

Corollary. *If* $\text{Der}_k A$ *is free on* $\partial/\partial y_1, \ldots, \partial/\partial y_n$ *then*
(i) $K(\Delta(A)) = n$, *and*
(ii) *if* gld $A < \infty$ *then* gld $\Delta(A) = n$.

Proof. Let $S_t = A[x_1; -\partial/\partial y_1] \cdots [x_t; -\partial/\partial y_t]$. Evidently this is a constructible k-algebra and so satisfies the Nullstellensatz, by 9.4.21. Thus 9.1.14 can be applied repeatedly.   □

Once again, the hypothesis that gld $A < \infty$ can be removed; see 3.9.

3.6 The results so far have dealt, essentially, with the localized rings $\Delta(A_c)$. There are corresponding global results concerning $\Delta(A)$.

Proposition. (i) $\Delta(A)$ *is an integral domain.*
(ii) $\text{GK}(\Delta(A)) = 2n$.
(iii) $n \leqslant K(\Delta(A)) \leqslant 2n$.

Proof. (i) In 1.25 it is shown that $\Delta(A) \hookrightarrow \Delta(A_c)$ which by 3.2, is an integral domain.

(ii) With c chosen as in 2.6, $\text{GK}(\Delta(A)) \leqslant \text{GK}(\Delta(A_c)) = 2n$. On the other hand, note that
$$\Delta(A_c) \supseteq A_n(k) = k[y_1, \ldots, y_n][x_1; -\partial/\partial y_1] \cdots [x_n; -\partial/\partial y_n],$$

where $y_1, \ldots, y_n \in A$. One can choose a power of c, t say, such that $tx_1, \ldots, tx_n \in \Delta(A)$. In the k-subalgebra S of $\Delta(A) \cap A_n(k)$ generated by $y_1, \ldots, y_n$, $tx_1, \ldots, tx_n$ one can check that the standard monomials are independent. (First consider the words of highest degree in the tx_i.) Hence $GK(S) \geq 2n$ and so $GK(\Delta(A)) \geq 2n$.

(iii) With c as before, 3.5, or 6.5.3 and 6.5.4, shows that

$$K(\Delta(A)) \geq K(\Delta(A_c)) \geq n.$$

On the other hand, 1.21 shows that $\Delta(A)$ is right finitely partitive and so, by 8.3.18,

$$K(\Delta(A)) \leq GK(\Delta(A)) = 2n.$$

(Alternatively 6.5.6, 8.2.14 and 8.3.20 provide the inequalities and equalities

$$K(\Delta(A)) \leq K(\mathrm{gr}\,\Delta(A)) = GK(\mathrm{gr}\,\Delta(A)) \leq GK(\Delta(A)).) \qquad \square$$

3.7 There remains global dimension. This, of course, concerns the case when A is regular. In this case one can improve upon 3.6.

Theorem. *Let A be regular. Then:*
(i) $\Delta(A)$ *is simple;*
(ii) $K(\Delta(A)) = n$;
(iii) $\mathrm{gld}\,\Delta(A) = n$.

Proof. The proofs all use the faithfully flat overring $\prod_{j=1}^{t} B_j$ provided by 2.13, where $B_j = \Delta(A_{c_j})$.

(i) If $0 \neq I \triangleleft \Delta(A)$ then $0 \neq I_{c_j} \triangleleft B_j$ for each j. Hence, by 3.2, $I_{c_j} = B_j$. The faithful flatness implies that $I = \Delta(A)$.

(ii) The combination of 3.2 and 6.5.3 gives

$$K(\Delta(A)) \leq \sup\{K(B_j)\} = n$$

and so, by 3.6, $K(\Delta(A)) = n$.

(iii) We know, by 3.5, that $\mathrm{r\,gld}\,B_j = n$. We also know that B_j, being a localization, is flat over $\Delta(A)$. (However, 7.2.6 cannot be applied here since it is not yet clear that $\mathrm{r\,gld}\,\Delta(A) < \infty$).

For each j and each module M, let M_j be the j-torsion submodule, that meaning the kernel of the map $M \to B_j \otimes M$. If $M = M_j$ we say M is j-torsion. Note that, since the c_j commute, if M is j-torsion then $B_i \otimes M$ also is j-torsion for each i.

Now suppose that there are modules having flat dimension greater than n. Amongst these we choose one, M say, to be j-torsion for as many j as possible, say for $j = 1, \ldots, s$. The faithful flatness ensures that $s < t$. Consider the short exact sequences

$$0 \to M_{s+1} \to M \to \bar{M} \to 0$$

and

$$0 \to \bar{M} \to B_{s+1} \otimes M \to (B_{s+1} \otimes M)/\bar{M} \to 0,$$

where $\bar{M} = M/M_{s+1}$. Note first that M_{s+1} and $(B_{s+1} \otimes M)/\bar{M}$ are both j-torsion for $j = 1, \ldots, s+1$ and hence have flat dimension n or less. Also

$$\mathrm{fd}_{\Delta(A)}(B_{s+1} \otimes M) = \mathrm{fd}_{B_{s+1}}(B_{s+1} \otimes M) \leq n$$

by 7.4.2. The second sequence then shows that $\mathrm{fd}\,\bar{M} \leq n$ and the first that $\mathrm{fd}\,M \leq n$, a contradiction.

We conclude that w gld $\Delta(A) \leq n$ and hence gld $\Delta(A) = n$. □

3.8 This provides further characterizations of regular rings.

Theorem. *The following conditions are equivalent*:
(i) *A is regular;*
(ii) *$\Delta(A)$ is simple;*
(iii) *A is simple as a $\Delta(A)$-module.*

Proof. (i)⇒(ii) Use 3.7.

(ii)⇒(iii) If A is not simple then it has a nonzero proper ideal I stable under Der A. It is clear that $\Delta(A)I = I\Delta(A)$ is then a proper nonzero ideal of $\Delta(A)$, a contradiction.

(iii)⇒(i) Suppose, to the contrary, that A is not regular. Let E be the ideal of A generated by all the $(m-n) \times (m-n)$ minors of the matrix β described in 2.14. By 2.14(iii), E must be contained in each maximal ideal P such that A_P is nonregular; thus E is proper. Moreover, 2.14(i) shows that E is invariant under Der A and 2.14(ii) shows that $E \neq 0$. This contradicts the hypothesis that A is simple over $\Delta(A)$. □

3.9 As a consequence we see that, in the situation dealt with earlier in 3.2, A must be regular.

Corollary. *If $\mathrm{Der}_k A$ is free on $\partial/\partial y_1, \ldots, \partial/\partial y_n$ then A is regular.*

Proof. By 3.2 $\Delta(A)$ is simple; so 3.8 applies. □

3.10 There are corresponding results for $\Delta(L)$.

Theorem. (i) $\Delta(L)$ *is simple.*
(ii) $\mathrm{GK}\,(\Delta(L)) = 2n$.
(iii) $K(\Delta(L)) = n$.
(iv) $\mathrm{gld}\,\Delta(L) = n$.

Proof. (i) $\Delta(L)$ is a localization of the simple algebra $\Delta(A_c)$, with c as in 2.6.

(ii) $\Delta(L)$ is a finitely generated module over $B_n(k)$; so $\mathrm{GK}(\Delta(L)) = \mathrm{GK}(B_n(k))$ by 8.2.9. However, any affine subalgebra of $B_n(k)$ is a subalgebra of $A_n(k)_c$ for some

544

$c \in k[y_1, \ldots, y_n]$; and 8.2.11 shows that

$$\mathrm{GK}\,(A_n(k)_c) = n + \mathrm{GK}\,(k[y_1, \ldots, y_n]_c) = 2n.$$

Hence $\mathrm{GK}\,(B_n(k)) = 2n$.

(iii) Of course $\mathcal{K}(\Delta(L)) \leqslant \mathcal{K}(\Delta(A_c)) = n$. To reverse the inequality, view $\Delta(L)$ as $L[x_1; -\partial/\partial y_1] \cdots [x_n; -\partial/\partial y_n]$. The x_i commute with each other. Thus, as in 6.5.10, it is easy to construct a chain of right ideals, with generators being monomials in the x_i, of ordinal type $(\omega^n)^{\mathrm{op}}$. Hence $\mathcal{K}(\Delta(L)) \geqslant n$.

(iv) A similar argument applies here, using the Koszul resolution; see 7.3.16. □

3.11 The following facts are now evident.

Corollary. (i) *If B is an algebra with $A \subseteq B \subseteq L$ then $\mathrm{GK}\,(\Delta(B)) = 2n$.*

(ii) *If A is regular and $\mathcal{S}$ is a multiplicatively closed subset of A then $\Delta(A_{\mathcal{S}})$ is simple Noetherian, $\mathrm{gld}\,\Delta(A_{\mathcal{S}}) = n$ and $\mathcal{K}(\Delta(A_{\mathcal{S}})) = n$.* □

3.12 The section ends with some illustrative examples. The first concerns the case when A is nonregular. By 3.8 and 2.12 we know that $\Delta(A)$ is not simple and that $\Omega(A)$ is not projective.

Example. *Let A be the coordinate ring of the cusp $y^2 = x^3$; so $A = k[x, y]/(x^3 - y^2)$. Then:*

(i) $\Omega(A)$ is not torsion-free;
(ii) Der A is not projective;
(iii) $\Delta(A)$ is not simple;
(iv) $\mathcal{K}(\Delta(A)) = 2 \neq \dim A$.

Proof. (i) By 1.17 (ii), $\Omega(A)$ is generated over A by dx and dy subject to the relation $3x^2\,dx - 2y\,dy = 0$. One can check that $2x\,dy - 3y\,dx \neq 0$; yet

$$y(2x\,dy - 3y\,dx) = 2xy\,dy - 3x^3\,dx = x(2y\,dy - 3x^2\,dx) = 0.$$

(ii) Note first that $A \simeq k[t^2, t^3] \subseteq k[t]$, the polynomial ring. Hence, by 1.24, each derivation of A is the restriction of one of $L = Q(A)$; and these all have the form

$$q\frac{d}{dt}, q \in L.$$

Now

$$q\frac{d}{dt}(t^2) \in A \text{ and } q\frac{d}{dt}(t^3) \in A \Leftrightarrow 2tq \in A \text{ and } 3t^2 q \in A.$$

It follows easily that this is equivalent to having $q \in At + At^2$. Hence

15.3.13   Dimensions   545

$$\operatorname{Der}_k A = At \, d/dt + At^2 \, d/dt \simeq At^2 + At^3 = I,$$

this being a maximal ideal of A. Now if I were projective then $\operatorname{rank}_k I/I^2$ would be 1; but in fact rank $I/I^2 = 2$.

(iii) This follows from 3.8; in fact I is stable under $\operatorname{Der} A$ and so I generates a proper ideal of $\Delta(A)$.

(iv) A is a cyclic torsion $\Delta(A)$-module, so $K(_{\Delta(A)}A) < K(\Delta(A))$. The chain of submodules $A \supset I \supset \cdots \supset I^m \supset \cdots$ shows that $K(_{\Delta(A)}A) \geqslant 1$; and then 3.6 shows that $K(\Delta(A)) = 2$. □

3.13 The next example concerns the regular case. It shows that it is not necessarily the case that $\operatorname{Der} A$ is freely generated by $\partial/\partial y_1, \ldots, \partial/\partial y_n$.

Example. *Let A be the coordinate ring of $\mathbb{S}^1$; so $A = \mathbb{R}[x, y]/(x^2 + y^2 - 1)$. Then:*
(i) *A is regular and its only units are in $\mathbb{R}\setminus\{0\}$;*
(ii) *$\Omega(A)$ is free of rank 1 on $x \, dy - y \, dx$;*
(iii) *$\operatorname{Der} A$ is free of rank 1 on $x \partial/\partial y - y \partial/\partial x$;*
(iv) *$\operatorname{Der} A$ contains no derivation of the form $\partial/\partial a$ with $a \in A$.*

Proof. (i) One knows from 7.8.14 that A is regular and

$$\mathbb{C} \otimes_\mathbb{R} A = \mathbb{C}[x, y]/(x^2 + y^2 - 1) \simeq \mathbb{C}[z, z^{-1}]$$

via $x + iy \mapsto z$. The units of $\mathbb{C} \otimes_\mathbb{R} A$ correspond to the elements αz^m, $0 \neq \alpha \in \mathbb{C}$, $m \in \mathbb{Z}$. Those fixed under complex conjugation are in $\mathbb{R}\setminus\{0\}$.

(ii) Of course $\Omega(A)$ is generated by dx and dy, and has the relation $x \, dx + y \, dy = 0$. If we let $w = x \, dy - y \, dx$ then

$$xw = x^2 \, dy - xy \, dx = (x^2 + y^2) \, dy = dy$$

and

$$yw = xy \, dy - y^2 \, dx = -(x^2 + y^2) \, dx = -dx.$$

Hence $\Omega(A) = Aw$. Regularity shows this must be free.

(iii) This follows from (ii) by duality.

(iv) Let $\delta = x \partial/\partial y - y \partial/\partial x$. If $\operatorname{Der} A$ contained a derivation $\partial/\partial a$ then, by (iii), $\delta(a)$ would be a unit; so by (i), $\delta(a) \in \mathbb{R}\setminus\{0\}$. We may therefore suppose $\delta(a) = 1$. Now a can be written uniquely in the form $a = f(x) + g(x)y$ with $f(x), g(x) \in \mathbb{R}[x]$. Thus

$$1 = \delta(a) = xg(x) - yf'(x) - y^2 g'(x)$$
$$= xg(x) + (x^2 - 1)g'(x) - yf'(x).$$

Consideration of the highest degree term in x shows this to be impossible. □

3.14 This example also sheds some light on the discussions in Chapter 14. For $\Delta(A) \simeq U(g)/(x^2 + y^2 - 1)$ with g being the three-dimensional solvable real Lie algebra on generators x, y, u with relations $[x, y] = 0$, $[u, x] = -y$ and $[u, y] = x$. It is easy to check that $\mathbb{C} \otimes_\mathbb{R} \Delta(A) \simeq A'_1(\mathbb{C})$; but $\Delta(A)$ itself is far from $A'_1(\mathbb{R})$ since its only units belong to $\mathbb{R}\setminus\{0\}$. Thus the classification of simple factors of $U(g)$ with g solvable will not be a straightforward extension of the classification for g completely solvable.

3.15 It is also possible to have $\Omega(A)$ not being free with A regular. For if A is the coordinate ring of $\mathbb{S}^2$, i.e. $A \simeq \mathbb{R}[x, y, z]/(x^2 + y^2 + z^2 - 1)$ then $\Omega(A)$ is generated by dx, dy, dz modulo the relation $xdx + ydy + zdz = 0$. As in 11.2.3 this is stably free but not free.

§4 Further Properties

4.1 The fixed notation of §2 is retained again. Here several further properties of $\Delta(A)$ are described, when A is regular. It is shown that, as with $A_n(k)$, there is a lower bound of n on the GK dimension of nonzero $\Delta(A)$-modules and that $K_0(\Delta(A)) \simeq K_0(A)$. It is also shown that $\Delta(A) = \{f \in \Delta(L) | f(A) \subseteq A\}$.

4.2 In 3.6 it was shown that $\mathrm{GK}(\Delta(A_c)) = \mathrm{GK}(\Delta(A))$. This is also an easy consequence of the next result which is needed in dealing with their modules.

Lemma. (i) *Let $0 \neq c \in A$ and let W be a finite dimensional subspace of $\Delta(A_c)$. Then there are a finite dimensional subspace V of $\Delta(A)$ and natural numbers p, q such that, for all s, $W^s \subseteq c^{-ps}V^{qs}$.*

(ii) *If M is a $\Delta(A)$-module and $0 \neq c \in A$, then $\mathrm{GK}(M_c) \leq \mathrm{GK}(M)$.*

Proof. (i) Of course $Wc^t \subseteq \Delta(A)$ for some t. Note that $[c, \Delta(A)_s] \subseteq \Delta(A)_{s-1}$ for each $s \in \mathbb{N}$ and so $\delta = [c, -]$ is a locally nilpotent derivation. Therefore we can choose a finite dimensional subspace V of $\Delta(A)$ containing 1 and c, such that $Wc^t \subseteq V$ and $\delta(V) \subseteq V$. Since δ is locally nilpotent, $\delta^q(V) = 0$ for some q.

For each $v \in V$ let $v_i = \delta^i(v)$; thus $v_q = 0$ and v_{q-1} commutes with c. Note that for each i.

$$cv_i = v_i c + v_{i+1}$$

and so

$$v_i c^{-1} = c^{-1} v_i + c^{-1} v_{i+1} c^{-1}.$$

It follows easily that

$$vc^{-1} = c^{-1}v + c^{-2}v_1 + \cdots + c^{-q}v_{q-1}.$$

Repetition of this process reveals that

$$Vc^{-h} \subseteq c^{-h}V + \cdots + c^{-(h+q-1)}V$$

15.4.5   *Further properties*

for all h. Hence, if $p = t + q - 1$, then

$$W^s \subseteq (Vc^{-t})^s \subseteq c^{-ts}V^s + \cdots + c^{-ps}V^s$$
$$\subseteq c^{-ps}(c^{(q-1)s}V^s + \cdots + V^s) \subseteq c^{-ps}V^{qs}$$

as required.

(ii) Let M_0 be a finite dimensional subspace of M and let V, W be as in (i). Then, for all s, $W^s M_0 \subseteq c^{-ps}V^{qs}M_0$ and so $\dim W^s M_0 \leq \dim V^{qs}M_0$. Hence $GK(M_c) \leq GK(M)$. □

4.3 We can now extend 8.5.5 from $A_n(k)$ to $\Delta(A)$.

Theorem. *If A is regular and M is a nonzero left $\Delta(A)$-module then $GK(M) \geq n$.*

Proof. It is enough to consider the case when M is simple. By 2.13 we can choose c as specified there so that $M_c \neq 0$. In that case $M \hookrightarrow M_c$ and so $GK(M) \leq GK(M_c)$. Hence $GK(M_c) = GK(M)$, by 4.2(ii).

However $\Delta(A_c)$ contains a copy of $A_n(k)$; so $GK(M_c) \geq n$ by 8.5.5. □

Once again the regularity of A is essential here. For if A is taken to be the coordinate ring of the cusp, and I is the maximal ideal described in 3.12, then the simple module A/I has GK dimension 0.

4.4 As seen in Chapter 12, the associated graded ring is useful in calculating K_0. In order to do so, one needs to check that it has finite global dimension. That we approach using the symmetric algebra $S_A(\text{Der } A)$.

Lemma. *Let B be a localization of A. Then:*
(i) *the maps $A \hookrightarrow B$, $\text{Der } A \hookrightarrow \text{Der } B$ induce an A-algebra homomorphism $\alpha: S_A(\text{Der } A) \to S_B(\text{Der } B)$;*
(ii) *$1 \otimes \alpha: B \otimes_A S_A(\text{Der } A) \to S_B(\text{Der } B)$ is a B-algebra isomorphism.*

Proof. (i) Clear.

(ii) By the universal property of symmetric algebras, there exist α, γ as shown below; and β exists by the universal property of localization.

$$\begin{array}{ccccc}
\text{Der } A & \longrightarrow & S_A(\text{Der } A) & \longrightarrow & B \otimes S_A(\text{Der } A) \\
\downarrow & & \downarrow \alpha & \overset{\beta}{\nearrow} & \\
\text{Der } B & \longrightarrow & S_B(\text{Der } B) & \overset{\gamma}{\swarrow} &
\end{array}$$

Evidently β, γ are inverse to each other and $\beta = 1 \otimes \alpha$. □

4.5 When A is a regular we can improve upon 1.19

Proposition. *If A is regular then $\pi_A : S_A(\operatorname{Der} A) \to \operatorname{gr} \Delta(A)$ is an isomorphism.*

Proof. Let c be one of the elements $c_1, \ldots, c_t$ provided by 2.13, let $B = A_c$, and let $\vartheta : \operatorname{gr} \Delta(A) \to \operatorname{gr} \Delta(B)$ be the homomorphism induced by the inclusion of $\Delta(A)$ in $\Delta(B)$. One can readily check that the diagram

$$\begin{array}{ccc} B \otimes_A S_A(\operatorname{Der} A) & \xrightarrow{\beta} & S_B(\operatorname{Der} B) \\ \downarrow {\scriptstyle 1 \otimes \pi_A} & & \downarrow {\scriptstyle \pi_B} \\ B \otimes_A \operatorname{gr} \Delta(A) & \xrightarrow{\psi} & \operatorname{gr} \Delta(B) \end{array}$$

in which π_A, π_B are the canonical surjections, β is as in 4.4 and ψ is defined by $\psi(b \otimes h) = b \vartheta(h)$, is a commutative diagram of B-algebra homomorphisms. However, π_B is an isomorphism, because $\operatorname{Der} B$ is free on $\partial/\partial y_1, \ldots, \partial/\partial y_n$, β is an isomorphism by 4.4 and $1 \otimes \pi_A$ is surjective. Therefore $1 \otimes \pi_A$ must be an isomorphism. Since this is so for each $c \in \{c_1, \ldots, c_t\}$ the faithful flatness given by 2.13 establishes that π_A is an isomorphism. $\square$

4.6 Proposition. *(a) If B is a commutative ring and M is a projective B-module then $S_B(M)$ is a projective B-module.*
(b) If, further, $\operatorname{gld} B < \infty$ and M is finitely generated then $\operatorname{gld} S_B(M) < \infty$.

Proof. (a) If N is chosen to make $M \oplus N$ free then $S_B(M \oplus N)$ is free over B. Now $S_B(M \oplus N) \simeq S_B(M) \otimes_B S_B(N)$ since they share the same universal property; and $S_B(N) \simeq B \oplus I$ as a B-module, with I being the augmentation ideal. Therefore $S_B(M)$ is a direct summand of $S_B(M \oplus N)$ as an $S_B(M)$-bimodule, and so, in particular, is projective over B.

(b) In this case N can be chosen to be finitely generated and then $S_B(M \oplus N) \simeq B[z_1, \ldots, z_s]$, where $s = \operatorname{rank} M \oplus N$. Since $S_B(M)$ is a bimodule direct summand, 7.2.8 applies. $\square$

4.7 One can now readily deduce the properties of $\operatorname{gr} \Delta(A)$ when A is regular.

Corollary. *If A is regular then $\operatorname{gr} \Delta(A)$ is a commutative k-affine integral domain and*

$$\operatorname{gld}(\operatorname{gr} \Delta(A)) = K(\operatorname{gr} \Delta(A)) = \operatorname{GK}(\operatorname{gr} \Delta(A)) = 2 \dim A.$$

Proof. By 4.5, $\operatorname{gr} \Delta(A) \simeq S_A(\operatorname{Der} A)$. Choose $c \in A$ such that $\operatorname{Der} A_c$ is free. Since $S_A(\operatorname{Der} A)$ is projective over A the map

$$S_A(\operatorname{Der} A) \longrightarrow A_c \otimes S_A(\operatorname{Der} A) \simeq S_{A_c}(\operatorname{Der} A_c)$$

is injective. However, $S_{A_c}(\operatorname{Der} A_c)$ is a commutative polynomial ring in n

15.4.11   *Further properties*   549

indeterminates over A_c and so is an integral domain of transcendence degree $2n$. It follows that $S_A(\text{Der } A)$ is an integral domain having the same quotient field and hence $\text{tr deg } S_A(\text{Der } A) = 2n$. The rest is clear from 7.1.15 and 8.2.14. □

4.8 It is now easy to describe $K_0(\Delta(A))$.

Theorem. *If A is regular then the inclusion $A \hookrightarrow \Delta(A)$ induces an isomorphism $K_0(A) \to K_0(\Delta(A))$.*

Proof. By 4.5, 4.6 and 4.7, $\text{gld}(\text{gr } \Delta(A)) < \infty$ and $\text{gr } \Delta(A)$ is a projective A-module. Of course $(\text{gr } \Delta(A))_0 = A$ and so 12.6.13 applies directly. □

Thus, for example $K_0(\Delta(A(\mathbb{S}^1))) \simeq \mathbb{Z} \oplus (\mathbb{Z}/2\mathbb{Z})$, by 12.1.6.

4.9 We note another consequence of 4.6.

Proposition. *If A is regular then $\Delta(A)$ is projective and faithfully flat over A.*

Proof. Since $\text{gr } \Delta(A)$ is projective over A so too is each homogeneous component. Therefore the short exact sequence

$$0 \to \Delta(A)_t \to \Delta(A)_{t+1} \to (\text{gr } \Delta(A))_{t+1} \to 0$$

splits over A. It follows easily that, as an A-module, each $\Delta(A)_t$ is projective and $\Delta(A) \simeq \bigoplus_t (\text{gr } \Delta(A))_t$ and so is projective. Of course $A \subseteq \Delta(A)$; so it is faithfully flat too. □

The proof shows that $\Delta(A) \simeq \text{gr } \Delta(A)$ as an A-module.

4.10 It is clear from 1.24 that

$$\text{Der } A \simeq \{\delta \in \text{Der } L \mid \delta(A) \subseteq A\}.$$

The next sequence of results proves a similar fact for $\Delta(A)$ when A is regular. The need to restrict to this case is easily shown. Let A be the (nonregular) coordinate ring of the cusp, as in 3.12. Let $\delta = td/dt$ and $\gamma = t^{-3}\delta(\delta - 2)(\delta - 4)$. One can verify that $\gamma(A) \subseteq A$. However, $\gamma(t^3) = -3$ and so $\gamma(I) \not\subseteq I$, where I is the $\Delta(A)$-stable ideal (t^2, t^3). Hence $\gamma \notin \Delta(A)$.

4.11 The results involve a detailed investigation of $\Delta(L)$. For this some notation is required. If $f \in \Delta(L)$ then the **order** of f is its total degree when written as a polynomial in $\partial/\partial y_1, \ldots, \partial/\partial y_n$; i.e. order $f = t$ if and only if $f \in \Delta(L)_t$. It is easy to see that if $s \in L$ then $fs - sf$ has smaller order than has f; indeed, the order of f is the maximal length of a sequence $s_1, s_2, \ldots, s_t \in L$ such that

$$[\cdots[[f, s_1]s_2], \ldots, s_t] \neq 0.$$

4.12 We start by considering the 'localized' case.

Proposition. *Suppose that* Der A *is a free A-module with basis* $\partial/\partial y_1, \ldots, \partial/\partial y_n$. *If* $f \in \Delta(L)_t$ *with* $f(A) \subseteq A$ *then* $f \in \Delta(A)_t$.

Proof. This is trivial if $t = 0$. We proceed by induction, supposing it true for operators of order less than t. Thus if we let $f_i = fy_i - y_i f$ for each i, then order $f_i <$ order f and so $f_i \in \Delta(A)_{t-1}$. By 3.2, f_i may be viewed as a polynomial over A in $x_1, \ldots, x_n$, of degree $\leq t-1$ and with coefficients from A on the left. Of course
$$[[f, y_i], y_j] = \partial f_i / \partial x_j$$
and so, since
$$[[f, y_i], y_j] = [[f, y_j], y_i],$$
then $\partial f_i/\partial x_j = \partial f_j/\partial x_i$ for each i, j. Hence there exists $F \in \Delta(A)_t$ such that $\partial F/\partial x_i = f_i$ for each i. Let $g = f - F$. Then $g \in \Delta(L)$, $g(A) \subseteq A$, and $[g, y_i] = 0$ for all i. It follows that order $g = 0$; so $g \in A$ and $f \in \Delta(A)_t$. □

4.13 The extension to the 'global' case involves the localizations A_c.

Lemma. *If* $f \in \Delta(L)$ *and* $f(A) \subseteq A$ *then* $f(A_c) \subseteq A_c$.

Proof. Say f has order m, and $a, s \in A$ with s a power of c. We claim that $f(s^{-1}a) \in s^{-m-1}A$. This is clear if $m = 0$. Induction on order shows it holds for $fs - sf$; so $(fs - sf)(s^{-1}a) \in s^{-m}A$. Hence $f(a) - sf(s^{-1}a) \in s^{-m}A$ and so $f(s^{-1}a) \in s^{-m-1}A$ as claimed. Therefore $f(A_c) \subseteq A_c$. □

4.14 Theorem. *Let A be regular. Then:*
(i) $\Delta(A)_t = \{f \in \Delta(L)_t | f(A) \subseteq A\}$ *for each* t;
(ii) $\Delta(A) = \{f \in \Delta(L) | f(A) \subseteq A\}$;
(iii) *the inclusion* $\Delta(A) \hookrightarrow \Delta(L)$ *is a strict filtered homomorphism.*

Proof. (i) Let $c_1, \ldots, c_t$ be as in 2.13 and let $f \in \Delta(L)_t$ satisfy $f(A) \subseteq A$. By 4.13, $f(A_{c_i}) \subseteq A_{c_i}$ and so, by 4.12, $f \in \Delta(A_{c_i})_t$ for each i. Note that $A_{c_i} \otimes_A \Delta(A)_t = \Delta(A_{c_i})_t$. Thus if $M = (Af + \Delta(A)_t)/\Delta(A)_t$ then $A_{c_i} \otimes M = 0$ for each i. The faithful flatness provided by 2.13 shows that $M = 0$.
(ii), (iii) Clear from (i). □

§5 Rings of Differential Operators

5.1 This section concerns an algebra B over a field k of characteristic 0. It starts by defining its **ring of differential operators** $\mathscr{D}(B)$. Applied to A, with the notation of §2, it will be shown that
$$\mathscr{D}(A) = \{f \in \Delta(L) | f(A) \subseteq A\}$$

which, by 4.14, equals $\Delta(A)$ if A is regular. Thus the results of the earlier sections apply to $\mathcal{D}(A)$ in that case. However, 4.10 shows that $\mathcal{D}(A)$ and $\Delta(A)$ are not always equal.

The section then gives a different description of $\mathcal{D}(B)$ involving idealizer subrings.

5.2 First, given a k-algebra B, we describe $\mathcal{D}(B)$. This is a filtered k-algebra in which

$$\mathcal{D}(B)_0 = \{f \in \mathrm{End}_k B \mid fb - bf = 0 \text{ for all } b \in B\} = \mathrm{End}_B B = B$$

and

$$\mathcal{D}(B)_p = \{f \in \mathrm{End}_k B \mid fb - bf \in \mathcal{D}(B)_{p-1} \text{ for all } b \in B\}.$$

Thus $\mathcal{D}(B) = \bigcup_p \mathcal{D}(B)_p$; and an element $f \in \mathcal{D}(B)_p \setminus \mathcal{D}(B)_{p-1}$ has **order** p.

5.3 Lemma. (i) $\mathcal{D}(B)_1 = B + \mathrm{Der}_k B$.
(ii) $\mathcal{D}(B)$ is a filtered ring which has $\Delta(B)$ as a filtered subring.

Proof. (i) Clearly $B + \mathrm{Der}_k B \subseteq \mathcal{D}(B)_1$. To prove the reverse inequality let $f \in \mathcal{D}(B)_1$. Replacing f by $f - f(1)$ we may suppose $f(1) = 0$. Then, for any $a, b \in B$,

$$f(ab) - af(b) = (fa - af)(b) = (fa - af)(b1)$$
$$= b(fa - af)(1) = bf(a)$$

as required.
(ii) Straightforward. □

5.4 The next few results describe in more detail the links between $\mathcal{D}(B)$ and $\Delta(B)$ when B is an integral domain.

Lemma. *Let B be an integral domain, $f \in \mathcal{D}(B)_m$ and $\mathcal{S}$ be a multiplicatively closed subset of B.*
(i) *If $a, b \in B$, $s, t \in \mathcal{S}$, and $at = bs$, then*

$$\sum_{p=0}^{m}(-1)^p s^{-p-1}((\mathrm{ad}\, s)^p(f))(a) = \sum_{p=0}^{m}(-1)^p t^{-p-1}((\mathrm{ad}\, t)^p(f))(b).$$

(ii) *f has a unique extension to an element of $\mathcal{D}(B_\mathcal{S})_m$.*

Proof. (i) Proceed by induction on m, applying the inductive hypothesis to $fs - sf$ (or see [Hart **83**]).
(ii) Define f^* by setting $f^*(s^{-1}a)$ equal to the left-hand side of the equation in (i). Then (i) shows f^* to be well defined, and the rest is easily checked. □

5.5 Theorem. *Let B be an integral domain of finite transcendence degree n over k. Let L be the quotient field of B and $\{y_1, \ldots, y_n\}$ be a transcendence basis for L over k. Then:*

(i) if $f \in \mathscr{D}(L)$ and $[f, y_j] = 0$ for each j then $f \in L$;
(ii) $\mathscr{D}(L) = \Delta(L)$;
(iii) $\mathscr{D}(B) = \{f \in \mathscr{D}(L) | f(B) \subseteq B\}$;
(iv) $\mathscr{D}(B)_t = \mathscr{D}(L)_t \cap \mathscr{D}(B)$.

Proof. (i) Suppose order $f \geqslant 1$ and $a \in L$. Note that $[f, a]$ has the same property as f but has smaller order. By taking commutators with appropriate elements we can reduce to the case when order $f = 1$. Then $f = b + \delta$, where $b \in L$ and $\delta \in \text{Der } L$. The hypothesis ensures that δ is zero on $k(y_1, \ldots, y_n)$. But $\text{Der } L = \sum L \partial / \partial y_i$ and therefore $\delta = 0$. Thus order $f = 0$ as required.

(ii) Arguing as in the proof of 4.12, one sees that if $f \in \mathscr{D}(L)$ then there exists $F \in \Delta(L)$ such that $[g, y_j] = 0$ for each j, where $g = f - F$. By (i), $g \in L$ and so $f \in \Delta(L)$.

(iii), (iv) Combine (ii) with 5.4 (ii). □

5.6 In particular, this applies to A, as in §2.

Corollary. *If A is regular then $\mathscr{D}(A) = \Delta(A)$ and their filtrations coincide.* □

5.7 The remaining results consider the case when B is not necessarily an integral domain but is affine. In this case there is a polynomial ring $R = k[y_1, \ldots, y_n]$, say, and a surjection $\vartheta : R \twoheadrightarrow B$. Let $V = k + \sum_{i=1}^{n} k y_i$, $I = \ker \vartheta$ and $\bar{r} = \vartheta(r)$ for $r \in R$. **This notation will be retained**.

5.8 Our aim is to connect $\mathscr{D}(B)$ with $\mathscr{D}(R)$; this requires some preliminaries.

Lemma. (i) *Let $\beta \in \text{End } R_k$. The equation $\bar{\beta}(\bar{r}) = \overline{\beta(r)}$ defines a map $\bar{\beta} \in \text{End } B_k$ if and only if $\beta(I) \subseteq I$. In that case $\bar{\beta} = 0$ if and only if $\beta(R) \subseteq I$.*
(ii) *If $\beta \in \mathscr{D}(R)$ and $\beta(I) \subseteq I$ then $\bar{\beta} \in \mathscr{D}(B)$ and order $\bar{\beta} \leqslant$ order β.*

Proof. Straightforward. □

5.9 Let $\boldsymbol{i}$ denote $(i_1, \ldots, i_n)$ and set $|\boldsymbol{i}| = i_1 + \cdots + i_n$,

$$y^{\boldsymbol{i}} = y_1^{i_1} \cdots y_n^{i_n}, \quad \partial^{\boldsymbol{i}} = (\partial/\partial y_1)^{i_1} \cdots (\partial/\partial y_n)^{i_n} \quad \text{and} \quad \boldsymbol{i}! = i_1! \cdots i_n!.$$

Proposition. *Let $\beta \in \mathscr{D}(R)$. Then:*
(i) $\beta(R) \subseteq I$ *if and only if* $\beta \in I \mathscr{D}(R)$;
(ii) $\beta(I) \subseteq I$ *if and only if* $\beta \in \mathbb{I}(I \mathscr{D}(R))$.

Proof. (i) ($\Leftarrow$). Clear.
($\Rightarrow$). Write $\beta = \sum r_i \partial^{\boldsymbol{i}}$. If some $r_i \notin I$ we choose one to minimize $|\boldsymbol{i}|$. Then $\beta(y^{\boldsymbol{i}}) \equiv \boldsymbol{i}! r_i$ modulo I and so $\beta(y^{\boldsymbol{i}}) \notin I$.

15.5.13   *Rings of differential operators*   553

(ii) ($\Leftarrow$). If $a \in I$ then $a \in I\mathcal{D}(R)$ and so $\beta(a) = \beta a(1_R) \in I\mathcal{D}(R)(1_R) \subseteq IR = I$.
($\Rightarrow$). If $a \in I$ and $r \in R$ then $\beta a(r) = \beta(ar) \in I$ and therefore $\beta a(R) \subseteq I$. Hence, by (i), $\beta a \in I\mathcal{D}(R)$ and thus $\beta \in \mathbb{I}(I\mathcal{D}(R))$. □

5.10 Corollary. *The map $\beta \mapsto \bar{\beta}$ induces an injective ring homomorphism $\varphi : \mathbb{I}(I\mathcal{D}(R))/I\mathcal{D}(R) \to \mathcal{D}(B)$.* □

5.11 The next few results serve to prove that φ is an isomorphism.

Lemma. *Let m be a positive integer, $\alpha \in \mathcal{D}(B)$ and $\beta \in \mathcal{D}(R)$ with order $\alpha \leq m$, order $\beta \leq m$ and suppose that $\alpha(\bar{r}) = \overline{\beta(r)}$ for all $r \in V^m$. Then $\bar{\beta} = \alpha$.*

Proof. We proceed by induction on m, the case $m = 0$ being clear. Choose any $v \in V$ and let $\alpha' = \alpha \bar{v} - \bar{v}\alpha$, $\beta' = \beta v - v\beta$. Then α' and β' each have order at most $m - 1$ and one can check that $\alpha'(\bar{r}) = \overline{\beta'(r)}$ for all $r \in V^{m-1}$. Therefore, by the induction hypothesis, $\alpha' = \bar{\beta}'$.

We now aim, using this, to demonstrate, by induction on n, that $\alpha(\bar{r}) = \overline{\beta(r)}$ for all $r \in V^n$, this giving the desired result. If $n \leq m$, it is obvious. For $n > m$, let $r = v_1 \cdots v_n$ with $v_1, \ldots, v_n \in V$. Then

$$\alpha(\bar{r}) = \alpha(\bar{v}_1 \cdots \bar{v}_n) = (\alpha \bar{v}_1 - \bar{v}_1 \alpha)(\bar{v}_2 \cdots \bar{v}_n) + \bar{v}_1 \alpha(\bar{v}_2 \cdots \bar{v}_n)$$

and

$$\beta(r) = \beta(v_1 \cdots v_n) = (\beta v_1 - v_1 \beta)(v_2 \cdots v_n) + v_1 \beta(v_2 \cdots v_n).$$

The argument above, together with the induction on n, ensures that $\alpha(\bar{r}) = \overline{\beta(r)}$ as required. □

5.12 Lemma. *Let m be a positive integer and $\lambda \in \mathrm{Hom}_k(V^m, R)$. There exists a unique $\beta \in \mathcal{D}(R)_m$ which, when restricted to V^m, equals λ.*

Proof. We construct $\beta = \sum_{|i| \leq m} r_i \partial^i$ by a recursive construction of

$$\beta_j = \sum_{|i| \leq j} r_i \partial^i \quad \text{for } j = 0, \ldots, m.$$

Set $\beta_0 = \lambda(1) = r_0$. Given β_{j-1}, for each i with $|i| = j$ choose r_i so that

$$\lambda(y^i) = \beta_{j-1}(y^i) + i! r_i$$

and set $\beta_j = \sum_{|i| \leq j} r_i \partial^i$. It is easy to check that β_j and λ are equal, when restricted to V^j. Hence $\beta = \beta_m$ is as required; and its uniqueness is clear. □

5.13 Theorem. $\mathcal{D}(B) \simeq \mathbb{I}(I\mathcal{D}(R))/I\mathcal{D}(R)$.

Proof. We need only show that the map φ of 5.10 is surjective. Let $\alpha \in \mathcal{D}(B)_m$ and let $z_k = \vartheta(y_k)$. For each i with $|i| \leq m$, choose $r_i \in R$ such that $\bar{r}_i = \alpha(z^i)$.

By 5.12, there exists $\beta \in \mathscr{D}(R)_m$ such that $\beta(y^i) = r_i$. Hence, for all $r \in V^m$, $\overline{\beta(r)} = \alpha(\bar{r})$ and so, by 5.11, $\alpha = \bar{\beta}$. □

We note that this isomorphism preserves the filtrations.

5.14 Corollary. $\mathscr{D}(B) \simeq \mathrm{End}\,(\mathscr{D}(R)/I\mathscr{D}(R)_{\mathscr{D}(R)})$. □

§6 Additional Remarks

6.0 Many of the results in this chapter can be found in [Grothendieck **67**], [Sweedler **75**] or [Bjork **79**] or deduced from results there. However, this approach provides an elementary introduction to this area.

6.1 (a) $\Delta(A)$ was discussed in [Hochschild, Kostant and Rosenberg **62**].
(b) This discussion of derivations and differentials is based on [Matsumura **80**, §26]. An alternative description of $\Omega(A)$ is given there. It is shown that $\Omega_K(A) \simeq J/J^2$, where J is the kernel of the multiplication map $A \otimes A \to A$, $a \otimes b \mapsto ab$.

6.2 No comment.

6.3 (a) The argument used in the proof of 3.7 comes from [Hodges and Smith **85**]. 3.6(ii) appears in [S. P. Smith **84**].
(b) We note, in connection with 3.9, the **Zariski–Lipman conjecture**: If $\mathrm{Der}_k A$ is free then A is regular. See [Lipman **65**].

6.4 (a) In [Bjork **79**, p. 31] it was asked if all simple $\Delta(A)$-modules are holonomic when A is as in 4.3. The example of [Stafford **85**], cited in 8.5.8, answers this in the negative.
(b) Let A be regular, $\Delta(A)$ have the filtration described in 1.19 and M be a finitely generated $\Delta(A)$-module with a good filtration. The results mentioned in 8.7.6(b), combined with 2.13 and 4.2, show that $\mathrm{GK}\,(M) = \mathrm{GK}\,(\mathrm{gr}\,M)$.
(c) The condition, in 4.5, that A be regular is not redundant. Indeed if A is the coordinate ring of the cusp then $\mathrm{gr}\,\Delta(A)$ is an integral domain, whereas $S_A(\mathrm{Der}\,A)$ is not (Stafford, unpublished).

6.5 (a) If K is a commutative ring and B is a K-algebra then the definition in 5.2, with k replaced by K, gives the ring $\mathscr{D}_K(B)$ of differential operators on B over K.
(b) This discussion of $\mathscr{D}(B)$ comes in part from [Hart **83**], where further details can be found.
(c) As with $\Omega(A)$ in 6.1(b), there is another description of $\mathscr{D}(A)$. For $n \geq 0$, let P^n denote the $A \otimes A$-module $(A \otimes A)/J^{n+1}$. Then

$$\mathscr{D}(A)_n \simeq \mathrm{Hom}_{A \otimes A}(P^n, \mathrm{End}_k A).$$

For details see [Grothendieck **67**, §16], [Heyneman and Sweedler **69**], [Sweedler **75**, Chapters 8, 13].
This description is used in the proof of 5.13 in [S. P. Smith and Stafford, **P**]. This appears to be its first proof in the literature although it was known earlier.
(d) An example in [Bernstein, Gelfand and Gelfand **72**] shows that $\mathscr{D}(A)$ need not be affine nor right or left Noetherian. [S. P. Smith and Stafford **P**] show that it can be

affine and right Noetherian yet not left Noetherian. This demonstrates the lack of symmetry in $\mathscr{D}(A)$.

(e) [Nakai **70**] conjectured that $\mathscr{D}(A) = \Delta(A)$, for A an affine domain, if and only if A is regular. Some progress is reported in [Mount and Villamayor **73**] and [Becker **78**].

6.6 Further references: [Beilinson and Bernstein **81**], [Bell and Goodearl **P**], [Bjork **P**], [Borho and Brylinski **82, 85**], [Brylinski and Kashiwara **81**], [Chamarie and Stafford **P**], [Hart and Smith **87**], [Levasseur **81**], [Muhasky **P**], [Musson **86**], [Pham **79**], [S. P. Smith **86, P**], [Borel **87**], [Coutinho and Holland **P**].

REFERENCES

The titles listed below are those cited in the text, the points of citation being noted after each. Browsing near these titles in the volumes of *Reviews in Ring Theory* (ed. L. W. Small, Amer. Math. Soc., Providence, RI) makes available a much fuller bibliography than would be possible here. The digits after the author's name indicate the year of publication, generally in this century.

Aljadeff, E., and Rosset, S.
   86 Global dimension of crossed products, *J. Pure Appl. Algebra* **40**, 103–13. (7.12.5)

Amitsur, S. A.
   52 An embedding of PI-rings, *Proc. Amer. Math. Soc.* **3**, 3–9. (13.11.4)
   53 The identities of PI-rings, *Proc. Amer. Math. Soc.* **4**, 27–34. (13.11.4)
   55 The T-ideals of the free ring, *J. London Math. Soc.* **30**, 470–5. (13.11.1)
   56 Algebras over infinite fields, *Proc. Amer. Math. Soc.* **7**, 35–48. (9.7.1)
  56a Radicals of polynomial rings, *Canad. J. Math.* **8**, 355–61. (13.11.2)
   71 Rings of quotients and Morita contexts, *J. Algebra* **17**, 273–98. (3.7.6)
   72 On central division algebras, *Israel J. Math.* **12**, 408–20. (13.11.1)
   74 Polynomial identities, *Israel J. Math.* **19**, 183–99. (13.11.0)
   80 On the characteristic polynomial of a sum of matrices, *Linear and Multilinear Algebra* **8**, 177–82. (13.11.9)

Amitsur, S. A., and Procesi, C.
   66 Jacobson rings and Hilbert algebras with polynomial identities, *Ann. Mat. Pura. Appl.* **71**, 61–72. (13.11.10)

Amitsur, S. A., and Small, L. W.
   78 Polynomials over division rings, *Israel J. Math.* **31**, 353–8. (9.7.6)
   80 Prime ideals in PI rings, *J. Algebra* **62**, 358–83. (13.11.8, 13.11.9)

Anderson, F. W., and Fuller, K. R.
   74 *Rings and Categories of Modules*, Graduate Texts in Mathematics Vol. 13, Springer-Verlag, New York–Berlin. (3.7.5, 4.3.2, 7.1.1)

Artin, E., and Tate, J. T.
   51 A note on finite ring extensions, *J. Math. Soc. Japan* **3**, 74–7. (13.11.9)

Artin, M.
   69 On Azumaya algebras and finite dimensional representations of rings, *J. Algebra* **11**, 532–63. (13.11.7)

Artin, M. and Schelter, W. F.
   81 Integral ring homomorphisms, *Adv. in Math.* **39**, 289–329. (13.11.8)

Asano, K.
   49 Zur Arithmetik in Schiefringen I, *Osaka J. Math.* **1**, 98–134. (5.8.0, 13.11.9)
  49a Uber die Quotientbildung von Schiefringen, *J. Math. Soc. Japan* **1**, 73–8. (2.4.1)

Atiyah, M. F., and Macdonald, I. G.

69 *Introduction to Commutative Algebra*, Addison-Wesley, Reading, Mass. (4.1.10, 13.7.15, 13.8.2, 13.8.15, 15.1.13)

Bass, H.
64 K-theory and stable algebra, *Inst. Hautes Etudes Sci. Publ. Math.* **22**, 489–544. (11.3.10, 11.8.0, 11.8.3, 11.8.4)
68 *Algebraic K-theory*, Benjamin, New York. (12.8.0, 12.8.2)
72 The degree of polynomial growth of finitely generated nilpotent groups, *Proc. London Math. Soc.* **25**, 603–14. (8.7.2)
74 *Introduction to Some Methods of Algebraic K-theory*, CBMS Monographs 20, Amer. Math. Soc., Providence, R.I. (12.8.0, 12.8.4)
75 Algebraic K-theory: a historical survey, *Proc. Int. Cong. Math.* (Vancouver 1974) Canadian Math. Congress, 277–83. (12.8.0, 12.8.3)

Becker, J.
78 Higher derivatives and integral closure, *Amer. J. Math.* **100**, 495–521. (15.6.5)

Beidar, K. I.
81 On radicals of finitely generated algebras, *Russian Math. Surveys* **36**(6), 171–2. (9.7.1)

Beilinson, A., and Bernstein, J.
81 Localisation de g-modules, *C. R. Acad. Sci., Paris* **292**, 15–18. (15.6.6)

Bell, A. D.
84 Goldie dimension of prime factors of polynomial and skew polynomial rings, *J. London Math. Soc.* **29**, 418–24. (4.7.3)
87 Localization and ideal theory in iterated differential operator rings, *J. Algebra* **106**, 376–402. (4.7.3)
87a Localization and ideal theory in Noetherian strongly group-graded rings, *J. Algebra* **105**, 76–115. (4.7.3)

Bell, A. D., and Goodearl, K. R.
P Uniform rank over differential operator rings and Poincaré–Birkhoff–Witt extensions, preprint, University of Utah. (15.6.6)

Bergen, J., Montgomery, S. and Passman, D. S.
P Radicals of Lie algebra smash products, preprint, University of Wisconsin, Madison. (9.7.1)

Bergman, G. M.
71 Groups acting on hereditary rings, *Proc. London Math. Soc.* **23**, 70–82; Corrigendum, **24** (1972), 192. (7.12.8)
81 Gelfand–Kirillov dimension can go up in extension modules, *Comm. Algebra* **9**, 1567–70. (8.3.4)
83 Sfields finitely right-generated over subrings, *Comm. Algebra* **11**, 1893–902. (10.0.0)
P Gelfand–Kirillov dimensions of factor rings, preprint, University of California, Berkeley. (8.7.3)
U Lifting prime ideals to extensions by centralizing elements, University of California, Berkeley, unpublished. (10.7.0)
Ua More on extensions by centralizing elements, University of California, Berkeley, unpublished. (10.7.0)
Ub On Jacobson radicals of rings, University of California, Berkeley, unpublished. (10.7.0)
Uc Homogeneous elements and prime ideals in $\mathbb{Z}$-graded rings, University of California, Berkeley, unpublished. (10.7.6)
Ud A note on growth functions of algebras and semigroups, University of California, Berkeley, unpublished. (8.1.18)

Bergman, G. M., and Isaacs, I. M.

73   Rings with fixed-point-free group actions, *Proc. London Math. Soc.* **27**, 69–87. (10.7.5)
Bergman, G. M., and Small, L. W.
75   PI-degrees and prime ideals, *J. Algebra* **33**, 435–62. (13.11.7)
Bernstein, I. N.
71   Modules over a ring of differential operators. Study of the fundamental solutions of equations with constant coefficients, *Funct. Anal. Appl.* **5**, 89–101. (8.7.4, 8.7.6)
72   The analytic continuation of generalized functions with respect to a parameter, *Funct. Anal. Appl.* **6**, 273–85. (8.7.4, 8.7.5)
Bernstein, I. N., Gelfand, I. M., and Gelfand, S. I.
72   Differential operators on a cubic cone, *Russian Mathematical Surveys* **27**, 169–74. (15.6.5)
Berrick, A. J.
82   *An Approach to Algebraic K-theory*, Research Notes in Mathematics 56, Pitman, London. (12.8.0)
Bhatwadekar, S. M.
76   On the global dimension of some filtered algebras, *J. London Math. Soc.* **13**, 239–48. (7.12.1)
Bit-David, J.
80   Normalizing extensions II, in *Ring Theory (Antwerp 1980)*, ed. F. Van Oystaeyen, Lecture Notes in Mathematics 825, Springer-Verlag, New York–Berlin, pp. 6–9. (10.7.1)
Bit-David, J., and Robson, J. C.
80   Normalizing extensions I, in *Ring Theory (Antwerp 1980)*, ed. F. Van Oystaeyen, Lecture Notes in Mathematics 825, Springer-Verlag, New York–Berlin, pp. 1–5. (10.7.1, 10.7.2)
Bjork, J-E.
72   The global homological dimension of some algebras of differential operators, *Invent. Math.* **17**, 67–78. (7.12.10)
79   *Rings of Differential Operators*, North-Holland Mathematics Library 21, Amsterdam. (8.7.6, 15.6.0, 15.6.4)
P   *Algebraic Analysis of Differential Systems*, in preparation. (15.6.6)
Blair, W. D.
73   Right Noetherian rings integral over their centers, *J. Algebra* **27**, 187–98. (13.11.8)
Blair, W. D., and Small, L. W.
P   Embeddings in Artinian rings and Sylvester rank functions, preprint, University of California, San Diego. (4.7.1)
Bollobas, B.
79   *Graph Theory*, Graduate Texts in Mathematics 63, Springer-Verlag, New York–Berlin. (13.3.3)
Bonang, F.
P   Thesis, University of Leeds. (11.8.2)
Borel, A.
87   *Algebraic D-modules*, Academic Press, London–New York. (15.6.6)
Borho, W.
82   On the Joseph–Small additivity principle for Goldie ranks, *Compositio Math.* **47**, 3–29. (4.7.5, 8.7.3)
82a Invariant dimension and restricted extension of noetherian rings, in *Séminaire Dubreil-Malliavin (1981)*, ed. M.-P. Malliavin, Lecture Notes in Mathematics 924, Springer-Verlag, New York–Berlin, pp. 57–71. (6.10.8)
86   A survey on enveloping algebras of semisimple Lie algebras I, in *Lie Algebras and*

Related Topics, ed. D. J. Britten, F. W. Lemire, and R. V. Moody, Canadian Math. Soc. Conf. Proc. Vol. 5, Amer. Math. Soc., Providence R. I., pp. 19–50. (14.10.0)

Borho, W., and Brylinski, J-L.
82 Differential operators on homogeneous spaces I, *Invent. Math.* **69**, 437–76. (15.6.6)
85 Differential operators on homogeneous spaces III, *Invent. Math.* **80**, 1–68. (15.6.6)

Borho, W., Gabriel, P., and Rentschler, R.
73 *Primideale in Einhüllenden auflösbarer Lie-Algebren,* Lecture Notes in Mathematics 357, Springer-Verlag, New York–Berlin. (14.7.2, 14.10.0, 14.10.2)

Borho, W., and Kraft, H.
76 Über die Gelfand–Kirillov Dimension, *Math. Ann.* **220**, 1–24. (8.7.0, 8.7.1, 8.7.2)

Bourbaki, N.
61–65 *Algèbre commutative,* Hermann, Paris. (5.1.10, 5.3.7, 13.9.14)

Braun, A.
84 The nilpotency of the radical in a finitely generated PI ring, *J. Algebra* **89**, 375–396. (13.11.10)

Braun, A., and Small, L. W.
86 Localization in prime noetherian PI rings, *Math. Z.* **193**, 323–30. (13.11.7)

Braun, A., and Warfield, R. B.
P Symmetry and localization in Noetherian prime PI rings, preprint, University of Washington, Seattle. (13.11.7)

Brown, K. A.
81 Module extensions over Noetherian rings, *J. Algebra* **69**, 247–60. (4.7.3)
82 The Nullstellensatz for certain group rings, *J. London Math. Soc.* **26**, 425–34. (9.7.4)
84 Ore sets in enveloping algebras, *Compositio Math.* **53**, 347–67. (4.7.3)
85 Ore sets in Noetherian rings, in *Séminaire Dubreil-Malliavin (1983–1984),* ed. M.-P. Malliavin, Lecture Notes in Mathematics 1146, Springer-Verlag, New York–Berlin, pp. 355–66. (4.7.3)
86 Localisation at cliques in group rings, *J. Pure Appl. Algebra* **41**, 9–16. (4.7.3)

Brown, K. A., and Du Cloux, F.
P On the representation of solvable Lie algebras, preprint, University of Glasgow. (4.7.3)

Brown, K. A., and Hajarnavis, C. R.
84 Homologically homogeneous rings, *Trans. Amer. Math. Soc.* **281**, 197–208. (7.12.0)

Brown, K. A., Hajarnavis, C. R., and MacEacharn, A. B.
82 Noetherian rings of finite global dimension, *Proc. London Math. Soc.* **44**, 349–71. (7.12.0)
83 Rings of finite global dimension integral over their centres, *Comm. Algebra* **11**, 67–93. (7.12.0)

Brown, K. A., Howie, J., and Lorenz, M.
P Induced resolutions and Grothendieck groups of polycyclic-by-finite groups, preprint, University of Glasgow. (12.8.5)

Brown, K. A., Lenagan, T. H., and Stafford, J. T.
80 Weak ideal invariance and localization, *J. London Math. Soc.* **21**, 53–61. (4.7.3, 6.10.8)
81 K-theory and stable structure of some Noetherian group rings, *Proc. London Math. Soc.* **42**, 193–230. (6.8.24)

Brown, K. A. and Smith, S. P.
85 Bimodules over a solvable algebraic Lie algebra, *Quart. J. Math. Oxford* **36**, 129–39. (14.10.8)

Brown, K. A. and Warfield, R. B.

P   The influence of ideal structure on representation theory, preprint, University of Washington, Seattle. (4.7.3)

Brylinski, J-L., and Kashiwara, M.
  81 Kazhdan–Lusztig conjecture and holonomic systems, *Invent. Math.* **64**, 387–410. (15.6.6)

Cartan, H. and Eilenberg, S.
  56 *Homological Algebra*, Princeton Univ. Press, Princeton. (7.3.16, 7.12.0, 12.4.8)

Cauchon, G.
  76 Les T-anneaux, la condition (H) de Gabriel et ses conséquences, *Comm. Algebra* **4**, 11–50. (6.10.4)
  76a Anneaux semi-premiers, Noethériens, à identités polynomiales, *Bull. Soc. Math. France* **104**, 99–111. (13.11.6)

Chakravarti, R. S.
  85 The basic element theorem for fully bounded rings and some applications, *Comm. Algebra* **13**, 259–83. (11.8.7)

Chamarie, M.
  80 Maximal orders applied to enveloping algebras, in *Ring Theory (Antwerp 1980)*, ed. F. Van Oystaeyen, Lecture Notes in Mathematics 825, Springer-Verlag, New York–Berlin, pp. 19–27. (5.8.1)
  81 Anneaux de Krull non commutatifs, *J. Algebra* **72**, 210–22. (5.8.1)
  83 Modules sur les anneaux de Krull non commutatifs, in *Séminaire Dubreil-Malliavin (1982)*, ed. M.-P. Malliavin, Lecture Notes in Mathematics 1029, Springer-Verlag, New York–Berlin, pp. 283–310. (5.8.1)

Chamarie, M., and Stafford, J. T.
  P   When rings of differential operators are maximal orders, preprint, University of Leeds. (15.6.6)

Chatters, A. W.
  71 The restricted minimum condition in Noetherian hereditary rings, *J. London Math. Soc.* **4**, 83–7. (5.8.4)
  72 A decomposition theorem for Noetherian hereditary rings, *Bull. London Math. Soc.* **4**, 125–6. (5.8.4)

Chatters, A. W., Goldie, A. W., Hajarnavis, C. R., and Lenagan, T. H.
  79 Reduced rank in Noetherian rings, *J. Algebra*, **61**, 582–9. (4.7.1)

Chatters, A. W., and Hajarnavis, C. R.
  80 *Rings with Chain Conditions*, Research Notes in Mathematics 44, Pitman, London. (3.7.2, 4.7.0, 4.7.1)
  P   Ideal arithmetic in Noetherian PI rings, preprint, University of Warwick. (13.11.9)

Chatters, A. W., Hajarnavis, C. R., and Norton, N. C.
  77 The Artin radical of a Noetherian ring, *J. Austral. Math. Soc.* **23**, 379–84. (4.7.1)

Chevalley, C.
  36 *L'Arithmétique dans les Algèbres de Matrices*, Act. Sci. Ind. 323, Hermann, Paris. (5.8.0)

Chin, W.
  87 Prime ideals in differential operator rings and crossed products of infinite groups, *J. Algebra* **106**, 78–104. (10.7.6)

Chin, W. and Quinn, D.
  P   Rings graded by polycyclic-by-finite groups, preprint, University of Texas, Austin. (10.7.6)

Cohen, M.
  75 Semiprime Goldie centralizers, *Israel J. Math.* **20**, 37–45; Addendum **24** (1976), 89–93. (10.7.5)

Cohen, M. and Montgomery, S.
- 75 Semisimple Artinian rings of fixed points, *Canad. Math. Bull.* **18**, 189–90. (10.7.5)
- 79 The normal closure of a semiprime ring, in *Ring Theory (Antwerp 1978)*, ed. F. Van Oystaeyen, Lecture Notes in Pure and Applied Mathematics 51, Dekker, New York, pp. 43–59. (10.3.15)

Cohn, H.
- 80 *Advanced Number Theory*, Dover, New York (first published as *A Second Course in Number Theory*, Wiley, New York, 1962). (12.5.6)

Cohn, P. M.
- 63 A remark on the Birkhoff Witt theorem, *J. London Math. Soc.* **38**, 197–203. (1.7.8)
- 66 On the structure of the GL_2 of a ring, *Inst. Hautes Etudes Sci. Publ. Math.* **33**, 365–413. (11.5.7)
- 74 *Algebra I*, Wiley, London–New York. (4.1.2)
- 77 *Algebra II*, Wiley, London–New York. (0.1.1, 3.5.7, 5.7.18, 13.1.21, 13.3.3, 13.8.11, 13.11.4, 15.2.7)
- 85 *Free Rings and their Relations* (second edition), London Math. Soc. Monographs 19, Academic Press, London. (11.3.9)

Cortzen, B.
- 82 Finitistic dimensions of ring extensions, *Comm. Algebra* **10**, 993–1001. (7.12.2)

Coutinho, S. C.
- 85 K-theoretic properties of generic matrix rings, *J. London Math. Soc.* **32**, 51–6. (13.11.1)
- 86 Generating modules efficiently over noncommutative Noetherian rings, Thesis, University of Leeds. (11.8.0, 11.8.6, 11.8.7)

Coutinho, S. C., and Holland, M. P.
- P Module structure of rings of differential operators, preprint, University of Leeds. (15.6.6)

Cozzens, J. H.
- 76 Maximal orders and reflexive modules, *Trans. Amer. Math. Soc.* **219**, 323–36. (5.8.1)

Cozzens, J. H., and Faith, C.
- 75 *Simple Noetherian Rings*, Cambridge Tracts in Mathematics 69, Cambridge Univ. Press, Cambridge (1.9.8)

Curtis, C. W., and Reiner, I.
- 81 *Methods of Representation Theory* Vol. I, Wiley–Interscience, J. Wiley, New York. (5.8.0)

Daintree, R. M.
- 84 Ore and Laurent extensions of infinite Krull dimension, Thesis, University of Leeds. (6.9.24, 6.10.1)

Dean, C., and Stafford, J. T.
- P A non-embeddable Noetherian ring, preprint, University of Leeds. (4.7.1)

De Meyer, F. and Ingraham E.
- 70 *Separable Algebras over Commutative Rings*, Lecture Notes in Mathematics 181, Springer-Verlag, New York–Berlin. (13.7.8)

Dirac, P. A. M.
- 26 On quantum algebra, *Proc. Camb. Phil. Soc.* **23**, 412–18. (1.9.3)

Dixmier, J.
- 77 *Enveloping Algebras*, North-Holland Mathematics Library 14, Amsterdam. (Translated from *Algèbres enveloppantes*, Cahiers Scientifiques 37, Gauthier-Villars 1974.) (1.7.5, 4.6.2, 8.5.12, 9.7.5, 14.10.0, 14.10.2, 14.10.4)

Duflo, M.
- 73 Certaines algèbres de type fini sont des algèbres de Jacobson, *J. Algebra* **27**, 358–65. (8.7.4, 9.7.0, 9.7.1, 9.7.2)

Eisenbud, D. and Evans, E. G.
  73 Generating modules efficiently: theorems from algebraic K-theory, *J. Algebra* **27**, 278–305. (11.8.0, 11.8.7)

Eisenbud, D., and Robson, J. C.
  70 Modules over Dedekind prime rings, *J. Algebra* **16**, 67–85. (5.8.6, 5.8.7, 6.10.7)
  70a Hereditary Noetherian prime rings, *J. Algebra* **16**, 86–104. (5.8.6, 5.8.7)

Ely, R. E.
  74 Multiple idealizers and hereditary Noetherian prime rings, *J. London Math. Soc.* **7**, 673–80. (5.8.5)

Evans, E. G.
  73 Krull–Schmidt and cancellation over local rings, *Pacific J. Math.* **46**, 115–21. (11.8.4)

Faith, C. and Utumi, Y.
  65 On Noetherian prime rings, *Trans. Amer. Math. Soc.* **114**, 53–60. (3.7.2)

Farkas, D. R., and Snider, R. L.
  81 Commuting rings of simple $A(k)$-modules, *J. Austral. Math. Soc. Ser. A* **31**, 142–5. (9.7.5)

Farrell, F. T. and Hsiang, W. C.
  70 A formula for $K_1 R_\alpha[T]$ in *Applications of Categorical Algebra*, ed. A. Heller, Proc. Symposia in Pure Math. 18, Amer. Math. Soc., Providence R. I., pp. 192–218. (7.12.7, 12.8.5)
  81 The Whitehead group of poly- (finite or cyclic) groups, *J. London Math. Soc.* **24**, 308–24. (12.8.5)

Feldman, G. L.
  83 Global dimension of rings of differential operators, *Trans. Moscow Math. Soc.* **41**, 123–47. (7.12.6)

Feller, E. H., and Swokowski, E. W.
  64 The ring of endomorphisms of a torsionfree module, *J. London Math. Soc.* **39**, 41–2. (3.7.4)

Ferrero, M., and Kishimoto, K.
  85 On differential rings and skew polynomials, *Comm. Algebra* **13**, 285–304. (9.7.1)

Fields, K. L.
  69 On the global dimension of skew polynomial rings, *J. Algebra* **13**, 1–4. (7.12.5)
  70 On the global dimension of residue rings, *Pacific J. Math.* **32**, 345–9. (7.12.3)

Fisher, J. W., and Montgomery, S.
  78 Semiprime skew group rings, *J. Algebra* **52**, 241–7. (10.7.5)

Fisher, J. W., and Osterburg, J.
  78 Semiprime ideals in rings with finite group actions, *J. Algebra* **50**, 488–502. (10.7.5)

Fitting, H.
  35 Primärkomponentenzerlegung in nichtkommutativen Ringen, *Math. Ann.* **111**, 19–41. (1.9.1)

Formanek, E.
  72 Central polynomials for matrix rings, *J. Algebra* **23**, 129–32. (13.11.5)
  74 Noetherian PI-rings, *Comm. Algebra* **1**, 79–86. (13.11.6)
  84 Invariants and the ring of generic matrices, *J. Algebra* **89**, 178–223. (13.11.1)

Formanek, E., and Jategaonkar, A. V.
  74 Subrings of Noetherian rings, *Proc. Amer. Math. Soc.* **46**, 181–6. (10.7.1)

Forster, O.
  64 Über die Anzahl der Erzeugenden eines Ideals in einem Noetherschen Ring, *Math. Z.* **84**, 80–7. (11.8.7)

Gabber, O.

P   Equidimensionalité de la variété caractéristique, Exposé de O. Gabber redigé par T. Levasseur, preprint, University of Paris VI. (14.10.2)

Gabel, M. R.
- **75** Lower bounds on the stable range of polynomial rings, *Pacific J. Math.* **61**, 117–20. (11.5.10)

Gabriel, P.
- **62** Des catégories abéliennes, *Bull. Soc. Math. France* **90**, 323–448. (6.10.0, 6.10.2, 6.10.4)
- **71** Représentations des algèbres de Lie résolubles, in *Séminaire Bourbaki 347*, Lecture Notes in Mathematics 179, Springer-Verlag, New York–Berlin, pp. 1–22. (14.10.5)

Gelfand, I. M., and Kirillov, A. A.
- **66** Sur les corps liés aux algèbres enveloppantes des algèbres de Lie, *Inst. Hautes Etudes Sci. Publ. Math.* **31**, 5–19. (6.10.6, 8.7.0)
- **66a** Fields associated with enveloping algebras of Lie algebras, *Soviet Math. Dokl.* **167**, 407–9. (8.7.0)

Ginn, S. M. and Moss, P. B.
- **77** A decomposition theorem for Noetherian orders in Artinian rings, *Bull. London Math. Soc.* **9**, 177–81. (4.7.1)

Goldie, A. W.
- **58** The structure of prime rings under ascending chain conditions, *Proc. London Math. Soc.* **8**, 589–608. (2.4.1, 2.4.2, 2.4.3)
- **60** Semi-prime rings with maximum condition, *Proc. London Math. Soc.* **10**, 201–20. (1.9.2, 2.4.3)
- **62** Noncommutative principal ideal rings, *Arch. Math.* **13**, 213–21. (3.7.4)
- **64** Torsionfree modules and rings, *J. Algebra* **1**, 268–87. (4.7.1)
- **67** Localization in non-commutative Noetherian rings, *J. Algebra* **5**, 89–105. (4.7.3)
- **69** Some aspects of ring theory, *Bull. London Math. Soc.* **1**, 129–54. (2.4.0)
- **78** Reduced rank on modules, in *Ring Theory (Antwerp 1977)*, ed. F. Van Oystaeyen, Lecture Notes in Pure and Applied Mathematics 40, Dekker, New York, pp. 45–8. (6.10.4)

Goldie, A. W. and Krause, G.
- **84** Strongly regular elements of Noetherian rings, *J. Algebra* **91**, 410–29. (4.7.4)

Goldie, A. W., and Michler, G. O.
- **74** Ore extensions and polycyclic group rings, *J. London Math. Soc.* **9**, 337–45. (9.7.1, 10.7.6)

Goldman, O.
- **51** Hilbert rings and the Hilbert Nullstellensatz, *Math. Z.* **54**, 136–40. (9.7.3)

Golod, E. S., and Shafarevich, I. R.
- **64** On towers of class fields, *Izv. Akad. Nauk. SSSR, Ser. Mat.* **28**, 261–72 or *Amer. Math. Soc. Trans.* (2) **48**, (1965) 91–102. (13.11.10)

Goodearl, K. R.
- **73** Idealizers and nonsingular rings, *Pacific J. Math.* **48**, 395–402. (7.12.5)
- **74** Global dimension of differential operator rings, *Proc. Amer. Math. Soc.* **45**, 315–22. (7.12.2)
- **74a** Localization and splitting in hereditary Noetherian prime rings, *Pacific J. Math.* **53**, 137–51. (5.8.6)
- **75** Global dimension of differential operator rings II, *Trans. Amer. Math. Soc.* **209**, 65–85. (7.12.1, 7.12.9)
- **78** Global dimension of differential operator rings III, *J. London Math. Soc.* **17**, 397–409. (7.12.10)
- **79** Simple Noetherian rings, the Zaleskii–Neroslavskii examples, in *Ring Theory*

(*Waterloo 1978*), ed. D. Handelman and J. Lawrence, Lecture Notes in Mathematics 734, Springer-Verlag, New York–Berlin, pp. 118–30. (12.8.7)
- 84 Simple Noetherian rings not isomorphic to matrix rings over domains, *Comm. Algebra* **12**, 1421–34. (12.8.7)
- 86 Patch-continuity of normalized ranks of modules over one-sided Noetherian rings, *Pacific J. Math.* **122**, 83–94. (4.7.6)

Goodearl, K. R., Hodges, T. J., and Lenagan, T. H.
- 84 Krull and global dimensions of Weyl algebras over division rings, *J. Algebra* **91**, 334–59. (7.5.9)

Goodearl, K. R. and Lenagan, T. H.
- 83 Krull dimension of differential operator rings III: Noncommutative coefficients, *Trans. Amer. Math. Soc.* **275**, 833–59. (6.10.6)
- 83a Krull dimension of differential operator rings IV: Multiple derivations, *Proc. London Math. Soc.* **47**, 306–36. (6.10.9)
- 84 Krull dimension of skew Laurent extensions, *Pacific J. Math.* **114**, 109–47. (6.10.9)

Goodearl, K. R., Lenagan, T. H., and Roberts, P. C.
- 84 Height plus differential dimension in commutative Noetherian rings, *J. London Math. Soc.* **30**, 15–20. (6.10.6)

Goodearl, K. R. and Schofield, A. H.
- 86 Non-artinian essential extensions of simple modules, *Proc. Amer. Math. Soc.* **97**, 233–6. (6.10.4)

Goodearl, K. R., and Warfield, R. B.
- 79 Simple modules over hereditary Noetherian prime rings, *J. Algebra* **57**, 82–100. (5.8.7)
- 81 State spaces of K_0 of Noetherian rings, *J. Algebra* **71**, 322–78. (12.8.1)
- 82 Krull dimension of differential operator rings, *Proc. London Math. Soc.* **45**, 49–70. (6.9.24, 6.10.9, 7.12.10)

Gordon, R.
- 74 Gabriel and Krull dimension, in *Ring Theory* (*Oklahoma 1973*), ed. B. R. McDonald, A. R. Magid and K. C. Smith, Lecture Notes in Pure and Applied Mathematics 7, Dekker, New York, pp. 241–95. (6.10.0, 6.10.2)
- 75 Artinian quotient rings of FBN rings, *J. Algebra* **35**, 304–7. (6.10.8)

Gordon, R. and Robson, J. C.
- 73 *Krull Dimension*, Mem. Amer. Math. Soc. 133, Amer. Math. Soc., Providence, R.I. (6.4.9, 6.10.0, 6.10.2, 6.10.3, 13.10.10)
- 73a Semiprime rings with Krull dimension are Goldie, *J. Algebra* **25**, 519–21. (6.10.3)
- 74 The Gabriel dimension of a module, *J. Algebra* **29**, 459–73. (6.10.0)

Gray, A. J.
- 84 A note on the invertible ideal theorem, *Glasgow Math. J.* **25**, 27–30. (4.7.1)

Grayson, D. R.
- 80 K-theory and localization of non-commutative rings, *J. Pure Appl. Algebra* **18**, 125–7. (12.8.6)

Gromov, M.
- 81 Groups of polynomial growth and expanding maps, *Inst. Hautes Etudes Sci. Publ. Math.* **53**, 53–73. (8.2.18, 8.7.2)

Grothendieck, A.
- 67 Eléments de géométrie algébrique IV, Etude locale des schémas et des morphismes de schémas, *Inst. Hautes Etudes Sci. Publ. Math.* **32**. (15.6.0, 15.6.5)

Hajarnavis, C. R.
- P Homological and Cohen–Macaulay properties in non-commutative Noetherian rings, preprint, University of Warwick. (7.12.0)

Hall, P.
- **54** Finiteness conditions for soluble groups, *Proc. London Math. Soc.* **4**, 419–36 (1.9.5)
- **59** On the finiteness of certain soluble groups, *Proc. London Math. Soc.* **9**, 595–622. (9.7.0)

Harada, M.
- **66** Hereditary semiprimary rings and triangular matrix rings, *Nagoya Math. J.* **27**, 463–84. (5.4.7)

Hart, R.
- **66** Endomorphisms of modules over semiprime rings, *J. Algebra* **4**, 46–51. (3.7.4)
- **67** Simple rings with uniform right ideals, *J. London Math. Soc.* **42**, 614–17. (12.8.7)
- **71** Krull dimension and global dimension of simple Ore extensions, *Math. Z.* **121**, 341–5. (1.9.8, 6.10.2, 6.10.6, 6.10.9, 7.12.5)
- **72** A note on the tensor product of algebras, *J. Algebra* **21**, 422–7. (7.12.5)
- **74** Derivations on commutative rings, *J. London Math. Soc.* **8**, 171–5. (15.2.14)
- **75** Derivations on regular local rings of finitely generated type, *J. London Math. Soc.* **10**, 292–4. (1.9.8)
- **80** Invertible 2×2 matrices over skew polynomial rings, in *Ring Theory (Antwerp 1980)*, ed. F. Van Oystaeyen, Lecture Notes in Mathematics 825, Springer-Verlag, New York–Berlin, pp. 59–62 (12.8.7)
- **83** Differential operators on affine algebras, *J. London Math. Soc.* **28**, 470–6. (15.5.4, 15.6.5)

Hart, R., and Robson, J. C.
- **70** Simple rings and rings Morita equivalent to Ore domains, *Proc. London Math. Soc.* **21**, 232–42. (1.9.3, 12.8.7)

Hart, R., and Smith, S. P.
- **87** Differential operators on some singular surfaces, *Bull. London Math. Soc.* **19**, 145–8. (15.6.6)

Heinicke, A. G.
- **81** Localization and completion at primes generated by normalizing sequences in right Noetherian rings, *Canad. J. Math.* **33**, 325–46. (4.7.3)
- **81a** On the Krull symmetry of enveloping algebras, *J. London Math. Soc.* **24**, 109–12. (14.10.8)

Heinicke, A. G., and Robson, J. C.
- **81** Normalizing extensions: prime ideals and incomparability, *J. Algebra* **72**, 237–68. (10.7.2, 10.7.4)
- **84** Intermediate normalizing extensions, *Trans. Amer. Math. Soc.* **282**, 645–67. (10.7.1, 10.7.4)

Herstein, I. N.
- **65** A counter-example in Noetherian rings, *Proc. Nat. Acad. Sci. (U.S.A)* **54**, 1036–7. (6.10.4)
- **68** *Noncommutative Rings*, Carus Mathematics Monographs 15, Wiley, New York. (10.4.8, 13.11.10)

Heyneman, R. G., and Sweedler, M.
- **69** Affine Hopf algebras I, *J. Algebra* **13**, 192–241. (15.6.5)

Hilbert, D.
- **1898** Über die Theorie der algebraischen Formen, *Math. Ann.* **36**, 473–534. (7.12.0)
- **03** *Grundlagen der Geometrie*, Teubner, Leipzig. (1.9.4)

Hochschild, G., Kostant, B., and Rosenberg, A.
- **62** Differential forms on regular affine algebras, *Trans. Amer. Math. Soc.* **102**, 383–408. (15.6.1)

Hochster, M.

- 69 Prime ideal structure in commutative rings, *Trans. Amer. Math. Soc.* **142**, 43–60. (4.7.6)

Hodges, T. J.
- 84 The Krull dimension of skew Laurent extensions of commutative Noetherian rings, *Comm. Algebra* **12**, 1301–10. (6.10.9, 7.12.9)
- P K-theory and right ideal class groups for HNP rings, preprint, University of Cincinnati. (5.8.7)

Hodges, T. J., and McConnell, J. C.
- 81 On Ore and skew Laurent extensions of Noetherian rings, *J. Algebra* **73**, 56–64. (6.10.6)

Hodges, T. J., and Osterburg, J.
- 87 A rank two indecomposable projective module over a Noetherian domain of Krull dimension one, *Bull. London Math. Soc.* **19**, 139–44. (11.8.2)

Hodges, T. J., and Smith, S. P.
- 85 On the global dimension of certain primitive factors of the enveloping algebra of a semi-simple Lie algebra, *J. London Math. Soc.* **32**, 411–18. (15.6.3)

Humphreys, J. E.
- 80 *Introduction to Lie Algebras and Representation Theory,* Graduate Texts in Mathematics 9, Springer-Verlag, New York–Berlin. (1.7.5, 8.5.12)

Irving, R. S.
- 79 Generic flatness and the Nullstellensatz for Ore extensions, *Comm. Algebra* **7**, 259–77. (9.7.0)
- 79a Prime ideals of Ore extensions over commutative rings, *J. Algebra* **56**, 315–42. (10.7.6)
- 79b Prime ideals of Ore extensions over commutative rings II, *J. Algebra* **58**, 399–423. (10.7.6)

Irving, R. S., and Small, L. W.
- 83 The Goldie conditions for algebras with bounded growth, *Bull. London Math. Soc.* **15**, 596–600. (8.7.1)

Jacobson, N.
- 43 *The Theory of Rings*, Amer. Math. Soc. Math. Surveys 1, New York. (5.7.17, 5.7.18, 5.7.19, 5.8.0)
- 50 Some remarks on onesided inverses, *Proc. Amer. Math. Soc.* **1**, 352–5. (2.4.1)
- 62 *Lie Algebras*, Wiley–Interscience, J. Wiley, New York. (1.7.1, 14.0.0)
- 64 *Structure of Rings* (revised edition), Amer. Math. Soc., Providence, RI (first edition 1956). (0.3.1, 2.3.7, 3.2.6)
- 75 *PI-Algebras, an Introduction*, Lecture Notes in Mathematics 441, Springer-Verlag, New York–Berlin. (13.1.21, 13.11.0, 13.11.1)
- 80 *Basic Algebra II*, Freeman, San Francisco. (0.1.1)

Jantzen, J. C.
- 83 *Einhullende Algebren halbeinfacher Lie-Algebren*, Ergebnisse der Mathematik and ihrer Grenzgebiete 3, Springer-Verlag, New York–Berlin. (8.5.1, 8.7.0, 14.10.0)

Jategaonkar, A. V.
- 68 Left principal ideal domains, *J. Algebra* **8**, 148–55. (5.8.7)
- 69 Ore domains and free algebras, *Bull. London Math. Soc.* **1**, 45–6. (8.7.1)
- 69a A counter-example in ring theory and homological algebra, *J. Algebra* **12**, 418–40. (6.10.4)
- 71 Endomorphism rings of torsionless modules, *Trans. Amer. Math. Soc.* **161**, 457–66. (3.7.4)
- 73 Injective modules and classical localization in Noetherian rings, *Bull. Amer. Math. Soc.* **79**, 152–7. (4.7.3)

- **74** Relative Krull dimension and prime ideals in right Noetherian rings, *Comm. Algebra* **2**, 429–68. (4.7.1)
- **74a** Jacobson's conjecture and modules over fully bounded Noetherian rings, *J. Algebra* **30**, 103–21. (4.7.3, 6.10.2, 6.10.4)
- **75** Principal ideal theorem for Noetherian PI rings, *J. Algebra* **35**, 17–22. (4.7.1)
- **82** Solvable Lie algebras, polycyclic by finite groups and bimodule Krull dimension, *Comm. Algebra* **10**, 361–6 (4.7.3)
- **86** *Localization in Noetherian Rings*, London Math. Soc. Lecture Note Series 98, Cambridge University Press, Cambridge. (4.3.14, 4.5.7, 4.6.18, 4.7.0, 4.7.2, 4.7.3, 4.7.6, 6.10.8)

Jategaonkar, V. A.
- **84** A multiplicative analog of the Weyl Algebra, *Comm. Algebra* **12**, 1669–88. (1.9.8)

Johnson, R. E.
- **51** The extended centralizer of a ring over a module, *Proc. Amer. Math. Soc.* **2**, 891–5. (2.4.2)

Jondrup, S.
- **81** On the centre of a hereditary PI algebra, *Comm. Algebra* **9**, 1673–9. (13.11.9)

Jordan, D. A.
- **75** Noetherian Ore extensions and Jacobson rings, *J. London Math. Soc.* **10**, 281–91. (9.7.1)
- **78** Primitive skew Laurent polynomial rings, *Glasgow Math. J.* **19**, 79–86. (10.7.6)
- **81** Differentially simple rings with no invertible derivatives, *Quart. J. Math. Oxford* **32**, (2), 417–24. (1.9.8)
- **P** Normal elements and completions of non-commutative Noetherian rings, preprint, University of Sheffield. (4.7.3)

Joseph, A.
- **77** A generalization of Quillen's Lemma and its applications to the Weyl algebras, *Israel J. Math.* **28**, 177–92. (9.7.3)
- **80** Dimension en algèbre non-commutative, Cours de troisième cycle, Université de Paris VI, mimeographed notes. (8.7.0)
- **81** Applications de la thèorie des anneaux aux algèbres enveloppantes, Cours de troisième cycle, Université de Paris VI, mimeographed notes. (14.10.0)
- **83** On the classification of primitive ideals in the enveloping algebra of a semisimple Lie algebra, in *Lie Group Representations I*, ed. R. Herb, S. Kudla, R. Lipsman and J. Rosenberg, Lecture Notes in Mathematics 1024, Springer-Verlag, New York–Berlin, pp. 30–76. (14.10.0)
- **84** Primitive ideals in enveloping algebras, in *Proc. Int. Cong. Math.* (*Warsaw 1983*), North-Holland, Amsterdam, pp. 403–14. (14.10.0)

Joseph, A., and Small, L. W.
- **78** An additivity principle for Goldie rank, *Israel J. Math.* **31**, 89–101. (4.7.5)

Kaplansky, I.
- **48** Rings with a polynomial identity, *Bull. Amer. Math. Soc.* **54**, 575–80. (13.11.0, 13.11.3)
- **70** *Commutative Rings*, Allyn and Bacon, Boston (revised edition, University of Chicago Press (1974)). (4.1.10, 5.4.1, 7.1.15, 13.7.15, 13.9.14, 13.10.12)
- **72** *Fields and Rings* (second edition), University of Chicago Press, Chicago. (7.1.1, 7.12.3)

Kerr, J. W.
- **79** An example of a Goldie ring whose matrix ring is not Goldie, *J. Algebra* **61**, 590–2. (3.1.5)

Kharchenko, V. K.

- 74 Galois extensions and quotient rings, *Algebra and Logic* **13**, 264–81. (10.7.5)
- 75 Generalized identities with automorphisms, *Algebra and Logic* **14**, 132–48. (10.7.3)

Kirkman, E. E., and Kuzmanovich, J. J.
- P Matrix subrings having finite global dimension, preprint, Wake Forest University, Winston-Salem, N.C. (7.12.0)

Kosevoi, E. G.
- 70 On certain associative algebras with transcendental relations, *Algebra and Logic* **9**, 313–19. (8.7.1)

Krause, G.
- 70 On the Krull dimension of left Noetherian left Matlis rings, *Math. Z.* **118**, 207–14. (6.10.0, 6.10.2)
- 72 On fully left bounded left Noetherian rings, *J. Algebra* **23**, 88–99. (6.10.4)

Krause, G., and Lenagan, T. H.
- 85 *Growth of Algebras and Gelfand–Kirillov Dimension*, Research Notes in Mathematics 116, Pitman, London. (8.0.0, 8.1.18, 8.2.6, 8.2.12, 8.2.13, 8.2.18, 8.3.4, 8.5.11, 8.7.0, 8.7.1, 8.7.6)

Krause, G., Lenagan, T. H., and Stafford, J. T.
- 78 Ideal invariance and Artinian quotient rings, *J. Algebra* **55**, 145–54. (6.10.8)

Krull, W.
- 38 Dimensiontheorie in Stellenringen, *J. Reine Angew. Math. (Crelle's J)* **179**, 204–26. (6.10.4)

Kunz, E.
- 85 *Introduction to Commutative Algebra and Algebraic Geometry*, Birkhäuser, Boston. (7.1.15, 11.8.7, 15.2.8)

Kuzmanovich, J. J.
- 72 Localizations of HNP rings, *Trans. Amer. Math. Soc.* **173**, 137–57. (5.8.6)

Lam, T. Y.
- 78 *Serre's Conjecture*, Lecture Notes in Mathematics 635, Springer-Verlag, New York–Berlin. (11.2.2, 11.8.0, 12.8.3)

Lam, T. Y., and Siu, M. K.
- 75 K_0 and K_1—An introduction to algebraic K-theory, *Amer. Math. Monthly* **82**, 329–64. (12.8.0)

Lambek, J.
- 66 *Lectures on Rings and Modules*, Blaisdell, Waltham (reprinted by Chelsea, New York, 1976). (0.1.1, 10.3.4)

Lanski, C.
- 80 Goldie conditions in finite normalizing extensions, *Proc. Amer. Math. Soc.* **79**, 515–19. (10.1.13, 10.7.1)

Le Bruyn, L.
- 87 *Trace Rings of Generic* 2×2 *Matrices*, Mem. Amer. Math. Soc. 363, Amer. Math. Soc., Providence, R.I. (13.11.1)

Lemonnier, B.
- 72 Déviation des ensembles et groupes abéliens totalement ordonnés, *Bull. Sci. Math.* **96**, 289–303. (6.10.1)
- 72a Sur une classe d'anneaux définie à partir de la déviation, *C. R. Acad. Sci. Paris* **274**, 297–9. (6.10.3)
- 78 Dimension de Krull et codéviation, Application au théorème d'Eakin, *Comm. Algebra* **6**, 1647–65. (10.7.1)
- 84 Dimension de Krull et dualité de Morita dans les extensions triangulaires, *Comm. Algebra* **12**, 3071–110. (10.7.1)

Lenagan, T. H.

73 The nil radical of a ring with Krull dimension, *Bull. London Math. Soc.* **5**, 307–11. (6.10.3)
73a Bounded hereditary Noetherian prime rings, *J. London Math. Soc.* **6**, 241–6. (5.8.6)
75 Artinian ideals in Noetherian rings, *Proc. Amer. Math. Soc.* **51**, 499–500. (4.7.1)
77 Noetherian rings with Krull dimension one, *J. London Math. Soc.* **15**, 41–7. (6.10.4)
80 Modules with Krull dimension, *Bull. London Math. Soc.* **12**, 39–40. (6.10.2)
81 Gelfand–Kirillov dimension and affine PI rings, *Comm. Algebra* **10**, 87–92. (8.7.3)
83 Krull dimension of differential operator rings II: The infinite case, *Rocky Mountain J. Math.* **13**, 475–80. (6.9.24)

Lesieur, L. and Croisot, R.
59 Sur les anneaux premiers Noethériens à gauche, *Ann. Sci. École Norm. Sup.* **76**, 161–83. (2.4.1)

Levasseur, T.
81 Anneaux d'operateurs differentiels, in *Séminaire Dubreil–Malliavin (1980)*, ed. M.-P. Malliavin, Lecture Notes in Mathematics 867, Springer-Verlag, New York–Berlin, pp. 157–73. (15.6.6)
82 Sur la dimension de Krull de l'algèbre enveloppante d'une algèbre de Lie semi-simple, in *Séminaire Dubreil-Malliavin (1981)*, ed. M.-P. Malliavin, Lecture Notes in Mathematics 924, Springer-Verlag, New York–Berlin, pp. 173–83. (8.7.5)
86 La dimension de Krull de $U(\mathrm{sl}(3))$, *J. Algebra* **102**, 39–59. (8.7.5)

Levitzki, J.
35 On automorphisms of certain rings, *Annals. of Math.* **36**, 984–92. (10.7.5)
45 Solution of a problem of G. Koethe, *Amer. J. Math.* **67**, 437–42. (2.4.3)
50 A theorem on polynomial identities, *Proc. Amer. Math. Soc.* **1**, 334–41. (13.11.2)

Levy, L. S.
63 Torsionfree and divisible modules over non-integral-domains, *Canad. J. Math.* **15**, 132–51. (3.7.4)

Lewin, J.
74 A matrix representation for associative algebras, I, *Trans. Amer. Math. Soc.* **188**, 293–308. (8.7.3)

Lipman, J.
65 Free derivation modules on algebraic varieties, *Amer. J. Math.* **87**, 874–98. (15.6.3)

Littlewood, D. E.
33 On the classification of algebras, *Proc. London Math. Soc.* **35**, 200–40. (1.9.3)

Lorenz, M.
81 Finite normalizing extensions of rings, *Math. Z.* **176**, 447–84. (10.7.0, 10.7.2, 10.7.4)
81a Chains of prime ideals in enveloping algebras of solvable Lie algebras, *J. London Math. Soc.* **24**, 205–10. (14.10.2)
82 On the Gelfand–Kirillov dimension of skew polynomial rings, *J. Algebra* **77**, 186–8. (8.7.2)
85 K_0 of skew group rings and simple Noetherian rings without idempotents, *J. London Math. Soc.* **32**, 41–50. (12.8.7)

Lorenz. M., and Passman, D. S.
79 Prime ideals in crossed products of finite groups, *Israel J. Math.* **33**, 89–132. (10.7.5)

Lorenz, M., and Small, L. W.
82 On the Gelfand–Kirillov Dimension of Noetherian PI-Algebras, in *Algebraists' Homage* (Yale 1981), ed. S. A. Amitsur, D. J. Saltman and G. B. Seligman, Contemporary Mathematics 13, Amer. Math. Soc., Providence, R.I., pp. 199–205. (8.7.3)

Ludgate, A. T.
72 A note on noncommutative Noetherian rings, *J. London Math. Soc.* **5**, 406–8. (2.4.1)

McConnell, J. C.
- **68** Localisation in enveloping rings, *J. London Math. Soc.* **43**, 421–8. (1.9.2, 4.7.2, 14.10.3)
- **69** The Noetherian property in complete rings and modules, *J. Algebra* **12**, 143–53. (4.7.3)
- **74** A note on the Weyl algebra A_n, *Proc. London Math. Soc.* **28**, 89–98. (7.5.9)
- **74a** Representations of solvable Lie algebras and the Gelfand–Kirillov conjecture, *Proc. London Math. Soc.* **29**, 453–84. (14.10.6)
- **75** Representations of solvable Lie algebras II. Twisted group rings, *Ann. Sci. École Norm. Sup.* **8**, 157–78. (14.10.8)
- **77** On the global dimension of some rings, *Math. Z.* **153**, 253–4. (7.12.2)
- **77a** Representations of solvable Lie algebras III. Cancellation theorems, *J. Algebra* **44**, 262–70. (14.10.8)
- **77b** Representations of solvable Lie algebras IV. An elementary proof of the $(U/P)_E$-structure theorem, *Proc. Amer. Math. Soc.* **64**, 8–12. (14.10.8)
- **78** On completions of noncommutative Noetherian rings, *Comm. Algebra* **6**, 1485–8. (4.7.3)
- **79** *I*-adic completions of noncommutative rings, *Israel J. Math.* **32**, 305–10. (4.7.3)
- **82** The Nullstellensatz and Jacobson properties for rings of differential operators, *J. London Math. Soc.* **26**, 37–42. (1.9.6, 9.7.0, 9.7.4)
- **82a** Representations of solvable Lie algebras, V. On the Gelfand–Kirillov dimension of simple modules, *J. Algebra* **76**, 489–93. (14.10.8)
- **84** On the Krull and global dimensions of Weyl algebras over affine coefficient rings, *J. London Math. Soc.* **29**, 249–53. (9.7.1, 9.7.4)
- **P** Amalgams of Weyl algebras and the $\mathscr{A}(V, \delta, \Gamma)$ conjecture, preprint, University of Leeds. (14.10.8)

McConnell, J. C., and Pettit, J. J.
- **P** Crossed products and multiplicative analogues of Weyl algebras, preprint, University of Leeds. (1.9.8)

McConnell, J. C., and Robson, J. C.
- **73** Homomorphisms and extensions of modules over certain differential polynomial rings, *J. Algebra* **26**, 319–42. (5.7.18)

McConnell, J. C., and Sweedler, M. E.
- **71** Simplicity of smash products, *Proc. London Math. Soc.* **23**, 251–66. (1.9.8)

McNaughton, J. M.
- **77** A note on cocycle twisted rings of differential operators, *J. London Math. Soc.* **15**, 409–14. (14.10.8)

Makar-Limanov, L.
- **83** The skew field of fractions of the Weyl algebra contains a free subalgebra, *Comm. Algebra* **11**, 2003–6. (8.2.13)

Malliavin, M.-P.
- **76** Dimension de Gelfand–Kirillov des algèbres à identités polynomiales, *C. R. Acad. Sci. Paris* **282**, 679–81. (13.11.10)
- **79** Caténarité et théorème d'intersection en algèbre non commutative, in *Séminaire Dubreil (1977–1978)*, ed. M.-P. Malliavin, Lecture Notes in Mathematics 740, Springer-Verlag, New York–Berlin, pp. 408–31. (14.10.2)

Markov, V. T.
- **73** On the dimension of noncommutative affine algebras, *Math. USSR Izvestija* **7**, 281–5. (13.11.10)

Martindale, W. S.
- **69** Prime rings satisfying a generalized polynomial identity, *J. Algebra* **12**, 576–84. (10.7.3)

Marubayashi, H.
75 Noncommutative Krull rings, *Osaka J. Math.* **12**, 703–14. (5.8.1)
Matsumura, H.
80 *Commutative Algebra*, Benjamin, Reading, Mass. (15.6.1)
Maury, G., and Raynaud, J.
80 *Ordres maximaux au sens de K. Asano*, Lecture Notes in Mathematics 808, Springer-Verlag, New York–Berlin. (5.8.1)
Mewborn, A. C. and Winton, C. N.
69 Orders in self-injective semiperfect rings, *J. Algebra* **13**, 5–9. (2.4.3)
Michler, G. O.
69 Asano orders, *Proc. London Math. Soc.* **19**, 421–43. (5.8.2)
Milnor, J.
68 A note on curvature and fundamental group, *J. Diff. Geom.* **2**, 1–7. (8.7.2)
68a Growth of finitely generated solvable groups, *J. Diff. Geom.* **2**, 447–9. (8.7.2)
71 *Introduction to Algebraic K-Theory*, Annals of Math. Studies No. 72, Princeton University Press, Princeton. (12.8.0)
78 Analytic proofs of the 'hairy ball theorem' and the Brouwer fixed point theorem, *Amer. Math. Monthly* **85**, 521–4. (11.2.3)
Montgomery, S.
79 Automorphism groups of rings with no nilpotent elements, *J. Algebra* **60**, 238–48. (10.7.3)
80 *Fixed Rings of Finite Automorphism Groups of Associative Rings*, Lecture Notes in Mathematics 818, Springer-Verlag, New York–Berlin. (7.12.8, 10.7.5)
81 Prime ideals in fixed rings, *Comm. Algebra* **9**, 423–49. (10.7.5)
P Prime ideals and group actions in noncommutative algebras, preprint, University of Southern California. (10.7.5)
Morita, K.
58 Duality for modules and its applications to the theory of rings with minimum condition, *Sci. Rep. Tokyo Kyoiku Daigaka*, Sect. A, **6**, 83–142. (3.7.5)
Mount, K. R., and Villamayor, O. E.
73 On a conjecture of Y. Nakai, *Osaka J. Math.* **10**, 325–7. (15.6.5)
Muhasky, J. L.
P The differential operator ring of an affine curve, preprint, University of Utah, Salt Lake City. (15.6.6)
Müller, B. J.
76 Localization in noncommutative Noetherian rings, *Canad. J. Math.* **28**, 600–10. (4.7.3)
76a Localization in fully bounded Noetherian rings, *Pacific J. Math.* **67**, 233–45. (4.7.3)
79 Ideal invariance and localization, *Comm. Algebra* **7**, 415–41. (4.7.3)
Musson, I. M.
86 Some rings of differential operators which are Morita equivalent to the Weyl algebra A_1, *Proc. Amer. Math. Soc.* **98**, 29–30. (15.6.6)
Nagata, M.
62 *Local Rings*, Interscience, New York. (6.10.4)
Nakai, Y.
70 High order derivations 1, *Osaka J. Math.* **7**, 1–27. (15.6.5)
Nastasescu, C., and Van Oystaeyen, F.
82 *Graded Ring Theory*, North-Holland Mathematics Library 28, North-Holland Amsterdam. (1.9.6)
Nicholson, W. K., and Watters, J. F.
79 Normal radicals and normal classes of rings, *J. Algebra* **59**, 5–15. (3.7.6)

Nouazé, Y., and Gabriel, P.
- 67 Idéaux premiers de l'algèbre enveloppante d'une algèbre de Lie nilpotente, *J. Algebra* **6**, 77–99. (4.7.2, 14.10.6)

Ojanguren, M., and Sridharan, R.
- 71 Cancellation of Azumaya algebras, *J. Algebra* **18**, 501–5. (11.8.2)

Ol'šanskiĭ, A. Ju.
- 79 Infinite groups with cyclic subgroups, *Soviet Math. Dokl.* **20**, 343–6. (1.9.5)

Ore, O.
- 32 Formal Theorie der linearen differential Gleichungen, *J. Reine Angew. Math. (Crelle's J.)* **168**, 233–52. (1.9.1)
- 33 Theory of noncommutative polynomials, *Ann. of Math.* **34**, 480–508. (1.9.2, 2.4.1)

Page, A.
- 84 On the centre of hereditary PI rings, *J. London Math. Soc.* **30**, 193–6. (13.11.9)

Paré, R., and Schelter, W. F.
- 78 Finite extensions are integral, *J. Algebra* **53**, 477–9. (13.11.8)

Passman, D. S.
- 77 *The Algebraic Structure of Group Rings*, Interscience, New York. (1.9.5)
- 81 Prime ideals in normalizing extensions, *J. Algebra* **73**, 556–72. (10.7.2, 13.11.2)
- 83 It's essentially Maschke's theorem, *Rocky Mountain J. Math.* **13**, 37–54. (10.7.5)
- 84 Group rings of polyclic groups, in *Group theory—Essays for Philip Hall*, ed. K. W. Gruenberg and J. E. Roseblade, Academic Press, London, pp. 207–56. (1.9.5)
- 86 *Group Rings, Crossed Products and Galois Theory*, CBMS Monographs 64, Amer. Math. Soc., Providence, R.I. (10.7.5)
- Pa Prime ideals in polycyclic crossed products, preprint, University of Wisconsin, Madison. (10.7.6)
- Pb Prime ideals in enveloping rings, preprint, University of Wisconsin, Madison. (14.10.2)

Pearson, K. R., and Stephenson, W.
- 77 A skew polynomial ring over a Jacobson ring need not be a Jacobson ring, *Comm. Algebra* **5**, 783–94. (9.7.1, 10.7.6)

Pearson, K. R., Stephenson, W., and Watters, J. F.
- 81 Skew polynomial rings and Jacobson rings, *Proc. London Math. Soc.* **42**, 559–76. (10.7.6)

Pham, F.
- 79 *Singularités des systèmes différentiels de Gauss-Manin*, Progress in Math. 2, Birkhäuser, Boston, Mass. (15.6.6)

Pierce, R. S.
- 82 *Associative Algebras*, Graduate Texts in Mathematics 88, Springer-Verlag, New York–Berlin. (13.11.1)

Posner, E. C.
- 60 Prime rings satisfying a polynomial identity, *Proc. Amer. Math. Soc.* **11**, 180–3. (13.11.6)

Procesi, C.
- 67 Noncommutative affine rings, *Atti. Accad. Naz. Lincei Mem. Cl. Sci. Fis. Mat. Natur. Sez. 1 (8)* **8**, 237–55. (13.11.10)
- 72 On a theorem of M. Artin, *J. Algebra* **22**, 309–15. (13.11.7)
- 73 *Rings with Polynomial Identities*, Dekker, New York. (13.11.0, 13.11.1, 13.11.10)

Procesi, C., and Small, L. W.
- 68 Endomorphism rings of modules over PI algebras, *Math. Z.* **106**, 178–80. (13.11.4)

Quebbemann, H-G.
- 79 Schiefkorper als Endomorphismenringe einfacher Moduln über einer Weyl-Algebra, *J. Algebra* **59**, 311–12. (9.7.5)

Quillen, D.
- **69** On the endomorphism ring of a simple module over an enveloping algebra, *Proc. Amer. Math. Soc.* **21**, 171–2. (9.7.0, 9.7.3)
- **73** Higher algebraic K-theory I, in *Algebraic K-theory I: Higher K-theories*, ed. H. Bass, Lecture Notes in Mathematics 341, Springer-Verlag, New York–Berlin, pp. 85–147. (12.8.5, 12.8.6)
- **75** Higher algebraic K-theory, *Proc. Int. Cong. Math.* (Vancouver 1974), Canadian Math. Congress, 171–6. (12.8.0)

Rainwater, J.
- **P** Global dimension of fully bounded Noetherian rings, preprint, University of Washington, Seattle. (7.12.1)

Raynaud, M.
- **68** Modules projectifs universels, *Invent. Math.* **6**, 1–26. (11.2.4)

Razmyslov, Ju. P.
- **73** A certain problem of Kaplansky, *Math. USSR Izv.* **7**, 479–96. (13.11.5)
- **74** The Jacobson radical in PI-algebras, *Algebra and Logic* **13**, 192–204. (13.11.9, 13.11.10)

Regev, A.
- **72** Existence of identities in $A \otimes B$, *Israel J. Math* **11**, 131–52. (13.4.10)

Reiner, I.
- **75** *Maximal Orders*, London Math. Soc. Monographs 5, Academic Press, London–New York. (5.8.0)

Rentschler, R. and Gabriel, P.
- **67** Sur la dimension des anneaux et ensembles ordonnés, *C. R. Acad. Sci. Paris*, **265**, 712–15. (6.10.0, 6.10.1, 6.10.6)

Resco, R.
- **79** Transcendental division algebras and simple Noetherian rings, *Israel J. Math.* **32**, 236–56. (6.10.6, 7.3.16, 9.7.6)
- **80** A dimension theorem for division rings, *Israel J. Math.* **35**, 215–21. (6.10.6, 7.12.5, 9.7.6)

Resco, R., Small, L. W., and Stafford, J. T.
- **82** Krull and global dimension of semiprime Noetherian PI rings, *Trans. Amer. Math. Soc.* **274**, 285–95. (7.12.9)

Resco, R., Small, L. W., and Wadsworth, A. R.
- **79** Tensor product of division rings and finite generation of subfields, *Proc. Amer. Math. Soc.* **77**, 7–10. (6.10.6)

Resco, R., Stafford, J. T., and Warfield, R. B.
- **86** Fully bounded G-rings, *Pacific J. Math.* **124**, 403–15. (9.7.6)

Rieffel, M. A.
- **83** Dimension and stable rank in the K-theory of C*-algebras, *Proc. London Math. Soc.* **46**, 301–33. (11.8.1)
- **83a** The cancellation theorem for projective modules over irrational rotation C*-algebras, *Proc. London Math. Soc.* **47**, 285–302. (11.8.7)
- **85** K-theory of crossed products of C*-algebras by discrete groups, in *Group Actions on Rings*, ed. S. Montgomery, Contemporary Mathematics 43, American Math. Soc., Providence, R.I., pp. 253–65. (12.8.0)

Rinehart, G. S.
- **62** Note on the global dimension of a certain ring, *Proc. Amer. Math. Soc.* **13**, 341–6. (7.12.5, 7.12.11)

Rinehart, G. S., and Rosenberg, A.
- **76** The global dimension of Ore extensions and Weyl algebras, in *Algebra, Topology and Category Theory*, ed. A. Heller and M. Tierney, Academic Press, New York, pp. 169–80 (7.12.9)

Robson, J. C.
- **67** Artinian quotient rings, *Proc. London Math. Soc.* **17**, 600–16. (3.7.3)
- **68** Non-commutative Dedekind rings, *J. Algebra* **9**, 249–65. (5.8.2)
- **71** A note on Dedekind prime rings, *Bull. London Math. Soc.* **3**, 42–6. (5.8.2, 5.8.7)
- **72** Idealizers and hereditary Noetherian prime rings, *J. Algebra* **22**, 45–81. (1.9.1, 5.8.5, 5.8.6, 7.12.5)
- **74** Decomposition of Noetherian rings, *Comm. Algebra* **1**, 345–9. (4.7.1)
- **75** Cyclic and faithful objects in quotient categories with applications to Noetherian simple or Asano rings, in *Noncommutative Ring Theory* (*Kent State, 1975*), ed. J. H. Cozzens and F. L. Sandomierski, Lecture Notes in Mathematics 545, Springer-Verlag, New York–Berlin, pp. 151–72. (6.10.7)
- **75a** The coincidence of idealizer subrings, *J. London Math. Soc.* **10**, 338–48. (5.8.5)
- **79** A unified approach to unity, *Comm. Algebra* **7**, 1245–55. (2.4.3)
- **85** Some constructions of rings of finite global dimension, *Glasgow Math. J.* **20**, 1–12. (7.12.0)

Robson, J. C., and Small, L. W.
- **74** Hereditary prime PI rings are classical hereditary orders, *J. London Math. Soc.* **8**, 499–503. (13.11.9)
- **76** Idempotent ideals in PI rings, *J. London Math. Soc.* **14**, 120–2. (13.11.9)
- **81** Liberal extensions, *Proc. London Math. Soc.* **42**, 87–103. (10.7.1)
- **86** Orders equivalent to the first Weyl algebra, *Quart. J. Math., Oxford*, **37**, 475–82. (9.7.5)
- **P** Another change of rings theorem, preprint, University of Leeds. (7.12.5)

Roggenkamp, K. W.
- **70** *Lattices over Orders II*, Lecture Notes in Mathematics 142, Springer-Verlag, New York–Berlin. (5.8.0)

Roggenkamp, K. W., and Huber-Dyson, V.
- **70** *Lattices over Orders I*, Lecture Notes in Mathematics 115, Springer-Verlag, New York–Berlin. (5.8.0)

Roos, J. E.
- **72** Détermination de la dimension homologique des algèbres de Weyl, *C. R. Acad. Sci. Paris* (*Ser. A*) **274**, 23–6. (7.12.2, 7.12.5)

Rosenberg, A. and Stafford, J. T.
- **76** Global dimension of Ore extensions, in *Algebra, Topology and Category Theory*, ed. A. Heller and M. Tierney, Academic Press, New York, pp. 181–8. (7.12.9)

Rotman, J.
- **79** *An Introduction to Homological Algebra*, Academic Press, New York–London. (7.1.1)

Rowen, L. H.
- **73** Some results on the center of a ring with polynomial identity, *Bull. Amer. Math. Soc.* **79**, 219–33. (13.11.6)
- **80** *Polynomial Identities in Ring Theory*, Academic Press, New York–London. (13.3.3, 13.4.10, 13.11.0, 13.11.1, 13.11.5, 13.11.7, 13.11.9)

Roy, A.
- **65** A note on filtered rings, *Arch. Math.* **16**, 421–7. (7.12.6)

Sands, A. D.
- **73** Radicals and Morita contexts, *J. Algebra* **24**, 335–45. (1.9.1)

Schelter, W. F.
- **76** Integral extensions of rings satisfying a polynomial identity, *J. Algebra* **40**, 245–57, Errata, **44** (1977), 576. (13.11.8)
- **78** Noncommutative affine PI rings are catenary, *J. Algebra* **51**, 12–18. (13.11.9, 13.11.10)

Schofield, A. H.
- **85** *Representations of Rings over Skew-fields*, London Math. Soc. Lecture Note Series 92, Cambridge Univ. Press, Cambridge. (4.7.1)
- **86** Stratiform simple Artinian rings, *Proc. London Math. Soc.* **53**, 267–87. (7.12.9)

Serre, J.-P.
- **58** Modules projectifs et espaces fibrés à fibre vectorielle, in *Séminaire Dubreil (1957/58)*, Secretariat Math., Paris. (7.12.6, 12.8.3)

Siebenmann, L. C.
- **70** A total Whitehead torsion obstruction to fibering over the circle, *Comm. Math. Helv.* **45**, 1–48. (12.8.5)

Sierpinski, W.
- **58** *Cardinal and Ordinal Numbers*, Polska Akad. Nauk Monog. Matem., Warsaw. (6.1.9)

Silvester, J. R.
- **81** *Introduction to Algebraic K-theory*, Chapman and Hall, London. (12.8.0)

Sirsov, A. I.
- **57** On rings with identity relations, *Mat. Sb. N.S.* **43** (85), 277–83 or *Amer. Math. Soc. Trans.* (2) **119**, (1983), 133–9. (13.11.8)

Small, L. W.
- **65** An example in Noetherian rings, *Proc. Nat. Acad. Sci. (U.S.A.)* **54**, 1035–6. (1.9.1)
- **66** Orders in Artinian rings, *J. Algebra* **4**, 13–41. (4.7.1)
- **66a** Hereditary rings, *Proc. Nat. Acad. Sci. (U.S.A.)* **55**, 25–7. (5.8.4)
- **67** Semi-hereditary rings, *Bull. Amer. Math. Soc.* **73**, 656–8. (5.8.4)
- **68** Orders in Artinian rings II, *J. Algebra* **9**, 266–73. (4.7.1)
- **68a** A change of rings theorem, *Proc. Amer. Math. Soc.* **19**, 662–6. (7.12.3)
- **71** Localization in PI-rings, *J. Algebra* **18**, 269–70. (13.11.7)
- **80** *Rings Satisfying a Polynomial Identity*, Lecture Notes, Universität Essen. (13.11.10)

Small, L. W., and Stafford, J. T.
- **81** Localization and completions of Noetherian PI algebras, *J. Algebra* **70**, 156–61. (4.7.3)
- **82** Regularity of zero divisors, *Proc. London Math. Soc.* **44**, 405–19. (4.4.12, 4.7.4)
- **85** Homological properties of generic matrix rings, *Israel J. Math.* **51**, 27–32. (13.11.1)

Small, L. W., Stafford, J. T., and Warfield, R. B.
- **85** Affine algebras of Gelfand–Kirillov dimension one are PI, *Math. Proc. Camb. Phil. Soc.* **97**, 407–14. (8.7.1)

Smith, M. K.
- **77** Growth of algebras, in *Ring Theory II (Oklahoma 1975)*, ed. B. R. McDonald and R. A. Morris, Lecture Notes in Pure and Applied Mathematics 26, Dekker, New York, pp. 247–59. (8.7.1)

Smith, P. F.
- **71** Localisation and the AR property, *Proc. London Math. Soc.* **22**, 39–68. (4.7.2)
- **72** On the dimension of group rings, *Proc. London Math. Soc.* **25**, 288–302; Corrigendum, **27** (1973), 766–8. (6.10.6)
- **82** The Artin–Rees property, in *Séminaire Dubreil-Malliavin (1981)*, ed. M.-P. Malliavin, Lecture Notes in Mathematics 924, Springer-Verlag, Berlin–New York, pp. 197–240. (4.7.2)

Smith, S. P.
- **81** Krull dimension of the enveloping algebra of sl (2, $\mathbb{C}$), *J. Algebra* **71**, 89–94. (8.7.3, 8.7.5)
- **84** Gelfand–Kirillov dimension of rings of formal differential operators on affine varieties, *Proc. Amer. Math. Soc.* **42**, 1–8. (15.6.3)

- **86** Differential operators on A^1 and P^1 in char $p > 0$, in *Séminaire Dubreil–Malliavin (1985)*, ed. M.-P. Malliavin, Lectures Notes in Mathematics 1220, Springer-Verlag, Berlin–New York, pp. 157–77. (15.6.6)
- **P** Curves, differential operators and finite dimensional algebras, in *Séminaire Dubreil–Malliavin (1986)* (to appear). (15.6.6)

Smith, S. P., and Stafford, J. T.
- **P** Differential operators on an affine curve, preprint, University of Leeds. (15.6.5)

Snider, R. L.
- **82** On primitive ideals in group rings, *Arch. Math. (Basel)* **38**, 423–5. (9.7.4)

Stafford, J. T.
- **76** Completely faithful modules and ideals of simple Noetherian rings, *Bull. London Math. Soc.* **8**, 168–73. (6.10.7)
- **77** Stable structure of noncommutative Noetherian rings, *J. Algebra* **47**, 244–67. (6.10.8, 11.8.0, 11.8.3)
- **78** Module structure of Weyl algebras, *J. London Math. Soc.* **18**, 429–42. (6.7.8, 11.2.15)
- **78a** Stable structure of noncommutative Noetherian rings II, *J. Algebra* **52**, 218–35. (11.8.0)
- **78b** A simple Noetherian ring not Morita equivalent to a domain, *Proc. Amer. Math. Soc.* **68**, 159–60. (12.8.7)
- **79** Cancellation for nonprojective modules, in *Module Theory (Seattle 1977)*, ed. C. Faith and S. Wiegand, Lecture Notes in Mathematics 700, Springer-Verlag, New York–Berlin, pp. 3–15. (11.8.7)
- **79a** Morita equivalence of simple Noetherian rings, *Proc. Amer. Math. Soc.* **74**, 212–14. (12.8.7)
- **80** Bounded numbers of generators of right ideals in polynomial rings, *Comm. Algebra* **8**, 1513–18. (6.10.7)
- **80a** Projective modules over polynomial extensions of division rings, *Invent. Math.* **59**, 105–17. (11.2.15)
- **81** Generating modules efficiently: algebraic K-theory for noncommutative Noetherian rings, *J. Algebra* **69**, 312–46; Corrigendum, **82** (1983), 294–6. (4.7.6, 11.0.0, 11.8.0, 11.8.3, 11.8.6, 11.8.7)
- **81a** On the stable range of right Noetherian rings, *Bull. London Math. Soc.* **13**, 39–41. (6.10.7)
- **82** Noetherian full quotient rings, *Proc. London Math. Soc.* **44**, 385–404. (4.7.1, 4.7.4)
- **82a** Generating modules efficiently over noncommutative rings, in *Séminaire Dubreil-Malliavin (1981)*, ed. M.-P. Malliavin, Lecture Notes in Mathematics 924, Springer-Verlag, New York–Berlin, pp. 72–88. (11.8.7)
- **82b** Homological properties of the enveloping algebra $U(Sl_2)$, *Math. Proc. Camb. Phil. Soc.* **91**, 29–37. (7.12.6)
- **83** Dimensions of division rings, *Israel J. Math.* **45**, 33–40. (6.10.6)
- **83a** Rings with a bounded number of generators for right ideals, *Quart. J. Math. Oxford* **34** (2), 107–14. (6.10.7)
- **85** Non-holonomic modules over Weyl algebras and enveloping algebras, *Invent. Math.* **79**, 619–38. (6.8.13, 8.5.8, 15.6.4)
- **85a** Stably free projective right ideals, *Compositio Math.* **54**, 63–78. (11.2.14, 11.8.2)
- **85b** Modules over prime Krull rings, *J. Algebra* **95**, 332–42. (5.8.1)
- **85c** On the ideals of a Noetherian ring, *Trans. Amer. Math. Soc.* **289**, 381–92. (4.7.3)
- **87** Endomorphisms of right ideals of the Weyl algebra, *Trans. Amer. Math. Soc.* **299**, 623–39. (12.8.0)
- **P** The Goldie rank of a module, in *Noetherian Rings and their Applications*, ed. L. W. Small, Amer. Math. Soc. Math. Surveys and Monographs (to appear). (4.6.18, 4.7.6)

Pa Global dimension of semiprime Noetherian rings, preprint, University of Leeds. (7.12.0)

Stafford, J. T. and Warfield, R. B.
85 Constructions of hereditary Noetherian rings and simple rings, *Proc. London Math. Soc.* **51**, 1–20. (5.6.13)

Stenström B.
75 *Rings of Quotients*, Springer-Verlag, New York–Berlin (2.4.1, 7.1.1, 10.3.4)

Stewart, P. N.
83 Nilpotence and normalizing extensions, *Comm. Algebra* **11**, 109–10. (10.7.0)

Suslin, A. A.
77 On the structure of the special linear group over polynomial rings, *Math. U.S.S.R. Izv.* **11**, 221–38. (11.5.6)
81 The cancellation problem for projective modules; in Eight lectures delivered at the ICM Helsinki 1978, *AMS Translations*, Series 2, **117**, 7–13. (11.5.8, 11.5.9, 11.8.5)

Swan, R. G.
62 Vector bundles and projective modules, *Trans. Amer. Math. Soc.* **105**, 264–77. (11.8.2)
67 The number of generators of a module, *Math. Z.* **102**, 318–22. (11.8.7)
68 *Algebraic K-theory*, Lecture Notes in Mathematics 76, Springer-Verlag, New York–Berlin. (7.12.7, 12.8.0)
77 Topological examples of projective modules, *Trans. Amer. Math. Soc.* **230**, 201–34. (11.1.15, 11.7.11)

Sweedler, M.
69 *Hopf Algebras*, Benjamin, New York. (1.9.7)
75 Groups of simple algebras, *Inst. Hautes Etudes Sci. Publ. Math.* **44**, 79–189. (15.6.0, 15.6.5)

Talintyre, T. D.
66 Quotient rings with minimum condition on right ideals, *J. London Math. Soc.* **41**, 141–4. (4.7.1)

Tauvel, P.
78 Sur les quotients premiers de l'algèbre enveloppante d'une algèbre de Lie resoluble, *Bull. Soc. Math. France* **106**, 177–205. (14.10.8)
82 Sur la dimension de Gelfand–Kirillov, *Comm. Algebra* **10**, 939–63. (8.7.3)

Utumi, Y.
56 A theorem of Levitzki, *Amer. Math. Monthly* **70**, 286. (2.4.3)

Van den Essen, A.
86 Algebraic micro-localization, *Comm. Algebra* **14**, 971–1000. (8.7.6)

Vaserstein, L. N.
71 Stable rank of rings and dimensionality of topological spaces, *Funct. Anal. and Appl.* **5**, 102–10. (11.8.3, 11.8.5)

Warfield, R. B.
79 Bezout rings and serial rings, *Comm. Algebra* **7**, 533–45. (4.7.1)
80 The number of generators of a module over a fully bounded ring, *J. Algebra* **66**, 425–47. (4.7.6, 11.8.0, 11.8.6)
80a Cancellation of modules and groups and stable range of endomorphism rings, *Pacific J. Math.* **91**, 457–85. (11.8.4, 11.8.5)
83 Prime ideals in ring extensions, *J. London Math. Soc.* **28**, 453–60. (4.5.6, 4.7.5)
84 The Gelfand–Kirillov dimension of a tensor product, *Math. Z.* **185**, 441–7. (8.2.6, 8.7.2)
86 Noncommutative localized rings, in *Séminaire Dubreil-Malliavin* (1985), ed. M.-P. Malliavin, Lecture Notes in Mathematics 1220, Springer-Verlag, New York–Berlin pp. 178–200. (4.7.3, 11.5.2)

Watters, J. F.
- 75 Polynomial extensions of Jacobson rings, *J. Algebra* **36**, 302–8. (9.7.1)

Webber, D. B.
- 70 Ideals and modules of simple Noetherian hereditary rings, *J. Algebra* **16**, 239–42. (7.12.8, 7.12.11)
- 71 Ideals and modules of simple Noetherian hereditary domains, Ph.D. Thesis, Leeds. (12.8.0)

Weyl, H.
- 28 *Gruppentheorie und Quantenmechanik*, Hirzel, Leipzig (English edition, Dover 1932). (1.9.3)

Whelan, E. A.
- 86 Finite subnormalizing extensions of rings, *J. Algebra* **101**, 418–32. (10.7.1)

Wolf, J.
- 68 Growth of finitely generated solvable groups and curvature of Riemannian manifolds, *J. Diff. Geom.* **2**, 421–46. (8.7.2)

Zalesskii, A. E. and Neroslavskii, O. M.
- 77 There exists a simple Noetherian ring with divisors of zero but without idempotents (Russian), *Comm. Algebra* **5**, 231–4. (12.8.7)

Zariski, O., and Samuel, P.
- 60 *Commutative Algebra II*, Van Nostrand, Princeton. (8.2.14, 13.10.12)

Zelmanowitz, J. M.
- 67 Endomorphism rings of torsionless modules, *J. Algebra* **5**, 325–41. (3.7.4)

INDEX OF NOTATION

Throughout this index, R denotes a ring, I an ideal, M a module, and X a set. The symbols $M_n(R)$, $Z(R)$, R^+, Hom, End, $\hookrightarrow$, $\twoheadrightarrow$, $\triangleleft$, $\mathscr{L}(\ )$, $\gg$, $\lceil\ \rceil$, $\mathbb{N}$, $\mathbb{Z}$, $\mathbb{Q}$, $\mathbb{R}$, $\mathbb{C}$, $\mathbb{Z}_{p\infty}$, appear in Notation (p xv), and so, too, does the convention that $\subset$ denotes strict inclusion.

Spec R	0.2.3	$A'_1(k)$	1.8.6
$N(R)$	0.2.4	$A'_n(k)$	1.8.6
$N(I)$	0.2.8	$P_{\lambda_1,\ldots,\lambda_n}(k)$	1.8.7
$J(R)$	0.3.8	$P_\lambda(k)$	1.8.7
R^n	1.1.1	$R * g$	1.9.9
M^*	1.1.5	$R_{\mathscr{S}}$	2.1.1
$\mathbb{I}(I)$	1.1.11	$\mathscr{C}_R(I)$	2.1.2
$R[x;\sigma,\delta]$	1.2.3	ass $\mathscr{S}$	2.1.3
$R[x;\delta]$	1.2.3	r ann X	2.1.10
$R[x;\sigma]$	1.2.3	l ann X	2.1.10
$A_n(k)$	1.3.1	$Q(R)$	2.1.14
$A_n(R)$	1.3.8	$M_{\mathscr{S}}$	2.1.17
$B_n(k)$	1.3.9	$\text{ass}_M \mathscr{S}$	2.1.17
$R[[x;\sigma]]$	1.4.1	$\triangleleft_e$	2.2.1
$R[x,x^{-1};\sigma]$	1.4.3	$\zeta(R)$	2.2.4
RG	1.5.1	u dim	2.2.10
$R\#G$	1.5.4	$\{e_{ij}\}$	2.2.11
$R*G$	1.5.8	R^1	2.3.9
$\bar{a}$	1.6.4	$R \sim R'$	3.1.9
gr R	1.6.4	$O_r(I)$	3.1.12
gr I	1.6.7	$O_l(I)$	3.1.12
[ ]	1.7.1	R-ideal	3.1.13
$U(g)$	1.7.2	$R \overset{M}{\sim} R'$	3.5.5
$\text{Der}_k(R)$	1.7.10	$\rho(M)$	4.1.2
$R\#U(g)$	1.7.10	$A(R)$	4.1.7
$R*U(g)$	1.7.12	$\mathscr{C}(X)$	4.3.1

Index of Notation

$E(M)$	4.3.2	Na^{-1}	10.1.6
$Q \rightsquigarrow P$	4.3.7	$b(N)$	10.1.7
ass M	4.3.9	$\mathscr{E}$	10.3.4
$\rho_P(M)$	4.6.7	$R_{\mathscr{F}}$	10.3.5
$\tau(M)$	4.6.7	N_G	10.3.13
I^*	5.1.8	R_G	10.3.17
A^{-1}	5.2.5	M_σ	10.4.6
dev A	6.1.2	M_G	10.4.6
deg α	6.1.9	$\tau(A)$	10.5.16
A'	6.1.15	FFR	11.1.6
$S_c(A)$	6.1.16	$U_c(n, R)$	11.1.8
$S_a(A)$	6.1.16	$U_r(n, R)$	11.1.8
a_∞	6.1.16	$Gl(n, R)$	11.1.11
$\mathcal{K}(M)$	6.2.2	glr R	11.1.14
$\tau_\alpha(M)$	6.2.18	sr R	11.3.4
dim R	6.4.4	$E(n, R)$	11.3.5
FBN	6.4.7	er R	11.3.10
$D_n(k)$	6.6.18	B_r/B_c	11.3.12
$\mathcal{K}_n(R)$	6.7.6	cr R	11.5.20
δ^*	6.8.9	$B(M, P)$	11.6.2
P^0	6.9.9	J-Spec	11.6.6
I_∞	6.9.12	$g(M, P)$	11.6.13
pd M	7.1.2	$b(M, P)$	11.7.2
id M	7.1.3	$\hat{g}(M, P)$	11.7.12
fd M	7.1.4	$P_f(R)$	12.1.2
gld R	7.1.8	$\langle P \rangle$	12.1.2
w gld R	7.1.9	$[P]$	12.1.2
Tor	7.1.10	$K_0(R)$	12.1.2
Ext	7.1.12	$C(R)$	12.1.3
M^σ	7.3.4	$\tilde{K}_0(R)$	12.1.5
MR	7.6.4	$\mathscr{I}(R)$	12.1.6
$M_{gr}R$	7.6.4	ϑ_*	12.1.8
$M_{fl}R$	7.6.8	$P_{gr}(R)$	12.2.4
R^G	7.8.1	$M_{fgr}(R)$	12.6.7
m'	7.9.5	s_n	13.1.3
M_∞	7.9.5	$g(R)^+$	13.5.2
$\gamma(f)$	8.1.5	c_n	13.5.5
$GK(R)$	8.1.11	∇	13.5.9
$GK(M)$	8.1.11	g_n	13.5.11
p_M	8.4.6	TR	13.9.2
$e(M)$	8.4.7	tr deg	13.10.5
$R_{[n]}$	8.6.2	ad	14.1.2
$I(w)$	9.4.16	$E(M, g)$	14.1.15

$E(M)$	14.1.15	V^δ	14.8.3
M^g	14.1.15	$S_\delta(V)$	14.8.3
M_λ	14.1.16	V^Γ	14.8.4
M^λ	14.1.19	$\mathscr{A}(V, \delta, \Gamma)$	14.8.4
$\hat{g}$	14.4.4	$\mathscr{A}$	14.8.4
$\hat{R}$	14.4.4	$A(V)$	15.0.0
$Sz(R)$	14.4.6	$\Delta(A)$	15.1.4
$\Lambda(R;g)$	14.4.8	$A[d]$	15.1.6
$\Gamma(R;g)$	14.4.8	$\Omega_K(A)$	15.1.8
$W(g)$	14.7.2	$S_A(M)$	15.1.18
$\Gamma(g)$	14.7.2	$\mathscr{D}(A)$	15.5.1

INDEX

This index does not include an index of authors. Instead the reference list indicates details of all citations.

$A_n(k), A'_n(k)$ (see Weyl algebra)
$\mathcal{A}(V, \delta, \Gamma)$ 14.8.1 ff
a.c.c. 0.1.2
AR ring 4.2.3
additive
 —function 12.1.7
 reduced rank being— 4.1.2
additivity 4.5.1 ff
 —principle 4.5.9, 14.0.0
adjoint representation 14.1.2
Ado and Iwasawa's Theorem 1.7.1
affiliated
 —chain 4.4.6
 —prime 4.3.4
 —sequence 4.4.6
affine
 —algebra 8.1.9
 —PI algebra 13.10.1 ff
algebra
 affine— 8.1.9
 almost commutative— 8.4.2
 Azumaya— 13.7.6, 13.7.13
 central separable— 13.7.8
 central simple— 5.3.4
 constructible— 9.4.12
 δ-symmetric— 14.8.3
 exterior— 13.4.5
 filtered— 9.4.9
 —of group 10.3.13, 10.4.6
 somewhat commutative— 8.6.9
 symmetric— 15.1.18
algebraic Lie algebra 14.7.2
almost centralizing
 derivation ring being— 15.1.20
 —extension 8.6.6
almost commutative algebra 8.4.2
 not— 14.3.9
almost normalizing extension 1.6.10
 generic flatness of— 9.4.11
 Noetherian— 1.6.14
Amitsur-Levitzki theorem 13.3.3
annihilator 2.1.10
 —ideal 2.2.15
 maximal— 2.3.2
arithmetic of ideals 5.2.9
Artinian
 —module 0.1.2
 —ring 0.1.6
Artinian quotient ring
 additivity and— 4.5.9
 affiliated primes and— 4.4.10
 —and largest stable ideal 6.9.10
 existence of— 4.1.4
 FBN and— 6.8.16
 hereditary ring has— 5.4.2
 ideal invariance and— 6.8.15
 Krull dimension and— 6.8.1 ff
Artin-Procesi theorem 13.7.14, 13.11.7
Artin radical 4.1.7
Artin-Rees property 4.1.10, 4.2.3
 —and localization 4.2.9 ff, 6.8.21 ff
Artin-Tate lemma 13.9.10
Artin-Wedderburn theory 0.1.9
Asano prime ring 5.2.7
 PI— 13.9.15
associated graded ring (of filtered ring) 1.6.1 ff
 —being generically flat 9.4.9

—being integral domain 1.6.6
—being Noetherian 1.6.9
—being non-Noetherian 8.3.9
—being prime 1.6.6
GK dimension of— 8.1.14, 8.3.20, 8.6.1 ff, 8.7.6
global dimension of— 7.6.18
K_0 of— 12.1.10, 12.3.5, 12.6.1 ff
Krull dimension of— 6.5.6
module theory of— 7.6.1 ff
—of derivation ring 15.4.5 ff
—of enveloping algebra 1.7.5
—of filtered ring 1.6.4
associated prime 4.3.9, 4.4.2, 4.4.4
augmentation ideal
—of enveloping algebra 14.3.8
—of graded ring 12.2.3
—of symmetric algebra 15.1.18
augmentation map 14.3.8
automorphism
 inner— 1.8.2
 outer— 7.8.12
 X-inner— 10.3.10, 10.6.15
Azumaya algebra 13.7.6, 13.7.13

$B_n(k)$ (see Weyl algebra)
basic
—composition series 11.6.2
—dimension 11.6.2, 11.8.6
—element 11.8.7
Bass' theorem 11.7.4, 11.7.13
Bernstein number 8.4.7
bimodule
—condition 4.5.7
 prime— 4.3.4
—question 6.4.11
bound 10.1.7
bounded 6.4.7
 fully— 6.4.7
bracket product 1.7.1
Brandt groupoid 5.2.14

c-integral 5.3.2
cancellation
—of modules 11.4.1 ff, 11.7.1 ff
—rank 11.5.20
Capelli polynomial 13.5.5
category
 admissible— 12.4.2
—equivalence 3.5.7, 3.7.5
 Grothendieck group of— 12.4.3

catenary property 13.10.13
—in $U(g)$ 14.10.2
central
—extension 13.1.11
—polynomial 13.5.2, 13.6.1
—separable algebra 13.7.8
central simple algebra 5.3.4, 13.3.1 ff
—having central polynomial 13.6.3
centralizer 3.2.6, 10.3.8
centralizing sequence 4.1.13
—in enveloping algebra 14.3.4
characteristic
—closure 13.9.2
—ideal 8.7.6
class group
 ideal— 12.1.6
 projective— 12.1.5
classical order 5.3.5
—having central polynomial 13.6.3
clique 4.3.7
closed
 centrally— 10.3.13
 G-— 10.3.13
 normally— 10.3.13
closure
 central— 10.3.13
 G-— 10.3.13
 normal— 10.3.13
codeviation 6.1.8
common multiple property 2.2.5
complement 2.2.3
completely faithful module 5.7.3
—and stable range 6.7.7
completely integrally closed 5.1.3
 characterization of— 5.3.3
completely prime ideal 0.2.2
—in enveloping algebra 14.2.11
completely solvable 7.5.7, 14.1.8
compressible 6.9.3
constructible algebra 9.4.12
 derivation ring being— 15.1.22
context
—equivalent 3.6.4
 Morita— 1.1.6
 prime— 3.6.5
coordinate ring
—of algebraic variety 15.0.0
—of cusp 15.3.12, 15.4.10
coordinate ring of $\mathbb{S}^1$
 derivations of— 15.3.13
—is Dedekind domain 7.8.14

K_0 of— 12.1.6
K_0 of derivation ring of— 15.4.8
overring of— 12.7.8
coordinate ring of $\mathbb{S}^2$
Der—is not free 15.3.15
—has stably free module 11.2.3
ranks of— 11.5.4
critical 6.2.9
—composition series 6.2.19
—for dimension function 6.8.25
—ideal 6.3.4
crossed product
$R*g$ 1.9.9
$R*G$ (see group ring)
$R*U(g)$ (see enveloping algebra)
cusp (see coordinate ring)
cutting down 10.2.4
—in enveloping algebra 14.2.5, 14.5.4
—in fixed ring 10.5.15
—in group ring 10.5.6

D_n 6.6.18 (see also Weyl algebra)
polynomials over— 9.6.12
subfields of— 6.6.18
d.c.c 0.1.2
Dedekind domain/prime ring 5.2.10
commutative— 5.2.8, 5.7.18
$1\frac{1}{2}$ generators in— 5.7.12
module theory of— 5.7.1 ff, 7.11.4 ff
PI— 13.9.14
simple— 7.11.1 ff
degree
—in filtered module 7.6.7
minimal— 13.2.2
—of function 8.1.7
—of ordinal 6.1.9
—of polynomial 1.2.8, 9.6.6
—of polynomial element 6.9.17
—of polynomial identity 13.1.2
PI— 13.3.6, 13.6.7
total— 9.4.15
transcendence— 8.2.14, 13.10.5
denominator set 2.1.13
dense
—right ideal 10.3.4
—subring 0.3.5
density theorem 0.3.6
derivation(s)
—and localization 14.2.2
—and prime ideals 14.2.3
inner— 1.8.2

k-— 1.7.10, 15.1.2 ff
k-—to module 15.1.7
locally nilpotent— 14.6.4
σ-— 1.2.1
universal— 15.1.8
universal bimodule of— 8.7.3
derivation ring 15.1.4 ff
associated graded ring of— 15.4.5 ff
deviation 6.1.2
differential operator(s) (see also derivation ring)
asymmetry of— 15.6.5
order of— 15.4.11, 15.5.2
ring of— 15.0.0 ff, 15.5.1 ff, 15.6.5
differentials (Kähler—) 15.1.8, 15.6.1
—of regular ring 15.2.12
dimension
basic— 11.6.2, 11.8.6
classical Krull— 6.4.4
exact—function 6.8.4
flat— 7.1.4
—function 6.8.4
GK— 8.1.11, 8.1.16
Gabriel— 6.10.0
Gelfand-Kirillov— 8.1.11, 8.1.16
global— 7.1.8
Goldie— 2.2.10
injective— 7.1.3
Kronecker— 11.5.9
Krull— 6.2.2
projective— 7.1.2
uniform— 2.2.10
weak— 7.1.4
weak global— 7.1.9
discriminant 13.8.11
dual
—basis lemma 3.5.2
—module 3.4.4

eigen
—ring 1.1.11
—value 14.1.16
—vector 14.1.15
Eisenbud-Evans theorem 11.7.4
element
automorphic— 10.1.3
c-integral— 5.3.2
—centralizes 10.1.3
homogeneous— 1.6.3, 7.6.3, 10.6.7
integral— 5.3.2
minimal— 9.4.18

nilpotent— 0.2.5
normal— 4.1.10
—normalizes 10.1.3
regular— 2.1.2
strongly nilpotent— 0.2.5
elementary
 —group 11.3.5
 —range 11.3.9
 —rank 11.3.10
endomorphism property
 —over k 9.1.4
 —over K 9.2.3
endomorphism ring
 algebraic— 9.1.4 ff
 —and matrices 8.2.8
 finite dimensional— 9.5.1 ff
 —of uniform ideal 3.3.5
 —over PI ring 13.4.9
 —over semiprime ring 3.4.1 ff
enveloping algebra (skew—, —crossed product) 1.7.1 ff, 14.0.0 ff
 associated graded ring of— 1.7.5
 —being almost centralizing 8.6.6
 —being almost commutative 8.4.1 ff
 —being almost normalizing 1.7.14
 —being constructible 9.4.12
 —being integral domain 1.7.5, 1.7.14
 —being maximal order 5.1.6
 —being Noetherian 1.7.4, 1.7.14
 —being prime 1.7.14
 —being regular 7.7.5
 —being somewhat commutative 8.6.10
 —crossed product 1.7.12
 —crossed product construction 1.9.7, 1.9.9
 example of— 15.3.14
 —having finite endomorphism ring 9.5.5
 —having projectives stably free 12.3.3
 GK dimension of— 8.1.15, 8.2.10
 global dimension of— 7.5.7, 7.6.10
 K_0 of— 12.6.14
 Krull dimension of— 6.5.7, 6.6.2
 not simple— 1.7.5
 Nullstellensatz for overrings of— 9.4.22
 Nullstellensatz in— 9.1.8, 14.4.1 ff
 —of $\mathbf{sl}(2,k)$ 8.5.12
 —of solvable Lie algebra 14.1.1 ff
 Poincaré-Birkhoff-Witt theorem for— 1.7.5, 1.9.7, 1.9.9
 prime ideals in— 14.2.1 ff, 14.3.1 ff
 primitive ideals in— 14.4.1 ff
 second layer condition in— 4.3.14
 semicentre of— 14.4.6
 simple— 1.8.3
 skew— 1.7.10
 —smash product 1.9.7
 stably free ideals of— 11.2.14
 standard monomials in— 1.7.5
 universal property of— 1.7.2
equivalent
 —orders 3.1.9
 right—orders 5.6.5
essential submodule 2.2.1
essential right ideal 2.2.1
 —contains regular element 2.3.5
 —has regular generators 3.3.7
essentiality 10.2.11
evaluation 13.5.2
exponential growth 8.1.8
Ext 7.1.12
extension 1.6.2
 almost centralizing— 8.6.6
 almost normalizing— 1.6.10
 automorphic— 10.1.3
 c-integral— 5.3.2
 central— 13.1.11
 centralizing— 10.1.3
 essential— 2.2.1
 finite— 10.1.2
 integral— 5.3.2, 10.7.0, 13.8.1 ff
 normalizing— 10.1.3
 Ore— 1.9.2
FBN 6.4.7
 Artinian quotient ring of— 6.8.16 ff
 characterization of— 6.10.4
 Jacobson conjecture for— 6.4.15
 —PI ring 13.6.6
 second layer condition for— 4.3.14
Faith-Utumi theorem 3.2.6
faithful module 0.3.2
 completely— 5.7.3
filtered
 —homomorphism 7.6.8
 —K-algebra 9.4.9
 —module 7.6.8
 —ring (see associated graded ring)
filtration(s)
 equivalent— 8.6.12
 finite dimensional— 8.1.9

good— 8.6.3
—having bounded difference 8.6.12
—of derivation ring 15.1.19
—of module 7.6.7
—of ring 1.6.1
standard— 1.6.2, 7.6.7, 8.1.9
translate of— 8.6.14
Fitting's lemma 2.3.2
fixed ring 7.8.1ff, 10.5.1ff
flat
  —dimension 7.1.4
  faithfully— 7.2.3, 15.2.13
  generically— 9.3.2
  —module 7.1.4
  (n,m)-generically— 9.3.12, 9.3.16
  —resolution 7.1.4
Forster-Swan theorem 11.7.4
fractional ideal 3.1.11
  order of— 3.1.12
  reflexive— 5.1.8
free
  —filtered 7.6.15
  —graded 7.6.5

G-
  —domain 9.3.9
  —ideal 9.3.9
  —ring 13.9.12, 14.5.6
GK dimension (see Gelfand-Kirillov dimension)
Gabriel dimension 6.10.0
Gabriel H-condition 6.10.4
Gelfand-Kirillov dimension 8.0.0ff, 8.1.11, 8.1.16
  —and transcendence degree 8.2.14
  —of $\mathscr{A}(V,\delta,\Gamma)$ 14.8.13
  —of derivation ring 14.3.2, 14.3.6
  —of gr $\Delta(A)$ 15.4.7
  —of modules over $\Delta(A)$ 15.4.1ff
  —of PI ring 13.10.6
general linear
  —group 11.1.11
  —range 11.1.12
  —rank 11.1.14
generating subspace
  —for module 8.1.9
  —for ring 8.1.9
generative 5.5.3
generator 3.5.3
generic
  —flatness 9.3.1ff

—regularity 4.6.11
global dimension 7.0.0ff, 7.1.8
  —of $\mathscr{A}(V,\delta,\Gamma)$ 14.8.13
  —of commutative rings 7.1.15, 15.2.8
  —of derivation ring 15.3.2, 15.3.7
  —of gr $\Delta(A)$ 15.4.7
  weak— 7.1.9
going down
  —in fixed ring 10.5.15
  —in group ring 10.5.12
  —in PI ring 13.8.13
going up
  —in fixed ring 10.5.15
  —in normalizing extension 10.2.10
  —in PI ring 13.8.14
Goldie
  —dimension 2.2.10
  normalizing extension being— 10.1.11
  —rank 14.0.0
  —rank polynomial 14.0.0
  —ring 2.3.1
  —Theorem 2.3.6
  —Theorem\1 3.1.7
graded
  —division ring 10.3.19
  —homomorphism 7.6.3, 12.2.3
  —simple 10.3.19
graded ring (see also associated—) 1.6.3
  graded right ideal of— 1.6.7
  group-— 10.3.18
  K_0 of— 12.1.10
  $\mathbb{Z}$-— 12.4.10
graded ring element
  degree of— 1.6.4
  homogeneous— 1.6.3, 10.3.19
  homogeneous component of— 1.6.7
  leading term of— 1.6.4
grading 7.6.2
  product— 12.2.5
  translate of— 7.6.2
  trivial— 12.2.2
Grothendieck group (K_0) 12.0.0ff
  —of category 12.4.3
  —of derivation ring 15.4.8
  —of filtered ring 12.3.5, 12.6.13
  —of graded ring 12.1.10
  —of ring 12.1.2
group ring (skew—, —crossed product) 1.5.1ff
  —being constructible 9.4.12
  —being Noetherian 1.5.11, 1.5.12

—being normalizing extension 10.1.4
—being regular 7.7.5
—being simple 1.8.3, 7.8.12
—crossed product 1.5.8
fixed ring of— 7.8.1 ff, 10.5.1 ff
GK dimension of— 8.2.9, 8.2.18
global dimension of— 7.5.6, 7.8.1 ff
—having finite endomorphism ring 9.5.5
—having projectives stably free 12.3.3
K_0 of— 12.5.1 ff, 12.7.1 ff
Krull dimension of— 6.5.5, 6.6.1
Nullstellensatz for— 9.1.8, 9.4.22
polycyclic by finite— 1.5.12
prime ideals in— 10.5.1 ff
second layer condition in— 4.3.14
skew— 1.5.4
stably free ideals of— 11.2.8
twisted— 1.5.8
universal property of— 1.5.2, 1.5.6

Heisenberg
—group 1.5.3, 1.5.10, 8.2.17
—Lie algebra 1.7.7
hereditary ring(s) 5.2.2
—and idealizers 5.6.1 ff
—and idempotents 7.8.9
—being simple 7.11.1 ff
examples of— 7.8.14
fixed ring of— 7.8.8
Krull dimension of— 6.2.8
prime PI— 13.9.16
structure of— 5.4.1 ff
Hilbert polynomial 8.4.6
—in somewhat commutative algebra 8.6.19
Hirsch number 6.5.5
holonomic module 8.5.8
—over derivation ring 15.6.4
homogeneous
—component 1.6.7, 7.6.3
—element 1.6.3, 7.6.3, 10.6.7
—module w.r.t. dimension function 6.8.8
homomorphism
filtered— 7.6.7
graded—of modules 7.6.3
graded—of rings 12.2.3
strict filtered— 7.6.12
Hopkins' theorem 0.1.13

ideal(s) (right—, left—)
annihilator— 2.1.10, 2.2.15
augmentation— 12.2.3, 14.3.8, 15.1.18
characteristic— 8.7.6
—class group 12.1.6
completely prime— 0.2.2
critical— 6.3.4
dense— 10.3.4
essential— 2.2.1
fractional— 3.1.11
G-stable— 10.5.4
g-stable— 14.2.4
generative— 5.5.3
graded— 1.6.7, 10.3.19
integral— 3.1.13
invariant— 6.8.13
invertible— 4.2.5, 5.2.5
isomaximal— 5.5.2
localizable— 4.0.0
nilpotent— 0.1.11
prime— 0.2.3
primitive— 0.3.4
semimaximal— 5.8.5
semiprime— 0.2.7
semiprimitive— 0.3.9
stable— 1.8.2
subisomorphic— 3.3.4
T— 13.1.4
trace— 10.5.16
ideal invariance 6.4.16, 6.8.13
—and GK dimension 8.3.16
idealizer ring 1.1.11, 5.5.1 ff
—and differential operators 15.5.13
chain conditions of— 1.1.12
example (in Weyl algebra) of— 1.3.10
global dimension of— 7.5.11 ff
Krull dimension of— 6.5.2
multiple— 5.8.5
identity
algebra polynomial— 13.1.15
alternating— 13.1.12
monic— 13.1.2
multilinear— 13.1.8
polynomial— 13.1.2
stable— 13.1.16
standard— 13.1.3
incomparability 10.2.13, 10.4.1 ff
—in enveloping algebra 14.2.8
—in PI ring 13.8.14
—in skew polynomial ring 10.6.6, 14.2.9
—theorem 10.4.15
index of nilpotency 3.2.7

Index

injective
— dimension 7.1.3
— envelope 4.3.2
— resolution 7.1.3
integral
  c-— 5.3.2
— domain 0.2.2
— element 5.3.2
— extension 10.7.0, 13.8.1 ff
— ideal 3.1.13
integrally closed 5.3.3
  completely— 5.1.3
intersection condition 4.3.17
invariant(s)
— basis property 11.1.2
— ideal 6.8.13
  ideal—ring 6.8.13
— in enveloping algebra 14.1.15
  ring of— (see fixed ring)
  weakly—ideal 6.8.13
invertible ideal 4.2.5, 5.2.5
irreducible
— module 0.1.2
— ring 6.8.17
isomaximal 5.5.2
isotypic 0.1.2

Jacobi identity 1.7.1
Jacobson
  derivation ring being— 15.1.22
— PI ring 13.10.3
— radical 0.3.8
— ring 9.1.2 ff, 9.7.1
Jacobson conjecture 6.10.4
  example of— 4.3.20, 4.7.3
— for FBN rings 6.4.15
Jordan-Hölder
— theorem 0.1.3
— weights 14.4.10

K_0-group (see Grothendieck group)
Kaplansky's theorem 13.3.8
Koszul resolution
— and global dimension 7.3.16
— and Krull dimension 6.5.9
Köthe conjecture 0.2.9
— for PI rings 13.2.5
Krull
— domain 5.1.10
— ring 5.8.1
Krull dimension 6.0.0 ff
— and GK dimension 8.3.18

— and monic localization 7.9.4
  classical— 6.4.4
— of $\mathscr{A}(V, \delta, \Gamma)$ 14.8.13
— of derivation ring 15.3.7
— of enveloping algebra 14.7.6
— of fixed rings 7.8.8
— of $\operatorname{gr}\Delta(A)$ 15.4.7
— of module 6.2.2
— of normalizing extension 10.1.10
— of ring 6.2.2, 6.3.1 ff
— of simple rings 7.11.1 ff
— of $U(\mathbf{sl}(2, k))$ 8.5.12 ff
  symmetry of— 6.4.10, 14.10.8
Krull socle 6.2.14
Kurosch problem 13.8.9

leading coefficient of polynomial 1.2.8, 7.9.5, 9.6.6
leading right ideal 1.2.9, 1.8.5, 9.4.16
length
— in skew Laurent polynomial ring 6.9.14, 10.6.7
— of chain (Notation)
Lie algebra(s) 1.7.1
  abelian— 14.1.5
— acting locally nilpotently 14.1.13
— acting nilpotently 14.1.13
— acting semisimply 14.1.12
  adjoint representation of— 14.1.2
  algebraic— 14.7.2
  completely solvable— 7.5.7, 14.1.8
  eigenvalue of— 14.1.16
  eigenvector of— 14.1.15
  Heisenberg— 1.7.7
  ideal of— 14.1.2
  invariant of— 14.1.16
  module over— 14.1.3
  nilpotent— 14.1.6
— over ring 1.7.8
  representation of— 1.7.1
  semidirect product of— 1.7.11
  semi-invariant of— 14.1.15
  semisimple— 14.1.12
  solvable— 6.6.2, 14.1.7
  triangular module over— 14.1.3
  universal enveloping algebra of— 1.7.2
Lie product 1.7.1
Lie's theorem 14.5.3
linearization 13.1.9
link
— graph 4.3.7

second layer— 4.3.7
local number of generators 11.6.13
localizable 4.0.0
localization (see also quotient ring) 2.1.4
 —and prime ideals 4.3.1 ff, 4.4.1 ff
 —at $\mathscr{F}$ 10.3.5
 —at set of ideals 10.3.3
 GK dimension of— 8.2.13
 global dimension of— 7.4.1 ff
 K_0 of— 12.1.12, 12.4.9
 Krull dimension of— 6.5.3
 monic— 7.9.3
 —of derivation ring 15.1.25
 stable range of— 11.5.2
locally
 —finite dimensional 8.1.17
 —nilpotent derivation 14.6.4
 —nilpotent ring 13.8.5
Loewy series 4.1.2
lying over 10.2.8
 —in enveloping algebra 14.2.5, 14.5.4
 —in fixed ring 10.5.15
 —in group ring 10.5.6
 —in normalizing extension 10.2.9
 —in PI ring 13.8.14

Martindale quotient ring 10.3.5
Maschke's theorem 7.5.6, 10.5.11
matrix ring(s) 1.1.1 ff
 —and Morita equivalence 3.5.5
 —being Goldie 3.1.5
 —being Noetherian 1.1.2
 embeddings in— 13.4.1 ff
 GK dimension of— 8.2.7
 generic— 13.1.9, 13.10.10
 global dimension of— 3.5.10
 identities of— 13.3.2
 K_0 of— 12.1.3
 Krull dimension of— 6.5.1
 ranks of— 11.5.11 ff
matrix ring, triangular (see triangular matrix ring)
matrix units
 set of— 3.2.6
 standard set of— 2.2.11
maximal
 —annihilator 2.3.2
 —condition 0.1.5
 —K-order 5.3.13
 —order 5.1.1 ff, 13.9.8
 —quotient ring 10.3.5

module
 α-torsion— 6.2.18
 ad g-— 14.3.2
 Artinian— 0.1.2
 complement— 2.2.3
 completely faithful— 5.7.3
 compressible— 6.9.3
 critical— 6.2.9
 dual— 3.4.4
 essential— 2.2.1
 faithful— 0.3.2
 filtered— 7.6.8
 flat— 7.1.4
 free-filtered— 7.6.15
 free-graded— 7.6.5
 $|G|$-torsionfree— 10.5.9
 g-— 14.1.3
 generator— 3.5.3
 graded— 7.6.3
 holonomic— 8.5.8
 irreducible— 0.1.2
 isotypic— 0.1.2
 locally finite dimensional— 8.1.17
 monoform— 4.6.4
 Morita invariants of— 3.5.8
 Noetherian— 0.1.2
 —of differentials 15.1.8, 15.6.1
 —of quotients 2.1.17
 prime— 4.3.4
 progenerator— 3.5.4
 projective— 3.5.2
 projective-graded— 7.6.6, 12.2.1
 reduced rank at P of— 4.6.7
 reduced rank of— 4.1.2
 reflexive— 5.1.7
 semisimple— 0.1.2
 simple— 0.1.2
 singular— 10.3.6
 stable range of— 6.7.2
 stably free— 11.1.1
 stably free-graded— 12.2.10
 strictly triangular g-— 14.1.13
 torsion— 2.1.17
 torsionfree— 3.4.2
 torsionless— 3.4.2
 triangular g-— 14.1.13
 uniform— 2.2.5
 $\mathbb{Z}$-graded— 12.4.10
monic
 —localization 7.9.3
 —polynomial identity 13.1.2

monoform module 4.6.4
Morita context 1.1.6
—and fixed rings 7.8.3
—being Noetherian 1.1.7
correspondence of primes in— 3.6.2
correspondence of primitives in—
  3.6.5
—of equivalent orders 3.1.14
prime— 3.6.5
ring of— 1.1.6
Morita equivalence/ent 3.5.5
—and category equivalence 3.5.7
characterization of— 3.5.6, 3.7.5
not—to domain 12.7.16
—to Dedekind domain 5.2.12
—to simple domain 12.7.5
Morita invariants
—of module 3.5.8
—of ring 3.5.10
multiplicatively closed (m.c.) set 2.1.1
multiplicity 8.4.7

Nakai conjecture 15.6.5
Nakayama's lemma 0.3.10
nil set 0.2.5
nilpotent
—action 14.1.13
—element 0.2.5
—ideal 0.1.11
—ideals in PI rings 13.2.1 ff
—index 3.2.7
—Lie algebra 14.1.6
locally—action 14.1.13
locally—derivation 14.6.4
locally—ring 13.8.5
nil subrings being— 2.3.7, 6.3.7
—prime radical 6.3.8, 13.10.9
Noetherian
—domain being Ore 2.1.15
—module 0.1.2
—ring 0.1.6
normal element 4.1.10
—in enveloping algebra 14.3.4
normalizer 10.3.8
normalizing extension 10.1.3
almost— 1.6.10
normalizing sequence 4.1.13
—in enveloping algebra 14.3.14
Nullstellensatz 9.0.0 ff
—for derivation rings 15.1.22
—for enveloping algebras 9.1.8, 9.4.22,
  14.4.1
—for group rings 9.1.8, 9.4.22
—for PI rings 13.10.3
—over k 9.1.4
—over K 9.2.3

orbit theory 14.10.2
order(s) 3.1.1 ff
classical— 5.3.5
equivalent— 3.1.9
—in quotient ring 3.1.2
K-— 5.3.6
maximal— 5.1.1, 13.9.1
maximal K-— 5.3.13
non-Noetherian centre of— 5.3.7
—of fractional ideal 3.1.12
right equivalent— 5.6.5
ordinal
degree of— 6.1.9
Ore
—domain 2.1.14
—extension 1.9.2
—set 2.1.13
Ore condition (see also localization) 2.1.6
—and AR property 6.8.21
—and ideal invariance 6.8.14
—for monic polynomials 7.9.3
outer automorphism(s) 7.8.12
group of— 7.8.12

$P_\lambda(k), P_{\lambda_1,\ldots,\lambda_n}(k)$ 1.8.7
—being Dedekind domain 5.2.11,
  7.11.3
—being simple Noetherian 1.8.7
global dimension of— 7.5.4
global dimension of overrings of—
  9.1.13
Krull dimension of— 6.5.4, 6.6.12
Krull dimension of overrings of—
  9.1.13
matrices over— 7.11.7
skew group ring over— 7.8.14, 12.7.16
PI ring 13.0.0 ff
characterization of— 5.3.10
GK dimension of— 8.7.6
generic matrix— 13.1.19
universal— 13.1.15
partitive, finitely 8.3.17, 8.7.3
derivation ring being— 15.1.21
enveloping algebra being— 8.4.9
patch

—closed   4.6.14, 11.6.6ff
—continuity   4.6.16
—open   4.6.14
—topology   4.6.14
permeated   11.4.2
—theorem   11.7.13
Poincaré-Birkhoff-Witt theorem   1.7.5, 1.9.7, 1.9.9
polycentral   4.1.13
polycyclic by finite group   1.5.12
polynomial(s) (see also skew—ring)
  algebra—identity   13.1.15
  alternating—   13.1.12, 13.5.4
  Capelli—   13.5.5
  central—   13.5.2
  —element   6.9.17
  evaluation of—   13.5.2
  Hilbert—   8.4.6
  —identity   13.1.2
  —identity ring   13.1.6
  monic—   13.1.2
  multilinear—   13.1.8
  —over division ring   9.6.1ff, 11.2.9, 11.2.15
  t-alternating—   13.5.4
polynormal   4.1.13
poset   6.0.0
  codeviation of—   6.1.8
  deviation of—   6.1.2
  factor of—   6.1.1
positive cone   12.1.3
Posner's theorem   13.6.5
pri-ring (see principal ideal ring)
prime
  —bimodule   4.3.4
  —module   4.3.4
  —radical   0.2.4, 0.2.8
  —spectrum   0.2.3
prime ideal(s)   0.2.3
  affiliated—   4.3.4
  affiliated sequence of—   4.4.6
  —and extension rings (see cutting down, going down, going up, incomparability, lying over)
  —and uniform dimension   4.6.1ff
  associated—   4.3.9, 4.4.2, 4.4.4
  G-—   10.5.4
  height of—   4.1.11
  —in enveloping algebras   14.2.1ff
  —in fixed rings   10.5.13ff
  —in Morita context   3.6.1ff
  —in polynomial rings   10.6.1ff
  J-—   11.6.6
  linked—   4.3.7
  minimal—   0.2.8
  minimal—in semiprime ring   2.2.14, 3.2.1ff
  regular—   13.7.3
  semistable—   6.9.8
  σ-cyclic—   10.6.11
  unstable—   6.9.8
prime ring   0.2.3
  Asano—   5.2.7
  Dedekind—   5.2.10
primitive
  —ideal   0.3.4
  —ideal in enveloping algebra   14.4.1ff
  —polynomial ring   9.6.11
  —property   9.2.3
  —ring   0.3.2
principal ideal ring(s)   0.1.8
  $B_1(k)$ being—   1.3.9
  decomposition of—   4.1.9
  matrix rings being—   7.11.7, 11.3.8
  matrix ring over—   3.4.10
  modules over—   5.7.19
  skew polynomials being—   11.2.12
  structure of semiprime—   3.4.9
principal ideal theorem   4.1.11
  generalized—   4.1.13
progenerator   3.5.4
projective
  —class group   12.1.5
  —dimension   7.1.2
  extended—module   12.2.11
  —module   3.5.2
  —resolution   7.1.2
projective-graded   7.6.6, 12.2.1
properness   10.2.14

Quillen
  —lemma   9.7.3
  —Suslin theorem   11.2.2
  —theorem   12.6.13
quotient division ring   2.1.14
quotient ring(s)   (see also Ore condition)   2.1.3, 3.1.1
  —and localization   6.8.1ff
  construction of—   2.1.12
  Martindale—   10.3.5
  maximal—   10.3.6

universal property of— 2.1.4

radical
  Artin— 4.1.7
  Jacobson— 0.3.8
  —of Artinian ring 0.1.12
  prime— 0.2.4, 0.2.8
  —property 9.1.2
rank
  cancellation— 11.5.20
  elementary— 11.3.10
  general linear— 11.1.14
  Goldie— 14.0.0
  H-— 11.3.13
  —of stably free module 11.1.1
  reduced— 4.1.2
  reduced—with respect to P 4.6.7, 11.6.14ff
  stable— 11.3.4
reflexive module 5.1.7
regular
  —element 2.1.2
  —local ring 15.2.8
  —prime 13.7.3
  —ring 7.7.1
resolution
  finite free— 11.1.6
  flat— 7.1.4
  injective— 7.1.3
  projective— 7.1.2
ring(s) (see also algebra)
  AR— 4.2.3
  adjoining 1 to— 2.3.8
  Artinian— 0.1.6
  Asano— 5.2.7
  bounded— 6.4.7
  context equivalent— 3.6.4
  Dedekind— 5.2.10
  derivation— 15.1.4
  —extension 1.6.2
  FBN— 6.4.7
  filtered— 1.6.1
  fixed— 7.8.1
  fully bounded— 6.4.7
  G-— 13.9.12, 14.5.6
  generic matrix— 13.1.19
  Goldie— 2.3.1
  graded— 1.6.3
  graded-division— 10.3.19
  graded-simple— 10.3.19
  group-graded— 10.3.18

hereditary— 5.2.2
ideal invariant— 6.8.13
irreducible— 6.8.17
Jacobson— 9.1.2
localization of— 2.1.4
locally nilpotent— 13.8.5
Morita equivalent— 3.5.5
Morita invariants of— 3.5.10
Noetherian— 0.1.6
—of differential operators 15.0.0ff, 15.5.1
—of Morita context 1.1.6
quotient— 2.1.3, 3.1.1
PI— 13.0.0ff, 13.1.6
pri-— 0.1.8
prime— 0.2.3
primitive— 0.3.2
principal ideal— 0.1.8
regular— 7.7.1
semiprime— 0.2.7
semiprimitive— 0.3.9
semisimple— 0.1.11
simple— 0.1.10
skew Laurent polynomial— 1.4.3
skew Laurent power series— 1.4.4
skew polynomial— 1.2.4
skew power series— 1.4.1
trace— 13.9.2, 13.9.4
weakly ideal invariant— 6.8.13
$\mathbb{Z}$-graded— 12.4.10

$\mathbb{S}^n$ (see coordinate ring)
$\mathbf{sl}(2,k)$
  dimensions of enveloping algebra of— 8.5.12ff
saturated chain 13.10.13
Schanuel's lemma 7.1.2
  long— 7.1.2
Schelter's theorem 13.10.12
Schur's lemma 0.1.9
second layer
  —condition 4.3.12
  —link 4.3.7
semicentre 14.4.6
semidirect product of groups 1.5.7
semi-invariants 14.1.15
semimaximal 5.8.5
semiprime
  —ideal 0.2.7
  —ring 0.2.7
  structure of—Goldie rings 3.0.0ff

semiprimitive
—ideal 0.3.9
—polynomial ring 13.2.7
—ring 0.3.9
semisimple
—action 14.1.12
—Lie algebra 14.1.12
—module 0.1.2
—ring 0.1.11
semistable 6.9.8
Serre
—conjecture 11.8.0
—theorem 11.7.4, 11.7.13
simple module 0.1.2
endomorphism ring of— 9.0.0 ff
simple ring 0.1.10
$A'_n(k)$ being— 1.8.6
$\mathscr{A}(V, \delta, \Gamma)$ being— 14.8.9
—and enveloping algebras 14.3.6, 14.3.7
$B_n(k)$ being— 1.3.9
—being Dedekind domain 7.11.1 ff
—being pri-ring 11.2.13
bounds on generators in— 6.7.8
centre of— 2.1.16
cyclic module over— 5.7.3
derivation ring being— 15.3.8
examples of— 1.8.1 ff, 1.9.8, 7.8.14
GK dimension of— 8.3.19
global dimension of— 7.5.5
ideals in tensor product of— 9.6.9
K_0 of— 12.7.1 ff
Krull dimension of— 6.6.7, 6.6.11, 6.9.24, 8.3.19
—not being matrix ring 12.7.7 ff
—not being $\overset{M}{\sim}$ domain 12.7.16
$P_\lambda(k)$ being— 1.8.7
polynomials over— 6.7.9
projective modules over— 11.7.14
skew group ring being— 7.8.12
Weyl algebra being— 1.3.5
singular ideal 2.2.4
—being nilpotent 2.3.4
$\mathscr{F}$-— 10.3.5
skew enveloping algebra (see enveloping algebra)
skew group ring (see group ring)
skew Laurent polynomial ring (see skew polynomial ring)
skew Laurent power series ring 1.4.4
skew polynomial ring (—Laurent—)
1.2.1 ff, 1.4.1 ff
—and derivation ring 15.3.2
—being integral domain 1.2.9, 1.4.5
—being Noetherian 1.2.9, 1.4.5
—being non-Noetherian 1.2.11
—being prime 1.2.9, 1.4.5
—being pri-ring 11.2.13
—being regular 7.7.5
—being simple 1.8.4, 1.8.5
GK dimension of— 8.2.11, 8.2.15, 8.2.16
global dimension of— 7.5.3, 7.9.1 ff, 7.10.1 ff, 9.1.14
—having projectives stably free 12.3.3
—in several variables 1.6.1 ff
incomparability in— 14.2.9
K_0 of— 12.5.1 ff, 12.6.1 ff
Krull dimension of— 6.5.4, 6.6.6, 6.6.10, 6.9.1 ff, 9.1.14
leading ideals of— 1.2.9, 1.8.5
—over commutative rings 6.9.1 ff
primes of— 10.6.1 ff
stably free ideals of— 11.2.7, 11.2.8
universal property of— 1.2.5
skew power series ring 1.4.1
global dimension of— 7.5.3
Small's theorem 4.1.4
smash product 1.9.7
socle 6.2.14
Krull— 6.2.14
Krull—sequence 6.2.14
solvable Lie algebra 6.6.2, 14.1.7
somewhat commutative 8.6.9
derivation ring being— 15.1.21
sparse 4.6.11
spectrum
J-— 11.6.6
prime— 0.2.3
σ-— 10.6.3
splitting field 13.3.6
stability 11.0.0 ff
stable
G-— 10.5.4
g-— 14.2.4
—generation 6.7.1 ff, 11.6.11 ff
—ideal 1.8.2
—identity 13.1.16
—isomorphism 12.1.4
—matrix 11.5.16
—range of module 6.7.2
—rank 11.3.4

—unimodular row   11.3.2
stably
 —free   11.1.1, 11.2.1
 —free-graded   12.2.10
 —isomorphic   12.1.4
standard
 —filtration   1.6.2, 7.6.7, 8.1.9
 —matrix units   2.2.11
 —monomials   1.7.5
state space   12.8.1
strict   7.6.12
strongly nilpotent element   0.2.5
subisomorphic   3.3.4
symmetric
 —algebra   15.1.18
 δ-—algebra   14.8.3
syzygy theorem   12.3.1 ff

topology
 patch—   4.6.14
 Zariski—   4.6.14
Tor   7.1.10
torsion
 α-—   6.2.18
 —free   3.4.2
 $|G|$-—free   10.5.9
 —less   3.4.2
 —module   2.1.17
 —theory   2.4.1
 —with respect to   4.3.2
trace
 —function   12.7.13
 —ideal   10.5.16
 —of group ring   7.8.6
 —of ring   12.7.13
 proper—   12.7.13
 —ring   13.9.2, 13.9.4
 universal—function   12.7.13
translate
 —of filtration   8.6.14
 —of grading   7.6.2
triangular matrix ring   0.1.4
 chain conditions in—   1.1.4, 1.1.9, 1.1.10
 global dimension of—   7.5.1

uniform
 —dimension   2.2.10
 —module   2.2.5
 —right ideals   3.3.2 ff
unimodular

—row/column   11.1.8
stable—row/column   11.3.2
universal enveloping algebra (see enveloping algebra)
unstable   6.9.8

weak
 —algorithm   11.3.9
 —dimension   7.1.4
 —global dimension   7.1.9
weight(s)
 Jordan-Hölder—   14.4.10
 —space   14.1.19
Weyl algebra (A_n, A'_n, B_n) 1.3.1 ff
 A'_n 1.8.6
 B_n 1.3.9
 D_n 6.6.18
 —and $\mathcal{A}(V, \delta, \Gamma)$ 14.8.1 ff
 —and algebraic g 14.7.1 ff
 —and Nullstellensatz   9.1.8, 9.4.22
 —and solvable g 14.9.1 ff
 —as derivation ring   15.1.5
 —as differential operators   1.3.6
 —as factor of enveloping algebra   14.6.1 ff
 —being almost commutative   8.4.3
 —being Dedekind   5.2.11, 7.11.3
 —being maximal order   5.1.6
 —being not a principal ideal ring   7.11.8
 —being simple Noetherian domain   1.3.5, 1.3.8, 1.8.6
 embeddings of—   6.6.19
 GK dimension of—   8.1.15, 8.2.7
 GK dimension of modules over—   8.5.3 ff
 general linear rank of first—   11.1.16
 generators of right ideals of—   6.7.8
 global dimension of—   7.5.8
 global dimension of overrings of—   9.1.12, 9.4.23
 —has finite endomorphism rings   9.5.5, 9.7.5
 —has projectives stably free   12.3.3
 —has stably free ideals   11.1.4, 11.2.11
 —in derivation ring   15.2.6, 15.3.2
 K_0 of—   12.1.6, 12.3.3
 K_0 of overring of—   12.3.3, 12.6.14, 12.7.7
 Krull dimension of—   6.5.8, 6.5.10, 6.6.15, 7.5.9

Krull dimension of overrings of— 9.1.12, 9.4.23
matrices over— 7.11.7
modules over— 5.7.18
—over $\mathbb{Z}$ 4.6.13, 11.7.8
quotient division ring of— 6.6.18, 9.6.12
skew group ring over— 7.8.14, 12.7.7
stable range of— 11.2.15
word 1.6.2

Zariski-Lipman conjecture 15.6.3
Zariski topology 4.6.14